AF595353

Natural Environment and Cultural Heritage in the City, A Sustainability Perspective

Natural Environment and Cultural Heritage in the City, A Sustainability Perspective

Editors

Jan K. Kazak
Katarzyna Hodor
Magdalena Wilkosz-Mamcarczyk

MDPI • Basel • Beijing • Wuhan • Barcelona • Belgrade • Manchester • Tokyo • Cluj • Tianjin

Editors

Jan K. Kazak
Wrocław University of Environmental and Life Sciences Poland

Katarzyna Hodor
Cracow University of Technology
Poland

Magdalena Wilkosz-Mamcarczyk
University of Agriculture in Krakow
Poland

Editorial Office
MDPI
St. Alban-Anlage 66
4052 Basel, Switzerland

This is a reprint of articles from the Special Issue published online in the open access journal *Sustainability* (ISSN 2071-1050) (available at: https://www.mdpi.com/journal/sustainability/special_issues/heritage_city_sus).

For citation purposes, cite each article independently as indicated on the article page online and as indicated below:

LastName, A.A.; LastName, B.B.; LastName, C.C. Article Title. *Journal Name* **Year**, *Volume Number*, Page Range.

ISBN 978-3-0365-1752-0 (Hbk)
ISBN 978-3-0365-1751-3 (PDF)

Contents

About the Editors

Jan K. Kazak is associate professor at the Institute of Spatial Management at the Wrocław University of Environmental and Life Sciences. He is a leader of the Leading Research Team: Sustainable Cities and Regions. In his research, he integrates social, environmental, and economic aspects of local and regional development. He is a Management Committee Member in COST action: Implementing nature-based solutions for creating a resourceful circular city.

Katarzyna Hodor is an architect and associate professor of the CUT. She has a post-doctoral degree in technical science in architecture and urban planning. She graduated from the Faculty of Architecture at the Cracow University of Technology. She conducts research on identity and cultural heritage associated with the sacred, the restoration of historical greenery, and the revitalisation of rural and small-town areas. Since 2013, she has been the chief organiser of an international conference on garden design and historical dendrology. She has been a member of the SPAK and the SAK since 2011 and a member of the ICOMOS commission on historical gardens since 2020. She participates in international projects and organises and co-organises a cyclical academic competition in addition to numerous workshops, student project exhibitions, and competitions.

Magdalena Wilkosz-Mamcarczyk graduated from the Cracow University of Technology and holds a PhD in technical sciences, architecture, and urban planning. Since 2013, her academic career has been with the University of Agriculture in Kraków. She participated in research placements in Lviv, Budapest, and Vilnius. She is the head of the science popularisation project International Architecture and Landscaping Competition under contract 514/P-DUN/2019 financed by the Ministry of Science and Higher Education. She participates in the NAVA and Trans-European Education for Landscape Architects projects. Her academic interests include semiurbanisation and revitalisation.

Preface to "Natural Environment and Cultural Heritage in the City, A Sustainability Perspective"

In view of the pace regarding transformations in urban landscapes due to excessive urbanization combined with climate change that has become apparent in recent decades, concern for the future of urban landscapes seems essential. The challenge for designers and specialists in the field of creating urban space is establishing new spaces and revitalizing existing ones. These activities, based on the elements of cultural heritage, ecological design, as well as respect for the existing natural landscape of cities (such as parks or gardens), are important elements influencing the quality of life of the inhabitants. Projects supported by scientific research in the field of landscape, its transformations, and further functioning will allow us to develop solutions consistent with the principles of sustainable development.

The effects of scientific research on landscapes have traditionally been disseminated annually during scientific meetings related to landscape architecture since 1994. The conferences organised by the founders of the Krakow School of Landscape Architecture (KSLA) have seen active participation by academics, designers, and representatives of local and regional authorities, gathering an international audience. Studies presented during the XXVII scientific conference (9th international edition 2020) were focused on the subject of Urban ecology and cultural heritage in the city. The presented research works outlined in this Special Issue refer to contemporary problems of cities in a global and local context and formulate methods for their solutions based on a balanced approach to the cultural and natural heritage of urban areas.

Jan K. Kazak, Katarzyna Hodor, Magdalena Wilkosz-Mamcarczyk
Editors

Editorial

Natural Environment and Cultural Heritage in the City, a Sustainability Perspective

Jan K. Kazak [1,*], Katarzyna Hodor [2] and Magdalena Wilkosz-Mamcarczyk [3]

1 Institute of Spatial Management, Wrocław University of Environmental and Life Sciences, Grunwaldzka 55, 50-375 Wrocław, Poland
2 Chair of Landscape Architecture, Faculty of Architecture, Cracow University of Technology, Warszawska 24, 31-155 Kraków, Poland; khodor@pk.edu.pl
3 Department of Spatial Management and Landscape Architecture, Faculty of Environmental Engineering and Land Surveying, University of Agriculture in Krakow, Balicka 253 c, 30-198 Kraków, Poland; magdalena.wilkosz-mamcarczyk@urk.edu.pl
* Correspondence: jan.kazak@upwr.edu.pl

Citation: Kazak, J.K.; Hodor, K.; Wilkosz-Mamcarczyk, M. Natural Environment and Cultural Heritage in the City, a Sustainability Perspective. *Sustainability* **2021**, *13*, 7850. https://doi.org/10.3390/su13147850

Received: 25 June 2021
Accepted: 5 July 2021
Published: 14 July 2021

Publisher's Note: MDPI stays neutral with regard to jurisdictional claims in published maps and institutional affiliations.

The 21st century urges us to analyze urban problems and revise its components. Cities in the past followed various evolutionary paths depending on the level of development of the society, its ideology, or financial factors. In the past, the key was self-reliant and polycentric production, services, commerce, and finance. What is essential today in more complex socio-environmental systems [1] is to strengthen efforts to improve the quality of life and support residents throughout the transformation is activities aimed at implementing changes to create a socially-friendly [2,3] and climate-neutral built environment [4,5], which in parallel, maintains achievements of previous generations and cultural heritage [6–8]. A place that is at the intersection of these two spheres is a city. It is the focal point of human life on our planet and where critical socio-ecological challenges lie [9]. Therefore, city is a space where adaptation to climate change action will have to take place [10], and integrated spatial sustainable development should be implemented [11–13].

All the above-mentioned issues require involvement and a lot of activities on different levels. One of the leading institutions is the International Council on Monuments and Sites (ICOMOS), which is a non-governmental international organization dedicated to the conservation of the world's monuments and sites [14]. ICOMOS has started working on the 'ICOMOS SDGs Policy Guidance' in 2018, which is a flagship initiative "aiming to provide a robust and versatile resource to all kinds of stakeholders, within and outside of the heritage community, on the role of cultural heritage in sustainable development" [14]. Following the concept of integration of cultural heritage and sustainable development, the scientific committee of the XXVII Conference in the Series of Garden Art and Historical Dendrology, IX International Edition titled "Urban Ecology and Cultural Heritage in the City" has proposed topical scope combining two basic, yet highly relevant, issues that affect the cities today, the natural and urban environment viewed from various angles and through different relations. This Special Issue gathers selected papers which, despite different scientific perspectives and methodological approaches applied, have similar denominators. Main themes appearing in all 22 articles published in this Special Issue refer to cities and urban areas, their development, green and public spaces, trees and water, visual and functional features of space, gardens and built environment, as well as the local character of discussed cases and their use over time (Figure 1).

Figure 1. Word cloud generated from 22 publications collected in the Special Issue on "Natural Environment and Cultural Heritage in the City, A Sustainability Perspective" in *Sustainability*.

In this Special Issue, you can find recent and multidisciplinary studies focusing on four main pillars, which are:

- natural and cultural heritage covering conservation activities in urban and suburban zones, urbanscape and heritage relationships [15–21],
- ecological solutions in urban development and management [22–24],
- urban and land use planning, urban composition and impact of historical conditions on the modern city development [25–30],
- gardens and parks, their maintenance and use, ecosystem services served [31–36].

We believe that all collected studies presenting such an interdisciplinary approach will result in added value by sensitizing specialists from many fields to the need to use a complex approach in creating sustainable cities and regions.

Author Contributions: All authors equally contributed to editorial preparation. All authors have read and agreed to the published version of the manuscript.

Funding: This research received no external funding.

Acknowledgments: The cooperation under XXVII Conference in the Series of Garden Art and Historical Dendrology, IX International Edition titled "Urban Ecology and Cultural Heritage in the City" and Special Issue on "Natural Environment and Cultural Heritage in the City, A Sustainability Perspective" in *Sustainability* was supported by scientific activity conducted within the Leading Research Group: Sustainable Cities and Regions at the Wrocław University of Environmental and Life Science, the Chair of Landscape Architecture at the Cracow University of Technology and the Department of Spatial Management and Landscape Architecture at the University of Agriculture in Krakow.

Conflicts of Interest: The authors declare no conflict of interest.

References

1. Szewrański, S.; Kazak, J.K. Socio-environmental vulnerability assessment for sustainable management. *Sustainability* **2020**, *12*, 7906. [CrossRef]
2. Kisiała, W.; Rącka, I. Spatial and Statistical Analysis of Urban Poverty for Sustainable City Development. *Sustainability* **2021**, *13*, 858. [CrossRef]
3. van Hoof, J.; Bennetts, H.; Hansen, A.; Kazak, J.K.; Soebarto, V. The living environment and thermal behaviours of older south australians: A multi-focus group study. *Int. J. Environ. Res. Public Health* **2019**, *16*, 935. [CrossRef]
4. Świąder, M.; Lin, D.; Szewrański, S.; Kazak, J.K.; Iha, K.; van Hoof, J.; Belčáková, I.; Altiok, S. The application of ecological footprint and biocapacity for environmental carrying capacity assessment: A new approach for European cities. *Environ. Sci. Policy* **2020**, *105*, 56–74. [CrossRef]
5. Salvia, M.; Reckien, D.; Pietrapertosa, F.; Eckersley, P.; Spyridaki, N.A.; Krook-Riekkola, A.; Olazabal, M.; De Gregorio Hurtado, S.; Simoes, S.G.; Geneletti, D.; et al. Will climate mitigation ambitions lead to carbon neutrality? An analysis of the local-level plans of 327 cities in the EU. *Renew. Sustain. Energy Rev.* **2021**, *135*, 110253. [CrossRef]
6. Hodor, K.; Warys, E. Development and Role of Sacred Structures in Gliwice before the End of the 20th Century. *IOP Conf. Ser. Mater. Sci. Eng.* **2019**, *471*, 082015. [CrossRef]
7. Wilkosz-Mamcarczyk, M.; Olczak, B.; Prus, B. Urban Features in Rural Landscape: A Case Study of the Municipality of Skawina. *Sustainability* **2020**, *12*, 4638. [CrossRef]
8. Klapa, P.; Mitka, B.; Zygmunt, M. Application of Integrated Photogrammetric and Terrestrial Laser Scanning Data to Cultural Heritage Surveying. *IOP Conf. Ser. Earth Environ. Sci.* **2017**, *95*, 032007. [CrossRef]
9. Zięba, Z.; Dąbrowska, J.; Marschalko, M.; Pinto, J.; Mrówczyńska, M.; Leśniak, A.; Petrovski, A.; Kazak, J.K. Built environment challenges due to climate change. *IOP Conf. Ser. Earth Environ. Sci.* **2020**, *609*, 012061. [CrossRef]
10. Kiełkowska, J.; Tokarczyk-Dorociak, K.; Kazak, J.; Szewrański, S.; van Hoof, J. Urban adaptation to climate change plans and policies—The conceptual framework of a methodological approach. *J. Ecol. Eng.* **2018**, *19*. [CrossRef]
11. Furmankiewicz, M.; Campbell, A. From Single-Use Community Facilities Support to Integrated Sustainable Development The Aims of Inter-Municipal Cooperation in Poland, 1990–2018. *Sustainability* **2019**, *11*, 5890. [CrossRef]
12. Derlukiewicz, N.; Mempel-Śnieżyk, A.; Mankowska, D.; Dyjakon, A.; Minta, S.; Pilawka, T. How do Clusters Foster Sustainable Development? An Analysis of EU Policies. *Sustainability* **2020**, *12*, 1297. [CrossRef]
13. Ziemianska, M.; Kalbarczyk, R.; Sobota, M. Replacement planting in the light of the polish law- A n effective or token tool of environmental compensation? *Acta Sci. Pol. Adm. Locorum* **2019**, *18*, 323–333. [CrossRef]
14. International Council on Monuments and Sites. Available online: https://www.icomos.org/en (accessed on 27 January 2021).
15. Łakomy, K. Site-Specific Determinants and Remains of Medieval City Fortifications as the Potential for Creating Urban Greenery Systems Based on the Example of Historical Towns of the Opole Voivodeship. *Sustainability* **2021**, *13*, 7032. [CrossRef]
16. Forczek-Brataniec, U. Assessment of Visual Values as a Tool Supporting the Design Decisions of the Cultural Park Protection Plan. The Case of Kazimierz and Stradom in Kraków. *Sustainability* **2021**, *13*, 6990. [CrossRef]
17. Szilágyi, K.; Lahmar, C.; Rosa, C.A.P.; Szabó, K. Living Heritage in the Urban Landscape. Case Study of the Budapest World Heritage Site Andrássy Avenue. *Sustainability* **2021**, *13*, 4699. [CrossRef]
18. Dudzińska, M.; Prus, B.; Cellmer, R.; Bacior, S.; Kocur-Bera, K.; Klimach, A.; Trystuła, A. The Impact of Flood Risk on the Activity of the Residential Land Market in a Polish Cultural Heritage Town. *Sustainability* **2020**, *12*, 10098. [CrossRef]
19. Trojanowska, M. Therapeutic Qualities and Sustainable Approach to Heritage of the City. The Coastal Strip in Gdańsk, Poland. *Sustainability* **2020**, *12*, 9243. [CrossRef]
20. Gyurkovich, M.; Gyurkovich, J. New Housing Complexes in Post-Industrial Areas in City Centres in Poland Versus Cultural and Natural Heritage Protection—With a Particular Focus on Cracow. *Sustainability* **2021**, *13*, 418. [CrossRef]
21. Merino-Aranda, A.; Castillejo-González, I.L.; Velo-Gala, A.; de Paula Montes-Tubío, F.; Mesas-Carrascosa, F.-J.; Triviño-Tarradas, P. Strengthening Efforts to Protect and Safeguard the Industrial Cultural Heritage in Montilla-Moriles (PDO). Characterisation of Historic Wineries. *Sustainability* **2021**, *13*, 5791. [CrossRef]
22. Dudek-Klimiuk, J.; Warzecha, B. Intelligent Urban Planning and Ecological Urbanscape Solutions for Sustainable Urban Development. Case Study of Wolfsburg. *Sustainability* **2021**, *13*, 4903. [CrossRef]
23. Wowrzeczka, B. City of Waste—Importance of Scale. *Sustainability* **2021**, *13*, 3909. [CrossRef]
24. Ptak-Wojciechowska, A.; Januchta-Szostak, A.; Gawlak, A.; Matuszewska, M. The Importance of Water and Climate-Related Aspects in the Quality of Urban Life Assessment. *Sustainability* **2021**, *13*, 6573. [CrossRef]
25. Szpakowska-Loranc, E. Multi-Attribute Analysis of Contemporary Cultural Buildings in the Historic Urban Fabric as Sustainable Spaces—Krakow Case Study. *Sustainability* **2021**, *13*, 6126. [CrossRef]
26. Gierko, A. Towards an Understanding of the Pre-War Landscape Transformations in the Face of Contemporary Urban Challenges on the Example of Gajowice in Wrocław. *Sustainability* **2021**, *13*, 5962. [CrossRef]
27. Gyurkovich, M.; Pieczara, M. Using Composition to Assess and Enhance Visual Values in Landscapes. *Sustainability* **2021**, *13*, 4185. [CrossRef]
28. Jaszczak, A.; Kristianova, K.; Pochodyła, E.; Kazak, J.K.; Młynarczyk, K. Revitalization of Public Spaces in Cittaslow Towns: Recent Urban Redevelopment in Central Europe. *Sustainability* **2021**, *13*, 2564. [CrossRef]

29. Dudzic-Gyurkovich, K. Urban Development and Population Pressure: The Case of Młynówka Królewska Park in Krakow, Poland. *Sustainability* **2021**, *13*, 1116. [CrossRef]
30. Khadour, N.; Al Basha, N.; Sárospataki, M.; Fekete, A. Correlation between Land Use and the Transformation of Rural Housing Model in the Coastal Region of Syria. *Sustainability* **2021**, *13*, 4357. [CrossRef]
31. Jagiełło, M. Do We Need a New Florence Charter? The Importance of Authenticity for the Maintenance of Historic Gardens and Other Historic Greenery Layouts in the Context of Source Research (Past) and Taking into Account the Implementation of the Sustainable Development Idea (Future). *Sustainability* **2021**, *13*, 4900. [CrossRef]
32. Ludwig, B. Baroque Origins of the Greenery of Urban Interiors in Lower Silesia and the Border Areas of the Former Neumark and Lusatia. *Sustainability* **2021**, *13*, 2623. [CrossRef]
33. Ludwig, B. The Greenery of Early Modernist Housing Estates: The 1919–1927 Wałbrzych Agglomeration. *Sustainability* **2021**, *13*, 3921. [CrossRef]
34. Makowska, B. Practical Functioning of a Sustainable Urban Complex with a Park—The Case Study of Stavros Niarchos Foundation Cultural Center in Athens. *Sustainability* **2021**, *13*, 5071. [CrossRef]
35. Krzeptowska-Moszkowicz, I.; Moszkowicz, Ł.; Porada, K. Evolution of the Concept of Sensory Gardens in the Generally Accessible Space of a Large City: Analysis of Multiple Cases from Kraków (Poland) Using the Therapeutic Space Attribute Rating Method. *Sustainability* **2021**, *13*, 5904. [CrossRef]
36. Hodor, K.; Przybylak, Ł.; Kuśmierski, J.; Wilkosz-Mamcarczyk, M. Identification and Analysis of Problems in Selected European Historic Gardens during the COVID-19 Pandemic. *Sustainability* **2021**, *13*, 1332. [CrossRef]

Article

Site-Specific Determinants and Remains of Medieval City Fortifications as the Potential for Creating Urban Greenery Systems Based on the Example of Historical Towns of the Opole Voivodeship

Katarzyna Łakomy

Chair of Landscape Architecture, Faculty of Architecture, Cracow University of Technology CUT, Warszawska 24, 31-155 Kraków, Poland; klakomy@pk.edu.pl

Abstract: The article discusses the natural and historic heritage of medieval towns in the Opole Silesia region in the context of their ability to take advantage of their potential for sustainable development, especially in tourism. The chosen environmental, urban, architectural, and landscape factors were compared through this aspect and subjected to a multidimensional comparative analysis. The research studies applied mostly archival materials, contemporary topographic maps, statistical data, and both landscape as well as urban field studies. As a result, the studies indicated that the natural conditions of the locations, the preservation level of the urban system along with its development trends, and the areas of the old fortifications with their accompanying greenery constitute these towns' very value and identity. As they combine elements of nature and culture, they may serve as the basis for development of tourism, which is likely to contribute to the social and economic revitalization of the region itself. What may play a major role in the quest for sustainable development are the urban greenery systems to be designed based on former fortification areas, city greenery, and natural environmental resources, which have been integral elements of these towns over many past centuries.

Keywords: cultural heritage; medieval town planning; city defence walls; monument conservation; historical greenery; urban greenery system; landscape ecology; Silesian city; Opole Voivodeship

Citation: Łakomy, K. Site-Specific Determinants and Remains of Medieval City Fortifications as the Potential for Creating Urban Greenery Systems Based on the Example of Historical Towns of the Opole Voivodeship. *Sustainability* **2021**, *13*, 7032. https://doi.org/10.3390/su13137032

Academic Editors: Jan K. Kazak, Katarzyna Hodor and Magdalena Wilkosz-Mamcarczyk

Received: 25 April 2021
Accepted: 17 June 2021
Published: 23 June 2021

1. Introduction

Cultural heritage management is a fundamental part of preserving its value for future generations. Apart from obvious values, it is a universal value that shapes the identity and awareness of society. It also defines its wealth and can become a basis for important initiatives as well as socio-economic transitions. What is key is that when faced with contemporary social, economic, and ecological problems of cities, it can become a part of sustainability strategies. However, the total sum of our heritage assets constantly fluctuates and many sites lose their values, while many others remain unidentified and unstudied [1] In the Polish system of monument protection, there is also a problem with specifying the rank of a monument on the scale of a municipality, voivodeship and the country overall, which is associated with equal allocation of funds for their renovation. On the other hand, Poland does not possess a unified system of valuating heritage elements, including architectural, archaeological, or garden design monuments [2]. Many scholars currently pursue the establishment of such a system, with ICOMOS Polska playing a leadership role [3,4].

In this context, a group of small, poorly explored towns located in the Opole Voivodeship was noted. And they became the main subject of research. Selected out of the total pool of 36 towns in the voivodeship, selection was based on several criteria: their medieval origins, featuring a regular quadrangle-like market square, an oval layout, as well as the presence of former brick fortification walls (total of 15 towns).

Their local character, along with the region's specificity, means this heritage is underappreciated and underused in the region's development and its socio-economic context. Initial landscape and urban planning investigation provided a general overview of these towns as a system of specific forms of cultural heritage linked with the hydrological network, natural habitats, and agricultural areas.

A variety of materials were applied in the research study; however, the most important ones used for comparative analysis were archival plans and city views in addition to contemporary topographic data, data collected during field research, as well as statistical data.

In the context of further research, our intent was also to determine the degree to which decisions associated with a historical founding of a town can carry over to its contemporary functioning and its development potential in the context of sustainable development. Therefore, during the research, these issues were characterized: wildlife factors that historically defined the founding of the towns under study during the Middle Ages, assess the impact on their spatial form and development, and contemporarily—to isolate place-based landscape assets and the potential to use them to create urban greenery systems. The primary objective was to create a framework of principles for the design of such systems based on a major element—historical post-fortress areas along with guidelines of their refurbishment and directions of further study.

Building linkages between nature and the manmade environment has long since been a field of investigation for scholars and designers. The genesis of introducing greenery into cities dates back to ancient times [5,6]. Certainly, we can point to nineteenth-century proposals of American parkways, followed by Howard's garden-cities [7], which transformed into concepts of continuous systems like greenways and green belts [6] as precursory efforts in treating greenery in a systemic manner. City greenery systems can be divided into: strip, wedge, radial, and mixed layouts [6,8]. Significant changes in creating urban greenery systems were affected by the concept of ecological patches and corridors by Forman and Gordon [9,10]. They assume a continuity of environmental linkages and their significance to environmental balance, biodiversity, and the self-renewal of ecosystems [11]. Green infrastructure is a contemporary concept that implies a strategically planned network of natural and semi-natural areas that perform social and economic function, but most importantly, they support biodiversity [12].

Creating urban greenery systems based on areas around fortifications became popular in the nineteenth and at the turn of the twentieth century, when, in the face of urban development and a desire to improve health conditions in cities, large-scale demolition of medieval and modern fortifications was carried out. The most prominent examples of this include: Vienna, Brno, Salzburg, Olomouc, or Cologne [13], while Polish cases feature Wrocław [14] or Gdańsk [8]. The work initiated in Kraków in 1822 was one of the first ever in Europe and was based on transforming the decayed areas of former fortifications into a public strolling park surrounding the Old Town, so-called Planty Park [15].

In the contemporary reality of Polish heritage conservation, there is a lack of comprehensive studies of the development of greenery in areas previously occupied by medieval city fortifications. There are also no guidelines as to their design and arrangement or maintenance and renovation. The role of fortifications as objects of high landscape values was first noted in Poland by Bogdanowski [16]. His research is continued by K. Wielgus, J. Środulska-Wielgus, and A. Staniewska of the Cracow University of Technology [17]. This landscape and fortification structures from the Middle Ages, as attractions of cultural tourism, were analysed by Armin Mikos von Rohrscheidt [18] and J. Środulska-Wielgus [19]. The subject of greenery at the city walls was taken up in a historical [20,21] or comparative [22,23] context. It is worth noting that many cities have studies on greenery and urban planning, created for the purposes of land development studies or the needs of conservation protection.

The multi-volume publication by Bimler [24], which was followed by Przyłęcki's work [25], are considered to be among seminal works on the history of the massive buildings of medieval cities of Silesia. Among the important publications in the field of architecture, urban planning, and art, Thullie [26] and Chrzanowski and Kornecki [27] can be mentioned. Adamska [28] dealt in detail with the town squares of Opole Silesia. Few scholars have so far become narrowly specialised in medieval city walls. A comparative analysis of such structures in Silesia and the Opole Voivodeship, combined with assessments of their state of preservation, was provided by Piechaczek [29] and Przybyłok [30]. Legendziewicz is carrying out extensive work on Medieval town gates (Głubczyce, Głuchołazy, Grodków, Namysłów, Prudnik, Głogówek [31–33]) and walls (among others Kluczbork [34] or Grodków [35]) in Silesia.

The conservation of fortifications that are a part of a town's urban fabric is an important matter, as is the pursuit of new functions for historical defensive structures and the potential to adapt them based on the development directions of modern cities. The role of fortifications, including greenery, in the space of the city was discussed in the multi-author book 'Fortyfikacje w przestrzeni miasta' [36], while the issues of the conservation and exhibition of fortifications in Lower Silesia were widely discussed by Przyłęcki [37]. The state of preservation and the problems of revitalising selected medieval towns of the Opole Voivodeship were explored by Adamska [38].

Here, we should also mention studies associated with ICOMOS Polska, which resulted in numerous academic conferences, monographs, and best practices handbooks [39]. They discussed research and adaptation methodologies [40], as well as management in the context of sustainability. Scholars note that the proper assignation of functions to such structures is a major contemporary challenge, and determines the scope of necessary procedures while allowing for tourist access [41].

The architectural and urban heritage of the Opole Voivodship, as noted by scholars [28], has been explored to a varying degree and there is a lack of comprehensive investigations. The state of research concerning individual towns presents itself better—Paczków [42,43], Otmuchów [43], Opole [44], Namysłów [45], Biała [46], Byczyna [47], Głogówek [48], Głubczyce [49], Kluczbork [50], Krapkowice [51], Ujazd [52]. This is, of course, an incomplete list. Pre-war publications and chronicles, unpublished documents, historical files of each town or historical and urban planning studies drafted for land development studies or architectural conservation are additional sources of knowledge [53]. Their precision and volume nevertheless exceed the bounds of this paper.

The number of military architecture specimens in Poland is small. According to data provided by the NID, monuments of this type constitute 1.56% of all structures entered into the monuments register in Polish territory. They include city walls and gates, castle and defensive manor fortifications, fortresses and their elements, bunkers, shelters, guardhouses, fortified positions, etc. This group includes 1105 structures dated to between the Middle Ages and the Second World War. They typically have considerable historical value and are associated with major events in the country's history [1].

The town of Paczków, located in the Opole Voivodeship, is one of the highest-class structures of this type in Poland (Figure 1). Both the fortification system and the urban layout that it surrounds have been entered in the monuments register. Labelled as 'Paczków—old-town complex with a medieval fortification system', it was also placed on the list of Monuments to History [54]. It is one of the best examples of medieval urban planning Silesian towns [43,55], which were present in the territory of historical Silesia [56,57] and the present-day Śląsk, Górny Śląsk, and Opole Voivodeship.

Figure 1. Case study—Paczków (Patschkau), (**a**) View of the city (ca. middle eighteenth century, W.B. Werner (source: https://www.bibliotekacyfrowa.pl/dlibra/publication/8092/edition/15414, accessed on 15 October 2020), (**b**) View of the city (1935–1940) (source: https://polska-org.pl/3478452,foto.html?idEntity=536159, accessed on 15 October 2020), (**c**) Current bird's eye view (source: https://zabytek.pl/pl/obiekty/paczkow-zespol-staromiejski-ze-sredniowiecznym-systemem-fortyfik, accessed on 12 April 2021).

The Opole Voivodeship is the smallest of Poland's sixteen voivodeships. It was established in 1950, and has been functioning in its current form since 1999. It is located between two strong voivodeships: the Silesian Voivodeship and the Lower Silesian Voivodeship, with which it is historically and culturally tied. From the south, it borders the Czech Republic. It has an area of 9412 km^2 and occupies only 3% of the country's territory. It has a population of 980,771 [58] and an agrarian character.

The Opole Voivodeship's authorities, as well as those of its individual municipalities and powiats, face numerous socio-economic issues: growing unemployment, waste management, a significant burden being placed on the environment by particulate matter and gas emissions, a transformation of its economic structure, planning decisions, and others (based on a GUS analysis). A GIOŚ report from 2020 clearly indicated that the greatest threat to the environment of this region still included the unsatisfactory quality of its waters, air pollution with particulate matter, benzo[a]pyrene, and excessive noise levels [59].

At present, local strategic plans—the Voivodeship *Development Strategy and the Environmental Protection Programme*—are at the drafting stage. In the context of the subject under study, previous documents featured clearly identified priority goals, which included both improving the condition of the natural environment and citizens' quality of life, as well as the region's economic activation, based on its existing predispositions, resources, and wildlife and landscape assets and their reserves [60].

In the context of efforts towards achieving the 17 Goals of Sustainable Development, *the Sustainable Development Report* 2020 placed Poland in 24th place [61]. However, as demonstrated by Kapera in a study on the implementation of sustainable development in the country published in 2018, the Opole Voivodeship ranked last [62].

Voivodeship tourism potential studies performed in 2008 ranked voivodeships in terms of a taxonomic tourist attractivity indicator [63]. In this ranking, the Opole Voivodeship placed last. In 2015, it also placed last in terms of the taxonomic development indicator score (*taksonomiczny miernik rozwoju*, TMR) [64]. Furthermore, in terms of tourism geography, it was observed that Opolszczyzna has a poor tourism and cultural potential [65].

In light of the above, my study could support strategic decisions for the region, whose developmental perspectives should be based on broad action and change in the fields of economic development and environmental protection. This action can be based on its strengths, which also act as its fundamental assets from the perspective of tourism potential [66,67]—its historical cultural landscape of medieval towns and their accompanying areas [68].

This is why among all the identifiers of the towns under study, I chose research categories linked to historical landscape elements, which are of key significance to the contemporary state of cultural and natural heritage: town placement determinants, urban layouts and linkage networks between them, city fortifications, landscaped and natural greenery, as well as the composition of their skylines.

It was demonstrated that site-specific defensive determinants affected the contemporary state of the towns' development, and hindering development conditions allowed for retaining reserves of land of high environmental potential.

The assessment of the state of preservation and the quality of individual town layout elements showed that they were assets of high tourism value (wildlife assets, cultural heritage assets, and a well-developed land and fluvial transport infrastructure). The remains of historical city fortifications were found to be a particularly significant element that can become not only an attractor, but also an element of a town's identification. These remains are not only walls, gates, or towers, but also the area of former embankments and moats—which are currently both landscaped areas and wastelands.

I determined that the state of preservation and quality of architectural and urban heritage, together with areas of natural and landscaped greenery can become a basis for creating elaborate greenery systems on the city scale and a linked ecological network on the regional scale. This could benefit not only the natural environment, but also the proper protection and exposition of cultural heritage and restore historical place-based identity. These can, in turn, lay the foundations for sustainable tourism.

2. Materials and Methods

2.1. Research Side—Geographical Location, Environmentally Valuable Areas, Heritage Conservation

The Opole Voivodeship is located in southwestern Poland at the point of contact of three geographical regions: the Silesian-Kraków Uplands, the Silesian Lowlands, and the Sudetes. The spatial form of these areas results in Opolszczyzna being a basin open towards the west, and whose central axis is the River Oder. It is the second-longest river in Poland. The histographic network is formed by its numerous confluences, of which the most significant ones are the Osobłoga, the Nysa Kłodzka, the Mała Panew, and the Stobrawa. The voivodeship's territory also features two large artificial water bodies: the Otmuchowskie and Turawskie lakes.

The terrain is shaped by the Racibórz Basin in the south-eastern section, the Sudetes Foothills in the southwest, the Chełm Mountain Chain and the Niemodlin Lowlands in the centre, and the Opole Lowlands in the remainder of the area. The tallest point in Opolszczyzna is Biskupia Kopa (890 m above sea level), which is located in the Opawskie Mountains [59].

In terms of landscape conservation areas, the voivodeship is ranked last in Poland [69]. There are no national parks here. The largest areas of protected greenery are landscape parks (with a combined area of 63 thousand ha) and protected landscape zones (196.3 thousand ha). However, in terms of percentage of occupied territory to total territory, this results in 6.7% for landscape parks and 20.8% for protected landscape zones, respectively. In total, areas protected under law account for 27.6% of the voivodeship's territory (based on GUS data, January 2021 [58]).

According to an NID report, the amount of items in the immovable monuments register of the voivodeship was 3152 (twelfth place in Poland), yet when calculated per 1000 km^2, it placed the voivodeship in third place (right behind the Lower Silesian and Lesser Poland voivodeships) and amounted to 335 [1]. When compared against population, this number amounted to 3.2, which also placed the voivodeship in third place (right behind the Lubusz and Warmian-Masurian voivodeships). Defensive structures form an extraordinarily valuable, yet small (1.56%) group of items in the country's monuments register—in the Opole Voivodeship, they account for only 48 entries out of the 878 sites in all of Poland. Urban layouts likewise are very few—in the Opole Voivodeship, there are 26 such listed heritage sites out of a total of 802 in Poland overall.

2.2. Methodology

My study began with initial field analyses intended to reconnoitre the region in terms of its landscape and socio-economic conditions, and verify its cultural heritage assets. Description (state of preservation, dating, etc.) and formal analysis of urban layouts of cities, including the nature of medieval fortifications and urban green ensemble, has been made.

The data used in the study consisted of three groups: (1) Archival materials—both cartographic and iconographic—from the collection of the State Archives of Opole and Wrocław, the Library in Berlin and open repositories. The chief of these materials included: city plans from Christian Fredrich von Wrede's *Kriegs-karte von Schlesien*, from the years 1747–1753 [70], perspective views and city plans by Friedrich Bernhard Wernher from *Silesia in Compendio seu Topographia* [71]; his city panoramas are included in *Scenographia urbum Silesiae* [72] and a series of topographic maps from the Prussian Royal Measurement Office, the so-called Messtischblätter, drawn to a scale of 1:25,000 that had been published since 1875 (open repository). The second group of materials comprised GIS system data—a numerical terrain model, a land cover map, hydrographic maps, and CIR (Colour Infrared) orthophotomaps from "Opolskie w Internecie system informacji przestrzennej i portal informacyjno-promocyjny Województwa Opolskiego OWI–OGIS Portal (http://maps.opolskie.pl/start/ accessed on 5 January 2021) along with supplementary elements from the Geoportal (www.geoportal.gov.pl, accessed on 5 January 2021). The third group consisted of material collected in the field—primarily photographic and graphical documentation.

Of the 36 cities and towns located within the voivodeship's administrative limits, 29 were determined to be of medieval origin. Among these, 25 were observed to be typical Silesian towns with oval outlines and a geometricised plan with a central, quadrate market square. Their town charters were issued in similar periods (the thirteenth and fourteenth centuries) and most (15) had city fortifications in the form of stone walls with gates and towers surrounded by an embankment or moat (Table 1, Figure 2).

Table 1. Preliminary elimination of towns of the Opole Voivodeship.

Total Number of Towns in the Opole Voivodeship	Towns of Medieval Origins	Silesian Town with a Central, Quadrate Market Square	Town Charters Issued in the Thirteenth and Fourteenth Centuries	Towns with Documented Medieval City Fortifications	Medieval Fortifications Absorbed by Modern Ones	Towns Subjected to Detailed Analysis
36	29	25	25	18	3	15

In order to make the evaluations of the towns, multidimensional comparative analysis was applied, based on the analysis of the results from the field studies, statistical data, the archival materials, and contemporary data obtained from the Geographical Information System (GIS). This allowed to develop methods for categorizing and evaluating the characteristics connected to the historic elements of the landscape, which, in turn, happen to be of key importance to today's cultural and natural heritage. What particularly matters is the context of applying the heritage potential to protection [73–76] and sustainable development.

Figure 2. Outline of the Opole Voivodeship with towns subjected to detailed analysis.

The values (landscape values) of the surveyed centers were determined in the categories (indicators): environmental site-specific determinants, urban layout, city morphology, city fortifications, panoramas, and city greenery (Table 2).

These characteristics are the most distinctive for the physiognomy of these towns, essential both today and in the past, and they can be compared with historical cartography and iconography. In the author's opinion, they define the towns' respective identities and are their greatest value. The table below presents an outline of the actions taken. Indicator—originally isolated characteristics for the towns under study; historical significance—features the function and significance that a given indicator had in the past; contemporary significance—what we currently perceive as the indicator's value; value—the element under assessment (Table 8); investigation—what was explored to determine the value of a given indicator.

On this basis, the results were obtained. Each town was analyzed against the criteria mentioned (a. Indicator) and then rated (d. Value) on a scale of 0–2, where 0 denoted no value, and 2—the highest value. The sum of individual ratings denoted a town's individual cultural landscape potential (point 3.6). Landscaped greenery analyzes were carried out on the basis of identified: semi-natural greenery in river valleys, city parks, green squares, and public gardens within historical layouts, post-fortification areas converted into public greenery, post-fortification areas acting as wastelands, cemeteries, allotment gardens, block greenery, and other types (monasteries, hospitals, sports areas). The results are presented in Table 7 and then in Table 8.

Table 2. Assessment indicators for historical towns of Opolian Silesia.

a. Indicator	b. Historical Significance	c. Contemporary Significance	d. Value	e. Investigation
Environmental site-specific determinants	Impact on site-specific determinants	Environmental and landscape assets	Diversity	- hydrographic network, wetlands, hypsometric profile
Urban layout	Socio-legal, organisational, functional structure	Fundamental characteristic of a layout	Legibility of urban layout and structure	- scope of register entry - comparative analysis of archival and contemporary maps - urban composition analyses
City morphology (development directions)	Socio-economic development	Determinant of traditional landscape retention	Continuity of the traditional development model	- empirical method of determining urban transformation trajectories
City fortifications	Defensibility, exposition	Elements of identity, exposition	State of preservation	- scope of register entry - % of outline preservation - additional elements (towers, gates, embankments)
Panoramas (skyline composition)	Informative symbol, spatial orientation, a sign of status and wealth	Landscape asset	Quality and state of preservation—legibility of historical skyline composition	- landscape analysis of passive exposition, identification of observation points and sequences, landmarks, harmonious and disharmonious landmarks, - analysis and comparison against iconographic records

3. Results

3.1. Site-Specific Determinants of Historical Location

Silesian towns were located on the basis of a developed model, e.g., the Magdeburg Law [28,56]. They were often founded *In cruda radice*, or their origins could be associated with the existence of earlier gords, villages, or market settlements, as well as ancient Roman and early medieval period settlement [56,77].

The primary determinants of selecting a site for settlement were: trade development potential, agricultural potential (trade and agricultural character), the location of centres of power or worship, and environmental determinants. This confirms the pan-European trends [78].

I found that the group of towns under study shared two primary founding factors, namely economic and environmental, which impacted each other. The first group includes trade—the network of pre-existing trade routes, whose genesis dates back to Roman times, and earlier, early medieval trade settlements. This was confirmed by contemporary studies of trade routes and the location of customs chambers in the region in the fourteenth, fifteenth, and sixteenth centuries [56,79]. Of key importance was the well-developed waterway network, as confirmed by historical maps of Silesia, i.e., by Helwig from 1561, wherein river transport acted as the main spatial linkage system.

Water, along with terrain typical of river valleys, was of fundamental significance to the shape and development of the towns under study. As many as 86% of them were founded near rivers—along the Oder and its confluences (the Nysa Kłodzka, the Biała, the Osobłoga, the Psina, the Prosna, the Stobrawa, the Ścinawa Niemodlińska, the Prudnik, the Kłodnica and the Struga Grodkowska) or in their immediate vicinity.

Most importantly, proximity to a river enhanced a settlement's defensibility, but was also of commercial and economic significance (fishing, fluvial transport, placement of water crossings, customs chambers, potential to build watermills, beer breweries, etc.). Placement on hills, slopes, on the foregrounds of river valleys and river basins, also enhanced defensibility while providing suitable exposition to the city, ideatively elevating its status and rank, while also facilitating spatial orientation for travelers (Table 3). I found no direct link between terrain conditions and founding in only two cases (Byczyna, Strzelce Opolskie).

Table 3. Functions—water and terrain.

Terrain	Defensibility	Hills, Slopes
	Exposition	
Waterways	Transport	River routes, crossings
	Economy	Commerce Fishing Others (e.g., mills)
	Defensibility	River bed Wetlands, floodlands
	Exposition	Vista foregrounds of river valleys and wetlands

3.2. *Historical Heritage of Towns—Urban Layout, Architecture, Skylines*

The environmental determinants discussed above, together with the formal and legal conditions of town charters [56,80], affected the spatial form of towns. Their key features include an oval plan with a geometric street grid filled with blocks of buildings, a centrally-placed square, and masonry city fortifications with a system of gates and towers. The quality of such a layout's composition was determined by the proportions of each part, visual linkages, legible compositional axes, and skylines. The application of metrological methods of measuring parcellation modules began already in the 1950s, and is continued today [55,81].

The towns under study, in comparisons to the period's significant urban centres throughout Silesia, for example, Wrocław, can be considered small and very small. Their town charters were issued mainly in the thirteenth century. The area inside the city walls was between 8.7 ha, as in Krapkowice, to 18.6 ha in Namysłów. They possessed all of the previously mentioned features of Silesian towns, yet their planning was not as thorough, and often applied only to the blocks adjacent to the market square. Their major compositional axes were often the result of the course of a trade route (e.g., Krapkowice).

Market squares were typically placed in the centre of a layout—at the crossing of trade routes, yet there were instances where this rule was not followed, i.e., in Głogówek (due to the location of the castle). Their size ranged from 0.6 ha (Byczyna) to 1.4 ha (Paczków), and their central spaces were occupied by a so-called market block that consists of the town hall or town hall and adjacent townhouses [28]. These complexes had a representative and commercial function. Their main element—the tower, was the dominant of the urban interior and an important element of the city panorama. Also, churches were important elements in compositional terms, with visual linkages with the Market Square (Głubczyce being an exception). Due to functional reasons and a town's size, it was often placed adjacent to the walls, and often had a docking defensive wall (Krapkowice, Strzelce Opolskie). Any castles and nearby structures were undeniably important elements, which could: be located inside an urban structure (Głogówek), be adjacent to it (Otmuchów) or be a separate element (Opole). The essential characteristics of the towns under study have been presented in Table 4.

Table 4. General characteristic of Silesian towns (own study based on [28,30,56,77,80]).

Town charter issuance	Biała 1225, Byczyna 1268, Głogówek 1275, Głubczyce przed 1253, Głuchołazy 1225, Grodków 1278, Kluczbork 1253, Krapkowice 1284, Opole 1217, Namysłów via 1270, Otmuchów via 1347, Paczków via 1254, Prudnik 1279, Strzelce Op., 1290, Ujazd 1223
Plan	Oval plan (boat/pseudo-oval shape, 'dwarf city'), enclosed, grid-based, rectangular layout, with a geometricised main square (Ring, Circus) in the centre, surrounded by fortifications on an oval-like plan. Layouts both regular and distorted by local conditions
Major composition elements	Module, axis -> market square, parcel, streets
Functional buildings	Fortifications, town hall, church, monastery, hospital, castle, others (synagogues)
Fortifications	Moats, embankments, walls—made of stone (curtain wall), 1.5–2.5 m thick, height without breastwork at 4.5–8.4 m), towers, gates.
Main identification (skyline) elements	City walls, towers, church, castle

The shape of the town was also adapted to local environmental conditions, which is why they differed from a perfect oval, taking on more organic forms, as in the cases of Głogówek, Głubczyce, Otmuchów, Krapkowice. The most regular layouts were found in Paczków, Grodków, Głuchołazy, and Byczyna.

Economic determinants contributed to the development of many of these towns remaining wooden up to as late as the nineteenth century (except for town halls, churches, residential buildings, and town walls). This often led to extensive fires, both due to external and internal causes (e.g., Hussite raids in the fifteenth century). The current urban structure largely corresponded to the one from the period of the towns' respective charters.

City walls were erected in later periods than the towns themselves, namely in the fourteenth, fifteenth, and even the sixteenth century. They were built mostly from stone or brick. At present, the reconstruction of their appearance is highly problematic due to their fragmentary preservation. Their height varied. In Krapkowice, it reached 5.4 m up to the wall walk, while in Paczków, the entire wall was 9 m high. Their thickness also varied, from 0.9 m in Biała to 2.5 m in Paczków. Towers and gates were additional elements. The natural terrain was of great significance in the fortification structure, and was put to use to great effect. Wherever the terrain was not difficult enough, additional embankments and moats were placed.

The images of towns in Wernher's historical iconography [71,72], and those of others, proved to be a great source of information on the appearance, construction principles, and forms of individual structures, as well as period landscapes. The distinguishing features of old skylines included a wide vista foreground, horizontal greenery, water or field and fortification layouts, as well as landmarks and minor landmarks in the form of town halls, churches, and castles. Gates, city towers, and chapels acted as accents (Figure 1).

3.3. Contemporary Towns and Cities—Development Directions

Comparing historical maps with the numerical terrain model and maps of areas in danger of flooding for the towns under study showed that the development model defined in the Middle Ages, limited by the terrain, the course of waterways and swamplands, had been maintained until the nineteenth century. The more difficult the conditions, the slower the development, with a stronger utilisation of pre-urbanised areas. Such specimens possessed the best-preserved spatial layouts and followed a continuity-based development model (Biała, Byczyna, Krapkowice, Otmuchów).

Of course, a beneficial economic situation, especially the development of industry in the nineteenth century, accelerated the region's dynamic transformation. Especially those localities that were given access to the railroad and had the necessary resources quickly began turning into chaotically planned layouts with a mixed functional structure (Opole, Prudnik, Głubczyce, Głuchołazy).

What is important, the environmental factors that had constrained development in the past, can currently be said to define the high environmental and landscape potential of the towns under study.

The town of Krapkowice is an example of this, as it is located on a high escarpment above the Oder River. Swamplands, visible on an archival map, occupy the same area as indicated on a contemporary range of areas in danger of flooding and thus excluded from development. This made it possible to create extensive recreational areas in the present. Furthermore, interesting visual linkages between two neighbouring medieval towns, Krapkowice and Otmęt, were preserved (at present they constitute one city). This is one of the more interesting cases of a surviving historical cultural landscape of a river valley, further enhanced by the layer formed by industrial buildings from the nineteenth and twentieth centuries (which are listed heritage sites) (Figure 3).

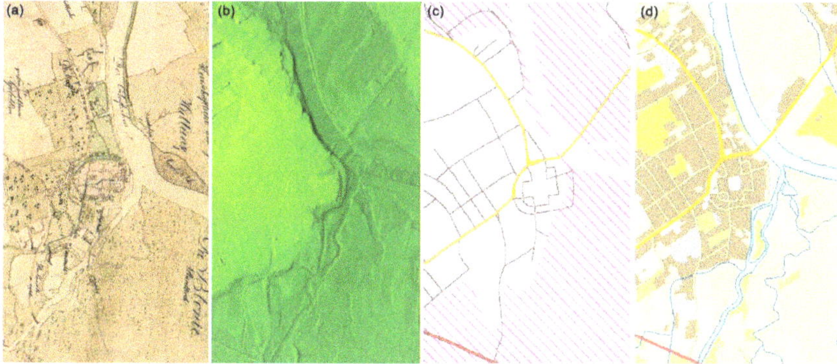

Figure 3. Case study—Krapkowice (Krappitz), (**a**) Town map from 1811 (Archiwum Państwowe w Opolu), (**b**) Numerical Terrain Model, (**c**) Areas in danger of flooding, (**d**) Contemporary development (own study based on OWI—OGIS Portal data).

In tourism geography studies, it is assumed that varied terrain enhances a given area's attractiveness to tourists [65,66]. In the context of this study, this provides additional advantages: (1) It enhances landscape attractiveness via distant visual linkages, interesting panoramas, and exposing heritage sites, (2) Surviving land reserves located mostly along river valleys (e.g., the Oder, the Osobłoga, the Nysa Kłodzka) are areas that are intriguing in terms of wildlife, (3) An urban form that has survived in a shape close to its historical version is an undeniable asset in urban heritage terms, (4) Surviving land reserves can currently be used to create recreational areas.

One example of this is the conversion of areas in danger of flooding or that were previously water reservoirs into allotment gardens (Prudnik), sports grounds (Grodków) or urban parks (e.g., Kluczbork) or keeping them as semi-natural areas of environmental value (e.g., Otmuchów).

The contemporary directions of development for each town should be devised based on the town's specific character and local conditions. Nevertheless, according to the studies above, the priority tasks should be: to increase the present housing density while adjusting to the height of the existing buildings, draw up landscape view protection zones for the landmark historic urban layouts, and mark natural environment protection zones to secure the preserved natural and semi-natural areas. New institutions should be located as remotely from the historical center as possible. They shall require a landscape view analysis as not to distort the town's historic silhouette. Their composition should be incorporated with the existing layout. Several cities shall require a reorganization of the wheel transportation system so that it by-passes the historic center.

3.4. Contemporary Town and Cities—State of Preservation of Urban Layouts, Fortifications, and Skylines

The value of the spatial layouts of the towns and cities under study, understood as the sum of the artistic, compositional, and historical value and their state of preservation, resulted in all of them being entered in the immovable monuments register of the Opole Voivodeship. However, the original urban layouts of many of them are blurred or deformed. This was caused by the Second World War and Polish post-war conservation theory.

Comparative analyses of historical von Wrede's plans and contemporary maps showed that all towns have a well-preserved urban layout. The surviving street network in all towns means that compositional axes and visual linkages can still be identified. Market squares, still surrounded by compact development, continue to play their formal functions. Even in the most transformed ones (Głubczyce, Otmuchów, and Strzelce Opolskie), these elements are still visible. Biała, Byczyna, Głogówek, Głuchołazy, Krapkowice, and Paczków have the best preserved town layout.

The market square, surrounded by densely developed buildings, still fulfils its representative functions. Town halls formed their main elements in 12 towns (the exceptions being Krapkowice, Ujazd, and Głuchołazy). They date back to the origins of the Middle Ages, but their forms show the difficult history of the region. They have gothic fragments (Opole), largely combining elements of Gothic and Renaissance (Namysłów), Renaissance (Głogówek), Renaissance and Baroque (Otmuchów), several from the 19th century (e.g., Grodków, Strzelce Opolskie). Despite a historicising form, the building in Głubczyce is from the twenty-first century.

As a result of wartime operations, the development of old-town complexes suffered varying degrees of damage. The towns that suffered the most included Nysa, Brzeg, Głogówek, Głubczyce, and Opole, which lost over half of their old-town buildings [38]. The spatial layout of Głubczyce was entered in the register immediately after the war (in 1949), but the development of its central area (100% of its fabric) in the form of multi-family housing blocks, that completely altered the character of its space, is from the 1960s.

Buildings of this type are in all market-side frontages of the towns under study, with the exception of Gluchołazy and Opole. These buildings formed between 20 and 100% of a given frontage's development. The majority of tenement houses have elevations shaped at the turn of the 19th and 20th centuries (e.g., Krapkowice), and in some cities also—Baroque (Głogówek, Namysłów).

The main religious buildings of medieval origin, remodelled and extended to varying degrees, have survived into the present. Even residential buildings continue to be major spatial elements, although their technical condition varies. In Prudnik and Opole, only a single tower has survived, while in Ujazd or Strzelce Opolskie, the buildings were found to be in a state of permanent ruin, while in other towns, they were used as museums (Namysłów), hotels (Otmuchów), or schools (Krapkowice).

The problem of the remains of town walls and their exhibition has, since the nineteenth century, been a conservation problem, which, depending on the local situation, was solved in different ways. In terms of conservation theory, reconstruction often saw use in the nineteenth century, while after the Second World War, there was a shift towards the concept of the permanent ruin [37]. The post-war situation in Lower Silesia meant that the remains of town walls were not sufficiently recognised and saw their gradual removal. A series of research and conservation projects that were mostly intended to secure them was performed as late as towards the end of the 1960s [37]. Contemporary revitalisation programmes (e.g., Brzeg, Paczków) approached their respective subjects in a much more comprehensive manner.

The state of preservation and conservation of fortifications proved to be highly differentiated (Table 5). Out of 15 towns, only 2 were found to have retained almost a complete ring of defensive walls with gates and towers (Paczków and Byczyna). Eight featured significant elements of the walls and singular towers and gates (22–56%), while in four cases, these were fragments at a level of 1–13% of circumference (or structures like towers and gatehouses, as in Głuchołazy and Prudnik). Kluczbork had a wooden and earthen rampart from the north [34]. Only Ujazd had no surviving wall fragments visible above grade. Of note is the fact that all surviving wall fragments, as well as the walls and towers, were either listed in the immovable monuments register individually or as a whole (the entire system).

Table 5. City walls—state of preservation (own study [30,31,34,35]).

		Period of Construction, Century	Approximate Wall Circumference	Surviving Sections [%]	Gatehaus	Tower	Turret	Listed Site
1	Biała	Fifteenth	1100	M	-	1	yes	yes
2	Byczyna	Fifteenth–sixteenth	1000	L	3	1	yes	yes
3	Głogówek	Fourteenth–fifteenth	1350	M	1	1	yes	yes
4	Głubczyce	Thirteenth, fourteenth–fifteenth	1500	M	-	-	yes	yes
5	Głuchołazy	Fourteenth, fifteenth	970	S	-	1	-	only a tower
6	Grodków	Fourteenth, sixteenth	1270	M	2	1	yes	yes
7	Kluczbork	Fifteenth–sixteenth	850 (1290)	M	-	1	-	yes
8	Krapkowice	Fourteenth	850	M	1	-	yes	yes
9	Namysłów	Fourteenth	1600	M	1	2	yes	yes
10	Opole	Fourteenth	1500	S	-	1	yes	yes
11	Otmuchów	Fourteenth	1200	S	-	1	no	yes
12	Paczków	Fourteenth	1200	L	1	3	yes	yes
13	Prudnik	Fourteenth	1100	S	-	1	yes	yes
14	Strzelce Opol.	Fourteenth	900	S	-	1	yes	yes
15	Ujazd	Fourteenth	1100	N	-	-	-	-

Surviving section in %: l—large part, m—medium, s—small, n—none.

In the context of the problem under study, it is not the masonry structures themselves that are significant, but their accompanying areas—sites that remained in places formerly occupied by demolished walls (that could still contain underground remains), moats, or embankments. Due to the shape of the towns, they form ring-like belts. In the nineteenth century, following European tendencies, they were converted into areas of landscaped public greenery (Byczyna). However, the densification of central development turned them into attractive sites for construction [37]. This is why they were used for public and private buildings—single-family (e.g., Grodków, Figure 4) as well as transport routes (Kluczbork) and car parks (Głogówek, Prudnik). The shape of the terrain and the non-uniform soil structure in the area made them into modern-day wastelands or overgrown gardens (Biała, Krapkowice).

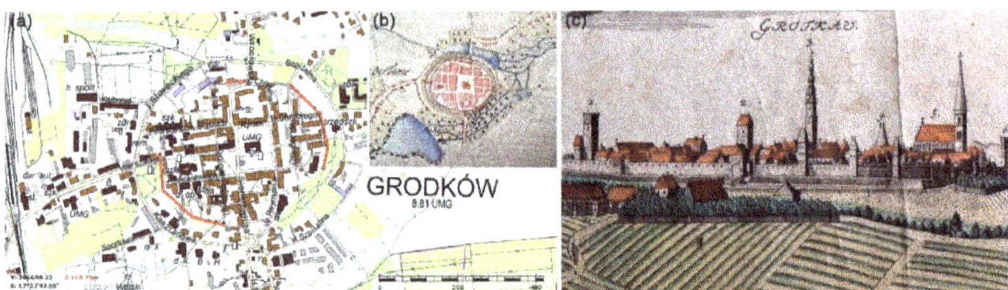

Figure 4. Case study—Grodków (Grotkau), (**a**) Topographic map (red line—surviving sections), (**b**) City plan from ca. 1750 (C.F. von Wrede [28]), (**c**) View of the city from the north (ca. middle eighteenth century, W.B. Wernher (source: https://medievalheritage.eu/en/main-page/heritage/poland/grodkow-city-defensive-walls/, accessed on 15 October 2020).

During my research, I identified the following major problems in areas of historical fortifications:

- a varied state of ownership;
- a varied state of preservation;
- a varied state of development;
- fragmentation, gaps in the defensive perimeter;
- lack of archaeological studies;
- landscaped accompanying greenery, in the form of typical city gardens and parks of low visual and functional quality;
- decisions to establish a green area not backed by a functional need in adjacent areas, made randomly and only to follow a general tendency;
- tall trees obscured walls (landscaped greenery);
- natural succession, which caused walls to be obscured from further observation points, erasure of skylines;
- low attractiveness and quality of nearby buildings—an element ignored in conservation and restoration (Figure 5).

Figure 5. City walls—Byczyna and Krapkowice.

The remains of medieval fortifications are an essential element of history and the present-day identification of the towns under study. This is why it is necessary to give them proper exposition—both regional and local, and the forgotten and unutilised exposition from water-level. It is an undeniable visual attractor that provides an opportunity to recreate medieval landscapes with visible important landmarks such as castles, churches, or fortification fragments. It is important to prevent the consequences of natural succession.

The degree of preservation of city walls gives them poor potential in individual terms, yet in connection with restoration and other elements, including accompanying greenery and an urban greenery system, it can result in an opportunity to enhance the attractiveness of these elements both visually and functionally. Attention should therefore be paid to their educational value and promotion, including as a system of linked cities. On the scale of a single town, we should merge the remaining fragments and connect them with tourist trails and other medieval structures. We should likewise provide correct exposition and accessibility, as creating a system of urban greenery based on fortified areas can bring benefits not only in the field of conservation, but also to society and ecology.

Skyline legibility analyses as observed from main road approaches allowed us to identify four towns with the highest skyline compositional value. These were towns that displayed the best state of preservation in terms of exposition, development directions, and legible functional historical structures—which acted as landscape landmarks and sub-landmarks. Despite an altered architectural fabric, primarily in terms of form and character of the cover of housing, key layout elements remained legible, and skylines retained their historical character (Biała, Głogówek, Otmychów, Paczków). In the case of Byczyna—poor exposition resulted from its terrain-related, site-specific conditions—it stands on a small hill—and partially from the presence of buffer strips and accompanying access roads.

3.5. Greenery in the Towns

Statistical data procured from the GUS included the following categories: rest and walking parks, urban gardens, street greenery, housing estate greenery, cemeteries, and municipal forests (Table 6).

Table 6. Landscaped greenery, data from the GUS (as given in 2020, accessed on 30 January 2021).

	Municipal Unit	Rest and Walking Parks [ha]	Urban Gardens [ha]	Street Greenery [ha]	Housing Estate Greenery [ha]	Cemeteries [ha]	Municipal Forests [ha]	Total [ha]
1	Biała	0.0	1.4	1.0	1.4	1.1	0.0	4.9
2	Byczyna	6.3	7.3	0.0	19.0	1.9	0.0	34.5
3	Głogówek	16.8	0.7	0.7	19.2	2.4	12.1	51.9
4	Głubczyce	11.3	7.2	6.8	31.3	5.6	0.0	62.2
5	Głuchołazy	0.0	4.3	0.5	20.3	8.6	178.6	212.3
6	Grodków	6.4	4.6	0.0	15.9	1.9	0.0	28.8
7	Kluczbork	70.0	4.6	4.7	107.2	7.2	128.6	322.3
8	Krapkowice	52.6	6.3	6.5	71.8	3.0	1.6	141.8
9	Namysłów	29.2	17.2	7.7	76.1	10.3	4.8	145.3
10	Opole	182.5	5.0	216.0	368.4	44.7	13.9	830.5
11	Otmuchów	0.0	8.5	0.3	8.5	3.0	109.6	129.9
12	Paczków	18.2	2.5	1.0	23.7	2.9	4.2	52.5
13	Prudnik	11.5	15.2	4.5	43.5	7.5	7.9	90.1
14	Strzelce Opolskie	64.6	18.0	3.3	104.3	7.6	2.3	200.1
15	Ujazd	5.4	0.2	0.2	5.8	1.5	0.0	13.1

According to statistical data, Biała, Ujazd, and Strzelce Opolskie had the least amount of greenery (0.3, 0.4, and 0.8% of their respective areas), while Głuchołazy and Kluczbork had the highest amount of greenery (31 and 26%, respectively). The average percentage of land covered by greenery in all towns was 7.7%, with a national average of 34.4%. This clearly points to such land having a very small share in the type of towns under study.

The data presented above concerns public greenery, with a broad range of area types ignored: single-family housing estates, greenery in religious complexes and special-purpose greenery—forming parts of schools, hospitals, sports areas, etc., in addition to allotment gardens, which are considered productive greenery. I could not find statistical data for environmentally valuable greenery in the central areas of the towns under study. I also observed that castle parks (Ujazd or Otmuchów) were not included in the statistics.

After an analysis of existing landscaped greenery, the fact that the towns analysed are small and located in urban-rural municipalities appeared worth noting. According to data provided by the GUS for 2019, urbanised areas amounted to only 6.2% of the Opole Voivodeship. The remaining land was used for agricultural or is semi-natural (forests, river valleys, wetlands). As such, the demand for city or estate parks is low, and landscaped greenery had a more formal function—in the form of ornamental city or front gardens.

Thus, assessment of urban greenery potential for shaping systems of greenery accounted for the diversity of present forms, semi-natural areas accompanying waterways in the immediate vicinity of central zones and preserved post-fortress areas (landscaped and non-landscaped, accompanying fully and partially preserved wall fragments) (Table 7).

Table 7. Landscaped greenery.

	Municipal Unit	Rv	Cp	Sq	Cm	Al	Pfg	Wl	He	Ot	Total
1	Biała	x	x	x	x	x	x	x	x	x	9
2	Byczyna	x	x	x	x	x	x	x	x	x	9
3	Głogówek	x	x		x	x		x	x		6
4	Głubczyce	x	x	x	x	x	x	x	x	x	9
5	Głuchołazy	x	x	x	x	x			x	x	7
6	Grodków	x	Pfg	x	x	x	x	x	x		7
7	Kluczbork	x	x		x	x	x	x	x		7
8	Krapkowice	x	x	x	x	x	x	x	x	x	9
9	Namysłów	x	x	x	x	x	x	x	x		8
10	Opole	x		x	x	x			x		5
11	Otmuchów	x	x		x	x		x	x	x	7
12	Paczków	x	Pfg		x	x	x	x	x	x	7
13	Prudnik	x	x		x	x			x	x	6
14	Strzelce Opolskie	x	x		x	x			x	x	6
15	Ujazd	x	x		x	x			x	x	6

Rv–river valley and wetland greenery, Cp—city parks, Sq—green squares inside the city walls, Cm—cemeteries, Al—allotment gardens, Pfg—Post-fortification greenery, Wl—waste lands next to city walls, He—housing estate greenery, Ot—other (in Table 8: 9-7 = 2p, 6-4 = 1p, 3-1 = 0p).

Table 8. Cultural landscape potential.

		Environmental Diversity of a Site	Urban Layout Legibility	Retained Development Trajectories	Surviving City Fortifications	Legibility of Skyline Composition	Landscaped Greenery	Individual Qualities and Cultural Landscape Potential
1	Biała	2	2	2	1	2	2	11
2	Byczyna	0	2	2	2	1	2	9
3	Głogówek	1	2	1, except for the east	1	2	1	8
4	Głubczyce	1	1	0	1	1	2	6
5	Głuchołazy	1	2	0	0	0	2	5
6	Grodków	1	1	2	1	1	2	8
7	Kluczbork	1	1	1	1	0	2	6
8	Krapkowice	2	2	2	1	1	2	10
9	Namysłów	1	1	1	1	1	2	7
10	Opole	1	1	0	0	1	1	4
11	Otmuchów	2	1	2	1	2	2	10
12	Paczków	1	2	1	2	2	2	10
13	Prudnik	1	1	0	0	1	1	4
14	Strzelce Opol.	0	1	1	0	1	1	4
15	Ujazd	1	1	1	0	1	1	5

In the first group, I examined all towns and identified: cemeteries, housing estate greenery, allotment gardens and, in most cases, city parks. River valley and wetland greenery was also found to be present in all towns.

Post-fortification areas were converted into landscaped greenery in eight towns (e.g., Paczków, Krapkowice). In two cases, even the role of the main city parks was taken over by Planty (Post-fortification greenery). It should be noted that only one case featured a complete fortification ring (Paczków), while in others, the fragments that were not developed became wastelands, wild gardens, and car parks. A possibility of their partial restoration was determined in nine cases. Historical greenery included mainly cemeteries and parks. Insofar as cemeteries had a stable form and function, the parks formed a diverse group. They originated from residential parks—palatial and castle parks (Głogówek, Strzelce Opolskie), and were founded in areas of former city ponds (Głubczyce, Byczyna, Kluczbork) or areas in danger of flooding (Głogówek). Numerous small squares were also found (Byczyna, Biała, Krapkowice), and typically accompanied transport and circulation or filled in gaps in the urban fabric. All cities have riverside greenery and large areas of allotment gardens.

In Opole Voivodeship's towns varied terrain and a diversity of existing plants, both in terms of form, genesis, function, state of development, and environmental and cultural assets, offers many possibilities for creating greenery systems. The greater this diversity, the more advanced can a system become, which is conducive to biodiversity and ecological stability. In combination with improving visual attractiveness and cultural identity, it can become a basis for creating sustainable cities [7,82,83]. The concepts of green infrastructure or even urban ecological networks can also be applied here [84,85].

They can be based on existing landscaped greenery fragments—preserved areas formerly occupied by fortifications, parks and gardens, suburban forests and areas with high environmental potential—along rivers, wetlands, buffer greenery, municipal forests.

Other forms will also play an important role here—partially accessible greenery—allotment gardens, cemeteries, monastery and residential gardens, as well as home gardens.

A high potential for urban greenery systems was detected in 8 out of the 15 cases under study. It was fully ruled out in only two cases due to highly dense development structures.

Accounting for the rating indicators of the towns under study, it can be observed that the systems would become a combination of ring (post-fortress area greenery), wedge (parks, promenades, street greenery), and linear (riverside greenery) systems. American riverside parkways [6] appear to be a proper reference point, yet on a slightly different scale. This is facilitated by the region's complex hydrographic system and existing regional and local ecological corridors outlined along the Oder and Osobłoga rivers.

3.6. Cultural Landscape Potential

As above, the following indicators were of key significance to the contemporary state of cultural and environmental heritage: site-specific environmental determinants, urban layout, urban morphology, city fortifications, skylines, and greenery. Their assessment allowed us to identify cities with the greatest landscape potential (Table 8).

The towns with the highest scores were those which managed to preserve their historical landscape on the highest level. These included: Biała, Byczyna, Krapkowice, Otmuchów, and Paczków. The towns with the lowest evaluation of the studied natural and cultural potential were: Głucholazy, Opole, Prudnik, Strzelec, and Ujazd.

4. Discussion

At present, the Opole Voivodeship features numerous cases of architecture and urban layouts from the period of the region's most dynamic development—the fourteenth and nineteenth centuries. They include the previously mentioned urban layouts of medieval towns. As observed by Eysymontt in relation to the area known since the sixteenth century as Lower Silesia, these towns did not make significant contributions to the development of western urbanization, and were of little value in terms of innovation [56]. This is a perspective from arts history studies and is essentially accurate, if not detrimental in the context of public awareness as to the attractiveness of such structures in the public conscious. However, this does not detract from the values that they possess and which have been detected, nor from their regional significance. In terms of other Silesian towns, their medieval urban planning structure and its key elements may be considered as well-preserved. However, while the city layout itself is clear, the post-war housing tissue that filled specific city blocks in such places as Strzelce Opolskie or Głubczyce has contributed to the loss of the historic landscape of the city center.

I compared environmental, urban, architectural, and landscape factors that were and continue to be significant to the shape of these towns. What is notable, contemporary GIS tools and source materials made them relatively simple to analyze (historical maps, GIS, field studies).

Several problems concerning data and the possibility of its distortion were noted during the course of this study. The first and very basic involved obtaining the archival city plans. In order to conduct a comparative analysis of a large group of towns with similar origins, it was crucial to secure the plans which originated in the same period or had the same author. This allowed them to make proper evaluation of the state of preservation of urban layouts during a similar time period or eliminate errors made within some markings on those plans. In this particular research study, the city plans used as basic were those by Christian Friedrich von Wrede. The studies with the least possible error factor in evaluating the original conditions were the studies conducted using the numeric maps of terrain and topographic maps. Interlaying the archival plans with the contemporary ones, it is easier to analyze and compare the development directions of the respective areas.

While granting points in the evaluation, there is always a possibility of a subjective evaluation, which results from the researcher's experience and his or her body of knowledge. For instance, when studying the level of preservation of the urban tissue, what can become a key element of evaluation are the quality and percentage of the housing tissue or that of the preserved market block. What was important in conducting this study was the preserved layout of the original city blocks and its key elements (e.g., town hall, castle, church, fortifications). On the other hand, when studying the panoramic views, it is crucial to analyze the sightseeing points and identify such ones that would showcase both the historic panoramas as well as the panoramic views, which would be successful in exposing the best contemporary views. Statistical data turned out to be unhelpful in studying greenery. What constitutes an additional element of the evaluation in this case may be the size of the area, not only its diversity.

All towns received positive evaluations in terms of the preservation urban layout legibility and green areas. They scored averagely with regard to environmental diversity of a site, retained development trajectories, and quality of the panoramas. The lowest evaluation was given to the visibility of the remains of city walls. It has to be noted though that the lowest scoring towns in fact feature strongly developed urban planning systems. On the other hand, the highest scoring towns are of similar size (with the exception of Krapkowice). They also feature city walls preserved at high or average level and are attractively located (with the exception of Byczyna) which—considering their rather not intensively developed urban housing tissue—results in quite clear panoramas.

In the light of some changes, which have affected the factors analyzed in the study from the time the cities were established to the present day, these factors may be divided into three categories: durable, medium-durable, and volatile/variable. The first category includes natural environmental conditions whereas the second one comprises those related to architectural and urban state of development. The third category pertains to city greenery. Some of the irreversible processes, such as the post-war architectural solutions, reshaping, and the chaotic city expansion, may be partially eliminated through appropriately designed green infrastructure.

Environmental conditions are particularly important (the hydrographic network and terrain) as they had a considerable impact on the spatial form and functioning of cities and towns in the Middle Ages and over subsequent development periods. I did not observe direct correlations in this respect in only two cases. It is notable that the more difficult and complicated a given site's conditions were, the more difficult it was for spatial development to take place, which also resulted in a stable and sustainable urban development. At present, this was found to carry over to the survival of considerable reserves of natural green areas of high environmental potential. These large areas of healthy and functioning ecosystems with minimal intervention required are core areas of green infrastructure.

Meanwhile, Poland has been following one major tendency when it came to the maintenance of post-fortress areas. We could characterize them as conservative. In the projects undertaken towards the end of the nineteenth and the beginning of the twentieth century, the goal was to create representative and richly designed city parks or promenades (Kraków, Chemno, Brzeg, Paczków). The post-war approached concentrated primarily on works within the city walls themselves while greenery was an added but not necessary element. In order to continue the old tradition, cities built alleys or small parks (Prudnik) or squares (Krapkowice). Several municipalities opted for grass covered areas (Byczyna). This leads to an important question on the possibility to match proper exposition with a diverse attractiveness and aesthetics as well the functional aspects of these areas.

In turn, the latest experience in creating greenery systems based on the training of medieval and modern fortifications has, among others, Krakow, Poznan and Wrocław [86–88]. However, they are all large metropolitan cities, which makes their experience difficult to carry over to the Opole Voivodeship's small towns.

Despite a wide range of research on historical walls by ICOMOS Polska [39,40], I found no studies or guidelines concerning landscaped greenery that should be placed near such walls. Historical and climate conditions do not quite allow for applying vegetation solutions from abroad [89,90]. Accepted conservation doctrines concerning historical ruins, including the ruins of city walls, recommend keeping them in a state of so-called permanent ruin. It is assumed that that the best course of action is their conservation, proofing, and making them accessible to tourists [41]. In the case of ruins of significant buildings—such as castles—this is not difficult, but for more modest wall fragments, this can prove a considerable challenge. The Historical Ruins Preservation Charter, adopted by Resolution of the General Assembly of Members of PKN ICOMOS on 4 December 2012, states: 'the preservation of historical ruins should be comprehensive—it should cover surviving ruins, earthen forms (which are relics of historical fortifications), rubble layers and the landscape (of which the ruins are elements)' [91]. This is a rather enigmatic guideline, which, based on the principles of historical landscape design and the renovation of historical gardens, can provide a basis for proper action within the field of city walls.

However, it should be noted that Trochonowicz mentioned the following as being factors responsible for wall degradation: biotic factors caused by microorganisms, fungi, mould, and plants which appear spontaneously in such structures [92]. Scholars note that the ingrowing of the roots of trees that grow in the immediate vicinity of an area can cause mechanical damage. Shade cast on wall surfaces by tree canopies is conducive to damp build-up and the growth of algae and bryophytes [93]. Tall and old trees can also pose a threat during strong winds.

Comparative analyses performed during field studies in countries like Ireland, Italy (San Leone, Monteriggioni, Montalcino, Pisa), Croatia (Mototvun, Rovini, Pula), the Czech Republic (Spišský Hrad), along with the general conservation guidelines in this field, prepared as a part of the INTEREG Central Europe RUINS project [94,95], revealed two universal tendencies: preserving monuments in their most authentic form and exposing them in a landscape that is as close to historical as possible.

5. Conclusions

Cultural heritage plays a major role in the sustainable development of every region acting as a strengthening or even creative factor of the local identity. Studies showed that despite pejorative evaluation of the Opole region, there exists a large group of small-sized towns with promising landscape potential. Indicated values provide biodiversity and multiple ecosystem service benefit as a system of green infrastructure.

I demonstrated that site-specific determinants affected the contemporary assets of a place when studied from the perspective of the geography of tourism (tourism assets), the landscape (retention of historical skylines), and cultural heritage (a harmonious landscape). The place-based assets I isolated included:

- Environmentally valuable, diverse areas around towns;
- Retained development trajectories—as determinants of preserving the traditional landscape;
- Preserved and legible urban layouts;
- Entirely or partially preserved medieval city fortifications;
- Preserved historical skyline composition;
- Diverse landscaped and semi-natural greenery in the town's structure.

This set of features of medieval origin towns located in the Opole Voivodeship makes them an interesting complex of cultural heritage linked with environmental determinants. As for the towns and surviving remains of fortifications, it is necessary to create a system of urban greenery, and for the complex of cities—creating a system of towns with similar features. My study demonstrated that despite their regional character, they can define the identity and uniqueness of the area under analysis. Their potential should be utilized in new Voivodeship development strategies, including sustainable tourism. This is immensely significant due to the region's observable stagnation as well as economic and environmental problems.

Sustainability meant the balance between society and environment is, in this case, very important [96]. And the main elements on which it can be based in this region are the natural environment and cultural heritage. Green infrastructure networks should also be based on them, in which individual cities have great potential to create.

Sustainable development of the region should also focus on determining site-specific tourism potential, primarily the material, technical and economic infrastructure, and social assets of each site [66]. In the long term, this could facilitate the implementation of precepts of sustainable development in the area's economic activation, improving the state of its environment. Local communities should play a major role in this process as they will be main future beneficiaries of the achieved improvements [97,98].

Development of a green infrastructure network should begin with identification of environmental and cultural resources, which give the best idea of the character of the very place. In the cities embedded in their history, these resources shall be relatively easily identified, studied, and subjected to comparative analysis. It allows to emphasize the uniqueness of the network and because of it to increase its attractiveness, also in the context of its touristic appeal. As demonstrated in several examples, what remains crucial is the retaining by cities of their authenticity and historical character. Interestingly enough, in the case of the Opole Voivodeship, what contributed to preservation (in the studied cities) the elements carrying modern potential were natural and economic limitations. Another step which needs to be undertaken while creating a greenery system in the cities of medieval origins involves identification of the most important elements of the system's grid. Consideration should be given to the diversity of the existing areas (in terms of functions and ecosystems). It is important to combine the elements typical for historical cities (residential parks and cemeteries) with natural greenery (riverbeds, wetlands).

The most recognizable will be the historic center surrounded by the city wall remains and the accompanied greenery. The greenery of the linear character should be treated with a great deal of care because it is associated with specific architectural structures, archeological relics, and ecosystems.

With regard to the above, as well as the conducted studies, general guidelines were established for the purpose of designing the areas of medieval fortifications. These guidelines, which can also be applied in other cases, have been divided into three stages: investigation, design, and maintenance.

- Investigation: historical and urban research, visibility and exposition analyses, archaeological research, geological and soil morphology testing,
- Design: it should expose the monument, highlight its form, and enhance its values via a visual effect and new accompanying functions, new technical infrastructure in a form that does not compete with the monument. It is essential to adapt the site to persons from different age groups and persons with disabilities, while ensuring night-time use and exposition. Greenery should also mask technical elements and those of poor aesthetic quality; it can also restrict access while not visually interfering with the monument.
- Maintenance (as well as existing structures): maintenance and plant replacement programme, ongoing tree, bush and lawn maintenance, seasonal plant planting plan; reducing the effects of destruction by biotic entities, and the effects of natural succession.

Further research is necessary to formulate plant species selection guidelines for post-fortification areas. These guidelines would depend on the character and size of the fortifications' remains, soil structure, and identifying areas for restoration and which should cover not only the historical remains themselves, but also accompanying areas. The principles above should also apply to areas already used as public greenery.

Funding: This research received no external funding.

Institutional Review Board Statement: Not applicable.

Informed Consent Statement: Not applicable.

Data Availability Statement: Publicly available datasets were analyzed in this study. They are all shown in the 'References' section.

Conflicts of Interest: The authors declare no conflict of interest.

References

1. *Raport o Stanie Zachowania Zabytków Nieruchomych w Polsce, Zabytki Wpisane do Rejestru Zabytków (Księgi Rejestru A i C)*; Narodowy Instytut Dziedzictwa: Warszawa, Poland, 2017. Available online: https://nid.pl/pl/Wydawnictwa/inne%20wydawnictwa/RAPORT%20O%20STANIE%20ZACHOWANIA%20ZABYTK%C3%93W%20NIERUCHOMYCH.pdf (accessed on 10 January 2021).
2. Szmygin, B. Teoria i kryteria wartościowania dziedzictwa jako podstawa jego ochrony. *J. Herit. Conserv.* **2015**, *43*, 44–52.
3. Szmygin, B. (Ed.) *Wartościowanie Zabytków Architektury*; Polski Komitet Narodowy ICOMOS, Muzeum Pałac w Wilanowie; Muzeum Pałac w Wilanowie: Warszawa, Poland, 2013.
4. Jarocka, A. (Ed.) *Heritage Value Assessment Systems—The Problems and the Current State of Research*; Lublin University of Technology; Polish National Committee of the International Council on Monuments and Sites ICOMOS: Lublin–Warsaw, Poland, 2015.
5. Hodor, K. Zieleń i ogrody w krajobrazach miast. *Tech. Trans. Archit.* **2012**, *6*, 7–15.
6. Zachariasz, A. *Zieleń Jako Współczesny Czynnik Miastotwórczy ze Szczególnym Uwzględnieniem Roli Parków Publicznych*; Wydawnictwo Politechniki Krakowskiej: Kraków, Poland, 2006; Volume 103, pp. 28–29.
7. Ignatieva, M.; Stewart, G.H.; Meurk, C. Planning and design of ecological networks in urban areas. *Landsc. Ecol. Eng.* **2011**, *7*, 17–25. [CrossRef]
8. Rozmarynowska, K. *Ogrody Odchodzące ... ? Z Dziejów Gdańskiej Zieleni Publicznej 1708–1945*; Słowo/Obraz Terytoria: Gdańsk, Poland, 2011; pp. 94–97.
9. Forman, R.T.T. Some general principles of landscape and regional ecology. *Landsc. Ecol.* **1995**, *10*, 133–142. [CrossRef]
10. Cook, E.A. Landscape structure indices for assessing urban ecological networks. *Landsc. Urban Plan.* **2002**, *58*, 269–280. [CrossRef]
11. Ružička, M.; Mišovičová, R. The general and special principles in landscape ecology. *Ekológia* **2009**, *28*, 1–6. [CrossRef]
12. John, H.; Neubert, M.; Marrs, C. (Eds.) *Green Infrastucture Handbook—Conceptual and Theoretical Background, Terms and Definition*; Technische Universität Dresden: Dresden, Germany, 2019. Available online: https://www.interreg-central.eu/Content.Node/MaGICLandscapes-Green-Infrastructure-Handbook.pdf (accessed on 5 January 2021).
13. Kieß, W. *Urbanismus im Industriezeitalter. Von der Klassizistischen Stadt zur Garden City*; Erst & Son: Berlin, Germany, 1991; pp. 200–205.
14. Bińkowska, I. *Natura i Miasto. Publiczna Zieleń Miejska we Wrocławiu od Schyłku XVIII do Początku XX Wieku*; Muzeum Architektury, Wydawnictwo Via Nova: Wrocław, Poland, 2011.
15. Zachariasz, A. *Zielony Kraków Dla Przyjemności I Pożytku Szanownej Publiczności*; Wydawnictwo Politechniki Krakowskiej: Kraków, Poland, 2016.
16. Bogdanowski, J. *Architektura Obronna w Krajobrazie Polski. Od Biskupina do Westerplatte*; PWN: Warszawa, Poland, 1996.
17. Wielgus, K.; Środulska-Wilelgus, J.; Staniewska, A. *Krajobraz Warowny Polski: Procesy Rewaloryzacji i Percepcji: Próba Syntezy*; Wydawnictwo Politechniki Krakowskiej: Kraków, Poland, 2019.
18. Von Rohrscheidt, A.M. Wykorzystanie średniowiecznych obiektów obronnych w Polsce w ramach różnych form turystyki kulturowej. *Tur. Kult.* **2010**, *6*, 4–25.
19. Środulska-Wilelgus, J. *Rola Turystyki Kulturowej w Ochronie i Udostępnieniu Krajobrazu Warownego*; Wydawnictwo Politechniki Krakowskiej: Kraków, Poland, 2016.
20. Adamska, M.E. Zabytkowe planty miejskie w Brzegu. *Przestrz. Urban. Archit.* **2015**, *1*, 45–58. Available online: https://www.ejournals.eu/PUA/2015/Volume-1/art/9542/ (accessed on 21 May 2021).
21. Adamska, W.E. The historic city parks of Opole Silesia. *Tech. Trans. Archit.* **2014**, *5-A*, 237–254.
22. Kimic, K. The location and role of old walls as green belts in shaping the landscape of towns in the 19th century. *Tech. Trans. Archit.* **2016**, *1*, 143–166. [CrossRef]
23. Hodor, K.; Fekete, A.; Matusik, A. The future of Planty Park in Cracow compared to other examples of city walls being transformed into urban parks. In Proceedings of the ECLAS Conference, Ghent, Belgium, 9–12 September 2018; Delarue, S., Dufour, R., Eds.; University College Ghent—School of Arts—Landscape & Garden Architecture and Landscape Development: Ghent, Belgium, 2018; pp. 411–416.
24. Bimler, K. *Die Schlesischen Massiven Wehrbauten, Bd. 1–5*; Komission Heydebrand-Verlag: Breslau, Poland, 1940–1944.
25. Przyłęcki, M. *Budowle i Zespoły Obronne na Śląsku*; Towarzystwo Opieki nad Zabytkami: Warszawa, Poland, 1998.
26. Thullie, C. *Zabytki Architektoniczne Ziemi Śląskiej na tle Rozwoju Architektury w Polsce*; Wydawnictwo "Śląsk": Katowice, Poland, 1965.
27. Chrzanowski, T.; Kornecki, M. *Sztuka Śląska Opolskiego, od Średniowiecza do Końca w. XIX*; Wydawnictwo Literackie: Kraków, Poland, 1974.
28. Adamska, M.E. *Transformacie Rynków Średniowiecznych Miast Śląska Opolskiego od XIII Wieku do Czasów Współczesnych. Przerwane Tradycje, Zachowane Dziedzictwo, Nowe Narracje*; Politechnika Opolska: Opole, Poland, 2019.
29. Piechaczek, B. Średniowieczne Kamienne Obwarowania Miast Opolszczyzny do Końca XV Wieku. Ph.D. Thesis, Katolicki Uniwersytet Lubelski Jana Pawła II, Lublin, Poland, 2006.

30. Przybyłok, A. Mury Miejskie na Górnym Śląsku w Późnym Średniowieczu. Ph.D. Thesis, Uniwersytet Łódzki, Łódź, Poland, 2014.
31. Legendziewicz, A. Selected city gates in Silesia—Research issues. *Technol. Trans.* **2019**, *3*, 71–103. [CrossRef]
32. Legendziewicz, A. Brama Grobnicka (zwana Klasztorną) w Głubczycach w świetle badań architektonicznych i ikonograficznych. In *Opolski Informator Konserwatorski 2015*; Wojewódzki Urząd Ochrony Zabytków w Opolu: Opole, Poland, 2015; pp. 103–112.
33. Legendziewicz, A. Architektura bram miejskich Głogówka od XIV do XX wieku. *Kwart. Opol.* **2018**, *4*, 105–120.
34. Legendziewicz, A. Mury miejskie Kluczborka—od średniowiecza do współczesności. In *Opolski Informator Konserwatorski 2015*; Wojewódzki Urząd Ochrony Zabytków w Opolu: Opole, Poland, 2015; pp. 113–121.
35. Legendziewicz, A. Badania architektury Grodkowa. Średniowieczne fortyfikacje miejskie. In *Z dziejów Grodkowa i ziemi grodkowskiej, p. 2*; Urząd Miejski w Grodkowie: Grodków, Poland, 2015; pp. 67–88.
36. Wilkaniec, A.; Wichrowski, M. (Eds.) *Fortyfikacje w Przestrzeni Miasta*; Uniwersytet Przyrodniczy w Poznaniu: Poznań, Poland, 2006.
37. Przyłęcki, M. *Miejskie Fortyfikacje Średniowieczne na Dolnym Śląsku. Ochrona, Konserwacja, Ekspozycja 1850–1980*; PP PKŻ: Warszawa, Poland, 1987.
38. Adamska, M.E. Średniowieczne układy urbanistyczne miast Śląska Opolskiego—Stan zachowania i rewitalizacja. *Przegląd Bud.* **2013**, *3*, 15–19.
39. Series of Publications Issued by PKN ICOMOS. Available online: http://www.icomos-poland.org/pl/publikacje.html (accessed on 15 March 2021).
40. Górski, A. (Ed.) Obwarowania miast. Problematyka ochrony, konserwacji, adaptacji i ekspozycji. In Proceedings of the International Scientific Conference, Kożuchów, Poland, 28–30 April 2010; Towarzystwo Przyjaciół Ziemi Kożuchowskiej: Kożuchów, Poland, 2010.
41. *Transnational Model of Sustainable Protection and Conversation of Historic Ruins: Best Practices Handbook*; Lublin University Of Technology: Lublin, Poland, 2020.
42. Cydzik, J. Przebudowa zabytkowego ośrodka staromiejskiego na przykładzie miasta Paczkowa. *Ochr. Zabytk.* **1961**, *14*, 75–92.
43. Steinborn, B. *Otmuchów i Paczków*; Zakład Narodowy im. Ossolińskich: Wrocław, Poland, 1961.
44. Dziewulski, W.; Hawranek, F. *Opole Monografia Miasta*; Instytut Śląski w Opolu: Opole, Poland, 1992.
45. Eysmonty, R.; Goliński, M. (Eds.) Namysłów. In *Atlas Historycznych Miast Polskich, t IV, Śląsk, Zeszyt 11*; Instytut Archeologii i Etnologii PAN: Wrocław, Poland, 2015.
46. Borkowski, M.; Czapliński, M.P. *Biała—Historia i Współczesność*; Instytut Śląski: Opole-Biała, Poland, 2011.
47. Meisner, J. (Ed.) *Byczyny Przeszłość i Dzień Dzisiejszy*; Instytut Śląski w Opolu: Opole, Poland, 1988.
48. Chrzanowski, T. *Głogówek*; Zakład Narodowy im. Ossolińskich: Wrocław, Poland, 1977.
49. Adamska, M. Rewaloryzacja starego miasta w Głubczycach. *Zesz. Nauk. Politech. Opol. Ser. Bud.* **1998**, *43*, 132–149.
50. Cimała, B. *Kluczbork-Dzieje Miasta*; Instytut Śląski w Opolu: Opole, Poland, 1992.
51. Chrząszcz, J. *Historia Miasta Krapkowice na Górnym Śląsku od 1914*; Sady: Krapkowice, Poland, 2011.
52. Heffner, K. *Ujazd—historia i Współczesność*; Instytut Śląski w Opolu: Opole, Poland, 1990.
53. Inwentarz Zespołu PP PKZ Oddział Wrocław, Zespół Pracowni Dokumentacji Naukowo-Historycznej PP PKZ. Available online: https://www.nid.pl/pl/Regiony/Dolnoslaskie/Wykaz%20dokumentacji/01_PP%20PKZ%20Wroc%C5%82aw%20Pracownia%20Dokumentacji%20Naukowo-Historycznej.pdf (accessed on 8 June 2021).
54. Dziennik Ustaw, Poz. 1240, Rozporządzenie Prezydenta Rzeczypospolitej Polskiej z Dnia 22 Października 2012 r. w Sprawie Uznania za Pomnik Historii Paczków—Zespół Staromiejski ze Średniowiecznym Systemem Fortyfikacji. Available online: https://nid.pl/pl/Informacje_ogolne/Zabytki_w_Polsce/Pomniki_historii/Lista_miejsc/Paczk%C3%B3w_rozporz%C4%85dzeenie.pdf (accessed on 15 October 2020).
55. Pudełko, J. *Zagadnienia Wielkości Powierzchni Średniowiecznych Miast Śląskich*; Zakład Narodowy im. Ossolińskich: Wrocław, Poland, 1967.
56. Eysymontt, R. *Kod Genetyczny Miasta. Średniowieczne Miasta Lokacyjne Dolnego Śląska na tle Urbanistyki Europejskiej*; Via Nova: Wrocław, Poland, 2009.
57. Bahlcke, J. Górny Śląsk—studium przypadku powstania: Regionów historycznych, wyobrażeń o obszarach kulturowych, historiograficznych koncepcji przestrzeni. In *Historia Górnego Śląska. Polityka, Gospodarka i Kultura Europejskiego Region*; Bahlcke, J., Gawrecki, D., Kaczmarek, R., Eds.; Dom Współpracy Polsko-Niemieckiej: Gliwice, Poland, 2011; pp. 17–37.
58. Główny Urząd Statystyczny (GUS)—Central Statistical Office of Poland; Local Data Bank. Available online: https://bdl.stat.gov.pl/BDL/start (accessed on 11 October 2020).
59. Raport o Stanie Środowiska Województwa Opolskiego 2020. Główny Inspektorat Ochrony Środowiska, Departament Monitoringu Środowiska, Regionalny Wydział Monitoringu Środowiska w Opolu: Opole, Poland, 2020. Available online: http://www.gios.gov.pl/images/dokumenty/pms/raporty/stan_srodowiska_2020_opolskie.pdf (accessed on 24 January 2021).
60. Program Ochrony Środowiska Województwa Opolskiego na Lata 2012–2015 z Perspektywą do Roku 2019. Available online: https://www.opolskie.pl/program-ochrony-srodowiska/ (accessed on 24 January 2021).
61. Sustainable Development Report 2020. Available online: https://dashboards.sdgindex.org/rankings (accessed on 24 January 2021).
62. Kapera, I. *Rozwój Zrównoważony Turystyki. Problemy Przyrodnicze, Społeczne i Gospodarcze na Przykładzie Polski*; Oficyna Wydawnicza AFM: Kraków, Poland, 2018.

63. Bąk, I.; Matlegiewicz, M. Przestrzenne zróżnicowanie atrakcyjności turystycznej w Polsce w 2008 roku. In *Zeszyty Naukowe Uniwersytetu Szczecińskiego. Ekonomiczne Problemy Usług 2010, 52 Potencjał Turystyczny. Zagadnienia Przestrzenne*; Wydawnictwo Naukowe Uniwersytetu Szczecińskiego: Szczecin, Poland, 2012; pp. 57–67.
64. Wiśniewska, A. Ocena rozwoju turystyki zrównoważonej w polskich regionach. *Gospodarka w Praktyce i Teorii* **2017**, *49*, 83–92. [CrossRef]
65. von Rohscheidt, A.M. *Turystyka Kulturowa. Fenomen, Potencjał, Perspektywa*, 3rd ed.; GWSHM: Gniezno, Poland, 2016.
66. Potocka, I. Atrakcyjność turystyczna i metody jej identyfikacji. In *Uwarunkowania i Plany Rozwoju Turystyki, Tom III, Walory i Atrakcje Turystyczne, Potencjał Turystyczny, Plany Rozwoju Turystyki*; Młynarczyk, S., Zajajdacz, A., Eds.; Wydawnictwo Naukowe Uniwersytetu im. Adama Mickiewicza w Poznaniu: Poznań, Poland, 2009; pp. 19–31.
67. Brezovec, T.; Bruce, D. Tourism Development: Issues for Historic Walled Towns. *Management* **2009**, *4*, 101–114.
68. Tülek, B.; Atik, M. Walled towns as defensive cultural landscapes: A case study of Alanya—A walled town in Turkey. *WIT Trans. Environ.* **2014**, *143*, 231–242. [CrossRef]
69. Analizy Statystyczne. *Ochrona Środowiska 2020/Statistical Analyses. Environment 2020*; Główny Urząd Statystyczny/Statistics: Warszawa, Poland, 2020. Available online: https://stat.gov.pl/obszary-tematyczne/srodowisko-energia/srodowisko/ochrona-srodowiska-2020,1,21.html (accessed on 25 January 2021).
70. Von Wrede, C.F. *Kriegs-Karte von Schlesien, 1747–1753*; Staatsbibliothek zu Berlin, Kartenabteilung, Sygn. 15060; Lengenfelder, H (Verlag): München, Germany, 1992.
71. Wernher, F.B. Silesia in Compendio seu Topographia das ist Praesentatio und Beschreibung des Herzogthums Schlesiens [...] Pars I. 1750–1800, BUWr Oddział Rękopisów. Available online: https://www.bibliotekacyfrowa.pl/dlibra/publication/8092/edition/15414 (accessed on 15 October 2020).
72. Wernher, F.B. Scenographia Urbium Silesiae, Nürnberg, 1737–1752, BUWr Oddział Rękopisów. Available online: https://bibliotekacyfrowa.pl/publication/3022 (accessed on 15 October 2020).
73. Sowińska-Świerkosz, B.N.; Michalik-Śnieżek, M. The Methodology of Landscape Quality (LQ) Indicators Analysis Based on Remote Sensing Data: Polish National Parks Case Study. *Sustainability* **2020**, *12*, 2810. [CrossRef]
74. Sowinska-Swierkosz, B.N.; Chmielewski, T.J. A new approach to the identification of Landscape Quality Objectives (LQOs) as a set of indicators. *J. Environ. Manag.* **2016**, *184*, 596–608. [CrossRef]
75. Myczkowski, Z. Kryteria waloryzacji krajobrazów Polski—propozycja systematyki. In *Identyfikacja i Waloryzacja Krajobrazów—Wdrażanie Europejskiej Konwencji Krajobrazowej*; Generalna Dyrekcja Ochrony Środowiska: Warszawa, Poland, 2020; pp. 64–73.
76. Fornal-Pieniak, B.; Żarska, B. Metody waloryzacji krajobrazowej na potrzeby turystyki i rekreacji. *Acta Sci. Pol.* **2014**, *13*, 3–9.
77. Czarnecki, W. *Planowanie Miast i Osiedli, T.I. Wiadomości Ogólne. Planowanie Przestrzenne*; PWN: Warszawa, Poland, 1965.
78. Jedwab, R.; Johnson, N.D.; Koyama, M. Medieval cities through the lens of urban economics. *Reg. Sci. Urban Econ.* **2020**. [CrossRef]
79. Syperka, J. Dzieje gospodarcze Górnego Śląska w średniowieczu. In *Historia Górnego Śląska. Polityka, Gospodarka i Kultura Europejskiego Regionu*; Bahlcke, J., Gawrecki, D., Kaczmarek, R., Eds.; Dom Współpracy Polsko-Niemieckiej: Gliwice, Poland, 2011; pp. 295–308.
80. Bogucka, M.; Samsonowicz, H. *Dzieje Miast i Mieszczaństwa w Polsce Przedrozbiorowej*; Zakład Narodowy Imienia Ossolińskich Wydawnictwo: Wrocław, Poland, 1989; pp. 45–104.
81. Malik, R. Czeladź. Ze studiów nad budową i kształtem miasta średniowiecznego. *Kwart. Arch. Urban.* **2010**, *55*, 3–12.
82. Guzman, P. Assessing the sustainable development of the historic urban landscape through local indicators. *Lessons from a Mexican World Heritage City. J. Cult. Herit.* **2020**, *46*, 320–327.
83. Christopher Tweed, Margaret Sutherland, Built cultural heritage and sustainable urban development. *Landsc. Urban Plan.* **2007**, *83*, 62–69. [CrossRef]
84. Aminzadeh, B.; Khansefid, M. A case study of urban ecological networks and a sustainablecity: Tehran's metropolitan area. *Urban Ecosyst.* **2010**, *13*, 23–36. [CrossRef]
85. Balestrieri, M.; Ganciu, A. Greenways and Ecological Networks: Concepts, Differences, Similarities. *Agric. Res. Technol.* **2017**, *12*, 0011–0013. [CrossRef]
86. Środulska-Wielgus, J.; Staniewska, A.; Łakomy, K.; Wielgus, K. The historical elements of Kraków's green system—protection challenges for landscape planning in the 21st century. In Proceedings of the 5th Fábos Conference on Landscape and Greenway Planning, Budapest, Hungary, 1 July 2016; Jombach, S., Ed.; Szent István University, Department of Landscape Planning and Regional Development: Budapest, Hungary, 2016; pp. 439–446.
87. Raszeja, E.; Gałecka-Drozda, A. Współczesne interpretacje poznańskiego systemu zieleni w kontekście strategii miasta zrównoważonego. *Stud. Miej.* **2015**, *19*, 75–86.
88. Niedźwiecka-Filipiak, I.; Rubaszek, J.; Potyrała, J.; Filipiak, P. The Method of Planning Green Infrastructure System with the Use of Landscape-Functional Units (Method LaFU) and its Implementation in the Wrocław Functional Area (Poland). *Sustainability* **2019**, *11*, 394. [CrossRef]
89. Xing, Y.; Brimblecombe, P. Traffic-derived noise, air pollution and urban park design. *J. Urban Des.* **2020**, *25*, 590–606. [CrossRef]
90. Wang, S.; Jiang, Y.; Xu, Y.; Zhang, L.; Li, X.; Zhu, L. Sustainability of Historical Heritage: The Conservation of the Xi'an City Wall. *Sustainability* **2019**, *11*, 740. [CrossRef]

91. Karta Ochrony Historycznych Ruin Przyjęta Uchwałą Walnego Zgromadzenia Członków PKN ICOMOS w Dniu 4 Grudnia 2012 r. Available online: http://www.icomos-poland.org/pl/dokumenty/uchwaly/130-karta-ochrony-historycznych-ruin.html (accessed on 29 January 2021).
92. Trochonowicz, M. Obiekty murowe w ruinie. Wpływ czynników degradujących na ich zachowanie. In *Ochrona, Konserwacja i Adaptacja Zabytkowych Murów: Trwała Ruina II: Problemy Utrzymania i Adaptacji*; Szmygin, B., Ed.; Lubelskie Towarzystwo Naukowe, Polski Komitet Narodowy ICOMOS; Politechnika Lubelska: Lublin, Poland, 2010; pp. 172–184.
93. Trochonowicz, M.; Klimek, B.; Lisiecki, D. Biological corrosion and vegetation in the aspect of permanent ruin. *Bud. Archit.* **2018**, *17*, 17–26. [CrossRef]
94. Report on Current State-of-Art of Use and Re-Use of Medieval Ruins; Interreg Central Europa, Version 2, 12/2017. Available online: https://www.interreg-central.eu/Content.Node/RUINS/D.T2.1.1---Report-on-the-current-state-of-art-on-contempo-1-pdf (accessed on 10 February 2021).
95. Ashworth, G.J.; Bruce, D.M. Town Walls, Walled Towns and Tourism: Paradoxes and paradigms. *J. Herit. Tour.* **2009**, *4*, 299–314. [CrossRef]
96. Transforming Our World. The 2030 Agenda for Sustainable Development. Available online: https://sdgs.un.org/2030agenda (accessed on 21 May 2020).
97. Walter, G.R. Economics, Ecology-Based Communities, and Sustainability. *Ecol. Econ.* **2002**, *42*, 81–87. [CrossRef]
98. Hampton, M.P. Heritage, local communities and economic development. *Ann. Tour. Res.* **2005**, *32*, 735–759. [CrossRef]

Article

Assessment of Visual Values as a Tool Supporting the Design Decisions of the Cultural Park Protection Plan. The Case of Kazimierz and Stradom in Kraków

Urszula Forczek-Brataniec

Chair of Landscape Architecture, Faculty of Architecture, Cracow University of Technology (CUT), Warszawska 24, 31-155 Kraków, Poland; uforczek-brataniec@pk.edu.pl

Abstract: Krakow is a city of high landscape values, which has found confirmation in the entry onto the UNESCO heritage list. Its cultural landscape requires protection and clarification within the context of intensive tourist use and a rapid pace of urban spatial development. For preservation protection and restoration of landscape values, the city authorities undertook work on the creation of a Cultural Park in the Stradom and Kazimierz districts, providing a comprehensive, sustainable, and multidisciplinary approach to natural, cultural and visual values of the urban structure. The article presents the application of the method of research on visual values in order to protect individual scenic resources of the historical urban structure. It is one of the analytical studies of a comprehensive protection plan project. This project defines the scope, framework and methods of development and management of a Cultural Park. The task of the visual analysis was to identify, characterize and evaluate the visual resources. It created a visual framework for further development of the historical district while preserving its local spatial identity. The studies resulted in a division into zones according to their nature and intensity of activities as well as outlining protection zones and intervention zones adjusted to individual characteristics of those places. An original method combining achievements of the method of landscape and visual assessment (LVIA) as well as achievements of the Krakow School of Landscape Architecture (KSLA) in terms of cultural landscape assessment was used for the research. The applied method provided guidelines to support sustainable project decisions regarding further development of the district for the preservation of local spatial identity. Its universal character creates possibilities for its application into the plans of other Krakow districts and is intended to be applicable to both urban and rural structures.

Keywords: heritage visual value; urban heritage protection; urban spatial structure; visual analysis; visual landscape

Citation: Forczek-Brataniec, U. Assessment of Visual Values as a Tool Supporting the Design Decisions of the Cultural Park Protection Plan. The Case of Kazimierz and Stradom in Kraków. *Sustainability* **2021**, *13*, 6990. https://doi.org/10.3390/su13136990

Academic Editor: Carmela Cucuzzella

Received: 30 April 2021
Accepted: 17 June 2021
Published: 22 June 2021

1. Introduction

Currently, historical and old city centres and districts remain influenced by a high number of factors impacting their space. Such issues as historical substance, intensive use and the necessity of adjusting to contemporary needs require a proper approach to design and management. At the same time, urban development outside the historical centre might impact their shape through borrowed views. In this context, townscape calls for multithreaded protection which allows for preserving the highest values while adjusting to the changing conditions of their use and needs related to the functioning of a modern city.

These problems are compounded in cities of significant historical value. They are mentioned and discussed in declarations and recommendations of UNESCO, whose approach to the city landscape is evolving [1]. On the basis of this evolving approach to UNESCO heritage in the context of Polish legal conditions, as early as 2003, a formula was created for the protection of the cultural landscape in the form of the Cultural Park. It assured a comprehensive approach to historic areas that requires taking into account natural, cultural and visual aspects.

A Cultural Park is a project that facilitates comprehensive protection of city landscape [2], a legally established form of integrated and area protection, covering a wide context. It is legally regulated by the Act on the Protection and Care of Monuments of 3 July 2003 [3].

Protection objectives are clearly defined in Art.16 point 1 of this Act that states the following: *the city council after having been advised by the provincial conservation officer, based on the Law, can create a Cultural Park for protection of cultural landscape and preservation of distinguished landscape areas containing immovable historical buildings characteristic for local building and settlement tradition. This Cultural Park is to serve for integrated protection of landscape urban, architectural, and natural values of the place while taking into account the needs for sustainable development and preservation of the character of functioning residential district while maintaining good life conditions and cultural specificity* [4].

The formula of a Cultural Park imposes the obligation to prepare a protection plan for its area and then to prepare a local spatial development plan for the area of a Cultural Park, which becomes an act of local law. The protection plan serves the realization of the protection objectives of the Cultural Park. The protection plan is a flexible formula that allows the plan scope to be adapted to the specificity of its resources. The plan includes the identification of the features of the cultural landscape, natural and social conditions, and the analysis of viewing conditions. An in-depth analysis of these issues leads to synthesis and further steps to develop and implement a Cultural Park protection plan.

This article presents one of the threads of an analysis which is a study of visual values. It presents the application of the research on visual resources on the basis of which the scenic values were identified and characterized. The purpose of the study was to outline a spatial framework of the development of cultural and natural substance thus allowing for preservation, protection or restoration of its most precious exposure. The visual values have become one of the components of the integrated protection of the cultural landscape.

The basis for the studies constituted guidelines on the preparation of protection plans, in which the survey of scenic resources is, apart from studies related to cultural and natural resources, an integral part of the Cultural Park protection plan project.

In 2000 landscape was defined as *an area, as perceived by people, whose character is the result of the action and interaction of natural and/or human factors* by the European Landscape Convention (ELC) [5]. This definition highlights the perceptive aspect which has become equivalent with natural and cultural resources while giving it its rightful place in managing landscape resources. The idea of a Cultural Park follows this approach to landscape. This holistic form allows us to develop a protection plan for the preservation of the most precious cultural and natural values in the process of further spatial planning. This form incorporates the idea of balanced integrated protection while taking into account cultural, natural and landscape and visual aspects. It uses the experience of the system of planning of protected areas, which has proved to be very successful [6,7].

2. State of the Art

The issue of the city view has a rich history of studies starting from works by K. Lynch [8] who discovered and endorsed visual perception of the city in planning, and such authors like G. Cullen [9] or J. A. Jackle [10] as well as S. Bell [11] who limited landscape perception to the alphabet of spatial forms. This line is continued in contemporary studies where special attention is paid to spatial analyses of Warsaw within the context of the language of patterns [10], an analysis of the curve of impressions evoked by contemporary urban space [12] which refers to the method by T. Wejchert [13,14]. The issues of identifying and assessing the city's views are developed by English Heritage, and the publication "Seeing the history on the View. A method of assessing heritage significance within Views" presents an approach that can be applied to any view that is significant in terms of its heritage values [15].

An important modern aspect of visual studies is the influence of the visual quality of landscape on the quality of life [16]. They are related to the approach to landscape initiated by Humbolt—landscape as a living environment [17]. Currently, humanistic geography, sociology and humanistic archeology continue this research by studying the relationships between man and the inhabited environment. The issues of the relationship between humans and space raised by Yi-Fu Tuana [18] are explored and developed in relation to the role of society in shaping space [19]. An important issue is also the question of the relationship between historical heritage and the way it is understood and perceived. In this respect, urban geography developed by David Lowenthal and his followers draws our attention to contemporary processes and problems expressed in the form of space [20,21] and the relationship between its community and the past [22].

Historical city centers struggle with numerous problems resulting from the condensation of sociological and spatial issues [7]. One of the most important dilemmas is the intensive development of city outskirts with a simultaneous decline in the number of permanent residents in historical centers [23]. One of the manifestations of this process is "gentrification", loss of inhabitants and, consequently, loss of the identity of the place [2]. When the focus on profit and tourist exploitation is added to this, the situation gives rise to a high number of negative effects manifested in the transformation of space. As a result, they undergo processes that pose a threat to both tangible and intangible values [2].

The effects of extensive exploitation take on different forms. These are both attempts to increase real property development intensity and introducing spatial objects that deform the urban structure [24] and accumulation of secondary elements related to use and deforming the city space defined as visual pollution [25].

In this context, UNESCO's declarations mean great opportunities in implementing protection measures based on increasing awareness of the local community and participation of residents in protection [2,22,26]. In the face of changes, a view can constitute a permanent base for creating and maintaining a bond with space. Particularly in the face of the changing reality, Kobayashi [27] demonstrated the significance of cultural forms in the landscape while Cosgrowe emphasized the importance of visual perception both in the creation and interpretation of landscapes [28]. Technical sciences and architectural and landscape approaches to shaping space come out against the humanities. As one of the elements of an integrated approach to heritage, they deal with the protection and shaping of a form that constitutes a cultural resource in the form of a place and the surrounding space. Their role is emphasized by Ti Fu Tuan, pointing to the need for a "creative and artistic" approach to space, which eludes "computational–statistical" methods [18]. The processes taking place in historic urban spaces are reflected in the evolution of documents related to the UNESCO heritage. Particularly noteworthy is the 1962 Declaration on the Beauty and Character of Landscapes and Places [29], the ICOMOS Xi'an declaration of 2005 emphasizing the importance of the building's surroundings [30], including visual interaction with the surrounding landscape. Contemporary threats and conservation objectives are clearly defined in the UNESCO Recommendations of 2011 [2]. They emphasize the definition of cultural values, including visual values, the need for integrated protection and the threats resulting from the rapid development and fragmentation of cities. On the other hand, the issue of the protection of the wide landscape as a component perceived of natural and cultural origins was extended in the ICOMOS Florence Declaration of 2014 [31], drawing attention to the need for creating a sustainable landscape of contemporary cities. Visual values play an important role in the definitions of subsequent UNESCO and ICOMOS documents while the existence of a protection zone in the form of a buffer zone plays not only a protective but also an educational role, highlighting the importance of the context as well as the visual connections between the site and space. The need to study the viewing context of cultural heritage meets the methods developed in the context of the protection of the open landscape.

They constitute the contemporary trend developed within the method which referred to the assumptions of the British Landscape Character Assessment (LCA) and the Landscape and Visual Impact Assessments (LVIA) conducted for the purposes of environmental impact assessments (EIA). They originate from visual quality studies and new investments impacting open landscape [6,32]. However, the main methodical scheme finds application in various landscape visual studies, which Gobster, Ribe and Palmer stress in an overview for application of this method [33]. Furthermore, the authors of the latest issue of the instruction for defining visual impact on landscape Guidelines for Landscape and Visual Impact Assessment (GLVIA) clearly indicate its significant contribution into studies of urban space while articulating the need for its application and the need to conduct new studies that would integrate the achievements in this respect [34]. The implementation of these methods has many uses [33].

One of them is an analysis of the impact of new investments on the urban structure. Due to the nature of the studies and application of modern digital tools, they impact both significant development of analysis methods and updating the guidelines adjusted to new possibilities. What deserves particular attention here is the development of the methods for studying viewsheds [35,36], the development of guidelines for regulating visualisations [37] as well as the development of the methods related to the use of digital analytical tools for studying urban and suburban zones [38,39].

At the same time, it applies the methods related to protected areas. The concept of monitoring points applied here is based on the effectiveness of the method called Red Flag Points (RFP) by R.B. Litton [40] and Palmer's [41] method of Key Observation Points (KOP) developed based on the previous method. On this basis, a network of monitoring points of urban space was created.

The Krakow School of Landscape Architecture (KSLA) reaches for early methods of landscape analysis. While it is based on analyses of architectural forms and their exposures [42], it also built its own approach which was later developed and improved through new experiences and challenges [43]. Their synthesis constitutes a comprehensive method by Professor Bogdanowski based on the comprehensive approach to landscape issues [44]. The starting point is the landscape form which constitutes the basis for landscape division into units or architectural and landscape enclosures depending on the analysis scale. This division allows for creating landscape separations which are later assessed and then zoned out. Toned down urban enclosures of WAK (architectural and landscape enclosures) at the outset define the spatial form as the framework for further studies in terms of the analysis and assessment of cultural resources [45]. The enclosure method was created to analyse cultural landscapes with clear-cut historical values. It focuses its attention on substance values and composition values. This approach became the focus for the development of this form of landscape cultural protection, i.e., the Cultural Park [46]. While it is supplemented with natural threads, it represents an integrated approach to urban space thus becoming the basis for planning in the spirit of sustained development. Here, the sustainable aspect refers to landscape management both with reference to its substance and the form perceived, which, as a result, leads to recognizing the view as a resource that is the subject of proper sustained management [47].

Landscape visual values, which constitute the value of a given place, require a specific approach so that this value would not be degraded. This approach will allow their preservation in the course of further development and use. In this case, the visual analysis takes into account individual landscape resources, including historical resources and spatial dependencies so as to create a framework for further development while being aware of their values. It aims at preserving the most precious views and spatial relations as well as finding space for activities responding to the needs of a modern city.

The approach that combines the achievements of *visual assessment* [34] and the Krakow School of Landscape Architecture (KSLA) allows for using and integrating achievements and experiences in this respect, which proved especially valuable and beneficial in historical urban space.

Expert studies related to visual aspects are conducted in a number of scientific centres. Most of them concern selective issues related to the location of particular objects in urban space [48] while focusing on their impact on the composition and analysing the possibilities for further development [49]. Special attention should be devoted to the studies of possibilities of the location of high-rise buildings [50] in the context of city panoramas as well as extensive research of visual absorbency [51]. At the same time, there are studies of historical substance and its perception [52].

The method presented in the article brings together the issues of valuation of cultural landscape specific to the Krakow School of Landscape Architecture with studies of spatial dependencies conducted with the use of contemporary methods in the field of *visual assessment* [34]. It refers to spatial substance and its exposure. While based on the method of division into landscape enclosures [43,44] it indicates their visual characteristics due to active and passive exposure [6]. In terms of active exposure, it determines the content of visual points and visual paths in space. In terms of passive exposure, it characterises urban structure with reference to its historical quality and composition. This approach allows us to bring out the individual spatial character of every enclosure as well as assess its visual value and conduct proper zoning out due to landscape protection. In the case of Stradom and Kazimierz, more significant criteria were added, namely the view pollution index. This criterion is related to the build-up of elements disturbing the perception of space as a result of negligence and lack of regulations in the use of space. Defining this factor was essential for the classification of interventions in particular zones. Enriching the KSLA method with *visual assessment* tools allowed for developing an individual method that makes it possible to identify visual values, their characteristics and assessment in order to then separate zones with specific guidelines. The study conducted in this way solidified the need to protect exposure and constituted the basis for the application of the guidelines in further planning and management of space.

3. Materials and Methods

The study area is located in Krakow, an early Medieval city located in southern Poland. The urban structure of the city is a layered system. It is composed of districts of different origins and times of creation [53]. The study areas of Stradom and Kazimierz are Medieval districts with preserved urban layout and substance in the form of historical objects. These include sacral buildings dating back to the 14th century, public buildings as well as housing. In the 19th century, the Medieval city walls were torn down while the old moats, as well as the Vistula riverbed, were filled up due to sanitary reasons and because of merging adjacent districts in the city [54]. It was then that new streets were marked out and the urban tissue was replenished. The Second World War affected the district greatly due to mass resettlements of the Jewish community from Kazimierz to the nearby ghetto. After the Second World War, this area gradually degraded, and underwent reconstruction and reinvestment. The significant redevelopment was connected with the winding-up of 19th-century industrial plants and replenishing real property development. As a result of many phases of development, the subject study area presents a complex spatial structure of different origins and different states of preservation [55]. It stands out in terms of picturesqueness and boasts an intriguing compositional arrangement built on valuable visual connections. At the same time, there are areas that are significantly polluted due to intensive tourist traffic and many years of negligence in terms of clearing out public space. In recent years, Krakow has been visited by an average of almost 10 million tourists a year [56]. Because of the value of historical substance, spatial diversity and many years of layered neglect, the city authorities decided to start the procedure of encompassing the area with a holistic protection plan in the form of a Cultural Park. On 12 June 2019, the Krakow City Council adopted a resolution on the announcement of the commencement of work on the creation of a Cultural Park Kazimierz and Stradom. In 2020, work began on the protection plan project, in 2021 project was completed.

One of the elements for developing this protection plan project was a visual analysis. Visual analysis was to identify and assess visual values. This then leads to formulating guidelines for preservation protection and restoration of visual values.

The method of visual landscape studies applied for the needs of the plan for Cultural Park protection combines the achievements of *visual assessment* and the Krakow School of Landscape Architecture. From *visual assessment,* it takes the procedure and it uses the advanced tools for visual analyses from the KSLA method it borrows a specific approach that is typical for protected landscapes as well as the issues of the visual analysis [46].

It consists of identifying and characterizing the resource, assessment of visual values and formulating guidelines for further stages of project works.

Identifying the resource occurs on the basis of desk study and field study [57]. This stage of the project relies first and foremost on obtaining the information necessary for further analyses. The preliminary works included analyses of archive materials, developing a model for defining visual relations of spatial elements as well as an initial analysis of the spatial structure. Since the study is a part of a multidisciplinary report, it is based on one common matrix. The basis of the study was a division into architectural and landscape enclosure [58] which all of the studies followed. The starting point was an enclosure while enclosures of similar characteristic features and origin constituted a group of enclosures.

Identification of visual resources of enclosures was based on the KSLA division into active and passive exposure [42,59]. Active exposure is related to landscape observation whereas passive exposure means exposing it in the form of frames, views, and panoramas. This distinction made it possible to distinguish functional dependencies of the exposure on the seen views. Active exposure elements comprise points, visual paths, visual surfaces and visual axes. Passive exposure elements comprise the following: dominants, subdominants, accents, foreground and background. The specifics of active exposure results from the location of the point, visual paths, etc., and it is connected to its management. Passive exposure is a view while a panorama is an image whose unveiling is dependant on the spatial structure and substance it is built on. Active exposure is also called view availability. Passive exposure is also called view substantion. Both exposure categories are closely related to each other and interdependent. However, their specification decides on a different approach.

Due to the nature of Stradom and Kazimierz, an additional category of identification was applied in the form of the degree of view pollution. Investment layering, lack of renovation activity, lack of successive cleaning activity with reference to public transport, advertisement and information space created a surface of clutter that calls for immediate intervention in terms of cleaning. Defining the degree of pollution and its identification was the source of significant data to define further activity for improving the visual values of the analysed object (Figure 1).

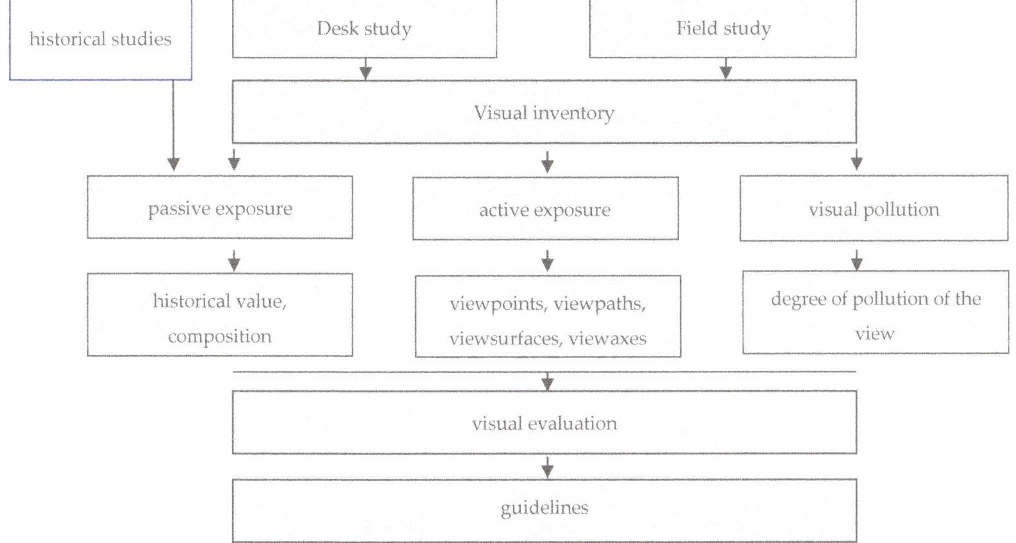

Figure 1. Visual analysis method diagram.

4. Results

4.1. Visual Inventory

Identification of the visual resource constitutes the basis work for assessment and then zoning. It comprises desk study and field study. Desk study means, first of all, collecting starting materials on both historical and spatial issues. Since the team responsible for the development of the Cultural Park protection plan comprises representatives of various fields, detailed historical research was conducted by historians and art historians. They provided selected materials required for the identification of visual resources. These included archival materials, historical pictures, maps as well as all types of source materials. Moreover, a team of historians made an in-depth analysis of the stages of the city's development and its transformations. The visual analysis was based on these findings. On the basis of iconographic materials, the selection was made and historical pictures were compared to the current state. This constituted the basis for selecting visual points and their hierarchy which was at a later stage verified during field study (Figure 2).

A desk study was also carried out in the field of spatial relations analyses. The maps were analyzed and preliminary visibility charts were prepared as well as crosscut sections for defining fundamental visual connections. Items significant for urban composition were pre-defined similarly to the visual ranges of the dominant objects. An important element of desk study was an analysis of visual ranges (Figure 3). Placing them made it possible to identify the areas where views intensified onto the key spatial objects.

A field study means, first and foremost, the verification of data and keeping inventory. The identification process takes place alongside the comparison of historical pictures to the real state. Another process is the verification of view ranges and the photographic inventory of views where view ranges are overlapping. Hierarchy of views also takes place.

As a result of combining desk and field studies, enclosure characterisation is defined in terms of active and passive exposure.

Figure 2. Comparison of archival and contemporary pictures as a basis for creating hierarchy of visual points.

Figure 3. Visibility maps for dominants and subdominants of Stradom and Kazimierz. From the top: an exemplary viewshed of the Corpus Christi Basilica; viewshed of the Wawel Cathedral: the bottom figure presents a compilation of the viewsheds of all the landscape dominants.

4.1.1. Active Exposure

Active exposure of an enclosure is defined by exposure elements in the form of points, visual paths, surfaces, and visual axes. Elements of active exposure are at the same time characterized by their significance. Key, characteristic and supplementary elements are distinguished. Another step involves the selection of monitoring points for particular enclosure groups.

The basis for defining elements of active exposure is field study supported by desk study. In this case, methodology achievements in the field of *visual assessment* are used [34]. In order to identify the points of the widest range, analyses of visibility are applied. An analysis of views from the highest and most attractive objects of the urban layout defined as dominants and subdominants makes it possible to locate the places from which they are seen as a single one or a group. Their exposure defines the spatial shape of the views as well as their attractiveness [60,61]. For studying this issue, the degree of visibility index is used [62]. Overlaying viewsheds becomes the basis for defining the number of visible dominants from a given place. These data take on the form of a degree of visibility map (Figure 4). The higher the degree, the more dominants are visible from a given place. The data obtained in digital analyses are verified in the field. As a result, combining the desk study and field study helps to define and create a hierarchy of elements of active exposure.

Figure 4. A degree of visibility map.

On the basis of studies of active exposure, the final version of the map of elements of active exposure is prepared (Figure 5). It is supplemented with panoramas of key characteristics and supplementary points. Additionally, a comparison table is prepared (Table 1). It contains elements of active exposure for every enclosure that then becomes the basis of its valuation in this respect, in terms of the content of places from which the views are most attractive.

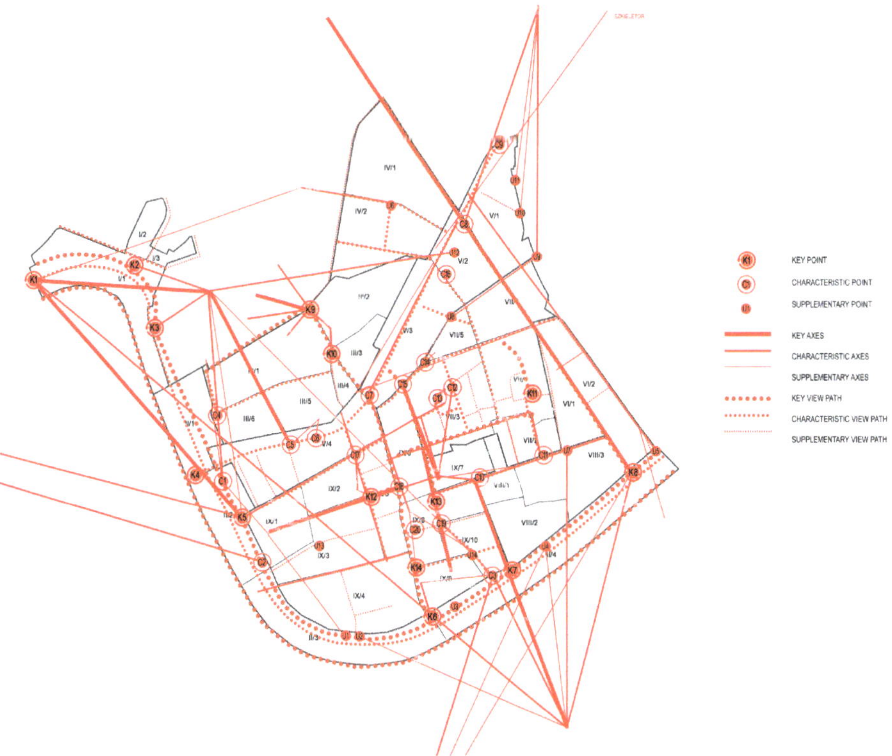

Figure 5. Map of active exposure.

Table 1. Comparison of elements of active exposure in particular enclosures (WAK). Letter symbols: K—key element; C—characteristic element; S—supplementary element. The table presents the number of active exhibition elements present in individual enclosures.

ZWAK	Wak	Viewpoints			Viewpaths			Viewaxis			Viewsurfaces		
		K	C	S	K	C	S	K	C	S	K	C	S
I/1	Wistula band	3			2	2		2	3	1	1		
I/2	Na groblach sqare												
I/3	Podzamcze						1				1		
II/1	Rybaki	1			2	1					1		
II/2	Bulwar Inflancki pn	1	2		2	1		1	3	2	1		
II/3	Bulwar Inflancki pd			2	2	1				2	1		
I/4	Bulwar Kurlandzki	3	1	3	2	1		1	4	3	1		

Table 1. *Cont.*

ZWAK	Wak	Viewpoints			Viewpaths			Viewaxis			Viewsurfaces		
		K	C	S	K	C	S	K	C	S	K	C	S
III.1	Bernardyni	1				2	1	1	6				
III.2	Misjonarze				1								
III.3	Bożogrobcy				1								
III.4	Piekiełko	1							1				
III.5	Koletek						1						
III.6	Nadwiślan		1				1						
IV.1	Starowiślna			1		1	1		1				
IV.2	Św. Sebastiana					2	1						
V.1	Wrzesińska		1	3		1	1		1	2			
V.2	Berka joselewicza		2	1	1	2			2				
V.3	Dietla pn		1		1	3			1				
V.4	Dietla pd		2		1	2	3	1	1	2			
VI.1	Dajwór				1		1				1		
VI.2	Halicka					1	4			2			
VII.1	Szeroka	1			1	1	3				1		
VII.2	Bawół		1		1	1	1						
VII.3	Plac nowy		3		2	3	4		1		1		
VII.4	Miodowa				1	2	1						
VII.5	Brzozowa		1	1		1	1			1			
VIII.1	Zajezdnia		1			1	1						
VIII.2	Gazownia				1		1						
VIII.3	Elektrownia				1				1				
IX.1	Skałka				1	1							
IX.2	Augustianie				1	2							
IX.3	Piekarska			1			1		1	1			
IX.4	Skawińska						3						
IX.5	Krakowska zach.	1	1		2	2	2			1			
IX.6	Krakowska wsch.		1		3	1		3					
IX.7	Boże ciało				1	1							
IX.8	Wolnica	1	2		2	1	1	2			1		
IX.9	Bonifratrzy	1			1	1							
IX.10	Mostowa			1					3	1			

The graphic presentation in the form of a map reveals a specific spatial composition of the analysed area as a synthesis. In the diagram of the visual axes pattern of Kazimierz and Stradom, fundamental features of visual connections can be observed. These characteristics demonstrate the individual nature of the analysed space. The graphics are in the form of a diagram. It exposes the compositional and spatial specifics of the townscape and reflects its internal arrangement as well as the degree of visual connection with other districts in the city (Figure 6).

4.1.2. Passive Exposure

The nature of passive exposure is defined by two factors, namely the quality of the substance which builds the space as well as the composition which creates the mutual pattern of particular elements.

The quality of substance was defined on the basis of works by a team of historians (Table 2). Every enclosure gained historical value. The approved criteria were based on the state of preservation, the clarity of the pattern as well as the diversity of the pattern.

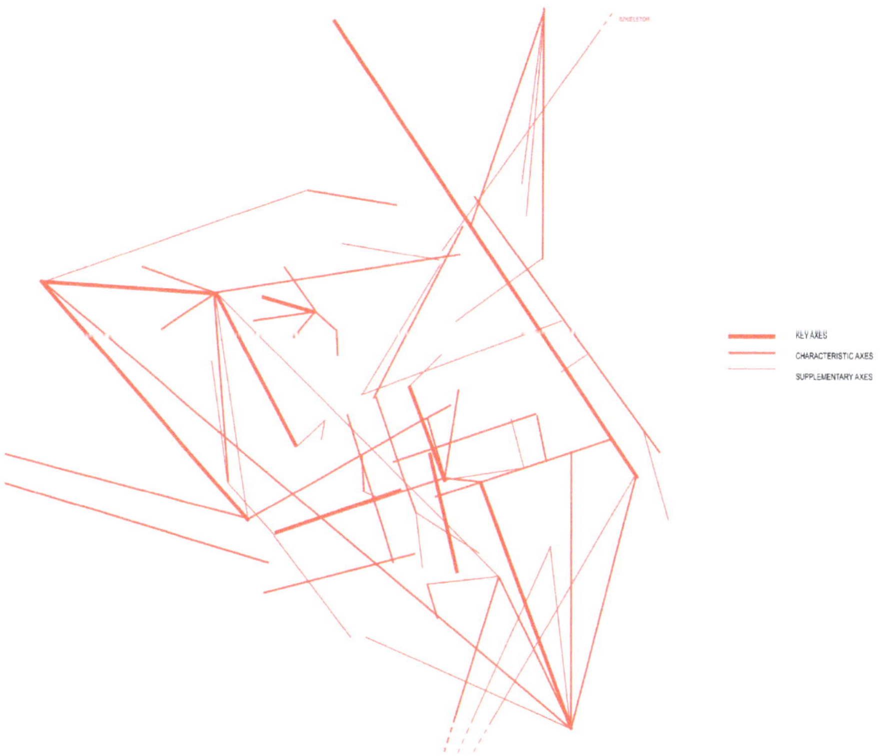

Figure 6. Map of visual axes—diagram of the compositional and spatial specifics of townscape.

Table 2. Historical values of the landscape substance (example ZWAK I and II).

ZWAK	WAK	Historical Values
I/1	Vistula band	Low
I/2	Na Groblach Square	Medium
I/3	Podzamcze	Medium
II/1	Rybaki	High
II/2	Bulwar Inflancki N	High
II/3	Bulwar Inflancki S	High
II/4	Bulwar Kurlandzki	High

The compositional nature was defined on the basis of visual features of the enclosure. The following view features were taken into account: depth, distance, width, multiplanning, and presence of borrowed views. Comparing compositional features of particular enclosures is presented in the table (Table 3). The material in the form of panoramas illustrates the occurrence of features characteristic for the composition (Figure 7). The table specifies the presence of particular features of the view while graduating them in three degrees: high, medium, and low. The presence of these features becomes the basis for the assessment of passive exposure.

Table 3. Composition characteristics of enclosures were made on the basis of the occurrence of specific visual features of the enclosure, such as: 1. Distance, 2. View depth, 3. Width, 4. Multiplanning, 5. Presence of borrowed views (example ZWAK I and II). Letter symbols refer to the significance of these features: H—high, M—medium, L—low.

ZWAK	WAK	Composition Elements				
		1	2	3	4	5
I/1	Vistula band	H	H	H	H	H
I/2	Na Groblach Square	L	L	L	L	L
I/3	Podzamcze	L	L	L	M	M
II/1	Rybaki	M	L	H	H	M
II/2	Bulwar Inflancki N	M	M	H	M	M
II/3	Bulwar Inflancki S	M	H	H	M	M
II/4	Bulwar Kurlandzki	M	H	H	M	M

Figure 7. Composition elements of the WAK II/2, 1. Distance (M), 2. View depth (M), 3. Width (H), 4. Multiplanning (M), 5. Presence of borrowed views (M).

4.1.3. Visual Pollution

Due to significant differences between the state of the substance, the composition value, and neglect of the public sphere due to intensive use, an additional aspect of the characteristics of the study area was identified, namely the visual pollution index. The assessed elements here are those that have become layered as a result of neglect in maintenance, elements that distinguish themselves in urban space or temporary elements. They do not constitute historical substance nor do they belong to the main urban structure, yet they decide about the visual reception of townscape, so they are treated as a separate category. They are presented as the degree of visual pollution of urban space. The level of pollution is impacted by elements observed in space and their intensity on the three-grade scale: high, medium, and low. Among those elements, we distinguish the following: 1. Neglected surfaces, 2. Neglected facades, 3. Chaotic street furniture, 4. Neglected and chaotic greenery, 5. Presence of outdoor advertisement, 6. Excess of road signs, 7. Burdensome presence of parked vehicles (Table 4).

Table 4. Visual pollution listed for particular enclosures 1. Neglected surfaces, 2. Neglected facades, 3. Chaotic street furniture, 4. Neglected and chaotic greenery, 5. Presence of outdoor advertisement, 6. Excess of road signs, 7. Burdensome presence of parked vehicles.

ZWAK	WAK	Elements							Pollution Degree
		1	2	3	4	5	6	7	
I/1	Wistula band	L				M			L
I/2	Na Groblach Sqare				L	L		M	L
I/3	Podzamcze							L	L

Table 4. *Cont.*

ZWAK	WAK	Elements							Pollution Degree
		1	2	3	4	5	6	7	
II/1	Rybaki	M				L			L
II/2	Bulwar Inflancki Pn	L				L			L
II/3	Bulwar Inflancki Pd	M				H+			L
II/4	Bulwar Kurlandzki	M				L	L		L
III.1	Bernardyni	L		L		L	M	M	M
III.2	Misjonarze								L
III.3	Bożogrobcy	L				M	l	M	M
III.4	Piekiełko	M	L	M	L	M	M		M
III.5	Koletki		L				M	L	L
III.6	Nadwiślan	L	H	H	L	H	M		H
IV.1	Starowiślna	M	L	H	L	M	L	H	H
IV.2	Świętego Sebastiana	M		L	L			M	L
V.1	Wrzesińska	M	L	H	H			M	H
V.2	Berka Joselewicza	H	M	H	H	H	L	H	H
V.3	Dietla Pn	r	M	M		H	M		M
V.4	Dietla Pd			L	L	M	H	M	M
VI.1	Dajwór	M	M			M	M	M	H
VI.2	Halicka	H	M	H	H			H	H
VII.1	Szeroka	M		H	M	H	M	H	H
VII.2	Bawół		M	M		M	M		M
VII.3	Plac Nowy	H	M	H	L	H	H	H	H
VII.4	Miodowa	H	M	H	H	H	M	M	H
VII.5	Brzozowa	H	M	H	H	H	M	H	H
VIII.1	Zajezdnia					M	1		L
VIII.2	Gazownia					L	L	M	L
VIII.3	Elektrownia						L	M	L
IX.1	Skałka	M						L	L
IX.2	Augustianie		H				L	H	L
IX.3	Piekarska	L						M	L
IX.4	Skawińska	L					M	H	M
IX.5	Krakowska Zach			L		M	L	M	L
IX.6	Krakowska Wsch	H	H	M		H	M	-	H
IX.7	Bożego Ciała	H	M	M		H	M	H	H
IX.8	Wolnica	H	L	M	M	M	M	H	H
IX.9	Bonifratrzy	L			L		L		L
IX.10	Mostowa				L	L	L	M	L

4.2. Assessment of Visual Values

Baseline analyses became the basis for active exposure assessment. The elements of active exposure became the criteria of assessment of active exposure. The elements' layout and layering indicated enclosures with the most precious views. Those views are the district's landmarks. Their location shapes tourist routes and decides the concentration of services in the public sphere while indicating the need for more intensive management. The assessment criteria were the number and rank of elements of active exposure. Depending on the number of elements, the active exposure value can range from the highest, through high, medium, low up to the lowest value.

The value of passive exposure is estimated on the basis of compositional character and the quality of historical substance which was decided upon by studied by historians. The accepted criteria of composition characteristics allowed us to define and assess them in the case of particular enclosures. Distance, depth, width, multiplanning, and borrowed views helped us to characterize the richness of passive exposure and define the enclosure value in this respect.

Listing the values of passive and active exposure became the basis for the assessment of exposure of particular enclosures (Table 5).

Table 5. Active and passive exposure evaluation summary.

		Active Exposure		
		High	**Medium**	**Low**
Passive Exposure	**High**	Highest	High	Medium
	Medium	High	Medium	Low
	Low	Medium	Low	Lowest

The studies allowed us to obtain information regarding the resources and characterize architectural and landscape enclosures (WAK). As a result, an assessment of active and passive exposure was made as well as a summary assessment which allowed for zoning out the study area (Figure 8).

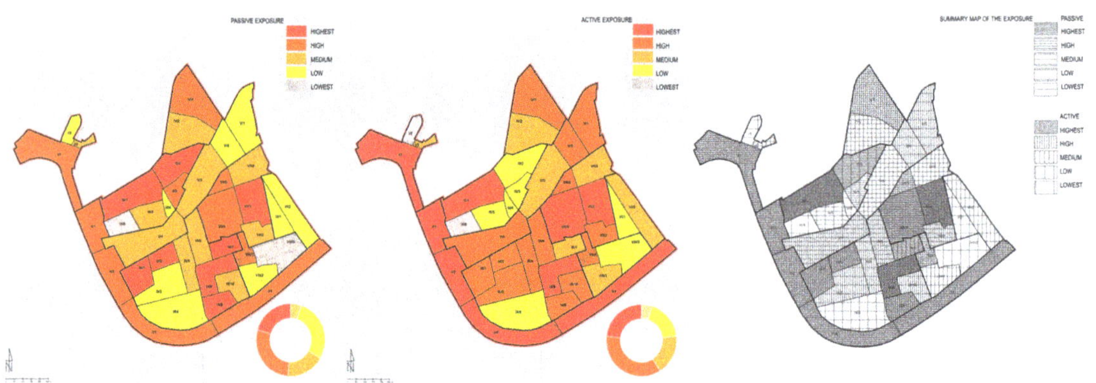

Figure 8. Assessment map of active and passive exposure and summary map.

The applied method allowed us to distinguish characteristic features of the urban structure. While basing on the key role of dominants and subdominants in the city [41,42] and defining their visibility degree [43], an analysis was made of the urban structure leading to creating a hierarchy of elements of active exposure. Comparing the analyses made on this basis for various spaces indicates how effective this method is and it allows us to distinguish characteristic features of the analyzed space and their assessment at a further stage. A graphic analysis of the elements of active exposure brought about some interesting results. The graphic form of the presentation showed visual characteristics of the area. It helped to define the character of internal visual connections, their range and intensity of occurrence. Moreover, it illustrated the openness of the analysed district towards neighbouring parts of the urban structure. In the case of Stradom and Kazimierz, the number of visual connections is high and it is determined by the presence of dominants and subdominants as well as the urban layout. This district also shows unique connections with neighbouring districts. This is due to the proximity to the Vistula riverbed and the shape of the surface of the river valley which provides for long views reaching the hills.

An analysis of passive and active exposure led to separating enclosures by value, which was shown in the form of the map, the table and the diagrams.

Valuing that takes into account the issues of active exposure as well as substance value and its visual realisation as passive exposure made it possible to divide the area of the

Cultural Park in accordance with five value categories. On this basis, we distinguished a zone of special exposure value with the highest values in terms of composition and substance as well as an abundance of sites of active exposure.

Further stages of analyses became the basis for formulating project guidelines. They were related to active and passive exposure and came down to preserving and protecting existing visual connections, clearing the vanishing or losing clarity views, subjecting foreground of the view and clearing particular view plans.

Based on this hierarchy of elements of active exposure, key views very important for historic urban space were indicated. Their identification was the basis for the Cultural Park managing project and for the preparation of impact assessments for new development proposals [6]. The research also indicated monitoring points for landscape changes [26]. For every group of enclosures, one monitoring point was indicated. Guidelines related to monitoring were applied in urban space while using the experience gained during studies of open protected areas.

As a result of the analysis of view, pollution information was obtained regarding the degree of pollution and its territorial layout (Figure 9). Comparing exposure assessment to the visual pollution degree provided us with additional data. Places with the highest visual pollution were identified. It was also possible to observe against which exposition values pollution takes place. It occurred that in several cases, the highest visual pollution affects enclosures with the highest exposition values. At the same time, little visual pollution covers enclosures that could boast medium to low visual values. These observations became the basis for marking out intervention zones within the studied area. There are zones of immediate intervention and moderate intervention. Immediate intervention zones are those which cover the highest degree of view pollution and the highest exposition values (Figure 10).

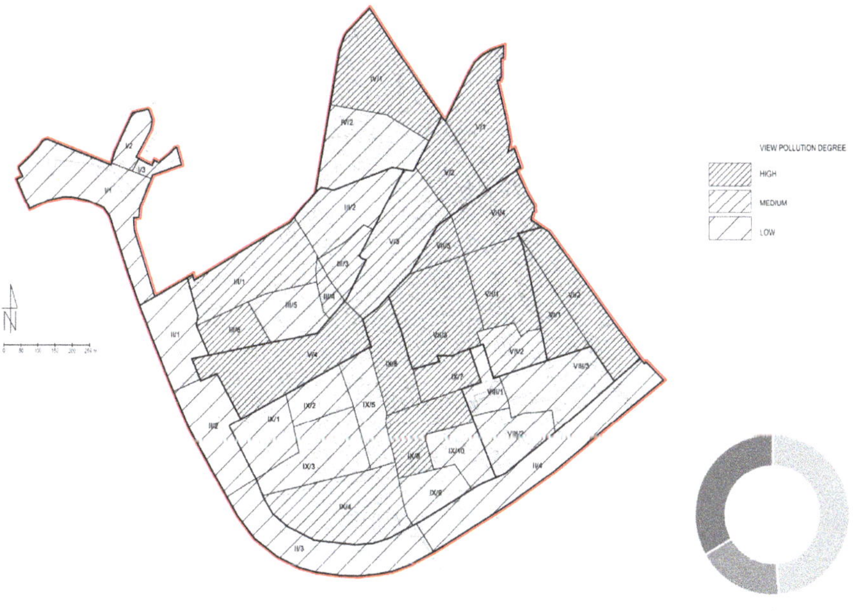

Figure 9. Degree of view pollution.

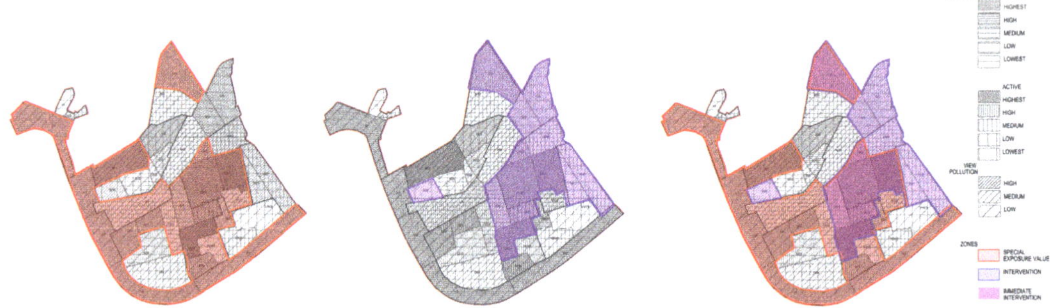

Figure 10. Zone with the highest exposure values, zone of intervention and zone of immediate intervention.

These observations distinguish the Cultural Park protection plan for Stradom and Kazimierz against the previously prepared protection plans. Individual characteristic features of the place indicated the need for developing a certain method and adjusting the tools to the nature of the place.

5. Discussion

Visual studies of urban substance are conducted with the use of various methods and to varying degrees. They use available digital tools and various types of analyses. Moreover, they present a different degree of detail. Advanced tools are used for in-depth studies of spatial phenomena and their impact on the functioning of the city [63] and open landscape [64]. The development of those methods is very dynamic and varied [19]. In Poland, such studies are usually conducted as expert opinions focusing on selected issues, such as urban panorama against new investments [65] or an influence of a high-rise building on spatial relations [50] as well as the degree of shading the area as a result of new investment [66]. They are related to commercial investment in real estate development or single buildings. Comprehensive analyses of whole districts or cities which could become a part of a local zoning plan are currently hardly ever carried out. According to Różycka [67], currently, we are dealing with a crisis of the modern planning system in Poland. Due to that fact, valuable urban substances require our special attention.

The form of protection as a Cultural Park and its implementation as a cultural park protection plan fills this void. It creates possibilities for development and comprehensive visual analysis of areas of high spatial values as indicated by Z. Myczkowski and others [3]. While being a starting document for a local zoning plan, it provides for the application of the results in the plan which then becomes an act of local law.

The presented method demonstrates the way of using possibilities that the protection plan offers in terms of visual values of urban structure. A comprehensive approach to visual analysis was presented as it combines contemporary achievements of visual analyses when it comes to *visual assessment* with KSLA assumptions where the landscape is treated as a spatial form. This fusion demonstrates an original approach to landscape represented by the field of landscape architecture. This work uses achievements in the field of analyses of urban structures [38]; it also creates a division into landscape units with reference to urban structure. At the same time, it applies the methods related to protected areas as suggested by Palmer that landscape observation and control are the basis for following processes and on their basis, effective actions can be undertaken [41].

Special attention should be given to a graphic interpretation of the results of the active exposure analysis. It acted as an exceptional comparative material to demonstrate the nature of visual connections of the analysed area, both internal connections and external ones. This form of graphic interpretation is appreciated by both designers and planners [68]. The one used in this case exposes composition and spatial specifics while reflecting uniformity or complexity of spatial structure. It visualises the degree of the visual connection with external structures. In this way, it shows critical directions and helps to adequately channel the course of action. The diagrams present a comparison of three Krakow districts, namely: Old Town, Nowa Huta, Stradom and Kazimierz (Figure 11).

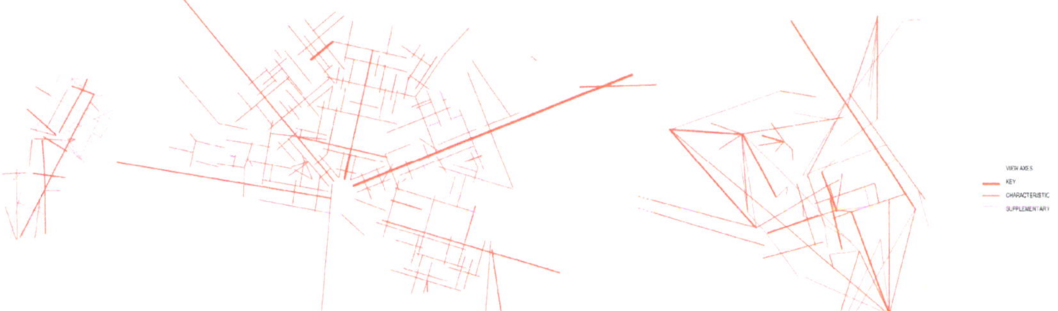

Figure 11. Comparison of the diagram of visual axes for Old Town, Nowa Huta, Stradom and Kazimierz.

The presented example of the visual analysis put together the experience of past visual analyses for Cultural Parks. On the other hand, it contributed to an evolution in the developed method. The source of this evolution was the recording and characterising of the place as well as the need to take those into account in the study process. In this case, it was a place of rich substance and compositional values, but also neglected and deprived of adequate management. Developed enlargement of the method creates new possibilities for its use. With reference to space of high values exposed to intensive use observations prove that combining these characteristics goes hand in hand with a lack of adequate regulations for use which poses a threat to landscape values and calls for immediate intervention.

The evolution of the method demonstrates its flexibility and that it can adjust to working conditions and individual characteristic features of the analysed area which is perceived as an advantage by the authors of the guidelines for the assessment of landscape impact (GLVIA). It is based on the cooperation of experts across different areas of expertise which ensures a contemporary multithreaded approach, which is crucial in the case of studying complex urban structures [69].

The need for these types of studies in a modern city is stressed by T. Brigen who indicates that the view as such is today a multi-aspect value. It has become a common quality whose value crosses the borders of not only the city but often the country which at the same time constitutes the self-portrait of its community. Proper legal protection of the view has a cultural and financial effect on a different scale. The view is that heritage constitutes not only one's cultural identity and personal aesthetic experience but also the popularity of the city, tourism and property management [70].

It is key in terms of intensive urban development and strong pro-investment lobby movements. The legitimacy of this type of analysis is proved in widely criticised cases of destroying visual axes in the cities, as in Warsaw in the case of the Saska axis. In this context, the issue of protecting the view as a cultural resource and developing transparent methods of identifying and assessing the value of a view becomes key [15].

The presented paper touches upon the issue of scenic values, whose role is crucial in a modern city. This is confirmed by the provisions of UNESCO's declarations and recommendations, which pay more and more attention to the context and the perceived landscape [2,28]. In Krakow, this problem is particularly important in view of very strong investment pressure and the generally poor condition of the planning system in Poland [71] in the context of enormous tourist traffic [56]. Krakow represents a city of very high historical value. On the one hand, this creates both economic and social opportunities; yet, on the other hand, it threatens the substance and its material and non-material values [72].

Overexploitation of tourism, investment pressure, and loss of permanent residents are processes still present and requiring regulation. These phenomena can be observed in Eastern European cities in, for example, Croatia, Romania and Bulgaria [73]. In turn, the problem of migration from cities, gentrification and loss of identity in the social dimension also affects other parts of Europe and the world requires more and more effective forms of protection [2,29]. In this context, urban landscape visibility constitutes a link between the future and the past well remembered [26,69,73]. While it evokes reactions at the social and economic level, it is unquestionably a landscape resource [2]. Moreover, visual balance is indicated in numerous sources as a picture of sustainable landscape management [1,46] and since it is a resource, it can be lost if inadequately managed.

6. Conclusions

The demonstrated example of a visual analysis prepared for the cultural park protection plan project presents a comprehensive study method.

Visual analysis in this case, while being based on the achievements of the Krakow School of Landscape Architecture as well as contemporary methods developed as a part of *visual assessment*, provided for adjusting tools to the needs of the analysis [34]. Meanwhile, combining knowledge regarding historical substance value and spatial dependencies with the use of contemporary study tools allows for a balanced approach that favourably connects the accessible methods.

The developed methodical scheme dedicated to the visual analysis of urban space presents subsequent stages and the scope of works. Its formula creates possibilities for adjusting detailed studies to the specifics of the analysed area. Its application allows for distinguishing characteristic visual features of particular spaces. These features have a significant meaning for further landscape development and management in the spirit of sustainable development which is to preserve the most precious visual resources for protection and securing diversity of cultural heritage and landscape.

It allows us to preserve individual visual features and preserve spatial diversity which leads to the creation of space for sustainable spatial development while keeping its visual identity. This is in line with the recommendations for UNESCO sites, allowing for the study of the broad spatial context of the most valuable urban structures.

It provides arguments for a discussion on specific cases, which is crucial in times of aggressive real estate development policy.

The case invoked proved the effectiveness of applying a chain of studies according to the accepted order, and at the same time, it pointed to the legitimacy of using visual analyses that support project decisions in protection plans. The evolution of the method indicates its versatility and usefulness in terms of analysing diverse urban spaces.

The applied form allows for further verification and development based on studies of space with individual features.

Funding: This research received no external funding.

Institutional Review Board Statement: Not applicable.

Informed Consent Statement: Not applicable.

Data Availability Statement: Not applicable.

Acknowledgments: The author would like to thank the entire team developing the project of Protection plan of the Stradom and Kazimierz Cultural Park for the cooperation. The author express gratitude to the reviewers, whose remarks and suggestions have helped to improve the work.

Conflicts of Interest: The author declares no conflict of interest.

References

1. UNESCO. *Recommendation on the Historic Urban Landscape*; UNESCO: Paris, France, 2011. Available online: https://whc.unesco.org/en/hul/ (accessed on 1 June 2021).
2. Myczkowski, Z.; Marcinek, R.; Siwek, A. *Park Kulturowy Jako Forma Ochrony Krajobrazu Kulturowego (en. Cultural Park as a Form of Cultural Landscape Protection)*; OT NID Press: Kraków, Poland, 2017.
3. Ustawa O Ochronie Zabytków I Opiece Nad Zabytkami Z dn. 3 Lipca 2003 Roku (en. Act on the Protection and Care of Monuments). 2003. Available online: http://isap.sejm.gov.pl/isap.nsf/DocDetails.xsp?id=WDu20030800717 (accessed on 20 May 2021).
4. *European Landscape Convention, European Treaty Series—No. 176*; Council of Europe: Florence, Italy, 2000. Available online: https://rm.coe.int/1680080621 (accessed on 26 December 2020).
5. Forczek-Brataniec, U. *Visible Space: A Visual Analysis in the Lanscape Planning and Designing*; Cracow University of Technology: Cracow, Poland, 2018.
6. González, A.P. *Cultural Parks and National Heritage Areas: Assembling Cultural Heritage, Development and Spatial Planning*; Cambridge Scholars Publishing: Newcastle, UK, 2013.
7. Lynch, K. *The Image of the City*, 1st ed.; The MIT Press: Cambridge, MA, USA, 1960.
8. Cullen, G. *Concise Townscape*, 1st ed.; Routledge Press: New York, NY, USA, 1961.
9. Jackle, J.A. *The Visual Elements of Landscape*, 1st ed.; The University of Massachusetts Press: Amherst, MA, USA, 1987.
10. Bell, S. *Elements of Visual Design in the Landscape*, 2nd ed.; Spon Press: London, UK, 2004.
11. Iwańczak, B.; Lewicka, M. Affective map of Warsaw: Testing Aleksander's pattern language theory in an urban landscape. *Landsc. Urban Plan.* **2020**, *204*, 1–16. [CrossRef]
12. Mordwa, S. Krzywa wrażeń dla ulicy Piotrowskiej w Łodzi (en. Piotrkowska Street In Łódź Sensations Curve). *Acta Univ. Lodz. Folia Geogr. Econ.* **2009**, *10*, 89–98.
13. Wejchert, K. *Elementy Kompozycji Urbanistycznej (en. Elements of Urban Composition)*; Arkady: Warsaw, Poland, 1984.
14. Green, S. *Seeing the History in the View, a Method for Assessing Heritage Significance Within Views*, 1st ed.; English Heritage: London, UK, 2011.
15. Sternberg, E.H. *Healing Spaces: The Science of Place and Well-Being*; Harvard University Press: Cambridge, MA, USA, 2010.
16. Antrop, M. Geography and landscape science. *Belgeo* **2000**, *30*, 9–36. [CrossRef]
17. Tuan, Y.F. *Space and Place: The Perspective of Experience*, 1st ed.; University of Minnesota Press: Minneapolis, MN, USA, 1977.
18. Vidal de La Blache, P. *Principes de Géographie Humaine: Publiés d'après les Manuscrits de l'auteur par Emmanuel de Martonne*; ENS Édition: Lyon, France, 2015.
19. Livingstone, D.N. Classics in human geography revisited: Commentary. *Prog. Hum. Geogr.* **1994**, *18*, 209–210. [CrossRef]
20. González, A.P. From a given to a construct: Heritage as a commons. *Cult. Stud.* **2014**, *28*, 359–390. [CrossRef]
21. Parga-Dans, E.; Gonzales, P.A.; Enriguez, R.O. The social value of heritage: Balancing the promotion-preservation relationship in the Altamira World Heritage Site, Spain. *J. Destin. Mark.* **2020**, *18*, 10499. [CrossRef]
22. Szpakowska-Loranc, E.; Matusik, A. Łódź—Towards a resilient city. *Cities* **2020**, *107*, 102936. [CrossRef]
23. Czyńska, K. *Kształtowanie Współczesnych Panoram Miast na Przykładzie Szczecina. Wykorzystanie Wirtualnych Modeli Miast w Monitoringu i Symulacji Panoram (en. Using Model of Virtual City for Research on Visibility Range of Panoramas of the City)*; Uniwersytet Wrocławski: Wrocław, Poland, 2006.
24. Sroczyńska, J. The Social Value of Architectural Monuments in the Light of Selected Documents of UNESCO Icomos, the Council of Europe, Shaping the Theory of Cultural Heritage Protection. *Wiadomości Konserw. J. Herit. Conserv.* **2021**, *65*, 7–19.
25. Kobayashi, A. A Critique of Dialectical Landscape. In *Remaking Human Geography (RLE Social & Cultural Geography)*, 1st ed.; Kobayashi, A., Mackenzie, S., Eds.; Routledge: London, UK, 1989.
26. Cosgrove, D.F. *Social Formation and Symbolic Landscape*, 2nd ed.; The University of Wisconsin Press: Madison, WI, USA, 1998.
27. UNESCO. *Recommendation Concerning the Safeguarding of Beauty and Character of Landscapes and Sites, Paris*; UNESCO: Paris, France, 1962. Available online: https://unesdoc.unesco.org/ark:/48223/pf0000114582.page=142 (accessed on 1 June 2021).
28. ICOMOS Xi'an Declaration on the Conservation of the Setting of Heritage Structures, Sites and Areas. Available online: http://orcp.hustoj.com/the-xian-declaration-on-the-conservation-of-the-setting-of-heritage-structures-sites-and-areas-2005/ (accessed on 1 June 2021).
29. ICOMOS. *Heritage and Landscape as Human Values, Florence*; ICOMOS: Florence, Italy, 2014. Available online: https://www.icomos.org/en/about-icomos/2016-11-17-13-14-08/minutes-of-the-general-assemblies (accessed on 1 June 2021).
30. Voronych, Y. Visual Pollution of urban space in Lviv. *Przestrz. Forma* **2013**, *20*, 309–314.
31. Zube, E.H.; Brush, R.O. *Landscape Assessment: Values, Perceptions and Resources*, 1st ed.; Zube, E.H., Brush, R.O., Fabos, J.G., Eds.; Dowden, Hutchinson & Ross: Stroudsburg, PA, USA, 1975.
32. Gobster, P.H.; Ribe Robert, G.; Palmer, J.F. Themes and trends in visual assessment research: Introduction to the "Landscape and Urban Planning special collection on the visual assessment of landscapes. *Landsc. Urban Plan.* **2019**, *191*, 1–7. [CrossRef]
33. Swanwick, C. *Guidelines for Landscape and Visual Impact Assessment*, 3rd ed.; Routledge: London, UK, 2013.
34. Ozimek, P.; Böhm, A.; Ozimek, A.; Wańkowicz, W. *Planowanie Przestrzeni o Wysokich Walorach Krajobrazowych Przy Użyciu Cyfrowych Analiz Terenu Wraz z Oceną Ekonomiczną (en. Planning Spaces with High Scenic Values by Means of Digital Terrain Analyses and Economic Evaluation)*; Cracow University of Technology: Cracow, Poland, 2013.

35. Ozimek, P.; Ozimek, A. Viewshed Analyses as Support for Objective Landscape Assessment. *J. Digit. Landsc. Archit.* **2017**, *2*, 190–197.
36. Photography and Photomontage in Landscape and Visual Impact Assessment. Available online: https://landscapewpstorage01.blob.core.windows.net/www-landscapeinstitute-org/2019/01/LIPhotographyAdviceNote01-11.pdf (accessed on 1 November 2019).
37. Szczepańska, A.; Pietrzyk, K.; Senetra, A. Analysis and Evaluation of Historical Public Spaces in Small Towns in the Polish Region of Warmia. *Sustainability* **2020**, *12*, 8356.
38. Górka, A. Visual Capacity Assessment of the Open Landscape in Terms of Protection and Shaping: Case Study of a Village in Poland. *Sustainability* **2020**, *12*, 6319. [CrossRef]
39. Litton, R.B. *Landscape Control Points: A Procedure for Predicting and Monitoring Visual Impacts*; Pacific Southwest Forest and Range Exp.Stn.: Berkeley, CA, USA, 1973.
40. Palmer James, F. The contribution of key observation point evaluation to a scientifically rigorous approach to visual impact assessment. *Landsc. Urban Plan.* **2019**, *183*, 100–110. [CrossRef]
41. Bogdanowski, J. *Kompozycja i Planowanie w Architekturze Krajobrazu (en. Composition and Planning in Landscape Architecture)*, 1st ed.; Zakład Narodowy Im. Ossolińskich: Wrocław, Poland, 1976.
42. Myczkowski, Z.; Forczek-Brataniec, U.; Chajdys, K.; Rymsza-Mazur, W.; Środulska-Wielgus, J.; Wielgus, K. Metoda jednostek i wnętrz architektoniczno—krajobrazowych (JARK–WAK) w relacjach XXI wieku (en. JARK–WAK 2000 + the method of landscape architectural units and interiors on the background of researches in XXI century). In *Metoda Architektury Krajobrazu*; Szulczewska, B., Szumański, M., Eds.; Wieś Jutra Press: Warszawa, Poland, 2010; pp. 7–15.
43. Bogdanowski, J. *Metoda Jednostek i Wnętrz Architektoniczno-Krajobrazowych (JARK–WAK) w Studiach i Projektowaniu (en. The Units and Rooms Method (JARK-WAK) in Studies and Design)*, 2nd ed.; Cracow University of Technology: Cracow, Poland, 1990.
44. Dąbrowska-Budziło, K. *Treść krajobrazu Kulturowego w Jego Kształtowaniu i Ochronie (en. Symbolism of the Cultural Landscape in Its Shaping and Protection)*; Cracow University of Technology: Cracow, Poland, 2002.
45. Myczkowski, Z.; Forczek-Brataniec, U. Landscape conservation in the research and development of the Krakow School of landscape architecture from 1970s to 2017—From Jurassic landscape parks to cultural parks in Krakow. *Landsc. Archit. Art* **2018**, *13*, 129–134.
46. Forczek-Brataniec, U. Zmiany w krajobrazie wokół zbiorników wodnych w Pieninach. *Monogr. Pienińskie* **2010**, *2*, 259–279.
47. Czyńska, K.; Rubinowicz, P. Sky Tower impact on the landscape of Wrocław—Analysing based on the VIS method. *Architectus* **2017**, *50*, 87–98.
48. Oleński, W. Postrzeganie Krajobrazu Miasta w Warunkach Wertykalizacji Zabudowy (en. The Perception of Cityscape in the Environment of Verticalization of Buildings). Biblioteka Cyfrowa Politechniki Krakowskiej. Available online: https://repozytorium.biblos.pk.edu.pl/resources/26294 (accessed on 1 November 2019).
49. Czyńska, K.; Rubinowicz, P. Application of 3D virtual city models in urban analyses of tall buildings—Today practice and future challenges. *Archtecurae Artibus* **2014**, *19*, 9–13.
50. Rubinowicz, P.; Czyńska, K. Study of City Landscape Heritage Using Lidar Data and 3d-City Models, The International Archives of the Photogrammetry. *Remote Sens. Spat. Form. Sci.* **2015**, *XL-7/W3*, 1395–1402.
51. Szczepańska, A.; Pietrzyk, K. Evaluation of public spaces in historical centers of small towns—Case study. *J. Urban Plan. Dev.* **2020**, *146*, 05019020. [CrossRef]
52. Adamczewski, J. *Mała Encyklopedia Krakowa (en. Little Encyclopedia of Krakow)*; Wanda Press: Kraków, Poland, 1997.
53. Marcinek, R.; Myczkowski, Z.; Bohm, A. *Bulwary Wiślane*; OKiCKN Press: Kraków, Poland, 2012.
54. Bieniarzówna, J.; Małecki, J.M. *Dzieje Krakowa. Tom 3. Kraków w latach 1796–1918 (The History of Krakow. Volume 3: Krakow in the years 1796–1918)*; Wydawnictwo Literackie: Krakow, Poland, 1994.
55. Kraków w Liczbach (Krakow in Figures). 2019. Available online: https://www.bip.krakow.pl/?mmi=6353 (accessed on 1 November 2020).
56. Palmer James, F. A Landscape Assessment Framework for Visual Impact Assessment in the USA. *J. Digit. Landsc. Archit.* **2016**, *1*, 10–17.
57. Myczkowski, Z.; Forczek-Brataniec, U.; Wielgus, K. Plan ochrony Parku Kulturowego Stare Miasto w Krakowie w kontekście ochrony i planowania krajobrazu miasta (en. Protection plan of Cultural Park of Downtown of Cracow in the context of conservation and planning of townscape). In *Planowanie Krajobrazu: Wybrane Zagadnienia*; Przesmycka, E., Ed.; Wydawnictwo Uniwersytetu Przyrodniczego w Lublinie: Lublin, Poland, 2013; pp. 38–60.
58. Ozimek, A. *Landscape Measure. Objectification of Views and Panoramas Assessment Supported by Computer Tools*, 1st ed.; Cracow University of Technology: Cracow, Poland, 2019.
59. Graczyk, R. *Identyfikacyjna Funkcja Dominanty Architektonicznej w Strukturze Małego Miasta (en. Identifying Function of an Architectural Dominant in the Structure of a Small Town)*, 1st ed.; Poznań University of Technology: Poznań, Poland, 2015; p. 202.
60. Ozimek, A. Landscape dominant ELEMENT—An attempt to parametrize the concept. *Tech. Trans.* **2019**, *116*, 35–62. [CrossRef]
61. Forczek-Brataniec, U. The Degree of Visibility as a Tool to Assess and Design of a Visual Corridor. In *Proceedings of the Fábos Conference on Landscape and Greenway Planning, Budapest, Hungary, 30 June–3 July 2016*; Valanszki, I., Jombach, S., Filep-Kovacs, K., Fabos, J.G., Ryan, R., Lindhult, R.M., Kollanyi, L., Eds.; Szent Istvan University: Budapest, Hungary, 2016; Volume 5, pp. 295–302.
62. Chmielewski, S. Chaos in Motion: Measuring visual pollution with Tangential View Metricks. *Land* **2020**, *9*, 515. [CrossRef]

63. Palmer, J.F. The contribution of a GIS-based landscape assessment model to a scientifically rigorous approach to visual impact assessment. *Landsc. Urban Plan.* **2019**, *189*, 80–90. [CrossRef]
64. Rubinowicz, P. Protection of the waterfront panoramas based on computational 3D-analysis. In *Architecture in the Age of the 4th Industrial Revolution*; eCAADe + SIGraDi: Porto, Portugal, 2019; pp. 325–332.
65. Rubinowicz, P. Application of the Visual Protection Surface Method (VPS) for Protectionof Landscape Interiors within a City. *IOP Conf. Ser. Mater. Sci. Eng.* **2019**, *471*, 092048. [CrossRef]
66. Różycka-Czas, R.; Czesak, B.; Cegielska, B. Towards Evaluation of Environmental Spatial Order of Natural Valuable Landscapes in Suburban Areas: Evidence from Poland. *Sustainability* **2019**, *11*, 6555. [CrossRef]
67. Zubel, H. Diagram Notation of a Concept in the Process of Architectural Design, Illustrated with the Example of Own Work, with Particular Focus on the Project of Community Centre in Grodzisk Mazowiecki. Available online: Htps://repolis.bg.polsl.pl/dlibra/publication/70083/edition/61663/content (accessed on 25 December 2020).
68. Hodor, K.; Fekete, A. The sacred in the landscape of the city. *Tech. Trans.* **2019**, *116*, 15–22. [CrossRef]
69. Bridgen, T. Value in the "View". In *Conserving Historic Urban View*; RIBA Publishing: London, UK, 2018.
70. Badach, J., Raszeja, E. Developing a Framework for the Implementation of Landscape and Greenspace Indicators in Sustainable Urban Planning. Waterfront Landscape Management: Case Studies in Gda´ nsk, Pozna´n and Bristol. *Sustainability* **2019**, *11*, 2291. [CrossRef]
71. Bucurescu, I. Managing tourism and cultural heritage in historic towns: Examples from Romania. *J. Herit. Tour.* **2015**, *10*, 248–262. [CrossRef]
72. González Santa-Cruz, G.; López-Guzmán, T. Culture, tourism and World Heritage Sites. *Tour. Manag. Perspect.* **2017**, *24*, 111–116. [CrossRef]
73. Woźniak, M. Ład przestrzenny jako paradygmat zrównoważonego gospodarowania przestrzenia (en. Spatial order as a paradigm of sustainable space). *Białostockie Stud. Prawnicze* **2015**, *18*, 167–182. [CrossRef]

Article

Living Heritage in the Urban Landscape. Case Study of the Budapest World Heritage Site Andrássy Avenue

Kinga Szilágyi [1,*], Chaima Lahmar [1], Camila Andressa Pereira Rosa [1] and Krisztina Szabó [2]

1 Doctoral School of Landscape Architecture and Landscape Ecology, Hungarian University of Agricultural and Life Sciences (MATE), 1118 Budapest, Hungary; chaimalahmer@gmail.com (C.L.); camilarosa.capr@gmail.com (C.A.P.R.)

2 Department of Garden and Open Space Design, Institute of Landscape Architecture and Urbanism, Hungarian University of Agricultural and Life Sciences (MATE), 1118 Budapest, Hungary; Szabo.Krisztina.dendro@uni-mate.hu

* Correspondence: proverde.53@gmail.com

Abstract: Historic allées and urban avenues reflect a far-sighted and forward-thinking design attitude. These compositions are the living witnesses of olden times, suggesting permanence. However, the 20th century's urban development severely damaged the environment, therefore hundred-year-old mature trees are relatively rare among city avenues' stands. Due to the deteriorated habitat conditions, replantation may be necessary from time to time. However, there are a large number of replanted allées and urban avenues considered historical monuments, according to the relevant international literature in urban and living heritage's preservation. The renewal often results in planting a different, urban tolerant taxon, as seen in several examples reviewed. Nevertheless, the allée remains an essential urban structural element, though often with a changed character. The Budapest Andrássy Avenue, a city and nature connection defined in the late 19th century's urban landscape planning, aimed to offer a splendid link between city core and nature in Városliget Public Park. The 19–20th century's history and urban development are well documented in Hungarian and several English publications, though current tree stock stand and linear urban green infrastructure as part of the urban landscape need a detailed survey. The site analyses ran in 2020–early 2021 created a basis for assessing the allées and the whole avenue as an urban ecosystem and a valuable case study of contemporary heritage protection problems. Andrassy Avenue, the unique urban fabric, architecture, and promenades have been a world heritage monument of cultural value since 2002. The allées became endangered despite reconstruction type maintenance efforts. The presented survey analyses the living heritage's former renewal programs and underlines the necessity of new reconstruction concepts in urban heritage protection. We hypothesize that urban green infrastructure development, the main issue in the 21st century to improve the urban ecological system and human liveability, may support heritage protection. The Budapest World Heritage Site is worthwhile for a complex renewal where the urban green ecosystem supply and liveable, pedestrian-friendly urban open space system are at the forefront to recall the once glorious, socially and aesthetically attractive avenue.

Keywords: allées; urban avenues; World Heritage Site; urban habitat; urban ecosystem supply; renewal method

Citation: Szilágyi, K.; Lahmar, C.; Rosa, C.A.P. ; Szabó, K. Living Heritage in the Urban Landscape. Case Study of the Budapest World Heritage Site Andrássy Avenue. *Sustainability* **2021**, *13*, 4699. https://doi.org/10.3390/su13094699

Academic Editors: Jan K. Kazak, Katarzyna Hodor and Magdalena Wilkosz-Mamcarczyk

Received: 9 March 2021
Accepted: 17 April 2021
Published: 22 April 2021

1. Introduction

Complete allée renewals occur in city avenues. Due to the ecological alteration of the urban habitat and the deterioration of its quality, it is often impossible to keep the mature trees and even the original tree species in the necessary renewal process. From the many examples, we highlight Andrássy Avenue's birth [1] and renewal in Budapest. A 19th-century development went on a staged, total renewal with the tree species' substitution in several phases during the 20th-century. Currently the avenue offers a different streetscape in each segment (regarding the taxon, the age and quality), somewhat adapting to the changing urban habitat conditions. However, the original compositional intention of urban open space design is now the only history, as the avenue's image has radically changed. A new approach would be needed to preserve the listed world heritage's living components and strengthen the urban green infrastructure element's ecosystem supply.

1.1. Urban Allées–from Garden Art to Urban Landscape Architecture—A Literature Review

According to the origin of urban allées, we have to distinguish the formal, functional and social aspects. The allée was a traditional garden art element in the late Renaissance. Then, allées in Baroque gardens and hunting parks stepped out into the open landscape with representation and directing functions. However, in private ownership gardens, they supplied only the privileged.

The London Hyde Park, originally a sizeable royal hunting park, was the first offered for restricted public use in 1637 [2]. The Berlin Tiergarten opened its gates for the public in 1659. The Austrian Empress, Maria Theresa, also allowed regular public use for Vienna's residents in the Prater (1766), the Augarten (1775) and the Belvedere (1776). These parks, or hunting fields, had a large-scaled garden character using perspective design and traditional garden elements like ponds and lakes, large meadows, roads, paths and architectural features [3]. They created the canopy by woods, clumps, and allées, even in the landscape garden period from the early 18th century. The Cracow Planty garden, built on the site of old town walls and towers in the early 19th century, is an excellent example of a residential activity that gave birth to a ring-form allée and garden system [4].

Trees and allées planned and planted in Paris in 1616 for a mere recreational carriage riding, served Marie de' Medici's pleasures. The allée run along the Seine beyond the Tuileries garden walls [5]. Later in the 17th century, tree allées "in the urban landscape were largely public spaces, to be enjoyed by many people, if not by all classes of society" [6]. A more than 1 km-long Unter den Linden in Berlin was planted with 2000 nut and linden trees in 1647. It ran from the city core to the open landscape, the hunting park, the Tiergarten, later a public park for the citizens of Berlin. In the 18th century, the city grew around the linden tree avenue. Hamburg had a similar double tree-lined avenue called the Jungfernstied in this period and later, in the 19th century, the famous Palmaille in Hamburg-Altona. Together with the Munich Maximilianstrasse, the Berlin Kurfürsterdamm, and the Düsseldorf Königsallée, these large-scaled green linear public open spaces perished during the World War II bombings. The avenues' reconstruction or renewal started only from the late 1950s and early 1960s, when tree nurseries' production could supply the cities with sufficient, good-quality saplings [7].

Urban expansion and the growing economic differences of the urban population in the 18th century led to severe tension in the society, among others, in environment quality and access to urban public open spaces. Along with the Enlightenment's philosophy, new ideas and principles for urban development formed the Embellishment movement. With four aims: order, hygiene, light, and cultural, aesthetic requirements, urban development could help solve social and urban environment problems. Urban embellishment utilised some large-scaled Baroque principles, the so-called French-style's garden design elements, as vues and allées. By the end of the century, with the superb French and British examples, most European towns created promenades, urban allées for public use [6].

The Paris Champs-Élysées might be one of the exemplary promenades and the Baroque garden style's living evidence of the origin of urban allées. It was André Le Notre, the royal garden designer of Louis XIV., who laid out the Tuileries Garden extension in 1667. The wide promenade with two rows of elm trees run from the royal palace to the Rond Point. When new sections were added, first in 1710 to the Place d'Étoile, then in 1765 to the Porte, the Le Notre style still ruled the design [8]. Since then, the Champs-Élysées, with its two double rows of trees and 70-m width, could accommodate both the urban traffic and the promenade function. However, the past decades' heavy urban car traffic seriously damaged both the environmental quality and social acceptance. According to

Paris chief major's statement, an exemplary renewal may come in the short future. Anne Hidalgo told the *Le Journal du Dimanche* on January 10, 2021, that she envisioned an extensional urban open space, a so-called linear garden where car traffic would be halved to give pedestrians space, intensive greening and liveable urban open space.

Besides Paris and the Champs-Élysées, we may refer to many other European examples, but search for previews to the Andrássy Avenue development in the Hungarian capital, Vienna, and even the German-speaking countries undoubtedly offer suitable models for urban allées. Until the mid-19th century, the Austrian capital offered urban recreation only in the suburban green areas and allées, for example, the Prater with its large-scaled main allée. Vienna maintained the town wall system and the old town centre's glacis where building was strictly prohibited. The glacis, crisscrossed by various allées linking the old town and the suburbs, gave place to the new Ringstrasse only after the wall demolition in the 1850s. As a result of a planning competition, the new, tree-lined boulevard served the necessary transport and traffic needs, besides organised urban squares, public gardens, parks, and cultural institutions, such as theatres, museums, and government buildings [6].

Owing to the industrial and economic development during the twentieth century, the world witnessed the rise of many metropolises representing the centre of political power and residential compaction. However, the mass urbanisation and the dense population have negatively impacted the cities' historic cores and urban environment. Due to the significant changes in the daily activities and needs of the citizens, it is unlikely for a traditional city to retain its original character; hence, the concept of urban conservation aimed to preserve historical cities as recognised heritage and help to maintain the sense of continuity and tradition [9].

In addition to the international urban conservation guidelines, regional and national regulations are mandatory to consider the treated region's local specificities, as a case in point, the example of the Five Avenues District in Tianjin, China. The region's British occupation profoundly impacted the district's development, resulting in numerous high-quality residential areas for the British concession, foreigners, and Chinese celebrities. On the other hand, a new neighbourhood borne with the development in the 1930s. However, the Five Avenues District's original function has changed throughout time to become a more mixed-use district. Nevertheless, the district has shown the capacity to meet the contemporary demands and save its authenticity owing to local conservation strategies. The scientifically based conservation plan launched by stakeholders aimed to preserve the urban and regional heritage by preventing high-rise buildings from the perspective view. Moreover, the plan aimed to buildings' categorisation based on historical values and requirements and applied a technical renovation for each category with strict control for non-historic buildings' facades. Lastly, stakeholders invited the community to the conservation procedure by creating online surveys and collecting data about citizens' contemporary demands. The municipality's new implementations align with the historic urban landscape paradigm and promote society's development [10].

1.2. Andrássy Avenue and Its Importance in the Urban Landscape

The story of Andrassy Avenue can be traced back to the times after the Austro-Hungarian Compromise of 1867. The establishment of an avenue connecting the city centre and the City Park (Városliget) was an essential element in Prime Minister Gyula And-rássy's urban development programme [1]. The avenue got its name after him in 1883, to be later renamed Stalin Avenue in 1950, Avenue of Hungarian Youth in 1956, Avenue of People's Republic in 1957 and finally, in 1990, the original and authentic name of Count Andrássy [11]. The long list of street names and political stress to rewrite the city-text reflects a Hungary tradition, similarly to several Central and Eastern European countries, although here in Budapest, the Andrássy Avenue has left the public memory [12].

As aforementioned, the renaming of important streets usually derived from cultural politics, especially in post-socialist eras. The cities played an essential role in socialist regimes [13], where the symbolic control of urban centres belonged to the socialist system [14]. Therefore, this can be interpreted as a reconfiguration of both the places and their history. An example of such is the Bulevardul Carol I, located in Bucharest and named after Romania's first king, which later became Bulevardul Republicii (Boulevard of the Republic) in the way of de-commemoration of the previous regime and commemorating the socialist era. After the revolution of 1989, the original name of the boulevard was reinstated [15]. Within a more contemporary scenario, the Romanian Timisoara city renamed several streets after martyrs, mainly in the 1990s and after 2010. This political approach proved to be not so welcomed by the population due to cultural and historical aspects and the economic and practical ones as well [13]. Another example is Capital Street (Ulica Stoteczna) in Poland. This artery was planned in 1919 to connect the first housing development of the Warsaw Housing Cooperative to the city centre. In 1934, it was renamed as Popiehiszko Street, after the chaplain Jerzy Popieluszko. Even though he was cherished by the community and played an essential role during the Solidarity movement, this decision was contested by some part of the populations, claiming that Capital Street had too great historical value to be renamed [16].

Moving back to the Hungarian scenario, the plan of the great Andrássy Avenue had been elaborated by 1876. The promenade is divided into three segments, with a broader space in each part, offering a pleasant and elegant open space suited equally for carriage, horse rides and walks [17] (Figure 1). The greenway consisted of double-tree lines in its first segment running from the city centre. Further, in the second and third sections running through the outskirts, the plantation formed 2 × 2 lines in the even broader cross-sections. This exquisite design symbolised the experience of arriving from Town to Nature; the latter represented by the first urban public park, the City Park of Budapest, designed by H. Nebbien in classical landscape garden style. The approximately 2km-long avenue offered the green corridor of almost 600 plane trees (*Platanus* × *hibrida*), the same species used in the Városliget's allées and clumps [18,19].

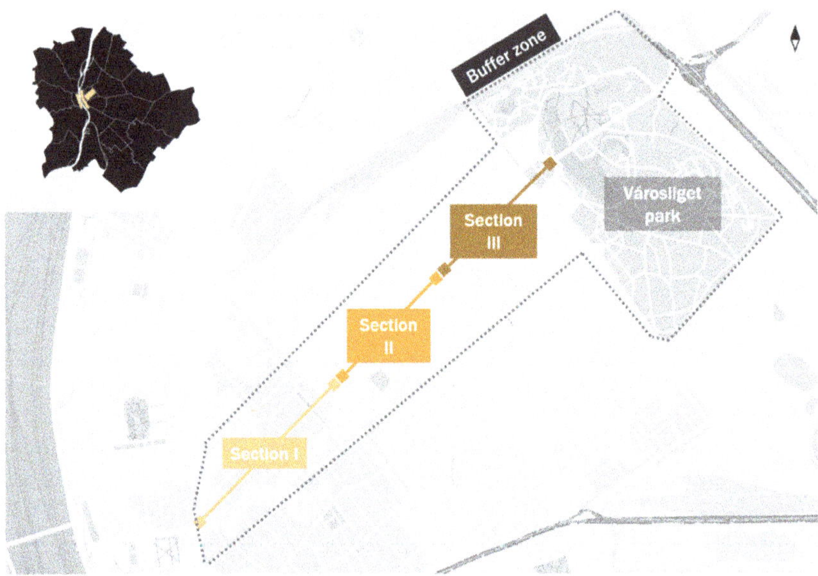

Figure 1. Map of the inner Pest region with the three sections. (edited by Rosa & Lahmar).

During the preliminaries to the millennium exhibitions and ceremonies of 1896, the first continental underground was built along Andrássy Avenue, emphasising the protection of the well-established twenty-year-old plane trees with lovely canopies at that time. Owing to the careful planning and execution, only 23 trees needed to be felled, primarily because of the stations' construction. Thanks to the Public Works Council's commitment and the City Park's entire community, the avenue's reconstruction reflected an ideal investment approach in terms of environmental ethics aspects (Figure 2).

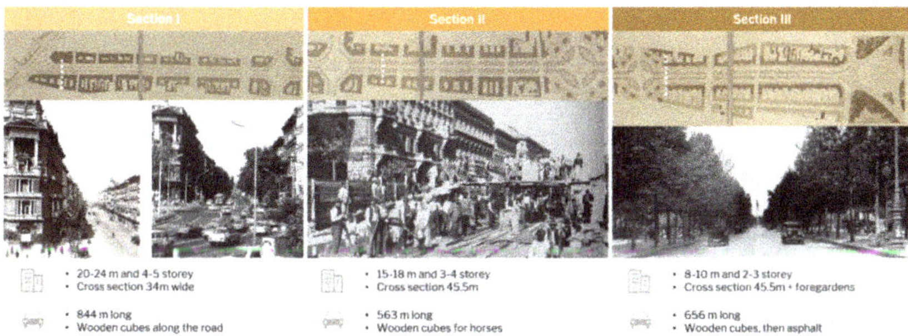

Figure 2. Character and cross-sections along Andrássy Avenue (19th–mid 20th century). (edited by Rosa & Lahmar).

However, the plane tree avenue as an urban habitat suffered its first damage due to the underground's structure that strongly inhibited the roots' spread. Building debris and stone caused severe deterioration in the natural soil system. Instead of the original wooden cube cover, the new cobblestone cover resulted in a decreased soil's ventilation.

The First World War caused severe destruction in Andrássy Avenue since, because of the overall coal shortage, people used the trees as firewood. As a result, in 1920, the whole avenue was replanted. This time, as if correcting the earlier mistake, the nettle tree (*Celtis occidentalis*) was used in the inner part where the living space is tighter and more shaded, thus not convenient for *Platanus* trees. The deterioration of the allée's habitat and quality accelerated in the 1930s when the block-pavement was replaced by asphalt. Because of the growing vehicle traffic and consequent road salting in winter, by the 1960s, the plane trees' decline became visible. Degraded trees cannot withstand the attack of bacteria, fungi and animal pests known in the literature as "weakness" pests; the Andrassy's plane trees suffered from the spread of *Apiognomonia veneta* disease. The spacious crown shape has overgrown the boulevard space over time. The dark foliage tunnel, however, in calm weather, might have reduced the vertical mixing and transmission of vehicle's pollutants, which led to a further problem because of the ever-increasing traffic [19].

Despite the deteriorating health of the trees, the avenue still showed an almost enclosed vaulted structure. According to Mihály Mőcsényi's professional opinion, the plane trees should have been rejuvenated in the early 1970s at the latest, when the signs of deterioration were clear. The tree's disease and destruction is the symptom of the complex "urban damage". Modern urban landscape planning principles and high quality technical and reconstruction methods may reduce the disadvantages [20].

Unfortunately, nothing happened to improve the urban habitat along Andrássy Avenue. The tree survey in 1987 recorded merely 36 healthy plane trees. The repair and preservation of the plane tree avenue became more and more desperate. In 1996 and finally, in 2005, a total renewal went on using ash trees (*Fraxinus excelsior*) in the middle and columnar ash trees (*Fraxinus excelsior* 'Westhof's Glorie') in the third section [19]. For a decade, the ash tree avenue showed promising growth. Still, the once supposed urban tolerant taxa seem to start suffering from the traffic burden and a new ash tree disease. In contrast, the columnar habit of ash trees presents an entirely different view from the original stand.

The total avenue renewal might be inevitable since there was no chance and no will to reduce traffic intensity. However, the magnificent, almost closed green vault disappeared with the plane trees, which was characteristic in the middle and outer part of the Andrássy Avenue and evoked 19th-century garden art. The new plantation character, the narrow conical canopy shape and the weaker branch system, and the smaller foliage are not ideal for the avenue's scale. The new plantation along the allée offers a far-too urban image with the canopy branching grown on a high stem (around 3.50 m). The result of the total renewal is a certainly lasting avenue ensemble that requires less tree maintenance but unfortunately distorts Andrássy Avenue's design philosophy, its character, and aesthetic value.

In 2002 the Andrássy Avenue with the buffer zone in the 6th-district, the Heroes square, and Városliget urban park in the 14th-district won World Heritage Site's honorary title as cultural heritage (Figure 3). Unfortunately, owing to excessive urban development in the protected zone, the UNESCO World Heritage Centre raised severe objections. It warned the Hungarian government that the Budapest site could be declared endangered [21]. The related Decision 43 COM 7B.84 of WHC/19/43.COM/18 severely criticised the potential impacts of extensive developments within the Városliget area and the Buda Castle Quarter. It noted, with regret, the state of property conservation and the overall approach to conservation and development. Unfortunately, the phenomenon is similar to post-socialist cities' actions that followed a path of economic growth and the idea of neoliberal competition in a regional context. Urban policy and development served city branding and the dynamism of economic growth in the end [22]. As a result, conservation considerations were relegated to the background. It can be seen in the properties' renewal of valuable public and private buildings in the listed areas and zones as along Andrássy Avenue.

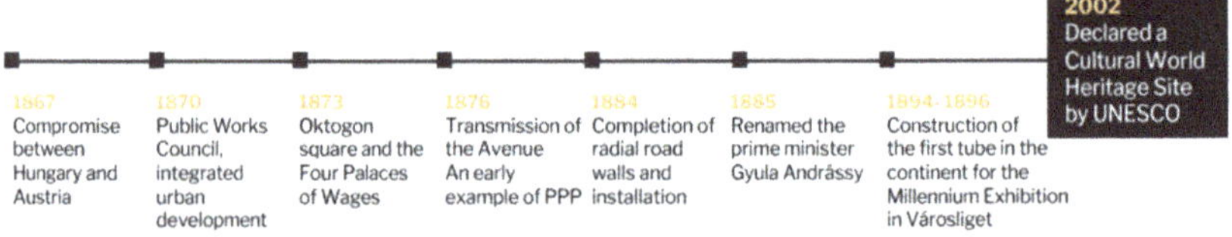

Figure 3. Andrássy Avenue development—Timeline. (edited by Rosa & Lahmar).

Along with our hypothesis that the contemporary European issue of urban green infrastructure should be a vital issue in heritage protection, we highlight the moments when the urban avenue's allées played a determining role in the construction and renewal process. In this way, the urban landscape aspect would call attention to the central issue in the 21st century's urban liveability and ecological matters. On the other hand, we will add new elements to the academic debate on the renewal possibilities and ways. Besides elementary urban ecological aspects, the social context and the visual-aesthetical embellishment are essential in planning and decision making. The debate about urban tolerant taxon and the diversity in planting design raises questions in landscape architectural harmony and sustainability aspects.

2. Materials and Methods

Andrássy Avenue's research reflects the general scientific method used in landscape architecture topics. First, the literature survey helped the historical introduction of urban allées and Andrássy Avenue development. Then, we run a detailed site analysis to understand the avenue's character, role and values in the urban structural–functional and ecosystem supply contexts. As for the greenery, the tree stock investigation consisted of dendrological and dendrometrical visual analyses in the same period of the year.

2.1. Site Survey of the Urban Avenue as a Linear Green Open Space

A site survey of the urban landscape went on in the autumn and winter of 2020–2021 to analyse Andrássy Avenue as an urban linear green space, part of the Budapest World Heritage site. This part of the research focuses on the urban landscape, the architectural values, the urban habitat, the environmental problems, public-use forms and intensity, and urban character. As aforementioned in this paper, Andrássy Avenue has three different sections with typical cross-sections and urban landscape characters; these sections create the logic of site analyses and evaluation. This part of the research focuses on Andrássy Avenue's development documented in urban planning maps and publications. During the site survey, we detected the public-use forms and the intensity. We examined the urban landscape's character in each section and subsection on trees' quality and uniformity, and urban habitat.

2.2. Dendrological and Dendrometrical Survey

The trees are the "aristocrats of the plant kingdom" [23], and in an urban setting, a specimen with a beautiful, healthy conspicuous crown can be a prominent element. Trees in the urban environment have a unique value [24], and with proper preservation, their outstanding value may be higher from year to year. Urban habitat can be stressful and far from ideal for urban tree planting; still, we expect healthy tree growth to fulfil the maximum in all aspects of the trees' ecosystem supply from environmental regulation to social and aesthetic functions. The urban conditions and factors (reflected, radiated heat, polluted air and soil, lack of intensive maintenance, traffic burden, social overuse, restricted root volume, soil compaction, etc.) cause severe stress to urban trees' visible signs quality [25,26]. The tree stock's measurement and evaluation along Andrassy Avenue used the Hungarian Association for Tree Management's [27] method and criteria, based on European visual tree assessment methods [28,29]. The dendrological survey is the overall tree recording, where the dendrometric characteristics express the full tree height, trunk height, trunk circumference size, trunk diameter, and crown diameter in numbers. Trunk diameters come in two ways; either with the average of two diagonals' measured values or with the calculation out of the trunk circumference size. The diameter's measurement or its calculation is essential for the definition of the tree's approximate age.

The trees' general data are as follows: survey number of trees, scientific and common name; Individual data: total height, trunk height, average trunk diameter at an altitude of 1 m, average crown diameter, breast height (130 cm), and trunk diameter in specimens tested instrumentally.

All trees are unique and form a physiologically coherent whole. A detailed examination is necessary because the habitat condition is often different from ideal, which affects the whole tree's life chances. During the visual inspection, the general state of the roots, trunk, and crown are essential. We applied the EU conform method developed by MFE (association of Hungarian arborists) [30], which uses five parts and values (value aspects details in Appendix A): A–Root system including roots and collars and the type and condition of tree's plantation site; B–Trunk condition; C–Crown condition, including the crown base and the full crown (branches, branchlets, twigs, and shoots with leaves); D–Assessment of viability; E–Degree of care, maintenance.

Visual examination of the root surface and habitat (A) is the most decisive in the root system survey without excavation. The shape of the root collar and its changes and injuries refer to the whole root system; therefore, examining these elements together gives the root system value. The trunk (B) condition fundamentally determines the tree's health; in the rotting case, the static state of wood deteriorates, limiting the nutrient and water transport. To visually inspect the trunk, it is also essential to consider the root collar and crown base, as their condition also affects the trunk. The crown base and root neck are the two most sensitive parts of the tree. We evaluated the crown structure (C) together with the crown base condition, during which the primary consideration is to determine the ratio of real to ideal foliage weight. Viability (D) depends on the crown's disease, the tree's health as a whole, and the root's and trunk's condition. The degree of tree-care (E) should be given to the ideal maintenance. Individuals who have their physiological needs, have a species-specific growth vigour, and have been cared for in timely and sound quality can be considered optimally groomed trees. With all these assessments, wood care work will become more predictable in the long run. The general condition's indicator referring to the overall tree value is a percentage derived from the above survey's values. The calculation formula: (A + B + C + D + E − 5)/20.

3. Results of Site Analyses and Tree Survey

The first section, or inner part, comprehends Bajcsy-Zsilinszky—Oktogon; the second, middle part, is from Oktogon to Kodály körönd; the third, outer part, extends from Kodály körönd to Heroes Squares. Within the main sections, we can differentiate subsections bordered by crossing streets and the two main squares—Oktogon and Kodály körönd—and the Heroes' Square at the City Park termination (Figure 4).

Figure 4. The three sections and their typical urban landscape character. (Section 1. *Celtis occidentalis*; Section 2. *Fraxinus excelsior*; Section 3. *Fraxinus excelsior* 'Westhof's Glorie'. (edited by Szilágyi, K.; source: https://www.nomadepicureans.com/europe/hungary/free-budapest-walking-tour-self-guided/ (accessed on 27 February 2021)).

3.1. *Andrássy's Urban Character and Functions*

Alongside the street, public and private buildings, cultural and educational institutes, and valuable monuments create the elegant urban space walls reflecting the most exemplary architecture of late classicism and eclectic style. Here is the famous Hungarian State Opera, for example. Moreover, urban design and architecture quality are equal to the 19th-century European urban development and renewal trends.

The research findings for the urban land use and functions show that Section 1 has an imposing metropolitan atmosphere, together with strong commercial and cultural values, high tourism activity and internationally famous brands' shopping stores (Figure 5a). With a significant profile change, Section 2 presents an increase in urban green character. Consisting of more residential and educational buildings and with a double tree-lined promenade on each side, this part still has a high tourism activity (Figure 5b). From Kodály körönd square to Városliget (City Park), in Section 3, the avenue reflects a more "natural" atmosphere. There, detached housing and villas' gardens resemble a significant increase of green coverage. Continuing with the doubled tree-lined promenade, many buildings here function nowadays as embassies, institutions, hotels, museums, and restaurants, maintaining the tourism activity (Figure 5c).

Regarding the allées traffic & functional connections, Andrássy can be considered an urban motorway with heavy traffic throughout. Starting at Bajcsy-Zsilinszky, the first section with a 34m width consists of the main road (2 + 2 traffic lanes) and a 1 + 1 bike lane, up to the Oktogon (Figure 5a). The 45.5 m wide second section progresses to Kodály körönd with the main road plus 1 + 1 service roads and 1 + 1 bike lane (Figure 5b). The width and traffic lane structure is similar in the third section, but the residential villas' front gardens add extra green to the urban landscape. (Figure 5c). Furthermore, there is no bus line, except for tourism buses going to and from Heroes square. Multiple stops of the metro line 1 are also present, connecting the city centre to the Városliget Park and beyond.

The degradation of Andrássy Avenue from an open and green urban linear space system into a busy urban motorway resulted in the loss of the original, glorious intention to create a large-scale link to the nature of Városliget. Especially in the third section, the new tree plantation with the compact crown form and the unnatural trunk height serve only the heavy traffic and the tourist buses.

Figure 5. Urban context, functions, monuments and character; (**a**) Section 1; (**b**) Section 2; (**c**) Section 3. (edited by Rosa & Lahmar; sources: https://pocketoz.com.au/suitcase/dest-budapest.html (accessed on 27 February 2021); https://www.budapesttips.co.uk/budapest-attractions/opera-budapest/ (accessed on 27 February 2021).

3.2. Green System and Habitat Analyses along Andrássy Avenue

All along Andrássy Avenue, the urban habitat and the trees' vulnerable condition are negatively impacted by the urban context and functions (Figure 6). The dense traffic, four-five-storey high buildings (typically in Section 1) causing sizeable shaded area together with compacted and polluted soil and artificially restricted place for the root system resulted in severe damages, like heavily tilted trunks and deformed crowns. The century-old hackberry trees need regular crown shaping maintenance to improve stability.

Figure 6. Stressful urban habitat along Andrássy Avenue, Section 2. and 1. (edited by Rosa & Lahmar; source: http://holpihenj.com/andrassy-ut/ (accessed on 27 February 2021)).

Due to the broader cross-section in the second part and the tree canopy's lower density, the crown conditions are slightly better than in Section 1. However, during the last maintenance procedure, the artificial grass layer disappeared, revealing the poor root state, causing trunk damages and diseases. The third section seems the best thanks to the even broader cross-section, the linear green area offering a bit healthier habitat for the trees and the protective planting design applied on both sides of the linear green space.

3.2.1. Urban Habitat and Planting Places

There are different tree-planting situations in the three sections, which strongly determine the life cycle and lifespan and their decorative character and health conditions (Figure 7). The examined trees were 816 in three sections and two squares among different states (Table 1). In terms of individual numbers, the data are as follows: Ten trees stand in the four green islands of Kodály körönd square; 13 trees are in Oktogon square along the pavement. The first section contains 182 trees in two rows (A- 95, B-87), the second section has 274 individual trees in four lines (A-74, B-68, C-59, D-73), and the third section has 337 trees in four lines (A-92, B-81, C-77, D-87). In Sections 2 and 3, the inner rows (B and C) are significantly weaker because of the traffic disturbance and have much fewer trees than the outside lines.

Figure 7. Urban allée's habitat in the three sections from the strictly urban to a moderate urban green space, 2021 January. (photo: Szabó, K., 30 January 2021).

The root growth is severely limited in the first section because of small tree pits (1 × 1 m). In the case of older individuals, the roots push into the tree grates and covers. Due to the lack of space and the buildings' shading, the trunks are oblique, and the crowns asymmetrical.

In the second section, the trees grow in narrow green stripes without protection against trampling, compaction, salting, etc. The roots of medium, adult, and old trees are visible on the surface due to maintenance problems. Despite the intensive pedestrian use, the maintenance insisted on creating and keeping up green surfaces; as all plantation and greening efforts failed, artificial grass cover supplied the green surface colour without supporting, regulating green ecosystem supply. The roots started to grow upwards in the absence of sufficient oxygen. The artificial irrigation system, keeping the perennials' plantation, also motivated the root system towards surface growth. The wide pathway between the two green stripes is not water permeable. The narrow plantation line has no protection against winter salting, causing severe damage along the traffic road lanes. Salting material, dangerous for young trees' root system, has a high chlorine emission into the absorption zone. Thus, the overall environmental condition is deficient in Section 2, resulting in many trees dying out or felling.

The third section's habitat seems much better than the two previous. There are broad green stripes, semipermeable cobble cover on pathways, linear shrub plantation for protection against winter salting, and less intensive use.

Table 1. Primary data of the Andrássy tree survey. (edited by authors).

Sections	Length (m)	Character	No. of Rows	No. of Trees	Species (Dominant in Bold)
I.	844	closed rows installation, no service road and walkway, trees in tree pits, opposite plantations	1 × 1	182	*Celtis australis*, ***Celtis occidentalis***
II.	563	closed rows installation, service road and walkway, trees in the green bar, alternate plantations (triple bond)	2 × 2	271	***Fraxinus excelsior***, *F. ex.* 'Westhof's Glorie', *F. pennsylvanica*, *Platanus × hispanica*, *Ulmus grabra*, *Celtis australis*
III.	656	closed rows installation, service road and walkway, trees in the green bar, alternate plantations (triple bond)	2 × 2	337	*F. ex.***'Westhof's Glorie'**
Oktogon		plantation in tree pits, along the four square borders	1 × 1	13	*Celtis australis*, ***Celtis occidentalis***, *Platanus × hispanica*
Kodály Köröndd		the plantation is in the green area on green islands	1 × 1	10	***Platanus × hispanica***, *Celtis occidentalis*

3.2.2. Species Uniformity/Diversity

Seven different taxa live nowadays in the allée owing to several reconstructions and renewals (Figure 8). There are still some old individuals, such as the sycamore trees (*Platanus × hispanica*) from the first plantation on Kodály körönd square, and in the second section from the replantation program after WW1. Old *Celtis* trees are still in Section 1, and one single old elm tree (*Ulmus glabra*) planted in 1949 stands in the middle area. Hackberries (*Celtis* spp.) grow in the squares too. The first and second sections show a great variety in the time dimension.

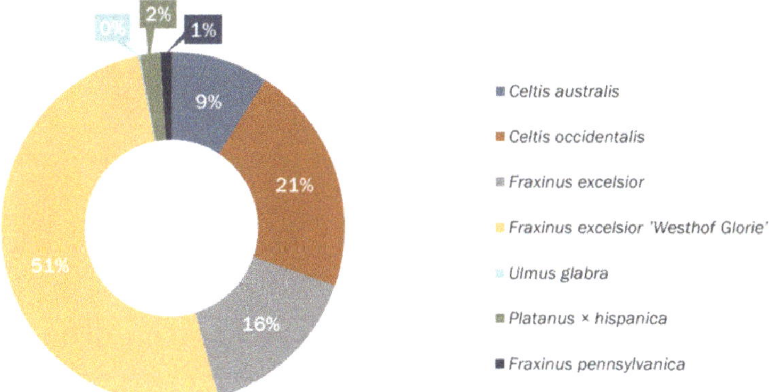

Figure 8. Taxa variety in Andrássy Avenue. (edited by Szabó, Rosa & Lahmar).

Due to the several replantation periods and the variable planting design, uniformity is only characteristic in the third section. The first part contains old and younger trees, where older ones are *Celtis occidentalis*, while the younger, newly replanted trees are *Celtis australis* and *C. occidentalis* mixed. The taxa confusion is more significant in the second section. Some old sycamores and a single *Ulmus glabra* stand there, with several *Fraxinus* taxa. The last year's replantation resulted in a new taxon, the *Celtis australis* appearing in the first and middle sections (Figure 9).

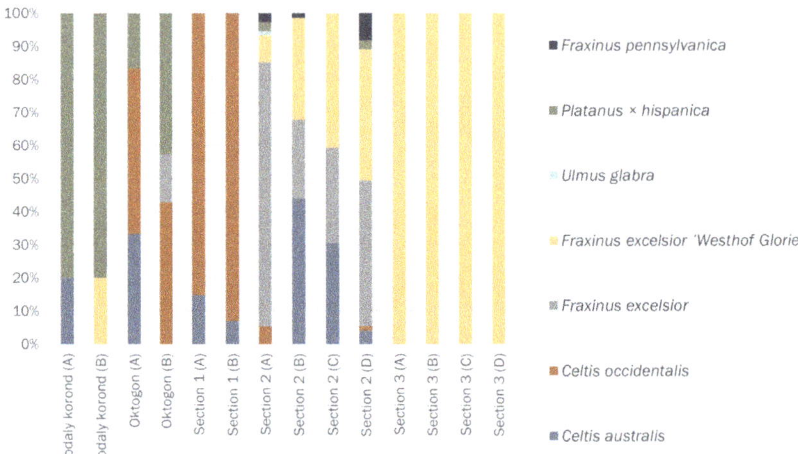

Figure 9. Tree taxa in the linear sections and the two main squares. (edited by Szabó, Rosa & Lahmar).

3.2.3. Age Uniformity

The oldest trees live in the Kodály körönd square, while the youngest individuals stand in the second section. The age difference between the oldest and the most immature trees is around 150 years. In many cases, the difference is about 70–80 years. Many trees had to be replanted in the past decade owing to severe environmental load and plant protection problems. The proportion of young or freshly planted trees is high in Section 2, line B and C (Figure 10a,b).

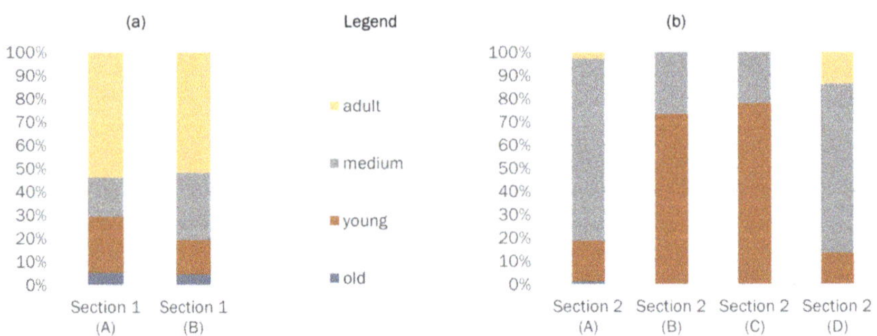

Figure 10. Tree's age variety; (**a**) Section 1; (**b**) Section 2. (edited by Szabó, Rosa & Lahmar).

3.2.4. Condition of Roots, Root Collars and Tree Pits

Based on Rado's value (see Appendix A), the relationships of roots, root collars and planting circumstances are different among places and sections. Individuals who did not show any abnormal symptoms (ground cracks, root congestion, girdling roots, damaged roots, bark, phloem, xylem injury, fungal infection, etc.) received excellent values. The highest value (5) is characteristic in the second section due to the many new replanted trees; good and medium values are typical in the first section thanks to healthy rooted individuals (Figure 11).

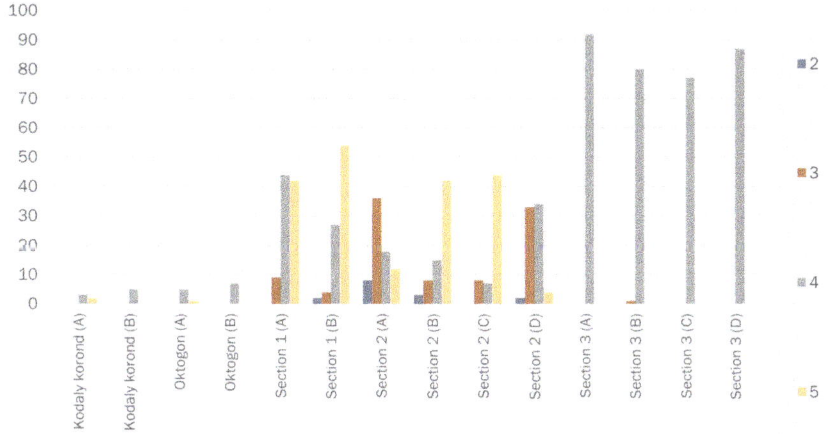

Figure 11. State of the root system in the studied area. (edited by Szabó, Rosa & Lahmar).

3.2.5. Trunks' Condition

The trunk condition shows relatively high values. Severe wounds, damages, and serious injuries are infrequent (Figure 12). Values B in Appendix A.

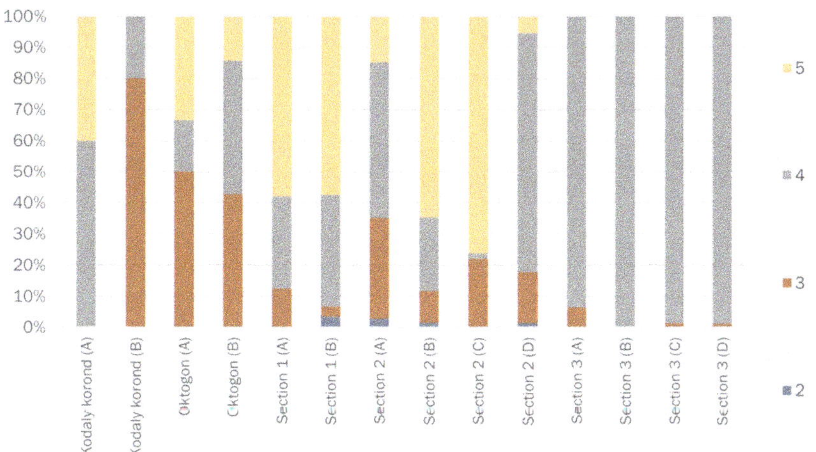

Figure 12. Correlations of the trunk values in the studied sections. (edited by Szabó & Rosa & Lahmar).

3.2.6. Crowns' Condition

The crown shape and size depends on location and retention condition. If root development is limited, water absorption and the nutrient transition is also limited, and the crown shows a less decorative value than the ideal. A value of 5 is scarce, almost only in the case of a new plantation as in the inner row C in Section 2 (see in Figure 13). Value 2 occurs in some individuals with foliage loss of more than 50% and severe crown damages. Value 4 reflects a good tree stand with a foliage loss of 11%–25%. We found many weak crowns in (values 3 and 2) Section 2 during the survey, where replantation is necessary within five years.

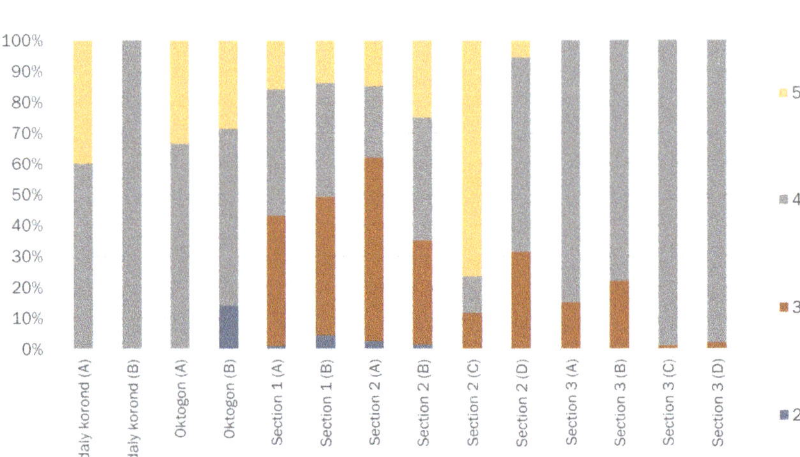

Figure 13. Correlations of crown values in the studied sections (value C in Appendix A). (edited by Szabó, Rosa & Lahmar).

3.2.7. Correlation of Vitality

The vitality is a complex value, and it is difficult to determine, as assumptions are needed regarding future habitat conditions. Trees planted in congested urban environments, especially when planted in a tree pit, cannot survive without proper maintenance. Thus, only in exceptionally optimal cases did the individuals receive a rating of 5 (excellent) for their vitality; otherwise, trees with good values received only 4 (it can reach maximum age by intervening). At the time of our study, only one individual needed urgent removal (values 1) (Figure 14).

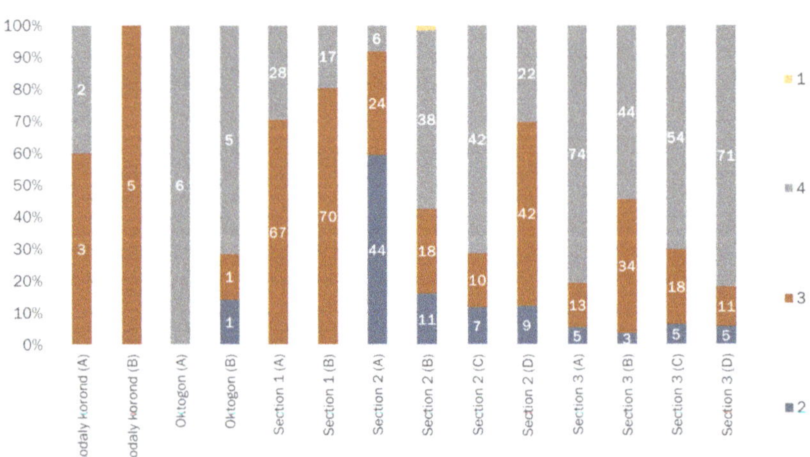

Figure 14. Correlations of vitality values in the studied sections (value D in Appendix A). (edited by Szabó, Rosa & Lahmar).

The professional maintenance experienced a 15-year decrease in life expectancy due to the polluted and disturbed urban environment. This is the second section's very situation so far (Figure 15).

Figure 15. Trees' damages, diseases and accidental falls. (edited by Szilágyi, Rosa & Lahmar; photo: Szilágyi, Szabó, Rosa, Lahmar, Szaller, 2010, and online source: https://www.fokert.hu/andrassyut/ (accessed on 15 October 2020).

4. Discussion

As a summary of the survey, the first statement made concerns the weak condition of the urban allée's trees due to the unfavourable habitat offered in the linear open space. The main problem is the continually deteriorating condition of the urban environment with pollution and overall environmental burden caused by the urban motorway, the intensive traffic generating an unhealthy urban microclimate, a severe heat island effect, air and soil pollution, lack of natural soil system, contamination, and physical damages due to all sorts of traffic. The lack of green surface and the restricted root system in Section 1 resulted in a weak tree condition and many old trees dying out. The direct car traffic's disturbance is responsible for both inner tree rows' inadequate status in Sections 2 and 3.

Trunk and crown deformation is an aesthetical problem, and the visual aspect is a critical element of the urban landscape. Due to the buildings' proximity, some older individuals bend towards the street or the sidewalk, posing a hazard to both properties (cars, bicycles, motorcycles, etc.) and passers-by. Diseases and fungal infections cause tree dying and increase the risk of branch or trunk rupture. The damaged and exposed root system, spreading and breaking the pathways, adds to the risk to pedestrians and ruins the paving materials and urban infrastructure.

After several reconstructions and renewals, we found seven taxa in the allée's rows that means a great variety according to the original urban landscape plan. The initial planting design offered the sycamore trees for the entire length of Andrássy Avenue. The choice had a strategic and aesthetical meaning and counted with the ecological needs of this taxon. The green link between the city and the nature of Városliget was a potent symbol in the 19th century's avenue program. However, another emblem appeared in the planting design. Palatine Joseph presented sycamore trees in large quantity from his tree nursery in Alcsút for the plantation of the new urban park, Városliget.

As for the ecological needs, the habitat seemed convenient for these trees as *Platanus* taxa need a direct connection to underground water sources. Since then, the urban development's growing intensity, construction of Metro 1, public works and tubes, and several underground garages in the close neighbourhood have lowered groundwater level [27]. Among other factors, the lack of groundwater supply played a determining role in the sycamores' unhealthy state. Nowadays, *Platanus* × *hispanica* represents only 2% in the whole area. However, we cannot deny that the cross-sectional features in the first section were not exactly conducive to the healthy development of *Platanus* trees.

The planted tree species are very different along Andrássy Avenue except for Section 3 (Figure 16). Still, there is no relationship among the number of individuals, tree species and various areas. The tree diversity is due to previous plantations' abandoned specimens, the extent and replacement of dead trees owing to environmental burden and conditions, and different professional considerations. Section 1 has a characteristic, mature *Celtis occidentalis* allée with more and more dying trees. Still, thanks to intensive maintenance, tree pits do not remain empty for a long time. In Section 1 and even in Section 2 with a new taxon, *Celtis australis*, the replantation went on here, hoping to maintain a healthy allée for a longer time. Section 2 seems the most confused and mixed in the number of taxa (six different taxa in two rows), the age and the quality context. Section 3 is homogenous according to the taxon and the age resulting from the last renewal program, but the condition is not so much promising. *Fraxinus excelsior* 'Westhof's Glorie' was considered urban tolerant in the early 2000s. Since then, the plantation's state has questioned this professional statement.

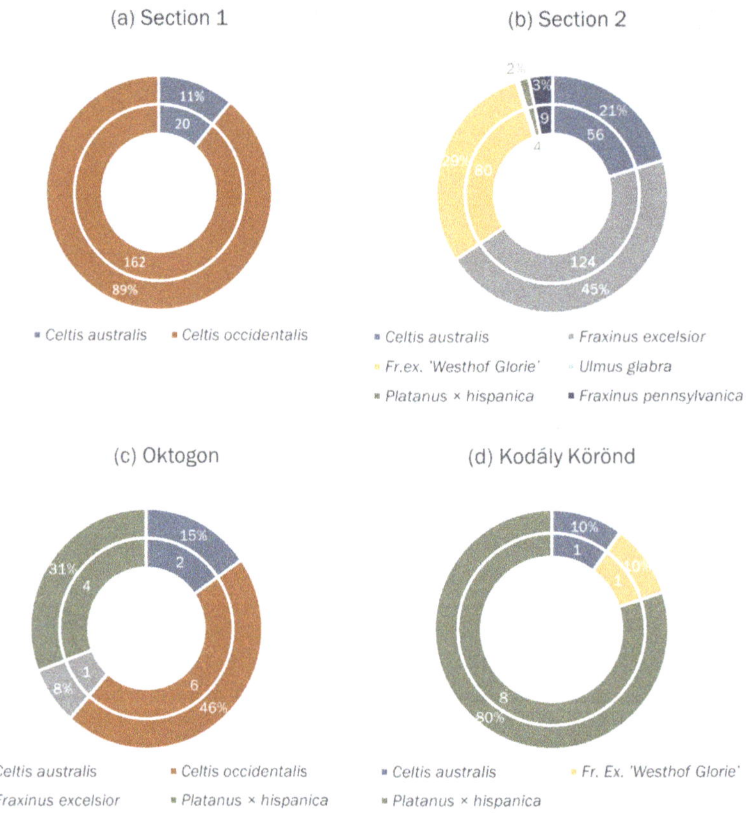

Figure 16. Tree's taxa variety (percentage and number of individuals); (**a**) Section 1; (**b**) Section 2; (**c**) Oktogon; (**d**) Kodály Köröndá. (edited by Szabó, Rosa & Lahmar).

Section 1 has a characteristic, mature *Celtis occidentalis* allée with more and more dying trees; thanks to the intensive maintenance, tree pits do not remain empty for a long time. In Section 1 and even in Section 2 with a new taxon, *Celtis australis*, the replantation went on here, hoping to maintain a healthy allée for a longer time. Section 2 seems the most confused and mixed in the number of taxa (six different taxa in two rows), the age and the quality context. Section 3 is homogenous according to the taxon and the age resulting from the last renewal program, but the condition is not so much promising. *Fraxinus excelsior* 'Westhof's Glorie' was considered urban tolerant in the early 2000s. Since then, the plantation's state questioned this professional statement.

The Andrássy Avenue tree survey confirmed the destructive effects of the urban environmental pressures on the trees' condition. Despite all maintenance efforts and even the recently replantation action, the allées and the green space obviously suffer in the given urban ecosystem. It is important to plan more feasible and efficient urban allée reconstruction strategies where successful precedent cases or current proposals may offer a new solution more effectively and sustainably. It would be possible to apply similar techniques, and approaches to the Andrássy Avenue case by precedent analyses. An excellent example of urban allée renewal is the ever-so-famous Paris avenue, the Champs Elysée.

The Paris Champs Elysée, the once model for the glorious urban avenue development in the 19th century, offers the new idea and model for a sustainable and livable urban renewal again. Three centuries after its official opening, the Paris' Champs Elysée avenue is still an outset of western modernity, inspiring several similar projects worldwide. The revolutionary design approach has changed the city's urban tissue and has created the impression of a city built with the prospect of infinite progress. Moreover, the unique charm and the symbolic dimensions make the Champs Elysée a site of national celebrations and major popular events, in addition to economic development and globalisation. However, since 1970 the avenue has lost its genuine aim and purpose, being a green open avenue for promenades and major festivities resulting from its significant economic, social, touristic, and political evolutions. The excesses of global over-tourism and intense automotive usage are negatively impacting the avenue's reputation and condition.

Throughout its history, the avenue had several renewal projects. In 1992, a renewal project launched by Paris city aimed to widen and raise the granite sidewalks, maintain the double rows of plane trees, place advertisement columns with telephones and tropical wood benches to improve the pedestrian experience. However, due to the intensity and type of use, renewal projects are necessary from time to time. The new project launched and communicated first in January 2021 by the official authorities' aims to transform the avenue into an urban demonstrator of sustainability and inclusivity.

The action plan has four aspects: 1. To reduce mobility nuisances by liberating more open space for pedestrians, and the installation of a silent pavement on the vehicular road by 2030, resulting in a drastic reduction in noise pollution; 2. Inventing new use forms by coordinating cultural and scientific events in parks, gardens, and green open spaces; 3. Installing free educational equipment for the younger generations while protecting and respecting the heritage and character of the avenue; 4. The project aims to rethink the ecosystem of the Champs Elysée by designing permeable pavement on the sidewalks, allowing water infiltration into retention basins, release the trees from their geometric shapes in favour of a free crown and root system expansion, increase the vegetation on the shrub level, and enrich the plantation in the gardens to increase the shaded areas and improve the evaporation for the benefit of biodiversity. Lastly, the project aims to collect data by installing sensors all along the avenue to monitor transportation flows, supply and evacuation of rainwater, benefit the economy, and environmental comfort in the urban ecosystem [31].

Unfortunately, Andrássy Avenue's green infrastructure's visual–aesthetical values are severely questioned due to the trees' health condition and the mixed, non-uniform appearance both in taxon and age contexts. The mixture of young, aged, mature, or old trees proves the constant efforts to maintain the allée on the one hand, but it may still disturb the overall character for an extended period. The taxon diversity is also an essential question, especially for a possible renewal program.

In climate change and global and urban warming, biodiversity in green infrastructure may be vital for ecosystem provision and stability. In Alnarp, Sweden, in 2014, the First

International Conference on Urban Tree Diversity offered a good understanding of the urban tree diversity issue [32]. The conference paper stressed the importance of tree diversity from socioeconomic status and cultural aspects too. Even the great variety of urban land use forms offers possibilities for the enlargement of assortment. "Urban ecosystem represents the most complex vegetative land cover, and multiple land uses of any landscape." [33]. As for taxon diversity in allées, the Morgenroth paper [32] notes that "species diversity may also be necessary for urban ecosystem stability", still "... increasing tree species diversity does not guarantee improved ecosystem function." For visual effects, still, the allée of only a single even-aged species may offer a designed urban character, even if the allée is a historical one. On the other hand, there are more possibilities to arrive at a higher tree diversity even in urban allées. "Santa Monica, USA, developed its urban tree diversity by planting single-species along individual streets, thus achieving diversity at the city-level, while also strengthening local distinctiveness [32].

On the contrary, MacDonagh's planting design idea is determined in the ecological and aesthetic values of tree diversity when asks: "let's have allées, avenues, boulevards, and groves of richly diverse tree species in our cities that are complementary in shape, form and colour." [34].

In Andrássy Avenue, moderate tree taxon diversity is a necessity because of the cross-sections variety. In the first section, we have to admit that due to the built-in urban system and the subsequent installation of metro 1, the sycamore trees did not have much chance of maintaining a healthy, beautiful crown. On the other hand, the two broad sections, Sections 2 and 3, may require different taxa to fit the large-scaled vue. The third section would be possible to find a way for the promenade's authentic reconstruction that fits into the urban landscape and attracts city dwellers and tourists again. Beautifully renovated city villas on both sides, the appearance of new features offering a tourist program also reinforces the idea of an authentic renovation.

As proven in our allée and tree survey, the inner rows along the busy traffic lanes in Sections 2 and 3 are, for the most part, in worthy condition owing to the direct environmental burden. If the present traffic system and intensity remain, the inner rows might have new, more tolerant taxon, as the chief landscape architect of Budapest Municipality envisions. We doubt the visual, aesthetic value of such diversity in the case of Andrássy Avenue. Our age's sustainability and liveability concepts need contemporary technical and urban planning solutions for urban ecological improvement and the decrement of urban traffic in inner-city areas.

5. Conclusions

Looking back into Andrássy avenue's one and a half-century-long history, we can observe all renewal and development projects used to support the urban traffic intensification. However, the allée's maintenance and renewal from time to time tried to keep pace with the environmental quality problems posed by the avenue, which had gone from a former promenade to a central urban road. Due to the severely damaged and polluted habitat along Andrássy Avenue, the trees stand cannot compete with the urban environment's disadvantages. Finally, in the 21st century, parallel with the green and friendly city idea and the liveability criteria, harmony and balance would be essential among the urban functions [35]. The listed World Heritage site cannot be an urban motorway anymore. The once glorious urban open space and its architecture and living heritage have more essential values and public life's opportunities for both the residents and tourist than driving, biking or walking along the running, hectic urban traffic lines. The only possible way to work out the renewal concept for the avenue might be resolving road traffic, reducing traffic lanes, breaking speeds, and prioritising pedestrian and other environmentally friendly transport. Time had arrived when such a brave decision should come. Here is Paris's bold, contemporary, and public-friendly plan as a possible example of city development and heritage protection.

However, even if the urban habitat's quality improves, the allée renewal requires a careful search for suitable tree species. Pests and pathogens that appear year after year, crossing countries very quickly, can cause a rapid collapse in plants. *Fraxinus excelsior* species seemed to be the best tolerant species 30 years ago, and today these individuals standing in urban allées dye out slowly. Could we take it for sure that the new plantation with *Celtis australis*, which looks excellent today, will not have the same fate in a decade or two?

Relying on the historical values, the appearance of the allée should be uniform, but we are aware of the new, contemporary planting design principles. Monoculture is always in great danger in urban planting, mainly because of pests' more comfortable spread. For all these reasons, an utterly uniform installation of more than 800 tree individuals is not necessarily recommended. It is worthwhile to improve the diversity, first of all by section-specific taxa, secondly by preserving some old witness trees, and depending on the new open space structure and use intensity, probably among the inner and outside rows.

The present Budapest Municipality, helped by a chief landscape architect, is committed to developing the city along with sustainable, liveable, and green principles and guidelines. Among others, the restriction of car traffic, the comprehensive urban trees' and allées' renewal program and the urban tree planting's technical solutions as the Stockholm method, the soil cells are at the forefront of urban landscape architecture development. For Andrássy Avenue renewal, the municipality will launch an open competition for a complex renewal where urban habitat improvement should be vital.

The present replacement works in the allées should be only temporary management until the long-term developments begin. Based on the successful implementation, two ways are possible at significantly different costs: forming the structural soil and the suspended pavement method. Both methods support the optimal urban precipitation management and the appropriate development in the subsurface media layer, enable the proper aeration, and thus the plants' healthy root development. The primary habitat improvements cover reducing urban traffic to give space for widening and demarcating spacious green lanes, reducing trampling and compaction. For the new plantation, it is essential to add biological formulations to help adaptation and healthy growing. Moreover, similarly to the German historic garden conservation technologies applied in a changing climate environment, urban climate and ecosystem data of exceptional urban living heritage would need permanent or at least regular digital control via sensor technologies of the tree stock, the soil system and the urban temperature distribution. [36]

Sustainability and liveability concepts mentioned for Andrássy Avenue's improvement and allée renewal played a determining role in the three years-long research and planning process to work out the new green vision for the capital. In 2019, the Budapest Capital Municipality accepted the Budapest Green System's Scientific Survey and Development Concept, in 2021, the Budapest Green Infrastructure Development and Management Plan (the so-called Dezső Radó Plan) [37].

The capital is committed to the value-preserving development of urban green infrastructure and modern construction technologies and maintenance methods. However, the capital's room for manoeuvre is limited in priority zones such as the World Heritage Site and its buffer zone, such as the City Park. These developments have been routinely brought under the authority of the government, and negotiations are tough. It is also a problem that there is still the lack of a management plan to preserve the whole cityscape, which would include strict rules on the impact assessment of projects in World Heritage sites.

We, dedicated landscape architects, are confident that urban communities and society will act more and more decisively for green spaces that improve urban life quality. If nothing else, but the covid epidemic has shown how vital everyday open space recreation is.

Civil societies certainly raise awareness of the importance of living heritage, green areas and liveable, residential, and environment-friendly open spaces. The most relevant socio-ecological surveys in open space use concluded that Budapest's residents love green places and enjoy nature. The renewed Andrássy Avenue is also likely to win lovers of urban open spaces while looking for community programs in pedestrian zones (Figure 17).

Figure 17. Functions, regular and occasional public use in Andrássy Avenue. (sources: https://timelord.blog.hu/2011/09/15/az_andrassy_ut_az_autosvilag_elott (accessed on 18 October 2020); https://antikterkep.hu/budapest-andrassy-ut-opera-fenykep-1900 (accessed on 18 October 2020)).

Author Contributions: The four named authors performed this research with no other substantial contribution. Conceptualisation: K.S. (Kinga Szilágyi); methodology: K.S. (Kinga Szilágyi) and K.S. (Krisztina Szabó); validation: K.S. (Kinga Szilágyi), K.S. (Krisztina Szabó), C.L. and C.A.P.R.; formal analysis: K.S. (Kinga Szilágyi), K.S. (Krisztina Szabó), C.L. and C.A.P.R.; investigation/site analyses: C.L., C.A.P.R. and K.S. (Krisztina Szabó); resources: K.S. (Krisztina Szabó), C.L. and C.A.P.R.; data curation: K.S. (Kinga Szilágyi), K.S. (Krisztina Szabó); writing—original draft preparation: K.S. (Kinga Szilágyi), K.S. (Krisztina Szabó), C.L. and C.A.P.R.; writing—review and editing: K.S. (Kinga Szilágyi), K.S. (Krisztina Szabó), C.L. and C.A.P.R.; visualisations: C.L. and C.A.P.R.; supervision: K.S. (Kinga Szilágyi) and K.S. (Krisztina Szabó); project administration: C.L., C.A.P.R. and K.S. (Krisztina Szabó); final editing: K.S. (Kinga Szilágyi) C.L. and C.A.P.R.; funding acquisition: Ormos Imre Foundation and Doctoral School of Landscape Architecture, MATE. All authors have read and agreed to the published version of the manuscript.

Funding: This research received financial research support from the ORMOS IMRE FOUNDATION (registration number "00 18 228421") and funds for publication cost from and the Doctoral School of Landscape Architecture, Hungarian University of Agricultural and Life Sciences.

Acknowledgments: We: Kinga SZILÁGYI, Chaima LAHMAR, Camila Andressa PEREIRA ROSA, Krisztina SZABÓ, the authors of this paper, would like to acknowledge and say thanks to the Ormos Imre Foundation and the Doctoral School of Landscape Architecture and Landscape Ecology (MATE) for the financial support given for the publication of this article in the MDPI journal sustainability.

Conflicts of Interest: The authors declare no conflict of interest. The funders had no role in the study's design, in the collection, analyses, or interpretation of data, in the writing of the manuscript, or in the decision to publish the results.

Appendix A

EU conform method developed by MFE:

Values A

5: "visibly developed root system, optimal place of production, intact root collar";

4: "root development slightly inhibited, place of production acceptable, root neck not damaged";

3: "minor wound rot on the roots/small defective site";
2: "strong surface damage, significantly unfavourable place of production";
1: "at least 50% damage to roots, poor place of production";
0; "dead root system, empty tree place"
Values B
5;"intact";
4;"small damage (some superficial wounds)";
3;"clear damage (surface wound, rot sites)";
2;"severe damage (more multiple large wounds, deep pitting)";
1;"advancerd damage, dead rotten strain (statically weakened, does not replenish nutrients";
0;" empty tree place"
Values C
5;"intact, max. 10% foliage loss";
4;" foliage loss 11–25%";
3;"significant foliage loss 26–50%";
2;"severe crown damage, more than 50%";
1;"dead crown, total foliage loss";
0;" empty tree place"
Values D
5; "excellent";
4; "it can reach maximum age by intervening";
3; "to be replaced before maximum age";
2; "to be replaced in 10 years";
1; "to be replaced urgently";
0;" empty tree place"
Values E
5;"optimal";
4;"small deficit";
3;"medium deficit";
2;"significant deficit";
1;"neglected";
0;"empty tree place"

References

1. László, S. *Hogyan épült Budapest? (1870-1930) (How Budapest's Building Went on?)*; Fővárosi Közmunkák Tanácsa (Council of Budapest Public Works): Budapest, Hungary, 1931; pp. 162–171.
2. Johnston, M. *Trees in Towns and Cities. A History of British Urban Arboriculture*; Oxbow Books: Oxford, UK, 2015; p. 41.
3. Benevolo, L. *A Város Európa Történetében. (La Citta Nella Storia D'Europa)*; Atlantisz Publisher: Budapest, Hungary, 1994; p. 165.
4. Zachariasz, A. Development of The System of the Green Areas of Krakow from The Nineteenth Century to The Present, In The Context of Model Solutions. *Proc. Iop Conf. Ser. Mater. Sci. Eng.* **2019**, *471*, 112097. [CrossRef]
5. Lawrence, H.W. From Private Allée to Public Shade Tree: Historic Roots of the Urban Forest. *Arnoldia* **1997**, *57*, 2. Available online: http://www.jstor.org/stable/42954514 (accessed on 14 April 2021).
6. Lawrence, H.W. *City Trees. A Historical Geography from the Renaissance through the Nineteenth Century*; University of Virginia Press: Charlottesville, VA, USA; London, UK, 2008; pp. 57–63.
7. Stilgenbauer, J.; McBride, J.R. Reconstruction of Urban Forests in Hamburg and Dresden after World War II. *Landsc. J.* **2010**, *29*, 144–160. Available online: www.jstor.org/stable/43323881 (accessed on 14 April 2021).
8. Jarrassé, D. *Grammaire des Jardins Parisiens*; Parigramme: Paris, France, 2007.
9. Bandarin, F.; van Oers, R. *The Historic Urban Landscape-Managing Heritage in an Urban Century*; John Wiley & Sons: Hoboken, NJ, USA, 2012; pp. 10–13.
10. Liu, T.; Butler, R.J.; Zhang, C. Evaluation of public perceptions of authenticity of urban heritage under the conservation paradigm of Historic Urban Landscape—A case study of the Five Avenues Historic District in Tianjin, China. *J. Archit. Conserv.* **2019**, *25*, 228–251. [CrossRef]

11. Andrássy, C.G. (1823-1890) Hungarian Statesman, Hungarian Prime Minister in the Austro-Hungarian Kingdom in 1867-1879. Lived in Paris after the Suppression of the 1848-49 Freedoms War, Where–among Others–the Gracious Development of the Champs Elysée Impressed him. In: Kovács Emőke: "A szép akasztott"–Id. Andrássy Gyula élete. Ujkor.hu 2017. 02. 07. Available online: https://ujkor.hu/content/a-szep-akasztott-id-andrassy-gyula-elete (accessed on 10 April 2021).
12. Palonen, E. The City-Text in Post-Communist Budapest: Street Names, Memorials, and the Politics of Commemoration. *GeoJournal* **2008**, *73*, 219–230. Available online: https://www.researchgate.net/publication/225587665_The_City-Text_in_Post-Communist_Budapest_Street_Names_Memorials_and_the_Politics_of_Commemoration (accessed on 7 April 2021). [CrossRef]
13. Harloe, M. Cities in the Transition. In *Cities After Socialism*; Andrusz, G., Harloe, M., Szelenyi, I., Eds.; Wiley-Blackwell: Oxford, UK, 1996; pp. 1–29.
14. Light, D. Street names in Bucharest, 1990–1997: Exploring the modern historical geographies of post-socialist change. *J. Hist. Geogr.* **2004**, *30*, 154–172. [CrossRef]
15. Creţan, R.; Matthews, W.P. Popular responses to city-text changes: Street naming and the politics of practicality in a post-socialist martyr city. *Area* **2016**, *48*, 92–102. [CrossRef]
16. Tucker, E.L. Renaming Capital Street: Competing Visions of the Past in Post-Communist Warsaw. *City Soc.* **1998**, *10*, 223–244. [CrossRef]
17. Preisich, G. *Budapest Városépítésének Története II. A Kiegyezéstől a Tanácsköztársaságig*; History of Budapest Urban Development; Műszaki Könyvkiadó: Budapest, Hungary, 1964; pp. 1867–1919.
18. Szilágyi, K. Ethical and Aesthetic Aspects in the Renewal of Historic Allées. *Hist. Environ. Policy Pract.* **2014**, *5*, 3–16. [CrossRef]
19. Szaller, V. Az Andrássy út; The Andrássy Avenue; FŐKERT Zrt. 2010. Available online: www.fokert.hu/dokumentumok/fasorok/andrassy.pdf (accessed on 19 April 2021).
20. Mőcsényi, M. *Band* II. kötet. Vélemény a Népköztársaság úti platánfákról (Professional opinion to the plane trees of Népköztársaság út). Ph.D. Thesis, Hungarian Academy of Sciences, Budapest, Hungary, 1993; pp. 483–487.
21. United Nations Educational, Scientific and Cultural Organisation. Budapest, including the Banks of the Danube, the Buda Castle Quarter and Andrássy Avenue. In *Convention Concerning the Protection of the World Cultural and Natural Heritage*; World Heritage Committee: Baku, Azerbaijan, 2019; Available online: https://whc.unesco.org/archive/2019/whc19-43com-18-en.pdf (accessed on 25 February 2021).
22. Vesalon, L.; Cretan, R. "Little Vienna" or "European Avant-Guard City"? Branding Narratives in a Romanian City. *J. Urban Reg. Anal.* **2019**, *11*, 19–34. [CrossRef]
23. Wilson, E.H. *Aristocrats of the Trees by Ernest Henry Wilson*; Dover Publications Inc.: Mineola, NY, USA, 1838; p. 1.
24. Sanders, J. Defining and Measuring Success in the Urban Forest. *Rutgers* **2012**. [CrossRef]
25. Goodwin, D. *The Urban Tree*; Routledge: Abingdon, UK, 2017; pp. 5–17. Available online: https://books.google.hu/books?id=ruyfDgAAQBAJ&printsec=frontcover&dq=duncan+goodwin&hl=en&sa=X&redir_esc=y#v=onepage&q=duncan%20goodwin&f=false (accessed on 5 March 2021).
26. Roloff, A. *Urban Tree Management: For the Sustainable Development of Green Cities*; John Wiley & Sons: Hoboken, NJ, USA, 2016; pp. 17–18, 36–45. Available online: https://books.google.hu/books?id=h4ibCgAAQBAJ&printsec=frontcover&dq=andreas+roloff&hl=en&sa=X&redir_esc=y#v=onepage&q=andreas%20roloff&f=false (accessed on 5 March 2021).
27. Szaller, V. *Útmutató a Vizuális és Műszeres Favizsgálatok Elvégzéséhez és Dokumentálásához*; Magyar Faápolók Egyesülete: Kiskunhalas, Hungary, 2017.
28. Mattheck, C. *Design in Nature: Learning from Trees*; Springer: Berlin/Heidelberg, Germany, 1998; pp. 1–4. [CrossRef]
29. Mattheck, C. *Updated Field Guide for Visual Tree Assessment*; Forschungszentrum Karlsruhe: Karlsruhe, Germany, 2007.
30. Lukács, Z. *Faápolás*; Garden Kft.: Budapest, Hungary, 2020; pp. 79–210.
31. Comité Champs-Élysées. Le Pavillon de l'Arsenal avec le Comité Champs-Élysées. Étude réalisée par Philippe Chiambaretta. In *Champs-Élysées Histoire & Perspectives*; PCA-STREAM: Février, France, 2020.
32. Morgenroth, J.; Ostberg, J.; van den Bosch, K.; Cecil, K. Urban Tree Diversity–Taking Stock and Looking Ahead. *Urban For. Urban Green.* **2016**, *15*, 1–5. [CrossRef]
33. Foresman, T.W.; Pickett, S.T.; Zipperer, W.C. Methods for spatial and temporal land use and 295 land cover assessment for urban ecosystems and application in the greater Baltimore-Chesapeake 296 region. *Urban Ecosyst.* **1997**, *1*, 201–216. [CrossRef]
34. MacDonagh, P. How "Clean and Simple" Becomes "Dead and Gone". Green Infrastructure for Your Community. February 16, 2015/L. Peter MacDonagh. Available online: https://www.deeproot.com/blog/blog-entries/how-clean-and-simple-becomes-dead-and-gone (accessed on 14 April 2021).
35. Marot, N.; Golobič, M.; Müller, B. Green Infrastructure in Central, Eastern and South Eastern Europe: A Universal Solution to Current Environmental and Spatial Challenges? *Urbani Izziv* **2015**, *26*, S1–S12. Available online: www.jstor.org/stable/24920943 (accessed on 14 April 2021).
36. Hüttl, R.F.; David, K.; Schneider, B.U. *Historic Gardens and Climate Change; Insight, Desiderata and Recommendations*; Open access, De Gruyter; Berlin-Brandenburgische Akademie der Wissenschaften: Berlin, Germany, 2019.
37. Radó, D.T. *Budapest Zöldinfrastruktúra Fejlesztési és Fenntartási Akcióterve (Budapest Green Infrastructure and Management Action Plan)*; Accepted by the Budapest Municipality General Meeting decision, 664/2021. (III.31.) Corresponding author was part of the professional research and authors' team; Budapest Municipality: Budapest, Hungary, 2021.

Article

The Impact of Flood Risk on the Activity of the Residential Land Market in a Polish Cultural Heritage Town

Małgorzata Dudzińska [1], Barbara Prus [2], Radosław Cellmer [3], Stanisław Bacior [4,*], Katarzyna Kocur-Bera [5], Anna Klimach [1] and Agnieszka Trystuła [1]

1 Institute of Geography and Land Management, Faculty of Geoengineering, University of Warmia and Mazury in Olsztyn, Prawocheńskiego str. 15, 10-719 Olsztyn, Poland; gosiadudzi@uwm.edu.pl (M.D.); anna.klimach@uwm.edu.pl (A.K.); agnieszka.trystula@uwm.edu.pl (A.T.)

2 Department of Land Management and Landscape Architecture, Faculty of Environmental Engineering and Land Surveying, University of Agriculture in Kraków, Balicka str. 253c, 30-149 Kraków, Poland; barbara.prus@urk.edu.pl

3 Department of Spatial Analysis and Real Estate Market, Faculty of Geoengineering, University of Warmia and Mazury in Olsztyn, Prawocheńskiego str. 15, 10-720 Olsztyn, Poland; rcellmer@uwm.edu.pl

4 Department of Geodesy, Cadastre and Photogrammetry, Faculty of Environmental Engineering and Land Surveying, University of Agriculture in Kraków, Balicka str. 253a, 30-149 Kraków, Poland

5 Institute of Geodesy and Civil Engineering, Faculty of Geoengineering, University of Warmia and Mazury in Olsztyn, Prawocheńskiego str. 15, 10-719 Olsztyn, Poland; katarzyna.kocur@uwm.edu.pl

* Correspondence: stanislaw.bacior@urk.edu.pl

Received: 25 October 2020; Accepted: 1 December 2020; Published: 3 December 2020

Abstract: The article attempts to determine the effect of perceived flood risk, based on identified flood hazard zones, on the level of activity in the market of land property designated for housing developments in the historical town of Sandomierz, Poland. The study employed graphical, analytical, quantitative methods, and spatial analyses with GIS tools. The proposed methodology, involving spatial interpolation of the phenomenon (Kernel Density Estimation (KDE) and Inverse Distance Weighting (IDW)) and an expert opinion survey, facilitates the assessment of the market activity in towns where transactions are scarce. Trade in property is lower in areas at risk of flooding than for the remaining parts of the town. The potential flood hazard zone affects both the activity of the property market and the average prices of land. The study demonstrated that both a flood and flood risk affect the levels of market activity and the prices of residential land. However, this impact differs at various times and locations and is greater immediately after a flood. Properties located in the most attractive location within an area are characterised by a greater sensitivity to this risk.

Keywords: flood risk zone; flood risk maps; housing development zone; GIS tools; sustainable development

---

1. Introduction

One of the goals of sustainable development is to provide better and safer living conditions for inhabitants. For many years, among many threats, flooding has been an extreme and unpredictable phenomenon [1]. However, with the advent of expensive forms of permanent land development, this dangerous element became to be recognised in economic terms, the losses it causes included [2]. The losses were the reason for searching for methods to estimate the probability of the occurrence of a flood, or the scale of the hazard it entails [3,4]. In many countries, measures have been taken to force relevant authorities to determine flood hazard areas through such means as flood hazard maps [5–7].

This policy, which involves the determination of the location of hazards and demarcation of restricted use areas, is aimed at highlighting the potential risk based on historical information of flood events. This phenomenon may be referred to as 'information pressure'. A multiplied and direct transfer of information may affect the awareness of various social groups [8]. The information pressure may be directed so that the individual's awareness changes through education and promotion of attitudes related to flood protection. Undoubtedly, information on flood hazard also shapes the awareness of businesses operating in the property market [9].

Floods are one of the most common natural disasters in the world [10]. The global frequency and severity of natural disasters have increased over the past several decades; the trend is expected to continue [11]. Flood hazard can be considered as the most serious threat. It is, however, possible to determine the scale of flood risk, which depends on the following factors: the location, hydrological conditions, meteorological conditions, topography, and past flood types [12]. A spatial economic assessment of the environmental flood risk is important because of decision-making regarding public and private projects building infrastructure to reduce the negative impact of high water [11,13].

At the European Union level, measures have been taken to prepare the Member States for controlling flood disasters. On 26 November 2007, Directive 2007/60/EC of the European Parliament and Council on the assessment and management of flood risks, commonly referred to as the Floods Directive, entered into force [14,15]. Based on the Floods Directive 2007/60/EC, each member of the European Union should design flood hazard maps using different probabilities of flooding. This would make the information more obvious to the local authorities and understandable to the public, providing valuable spatial information regarding the degree of flood hazard [16,17].

The methodology for mapping flood hazard in Poland was set out in the regulation of 21 December 2012, on the preparation of flood hazard maps and flood risk maps, which implements the requirements of the Floods Directive. Flood hazard maps show the spatial range of floods, the water depth, and, where possible, the rate of water flow [18]. Flood hazard maps contain three scenarios of the range of flood. These maps show the spatial range of floodwater with three classes of probability: ten-year flood (Q10%, high probability), hundred-year flood (Q1%, medium probability), and five-hundred-year flood (Q0.2%, low probability). For example, a 100-year floodplain is an area where a flood has a 1% chance of occurring in any given year [19].

Moreover, in areas with flood protection infrastructure, zones at risk of being flooded in the event of a breach of a flood embankment, water overflow over the embankment crest, or damage to or destruction of damming structures have been designated [20]. Porter and Demeritt [14] report that maps are instruments not only for defining and communicating flood risks, but also for regulating them and for rationalizing the inevitable limits and failures of those controls. The purpose of flood maps is more than just providing information with which to assess and thereby reduce the probability and repercussions of flooding.

The common assumption of these policies is that the publication of geographic locations of hazards should increase people's risk perception (perceived personal risk) so that they will take preventive measures to avoid risks. In the residential housing market, this means that residential properties in risk-prone areas would be less desirable and thus have lower values than equivalent units located elsewhere [21].

Many researchers investigated the flood hazard impact on house prices [21–25]. Zhang investigated heterogeneity in the flood hazard impact on the full conditional distribution of prices and found that the flood hazard impact on house prices varied across the conditional distribution of house prices. On the other hand, conditional lower-priced houses are more prone to be affected by flood risk. The price discount shows the owner's willingness to pay to reduce the cost of flooding, which means the difference between the market value of a house located within a floodplain and the house located outside a floodplain [19]. Harrison et al. [25] examined the effect of flood risk on the value of houses in Florida. The results showed that the location within the floodplain lowers the value of a house. Similarly, according to Atreya et al. [22], prices of houses in Georgia (United States) located within a

100-year floodplain fell significantly after a major flood. Still, the effect disappeared between four and nine years after the flood. Flood risk trends can disappear over time, particularly when flooding events are infrequent [9]. Beltrán et al. [26] proposed similar conclusions. They noted that an immediate cut in property prices after flooding did not herald a permanent reduction. They employed the repeat-sales model, which investigates the sale price of properties sold multiple times.

Rajapaksa et al. [10] investigated the effects floods had on values of properties following a flood event in Brisbane, Australia. It was the first analysis of the effect a flood risk map publication had on a property market. The impact was then compared to the effect of an actual flood event on the market. It caused a significant decrease in property prices in the flood zone. Rajapaksa et al. [10] estimated the effect of temporal variations in flooding on property values. The results indicated that property values in affluent suburbs recovered faster than in poorer areas. This research also highlighted the importance of suburb characteristics for property values and the pace of recovery of property values after a flood event. Atreya and Czajkowski [27] assessed the flood risk in Galveston County, Texas, United States. Their results showed that properties located in high-risk flood areas command a price premium, compared with those located elsewhere. They found that housing market properties located in the highest-risk flood area, up to almost 400 metres from the nearest coastline, actually offered a price premium.

Most studies found that flooding and floodplain locations had a negative impact on residential properties [28]. The findings are inconsistent, but most studies suggest that natural hazards, such as flood events, have a negative effect [19,29], while others suggest no effect [30]. The occurrence of flood hazard areas also affects the level of activity of the property market and the behaviour in the property market—whether a developer's, seller's, or buyer's. Eves and Wilkinson [31] approached the issue of flood impact on residential property markets by analysing the short-term behaviour of residential property market participants immediately after a flood event. This was achieved by assessing the change in the number of residential properties listed for sale or rent immediately before a major flood event and for 12 months following that event. Akbar et al. [32] also analysed property market activity before and after a flood. This study focused on identifying market vulnerability by comparing segments of the property market, that is, the number of total house (TH) sales (old and new houses), new house and land (HL) package sales, and land only (LO) sales before and after the 2011 floods. They demonstrated a lower level of the activity of the residential property market in areas that were flooded. The study was conducted immediately after the flood.

The objective of this paper is to determine the impact of perceived flood hazard (based on delineated flood risk zones) on the activity of the residential land market in Sandomierz. The authors determine the extent to which the awareness of flood risk areas shapes the residential property market and prices in the entire town. To this end, they developed a methodology to investigate this phenomenon in towns where the number of transactions is relatively small. The analysis was based on geoprocessing. It yielded thematic maps of the activity of the property market in the town. Neither literature nor practical guides propose this approach to space assessment, especially regarding Polish urban areas.

2. Materials and Methods

2.1. Study Area

Sandomierz is one of the oldest, and, historically, most important Polish towns located on the Vistula River (Figure 1). It's stretched between 50°38′33″ N and 50°43′27″ N and 21°41′56″ E and 21°47′19″ E. The total area of the town is 28.69 km^2, and the number of inhabitants is 24,600. The town is famous for its rich traditions and culture. The Vistula River divides the town in two: the northern part (on the left bank) is located on the Kielce–Sandomierz Upland, and the southern side (on the right bank) is situated in the area of the Sandomierz Valley. Each differs from the other in its nature and functions. The right bank part is of industrial character with single- and multi-family buildings.

It has a glass factory, a manufacturer of fodder concentrates, and numerous warehouses and storages. The left bank part of Sandomierz has a historical, residential, and administrative services character. It has the majority of the town's multi-family buildings. Single-family houses are located in peripheral districts and also in enclaves between blocks of flats. Only the public utility, a dairy, and a farmer's market are situated in the areas at the foot of a slope.

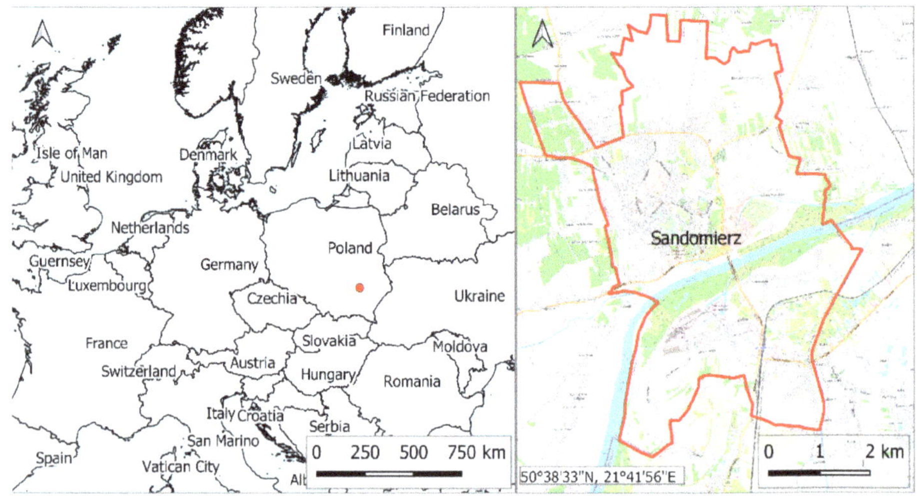

Figure 1. Poland within Europe and the location of Sandomierz, EPSG 2180.

The balance of areas with different functional and spatial use types in the town of Sandomierz indicates that 30% of the land is built-up areas. Approximately 50% of the surface area of this zone include housing developments, 25% is industrial, warehouse, and production facilities, and 15% is public service areas with housing and administration developments. In regard to the areas with developments, approximately 80% of them are single-family houses, and 20% of them are multi-family buildings. In 2010, the town was hit by a great flood called the Millennium Flood. It covered its entire southern part. Areas marked as floodplains were inundated.

2.2. Research Methodology

The article shows the integrated results. The historical-statistical method, as well as the GIS spatial analysis (graphical and analytical) of hydrological risk maps, were used to analyse the real estate market in Sandomierz. The hedonic pricing price analysis, which is a widely used method to estimate the indirect flood risks [10], requires a very large sample of transactions in land properties. It is impossible to carry out such an analysis for smaller areas. The research on the town of Sandomierz was thus based on spatial analyses using GIS methods and technology, including Kernel Density Estimation (KDE), Inverse Distance Weighting (IDW), quantitative methods, and hierarchical linear modelling (HLM). Kernel Density Estimation (KDE) is a widely-used nonparametric method for density estimation. According to Cellmer et al. [33], the results of a spatial analysis of market processes, with their specific spatial intensity, are crucial for understanding the conditions of local property markets. They can be presented on a map with such information as the activity of the market expressed as a number of transactions. Kernel estimation can be used to determine the density. It facilitates taking spatial resolution into account. Both the Inverse Distance Weighting method and kriging are widely applied in spatial variation analyses of market pricing. Morillo et al. [34] noted that while both interpolators produced similar results, IDW turned out to be better at predicting variation in "unit prices".

Hierarchical linear modelling (HLM) is a particular regression model designed to consider a hierarchical or nested structure of data. HLM is also known as the multi-level model, the linear mixed-effects model, or the covariance components model [35,36]. From a conceptual perspective, the hierarchical linear model (HLM) is a set of simultaneous equations describing dependent variables on consecutive levels of the hierarchy, which can be expressed as follows for a two-tier model [37,38]:

$$Y_{ij} = \beta_{0j} + \beta_{1j}X_{ij} + \varepsilon_{ij} \tag{1}$$

$$\beta_{0j} = \gamma_{00} + \gamma_{01}Z_j + u_{0j},\ \beta_{1j} = \gamma_{10} + \gamma_{11}Z_j + u_{1j} \tag{2}$$

$$\varepsilon_{ij} \sim N\left(0,\ \sigma_{\varepsilon}^2\right),\ u_{0j} \sim N\left(0,\ \sigma_{u0}^2\right),\ u_{1j} \sim N\left(0,\ \sigma_{u1}^2\right) \tag{3}$$

or, after substitution and rearrangement, as follows:

$$Y_{ij} = \gamma_{00} + \gamma_{10}X_{ij} + \gamma_{01}Z_j + \gamma_{11}Z_jX_{ij} + u_{1j}X_{ij} + u_{0j} + \varepsilon_{ij} \tag{4}$$

where γ_{00} is the global mean, γ_{10}, γ_{01}, γ_{11} is the regression coefficient, X_{ij} is the value of an individual-tier variable X for the ith object from the jth group, Z_j is the value of a group-tier variable Z for the jth group, and expression $Y_{ij} = \gamma_{00} + \gamma_{10}X_{ij} + \gamma_{01}Z_j + \gamma_{11}Z_jX_{ij}$ is the intercept of the equation, while expression $u_{1j}X_{ij} + u_{0j} + \varepsilon_{ij}$ is the random component.

Parameters of the model are estimated with the Maximum Likelihood Estimation method or optionally with the RML method (Restricted Maximum Likelihood). Parameter significance γ is assessed with the Wald chi-square test or the likelihood-ratio test [39]. Details of the methods for estimating and testing mixed models are available in the literature (such as [37,39]).

The study was divided into the following stages. At Stage 1, based on flood hazard maps, the graphical and analytical method was used to identify areas with a specific probability of flood occurrence. Here, floodplains with different return periods were delineated (high—ten-year flood, Q10%; medium—hundred-year flood, Q1%; and low—five-hundred-year flood, Q0.2%) and areas at risk of being flooded in the event of a breach of flood embankments with a probability of Q1% were pinpointed.

At Stage 2, the level of activity of the residential land market in the area of the town of Sandomierz was determined. The analysis was based on transactions recorded in the local land property market from 2009 to 2019. Kernel Density Estimation was applied to assess the activity of the property market, expressed as the number of transactions in spatial terms. Kernel Density Estimation, devised for estimating a smooth empirical probability function [40], is now a commonly applied spatial analysis technique to transform a geographically distributed set of points into a density surface in a GIS environment [41,42].

The authors decided that the spatial density of transactions should be considered only for the part of the town with low-rise residential buildings. The spatial distribution of low-rise residential buildings was determined with an urban planning document, Zoning Conditions and Directions for the Town of Sandomierz (Resolution of the Town Council No. XXII/236/2012, as amended) (Figure 2). The authors delineated potential low-intensity residential areas.

A map of unit prices of land property designated for housing developments was compiled with the Inverse Distance Weighting (IDW) method. It is one of the most frequently used deterministic models in spatial interpolation. Its general idea is based on the assumption that the attribute value of an unsampled point is a weighted average of known values in its neighbourhood [43]. This involves the process of assigning values to unknown points using values from a scattered set of known points. The value at the unknown point is a weighted sum of the values of N known points [44].

Stage 3 was to assess the impact of flood hazard zones on the activity of the property market in the town. The first step involved the delineation of uniform town zones in Sandomierz (categories A, B, and C) with an expert opinion survey. The next investigated aspect was the attitude of potential buyers. The authors looked into the level of interest in the area free from flood risk and parts of the town at risk of flooding. An indirect index of the number of issued building permits was employed. The selection of locations of residential projects is the primary gauge of the attitude of market participants.

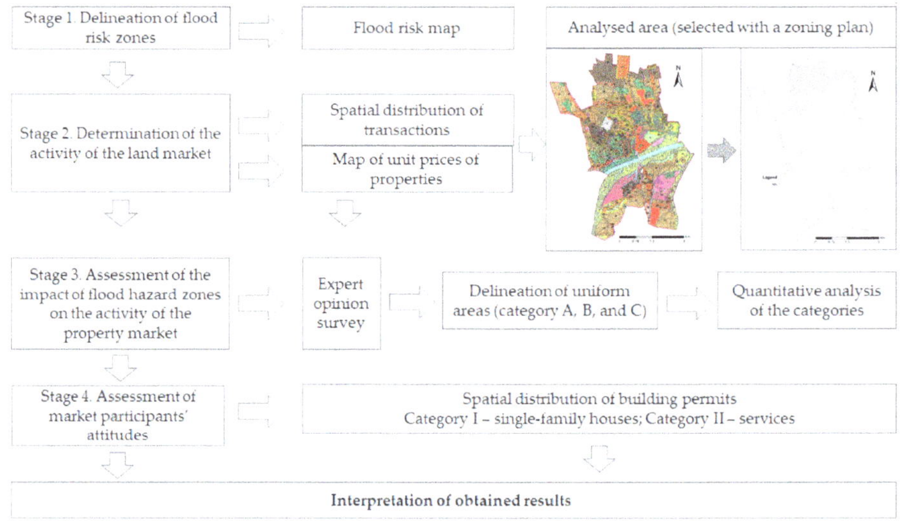

Figure 2. Research methodology diagram.

3. Results

3.1. Stage 1: Identification and Location of Flood Hazard Areas in the Town of Sandomierz

Flood maps are indispensable sources of information about hazards, vulnerabilities, and risks in a particular area [45]. Flood hazard maps include crucial information for flood management, such as flood extent, water levels, and flow velocity, that provide the basis for flood risk management plans [16]. The present study determined the spatial range of the flood based on flood risk maps compiled by the National Water Management Authority. The town of Sandomierz was represented on four sheets of 1:10,000 flood hazard maps with markings, as in Figure 3 (M-34-44-C-d-4, M-34-44-D-c-3, M-34-56-A-b-2, M-34-56-B-a-1). Flood hazard areas with the probabilities of Q10%, Q1%, and Q0.2% have the same surface area for the town of Sandomierz (Figure 3).

3.2. Stage 2: The Level of Activity in the Market of Land Property Designated for Low-Rise Housing Development in the Town of Sandomierz

The analysis of trade in undeveloped land properties designated for low-rise housing development in Sandomierz was based on materials originating from the Register of Property Prices and Values, provided by the County Office in Sandomierz. The data set included 474 market transactions concluded between 1 January 2006 and 31 December 2019, covering an area of 86.92 ha. A total of 647 plots were sold, with transactions concerning properties comprising from 1 to 11 plots. Most of the transactions concerned properties designated for low-rise housing developments. A total of 339 transactions were included in the analysis. The unit prices of properties designated for low-rise housing development ranged from 0.24 PLN (0.06 USD)/m^2 to 504.41 PLN (132.51 USD)/m^2. An average surface area per transaction was 0.1997 ha, while the average unit price during the analysed period was 52.58 PLN (13.81 USD)/m^2. Most of the sale contracts were executed in the northern, historical part of the town. The smallest number of transactions, 14%, were done in the southern part of the town at risk of flooding. Natural persons made up 99% of the buyers of such properties. The highest density of transactions (the number of transactions per km^2) was found in the northern part of the town and reached the level of 60 transactions per km^2.

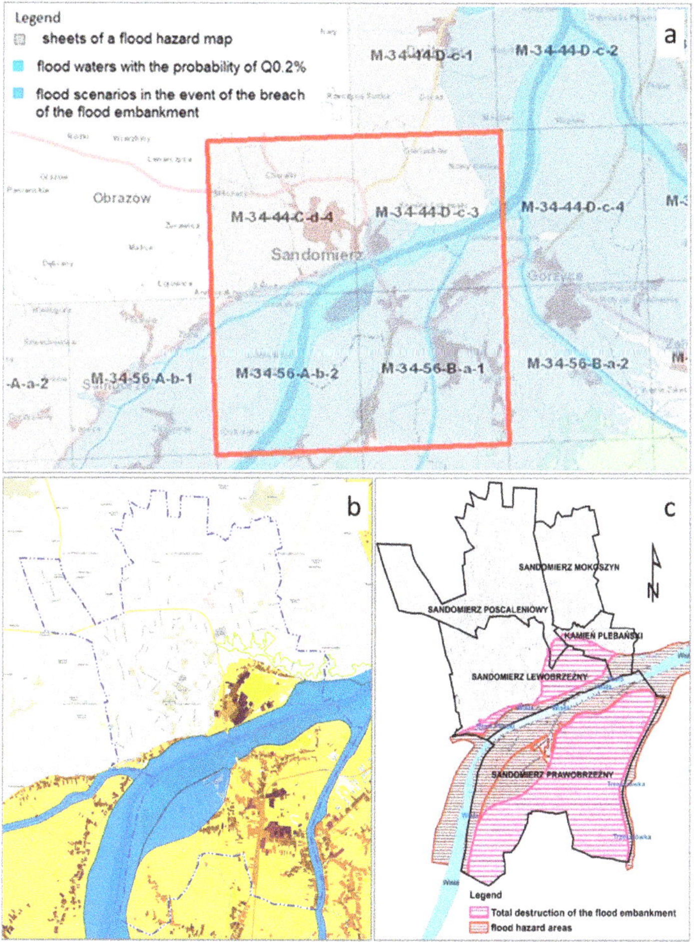

Figure 3. (**a**) The flood hazard map sheets for the area of Sandomierz. (**b**) Flood hazard areas in Sandomierz; (**c**) visual presentation of hazard areas in the town, purple areas are total destruction of the flood embankment areas, red areas are flood hazard areas. Source: our work is based on http://mapy.isok.gov.pl/imap/, which is a water data portal publishing flood hazard maps and flood risk maps.

The activity of the market in the investigated intervals (2006–2010, 2010–2014, and 2015–2019) was the highest in the northern part of the town free of flood risk. It was the lowest in the southern part where hazard was identified (Figure 4). Using the transaction data and the Inverse Distance Weighting method, the authors created a map forecasting the unit prices of land properties designated for low-rise housing developments.

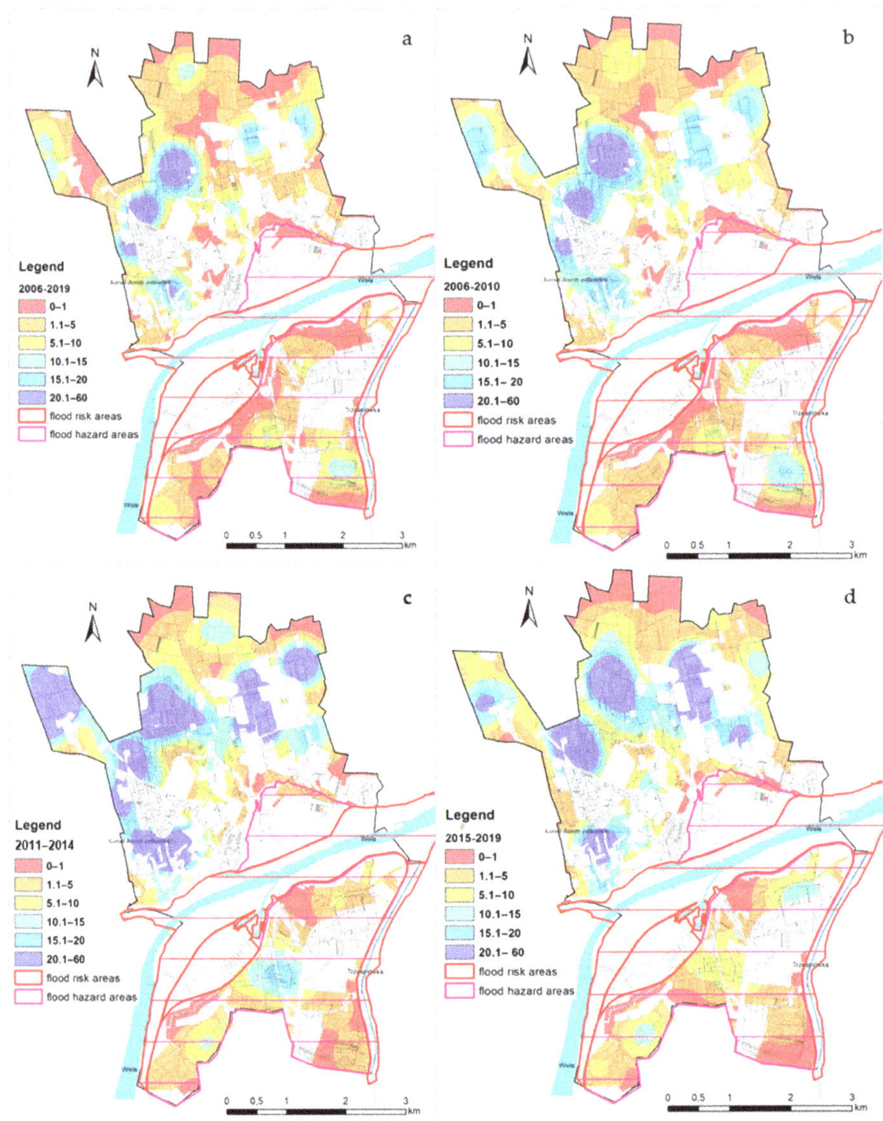

Figure 4. A map of activity in the land property market (the number of transactions per km^2) in the town of Sandomierz (**a**) from 2006 to 2019; (**b**) from 2006 to 2010; (**c**) from 2011 to 2014; (**d**) from 2015 to 2019.

The lowest property prices were noted in the southern part of Sandomierz (the district on the right bank) and at the northern edges of the town, at a considerable distance from the centre. The highest prices of land properties were noted in the central part of the town, and they reached up to around PLN 300 (USD 80) per 1 m^2. The spatial distribution of the prices was similar for each period, but their values were particularly stable in the southwestern part of the town (Figure 5).

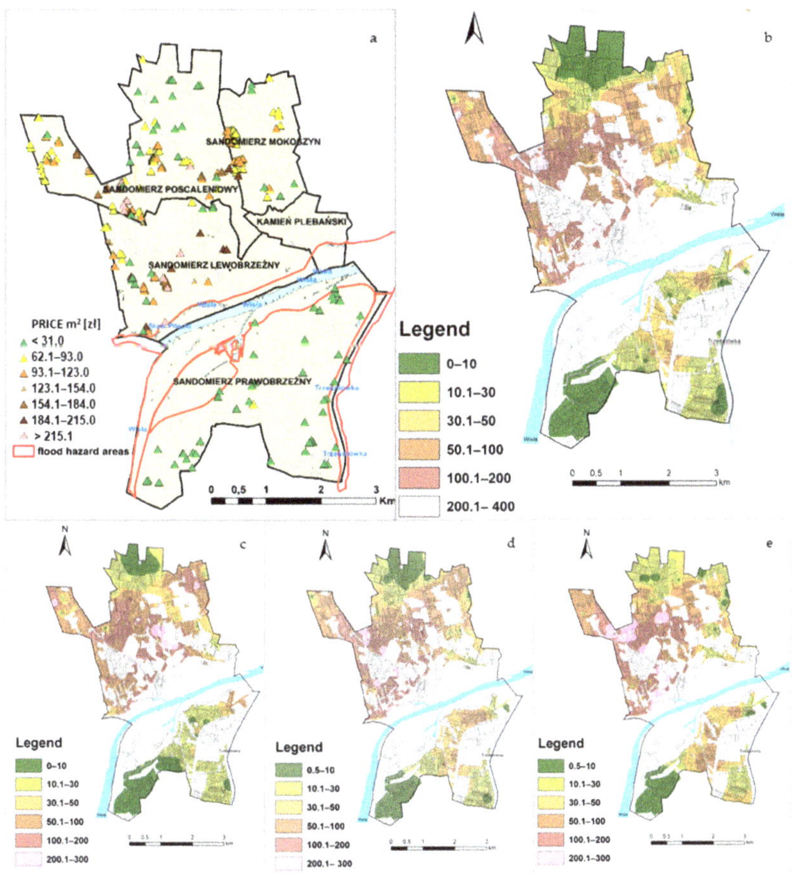

Figure 5. (**a**) The distribution of land properties designated for housing developments, sold from 2006 to 2019; (**b**) a map forecasting unit prices of land properties designated for housing developments from 2006 to 2019, estimated with the IDW method; (**c**) forecast of unit prices based on prices from 2006 to 2010; (**d**) from 2011 to 2014; (**e**) from 2016 to 2019.

3.3. *Stage 3: Assessment of the Effect of Flood Hazard Areas on the Activity of the Land Property Market, and on Prices*

The spatial variability of the market activity is primarily related to socio-economic factors affecting the quality of life, income, or the condition of the local economy. In local markets, spatial variability is caused by such local factors as spatial and planning determinants as well as localisation aspects associated with the temporary trends, preferences, safety, and the image of a particular estate or district [46]. As Usman et al. [47] noted, a single method for segmenting a property market into submarkets that are internally homogeneous and heterogeneous among the submarkets is yet to be proposed and generally accepted. The property market is generally subdivided into two classes. They are segmentation-based on a priori knowledge of the submarkets and data-driven submarket classification. This recognised and popular research approach can be found, for example, in [48–50].

Given the spatial variability in particular parts (districts) of Sandomierz, the areas were classified based on the results of a field survey among local experts, and spatial data. Real estate surveyors and real estate brokers were accepted as experts to assess the homogeneous features of particular areas of

Sandomierz. Only those experts who had at least five years of professional experience in real estate in Sandomierz were invited to participate. Five experts conforming to these requirements were found. The experts were shown the market activity and price forecast maps (Stage 2) as auxiliary materials for delineating categories of areas. Each of the interviewees was given a town map to mark categories of areas and characterise them. The survey divided the town into three categories of areas (Zones A, B, and C) with land property designated for low-rise residential buildings:

1. Very favourable—areas most frequently formed through the intensification (compacting) of the existing housing estates (Zone A)
2. Favourable—most commonly formed from rural settlement units (Zone B)
3. Moderate—including areas less favourable due to the neighbourhood, restricted access to services, etc. (Zone C).

The authors then determined the market activity in the categories of areas (Figures 6 and 7) and delineated submarket categories based on spatial data. The subdivision into the three zones was based on three main criteria: development intensity, distance to the Vistula, and mean unit transaction prices. To determine the classification, we produced raster spatial distribution maps for the criteria. The development intensity map was drawn using the Kernel Density Estimation method (KDE), taking into account the total building area. The mean unit price map was developed with Inverse Distance Weighting (IDW). The problem with simultaneous presentation of the criteria was different units and orders of magnitude. The values, therefore, were standardised for the analysis. The spatial distribution of the standardised values for the criteria is shown in Figure 6.

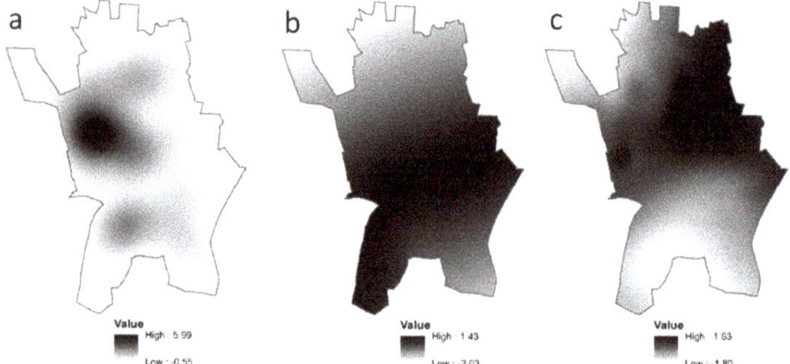

Figure 6. Spatial distribution maps for standardised values of the criteria for delineating market zones: (**a**) development intensity, (**b**) distance to the Vistula River, and (**c**) mean unit prices.

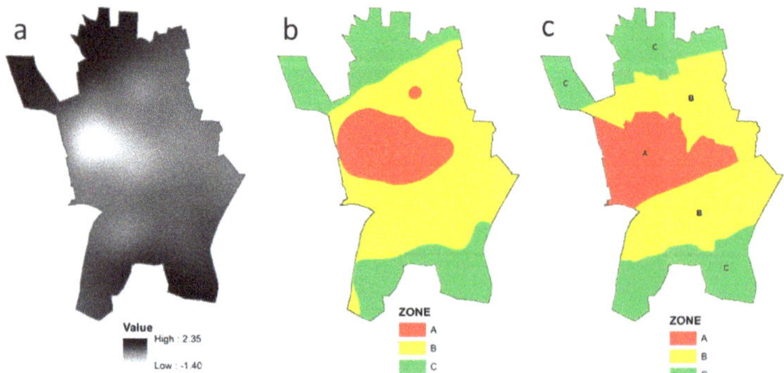

Figure 7. Delineation of market zones: (**a**) the spatial distribution of the mean standardised value, (**b**) raster reclassification results, (**c**) the final boundaries of the market zones.

The mean standardised value of the criteria was used to identify boundaries between the zones (Figure 7). The zone boundaries were approximated using the results of the final raster reclassification.

The results of both the methods were consistent, so they were used to determine the final boundaries between property market zones. The analysis was carried out separately for the parts exposed to flood hazard. To remove the influence of the differences in size of the area categories, the index of transaction density (a measure assigning transactions to areas) was employed. The transaction density (GT) index was proposed as the number of transactions concluded in a time unit per km^2, according to the following formula (5):

$$\mathrm{GT} = \frac{LT}{P} \tag{5}$$

where LT is the number of transactions, and P is the area in km^2.

Areas the experts believed to be the most valuable (Zone A) were located in parts of the town free from the risk of flooding. Real property in this category was the most expensive in the town with average levels of 114.36 PLN (30.04 USD), 123.25 PLN (32.38 USD), and 156.32 PLN (41.07 USD) per m^2 in the intervals of 2006–2010, 2011–2014, and 2015–2019, respectively. The market activity was relatively constant in this category, at about 25 transactions per km^2. This group takes up 24% of all transactions on the market (Figure 8).

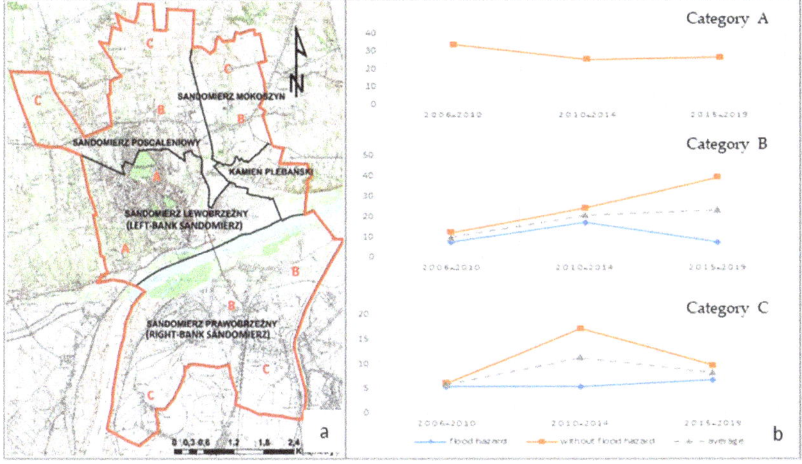

Figure 8. (**a**) Study area with categories of areas overlaid; (**b**) the activity of the market of residential land property for Zones A, B, and C; on the Y-axis, the value of the density index for areas in Zones A, B, and C. The diagrams juxtapose areas with flood hazard and areas without flood hazard.

The highest turnover was identified in Zone B with favourable conditions; 51% of transactions were done here. The average price in the area was 64.51 PLN (16.95 USD), 65.04 PLN (17.09 USD), and 76.28 PLN (20.04 USD) for respective intervals. There was a distinctive difference in prices for land situated in flood hazard zones versus outside of them; the largest was from 2006 to 2010, with the mean price of risk-free land being 87% higher. From 2011 to 2014, it was 62% higher, while from 2015 to 2019, it was 53% higher (Figure 9). A drop in market activity concerning areas at risk of flooding after 2010, when the Millennium Flood occurred, is apparent.

Figure 9. Average prices of land property designated for residential buildings in Zones A, B, and C.

Real property in Zone C comprised 26% of all transactions on the market and exhibited a significant price variation for areas at risk of flooding and those free of risk. The difference was 72% to 85%. The authors note that the market activity in this category was similar for areas with and without the risk in two periods: from 2006 to 2009, and from 2014 to 2019. The activity clearly increased in risk-free areas in the period from 2010 to 2014. This may be related to the flood of 2010. Land in this area exhibited the lowest mean price, which was 34.33 PLN (9.02 USD), 36.00 PLN, and 30.94 PLN for the intervals, respectively.

The authors attempted to determine the impact of flood risk on property prices using statistical methods. To this end, they employed a particular model of regression, Hierarchical Linear Model (HLM). Zone A is free of flood risk. Zones B and C are at risk. Therefore, the study disregarded transactions in Zone A. The employed model had two levels with a grouping variable determining the assignment to the individual zone (B or C). Explanatory variables were transaction date, flood risk, utilities, area, and location qualities. It was assumed that random effects would affect the intercept and flood risk in the HLM. Results of the HLM estimation are presented in Tables 1 and 2.

Table 1. HLM estimation results with fixed effects for the intercept and flood risk.

Variable	Effect	Standard Error	*t*-Value
Date	0.098	0.047	2.116
Utilities	1.050	1.913	0.549
Area	−0.774	2.226	−0.348
Location	2.683	3.315	0.809

Table 2. HLM estimation results with random effects for the intercept and flood risk.

Variable	Zone B	Zone C	Variance	Standard Deviation
Intercept	77.267	52.957	4540.80	67.385
Risk	−10.338	5.136	90.380	9.507

Fixed effects turned out to be insignificant with random effects being of key importance. The authors conducted a likelihood-ratio test for the likelihood functions for a classic multiple regression model and the HLM to investigate the significance of random effects. The difference between the likelihood function logarithms was −2.978. Assuming that the likelihood ratio conforms to the chi-square distribution (χ_2^2), the authors found random effects to be significant at the level of significance of $p = 0.05$. The intercept for Zone B was 77.27 and for Zone C, 52.96. This confirms the proposition that property in Zone B commanded a premium compared to property in Zone C. The random effect representing the risk of flood was −10.34 for Zone B. This means that a risk of flood reduces the unit price by this amount (in PLN).

3.4. Stage 4: Buyers' Attitude in the Time of Flood Information Pressure

An increasing number of studies shows that private buyers could probably take into consideration material risk when making investment decisions, leading to reduced capital in high-risk areas [51–54]. In their search for safe capital allocation possibilities, potential project owners assess the growth potential, risk levels, and market transparency related to it. Project owners should also evaluate the profitability of their projects by determining the profitability index (rate of return), taking into account all project costs, including (urban) planning fees or betterment levies for technical infrastructure [55], as well as tax on property function change, such as a conversion from agricultural to building parcel [56]. The present study looked into the willingness of buyers to invest in various locations in the town. The analysis focused on land designated for low-rise residential developments. The input data for this stage was the number and distribution of building permits in the area. This analysis focused on the period of 2016 to 2019. The data were acquired from the Chief Construction Supervision Authority of Poland. The authors reviewed the information and kept those permits that applied to the delineated part of the town (Figure 10). They were then classified as Category I or II.

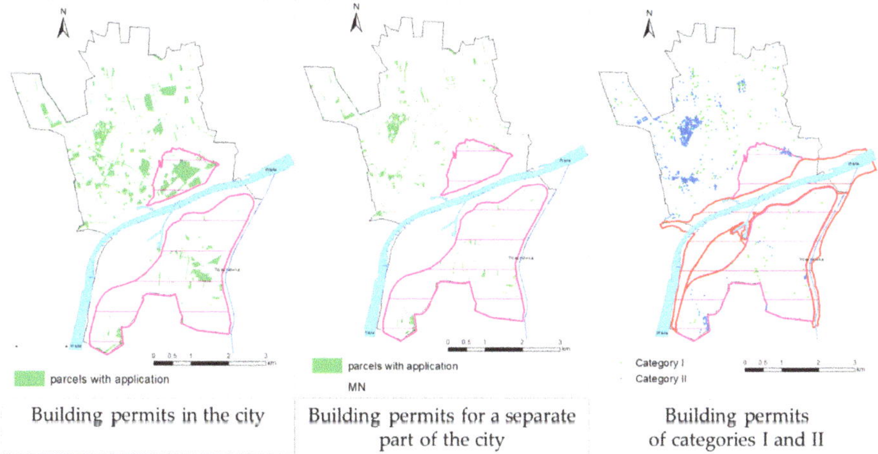

Figure 10. Building permits—spatial distribution on the left and right bank of the Vistula; Category I—construction or alteration of a residential building; Category II—improvement of residential developments, construction and upgrade of services such as water supply, sewerage system, gas, etc.

The authors determined the density of these projects in the town. The division of permits into two categories represents two groups of project owners in the market. The first one consisted mainly of natural persons seeking to build or purchase single-family houses. In Polish towns, it is most commonly the future owners who build low-rise residential buildings. The other group were stakeholders that provide services, such institutions as the municipality, gasworks, power supplier, and so forth.

The largest number of building permits was issued for properties situated in the risk-free zone (Figure 11). It was true for both the construction of low-rise residential buildings and services. The smallest number of permits was issued for the part of the town at risk of flooding. Surprisingly, a relatively large number of building permits for services was issued for the southwestern part of the town. This fact could be indicative of a local government scheme to stimulate the development of the area [57] and a future increase in market activity there [58]; and [59]. The investigated area exhibited a link between the transaction density index and the number of low-rise residential building permits. More permits were issued for areas with higher prices.

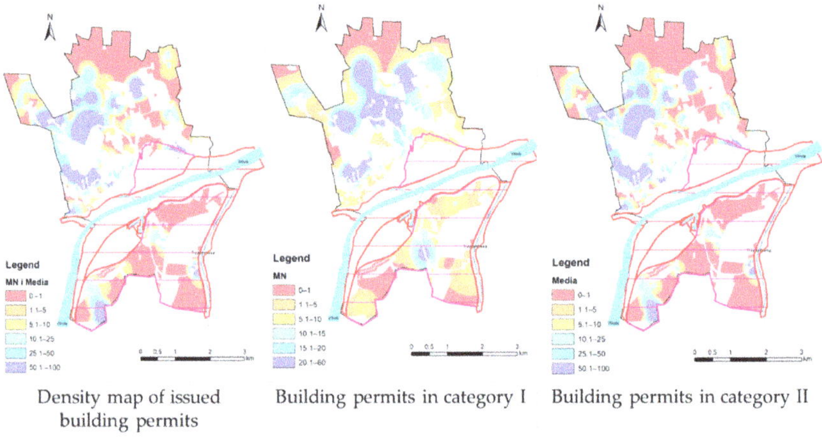

Figure 11. Building permits in Categories I and II—a density map.

4. Discussion

Urban structures shaped by historical factors face various challenges, including natural ones such as floods. The appreciation of the threats is evident in the way property markets operate, among other things. Basic property market data, such as the number of transactions and their spatial distribution, market prices, and attitudes of market participants, can be used to determine the property market activity. Historical market activity levels were determined with transaction volume and price data [32]. Market growth was inferred from participant attitudes. The market activity and residential property prices were lower in areas with flood risk than in areas free of risk. The authors are aware that property prices are affected by numerous factors [60,61], not only location or natural disasters. The results are consistent with the research by Rabassa and Zoloa [29], Samarasinghe and Sharp [62], and Bin and Landry [9]. They were convinced about the significant price discounting for properties in flood-prone areas. However, the findings vary due to the use of different methodologies and the nature of impacts [63], and discounted property values vary according to the country, location, and the type of analysed property.

Areas marked as Zone C exhibited an unambiguous and lasting price change. It reached over 72% for areas free of flood risk and those burdened with flood risk. Areas marked as Zone B also had a price difference, but it was not as spectacular as in Zone C. This was confirmed by Akbar et al. [32] who claim that there is potential for longer-term impacts of floods on housing markets, as residents reassess flood risks and the personal and economic disruption that they cause. It is not insignificant to note that Sandomierz is built up only to approximately 30%, and therefore no significant problems are encountered with finding land properties designated for residential purposes within its area. Our findings are not consistent with the research by Rajapaksa et al. [10], who demonstrated that the buy-to-sell period is shorter in affluent suburbs than in poorer ones.

The study shows that property prices decreased after the 2010 flood in areas at risk of flooding in Zone C. Many authors suggest that the price response is stronger immediately after a flood in the area [9,10,22,64,65]. According to Jung [66], results of analyses indicated that the impact of natural disasters on property prices could vary in time. This influences the sustainable development of areas affected by this phenomenon. The Millennium Flood of 2010 affected the market activity in the town. The property market in the northern part of the town, which is free from flood risk, was much more active, especially after the flood. The transaction density was considered very high for some locations. Local governments should make an effort to promote these areas and reduce the impact of the risk on the market to minimise the duration of the negative influence of disasters on the residential property market. Examples of such successful attempts can be found in Europe [2] and around the world [58]. The analysis has demonstrated that the promotion of buildable areas and stimulation of the land market through the construction of service networks has been implemented in a part of the southwestern district of the town. The authors confirm that flood risk maps are an irreplaceable tool for identifying areas and the potential level of impact of flooding on various elements distributed in the territory [67]. This information helps to analyse and make decisions related to flood risk management and planning, and also helps market participants to determine the risk of a project.

5. Conclusions

The occurrence of flood risk areas affects sustainable development of the town Sandomierz, because it influences, for example, residential land market activity. The impact varies by location: Zones B and C (with flood risk) exhibited lower transaction activity and property prices. The trade in these zones did not stop immediately after the flood when land prices were the lowest. The market activity and prices in this area changed over time. However, they did not reach the level for comparable areas that are not at flood risk. The study demonstrated that both a flood and a flood risk affect the levels of activity and the prices of residential land. However, this impact differs at various times and locations and is greater immediately after a flood.

The analyses have confirmed that the awareness of the risk of flood is great in the town of Sandomierz, and affects the level of market activity and the level of prices. The proposed methodology, involving spatial interpolation of the phenomenon (KDE and IDW) and an expert opinion survey, facilitates the assessment of the market activity in towns where transactions are scarce. Moreover, it is rather simple to apply, and very cost-effective (economically justifiable). An in-depth investigation with ArcGIS reveals the variability in market activity. Such studies help investors understand market patterns both regarding leasing and investment projects. This knowledge can be applied in urban planning to affect urban development policies, and in the sustainable management of natural disasters and urban resilience.

Author Contributions: Conceptualization, M.D. and B.P.; methodology, M.D.; software, M.D.; validation, M.D., B.P. and S.B.; formal analysis, M.D.; investigation, M.D., R.C.; resources, M.D. and B.P.; data curation, M.D.; writing—original draft preparation, M.D. and B.P.; writing—review and editing, B.P., M.D., A.K., K.K.-B., A.T.; visualization, M.D., R.C.; supervision, B.P.; project administration, S.B.; funding acquisition, B.P. All authors have read and agreed to the published version of the manuscript.

Funding: This research was funded by the Ministry of Science and Higher Education for the University of Agriculture in Kraków for 2020; Civil engineering and transport 030008-D018/KGPiAK/2020.

Acknowledgments: The authors would like to thank the anonymous reviewers for their thorough work with the manuscript and for providing constructive and insightful comments on this paper.

Conflicts of Interest: The authors declare no conflict of interest.

References

1. Echendu, A.J. The impact of flooding on Nigeria's sustainable development goals (SDGs). *Ecosyst. Health Sustain.* **2020**, *6*, 1–13. [CrossRef]

2. Jonkman, S.N.; Bočkarjova, M.; Kok, M.; Bernardini, P. Integrated hydrodynamic and economic modelling of flood damage in the Netherlands. *Ecol. Econ.* **2008**, *66*, 77–90. [CrossRef]
3. Smith, M.J. Sustainable development goals: Genuine global change requires genuine measures of efficacy. *J. Maps.* **2020**, *16*, 1–4. [CrossRef]
4. Merz, A.; Kreibich, H.; Thieken, A.; Schmidtke, R. Estimation uncertainty of direct monetary flood damage to buildings. *Nat. Hazards Earth Syst. Sci.* **2004**, *4*, 153–163. [CrossRef]
5. Dottori, F.; Salamon, P.; Bianchi, A.; Alfieri, L.; Hirpa, F.A.; Feyen, L. Development and evaluation of a framework for global flood hazard mapping. *Adv. Water Resour.* **2016**, *94*, 87–102. [CrossRef]
6. Alfieri, L.; Salamon, P.; Bianchi, A.; Neal, J.; Bates, P.; Feyen, L. Advances in pan-European flood hazard mapping. *Hydrol. Process.* **2014**, *28*, 4067–4077. [CrossRef]
7. Pappenberger, F.; Dutra, E.; Wetterhall, F.; Cloke, H.L. Deriving global flood hazard maps of fluvial floods through a physical model cascade. *Hydrol. Earth Syst. Sci.* **2012**, *16*, 4143–4156. [CrossRef]
8. Jongman, B.; Hochrainer-Stigler, S.; Feyen, L.; Aerts, J.C.; Mechler, R.; Botzen, W.J.; Bouwer, L.M.; Pflug, G.; Rojas, R.; Ward, P.J. Increasing stress on disaster-risk finance due to large floods. *Nat. Clim. Chang.* **2014**, *4*, 264–268. [CrossRef]
9. Bin, O.; Landry, C.E. Changes in implicit flood risk premiums. empirical evidence from the housing market. *J. Environ. Econ. Manag.* **2013**, *65*, 361–376. [CrossRef]
10. Rajapaksa, D.; Wilson, C.; Shunsuke, M.; Hoang, V.; Lee, B. Flood risk information, actual floods and property values: A quasi-experimental analysis. *Econ. Rec.* **2016**, *92*, 52–67. [CrossRef]
11. Daniel, V.E.; Florax, R.J.; Rietveld, P. Flooding risk and housing value: An economic assessment of environmental hazard. *Ecol. Econ.* **2009**, *69*, 355–365. [CrossRef]
12. Ji, Z.; Li, N.; Xie, W.; Wu, J.; Zhou, Y. Comprehensive assessment of flood risk using the classification and regression three method. *Stoch. Environ. Res. Risk Assess.* **2013**, *27*, 1815–1828. [CrossRef]
13. Halama, A. Zrównoważony rozwój małych miast w aspekcie zagrożenia powodziowego (Sustainable development of small towns in the aspect of flooding). *Acta Univ. Lodz. Folia Geogr. Socio-Oecon.* **2013**, *15*, 255–269.
14. Porter, J.; Demeritt, D. Flood-risk management, mapping, and planning: The institutional politics of decision support in England. *Environ. Plan. A* **2012**, *44*, 2359–2378. [CrossRef]
15. Kurczyński, Z. Mapy zagrożenia powodziowego i katalogów powodziowych, a Dyrektywa Powodziowa (Flood hazard maps and flood catalogs, according to the Floods Directive). *Arch. Fotogram. Kartogr. I Teledetekcji* **2012**, *23*, 209–217.
16. Yannopoulos, S.; Eleftheriadou, E.; Mpouri, S.; Io, G. Implementing the requirements of the European flood directive: The Case of ungauged and poorly gauged watersheds. *Environ. Process.* **2015**, *2*, 191–207. [CrossRef]
17. Kourgialas, N.N.; Karatzas, G.P. A national scale flood hazard mapping methodology: The case of Greece–Protection and adaptation policy approaches. *Sci. Total. Environ.* **2017**, *601602*, 441–452. [CrossRef]
18. Riegert, A. Doraźne metody ochrony przed powodzią (Ad hoc methods of flood protection). *Bezp. I Technika Pożar.* **2014**, *3*, 139–147.
19. Zhang, L. Flood hazards impact on neighborhood house prices: A spatial quantile regression analysis. *Reg. Sci. Urban Econ.* **2016**, *60*, 12–19. [CrossRef]
20. Franczak, P.; Listwan-Franczak, K.; Działek, J.; Biernacki, W. Planowanie przestrzenne na obszarach zalewowych w zlewniach górskich różnego rzędu w dorzeczu górnej Wisły oraz górnej i środkowej Odry. *Pr. I Stud. Geogr.* **2016**, *61*, 24–45.
21. Zhang, Y.; Hwang, S.N.; Lindell, M.K. Hazard proximity or risk perception? Evaluating effects of natural and technological hazards on housing values. *Environ. Behav.* **2010**, *42*, 597–624. [CrossRef]
22. Atreya, A.; Ferreira, S.; Kriesel, W. Forgetting the flood? An analysis of the flood risk discount over time. *Land. Econ.* **2013**, *89*, 577–596. [CrossRef]
23. Bin, O.; Kruse, J.B. Real estate market response to coastal flood hazards. *Nat. Hazards Rev.* **2006**, *7*, 137–144. [CrossRef]
24. Eves, C. The long term impacts of flooding on residential property values. *Prop. Manag.* **2002**, *20*, 214–227. [CrossRef]
25. Harrison, D.M.; Smerh, G.T.; Schwartz, A.L. Environmental determinants of housingprices: The impacts of flood zone status. *J. Real Estate Res.* **2001**, *21*, 3–20.

26. Beltrán, A.; Maddison, D.; Elliott, R. The impact of flooding on property prices: A repeat-sales approach. *J. Environ. Econ. Manag.* **2019**, *95*, 62–86. [CrossRef]
27. Atreya, A.; Czajkowski, J. Graduated flood risks and property prices in Galveston county. *Real Estate Econ.* **2019**, *47*, 807–844. [CrossRef]
28. Rajapaksa, D.; Zhou, M.; Lee, B.; Hoang, V.; Wilson, C.; Managi, S. The impact of flood dynamics on property values. *Land Use Policy* **2017**, *69*, 317–325. [CrossRef]
29. Rabassa, M.J.; Zoloa, J.I. Flooding risks and housing markets: A spatial hedonic analysis for La Plata City. *Environ. Dev. Econ.* **2016**, *21*, 464–489. [CrossRef]
30. Cupal, M. Flood risk as a price-setting factor in the market value of real property. *Procedia Econ. Financ.* **2015**, *23*, 658–664. [CrossRef]
31. Eves, C.; Wilkinson, S. Assessing the immediate and short-term impact of flooding on residential property participant behaviour. *Nat. Hazards* **2014**, *71*, 1519–1536. [CrossRef]
32. Akbar, D.; Rolfe, J.; Small, G.; Hossain, R. Assessing flood impacts on the regional property markets in Queensland, Australia. *Australas. J. Reg. Stud.* **2015**, *21*, 160–177.
33. Cellmer, R.; Trojanek, R. Towards increasing residential market transparency: Mapping local housing prices and dynamics. *ISPRS Int. J. Geo-Inf.* **2020**, *9*, 2. [CrossRef]
34. Morillo, M.C.; García-Cepeda, F.; Martínez-Cuevas, S.; Molina, I.; García-Aranda, C. Geostatistical study of the rural property market applicable to the region of Murcia (Spain) by M. carmen morillo1 et al. *Appl. Spat. Anal. Policy* **2017**, *10*, 585–607. [CrossRef]
35. Leyland, A.H.; Goldstein, H. (Eds.) *Multilevel Modelling of Health Statistics*; John Wiley & Sons: Hoboken, NJ, USA; Chichester, UK, 2001.
36. Matsuyama, Y. Hierarchical linear modeling (HLM). In *Encyclopedia of Behavioral Medicine*; Gellman, M.D., Turner, J.R., Eds.; Springer: Berlin/Heidelberg, Germany; New York, NY, USA, 2013. [CrossRef]
37. Goldstein, H. *Multilevel Statistical Models*, 4th ed.; Wiley: Chichester, UK; Hoboken, NJ, USA, 2011.
38. Cellmer, R.; Cichulska, A.; Renigier-Biłozor, M.; Biłozor, A. Application of mixed-effect regression models for compiling a land value map. In Proceedings of the Baltic Geodetic Congress, Olsztyn, Poland, 21–23 June 2018. [CrossRef]
39. Hox, J.J.; Roberts, J.K. *Handbook of Advanced Multilevel Analysis*; Routledge, Talyor & Francis: New York, NY, USA, 2011.
40. Silverman, B.W. *Density Estimation for Statistics and Data Analysis*; Chapman & Hall: London, UK, 1986.
41. Shultz, S.D.; Fridgen, P.M. Floodplains and property values: Implications for flood mitigation projects. *J. Am. Water Resour. Assoc.* **2001**, *37*, 595–603. [CrossRef]
42. Nakaya, T.; Yano, K. Visualising crime clusters in a space–time cube: An exploratory data-analysis approach using space–time kernel density estimation and scan statistics. *Trans. GIS* **2010**, *14*, 223–239. [CrossRef]
43. Lu, Y.; Wong, D.W. An adaptive inverse-distance weighting spatial interpolation technique. *Comput. Geosci.* **2008**, *34*, 1044–1055. [CrossRef]
44. Chen, F.-W.; Liu, C.-W. Estimation of the spatial rainfall distribution using inverse distance weighting (IDW) in the middle of Taiwan. *Paddy Water Environ.* **2012**, *10*, 209–222. [CrossRef]
45. Handbook on Good Practices for Flood Mapping in Europe. Available online: https://ec.europa.eu/environment/water/flood_risk/flood_atlas/pdf/handbook_goodpractice.pdf (accessed on 21 September 2020).
46. Cellmer, R. *Modelowanie Przestrzenne w Procesie Opracowywania Map Wartości Gruntów (Spatial Modeling in the Process of Preparing Land Value Maps)*, University of Warmia and Mazury: Olsztyn, Poland, 2014.
47. Usman, H.; Lizam, M.; Adekunle, M.U. Property price modelling, market segmentation and submarket classifications: A review. *Real Estate Manag. Valuat.* **2020**, *28*, 24–35. [CrossRef]
48. Ling, Z.; Hui, E.C.M. Structural change in housing submarkets in burgeoning real estate market: A case of Hangzhou, China. *Habitat Int.* **2013**, *39*, 214–223. [CrossRef]
49. Hwang, S. Residential segregation, housing submarkets, and spatial analysis: St. Louis and Cincinnati as a case study. *Hous. Policy Debate.* **2015**, *25*, 91–115. [CrossRef]
50. Levkovich, O.; Rouwendal, J.; Brugman, L. Spatial planning and segmentation of the land market: The case of the Netherlands. *Land Econ.* **2018**, *94*, 137–154. [CrossRef]
51. Kousky, C.; Luttmer, E.F.; Zeckhauser, R.J. Private investment and government protection. *J. Risk Uncertain.* **2006**, *33*, 73–100. [CrossRef]

52. Balvers, R.; Du, D.; Zhao, X. *What Financial Markets Reveal About Global Warming?* Department of Economics, West Virginia University: Morgantown, WV, USA, 2009.
53. Hallegatte, S. How economic growth and rational decisions can make disaster losses grow faster than wealth. *Policy Res. Work. Pap. Ser.* **2011**, 5617. [CrossRef]
54. Husby, T.G.; Mechler, R.; Jongman, B. What if Dutch investors started worrying about flood risk? Implications for disaster risk reduction. *Reg. Environ. Chang.* **2016**, *16*, 565–574. [CrossRef]
55. Battisti, F.; Campo, O. A methodology for determining the profitability index of real estate initiatives involving public-private partnerships. A case study: The integrated intervention programs in Rome. *Sustainability* **2019**, *11*, 1371. [CrossRef]
56. Battisti, F.; Campo, O.; Forte, F. A methodological approach for the assessment of potentially buildable land for tax purposes: The Italian case study. *Land* **2020**, *9*, 8. [CrossRef]
57. Malkowska, A.; Gluszak, M. Pro-investment local policies in the area of real estate economics-similarities and differences in the strategies used by communes. *Oecon. Copernic.* **2008**, *7*, 269–283. [CrossRef]
58. Steinberg, F. Strategic urban planning in Latin America: Experiences of building and managing the future. *Habitat Int.* **2005**, *29*, 69–93. [CrossRef]
59. Gaile, L. Improving rural-urban linkages through small town market-based development. *Third World Plan. Rev.* **1992**, *14*, 131–148. [CrossRef]
60. Sklenicka, P.; Molnarowa, K.; Pixova, K.C. Factors affecting farmland prices in the Czech Republic. *Land Use Policy* **2013**, *30*, 130–136. [CrossRef]
61. Wang, S.; Yang, Z.; Liu, H. Impact of urban economic openness on real estate prices: Evidence from thirty-five cities in China. *China Econ. Rev.* **2011**, *22*, 42–54. [CrossRef]
62. Samarashinghe, O.; Sharp, B. Flood prone risk and amenity values: A spatial hedonic analysis. *Aust. J. Agric. Resour. Econ.* **2010**, *54*, 457–475. [CrossRef]
63. Lamond, I. *The Impact of Flooding on the Value of Residential Property in the UK*; School of Engineering and the Built Environment, University of Wolverhampton: Wolverhampton, UK, 2008.
64. Kousky, C. Learning from extreme events: Risk perceptions after the flood. *Land Econ.* **2010**, *86*, 395–422. [CrossRef]
65. Naoi, M.; Seko, M.; Sumita, K. Earthquake risk and housing prices in Japan: Evidence before and after massive earthquakes. *Reg. Sci. Urban Econ.* **2009**, *39*, 658–669. [CrossRef]
66. Jung, E.; Yoon, H. Is flood risk capitalized into real estate market value? A mahalanobis-metric matching approach to the housing market in Gyeonggi, South Korea. *Sustainability* **2018**, *10*, 4008. [CrossRef]
67. Morales, P. Current status of the mapping of flood risk and its application in the land. The case of the region of Murcia. *Bol. Asoc. Geógr. Esp.* **2012**, *58*, 443–448.

Publisher's Note: MDPI stays neutral with regard to jurisdictional claims in published maps and institutional affiliations.

Article

Therapeutic Qualities and Sustainable Approach to Heritage of the City. The Coastal Strip in Gdańsk, Poland

Monika Trojanowska

Faculty of Civil and Environmental Engineering and Architecture, UTP University of Science and Technology, 85-796 Bydgoszcz, Poland; monika.trojanowska@utp.edu.pl

Received: 10 October 2020; Accepted: 3 November 2020; Published: 6 November 2020

Abstract: In this paper, the case of the Coastal Strip in Gdańsk is presented. Gdańsk has natural and cultural heritage of great value and is included on Tentative list of UNESCO World Heritage List as "Gdansk—Town of Memory and Freedom". The Coastal Strip is a rare example of landscape with natural dunes located within walking distance from densely populated residential districts. Therefore, the economic pressure for urban development is extremely strong. At the same time, the rise of social awareness about potential consequences of urbanization brought numerous efforts by local activists and researchers to prevent further development of the Coastal Strip. This study consisted of assessment of therapeutic qualities of The Coastal Strip using a conceptual framework—a universal standard for health promoting places. The results demonstrated that The Coastal Strip is a health-promoting place, thus adding new research-based evidence against plans for urban development of that area.

Keywords: Gdańsk; The Coastal Strip; environmental justice; preservation of heritage; therapeutic landscapes; public health; urban sustainability

1. Introduction

In this study, the case of The Coastal Strip in Gdańsk, Northern Poland is presented [1,2]. The Coastal Strip is a rare example of coastal landscape with natural dunes within walking distance from densely populated residential districts. It is a favorite recreational destination for not only local inhabitants but also tourists. Therefore, the economic pressure for urban development is extremely strong, as the market prices of apartments in this area are among the highest in the country. At the same time, people are increasingly aware of health promoting values of natural landscape and cultural heritage of the city. Thus, the plans for urban development of the Coastal Strip are raising social disapproval. Numerous studies concerning ecological and social values of that space were published. The Civic Project Development of the Coastal Strip in Gdansk was proposed in 2010 [1,2].

This case report investigates the therapeutic qualities of Gdańsk Coastal Strip. The assessment of therapeutic qualities was conducted with the conceptual framework of the universal standard of health-promoting places created by author. This case report can be used to justify the demand to protect the Coastal Strip from further development.

Literature Review

There is a plenty of evidence that environment affects human health [3,4]. Environment could be health- promoting and even therapeutic [5–9]. Gesler defined *therapeutic landscapes* as places where *"physical and built environments, social conditions and human perceptions combine to produce an atmosphere which is conducive to healing"* [6]. Researchers provided examples of places that have potential to promote healing, for examples Lourdes in France, Epidaurus in Greece, or Bama in China [5–7,9].

At the same time, Gesler and Conradson draw attention to the fact that the perception of therapeutic properties of the landscape is highly subjective and depends on the social context of the place [7,10].

Numerous researchers from various fields, e.g., environmental psychology, medicine, sociology, architecture, and urban planning, have described the main qualities of *therapeutic landscapes* [3,4,6,8]. However, a need for implementation science was determined. The universal standard for health-promoting places, developed by the author, could be used to evaluate the therapeutic qualities of any open green space [11].

Carlson recalls the *"four-level structure of theory"* proposed by Moore, Tuttle, and Howell (1982) which involved theoretical orientations, organizational frameworks, conceptual models, and explanatory hypotheses [12]. Referring to the *"four-level structure of theory"*, the universal standard is the conceptual model, based on theoretical orientations, organizational frameworks, and explanatory hypotheses captured during the literature review and field research.

As people are becoming aware of the benefits of everyday contact with nature, researchers from various countries are pointing to the uneven distribution of access to green spaces. [13,14]. Urban green space can provide multiple health, environmental, social, and economic benefits, but people with lower socio-economic status are often deprived of possibilities of every day contact with nature within walking distance [15]. Many researchers revealed the disparities regarding urban greenery and socio-economic status and called for access to natural landscapes for everyone [16]. People are increasingly demanding the right to the city and environmental justice [17].

The study reported in this paper was influenced by efforts to stop the urban development. The risk of deterioration of large scale open green space of the Coastal Strip Gdańsk should become important factor in decision making.

The therapeutic qualities are difficult to evidence. Therefore the universal standard for health-promoting places may serve as justification tool according to Lincoln and Guba evaluative criteria of credibility, generalizability, reliability, and confirmability [18]. Local community could use it to justify the call for protection of health promoting places endangered by urban sprawl.

2. Materials and Methods

2.1. Case Study. The Coastal Strip in Gdańsk

The coastline of Gdańsk (The Coastal Strip) is selected as the case study (Figure 1) for the following reasons. Gdańsk is a historic city and its coastline is perceived as cultural and natural heritage. Gdansk—Town of Memory and Freedom is included on Tentative list of UNESCO World Heritage List. [19] The Coastal Strip has important natural, cultural, and social qualities. Sandy beaches and natural dunes—younger 'yellow' coastal dunes and older 'grey' dunes overgrown by seaside forest are located next to densely inhabited popular districts. (Figure 1) The dunes are populated by surprising richness of species. Many of them are rare and endangered (e.g., Arhenia Spathulata and Epipactis Atrorubens (Dark Red Helleborine) [1,2]. Moreover, small patches of well-preserved natural habitats mentioned in the Annexes of the European Union Habitats Directive, e.g., psammophilic vegetation with Koeleria Glauca, can be found there [1,2].

There are numerous places of cultural heritage—historic seaside parks: Brzeźno Park and Jelitkowo Park, relicts of fisherman villages, cemeteries, historical buildings, as well as remains of coastal artillery.

The coastline is a favorite place for recreational activities and is highly popular among holiday seekers during summer season. The pedestrian and bicycle promenade along the beach stretches from Gdańsk to Sopot and even further to Gdynia. It is a favorite path for family walks.

It is a place for physical activity, mental regeneration, and organizing social events. Sadly, today this place is endangered by sprawling urban development.

Figure 1. Outlines of Gdańsk Coastal zone. Source: Google maps, 2020.

The coastal zone is a subject of strong economic pressure for further development [1,2,20–22]. As it is one of the most demanded locations (Figure 2), plans for new development, upscale apartments, or hotels are created. On the other side, numerous efforts to prevent urban development of the Coastal Strip were undertaken by local activists and researchers from Gdańsk universities and institutions [1,2,21]. The Civic Project Development of the Coastal Strip in Gdansk presented sustainable proposals for development [2]. The social movement is very active, with social media presence (e.g., www page, Facebook, etc.) The local spatial development plans are being constantly monitored [2].

2.2. Method and Data Collection

The main objective of this study was to find additional evidence to justify the social demand to protect the open green areas of The Coastal Strip from further urban development. The research question was whether The Coastal Strip is a health-promoting place? To answer this question, a conceptual framework for a universal standard of health-promoting places (Table 1) was used. It was created to consolidate the therapeutic attributes described by researchers and unify evidence into a ready-to-use tool. The tool organizes therapeutic attributes and allows for assessment and comparison.

The health-promoting qualities were divided into five categories: Sustainability, accessibility, amenities, design, and placemaking [11].

Table 1. A universal standard for health-promoting urban places. Source: Author.

1. Sustainability	2. Accessibility	3. Amenities	4. Design	5. Placemaking
1.1 Place Area Location Surrounding urban pattern 1.2 Environmental characteristics Soil quality Water quality Air quality Noise level Forms of natural protection Green and Blue Infrastructure 1.3 Biodiversity protection Parts of open green space not available to visitors Native plants Native animals Natural maintenance methods 1.4 Sustainable water management Rainwater infiltration Irrigation with non-potable water 1.5 Parks of Second (New) Generation 1.6 Urban metabolism 1.7 Ecological energy sources	2.1 Distance to park 2.2 Sidewalk Infrastructure- Width of sidewalk Evenness of surface Lack of obstructions Slope Sufficient drainage 2.3 General conditions of walkways Maintenance Overall aesthetics Street art Sufficient seating Perceived safety Buffering from traffic Street activities Vacant lots 2.4 Traffic Speed Volume Number and safety of crossings Stop signs On-street parking 2.5 User Experience Air quality Noise level Sufficient lighting Sunshine and shade Visibility of nearby building 2.6 Public transport stops 2.7 Sufficient Parking	3.1. Psychological and physical rejuvenation Natural Landscapes Green open space Presence of water Places to rest in the sun shade Places to rest in quiet and solitude 3.2. Promotion of Physical Activities Sports and recreational infrastructure Community gardens Addressing the needs of people with disabilities 3.3. Catering for basic needs Safety and security (presence of guards, cleanliness, maintenance, etc.) Places to sit and rest Shelter Restrooms Drinking water Food (possibility to buy food in the park or in the closest vicinities)	4.1. Architectural design Human scale Focal points and landmarks Structure of interior connections Framed views Long vistas (Extent) Pathways with views Invisible parts of the scenery (Vistas which engage the imagination) Possibility to watch other people Possibility to see wildlife 4.2. Salutogenic design Optimal levels of complexity Engaging features Risk Mystery/Fascination Movement 4.3. Sensory stimuli design Sensory stimuli: Sight Sensory stimuli: Hearing Sensory stimuli: Smell Sensory stimuli: Touch Sensory stimuli: Taste Sensory path	5.1 Enhancement of Social Contacts Organization of events Meeting places for groups 5.2 Human perception -spiritual & symbolic Sacred places Works of Art Monuments Culture and connections to the past Thematic gardens Personalization

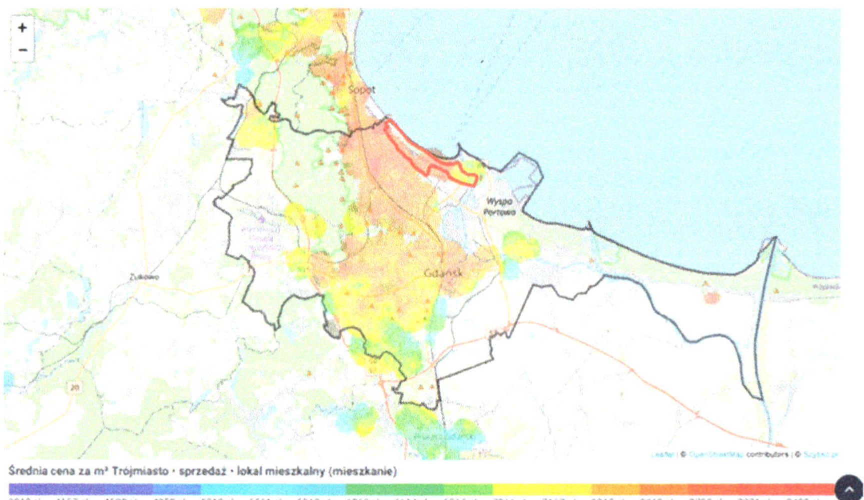

Figure 2. Spatial distribution of market retail value in Gdańsk. Average price per square m of apartments in Gdańsk [zł = polish zloty, pln]. The Coastal Strip, marked in red, is among the most expensive zones. Source: [22].

Three methods of research were used to develop the conceptual framework of the universal standard (Figure 3). The first method was a literature review in search of space characteristics linked to therapeutic qualities. Based on this research the draft for the universal standard was developed.

The second method was on-site field observation in selected parks in Europe and USA. Over 100 parks were studied. The third method—theory triangulation—was used to integrate the synthesized results of the literature review and field research in order to develop the final ready-to-use version of the universal standard.

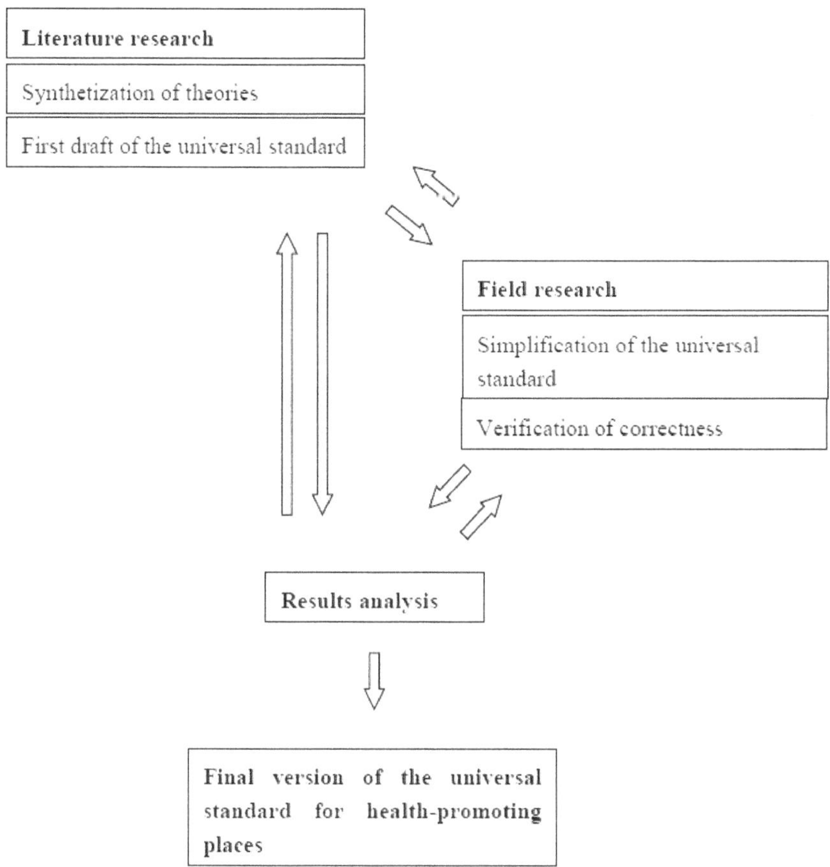

Figure 3. Scheme of the theory triangulation. Source: Author.

The universal standard was used to assess The Coastal Strip in Gdańsk therapeutic qualities. The evaluation was performed by the author—professional researcher—with a degree in architecture and urban design. The review of existing literature about the Coastal Strip was the first step, followed by numerous site visits, conducted over a couple of years (2013–2019). During those visits, observation, mapping the presence of visitors, and unstructured interviews were conducted.

In this study, The Coastal Strip in Gdańsk was treated as a large-scale urban park of new generation, encompassing smaller public parks and gardens in seaside districts of Gdańsk: Jelitkowo, Brzeźno, and Nowy Port. The park was evaluated using the universal standard (Table 1). Both a thick binary and a detailed assessment were performed.

The thick binary assessment has 2 categories (0, 1):

No, not observed-0

Yes, satisfactory-1

3. Results

The detailed assessment required a written explanation of why the researcher thought that the attribute was present. For better clarity, the results of the assessment are grouped into five tables representing five sections of the universal standard.

3.1. Sustainability

The assessment of sustainability (Table 2) demonstrated that The Coastal Strip can be treated as a modern park of new generation. Those parks are places where the sustainable development goes hand in hand with new forms of biodiversity protection. Moreover, the park is an important place for the local community. (Figures 4 and 5) It has no boundaries and spreads to transform the entire neighborhood into the grid of blue and green infrastructure [23–25].

Table 2. Assessment of Coastal Zone, Gdańsk—part 1, source: Author.

		POINTS
1. SUSTAINABILITY		**12/15**
1.1 Place		-/-
Area	Approx. 500 ha	
Location	Gdańsk coast line, a strip of coastline approx. 800 m wide and 6 km long	
Surrounding urban pattern	High density residential urban tissue, large scale residential districts	
1.2 Environmental Characteristics		6/6
Soil quality	Sufficient for recreational use. No visible traces of pollution	1
Water quality	The quality of water of Baltic Sea is sufficient for recreational use, e.g., bathing, swimming, etc. (Figure 4) Numerous potable water works which serve the city of Gdańsk area are located in that area.	1
Air quality	Very good [26]	1
Noise level	Low to moderate noise level in areas close to traffic routes [27]	1
Forms of natural protection	No specific legal protection apart from conditions of development specified in local spatial development plans, parts of dune are fenced to discourage cross passing and devastation of fragile ecosystems	1
Green and Blue Infrastructure	Important part of the green and blue infrastructure. The coastal line is composed of the complete dune system with natural fore dunes. The established dunes are covered by coniferous forest (Figures 4 and 5)	1
1.3 Biodiversity Protection		3/4
Parts of open green space not available to visitors	Parts of yellow dunes are fenced for biodiversity protection. The areas of water intake works are fenced.	1
Native plants	Planting is a combination of native and non-native species	1
Native animals	Both native and foreign species were observed	1
Natural maintenance methods	Data n/a	0
1.4 Sustainable Water Management		1/2
Rainwater infiltration	Porous, permeable surfaces	1
Irrigation with non-potable water	Data n/a	
1.5 Parks of Second (New) Generation		1/1
	Yes, The Coastal Strip can be regarded as park of new generation	1
1.6 Urban Metabolism		1/1
	Waste segregation and collection	1
1.7 Ecological Energy Sources		0/1
	Data n/a	

Figure 4. Coastal Zone, Gdańsk, Poland. Sandy beaches are full of holiday makers during summer. Source: Author.

Figure 5. Coastal Zone, Gdańsk, Poland. Open green space with play areas. Source: Author.

The park was evaluated as sustainable open green and blue area. The result was 12 out of 15 points, because there was no evidence about natural maintenance methods, irrigation with non-potable water and ecological energy sources. However, those issues can be resolved by local authorities and park management.

3.2. Accessibility

The park was assessed as universally accessible to people of every age and special needs.

Maximum of available points 26/26 were awarded. (Table 3) One of the major advantages of The Coastal Strip are attractive sandy beaches and open green space surrounded by forest. (Figures 4 and 5). The paths are comfortable, have even surfaces, and provide interesting views with long vistas (Figures 6–8).

Table 3. Assessment of Coastal Zone, Gdańsk, Poland—part 2, source: author.

		POINTS
2. ACCESSIBILITY		**26/26**
2.1 Distance to Park		1/1
	A majority of users walk to The Coastline Strip. Sufficient public transportation with local bus and tramway stops	1
2.2 Sidewalk Infrastructure (Figures 6–8)		5/5
Width of sidewalk	Sufficient	1
Evenness of surface	Good	1
Lack of obstructions	Majority of terrain is accessible, but there are fenced residential complexes and fenced areas around municipal infrastructure	1
Slope	Flat, no significant slope	1
Sufficient drainage	Sufficient	1
2.3 General Conditions of Walkways		8/8
Maintenance	The park is perceived as clean. No visible traces of litter.	1
Overall aesthetics	Good	1
Street art	None	1
Sufficient seating	Yes, multiple benches in popular places, but may be perceived as insufficient in other areas	1
Perceived safety	The park is perceived as a safe place	1
Buffering from traffic	Sufficient	1
Street activities	Yes, occasional events, both organized and spontaneous	1
Vacant lots	Yes, fenced areas around municipal infrastructure	1
2.4 Traffic		5/5
Speed	Slow	1
Volume	Moderate	1
Number and safety of crossings	Numerous possibilities for safe crossing of the street	1
Stop signs	Yes	1
On-street parking	Yes	1
2.5 User Experience		5/5
Air quality	Good	1
Noise level	Low to moderate in places close to traffic lines and crowded beaches	1
Sufficient lighting	Numerous lamps and sufficient lighting along the walking routes for pedestrians and roads, but there are parts of the area that are dark after dusk	1
Sunshine and shade	Yes, sunny open spaces surrounded by trees providing shade (Figures 4 and 5)	1
Visibility of nearby buildings	Part of area adjacent to residential districts has good visibility of nearby buildings, but there are part of the green space (e.g., forest, beaches) with no view of the city (Figures 4–8)	1
2.6 Public Transports Stops		1/1
	There are bus and tramway stops within walking distances	1
2.7 Sufficient Parking		1/1
	Yes, there are numerous parking spots, as well as on-street parking	1

Figure 6. Coastal Zone, Gdańsk, Poland. Vista engaging the imagination. Source: Author.

Figure 7. Coastal Zone, Gdańsk, Poland. Wooden pier over wetlands. Source: Author.

Figure 8. Coastal Zone, Gdańsk, Poland. Wooden pier in Brzeźno. Source: Author.

3.3. *Amenities*

The recreational infrastructure was assessed as satisfactory. (Table 4) The basic needs of users are satisfied by park infrastructure (drinking fountains, cafes, restaurants, food stands, etc.)—(Figures 9 and 10) Therefore 14 out of 15 points were awarded. One point which was missing were community gardens. However, it is debatable whether they should be introduced.

Table 4. Assessment of Coastal Zone, Gdańsk, Poland – part 3, source: author.

		POINTS
3. AMENITIES		**14/15**
3.1. Psychological and Physical Rejuvenation		5/5
Natural Landscapes	Coniferous forest, dunes and sandy beaches give an impression of a pristine natural landscape	1
Green open space	Numerous extensive grass-covered grounds (Figures 4 and 5)	1
Presence of water	Coast of Baltic sea, rain collectors, ponds (Figure 5)	1
Places to rest in the sun and shade	Multiple places including picnic and play areas	1
Places to rest in quiet and solitude	Multiple places to rest in quiet and solitude	1
3.2. Physical Activity Promotion		3/4
Sports infrastructure	Running loops, bicycle paths, cross-fit stations, boules pitch, skate park	1
Recreational infrastructure	Recreational infrastructure for all age groups	1
Community gardens	No	0
Addressing the needs of people with disabilities	Pathways are wide and even, the majority of the park area is accessible	1
3.3. Catering for Basic Needs		6/6
Safety and security (presence of guards, cleanliness, maintenance, etc.)	Assessed as a safe place during daytime, but difficult to provide sufficient surveillance after dusk	1
Places to sit and rest	Numerous benches in most popular areas	1
Shelter	Visitors may find shelter under tree canopies or inside buildings, e.g., scattered restaurants, cafes, etc. Provisional temporary structures provide shelter during organized events	1
Restrooms	Yes	1
Drinking water	Yes, drinking fountains, refreshment stands	1
Food (possibility to buy food in the park or close vicinity)	Snack bars, food stands, restaurants (Figures 9 and 10)	1

Figure 9. Coastal Zone, Gdańsk, Poland. Summer restaurants. Source: Author.

Figure 10. Coastal Zone, Gdańsk, Poland. Summer stands with food and souvenirs. Source: Author.

3.4. Design

The results of the DESIGN section clearly indicate that any type of urban development would hinder the health-promoting qualities of The Coastal Strip (e.g., long vistas, framed views, sensory stimuli, engaging features, etc. (Table 5) (Figures 4–11). Therefore, 19 out of 20 points were awarded. One point was missing-sensory path, which could enhance the sensory experience.

Table 5. Assessment of Coastal Zone, Gdańsk, Poland—part 4, source: Author.

		POINTS
4. DESIGN		**19/20**
4.1. Architectural Design		9/9
Human scale	Park offers various landscapes, ranging from open space of sandy beaches to cozy places inside the forest	1
Focal points and landmarks	Recognizable landmarks, monuments, sculptures and buildings.	1
Structure of interior connections	A clear structure of interior connections	1
Framed views	Natural frames are created by mature trees	1
Long vistas (Extent)	Park offers numerous extensive vistas	1
Pathways with views	Many paths offer interesting views	1
Invisible parts of the scenery (Vistas which engage the imagination)	Numerous designed vistas which engage the imagination	1
Possibility to observe other people	Plenty of places to watch the activities of other people from a distance.	1
Possibility to observe animals	Plenty of places to see wildlife from a distance.	1
4.2. Salutogenic Design		5/5
Optimal levels of complexity	Yes, the composition of the park is legible, yet offers optimal levels of complexity	1
Engaging features	There are multiple elements which attract attention (Figure 5)	1
Controlled Risk	Several elements offer a subjective feeling of overcoming controlled risk, e.g., wooden platforms over wetlands (Figure 7, wooden pier in Brzeźno (Figure 8)	1

Table 5. *Cont.*

Mystery/Fascination	Presence of sculptures and monuments draw the attention of users (Figure 11)	1
Movement	Baltic Sea waves, shimmering greenery	1
4.3. Sensory Stimuli Design		5/6
Sensory stimuli: Sight	Colorful leaves in the autumn, flowering trees in the spring	1
Sensory stimuli: Hearing	Sound of sea waves	1
Sensory stimuli: Smell	Flowering trees in the spring	1
Sensory stimuli: Touch	Trees, water.	1
Sensory stimuli: Taste	Refreshment stands, Snack bar (Figures 7 and 8)	1
Sensory path	No	0

Figure 11. Coastal Zone, Gdańsk, Poland. Monument of Arthur Schopenhauer in the Ronald Reagan park. Source: Author.

3.5. *Placemaking*

The aspect of placemaking is very important. The Coastal Strip has a long tradition of being a favorite holiday destination. The historic parks, hotels, restaurants and wooden pier served visitors for decades, if not centuries. Today, that tradition is being reinvigorated with organization of cultural and sport events, e.g., Festival of Arts, Park Runs, etc. Therefore, 7 out of 8 points were awarded, as only thematic gardens could be added to reinforce the place identity. (Table 6) There are numerous points of interest to facilitate the creation of mental maps and wayfinding (Figures 8–11).

Table 6. Assessment of Coastal Zone, Gdańsk, Poland—part 5, source: Author.

		POINTS
5. PLACEMAKING		**7/8**
5.1 Social Contact Enhancement		2/2
Organization of events	Multiple events, sport challenges, etc.	1
Meeting places for groups	Numerous picnic areas	1
5.2 Human Perception-Spiritual & Symbolic		4/6
Sacred places	Nearby churches	1
Works of Art	Sculptures created during festivals of art	1
Monuments	Monuments of Saint John Paul the II and Ronald Reagan, monument of Arthur Schopenhauer (Figure 6)	1

Table 6. *Cont.*

Culture and connections to the past	Multiple, monuments, works of art, historic parks, fisherman village, cemeteries, relicts of coastal battery	1
Thematic gardens	No	0
Personalization	No	1

4. Discussion

4.1. The Main Findings of the Research

The binary assessment demonstrated that the coastal zone scored 90%—76 out of maximum 84 points, which is a very good result. This result confirmed that The Coastal Strip is a health-promoting place, important for public health. The loss of such a place would be difficult to compensate in a densely populated area. The scores are high in all five categories. The missing points are limited and could be easily amended. The result indicates that any development of The Coastal Strip should be carefully analyzed. Any fragmentation may hinder the health-promoting qualities.

The results of assessment of therapeutic qualities are confirmed by existing research on coastal landscapes. Coastal communities may attain better physical health due to leisure time spent near the sea, as blue settings offer numerous therapeutic qualities [25–29].

4.2. Future Research Directions

The limitation of the universal standard comes from the subjectivity of individual perception. While the majority of therapeutic attributes can be assessed objectively, some are subjective. The precise methods of comparison cannot be used, as it is impossible to evaluate and compare some therapeutic attributes: i.e., Sensory stimuli, Mystery, Fascination, Risk/Peril, etc. Moreover, the therapeutic experience of green area can vary among individuals. The subjectivisms of assessment could be mitigated only with a more detailed description. Detailed studies, conducted by a team of researchers who discuss and compare the results, would be recommended. It would also be important to repeat the assessment to monitor the health-promoting potential of The Coastal Strip. The results could indicate areas for potential improvement.

The universal standard for health-promoting places could be used to assess the therapeutic qualities of urban open green space in any city.

5. Conclusions

A universal standard for health-promoting places is a conceptual framework to evaluate open green space. It was developed by the author after many years of research. In this study, it was used to evaluate The Coastal Strip in Gdańsk. The results, 76 out of maximum 84 points (90%), demonstrated that The Coastal Strip is a health-promoting place, invaluable for health promotion of local inhabitants and occasional visitors.

The therapeutic qualities are perceived as subjective phenomena and therefore any claims funded on lost therapeutic values are often rebutted as ungrounded or unimportant. The assessment of therapeutic qualities could be used to justify the social demand for limiting plans of further urban development of The Coastal Strip. The protection of natural heritage and public health promotion is a good reason to turn down the prospective short-term economic gains resulting from urban development. The therapeutic qualities could be lost and are difficult, if not impossible, to compensate.

The recognized limitation of this study is evaluation by only one researcher. Although the majority of awarded points are invariable and would not change if the evaluation was performed by a team of researchers, there are still some subjective points. Therefore, further, more detailed studies may require evaluation by a team of researchers and comparison of results. It would be also recommended to repeat the assessment to monitor the health-promoting features.

In this study, The Coastal Strip was chosen, but the conceptual framework of the universal standard for health-promoting urban places could be used to evaluate the qualities of any open public green space.

Funding: This research received no external funding.

Conflicts of Interest: The authors declare no conflict of interest.

References

1. Rembeza, M.; Rozwadowski, T. Public Participation practice versus economic pressure: Transformation of the Gdansk Coastal Strip. *Studia Komitetu Przestrzennego Zagospodarowania Kraju PAN* **2016**, *264*, 147–168.
2. Pas Nadmorski. Available online: http://gdanskpasnadmorski.vgh.pl/viewpage.php?page_id=6 (accessed on 9 October 2020).
3. Largo-Wight, E. Cultivating healthy places and communities: Evidenced-based nature contact recommendations. *Int. J. Environ. Health Res.* **2011**, *21*, 41–61. [CrossRef] [PubMed]
4. Frumkin, H. Beyond toxicity: Human health and the natural environment. *Am. J. Prev. Med.* **2001**, *20*, 234–240. [CrossRef]
5. Huang, L.; Xu, H. Therapeutic landscapes and longevity: Wellness tourism in Bama. *Soc. Sci. Med.* **2018**, *197*, 24–32. [CrossRef]
6. Gesler, W. Lourdes: Healing in a place of pilgrimage. *Health Place* **1996**, *2*, 95–105. [CrossRef]
7. Gesler, W. Therapeutic Landscapes: An evolving theme. *Health Place* **2005**, *11*, 295–297. [CrossRef] [PubMed]
8. Cooper-Marcus, C.; Sachs, N. *Therapeutic Landscapes. An Evidence-Based Approach to Designing Healing Gardens and Restorative Outdoor Spaces*; John Wiley & Sons, Inc.: Hoboken, NJ, USA, 2014; pp. 14–35.
9. Williams, A. Spiritual therapeutic landscapes and healing: A case study of St. Anne de Beaupre, Quebec, Canada. *Soc. Sci. Med.* **2010**, *70*, 1633–1640. [CrossRef] [PubMed]
10. Conradson, D. Landscape, care and the relational self: Therapeutic encounters in rural England. *Health Place* **2005**, *11*, 337–348. [CrossRef] [PubMed]
11. Trojanowska, M. *Parki i Ogrody Terapeutyczne*; Wydawnictwo Naukowe PWN: Warszawa, Poland, 2017.
12. Carlson, A. On the Theoretical Vacuum in Landscape Assessment. *Landsc. J.* **1993**, *12*, 51–56. [CrossRef]
13. Maas, J.; Verheij, R.; De Vries, S.; Spreeuwenberg, P.; Schellevis, F.G.; Groenewegen, P.P. Morbidity is related to a green living environment. *J. Epidemiol. Community Health* **2009**, *63*, 967–973. [CrossRef] [PubMed]
14. De Vries, S.; Buijs, A.E.; Snep, R.P.H. Environmental Justice in The Netherlands: Presence and Quality of Greenspace Differ by Socioeconomic Status of Neighbourhoods. *Sustainability* **2020**, *12*, 5889. [CrossRef]
15. Choi, D.-A.; Park, K.; Rigolon, A. From XS to XL Urban Nature: Examining Access to Different Types of Green Space Using a 'Just Sustainabilities' Framework. *Sustainability* **2020**, *12*, 6998. [CrossRef]
16. Kuo, F.E. Coping with poverty: Impacts of environment and attention in the inner city. *Environ. Behav.* **2001**, *33*, 5–34. [CrossRef]
17. Lefebvre, H. Le droit à la ville. *L'Homme et la société* **1967**, *6*, 29–35. [CrossRef]
18. Lincoln, Y.S.; Guba, E.G.; Pilotta, J.J. Naturalistic inquiry. *Int. J. Intercult. Relat.* **1985**, *9*, 438–439. [CrossRef]
19. Gdansk—Town of Memory and Freedom. Available online: https://whc.unesco.org/en/tentativelists/530/ (accessed on 27 October 2020).
20. Trojanowska, M. The Health impact of parks along waterways. In Proceedings of the 50th ISOCARP Congress, Gdynia, Poland, 23–26 September 2014; pp. 716–725.
21. Sas-Bojarska, A. The green waterfront of a city—Where are the limits of good planning? In Proceedings of the Gdansk case 49th ISOCARP Congress, Brisbane, Australia, 1–4 October 2013.
22. Ceny.szybko.pl. Available online: www.ceny.szybko.pl; https://ceny.szybko.pl/Gda%C5%84sk-ceny-mieszkan.html (accessed on 8 October 2020).
23. Völker, S.; Kistemann, T. "I'm always entirely happy when I'm here!" Urban blue enhancing human health and well-being in Cologne and Düsseldorf, Germany. *Soc. Sci. Med.* **2013**, *78*, 113–124. [CrossRef] [PubMed]
24. White, M.; Smith, A.; Humphryes, K.; Pahl, S.; Snelling, D.; Depledge, M. Blue space: The importance of water for preference, affect, and restorativeness ratings of natural and built scenes. *J. Environ. Psychol.* **2010**, *30*, 482–493. [CrossRef]

25. Foley, R.; Kistemann, T. Blue space geographies: Enabling health in place. *Health Place* **2015**, *35*, 157–165. [CrossRef] [PubMed]
26. Fundacja ARMAAG. Available online: https://www.armaag.gda.pl/indeks_jakosci_powietrza.htm (accessed on 8 October 2020).
27. Portal Stałego Monitoring Hałasu Miasta Gdańska. Available online: https://mag.bmt.com.pl/VisMap/apps/gdansk/public/index.html (accessed on 8 October 2020).
28. Wheeler, B.W.; White, M.; Stahl-Timmins, W.; Depledge, M.H. Does living by the coast improve health and wellbeing? *Health Place* **2012**, *18*, 1198–1201. [CrossRef] [PubMed]
29. White, M.P.; Alcock, I.; Wheeler, B.W.; Depledge, M.H. Coastal proximity and health: A fixed effects analysis of longitudinal panel data. *Health Place* **2013**, *23*, 97–103. [CrossRef]

Article

New Housing Complexes in Post-Industrial Areas in City Centres in Poland Versus Cultural and Natural Heritage Protection—With a Particular Focus on Cracow

Mateusz Gyurkovich * **and Jacek Gyurkovich**

Chair of Urbanism and City Structure Architecture, Faculty of Architecture, Cracow University of Technology, 31-155 Cracow, Poland; jgyurkovich@pk.edu.pl
* Correspondence: mateusz.gyurkovich@pk.edu.pl; Tel.: +48-606-605-272

Abstract: The cityscape changes constantly, reflecting the socio-economic conditions of a given urbanised area—both globally and in any given country. Post-industrial buildings and complexes have been its important elements since the nineteenth century. At present, many of them are undergoing adaptive reuse. The oldest, which are parts of post-industrial heritage and define the local identity, are now located in city centres. Some are revitalised and often adapted into multi-family housing. This paper fills a gap in the research on revitalised areas in Polish city centres, especially the ones converted into housing. It notes the links between these projects with elements of urban green-blue infrastructure, as well as the methods of protection of the reused postindustrial heritage. Studies from 2000–2020 on Polish multi-family housing architecture prove that the quality of buildings and semi-public green spaces is becoming increasingly important to developers and buyers. Properly used and exposed post-industrial heritage can contribute to raising the attractiveness of such spaces. In combination with city greenery systems, they can form attractive townscape sequences, as proven by Cracow cases. The paper's conclusions indicate that the preservation and exposition of post-industrial heritage in newly built housing complexes is affected by numerous factors. The most important of these are legal determinants based on both state-level and local law. Economic factors also play a major role, as they directly affect projects. The skills and talent of designers who can create unique proposals that expose surviving relicts and a given place's genius loci even in the most restrictive of economic and legal conditions, are also not without significance.

Keywords: urban renewal; post-industrial heritage; multi-family housing; revitalisation of city centres; protection and preservation of built heritage; blue-green infrastructure of a city; Polish major cities; Cracow

Citation: Gyurkovich, M.; Gyurkovich, J. New Housing Complexes in Post-Industrial Areas in City Centres in Poland Versus Cultural and Natural Heritage Protection—With a Particular Focus on Cracow. *Sustainability* **2021**, *13*, 418. http://doi.org/10.3390/su13010418

Received: 17 November 2020
Accepted: 29 December 2020
Published: 5 January 2021

1. Introduction

The need for living and residing in confined and safe areas has been a part of human nature for millennia. The first cities to be documented by archaeological studies appeared seven thousand years ago [1]. Some have survived to the present, yet those that had been destroyed were continuously being replaced by new ones. Humans settled in new, previously virgin areas and continue to do so. This process appears to be increasingly dynamic and thus generates negative consequences for the natural environment. The spatial structures of cities have changed over time, along with new needs and ideas generated by increasingly complex societies. Better and more complex urban forms emerged as a response to these changes [2]. Over the centuries, the city became the natural environment of human life [3], and the twentieth century specifically was proclaimed the century of cities. At present, already half of the world's population lives in cities, as demonstrated by numerous academic reports and statistical studies. It is projected that 75% of the world's population will live in cities by 2050. Forecasts also note the global phenomenon of shrinking cities as clearly regressing, while over 90% of the one thousand of the world's largest cities continue to increase their populations [4,5]. Despite European cities ranking increasingly lower in these types of lists (the first fifty positions for 2020 only included four European metropolises: Istanbul, Moscow, Paris and London, while four Polish cities were listed in the first thousand: Warsaw, Cracow, Łódź and Wrocław), the share of city dwellers in the continent's population is much higher. A European Commission report from 2006 stated that, at the time of its publication, as much as 75% of the European Union's population inhabited urbanised areas: cities and urban functional areas, including metropolitan ones [6]. Thus, the primary function of cities appears to be housing, yet they would not be able to function correctly without other programmatic elements which attract new residents. Cities can also be treated as systems that comprise numerous subsystems (natural and man-made) that either are or can turn out to be some of the most perfect known achievements of human civilisation [7]. The city and the phenomena that take place within them remain in the field of interest of many specialists who represent disciplines like psychology, sociology, architecture, history, geography or traffic engineering. Urban design is an interdisciplinary field [8], which provides multi-track opportunities for research [9].

In the discussion engaged in the paper, we use fundamental notions of cultural heritage and natural heritage which, as argued by Lowenthal [10], form humanity's legacy. The former is typically understood as the entirety of humanity's tangible and intangible achievements [11]. Urban heritage is a specific case of this and is of interest to the presented work [12]. Elements of architectural and urban heritage contribute to the urban form of the contemporary city, along with new additions [13]. Despite numerous scholars' understanding of natural heritage solely as nature unaltered by man, it is difficult to find truly wild areas in Europe. Therefore, it is assumed that areas that are partially landscaped or created by humans should be acknowledged as elements of natural heritage [10,14]. This reasoning is justified by the fact that even in city parks, created by humans, and often in post-industrial areas, separate, even niche ecosystems can develop [15]. They can contribute to ecological linkages on the regional and sometimes even the supraregional scale [16]. They also cross urban and metropolitan areas, including the area discussed in this paper [17].

This research includes studies of the development and transformation of housing areas in cities. These studies focus both on their design in a compositional harmony with the urban spatial layout [18,19] and on the context of the optimal use of natural resources and the reclamation of already developed areas [20–22]. Experience collected in the twentieth century demonstrated that large, monofunctional housing estates were often erected without the necessary educational, service or transport infrastructure [23]. They were built all over Europe after the Second World War as an answer to the demand caused by the new social and political situation [24]. The long-term consequences of these projects led to social and spatial pathologies, which have been identified and investigated only relatively recently [25,26], particularly in the countries of Central and Eastern Europe. Numerous measures have been taken to improve this situation and create multi-functional areas that can satisfy the needs of local communities [27]. The findings of studies that have been published for several decades all over the world, by presenting diagnoses and proposals of solutions, can contribute to a more sustainable urban development and the elimination of adverse phenomena such as urban shrinkage or urban sprawl [28–32]. In this context, efforts must be made to reclaim and reuse already developed areas inside cities, especially brownfields [33], by both developing and enhancing the network of blue-green linkages in these areas wherever possible [34].

This paper is a result of many years of study of the built environment, including multi-family housing complexes. The authors have continuously studied cities and urban housing areas in Poland and Western Europe for several decades. One of the foci of this research were areas built in post-industrial areas. In the years 2000–2020, comparative analyses of several dozens of new multi-family housing developments were performed in the five largest urban centres and metropolitan areas of Poland (Warsaw, Cracow, Łódź, Wrocław, the Tri-city: Gdańsk–Sopot–Gdynia). Among the projects that were analysed—ranging from singular multi-family residential buildings to complexes of varying size—those built in post-industrial areas located in city centres deserve particular attention. This paper fills a gap in the state of the art concerning revitalised areas in the centres of Polish cities, focusing on the linkages between these projects and elements of urban green-blue infrastructure [35], as well as the degree of protection and exposure of post-industrial heritage elements. To keep the text short, only cases from Cracow were discussed in this particular paper.

2. Materials and Methods

The methodology used in the study included the follow tools: an analysis of the literature (including printed and online sources), archival queries and analyses of the city's reports and planning documents, including documents associated with the protection of natural and cultural heritage. The study also covered historical texts and reports, applicable documents and policies, as well as drafts of documents available for public and expert consultation. Urban analysis, performed in the form of numerous site visits, was the primary research tool. The confrontation of the findings from site visits with applicable planning documents and the results of previous global and domestic studies, was crucial in formulating the study's findings.

The study presented here covered new residential buildings and complexes that are characterised by: a site with a post-industrial past, proximity to historical urban tissue and to elements of the city's blue-green infrastructure. Each of the analysed buildings and complexes was examined in situ. This enabled the collection of photographic documentation, which proved useful in the analyses that followed. Furthermore, based on available maps and surveying materials (procured online, from municipal institutions and design firms) the urban analyses of development structure were performed (based on a figure-ground plan): development height, public space structure, the green space system, the hydrological system (surface waters) and the vehicular and public transport layout. The composition of the layout was carefully analysed. Focused primarily on morphological changes in reference to the form of the city prior to and after the transformation, both in regard to the fragments under study and the entire structure. The focus was also placed on the formation of proper compositional and functional linkages with the spatial context including—wherever possible—green areas of significance to ensuring a healthy housing environment [14]. These areas—both in a landscaped and natural state—often constitute elements of natural heritage of the city, that are important on the urban or regional scale.

One of the goals of the study was to determine how elements of cultural heritage, most importantly those associated with previous land use (industrial or infrastructural) affected the character of architecture and land development in the projects under study. While performing this task, we studied maps and archival photographs, available both online [36] and in archives. We surveyed the literature: academic and popularising texts as well as source documents. We examined multi-family residential buildings and complexes built in post-industrial areas in the centre of Cracow. The analysis of applicable and historical planning documents played a major part in the study. We analysed applicable documents and listings concerning the city's heritage. Some of these elements were also discussed in the literature [37–41].

The literature that we based our study on and cited in the paper discusses the development of urban structures and urban composition, with a particular emphasis on European and Polish cities, as well as Cracow. Many items discussed the transformation of post-industrial areas, both in Europe and Poland, as well as housing buildings and complexes and their relationship with blue-green infrastructure. Some of the cited items reference definitions of fundamental notions linked with the subject under study (cultural heritage, natural heritage, architectural and urban heritage, blue-green infrastructure and housing environment). Others discuss the cases under study directly. In this context, we also cited applicable legal acts, including acts of local law (local spatial development plans—MPZP) as well as published planning documents concerning Cracow that we used in the study, and which were mostly available online. We have referenced these documents throughout the entirety of this paper.

3. Post-Industrial Areas in the Inner-Cities

3.1. Potential of Post-Industrial Areas in European Cities

Civilisational changes and the associated processes of urbanisation have altered the landscape of Europe—particularly its cityscapes and townscapes—several times over the past two centuries [42,43]. The European city, which has been evolving for several thousand years, has failed to produce a single commonly applicable spatial model, which appears to be both its strength and distinguishing feature [44–46]. However, it has created a certain set of characteristics and a hierarchical system of public spaces that make it recognisable [47,48]. These are the public buildings and housing tissue that act as its fundamental components along with the public spaces that bind them—and which have a varied typology and purpose. In addition, in many cities it is industrial and storage areas, primarily nineteenth- and twentieth-century factory and storage yard grounds, that are the most valuable elements of cultural heritage. Due to socio-economic changes, including the global pursuit of cheaper production solutions or the shutting down of manufacturing in favour of the creative sector [49], these areas became abandoned. They freed up areas seen as attractive for development and that were relatively close to historical city centres—areas that should have been reclaimed for cities [50]. In comparison to dynamically developing western countries, in Poland, these changes typically play out with a certain delay and do so at a slower pace, reflecting the general political and historical situation.

In the nineteenth century, it was the partitions that exerted a major impact on the uneven urbanisation of the present-day territory of Poland and the different development patterns of each city. During this period, what is now Poland was divided between foreign powers—Prussia, Russia and Austria (in the years 1795–1918, although this process began in 1772). In the twentieth century, the destruction of the urban fabric was caused by the fronts of the two world wars that swept across the country. This was combined with the toilsome reconstruction that came afterwards. In addition, it should be remembered that, as a result of the peace treaties that ended the Second World War, Poland lost over half of its territory to the Soviet Union [51], gaining a portion of highly urbanised German lands—in Silesia, Pomerania and Masuria. These territories were heavily damaged at the start of the war, but after its conclusion they became (and often reprised their role as) Poland's western and northern provinces [52]. Understanding these historical processes and their associated economic and social changes explains why uneven industrialisation exerted different types of impacts on the urban layouts of contemporary Polish cities. The totalitarian communist regime that governed the country up to 1989 (dependent on the USSR similarly as in the remainder of Central Europe) promoted industry at the cost of other branches of the economy. Due to this fact, the liquidation of large industrial plants in cities and the associated economic and spatial changes in Poland [53] started to take place during the final years of the twentieth century and continue to this day [54]. In Western Europe, these changes began earlier, in the 1980s [55].

As stipulated by the principles of sustainable development, confirmed both in urban planning manifestoes [56] and numerous international [57], national and local acts of law [58], reports [59] and planning documents, and stated in the findings of academic studies, the reuse of brownfields and their reincorporation into urban structures is desirable [60]. It appears to limit the negative consequences of exurbanisation. It can lead to the renewal of districts located closer to city centres and an inwards-oriented urban development as opposed to urban sprawl. Preceded by numerous analyses and multidisciplinary studies, supported by elaborate and sophisticated legal and planning systems that appear to guard the public interest and aid the development of cities and regions, European cases of successful post-industrial area revitalisation processes [61] show a broad range of possibilities [62].

3.2. Brownfields in Polish Cities—Problems with Spatial Development

Successful cases of revitalising urban brownfields in Western European countries as discussed in the literature [63–65] can in many cases act as models for strategies for transforming similar areas in Poland [66,67]. Especially as their careful analysis could aid in avoiding numerous mistakes or unneeded procedures that would only extend their transformation. Therefore, apart from transforming the physical urban structure, they could, based on interdisciplinary studies, also contribute to modifying the legal system and the manner of formulating planning documentation provisions [68].

Sometimes this is actually the case, and Gdańsk-based projects can act as model examples for similar projects in Poland, especially the still ongoing revitalisation of post-shipyard areas as a part of the 'Young City' and the almost-complete measures at Wyspa Spichrzów [69]. Urban renewal was preceded there by many years of studies, workshops and reports prepared with the participation of urbanists and planners from the Faculty of Architecture of the Gdańsk University of Technology (under the supervision of M. Kochanowski and later P. Lorens). They were associated with public participation and were conducted with a great number of stakeholders, which enabled the best possible preparation of similar projects under Polish legal conditions. In the case of Gdańsk, the measures and preparation of research and planning documents, together with multi-alternative studies, urban and architectural design competitions and even masterly workshops conducted by world-famous architects, were undertaken long before the passing of the Revitalisation Act in Poland in 2015. The Act stresses social and economic matters and greater participation of all stakeholders in decision-making via the legal strengthening of public participation measures [57]. Despite the final outcome often being different than stipulated in the ideal model assumptions from reports from many years ago, they have made their mark on the city's current planning documents. Over a decade later, the measures taken to transform Gdańsk's central, post-industrial zone have their own dedicated and perfect academic documentation [70–72].

However, despite the new Act being in force for five years, in light of the crisis of Poland's spatial planning system that shall be briefly discussed later, revitalisation measures targeting areas abandoned by industry and the military and the need to provide them with infrastructure leave much to be desired. This is partly to blame on legal imperfections and the pioneering nature of the relevant solutions. For over a decade, considerable focus was placed primarily on aesthetic and construction-related aspects [73], which led and still leads to the gentrification of areas subjected to revitalisation [74]. Revitalisation has often been performed as if played by ear, often based on good foreign models (which cannot always be directly transplanted to a different spatial and socio-economic setting), and sometimes without paying them any mind [75]. However, the outcome of brownfield transformation projects in Poland is primarily the result of low social awareness among decision-makers, real estate developers and future buyers of buildings and spaces created through the process in question [76]. The problem applies to the principles of shaping spatial structures in cities. This can also be associated with a very low level of architectural and urban education in the country at the primary- and high-school level [77]. Likewise, the country's liberal brand of capitalism is also to blame, as it strives to maximise profits with disregard for social and spatial costs of its decisions and is an understandable and natural social reaction to decades of communist rule [27].

In the country's largest cities and metropolises (Warsaw, Cracow, the Gdańsk–Sopot-Gdynia Tri-city, Wrocław, Poznań, Łódź or the Upper Silesia–Zagłębie Metropolis that was established in 2018 and consists of 41 cities and communities in Upper Silesia), there has been increasing development pressure on areas freed up by industry and on the infrastructure necessary for them to operate. This increase has had a different pace in different areas [78]. Especially that once-peripheral areas have currently found themselves in city centres, such as the areas of 'Młode Miasto', Wyspa Spichrzów, Dolny Wrzeszcz and Garnizon in Gdańsk, or Zabłocie, Podgórze and Grzegórzki in Cracow, Breweries in Cracow and Warsaw, Praski Port in Warsaw. The same has happened with numerous post-factory areas in Łódź [79], cities of the Upper Silesia–Zagłębie Metropolis and many others [40,80–83].

After 2003, Poland has entered into a crisis of spatial planning as previously enforced spatial development plans (acts of local law) were voided by a new Spatial Planning and Development Act [57]. This was combined with the obligation placed on municipalities to enact new spatial development conditions and directions studies. These studies did not have the status of local law and were merely intended as guidelines, to provide the basis for drafting new plans. Many municipal governments—including Poland's largest cities—have not fulfilled this obligation to date (October 2020). Some of them have done so only partially, enacting local plans only for fragments of their territories. This situation has led to land speculation that is difficult to control and the widespread practice of using legal loopholes during the issuing of construction permits. They are based on administrative decisions that are permitted in situations where there is no development plan in place for a given area, despite the fact that legislation that has been amended several times after 1989 [84].

This has also contributed to an excessive densification of new, functionally varied development (primarily commercial, office and residential) on brownfields obtained for real estate projects, primarily those located in city centres. This densification is typically motivated solely by short-term profit instead of correct relationships between urban structures and spatial order. It typically takes place without ensuring proper transportation, educational, cultural or blue-green infrastructure. This problem, apart from periodically generating media outrage and public protests, has also become the subject of numerous academic studies in Poland [85,86].

The irregularities and flaws of this form of shaping the city are particularly visible in multi-family housing complexes of varying size, which are primarily built by real estate development companies that offer apartments for sale [87]. Housing shortage, caused by years of underfunding and crises during the period of communist rule, continues to be a problem in Poland, as indicated by a report on the housing situation from March 2020 [88]. The report points to a state-wide apartment shortage of around 650 thousand units. Demographic change, as well as treating apartments and houses as a form of capital investment, have led to an overproduction of expensive, privately-owned apartments, built primarily by large real estate development companies and non-institutional developers [89].

The desire to maximise sales profits caused housing environment quality in complexes built in the years 1999–2020 to typically leave much to be desired [90]—even despite the observably better visual attractiveness of their architecture in most cases, although this cannot always be said of apartment functionality. However, it appears that the market for privately-owned housing associated with real estate development company projects is slowly beginning to exhaust itself—particularly in large cities, where real estate prices have been increasing along with the number of completed projects for over a decade. Perhaps the current crisis associated with the COVID-19 pandemic and lower demand for dwellings and office spaces for rent, caused by numerous associated factors [91] is one of the reasons behind this. For comparison, the number of social or municipally owned housing handed over for use has been declining and in 2019 amounted to only 2.2% of all construction projects [88].

3.3. Post-Industrial Areas—Overview and Specificity of Cracow's Downtown Area

The beginnings of Cracow date back to the end of the ninth and the beginning of the tenth century and are associated with a fortified gord on Wawel Hill (which is now the site of Poland's largest castle and cathedral complex). It towered above a stretch of swampland in the bend of Poland's largest river—the Vistula, which has been a major circulatory and commercial thoroughfare for over 900 years, ensuring prosperity and economic development to cities and settlements erected along its course [92]. Cracow is currently Poland's second-largest city (with around 769,000 residents in 2020) and a former state capital (between the eleventh and sixteenth century). The city is of immense symbolic and cultural significance to Polish national identity, a city of culture and science that attracts thousands of students and researchers and millions of tourists and pilgrims. In 1978, the urban layout of Cracow's Old Town (from 1257) together with the suburbs of Stradom and Kazimierz (from 1335), as well as Wawel Hill, was placed on the UNESCO World Heritage List. In 2010, a large portion of the city centre, within the so-called second ring road [93], which corresponds to the size of the city's urban area towards the end of the nineteenth century [94], was also added to the UNESCO List as a buffer zone.

However, like every city, Cracow's history features periods of prosperity and decline. Towards the end of the eighteenth century, after the third partition of Poland in 1795, after suffering damage in almost two centuries of wars and epidemics and a population decrease to around 10,000 [37], Cracow found itself under Austrian rule. From among the Polish cities incorporated into Austria, it was Lviv (currently in Ukraine) that developed much more dynamically and became the capital of Galicia, a new province created by the occupying power. With a peripheral location, close to a new border with Prussia and Russia that had been created after the partition of Poland among the European powers, Cracow became a distant backwater to the Austrian Empire, without major significance to the state's economic development. The rise and fall of Napoleon I and the associated geopolitical events, primarily the establishment of the Cracow Republic facade state in 1815, enabled demographic growth. In 1843, the city had a population of 43,000, with areas located on the southern shore of the Vistula—including the district of Podgórze—not counted, as formally Austrian territory. Due to its location at the border between three major powers—Russia, Austria and Prussia—the city gained a certain stabilisation and moderate economic development, primarily based on trade.

The development of industry in Cracow was largely confined to milling and alcohol distillation—no attempts at establishing a modern manufacturing industry were recorded at the time, although it did start to develop in other areas of contemporary Poland (Warsaw, Wrocław, cities of Upper Silesia or Poznań). The reincorporation of Cracow into Austria in 1846 did not change the situation, especially as the city's development was limited and subjected to military regulations. In the middle of the nineteenth century, Cracow was labelled a fortress city (Festung Krakau) and as such (with a later loosening of rigorous regulations) survived until the First World War and the end of the Austro-Hungarian monarchy [38]. It was the outline of the fortress's core (noyau), the first of three liens of defence, that determined Cracow's spatial development up to the first years of the twentieth century [95] and it is this area that was inscribed onto the UNESCO World Heritage List as a buffer zone (Figure 1). However, in the second half of the nineteenth century (following the Austrian act on industrialisation from December 1859), factories appeared in the city. They served not only its needs, but also the Austrian internal market and the production of export goods, along with increasingly modern industrial plants necessary to service the city and its infrastructure, such as waterworks, a gas plant and several power plants [37].

Figure 1. Centre of Cracow—UNESCO protected area from 1978 (marked with orange line) and its extension—the buffer zone from 2010 (marked with green line), source: http://whc.unesco.org/en/list/29/multiple=1&unique_number=1739 (public domain).

The first modern planning document—the Greater Cracow Regulation Plan—was drafted on the basis of a winning conceptual proposal (Figure 2) selected in a competition organised by the city in 1910. It sanctioned and regulated the placement of industrial areas in the eastern and southern part of the city, primarily in the contemporary districts of Wesoła, Grzegórzki, Kazimierz, Podgórze and Zabłocie [41]. Due to the previously mentioned military constraints, newly erected industrial plants found themselves in the immediate vicinity of the historic city centre. Some of these plants survived the interwar period and the Second World War and even continued to develop during the People's Republic of Poland (until 1989). Of these, some continue operating today, and the traces of others are a post-industrial heritage that is important to the city.

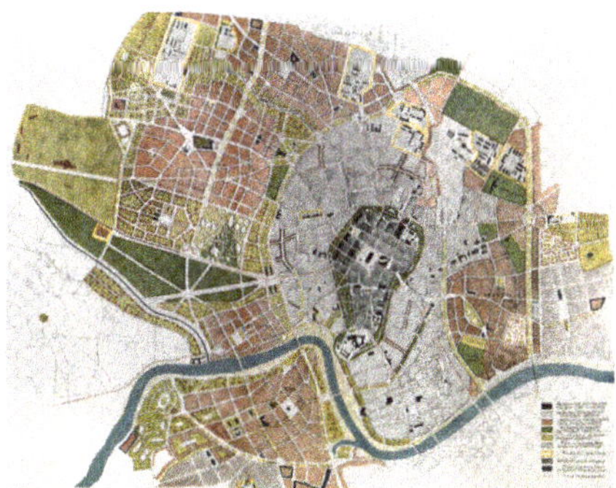

Figure 2. The Greater Cracow Regulation Plan—the winning competition entry by: Józef Czajkowski, Władysław Ekielski, Tadeusz Stryjeński, Ludwik Wojtyczko and Kazimierz Wyczyński, 1910, The creation of the urban form, extended beyond former fortifications of the inner city (everything that is marked differently than dark and light grey). The proposed layout, although never completed in that form, has been in many points repeated by the following urban plans of the city. source: reproduction in the collection of the Chair of Urbanism and City Structure Architecture of the CUT FA, photo by M. Nowak.

In the years 1945–1989, similarly as in other cities in the country, Cracow again saw an intensive wave of industrialisation, with several dozen factories and industrial plants of varying size and significance distributed across the entire city. They were located primarily within the territory outlined in the pre-war plans of the city—to the east of the centre and along the Vistula, as a suitable waterway. The infrastructure serving industry was extended as well—primarily in the form of roads and railways. These projects and their inconvenience over decades of operation have affected their immediate surroundings, the cityscape and the environment in various ways [96].

The peak manifestation of Cracow's industrialisation, as well as a form of political manifesto by the new, pro-Soviet government, was the construction of an enormous metallurgy plant in the years 1949–1954—the Vladimir Lenin Steelworks (now owned by ArcelorMittal Poland, which announced in October 2020 that it will cease to produce steel). The factory was established to the east of the existing urban structure of the city, on the northern shore of the Vistula. This metallurgy plant became the largest industrial complex in the city and one of the largest in the country. It was accompanied by the foundation of a Socialist Realist ideal city—Nowa Huta—on the territory of the existing villages of Mogiła, Bieńczyce and Krzesławice. These areas were incorporated into Cracow already in 1951, and over time became increasingly well-connected with it in physical terms, via its development structure, transportation system and a layout of green recreational areas that are now being increasingly damaged by new development projects [97]. This evolution can be recreated based on historical photographs, maps and planning documents [39] (Figure 3).

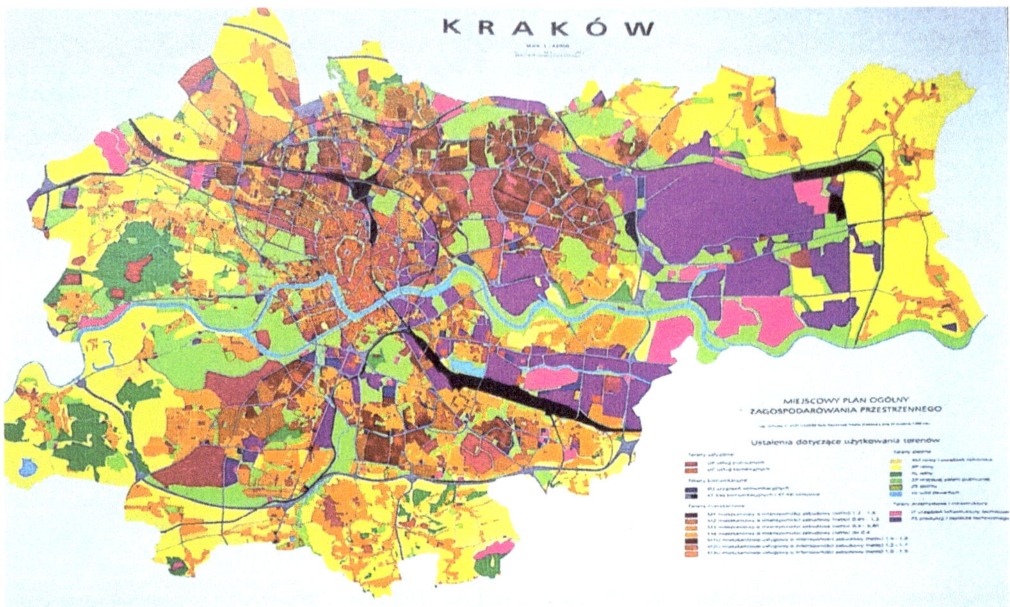

Figure 3. Drawing of the Local General Spatial Development Plan of the City of Cracow, drafted by a team under the supervision of Z. Ziobrowski in the years 1988–1994, shows a significant concentration of industrial and infrastructural uses (violet, pink) and their supporting railway infrastructure (black) in the eastern part of the city (around 70% of build-up areas) in relation to the western part of the city (around 12% of build-up areas). However, it is visible that the plan maintained the historically grounded industrial and infrastructural use even in central areas, particularly located near the Vistula River and along railway lines. This plan was in effect up to 2003. Source: reproduction in the collection of the Chair of Urbanism and City Structure Architecture of the CUT FA, photo by M. Nowak.

4. Results

4.1. Selected Cracow-Based Cases of Contemporary Housing Projects in City-Centre Revitalised Brownfields

The detailed study presented in this paper concerns cases of housing and mixed-use housing and service projects built in the years 2000–2020 on sites formerly occupied by industrial and infrastructural plants and storage complexes. They are located in the city centre of Cracow and as such primarily within and in the immediate vicinity of the UNESCO buffer zone and even inside the strict conservation zone established in 1978. Not all these areas were identified in applicable planning documents as intended for revitalisation (Figure 4). However, their transformation can be seen as offering a new spatial quality in zones that were excluded and inaccessible to residents for many years. Sometimes, it even forms a desirable continuation of generally accessible urban fabric, with a particular focus on the system of green public spaces.

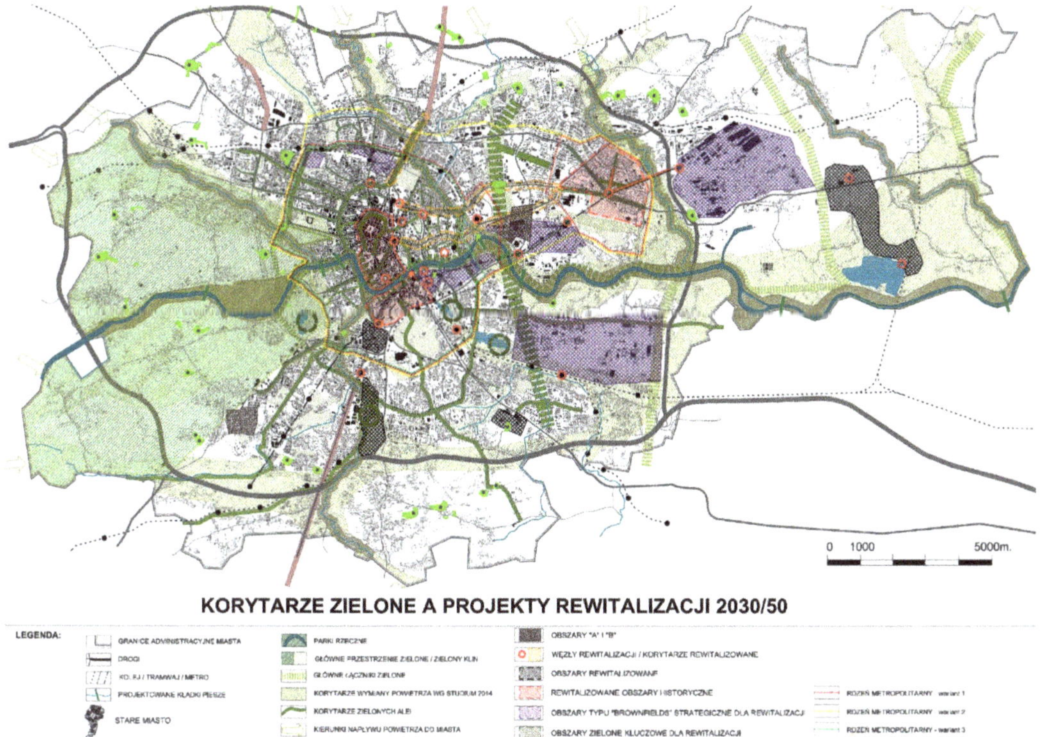

Figure 4. Areas planned for revitalisation and the network of the city's main green corridors—vision 2030–50; post-industrial areas covered by revitalisation plans in 2017 (black grid) and key to revitalisation (violet grid); by M. Gyurkovich, A. Szarata, Z. Zuziak and team, drafted by A. Derlatka 2017, source: [58], available by the courtesy of the Spatial Planning Bureau of the Office of the City of Cracow.

Citywide blue-green infrastructure is crucial for creating a healthy housing environment. In Cracow, it is primarily formed by a network of city parks and riverside green areas (called 'river parks' in planning documents), both landscaped and left in their natural state. They are supplemented by large greenery complexes—those of Las Wolski and the protected areas of several landscape parks in the western part of the city and large green areas, typically flood plains or farmland, in the northern and eastern edges of the city (some, including the Nowa Huta Meadows, labelled as ecological use or Natura 2000 areas). This system, although it is supplemented by greenery that accompanies the road and path network, is not continuous. The most important citywide connector is the Vistula river park [98], as the river flows through the entire city from west to east. Many areas located along the Vistula possess unique landscape, recreational or ecological assets (Figure 5).

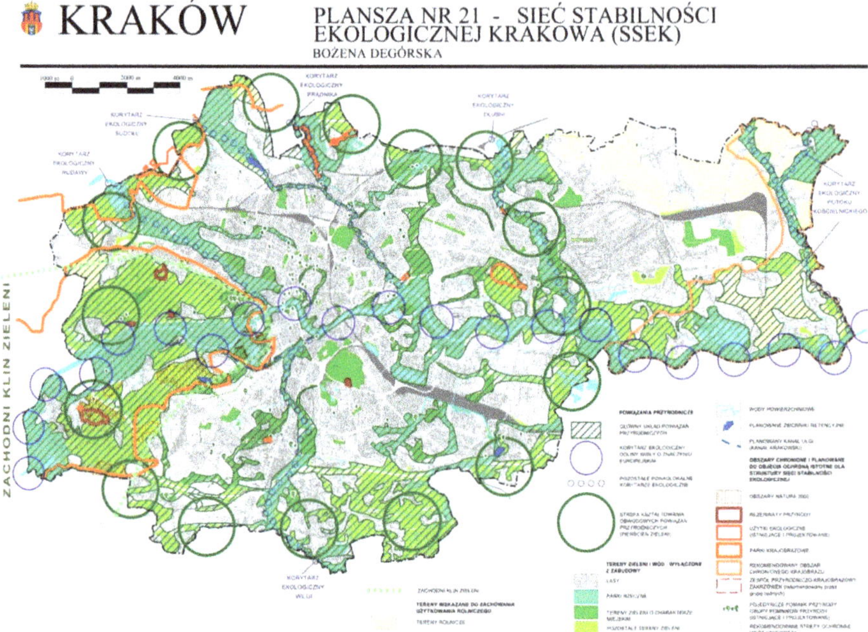

Figure 5. Cracow Ecological Stability Network—Sieć Stabilności Ekologicznej Krakowa (SSEK), by B. Degórska, source: [17]. Rivers and its surroundings marked cyan, forests -dark green, lawns—light green. Different forms of protected landscapes marked with red, pink and orange outlines. The ecological corridor of Vistula River Valley (crucial in the scale of entire Europe) marked with large dark-blue circles, smaller blue circles along the river valleys shows the ecological corridors of supra-local importance. Dark-green circles shows the range of "green ring" around the city.

Four completed housing projects that were built in post-industrial central plots and possess linkages with the city's blue-green ecological system were selected for further in-depth analysis. This was the most important criterion for case selection, apart from all of them having a dominant housing function. Three of the projects were built as singular, large endeavours, adaptations or as a result of remodelling small industrial plant buildings: Browar Lubicz at Lubicz Street, Wawrzyńca 21 in the Kazimierz district and a mixed-use service and residential building at Nadwiślańska Street, in the historical area of Podgórze. The study also covered two large industrial and storage areas that are currently being adapted into exclusive housing areas. These areas are located in the immediate vicinity of the Vistula River Valley—in the districts of Grzegórzki and Zabłocie, east of the historic city core that features the most precious elements of urban and architectural heritage (Scheme 1). These areas functioned almost up to the start of the twenty-first century. They comprised numerous manufacturing plants and institutions that were founded between the middle of the nineteenth century and 1989, and their multi-stage remodelling is still ongoing, with different real estate developers participating (Figure 6).

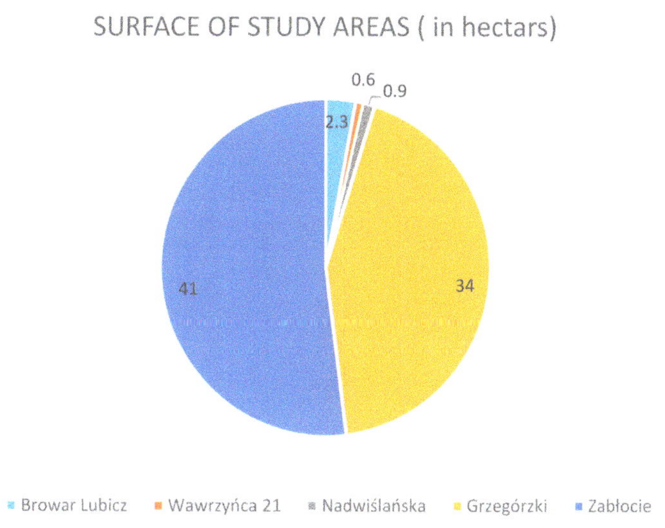

Scheme 1. The comparison of the surface of the areas under study.

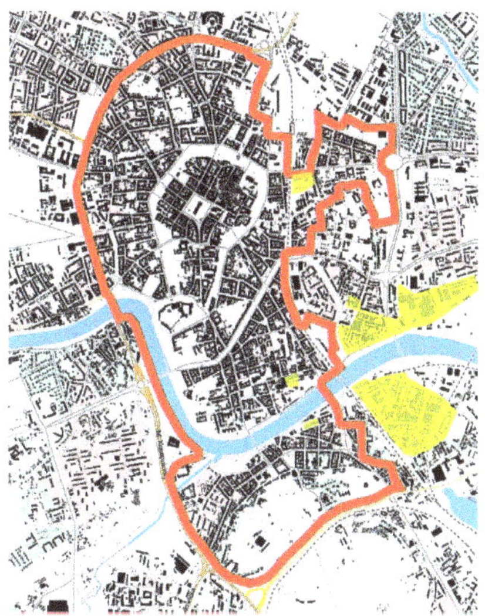

Figure 6. City centre of Cracow—territory covered by the Local Old Town Revitalisation Plan from 2008 (red), the present-day Historical Urban Complex, Monument to History (https://msip.um.krakow.pl). The areas discussed in the paper are shown in yellow—on some sites one can still see the industrial structure present at the time—based on: [99] https://rewitalizacja.krakow.pl/.

4.1.1. Small Interventions within the Historic Urban Tissue of Cracow

The Browar Lubicz project is one of the first and most successful examples of revitalising post-industrial areas in Cracow's historic urban centre. The small brewery, which belonged to the Goetz family of Galician market potentates, was built in 1840 on the border of the suburbs of Wesoła and Lubicz [37] and was afterwards successively extended and modernised. It operated up to the early 1990s. The site is located near a railway track, in the immediate vicinity of the Main Train Station, a Jesuit monastery and a complex of clinical hospitals that expanded throughout the entirety of the twentieth century [100]. The most valuable buildings were placed in the municipal register of monuments in 1995 [101], while many later buildings, primarily from the 1960s and 70s (Figure 7), of low architectural and aesthetic value, were demolished. Changes in designers and developers, as well as precise and restrictive local development plan provisions (that only covered the project site) led to the preparation of a series of highly similar design proposals. Ultimately, the design was prepared by the Cracow-based MOFO architectural firm [102].

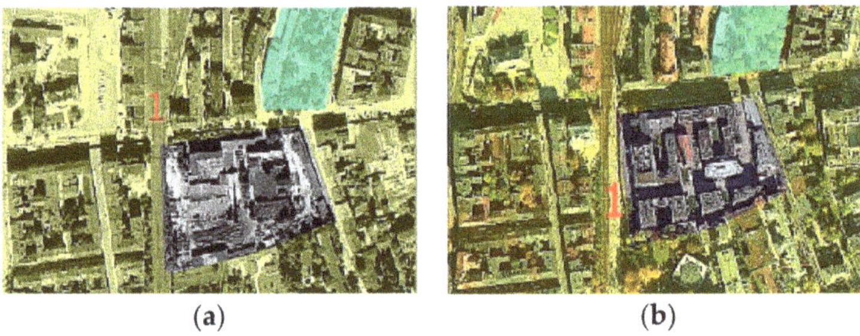

(a) (b)

Figure 7. Comparison of orthophotomaps of the selected fragment of Cracow's urban fabric under study (1. Browar Lubicz) from 1970 (**a**) and 2019 (**b**) against public greenery (cyan). There is an observable change in urban fabric typology and density. Based on aerial photography from https://msip.krakow.pl.

A housing and office complex was built at the site of the former production plant, and the former gates to the brewery are now open, inviting pedestrians into its interior. These openings are a continuation of locally significant compositional axes, primarily the longer axis of Park Strzelecki (Marksmen's Park). 'The new massings were combined with existing ones in a manner that references the layout of the brewery's historic development' [103]. A tall, slender 'factory' brick smokestack continues to act as the complex's dominant element, which, together with a restored boiler building, became the centre of a small recreational space. The architecture of the complex's housing buildings primarily features brick and steel, as well as wood, as finishing materials, blending new facades with the restored walls of historic sections, adapted to new functions—primarily commercial ones (offices, restaurant, shops). The brewery's old equipment further enhances the attractiveness of the complex's partially green social spaces. The proximity of the historic centre and the city's main transportation node, along with numerous public institutions and corporation offices, has contributed to the commercial success of the project, whose buildings were built to a very high finishing standard (Figure 8).

(a) (b) (c)

Figure 8. Browar Lubicz housing complex—as seen in 2019. Photo by M.Gyurkovich (**a**) view from the complex towards the park; (**b**) internal street with the former palace of the brewery owner on the left; (**c**) chiminey left as a symbol.

The course of Wawrzyńca Street is a trace of the outline of Kazimierz's medieval city walls. It is here that, on previously undeveloped land, the first infrastructural plants that provided amenities to the city began to be built in the second half of the nineteenth century: a gas plant, a power plant and a tram terminus [37]. Some of the historic buildings, including 'gasometres', disappeared, while others, which remained in the hands of energy and gas companies, act as administrative and office buildings (Figure 9). The half-timbered buildings of the terminus now house the Municipal Engineering Museum and gastronomic establishments. Other historic buildings were taken over by private real estate developers. This was the case for one of the buildings that once belonged to the municipal power plant—at 21 Wawrzyńca Street. Detailed architectural and conservatorial documentation and a series of conceptual design alternatives led to the construction of the Wawrzyńca 21 apartment complex. The complex, built to a design by architects from B2 Studio, consists of two city blocks (Figure 10a,b). In the frontal, northern block, historic buildings were extended and connected to new sections, whose neutral detail, maintained in a 'post-industrial' aesthetic, highlights the remains of former structures (Figure 11). The southern block, in the shape of the letter U, opens towards the Vistula, the green Krzemionki Hill on the river's southern bank and, in its immediate vicinity, the site of the former gasometres, commemorated in the form of a small green square. This complex, due to its ownership situation, does not have direct linkages (apart from visual ones) with the green square or the riverside boulevards, which are separated from it by a street. However, the development's central location, unique views and a particular attention to the post-industrial place-based heritage remain its key assets.

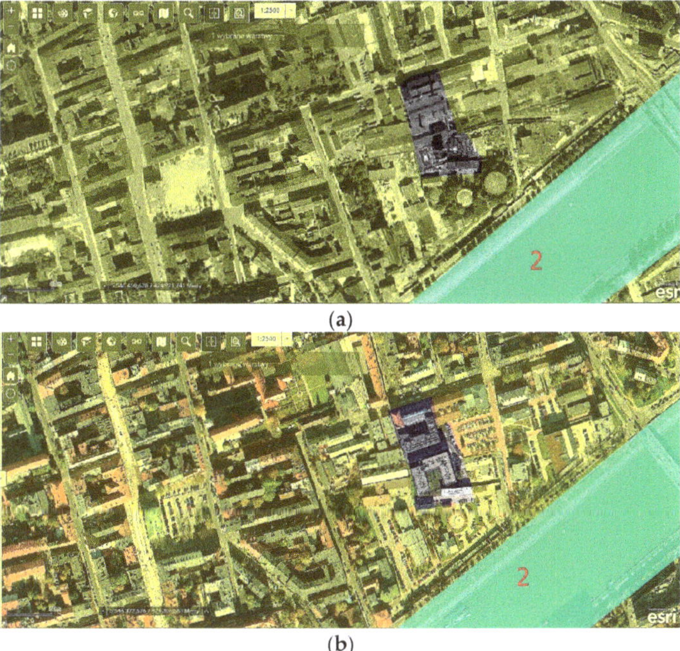

Figure 9. Comparison of orthophotomaps of the selected fragment of Cracow's urban fabric under study (2. Wawrzyńca 21) from 1970 (**a**) and 2019 (**b**) against public greenery (cyan). There is a visible change in urban fabric typology and density—based on aerial photography from: https://msip.krakow.pl.

(a)

(b)

Figure 10. Exclusive housing complex on the site of a former power plant and gas plant in Kazimierz—Wawrzyńca 21—as seen in 2020—view towards the south-west (**a**) and towards the east (**b**); Photo by P. Krajewski, courtesy of J. Białasik, source: http://b2studio.com.pl/projekt/apartamenty-wawrzynca-21/.

Figure 11. Details and materials inspired by the post-industrial *genius loci*—Wawrzyńca 21—as seen in 2020; Photo by P. Krajewski, courtesy of J. Białasik, source: http://b2studio.com.pl/projekt/apartamenty-wawrzynca-21/.

The *Nadwiślańska Apartments* complex of residential and service buildings in the district of Podgórze, located in a narrow, elongated city block between Nadwiślańska, Piwna and Krakusa streets, was built in the years 2011–2014 almost at the Vistula's shore. This area is located in the vicinity of Podgórze's former nineteenth-century power plant, built near a railway line that ran along the river. The power plant was redeveloped and adapted in the years 2006–2014 into a cultural facility—one of the city's most iconic buildings, the Cricoteka [92].

In the nineteenth century, the plot in question was occupied by small storage buildings and industrial plants. After the Second World War, a complex of buildings of the Vistula textile factory was built here, together with a tall and characterless sewing hall (Figure 12a). After the complete relocation of the plant outside of Cracow in the beginning of this century (to the town of Myślenice, located 35 km away), the plot was sold, and the buildings were demolished. The new project continues or perhaps even goes beyond the scale of the former sewing hall, referencing a nearby hotel from the 1990s and an office building at neighbouring Bohaterów Getta Square.

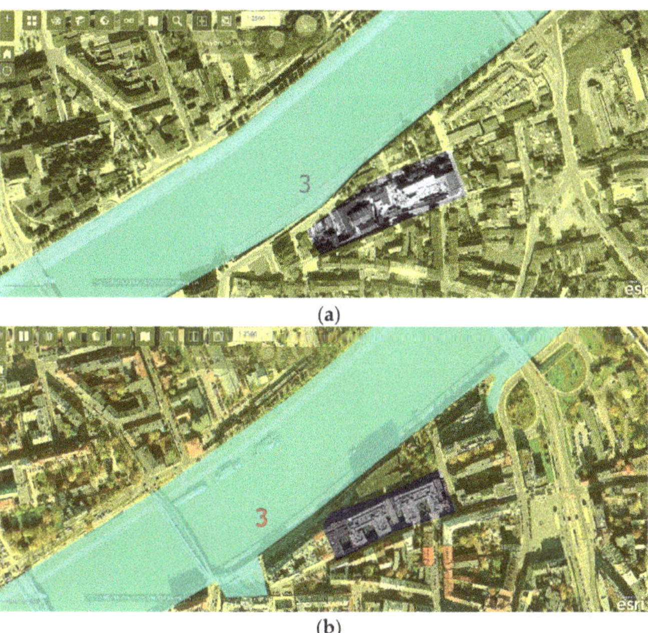

(a)

(b)

Figure 12. Comparison of orthophotomaps of the selected fragment of Cracow's urban fabric under study (3. *Nadwiślańska Apts.*) from 1970 (**a**) and 2019 (**b**) against public greenery (cyan). There is a visible change in the urban fabric's typology and density, based on aerial photography from: https://msip.krakow.pl.

It is one of the tallest buildings in Old Podgórze, many times taller than the nearby historical development and even the later urban infills built between the 1960s and 1990s (Figure 12b). Due to the sculptural form of its massing, which creates an illusion of a division into a sequence of several townhouses, the mass is not excessively overwhelming, and the scale appears adequate to building the frontage of a river valley in the city centre (Figure 13).

Figure 13. The Vistula River boulevards as seen from a footbridge, with Cricoteka and the project under study (Nadwiślańska Apartments) in the background—as seen in 2019. Photo by M.Gyurkovich.

Primarily featuring dark, industrial brick in several shades of colour, along with large, glazed surfaces and steel elements, the project by Saran Architekci is a neutral backdrop for the unique architectural form of the Cricoteka (Figure 14a). Service premises on the ground floor and the presence of new residents, mostly young professionals, have made this once-excluded area a living fragment of Podgórze's urban fabric, which is undergoing gentrification (Figure 14b). Apartments in the building offer unique views of the Vistula River Valley's panorama, the historic area of Cracow towards the north, and the green strip of Krzemionki Hill, which compositionally encloses from the south the historic urban layout of Cracow, developed for 9 centuries along the north–south axis [104].

(a)

(b)

Figure 14. Nadwiślańska residential and service complex- view from across the river (**a**) and from the south- opposite side (**b**). Photo by authors.

4.1.2. Transformation of Large Post-Industrial Areas in the City Centre near the Vistula River

The Grzegórzki district, located on the left bank of the Vistula, to the east of the historic city centre, together with Zabłocie, located on the opposite side of the river, underwent rapid industrial development at the turn of the twentieth century. Since the Middle Ages, the river had been the major route for transporting goods [96], and the construction of railway lines eastwards from Cracow's centre enabled the development of additional factories. The combination of newly erected plants and riverside storage areas with a system of railway tracks running along shores and connected with train stations enabled the plants to survive and develop at the site. They operated up to the end of the People's Republic of Poland in 1989. It was then that the global economic situation began to affect Poland's economy and thus the spatial form of Polish cities.

Unprofitable plants that typically occupied attractive land with good spatial linkages in city centres became increasingly attractive to real estate developers. These changes were noted by city planners and decision-makers, contributing to the planning and construction of significant infrastructural projects in this area. The most important ones included a new river crossing—Kotlarski Bridge (competition 1999; construction 2001). Along with the modernisation and extension of streets that led to it, it linked Grzegórzki and Zabłocie (by car, tram, bus, on foot and bicycle). It also led to the closing of the loop of the second ring-road of the city, which roughly corresponded to the line of the core of nineteenth-century Austrian fortifications [38]. The project clearly sped up the revitalisation of Zabłocie. It began several years earlier on the 'better' shore, in Grzegórzki, closer to the city centre.

The fall of large industrial plants in Grzegórzki towards the end of the 1980s, which affected, among others, the Zieleniewscy Machine Industry Plant (which was the largest factory in Cracow up to the 1940s) or the relocation of others—the municipal slaughter-house or the neighbouring chocolate factory (Figure 15a)—freed up large areas for new development. In the western part of the district, on the site of the former slaughterhouse, the Galeria Kazimierz commercial ad service centre was built in the years 2004–2005 [105]. It was soon surrounded by office and hotel buildings and later several smaller housing complexes (built in stages up to 2020). Apart from fragments of the road layout and three heritage-listed buildings of the former slaughterhouse (from 1871), the post-industrial heritage of the site is completely imperceptible. This is felt even more in the eastern part of the district, outside of Kotlarski Bridge, where no local spatial development plan was enacted between 2003 and October 2020 (Figure 15b).

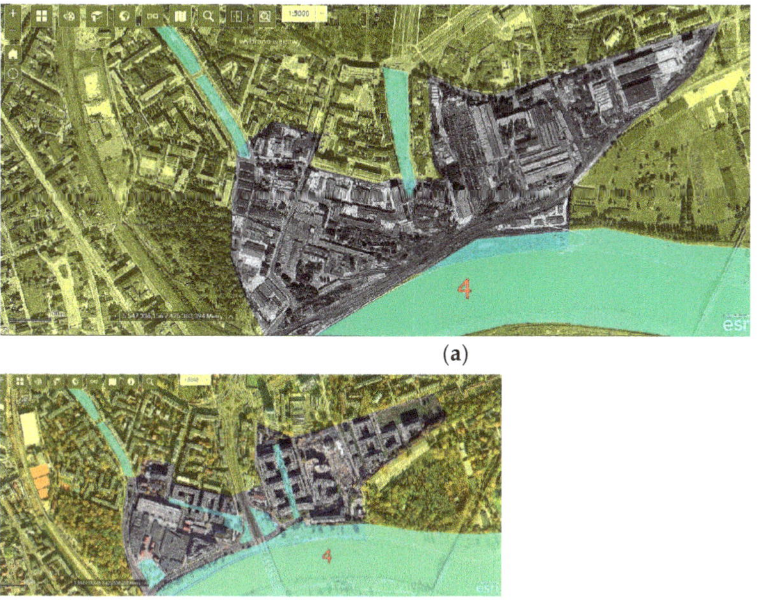

(**a**)

(**b**)

Figure 15. Comparison of orthophotomaps of the selected fragment of Cracow's urban fabric under study (4. Grzegórzki) from 1970 (**a**) and 2019 (**b**) against public greenery (cyan). There is a visible change in the urban fabric's typology and density, based on aerial photography from: https://msip.krakow.pl.

These areas are converted into housing estates without proper service and educational infrastructure. In these departments, they are based on existing neighbouring institutions from the 1970s and 1980s. Sometimes new educational functions appear as secondary adaptations of the ground floors of existing or even newly designed buildings. The second ones typically feature a certain number of commercial premises for rent. Nearly all buildings and infrastructural elements that signified the district's post-industrial heritage have been demolished (including two factory halls which continue to be listed in the municipal monuments register!). The only building that is under protection is a Socialist Realist cinema theatre (currently a musical theatre). In architectural terms, the buildings and complexes built in this part of the city are characterised by high diversity of styles and sizes, but also by formal attractiveness. However, their architecture does not reference local tradition and could be built in any place in the world looking equally as attractive (Figure 16a,b). Some complexes (like Wiślane Tarasy I) could be characterised by attention to and a quality of the housing environment, along with pleasantly composed semi-private and public greenery. They supplement the municipal system of blue-green infrastructure (Figure 17a,b).

Figure 16. Grzegórzki— "international style" architecture of housing complexes with no connection to the local identity—as seen in 2020. Photo by authors. (**a**) view towards Wiślane Tarasy I complex; (**b**) Wiślane Tarasy II complex looking more like some resort in the Mediteranean, than a housing complex in Cracow.

Figure 17. Grzegórzki- public greenery around and within housing complexes—as seen in 2020. Photo by authors. (**a**) internal public passage with greenery; (**b**) publicly accessible greenery on the edge of one of the housing complexes.

The transformation of post-industrial areas in Zabłocie began almost simultaneously. However, it initially progressed at a slower pace [86], also because production and other economic activity, often relocated from other areas of the city, was still ongoing in many buildings that existed at the time. Another reason was the isolation of the district, which has the shape of the letter V, surrounded on both sides by railway escarpments and by the Vistula River from the top (Figure 18). This area was connected with the remainder of the city's urban fabric by only three narrow streets. Only the construction of the previously mentioned bridge and tram line across the Vistula to Grzegórzki brought this area closer to the city, both mentally and functionally. Paradoxically, the resulting offer was first taken up by educational and cultural institutions, with housing complexes built primarily in the last five years. The district is the home of one of the largest private universities in the city—the Andrzej Frycz-Modrzewski Krakow University (since 2000) whose campus (2000–2020) is currently located near the river (initially, the university operated in office and post-industrial buildings scattered across the district). The adapted buildings of the former 'Rekord' Enamelware and Tin Products Factory, known from Steven Spielberg's film Schindler's List, became two important cultural institutions: a branch of the historical museum of the city of Cracow which relays the history of the area [106] and MOCAK—the Museum of Contemporary Art Kraków (since 2010).

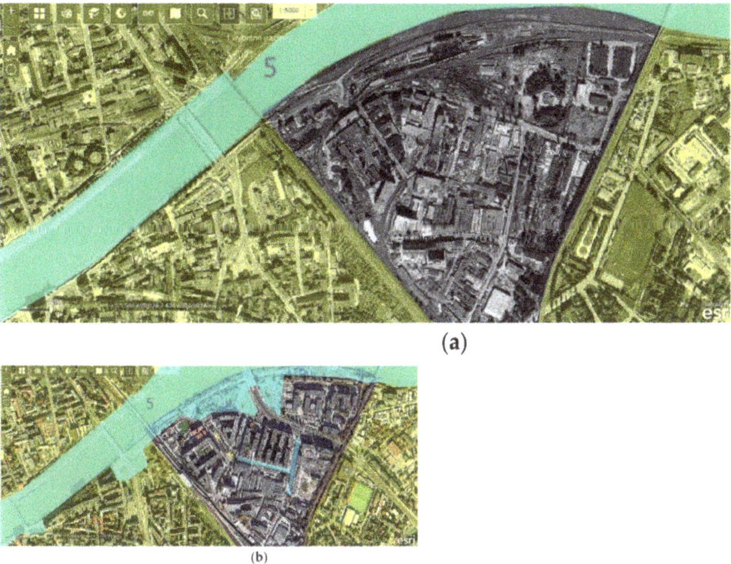

Figure 18. Comparison of orthophotomaps of the selected fragment of Cracow's urban fabric under study (5. Zabłocie) from 1970 (**a**) and 2019 (**b**) against public greenery (cyan). There is a visible change in the urban fabric's typology and density, based on aerial photography from: https://msip.krakow.pl.

In 2010, a local spatial development plan entered into force for the territory of Zabłocie. Its provisions, which are applicable to this day, have completely transformed the image of the district. Apart from a number of buildings in the municipal historic monuments record, the district's post-industrial architectural heritage was not placed under any form of conservation. Similarly, due to soil degradation, natural heritage was practically non-existent. The plan introduced a regular grid with narrow blocks allowing high-density development and mixed-use housing and commercial or service buildings [107]. Two short pedestrian and recreational axes were delineated, with a small amount of greenery—at the back of the MOCAK and along one of the area's streets (Romanowicza).

The site is not connected with the urban greenery system. New buildings and complexes are consistently built by different real estate developers, based on designs by numerous architectural firms. Thousands of new apartments and workplaces are offered (Figure 19a,b). Many of them, due to the use of attractive details that bring industrial aesthetics to mind—with steel and brick or raw concrete—have managed to create an illusion of place-based atmosphere, different from the situation in Grzegórzki (Figure 20a). Certain post-industrial buildings, such as a mill from the start of the twentieth century and a six-storey factory hall from the 1970s (Figure 20b), were subjected to adaptive reuse, but most were demolished. At present, a rapid municipal railway stop has been modernised nearby. As is often the case, the city did not develop essential buildings—primary schools, kindergartens or public healthcare centres—that contribute to housing comfort. The private sector is trying to fill this gap, but so far this has proved insufficient.

(a) (b)

Figure 19. Zabłocie- selected housing projects—as seen in 2020. Photo by authors. (**a**) difference of scale and density between existing and new architecture; (**b**) internal green courtyards are accessible only for the inhabitants.

(a) (b)

Figure 20. Zabłocie- selected housing projects—as seen in 2020. Photo by authors. (**a**) newly formulated street corner with shops in the ground floor; (**b**) public space of Romanowicza street with former factory adopted to apartment use (left).

In terms of conservation and offering access to natural heritage, the most important change in Grzegórzki was reconnecting this district with the river. The elimination of railway infrastructure that served industry began in the 1980s and led to the construction of pedestrian paths surrounded by greenery along the Vistula. They connect the existing boulevards at the level of Kazimierz with areas that are shaped more naturally and surround the Vistula in the eastern part of the city (also indirectly—with the Nowa Huta Meadows). The boulevards are modernised and equipped with new sports and recreation infrastructure elements from time to time (most recently—a beach in the eastern side of Kotlarski Bridge has been added). Zabłocie is still waiting for major projects near the Vistula boulevards, yet, similarly as in Grzegórzki, the railway lines that separated the district from the river have been demolished. A small green square called Stacja Wisła has been established in the western part of the area, and the banks of the river are currently more or less skilfully renaturalised [108].

4.2. Main Findings

The study was performed in a multi-track manner. The main criteria for evaluation were the number of elements of post-industrial heritage preserved; the preservation of the pre-existing *genius loci* in contemporary architectural interventions—such as forms, volumes of materials; the accessibility of semi-public or public spaces within the complex and access to elements of public greenery form the complex. Attention was mainly focused on functional and compositional linkages between the projects under study and elements of Cracow's blue-green infrastructure (Figure 21), understood as the natural heritage of the city. The distance to green areas and compositional linkages between the building/complex and the closest nearby park or river park were accounted for. It was also determined whether green areas were provided as a part of the housing complex projects as required by Polish construction law, as they can be found to constitute an attractive social space. It was also investigated whether they supplemented the urban system: providing passages, access paths or acting as continuations of the city's network understood as publicly accessible blue-green infrastructure. The elements of this infrastructure were divided into: crucial, moderate and local, due to the role which they play in forming the system (Table 1). The crucial elements are the ones, which are perceived as important in urban composition of the city, which has been also confirmed in many planning documents [58,98] and publications [37,96,97]. They are used by all the inhabitants of the city and tourists as recreational zones. The "moderate" ones are usually used by the inhabitants and users of a particular district of the city (e.g., small park). The "local" ones are the smallest and the less important elements of the entire blue-green infrastructure of Cracow. Usually, they are used by inhabitants of one or few urban blocks (e.g., something like a local green playground or small garden).

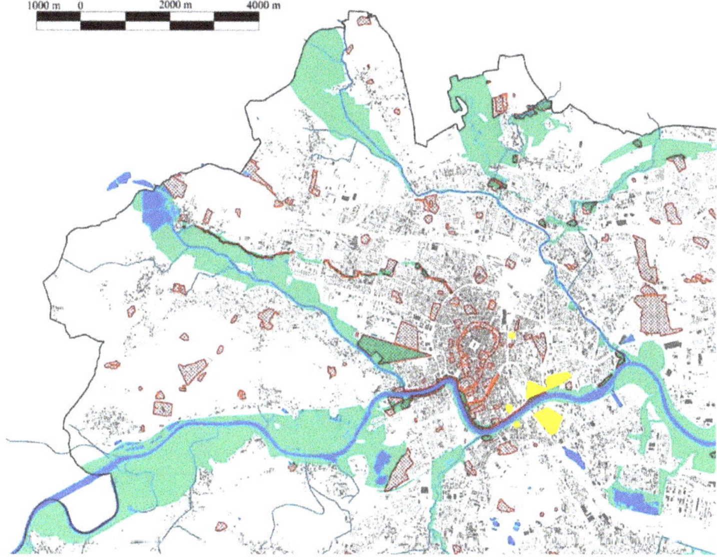

Figure 21. Planned and existing city parks (red) and river parks (green and blue) in Cracow, set against housing projects in post-industrial areas under discussion (yellow)—based on: [17] www.bip.krakow.pl.

Table 1. Areas under study and green areas (natural heritage) of the city.

Name of Area	Publicly Accessible Green Areas Featured within the Project	Compositionall /Functiona Linkages with Urban Greenery	Number of Public Green Areas in Walking Distance	Name of the Nearest Public Green Area	Distance	the Role of the Nearest Public Green Area in the Blue-Green System of the City
Browar Lubicz	partially	yes	3	Park Strzelecki	25 m/across the street	moderate
Wawrzyńca 21	no	no	1	Kurlandzki Boulevard (part of the Vistula River Park)	450 m	crucial
Nadwiślańska Apartments	no	no	2	Podolski Boulevard (part of the Vistula River Park)	12–50 m/across the street	crucial
Grzegórzki	partially	yes	2	Kurlandzki Boulevard (part of the Vistula River Park)	12–20 m/on the southern edge of the entire complex	crucial
Zabłocie	yes	partially	2	Stacja Wisła Park and Podolski Boulevard (part of the Vistula River Park)	12–250 m/on the northern edge of the entire complex	crucial (Boulevard) and local (park)

The smaller projects (Wawrzyńca 21 and Nadwiślańska Apartments) were found to offer no publicly accessible green areas on their grounds due to their size, composition and functional layout. They did not have compositional or functional linkages (except of visual) with the neighbouring elements of public greenery, because of the given location within the urban fabric. However, the bigger developments were found to feature such elements. The ones taken under consideration in this paper (Browar Lubicz or entire new parts of Grzegórzki and Zabłocie districts) are in fact multifunctional, with a predominant multi-family housing function. All of them are located close to elements of public greenery, which play different roles in the entire blue-green infrastructure system of Cracow (Scheme 2). The small complex within the adapted brewery offers partially accessible semi-public green squares within and opens towards Strzelecki Park with one of its visual and functional axes. Astonishingly, the two large multifunctional developments at Zabłocie and Grzegórzki districts, which are located almost at river embankments (river park) offered very little semi-public greenery to its users. In the case of Grzegórzki, only one public green passage was found (it is closed at night) which runs through one of the housing estates linking it to Kurlandzki Boulevard and the Vistula River. In Zabłocie, only two public passages with greenery were built to date (October 2020). In both cases, only the first line of the buildings forms the facade of river boulevards and have visual and compositional connections with the river and the river park.

The evaluation of the degree of preservation and exposure of post-industrial place-based heritage in the new projects was also an important element of the study. Apart from direct adaptations and preservation of existing build-up heritage elements, the form of inspiration for new architectural and material solutions was also taken into account. In heritage protection, including the protection of post-industrial heritage, Polish legislation provides useful tools in the listing of buildings and their complexes in historic monuments records and registers, especially the register of historical monuments (either municipal or voivodeship) [101,109], the establishment of conservation zones and detailed provisions in local spatial development plan (MPZP) texts. The last of the abovementioned tools can define a different method of protecting or maintaining characteristics that are valuable to local identity [110].

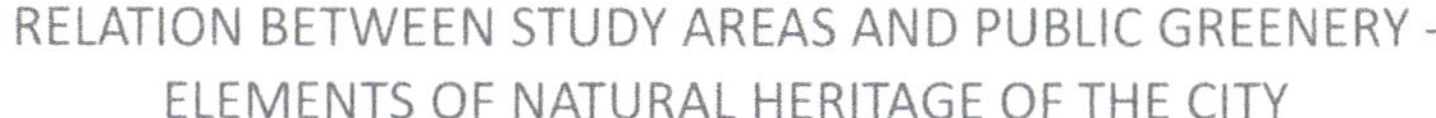

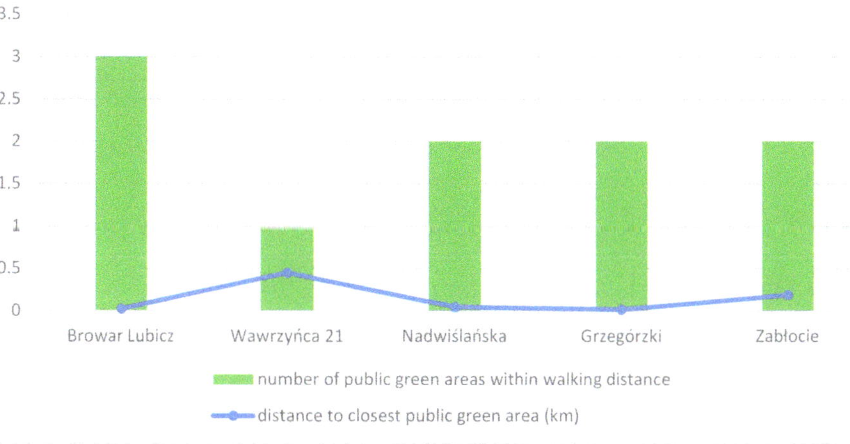

Scheme 2. The relation between areas under study and the elements of natural heritage.

Examining whether legal heritage protection instruments, together with those included in municipal planning documents, had an actual impact on the preservation and exposure of post-industrial relics in the areas under study, was also an important task. This is crucial, as Cracow, despite its long history and extensive heritage protection systems, possesses relatively few of such buildings and complexes in comparison to other large Polish cities.

In most of the areas under study, at least couple of instruments of heritage protection were incorporated. Browar Lubicz, Wawrzyńca 21, Nadwiślańska Apartments, the western part of Grzegórzki and most of Zabłocie districts are located within UNESCO protected areas of the city (the strictly protected zone or the buffer zone). Beyond these areas, the urban layouts of some of the parts of the city in which three of the analysed projects are located are placed in the municipal monuments register [93]. They include urban layouts of: the Wesoła district (no. A-650; since 16 February 1984)—for Browar Lubicz, where four buildings were registered separately (no. A-998; since November 1995); the Kazimierz district within the 'new town' with historical suburbs: Łąka Św. Sebastiana (St. Sebastian Meadow), Podbrzezie and Pola Kazimierzowskie, placed in the monuments register (no. A-1273/M; since 18 July 2011)—for Wawrzyńca 21; and the former town of Podgórze (no. A-608; since 26 October 1981)—for Nadwiślańska Apartments. The weakest protection of built-up heritage could be observed in the districts of Grzegórzki and Zabłocie, as analysed in the paper. Both were associated mainly with industrial production and storage up to the end of the twentieth century. As such, they were not a subject of interest to conservation offices at the time, especially that numerous buildings and structures dated back to no earlier than the beginning of the twentieth century. As such, and because of their manufacturing functions, they were not sufficiently protected. The historical monuments register has only three entries located in Grzegórzki. This number includes the post-Austrian fort complex (no. A-1048, since 20 February 1998), the three buildings of the former slaughterhouse (no. A-936, since 28 January 1993), which blend well with the commercial complex of Galeria Kazimierz, and the Socialist Realist building of the present-day Variete Theatre (no. A-1359/M, since 21 January 2014). Two factory halls from the first half of the twentieth century are still listed in the municipal monuments record [109], despite having been demolished during the construction of housing complexes in the early 2000s. None of the remains of the post-industrial past of Zabłocie district are listed in monuments register [101], despite the fact that almost the entire area has been protected by UNESCO as a buffer zone since 2010. Five factory complexes from 1899–1939, which were remodelled already in the interwar period and were listed the municipal monuments record [109]—and are currently also being remodelled and adapted to new uses. It includes two museums at Lipowa Street, apartments and offices in the mill at Zabłocie Street. The comparisons below demonstrate just how small the fragment of post-industrial heritage is that is under conservation in the analysed areas (Table 2, Scheme 3). The reference to the post-industrial heritage in new architecture were validated by the appearance of new buildings [1,13,22,102,103]. The use of materials, colours of the finishing of the facades, as well as details characteristic for industrial architecture in new structures was validated as "strong". The use of some (at least 2) of abovementioned elements in new buildings was validated as "medium". The lack of the references or small percentage of it (1 element) was evaluated as "weak".

Table 2. Areas under study analysed in terms of the protection of post-industrial architectural and urban heritage.

Name of Area	Degree of Heritage Preservation and Exposure	Number of Buildings/Complexes under Conservatory Protection or with the Status of Heritage Structures in Planning Documents	References to Post-Industrial Heritage in New Architecture
Browar Lubicz	high	4 buildings of the former brewery, Urban layout of Wesoła district	strong
Wawrzyńca 21	high	Urban layout of Kazimierz district	strong
Nadwiślańska Apartments	none	Urban layout of Podgórze	strong
Grzegórzki	none	3 entries in the municipal monuments register	weak
Zabłocie	low	5 factory complexes listed the municipal monuments record	medium

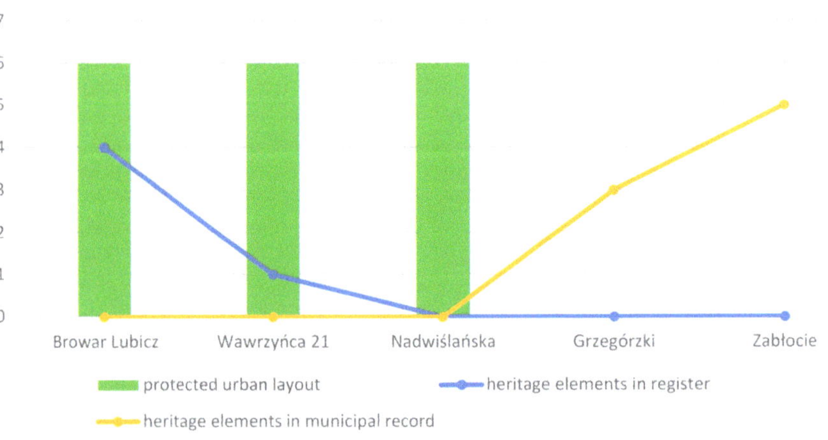

Scheme 3. The relation between the elements of protected cultural heritage areas under study.

The territory of Cracow is covered by a Spatial Development Study that has been in force since 2003 and was updated in 2014 [98]. Another amendment to the Study has been in preparation since 2019. However, many areas of the city do not have local spatial development plans in place (or they have been voided). This leads to growing spatial chaos, as projects are approved on the basis of mutually uncoordinated planning permits. Most of the projects discussed in this paper were built in areas covered by local spatial development plans (Table 3). In some cases, these plans were introduced or changed after the beginning of 'revitalisation' works and merely sanctioned ongoing projects that were being built based on independent administrative proceedings.

Table 3. Areas under study and local spatial developments plans (MPZP).

Name of Area	Name of the Local Spatial Development Plan (MPZP)	Date of Plan Entering into Force	Area Covered by Local Spatial Plan (in Hectares)	Area under Study (in Hectares)	Duration of Design/Design and Construction
Browar Lubicz	MPZP 'Browar Lubicz'	24 Oct.2007	2.23	2.23	2004–2011 2011–2013
Wawrzyńca 21	MPZP 'Bulwary Wisły'	8 Oct. 2013	168.24	0.6	2006–2017
Nadwiślańska Apartments	MPZP 'Bulwary Wisły'	8 Oct. 2013	168.24	0.9	2011–2014
Grzegórzki	I. Western part of the area- MPZP 'Rejon Al. Daszyńskiego' II. No local plan for eastern part	I. 22 Nov. 2018	I. 45.7	I. 12 II. 22	1994 and still ongoing
Zabłocie	MPZP 'Zabłocie'	26 June 2006	175	41	2000 and still ongoing

The areas in the eastern part of the Grzegórzki district, which are located outside of the conservation zone associated with the Monument to History and UNESCO, are an exception. They were the only areas from among those under study not to have a local spatial development plan in place as of the time of writing of this paper (October 2020). Despite heavy development in the district. The lack of detailed planning documents and the often-vague provisions of the Study [98] caused the projects under analysis to contribute to the protection of cultural and natural heritage and the preservation of the genius loci to differing degrees. Most of the projects were initiated prior to the passing of current spatial planning and revitalisation acts and were designed and built under different legal conditions.

It should also be remembered the local spatial development plans [111,112] and municipal revitalisation documents [99,113] change every couple of years and are not always mutually coherent. In terms of the protection of natural heritage, apart from the provisions of the study and local plans, a physiographic report prepared almost a decade ago can be a point of reference [17]. Therefore, numerous projects from the last two decades, even those in neighbouring areas, were built under different legal and spatial conditions—which applies particularly to the latest buildings, built in the context of previously completed projects (primarily in Zabłocie and Grzegórzki).

5. Discussion

Civilisational progress and technological development have always enforced changes in the use of urban space, primarily including architectural heritage, which forms the urban fabric [114]. It seems that it is widely accepted that historic structures in European cities and towns should be adapted to new functions, instead of being destroyed and demolished. In Poland, this way of thinking is not always associated with post-industrial heritage. As a result of global economic processes, during the last decades we could observe a disappearance of industrial production which used to determine the power and strong position of cities and region [37]. It has been transferred to other areas of the world.

Thus, as stated at the beginning of the paper, many abandoned and unused complexes of buildings and engineering structures were left within the central districts of Polish major cities. They are evidence of the economic development of cities, mostly in nineteenth and twentieth century. Many monuments of post-industrial heritage frequently possess not merely historic but also considerable artistic value. Such complexes and singular buildings, which no longer serve their original functions, are adapted to different needs [115]. Multi-family housing is one of them [61–65]. This way of thinking is connected to guidelines resulting from the needs of sustainable development [35]. They advocate the necessity to reuse already urbanised brownfields in order to stop urban sprawl.

The research presented in this paper was focused on multi-family housing projects of different size, built on central, post-industrial sites. Cracow was chosen as a case study, as the authors have been conducting many research projects focused on it over the previous decades. Furthermore, various forms of conservation of built-up and natural heritage were applied in the city. Therefore, the authors were interested in investigating how they affected the character of architecture and land development in the renewed post-industrial areas under study.

In the case of listing the entirety of a city's fabric as a monument to history (as with the historic centre of Cracow), every new project must be approved by proper administrative organs (including conservation services). The provisions concerning conservation and exposure of surviving post-industrial heritage are typically quite restrictive in the areas under study, although in some cases they were introduced after the commencement or even the conclusion of new projects.

They are defined by provisions of local law (local spatial development plans) or listings of individual buildings and complexes in applicable monuments registers (at the municipal or voivodeship level). The revitalisation of post-industrial sites must nevertheless allow for a high degree of interference with and the transformation of surviving tissue [49,52,54]—especially as not all of its elements are under protection, since many of them are twentieth-century factory and storage halls of negligible spatial and architectural value. However, it is also the awareness of developers, decision-makers and the talent of designers that define the degree to which the often-unique post-industrial character of these areas has been either preserved or lost. The findings of the study also showed that not all legal forms of heritage protection that are present in the Polish system are effective. Surprisingly, as proven in the study, only the combination of two legal instruments can effectively protect post-industrial and natural heritage: an entry in the monument register [101] together with an enacted local spatial development plan (Scheme 4). This holds true even despite the existence of UNESCO protection zones [93], which apply to large territories of the city centre of Cracow, so it seems to be equally protected. Practice has demonstrated that it is not the case (Table 4). The findings of the study open the way to further investigation. The COVID-19 pandemic has already changed the perspective and the way we use our cities [116,117]. The public sector and commercial businesses, which were the main actors in the refurbishment of post-industrial sites, will not need so many square metres within cities anymore. Most activity has already been switched to remote work. The question as to whether multi-family housing in converted, post-industrial sites can become a more efficient way of protecting this kind of heritage for the years to come, instead of adapting them to other functions, remains open.

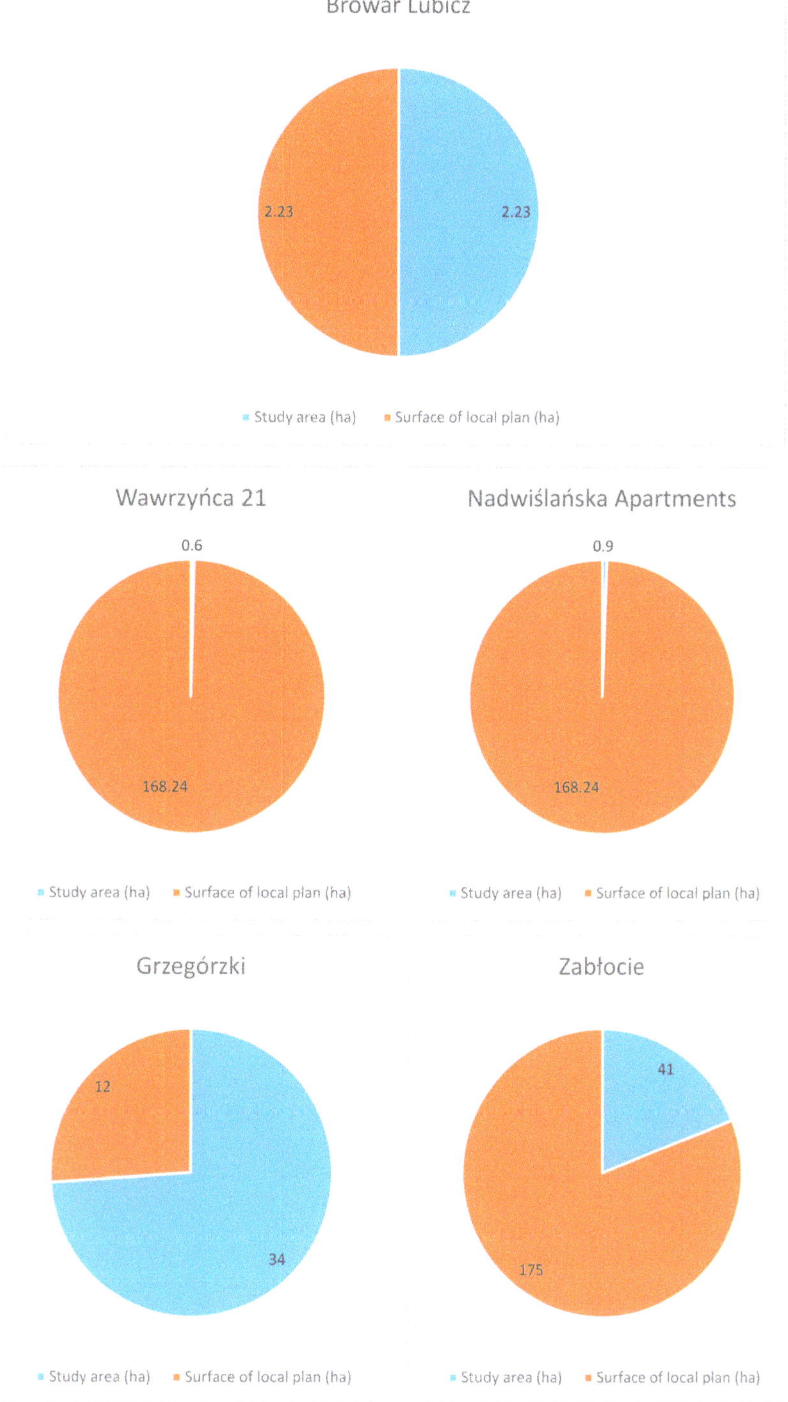

Scheme 4. The relation between the local spatial development plans areas and the areas under study.

Table 4. Summary of findings.

Name of Area	Degree of Cultural Heritage Preservation and Exposure	Degree of Linking of the Building/Complex with Natural Heritage	Role of Local Spatial Plan Provisions and Conservation Documents in the Final Project Outcome
Browar Lubicz	High	High	Significant
Wawrzyńca 21	High	None	Significant
Nadwiślańska Apartments	None	None	Average
Grzegórzki	Low	Average	Insufficient
Zabłocie	Low	Average	Average

The paper also focused on the formation of proper compositional and functional links between new projects and their spatial context [18,46], especially the city's publicly accessible green areas [98]. Therefore, this paper fills a gap in research on revitalised areas in the centres of Polish cities, primarily focusing on the links between the areas under study and elements of urban green-blue infrastructure of the city of Cracow [17], which constitutes its natural heritage in central areas. The results show that those linkages are connected primarily by the position of new projects, but also by properly drafted spatial development plans (Table 4). The case of new district in post-industrial areas of Zabłocie proves this statement. After demolishing most of pre-existing structures, the new district was built on an all but empty plot in last two decades. The potential to create good green public connectors between any given point of the site and the crucial element of blue-green infrastructure of the Cracow (Vistula River Park), neighbouring to the area, was squandered by deficient local development plan provisions [107].

6. Conclusions

Some of the residential buildings and complexes built in the twenty-first century in post-industrial areas in Cracow are located in dense urban fabric and do not feature any spatial, visual and functional linkages with the city's blue-green ecological system. It should therefore be stated that, based on the authors' studies and the literature, they were not found to offer a suitable housing environment. Detailed research presented in the paper on five post-industrial areas located in Cracow's city centre and where new projects with primarily multi-family residential functions were built, was as a part of a larger study conducted in five major Polish metropolises. One of the main criteria of the selection of these areas was their placement in close proximity to the city's blue-green infrastructure. Water and greenery are key elements of good quality housing environment [118]. The study found that despite this proximity, not all complexes were observed to have compositional and functional linkages with nearby public green areas. In two cases, these were found to be merely visual linkages, and in the third—only a part of the complex had these linkages and offered easy access to the elements of Cracow's natural heritage (Table 1). Green spaces in two complexes can be considered a continuation of urban greenery. In addition, the analysis of applicable planning documents (Table 3) demonstrated that local spatial development plans do not always require such linkages.

Examining the manner of post-industrial heritage protection was also an important aspect, focusing on architectural, urban and infrastructural heritage. The analyses of documents and an on-site urban analysis of the complexes under study found that not all forms of conservation proved effective. Only a listing in the historical monuments register, supported by spatial development plan provisions, guarantees the preservation of elements that are important to place-based tradition (Table 2andTable 3). They also become an inspiration for new architectural forms, which is visible in two of the examples under discussion (Browar Lubicz, Wawrzyńca 21). In two others (Nadwiślańska Apartments, Zabłocie), post-industrial heritage became an inspiration for architectural material solutions, but not necessarily formal ones or for the scale of the buildings and complexes.

The complete loss of the spirit of the place in the eastern part of the district of Grzegórzki is a result of the weakness of conservation instruments that were to protect the area's post-industrial structures (Table 2). This also stems from the fact that this area, as the only one among the sites under study, was never covered by a local spatial development plan and all projects were erected based on construction permits issued on the basis of individual, uncoordinated administrative decisions. This led to a series of pathologies, e.g., the construction of a modern luxury building practically on top of the Vistula River's flood embankments, the deregulation of buildings lines and their height, or the predominance of paved areas over green ones in some complexes. This final aspect, this time legally sanctioned by local plans, is unfortunately visible in numerous projects in Zabłocie.

Elements of natural and cultural heritage that could aid in the correct shaping of the housing environment have only recently gained in significance in Poland. It is the talent of designers and will of real estate developers that largely determine whether they shall be put to proper use. The original research performed in the years 2000–2020 on Polish multi-family residential architecture proves that developers have started to pay greater attention to the quality of buildings and their surrounding spaces in recent years. The examples from Cracow presented above appear to confirm this statement. However, it should be remembered that the projects that were analysed are located in exceptional locations, where developers, expecting high profits [119], can provide additional, non-commercial value that stems from the exposure and conservation of heritage. This trend can be considered satisfactory.

The study and findings presented here concerned local cases from Cracow. Many elements of the development of post-industrial areas in Cracow (and other Polish cities) and, by extension, the outcome of the study, were affected by local legal conditions and the country's socio-economic situation. However, the urban analysis methodology and tools used in the study can be used to assess similar buildings and complexes all over the world [120]. They were often used by many scholars, including the authors themselves, for studies on other cities [23,24,62,63].

Future research on Polish cases of introducing housing functions into post-industrial areas should account for the impact of new legal regulations at both state-level, such as the revitalisation act [69] and local level—new spatial development studies and local spatial development plans—on the transformation of other industrial areas in Polish cities. It should also focus on the manner of introducing housing functions is such areas. Interesting problems of the potential for creating a healthy housing environment in areas polluted by industry could result in interdisciplinary studies. This section of future research could go beyond Polish conditions and cover a wider, global context. Comparative studies of legal system impact on industrial heritage preservation in cities of Western Europe and Central and Eastern Europe could prove highly interesting. This is especially true when cities in countries with similar architectural traditions, common history and similar legal and political conditions in the nineteenth and twentieth century are accounted for (Poland, Czech Republic, Slovakia, Hungary, Austria, Eastern part of Germany, the Baltic States, Romania, Ukraine).

Author Contributions: Conceptualization, J.G. and M.G.; methodology, J.G.; validation, J.G. and M.G.; formal analysis, J.G. and M.G.; investigation, J.G. and M.G.; resources, J.G. and M.G.; data curation, M.G.; writing—original draft preparation, J.G. and M.G.; writing—review and editing, M.G.; visualization, M.G.; supervision, J.G. and M.G.; funding acquisition—J.G. and M.G. Both authors have read and agreed to the published version of the manuscript.

Funding: This publication has received financial support from the Polish Ministry of Science and Higher Education under subsidy for science for the FA CUT in 2019 and 2020.

Institutional Review Board Statement: Not applicable.

Informed Consent Statement: Not applicable.

Data Availability Statement: Data available in a publicly accessible repositories of the different branches of the Office of the City of Cracow, as well as Małopolska Voivodship. They are all showed in "References" section.

Acknowledgments: The authors would like to thank the Spatial Planning Bureau of the Office of the City of Cracow for repeated access to illustrative material (Figure 4) from the project *Model przestrzennej struktury miasta-Perspektwa planistyczna*, prepared by M. Gyurkovich and associates in the years 2016–2017, which remains the property of the Office of the City of Cracow; We would also like to thank Józef Białasik of B2 Studio in Cracow, for sharing illustrations of the *Wawrzyńca 21* project in Cracow (Figure 12, Figure 13, Figure 14). The authors would like to express their gratitude to Krzysztof Barnaś for the translation and proof-reading of the entire text on different stages of the paper's preparation and evaluation.

Conflicts of Interest: The authors declare no conflict of interest.

References

1. Rykwert, J. *The Seduction of Place. The History and Future of Cities*; Vintage Books: New York, NY, USA, 2002.
2. Mumford, L. *The Culture of Cities, A Harvest Book Series*; Harcourt Brace & Company: San Diego, CA, USA, 1996.
3. Busquets, J. *Barcelona, the Urban Evolution of a Compact City*; Nicolodi; Harvard University Graduate School of Design: Cambridge, MA, USA, 2005.
4. Sumara, A. Największe Miasta Świata-Ranking 2019, W Których Miastach Mieszka Więcej Niż 10 Mln Osób? Published on 17 March 2019. Available online: https://inzynieria.com/budownictwo/rankingi/48810,najwieksze-miasta-swiata-ranking-2019 (accessed on 16 August 2020).
5. World City Populations 2020. World Population Review. Available online: https://worldpopulationreview.com/world-cities (accessed on 16 August 2020).
6. European Commission, Directorate-General, Joint Research Centre, European Environment Agency. Urban Sprawl in Europe, The Ignored Challenge, (2006), EEA Report no. 10/2006. Available online: https://www.eea.europa.eu/publications/eea_report_2006_10 (accessed on 14 August 2007).
7. Rose, J.F.P. *The Well-Tempered City. What Modern Science, Ancient Civilizations and Human Nature Teach Us about the Future of Urban Life*; Harper Collins Publishers: New York, NY, USA, 2017.
8. Salingaros, N. *Principles of Urban Structure*; Techne: Amsterdam, The Netherlands, 2005.
9. Nan, E. *Good Urbanism. Six Steps to Creating Prosperous Places*; Springer: New York, NY, USA, 2013.
10. Lowenthal, D. Natural and cultural heritage. *Int. J. Herit. Stud.* **2005**, *11*, 81–92. [CrossRef]
11. Pruszyński, J. *Dziedzictwo Kultury Polski, Jego Straty i Ochrona Prawna*; t. I–III; Kantor Wydawniczy Zakamycze: Kraków, Poland, 2001.
12. Veldpaus, L.; Pereira Roders, A.R.; Colenbrander, B.J.F. Urban Heritage: Putting the Past into the Future. *Hist. Environ. Policy Pract.* **2013**, *4*, 3–18. [CrossRef]
13. Kadłuczka, A. The Idea of Historic City. Heritage and Creation. *Tech. Trans. Ser. Archit.* **2015**, *112*, 29–39.
14. Schneider-Skalska, G. Healthy housing environment in sustainable design. *IOP Conf. Ser. Mater. Sci. Eng.* **2019**, *471*, 1–10. [CrossRef]
15. Kovarik, I. Novel urban ecosystems, biodiversity and conservation. *Environ. Pollut.* **2011**, *159*, 1974–1983. [CrossRef]
16. Smets, J.; De Blust, G.; Verheyden, W.; Wanner, S.; Van Acker, M.; Turkelboom, F. Starting a Participative Approach to Develop Local Green Infrastructure; from Boundary Concept to Collective Action. *Sustainability* **2020**, *12*, 107. [CrossRef]
17. Degórska, B. (Ed.) Opracowanie Ekofizjograficzne Miasta Krakowa; Urząd Miasta Krakowa, 2010–2011. Available online: www.bip.krakow.pl (accessed on 12 June 2017).
18. Lynch, K. *Good City Form*; The MIT Press: Cambridge, MA, USA, 1984.
19. Rossi, A. *The Architecture of the City*; Opposition Books; MIT Press: Cambridge, MA, USA, 1982.
20. Kriken, J.L.; Enquist, P.; Rapaport, R. *City Building. Nine Planning Principles for the Twenty-First Century*; Princeton Architectural Press: New York, NY, USA, 2010.
21. Schneider-Skalska, G.; Kusińska, E. (Eds.) *Urban Housing Environment*; Wydawnictwo Politechniki Krakowskiej: Kraków, Poland, 2017.
22. Gyurkovich, J. *Architektura w Przestrzeni Miasta.Wybrane Problemy*; Wydawnictwo Politechniki Krakowskiej: Kraków, Poland, 2010.
23. Gyurkovich, M. (Ed.) *Future of the City: Mass HOUSING Estates or Multifamily Housing Complexes? Eco Rehab 3 Cracow 2012*; Cracow University of Technology: Cracow, Poland, 2012.
24. Sotoca, A. (Ed.) *After the Project—Updating Mass Housing Estates*; II EcoRehab, Barcelona 2011; UPC: Barcelona, Spain, 2012.
25. Ilmurzyńska, K. *Ewolucja przestrzenna Ursynowa Północnego. PUA Przestrzeń Urbanistyka Architektura* **2018**, *2018*, 203–218. [CrossRef]
26. Szczerek, E. Loss of potential: Large-panel housing estates—Czyżyny case. *IOP Conf. Ser. Mater. Sci. Eng.* **2019**, *471*, 1–10. [CrossRef]
27. Klumpner, H.; Twardoch, A.; Heciak, J. Miasto dobrze wymieszane. *Archit. Murator* **2017**, *10*, 32–40.

28. Bounds, M.; Morris, A. Second-wave gentrification in inner-city Sydney. *Cities* **2006**, *23*, 99–108. [CrossRef]
29. Haase, A.; Bontje, M.; Couch, C.; Marcinczak, S.; Rink, D.; Rumpel, P.; Wolf, M. Factors driving the regrowth of European Cities and the role of local and contextual impacts: A contrasting analysis of regrowing and shrinking cities. *Cities* **2021**, *108*, 1–11. [CrossRef]
30. Peterek, M.; Restrepo Rico, S.; Hebbo, Y.; Reichhardt, U.; Guerra Bustani, C. A flexible system for localised sustainable development. *Tech. Trans.* **2018**, *9*, 33–48. [CrossRef]
31. Peterek, M.; Restrepo Rico, S.; Hebbo, Y.; Reichhardt, U. Collaborative Planning for Sustainable Urban Infrastructure in Frankfurt am Main. *Tech. Trans.* **2019**, *8*, 31–50. [CrossRef]
32. Roca Cladera, J.; Biere Arenas, R.; Gyurkovich, M. (Eds.) *Back to the Sense of the City: 11th VCT International Monograph Book*; UPC CPSV: Barcelona, Spain, 2016; Available online: www.upcommons.upc.edu/handle/2117/90350 (accessed on 10 November 2016).
33. Zuziak, Z.; Grzybowski, A. (Eds.) *Centra Miast Metropolitalnych w Polsce. Urbanistyka a Polityka Przestrzenna*; Wyższa Szkoła Techniczna w Katowicach: Katowice, Poland, 2018.
34. Węcławowicz-Bilska, E. Recipe for a city. *IOP Conf. Ser. Mater. Sci. Eng.* **2019**, *471*, 1–10. [CrossRef]
35. Benedict, M.A.; McMahon, E.T. Green Infrastructure. Smart Conservation for 21st Century. *Renew. Resour. J.* **2002**, *20*, 12–17.
36. Miejski System Informacji Przestrzennej Krakowa (City System of Spatial Information of Cracow). Available online: https://msip.um.krakow.pl (accessed on 31 March 2020).
37. *Encyklopedia Krakowa*; PWN: Warszawa-Kraków, Poland, 2000.
38. Bogdanowski, J. *Warownie i zieleń Twierdzy Kraków*; Wydawnictwo Literackie: Kraków, Poland, 1979.
39. Sykta, I. Spacial Plans of the City of Krakow. *Tech. Trans. Ser. Archit.* **2008**, *104*, 51–67.
40. Gyurkovich, J.; Matusik, A.; Suchoń, F. (Eds.) *Cracow: Selected Problems of the Urban Structure Evolution*; Wydawnictwo Politechniki Krakowskiej: Kraków, Poland, 2016.
41. Seibert, K. *Plan Wielkiego Krakowa*; Wydawnictwo Literackie: Kraków, Poland, 1983.
42. Cullen, G. *The Concise Townscape*; The Architectural Press: London, UK, 1986.
43. Łakomy, K.; Hodor, K. About the changeability of the beauty criterion using selected industrial and sacral landscapes (18th-21st Centuries). *Space Form* **2016**, *26*, 349–360. [CrossRef]
44. Paszkowski, Z. *Miasto Idealne w Perspektywie Europejskiej i Jego Związki z Urbanistyką*; Universitas: Kraków, Poland, 2011.
45. Norberg-Schultz, C. *Genius Loci-Towards a Phenomenology of Architecture*; Rizzoli: New York, NY, USA, 2000.
46. Gyurkovich, J. Urbanity of a city—The dream of the past. *Tech. Trans. Ser. Archit.* **2015**, *112*, 97–114.
47. Sudjic, D. *The Language of Cities*; Penguin: London, UK, 2017.
48. Smętkowski, M.; Celińska-Janowicz, D.; Wojnar, A. Location patterns of advanced producer service firms in Warsaw: A tale of agglomeration in the era of creativity. *Cities* **2021**, *108*, 1–14. [CrossRef]
49. Eisinger, A.; Seifert, J. (Eds.) *Urban RESET: How to Activate Immanent Potentials of Urban Spaces*; Birkhäuser: Basel, Switzerland, 2012.
50. Davies, N. *God's Playground: A History of Poland; Volume 1: The Origins to 1795; Volume 2: 1795 to the Present*; Oxford University Press: Oxford, UK, 1981.
51. Racoń-Leja, K. Impact of wartime destruction and post-war politics on the social reconstruction of a modern city—on the example of Magdeburg. *IOP Conf. Ser. Mater. Sci. Eng.* **2019**, *471*, 1–10. [CrossRef]
52. Kaczmarek, S. *Rewitalizacja Terenów Poprzemysłowych: Nowy Wymiar w Rozwoju Miast*; Wydawnictwo Uniwersytetu Łódzkiego: Łódź, Poland, 2001.
53. Wyżykowski, A.; Kwiatkowski, K.; Racoń-Leja, K. (Eds.) *The Transformation of the City Space on the Background of Political-economic Changes*; Institute of Urban Design Cracow University of Technology: Cracow, Poland, 2002.
54. Zuziak, Z.K. Revitalization of cities and the theory of urbanism. *Tech. Trans. Ser. Archit.* **2012**, *109*, 7–18.
55. Lorens, P.; Mironowicz, I. (Eds.) *Wybrane Teorie Współczesnej Urbanistyki*; Miasto-Metropolia-Region, Akapit-DTP: Gdańsk, Poland, 2013.
56. Urban Agenda for the EU 'Pact of Amsterdam'; Agreed at the Informal Meeting of EU Ministers Responsible for Urban Matters on 30 May 2016 in Amsterdam, The Netherlands. Available online: https://ec.europa.eu/regional_policy/sources/policy/themes/urban-development/agenda/pact-of-amsterdam.pdf (accessed on 2 August 2020).
57. Ustawa z dn. 27 Marca 2003 Roku o Planowaniu I Zagospodarowaniu Przestrzennym z Późniejszymi Zmianami. Tekst Jednolity Dz. U. z 2020 r., poz. 293,471,782,1086 (Spatial Planning and Development Act of 27 March 2003, as Amended; Codified Text, Dz. U. 2020, item 293,471,782,1086). Available online: http://isap.sejm.gov.pl/isap.nsf/download.xsp/WDU20200000293/U/D20200293Lj.pdf (accessed on 2 August 2020).
58. Gyurkovich, M.; Szarata, A.; Zuziak, Z.K.; Ogrodnik, D.; Faron, A.; Ziobro, A.; Poklewski-Koziełł, D.; Suchoń, F.; Tota, P.; Sotoca, A.; et al. *Model Przestrzennej Struktury Krakowa. Perspektwa Planistyczna. Etap IIZdefiniowanie Modelu Przestrzennej Struktury Krakowa*; Reaserch project for BPP UM Kraków, Instytut Projektowania Urbanistycznego, Wydział Architektury; Politechnika Krakowska: Kraków, Poland, 2017.
59. Orlenko, M.; Ivashko, Y.; Kobylarczyk, J.; Kuśnierz-Krupa, D. Ways of revitalization with the restoration of historical industrial facilities in large cities. The experience of Ukraine and Poland. *Int. J. Conserv. Sci.* **2020**, *11*, 433–450.
60. Franta, A. The Role of the restructuring of postindustrial areas in the creation of new kinds of metropolitan public spaces—The stimulating function of regulations. *Tech. Trans. Ser. Archit.* **2007**, *104*, 35–43.

61. Gyurkovich, M. 22@Barcelona—The City of Knowledge Civilization. *Tech. Trans. Ser. Archit.* **2012**, *109*, 25–56.
62. Gyurkovich, J. Cologne—Revitalization of the former port of Rheinauhafen. *Tech. Trans. Ser. Archit.* **2011**, *108*, 25–46.
63. Gyurkovich, J. The livable place: Messestadt Riem—A new district of Munich. *Hous. Environ.* **2012**, *10*, 68–73.
64. Gyurkovich, J. Restoring destroyed urban structures. In *Reuso. III Congreso Internacional sobre Documentación, Conservación, y Reutilización del Patrimonio Arquitectonico y Paisajistico*; Palmero Iglesias, L.M., Ed.; UPV: Valencia, Spain, 2015; Available online: https://riunet.upv.es/handle/10251/56060# (accessed on 16 January 2016).
65. Franta, A.; Bojarowicz, A. From slums to a model example of revitalisation: Overcoming a difficult identity in the renewal process of the district of Hulme in Manchester. *Tech. Trans.* **2018**, *115*, 23–44. [CrossRef]
66. Poklewski-Koziełł, D. Role of waterfront areas and pocket parks in shaping of housing environment on the example of the Nordhavnen Housing Estate in Copenhagen. *Hous. Environ.* **2018**, *24*, 74–81. [CrossRef]
67. Korbel, W. Institution of land development plan—Valuation and expectations of changes in the eyes of the communal authorities of Poland. *Pr. Studia Geogr.* **2018**, *63*, 23–39.
68. Gyurkovich, J. Contemporary housing environment—In search for urban climate. *Urbanity Archit. Files* **2018**, *46*, 513–524.
69. Ustawa z dn.9 Października 2015 Roku o Rewitalizacji z Późniejszymi Zmianami; Tekst Jednolity Dz.U. z 2020 r., poz.802, 1086 (Revitalisation Act of 9 October 2015, as Amended; Codified Text Dz.U. 2020, Item 802, 1086). Available online: http://isap.sejm.gov.pl/isap.nsf/download.xsp/WDU20200000802/U/D20200802Lj.pdf (accessed on 2 August 2020).
70. Lipiński, J.; Lorens, P. *Młode Miasto Gdańsk: Laboratorium Miejskich Procesów Rozwojowych*; Monoplan: Warszawa, Poland, 2016.
71. Lorens, P. Shaping the new face of the Imperial Shipyard in Gdańsk. *Space Form* **2019**, *40*, 151–170.
72. Matusik, A.; Racoń-Leja, K.; Gyurkovich, M.; Dudzic-Gyurkovich, K. Hydrourban spatial development model for a resilient inner-city. The example of Gdańsk. *ACE Archit. City Environ.* **2020**, *15*, 1–21. [CrossRef]
73. Żylski, T. Operacja rewitalizacja. *Archit. Murator* **2016**, *1*, 22–25.
74. Bardzińska-Bonenberg, T. On gentrification of historical districts in Poznan. *Tech. Trans. Ser. Archit.* **2012**, *109*, 43–52.
75. Lorens, P. (Ed.) *Rewitalizacja miast w Polsce: Pierwsze Doświadczenia*; Urbanista: Warszawa, Poland, 2007.
76. Konior, A.; Pokojska, W. Management of Postindustrial Heritage in Urban Revitalization Processes. *Sustainability* **2020**, *12*, 5034. [CrossRef]
77. Haupt, P. Architectural education for primary school students mentored by university students. *World Trans. Eng. Technol. Educ.* **2019**, *17*, 454–458.
78. Rocznik Statystyczny Rzeczpospolitej Polskiej 2019, GUS, Opublikowano 23.12.2019, (Statistical Yearly of the Republic of Poland 2019, GUS, Published on 23.12.2019). Available online: https://stat.gov.pl/obszary-tematyczne/roczniki-statystyczne/roczniki-statystyczne/rocznik-statystyczny-rzeczypospolitej-polskiej-2019,2,19.html (accessed on 16 August 2020).
79. Szpakowska-Loranc, E.; Matusik, A. Łódź—Towards a resilient city. *Cities* **2020**, *107*, 1–14. [CrossRef]
80. Kozłowski, S.; Wojnarowska, A. *Rewitalizacja Zdegradowanych Obszarów Miejskich. Zagadnienia Teoretyczne*; Wydawnictwo Uniwersytetu Łódzkiego: Łódź, Poland, 2011.
81. Załuski, D.; Waloryszak, K.; Żylski, T. Trzy dzielnice, trzy rewitalizacje. *Archit. Murator* **2016**, *1*, 26–38.
82. Trzepacz, P.; Warchalska-Troll, A. (Eds.) *Rewitalizacja Miast: Teoria, Narzędzia, Doświadczenia*; IRM: Kraków, Poland, 2017.
83. Gyurkovich, M.; Sotoca, A. Quality of Social Space in Selected Contemporary Multifamily Housing Complexes in Poland's Three Biggest Cities. *IOP Conf. Ser. Mater. Sci. Eng.* **2019**, *471*, 1–11. [CrossRef]
84. Ustawa z dn. 13 Lutego 2020 o Zmianie Ustawy "Prawo Budowlane"; Tekst Jednolity Dz. U. z 2020 r. poz. 471, 695,782. (Act of 13 February 2020 on the Amendment of the Construction Law Act; Codified Text, Dz. U. 2020 item 471, 695,782). Available online: https://isap.sejm.gov.pl/isap.nsf/download.xsp/WDU20200000471/U/D20200471Lj.pdf (accessed on 2 August 2020).
85. Bartkowicz, B. The role of spatial planning in the realisation of tasks of strategic transformations and in the activation of space. *Tech. Trans. Ser. Archit.* **2011**, *108*, 7–11.
86. Dudzic-Gyurkovich, K. Effect of accessibility in housing complexes on shaping of beauty in the urban environment. Selected examples from Cracow. In *Teka Commission of Architecture, Urban Planning and Landscape Studies*; Polish Academy of Science Branch Lublin: Lublin, Poland, 2018; Volume XIV, pp. 116–125.
87. Happach, M.; Żylski, T. Kto dziś kształtuje miasto? *Architektura Murator* **2016**, *5*, 32–40.
88. Raport—Stan Mieszkalnictwa w Polsce; Ministerstwo Rozwoju; Marzec 2020 (Report—The State of Housing in Poland; Ministry of Development; March 2020). Available online: https://www.gov.pl/web/rozwoj/raport-o-stanie-mieszkalnictwa (accessed on 2 August 2020).
89. Leśniak-Rychlak, D. *Jesteśmy Wreszcie We Własnym Domu*; Instytut Architektury: Kraków, Poland, 2019.
90. Stiasny, G. Typowo polska architektura mieszkaniowa. *Architektura Murator* **2014**, *1*, 30–39.
91. Ogłoszenia o Wynajem Mieszkania—15.08.2020 (Apartment Rental Advertisements—15.08.2020). Available online: http://info.wyborcza.pl/temat/wyborcza/og%C5%82oszenia+o+wynajem+mieszkania (accessed on 15 August 2020).
92. Gyurkovich, M.; Dudzic-Gyurkovich, K.; Matusik, A.; Racoń-Leja, K. Vistula River Valley in Cracow: From an urban barrier to a new axis of culture in the scale of the city. *ACE Archit. City Environ.* **2020**, *15*, 9913. [CrossRef]
93. Lista Światowego Dziedzictwa UNESCO (UNESCO World Heritage List). Available online: https://www.unesco.pl/kultura/dziedzictwo-kulturowe/swiatowe-dziedzictwo/lista-swiatowego-dziedzictwa/ (accessed on 12 August 2020).
94. Porębska, A.; Godyń, I.; Radzicki, K.; Nachlik, E.; Rizzi, P. Built heritage, sustainable development, and natural hazards: Flood protection and UNESCO world heritage site protection strategies in Krakow, Poland. *Sustainability* **2019**, *11*, 4886. [CrossRef]

95. Suchoń, F. "The iron ring". The fortification system as the constituting factor in the hybrid structure of the city-fortress. In *Hybrid Urban Structures*; Gyurkovich, M., Ed.; Wydawnictwo Politechniki Krakowskiej: Kraków, Poland, 2016; pp. 183–205.
96. Izdebski, A.; Szmytka, R. (Eds.) *Ekobiografia Krakowa*; ZNAK Horyzont: Kraków, Poland, 2018.
97. Zachariasz, A. Development of the system of the green areas of Krakow from the nineteenth century to the present, in the context of model solutions. *IOP Conf. Ser. Mater. Sci. Eng.* **2019**, *471*, 1–11. [CrossRef]
98. Studium Uwarunkowań i Kierunków Zagospodarowania Przestrzennego Miasta Krakowa (2003), Uchwała RMK Nr XII/87/03 z dnia 16 kwietnia 2003 r. Zmiana Studium Uwarunkowań i Kierunków Zagospodarowania Przestrzennego Miasta Krakowa, t. I–III, załącznik do Uchwały nr CXII/1700/14 Rady Miasta Krakowa z 9 lipca 2014r., Biuro Planowania Przestrzennego UM Krakowa, Kraków 2014 (2003 Spatial Development Study, Resolution of the Cracow City Council of 16 April 2003 No. XII/87/03 together with the update vol. I–III dated on 9 July 2014). Available online: https://www.bip.krakow.pl/?bip_id=1&mmi=48 (accessed on 31 July 2020).
99. Adamczyk –Arns, G.; Wojnarowska, A.; Feresztyn, E.; Hultsch, F.; Hultsch, F. Lokalny Program Rewitalizacji Starego Miasta, zleceniodawca: Urząd Miasta Krakowa, (Old Town Local Revitalisation Programme, commissioned by: Office of the City of Cracow) BIG-STADTEBAU GmbH, Kronshagen, Germany, June 2008. Available online: https://www.google.com/search?q=lokalny+program+rewitalizacji+Stare+Miasto+Krak%C3%B3w+2008&source=lnms&tbm=isch&sa=X&ved=2ahUKEwjB5q_Lwp_rAhVK-aQKHQW7BpkQ_AUoAXoECAwQAw&biw=1366&bih=625 (accessed on 16 August 2020).
100. Waszczyszyn, E. Hospital in "the heart of the city". *Tech. Trans. Ser. Archit.* **2008**, *105*, 213–221.
101. Zespoły i Obiekty z Terenu miasta Krakowa Wpisane do Rejestru Zabytków (stan na lipiec 2020 r.) (Buildings and Complexes in Cracow Listed in the Register of Monuments—for July 2020). Available online: https://www.wuoz.malopolska.pl/rejestrzabytkow (accessed on 20 September 2020).
102. Suchoń, F. The revitalization of a postindustrial structure in the centre of historical city. Case study. *Hous. Environ.* **2013**, *12*, 154–158.
103. Czapnik, W.; Wużyk, K. Browar Lubicz w Krakowie. *Architektura-Murator* **2014**, *11*, 74–82.
104. Gyurkovich, M. In Search of Urban Composition of Sub-Centres in Polycentric European Metropolises. *ACE Archit. City Environ.* **2012**, *6*, 251–264.
105. Gilewicz, W.; Lisowski, M. "Galeria Kazimierz"- Shopping and Commercial Center in postindustrial city fabric of Krakow. *Tech. Trans. Ser. Archit.* **2006**, *103*, 323–330.
106. Museum of Cracow—Oskar Schindler's Enamel Factory. Available online: http://www.mhk.pl/oddzialy/fabryka-schindlera (accessed on 20 May 2016).
107. Uchwała NR CXIII/1156/06 Rady Miasta Krakowa z Dnia 28 Czerwca 2006 r. w Sprawie Uchwalenia Miejscowego Planu Zagospodarowania Przestrzennego "Obszaru Zabłocie"—Ogłoszona w Dzienniku Urzędowym Województwa Małopolskiego NR 559, poz. 3534 z Dnia 19 Września 2006 r. (Resolution No. CXIII/1156/06 of the Cracow City Council of 28 June 2006 on the Passing of the Local Spatial Development Plan for the 'ZABŁOCIE' area, Announced in the Official Journal of the Lesser Poland Voivodeship No. 559, item 3534 of 19 September 2006. The Plan Has Been in Effect Since 20 October 2006). Available online: https://www.bip.krakow.pl/?dok_id=14330 (accessed on 1 August 2020).
108. Blazy, R. Revitalization of riverside boulevards in Poland—A case study on the background of the European implementation. *IOP Conf. Ser. Mater. Sci. Eng.* **2019**, *603*, 1–6. [CrossRef]
109. Kraków-Gminna Ewidencja Zabytków (Cracow—Municipal Historic Monuments Record). Available online: https://www.bip.krakow.pl/zalaczniki/dokumenty/n/281325/karta (accessed on 10 October 2020).
110. Dobosz, P. Paradigm of future changes in the system of legal protection of monuments and care over them in Poland. *Wiad. Konserw. J. Herit. Conserv.* **2015**, *44*, 78–84. [CrossRef]
111. Uchwała NR XXIV/292/07 Rady Miasta Krakowa z Dnia 24 Października 2007 r. w Sprawie Uchwalenia Miejscowego Planu Zagospodarowania Przestrzennego Obszaru "Browar Lubicz", Ogłoszona w Dzienniku Urzędowym Województwa Małopolskiego NR 827, poz. 5405 z Dnia 28 Listopada 2007 r. (Resolution No. XXIV/292/07 of the Cracow City Council of 24 October 2007 on the Passing of the Local Spatial Development Plan for the "Browar Lubicz" Area, Announced in the Official Journal of the Lesser Poland Voivodeship No. 827 item 54505 of 28 November 2007. The Plan Has Been in Effect Since 27 December 2007). Available online: https://www.bip.krakow.pl/?dok_id=19431 (accessed on 1 August 2020).
112. Uchwała NR LXXXI/1240/13 Rady Miasta Krakowa z Dnia 11 Września 2013 r. w Sprawie Uchwalenia Miejscowego Planu Zagospodarowania Przestrzennego Obszaru "Bulwary Wisły"—Ogłoszona w Dzienniku Urzędowym Województwa Małopolskiego z Dnia 27 Września 2013 r., poz. 5685. Plan Obowiązuje od Dnia 28 Października 2013 r. (Resolution No. LXXXI/1240/13 of the Cracow City Council of 11 September 2013 on the Passing of a Local Spatial Development Plan for the "Bulwary Wisły" Area, Announced in the Official Journal of the Lesser Poland Voivodeship of 27 September 2013, Item 5685. The Plan Has Been in Effect Since 28 October 2013). Available online: https://www.bip.krakow.pl/?dok_id=58413 (accessed on 1 August 2020).

113. Uchwała NR XXXVI/929/20 Rady Miasta Krakowa z Dnia 26 Lutego 2020 r. Zmieniająca Uchwałę nr LIX/1288/16 w Sprawie Przyjęcia Aktualizacji Miejskiego Programu Rewitalizacji Krakowa (Resolution No. XXXVI/929/20 of the Cracow City Council of 26 February 2020, on the Amendment of RESOLUTION No. LIX/1288/16 on Approving an Updated Version of the Cracow Municipal Revitalisation Programme). Available online: https://www.bip.krakow.pl/?dok_id=167&sub_dok_id=167&sub=uchwala&query=id%3D24824%26typ%3Du&_ga=2.186729386.364313227.1597573055-787868020.1597573054 (accessed on 16 August 2020).
114. Kostof, S. *The City Shaped: Urban Patterns and Meanings Through History*; Thames & Hudson: London, UK; New York, NY, USA, 1999.
115. Gyurkovich, M. Adaptations of historic structures for new functions—Part 1—Selected examples from Madrid. *J. Herit. Conserv.* **2017**, *50*, 30–43. [CrossRef]
116. Hanzl, M. Urban forms and green infrastructure- the implications for public health during the COVID-19 pandemic. *Cities Health* **2020**, *4*, 1–6. [CrossRef]
117. Jasiński, A. Public space or safe space—Remarks during the COVID-19 pandemic. *Tech. Trans.* **2020**, *117*, 1–10. [CrossRef]
118. Seruga, W. Greenery and water in contemporary architectural and urban complexes. *Hous. Environ.* **2018**, *22*, 174–249. [CrossRef]
119. Nieruchomości-Online-pl—(Real Estate Sale Offers). Available online: https://krakow.nieruchomosci-online.pl/ (accessed on 10 October 2020).
120. Bojanowski, K.; Lewicki, P.; Moya Gonzàlez, L.; Palej, A.; Spaziante, A.; Wicher, W. *Elementy Analizy Urbanistycznej*; Program Tempus JEN-3533; Wydawnictwo Politechniki Krakowskiej: Kraków, Poland, 1998.

Article

Strengthening Efforts to Protect and Safeguard the Industrial Cultural Heritage in Montilla-Moriles (PDO). Characterisation of Historic Wineries

Antonia Merino-Aranda [1], Isabel Luisa Castillejo-González [1], Almudena Velo-Gala [2], Francisco de Paula Montes-Tubío [1], Francisco-Javier Mesas-Carrascosa [1] and Paula Triviño-Tarradas [1,*]

[1] Department of Graphic Engineering and Geomatics, University of Cordoba, 14071 Córdoba, Spain; am.antoniamerino@gmail.com (A.M.-A.); ma2cagoi@uco.es (I.L.C.-G.); ir1motuf@uco.es (F.d.P.M.-T.); fjmesas@uco.es (F.-J.M.-C.)

[2] Department of Prehistory, Archaeology, Ancient History, Medieval History and Historiographic Sciences and Techniques, University of Murcia, 30001 Murcia, Spain; almudena.v.g.@um.es

* Correspondence: ig2trtap@uco.es

Abstract: Industrial heritage is linked to the cultural processes that human society sets through the traces from the past. The conservation and dissemination of this industrial–cultural heritage are crucial for sustainable urban development, and positively influences the transition to resilient and sustainable cities. The wine industry around Montilla has suffered as a result of a sharp reduction of the vineyard area in the last 25 years. Wineries, as one of the historic typologies of wine-making facilities in the Montilla-Moriles Protected Designation of Origin (PDO), as well as their materials and construction techniques, are a reference in the agricultural landscape of Montilla. Many historic wineries are the result of the abandonment and cessation of the wine industry. These buildings are linked to the agrarian activity in this area, mostly wine-making, although in some cases, they coexist with similar production processes, such as milling the fruit of the olive grove. This research characterises and analyses four historic wineries in the Montilla-Moriles PDO, which represent an example of architecture in the wine-making transformation during the 19th–20th centuries. This manuscript contributes to the attainment of some objectives set in one of the Sustainable Development Goals (SDGs), protecting and disseminating the industrial cultural heritage in Montilla-Moriles.

Keywords: sustainable development; industrial–cultural heritage; historic wineries; wine industry; Montilla-Moriles PDO

Citation: Merino-Aranda, A.; Castillejo-González, I.L.; Velo-Gala, A.; de Paula Montes-Tubío, F.; Mesas-Carrascosa, F.-J.; Triviño-Tarradas, P. Strengthening Efforts to Protect and Safeguard the Industrial Cultural Heritage in Montilla-Moriles (PDO). Characterisation of Historic Wineries. *Sustainability* **2021**, *13*, 5791. https://doi.org/10.3390/su13115791

Academic Editors: Jan K. Kazak, Katarzyna Hodor and Magdalena Wilkosz-Mamcarczyk

Received: 30 April 2021
Accepted: 19 May 2021
Published: 21 May 2021

1. Introduction

1.1. 2030. Agenda: Focus on the Sustainable Development Goals

The 2030 Agenda for Sustainable Development, adopted by the United Nations General Assembly (UNGA), approved, in the year 2015, 17 Sustainable Development Goals (SDGs) for the next 13 years, [1] which are essentially focused on transforming the world. In this context, industrial agricultural heritage, including its procedures and agricultural techniques, as well as the ways of rural building, emerges as a repository of cultural resources endowed with vast power and visibility, acting as a structuring lynchpin for research. The National Plan for Industrial Heritage indicates that this type of heritage (i.e., industrial heritage) has vulnerable and occasionally misunderstood elements, which should be viewed as new cultural assets represented and interpreted through an updated scientific reading [2]. The economic, social and environmental aspects of sustainable development contribute to safeguarding the cultural heritage and nurturing creativity [3]. When the SDGs are grouped according to the three basic pillars of sustainable development (i.e., the economic, social and environmental dimensions), it is observed that culture plays a transversal role in all of them. Indeed, culture must be seen as a driver of sustainable development [4].

In Montilla-Moriles PDO, one of the most relevant pieces of cultural heritage is the existing wine industry in the area and its varied typology in terms of the buildings required for it, depending on the time of operation. In this context, the enhancement and dissemination of the agrarian industrial heritage in Montilla-Moriles tackles most of the SDGs, with greater influence on SDG 11 ("Sustainable Cities and Communities") in its fourth objective ("to strengthen efforts to protect and safeguard the world's cultural and natural heritage"). In our case, conserving the industrial cultural heritage in Montilla-Moriles (PDO) would help the UNGA in its adoption of the "Policy for the integration of a sustainable development perspective into the processes of the World Heritage Convention". Heritage conservation contributes to making resilient and sustainable societies. The overall goal of this strategy is to guide society through appropriate support, harness the potential of heritage in general and to contribute to sustainable development [4]. While authors like Calvo-Serrano et al. [5] have shown the history of a building through its architectural sustainability level, Gullino et al. [6] have monitored the sustainability worldwide in rural heritage sites according to the architecture related to agricultural activity, to contribute and advance toward future UNESCO cultural heritage. Several of these studied sites concerned vineyards, although none of them were located in Spain. While Cano et al. [7] have supported the conservation of rural buildings as a matter of cultural tourism in Southern Spain, the industrial wine-making activity in the area has only been shown through its exhibition in museums, such as the "Wine Museum in Montilla-Moriles" or throughout the wine tourism route in Montilla-Moriles (PDO). However, such heritage has not been widely published and disseminated. So far, research related to the development of the wine industry has been limited to the technical study of production and economy, leaving aside the characterisation and defense of the architecture linked to agricultural production in the area.

This work approaches the connections between material cultural aspects and their relationship with the environment. Therefore, the aim of this study is to provide a better understanding of how this traditional architecture reflects a society whose fundamental activity was, and continues to be today, agriculture, and how it was influenced by cultural sustainability. Focusing on its protection and valorisation, despite the numerous letters and plans of protection at the national and regional level, this architecture is still unprotected, unlike the wines that are produced in this territory. The architecture of these buildings tells us, among other things, about the history and environment of this region, and reflects its society and secular tradition.

1.2. Industrial Facilities for Wine-Making in the PDO Montilla-Moriles

The rural space in the agrarian region of the high countryside of Córdoba is constituted by traditional buildings such as wineries, farmhouses and estates, considered in the rural scope as production units. These architectural ensembles are spatially and architecturally rich. Throughout history, in the region of Montilla-Moriles, there have existed different industrial facilities for wine-making: (i) industrial cellars; (ii) wineries, (iii) lagaretas (winepresses), and (iiii) wine-making cooperatives.

Industrial cellars were dedicated to the making and maturation of local wines. The most representative industrial cellar is the primitive cellar of Alvear, built in 1729 in the middle of the urban area of Montilla. Wineries, as defined by Naranjo-Ramírez [8], are rural houses generated by the cultivation of the vineyard, which not only attend to the industrial operations of grape grinding and fermentation of musts, but also provide accommodation to the workers of this industry, and even serve as a second home for the owner of the agricultural holding (Figure 1). Lagaretas (winepresses) are small wineries located in the urban area, unlike wineries, which are located in the agricultural holdings. The anticipated introduction of the power grid in the cities led to the integration of this type of facility in the very homes of winegrowers. Lastly, wine-making cooperatives pose a great advantage to farmers in terms of supply concentration. These were created by farmers in Moriles (1955) and Montilla (1959) [8].

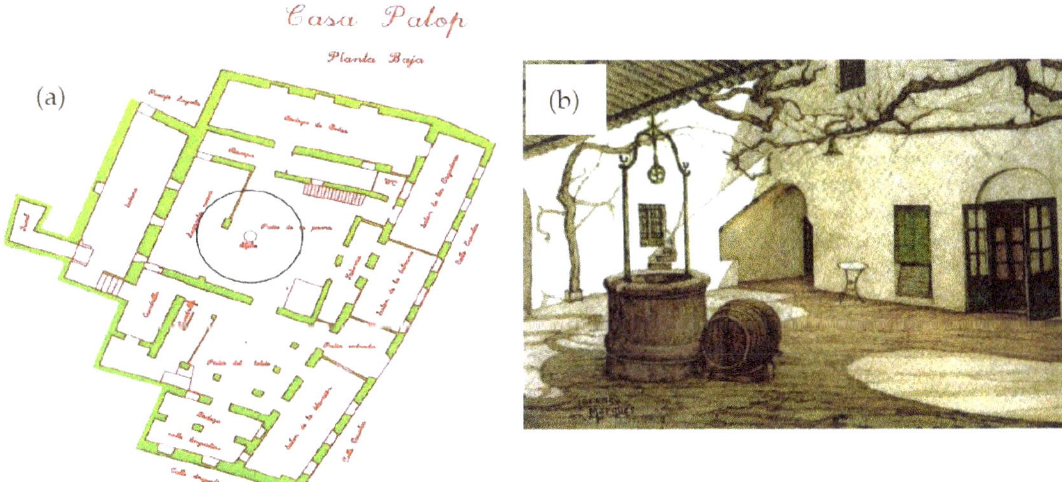

Figure 1. (**a**) Ground floor of Casa Palop (Palop House) in the centre of Montilla (Córdoba) [9]; (**b**) Drawing by Lorenzo Marqués of the current state of the patio (yard) of Casa Palop [9].

According to data from the Montilla-Moriles Regulatory Council, there are currently 40 wineries recognized as industrial heritage properties in the PDO. This industry typology demonstrates the successful integration of viticulture and sustainable development activities in this area. In this context, and according to the definition of sustainability applied to the wine-making sector, this activity can be maintained over a long period of time without depleting natural resources or causing serious damage to the environment. Regarding the constructions required for wine production, sustainability includes the use of materials present in the near environment and the use of available natural resources, thus integrating into the surrounding landscape. At the same time, the safeguarding and enhancement of this industrial heritage aim to guarantee and preserve these constructions and their territorial uniqueness and values.

1.3. Background: First News about Wineries in Montilla

As stated by Gullino et al. [6], rural properties, such as the abovementioned industrial facilities, are non-static features and are continuously evolving. Social, economic, and environmental changes are the main factors that impact land uses, agricultural practices and agricultural needs [10]. Indeed, rural landscapes underwent major transformations [11–13], and the Montilla-Moriles area is one example of this. Their protection and the management of PDO areas are crucial for the livelihood of the local populations and for the preservation of the traditional cultural heritage [14].

The first references about the wineries of Montilla-Moriles (Córdoba) date from the 16th, as a distinctive agricultural settlement in the area. In the year 1513, commissioned by Cardinal Cisneros, agronomist and treatise writer Gabriel Alonso de Herrera wrote the *Libro de Agricultura que es de la labrança y criança, y de muchas otras particularidades y provechos del campo* (Book of agriculture, concerning farming and maturing and many other particularities and uses of the field) (Valladolid: Francisco Fernández de Córdoba, 1513) (Figure 2), which was later known as *Agricultura general* (General agriculture) [15]. Likewise, in the book of Alonso de Herrera [16], there is a reference to the wineries of Córdoba province:

> *"There are three ways of harvesting. As is done in Cordova, they have their homes in the vineyards, which they call wineries, with their cellars and winepresses, and there they make their wine, and they cook it, and they let it settle, and at the time of racking they*

bring it home clean, and if there are good errands there, let them be done well and clean, this is the best thing . . . ".

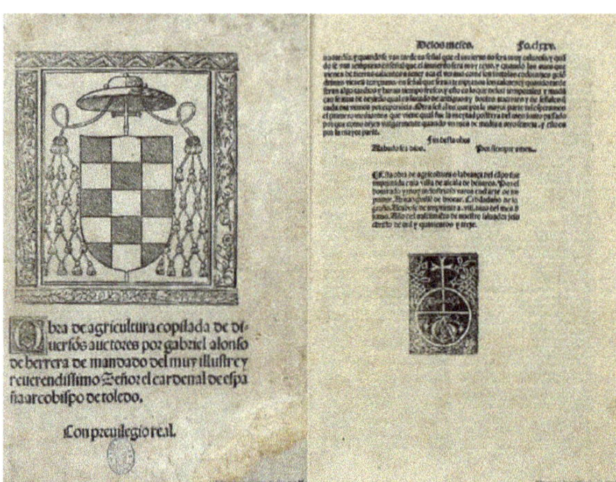

Figure 2. First edition of 1513 of the Libro de Agricultura que es de la labrança y criança, y de muchas otras particularidades y provechos del campo (Book of agriculture, concerning farming and maturing and many other particularities and uses of the field). Cover and colophon. Alcalá de Henares [17].

In that year, the lordship of Montilla was ruled by the Marquis of Priego, Mr. Pedro Fernández de Córdoba y Pacheco (1470–1517), which included the lordship of Aguilar and, in 1711, became part of the Duchy of Medinaceli (Figures 3 and 4). At the end of the lordships in 1873, in the times of María Cristina, and during the confiscation process in the mid-19th century, there was a considerable expansion of the cultivation of vineyards and the wineries associated with it [18]. Specifically, in 1845, P. Madoz [19] stated that the municipality of Montilla had 107 wineries, 28 farmhouses and 67 olive grove houses, although he did not identify the typology of such rural buildings. On his part, Ramírez de las Casas Deza [20] mentioned that Montilla had more than one hundred wineries with a good homestead.

During the second half of the 18th century, there was an increase in the litigation of small local producers against winery owners, who were mostly nobles and members of the Church [22]. This period is considered to be the starting point of the production cycle of wine in the region, where an entrepreneur agricultural middle class emerged. From the year 1860, an important wine-making industry thrived in the area of Montilla-Moriles, led by Francisco de Alvear and Gómez de la Cortina [23], Count of La Cortina, who protected the commercial work of the small farmers of the area. This was the beginning that set the foundation for the main wealth of the municipalities and their surroundings [24]. Between the years 1866 and 1888, the greatest levels of wine production were reached, due to the phylloxera plague crisis, which affected France and Italy [25,26]. However, the history of the vineyard in Cordoba, as well as in the whole of Spain and Europe, was dramatically truncated by this plague in the late 19th century [27]. After the phylloxera plague crisis, in the 20th century, under the protection of the rise of the wine production [8] in the entire mountainous region, numerous wineries and wine factories were built, which combined their agro-industrial nature with accommodation, constituting part of the traditional architectural heritage of the area [28].

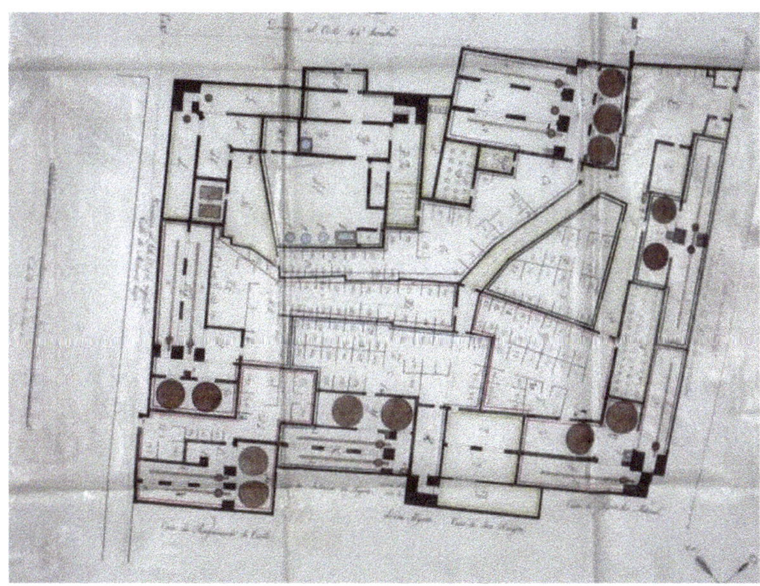

Figure 3. Ground plan of an oil mill, property of the Duke of Medinaceli in Montilla, by Benito de Mora, 1860. Priego Section, Ducal Archive of Medinaceli, Seville [18].

Figure 4. Engraving of "Bodega de La Tercia" (cellar) in Montilla, property of the Duke of Medinaceli, 1860 [21].

From the year 1960, the historic wineries were abandoned as a consequence of greater development of the industry and the emergence of the first wine-making cooperatives [24]. Similarly, another determining factor in the disappearance or cessation of the activity of the wineries was the decrease of land occupied by vineyards, by 70% in the last 30 years [28].

These reductions in the vineyard surface area, first in the 19th century with the phylloxera plague [27], and then in the 20th century with the adaptation of the law on cooperatives to the European regulations [24], were the main factors in the abandonment of this type of agricultural facilities [29].

In the Catalonian area, authors such as Llorens [26] identified the constructive features of the so-called "wine cathedrals", whereas, in Montilla-Moriles (PDO), there are few studies concerning the traditional constructions used for the industrialisation process of the wines produced. Therefore, the aim of this paper was to identify the industrialisation of the Montilla-Moriles wine production throughout history from a sustainable point of view and characterise four of the most representative wineries, as symbols of the agrarian industrial heritage in the area. Within this research, new approaches were considered to complement the scope of the industrial cultural heritage in Montilla-Moriles (PDO), from the architectural point of view. These constructions have not only contributed to economic and cultural development, but they are also a living memory of traditions, which differentiate us from other territories due to our cultural and architectural uniqueness. In terms of living culture, tradition and cultural uniqueness are reflected in the celebration of the Grape Harvest Festival, which has been held since 1816. It was King Ferdinand VII who authorised the celebration of this Fair during the first three days of September [30]. Nowadays, this cultural sustainability activity is the responsibility of the Denomination of Origin board, which is in charge of the representation, research, defense, guarantee, development and promotion of the wines produced in this region. The Montilla-Moriles Wine Route and the celebration of activities such as "Patios de bodega" offer a combination of culture and heritage in the area, in which one of its most emblematic resources is highlighted: its wineries.

2. Materials and Methods

The methodology followed in this research was dual. Firstly, an in-situ phase was conducted, visiting libraries and municipal archives, such as the Municipal Archive of Montilla, and the Archive Manuel Ruiz Luque, which was crucial for the data collection process. In this first phase, an exhaustive record of the historic wineries through the historical documents was carried out. Secondly, some relevant winery-owning families, such as the Alvear family, were interviewed. Thirdly, the protection figures for the safeguarding of this architectural heritage and of the territory were investigated, analysing the protection charters, in order to reflect on the current situation of protection present in the territory. Finally, each of the wineries was described and the situation of their state of conservation was analysed. The data were contrasted with previous studies on the phylloxera crisis, and also with the origins of the vineyard in Montilla-Moriles and the analysis of the technical aspects of the wine-making process in the area. All of them complement and enrich the present work, together with the rest of the literature consulted, allowing a global understanding of the architectural uniqueness and sustainability of the territory in the development of the wine industry in the Montilla-Moriles area.

In this study, we focused on a limited number of wineries, since scrutinising all the rural wine settlements in the area would go beyond the scope of this investigation.

3. Results and Discussion

In order to propose general guidelines and recommendations on the concept of industrial heritage, there have been numerous heritage charters written since the beginning of the 20th century, all aimed at proposing a conceptualisation, standardisation and methodology to preserve the heritage. In addition, wine tourism is currently developing this concept of architecture and landscape, including other activities such as grape harvest and gastronomic festivals [31]. As a result of this integration between architecture and the environment in which they are located, the European wine tourism charter was born. In 2003, the Nizhny Tagil Charter was drafted, where, for the first time, the fundamental importance of buildings and structures built for industrial activities were highlighted, as well as the territory where they are located, proposing the cataloguing, protection and maintenance according to the Charter of Venice, for their conservation. Finally, the Seville Charter of Industrial Heritage was presented in 2019, as a result of the inaction and destruction of industrial heritage, in order to update the fundamental aspects on practices regarding its protection.

Considering these aspects and values, the registration and analysis of industrial constructions become very useful tools for understanding the history of typological innovation processes and the economic structure of a territory [31].

3.1. *Conservation of Industrial Heritage*

The International Committee for the Conservation of Industrial Heritage (TICCIH), created in 1978, aims to promote international cooperation in the field of preservation, conservation, location, research, history, documentation, archaeology and revaluation of industrial heritage. There is extensive experience in the study and conservation of industrial heritage and landscapes, together with plans to enhance them, such as museums, interpretation centers and parks, as well as in the reuse of industrial buildings for other uses. The conservation of industrial heritage must involve public administrations, owners and social agents.

The examples regarding the experiences in industrial heritage conservation at the international level are varied. Some of them are: New Lanark (Scotland), Saltaire and the factories of the Derwent valley (England) and Völklingen (Germany) [32]. In Spain, among all the experiences in industrial heritage conservation, we must highlight the Almadén Mines, in Ciudad Real (Spain), registered in the World Heritage list in 2012 and, recently, adhered to the European Route of Industrial Heritage (ERIH), [33]. This case should be taken as a management and recovery model to safeguard the memory of the industrial past where landscape, buildings and machinery have been preserved and partially preserved [34].

3.2. *Heritage Importance, Heritage Values and Authenticity Attributes*

The wine-making tradition of the area is reflected in its landscape and buildings; these constitute a valuable material testimony of our history and our culture, acting as a link between the past and the present. The profound changes in agriculture and ways of life have motivated their abandonment, deterioration and, in some cases, their disappearance [35]. The industrial heritage of the Montilla-Moriles area is still little known and poorly appreciated. The own regulation of the Denomination of Origin board involves the characteristic natural factors that recognise them as unique compared to the rest of the wine-making regions. The value and authenticity of these industrial wineries are still in time to be safeguarded, jointly with the machinery and secondary elements to be preserved.

3.3. *Possible Strategies and Approaches for the Conservation of Wineries*

The functional obsolescence of these buildings and the crisis in the wine sector have led to the disuse of many wineries in recent decades. Regarding preservation, they have not had any protection figure and, after an initial period of demolitions, in recent years they have opted to give them a new use and thereby guarantee their heritage preservation. In general, heritage rehabilitation is carried out by the wine-makers themselves, who in recent times have seen the potential of wine tourism [36].

Historical precedents for the reuse and conservation of these spaces already existed: barracks in the 18th and 19th centuries in El Puerto and Jerez, schools in Jerez and social houses in Moguer at the beginning of the 20th and even bourgeois houses. The preservation of the values of this architectural heritage depends on its successful reuse [37].

The models of conservation and management of industrial heritage have a particular casuistry [38]. Institutional initiatives on winery rehabilitation are still scarce. Among the recent investments made by the General Directorate of Fine Arts and Cultural Assets of the Spanish Ministry of Culture, the project "Conservation of ethnographic heritage" affects five wineries [36]. Considering industrial heritage as a common thread for interventions in open spaces can contribute to combining industry, culture and nature in the creation of new avenues of wealth and well-being [38]. To this end, the institutions must be involved, so that the conservation and preservation of the wineries does not depend on the will of the wine-makers themselves. The control of the institutions over the inventories, restoration and management of assets considered as industrial heritage requires the collaboration and assistance of professionals in conservation and cataloguing tasks [36].

3.4. The Architecture of the Wineries

In this study, we focused on the architecture of the wineries of the study area, as one of the types of facilities used for wine production, linked to the traditional agricultural activity of the region, which was mainly based on cereals, olive groves and vineyards, and to the farmhouses and estates [39]. The latter had a stately home and accommodation for the workers, along with facilities in which the different production processes were carried out. In the mid-19th century, these buildings became exclusively wineries [8]. Some of them combined oil mills and areas for the transformation of wine, as is the case of "Lagar de la Capellanía" and "Lagar de las Monjas", and others that were exclusively dedicated to wine production, such as "Lagar de la Inglesa".

The studies conducted by [8] about these buildings specify that the wineries were within the rural houses of the Cordovan vineyard areas, and they were dedicated to different basic purposed: "*To organise the agrarian activities of a vineyard exploitation; to attend to the industrial operations of grape maturation and fermentation of musts; to provide accommodation to the workers of this industry, as well as, at certain times of the year, to serve as secondary home for the owner and his family; and other, different complementary activities*", among others. Construction habits are to some extent a stable element in any culture, thus it has an important impact on the shaping and sustaining of cultural landscapes [40]. Authors such as Naranjo Ramírez [8] identify the house-winery as a specific element that contributes to the agrarian landscape of Montilla with high scenic value. Regarding the building techniques, the wineries were initially predominated by whitewashed stucco factories with tiled roof on a structure of logs and cane matting. With the different renovations that took place in the 20th century, there was a combination of masonry, stucco and brick factories with tiled roofs on wooden or metal structures. These are recognisable external elements in the agrarian landscape of Montilla (Figure 5), defined by some researchers as a "*fundamental reference of the human presence in the landscape and even a reference of the property and exploitation of the land*" [41].

Figure 5. Landscape of Sierra de Montilla (Córdoba) with the "Lagar de los Raigones" (Montilla-Moriles wine route).

With respect to the interior structure of the buildings, and as reported by some authors [8], the wineries are organised as blockhouses around a central yard. This yard is connected to the rest of the rooms, which were used for either agricultural activities or accommodation, providing them with daylight and ventilation. Among the rooms dedicated to industrial purposes, there were winepresses for the reception, stomping and pressing of the grapes, as well as spaces used as cellars. These particularities were observed and described in the four examples studied in this investigation.

3.5. Historic Wineries in Montilla

3.5.1. Case Study: Lagar de la Capellanía (Winery of the Chaplaincy)

The first reference to this winery is in a document written by Mrs. Luisa Ward of Alvear (Figure 6) in 1834, in which she declared to be, along with her brother-in-law Manuel (clergyman), the owner of the first hydraulic press installed in their oil press of El Carril, in the municipality of Montilla. In this text, she also mentions another oil press/winery of their own located in the plain of El Mesto, known as the Chaplaincy. The project for the construction of the Chaplaincy (Figure 7) has an oil mill and a winery in the Chaplaincy of Alvear, which was constituted by: (a) a front façade; (b) an entrance to the sections; (c) the sections (rooms first and then cellars); (d) an old section; (e) a well; (f) stables; (g) a stomping winepress; (j) presses with their regaifas (circular stone slabs for milling); (k) a common well; (h) solid towers; (y) a well for musts; and (l) a lever winch. Currently, the "Lagar de la Capellanía" maintains its original structure and the yard connected to the rooms, although it has undergone different expansions and the use of its spaces has changed.

Sello 4º 40 mrs
Año de 1834
Valga para el Reynado de S. M. la Señora Doña Isabel II.

Figure 6. (**a**) Document of the declaration of property of the first hydraulic press installed in the oil press of El Carril, in the municipality of Montilla (Córdoba), written by Mrs. L. Ward of Alvear and her brother-in-law Manuel (clergyman) in 1834. Archive of the Alvear family [42]; (**b**) portrait of Mrs L. Ward of Alvear (photograph provided by Mª J. Jiménez Alvear).

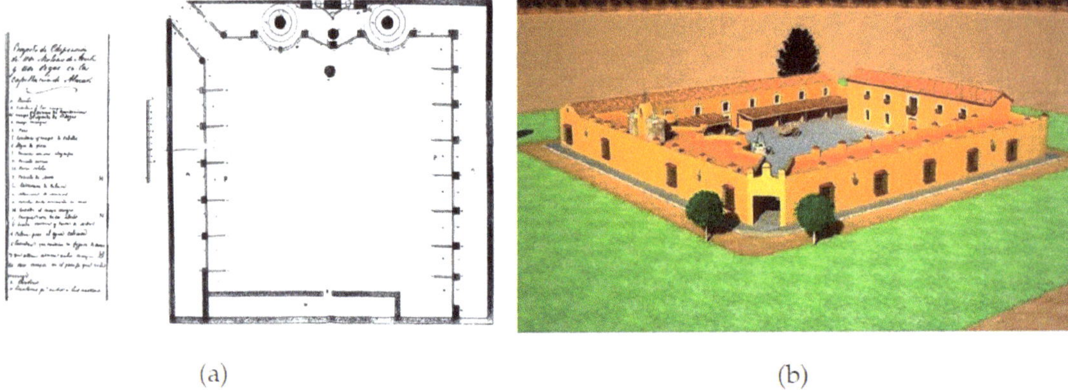

(a) (b)

Figure 7. (**a**) Construction project of the "Molino y Lagar de la Capellanía" (mill and winery of the chaplaincy). Archive of the Alverar family [42]; (**b**) reconstruction of the "Lagar de la Capellanía" [42].

3.5.2. Case Study: Lagar "el Parador de Montilla" (The Montilla Inn Winery)

During the reign of Elizabeth II, specifically on 20 July 1856, Montillian landowner Francisco Solano Rioboó requested the construction of a winery. Interested in agriculture, he sent a report to the sub-delegation of Córdoba in 1834, describing the economic utility of this type of buildings in olive milling [43], as well as the use of a machine brought from London by Mr. Diego of Alvear. A year later, the construction of the so-called Montilla Inn began.

In the documents analysed for the present study, provided by the Municipal Archive of Montilla, Mr. Francisco Rioboó informed the mayor of Montilla about the compliance with the road ordinances and the delineation performed by the overseer, Mr. R. Arjona (Figure 8). The reply of the mayor, Mr. Juan Mariano Algaba, highlighted the social interest of this investment and the granting of the license for the construction of the "Casa Lagar" (winery house), after the verifications and recommendations of the district engineer, the assistant of the Córdoba-Málaga road and the overseer. One year after its construction, this building had two bodies: one originally designed and used for accommodation, where the landlords lived and travelers were received, and another one reserved for the stable, the haystack, the winepress for grape stomping and a cellar with large earthenware jars. Both bodies were connected by another two bodies located to the right and to the left of the main bodies, forming a central square yard between them.

Some years after the events that occurred in Montilla as a result of the proclamation of the First Spanish Republic, which ended with the death of the wealthiest landowner of the region, Francisco Solano Rioboó, in 1877, "Casa Lagar" was already known as "Parador de la Concepción", and was inherited by devotee Mª Encarnación Rioboó y Ortiz.

The name "El Parador" comes from its primitive use, which was made compatible with that of the winery, as it was a stop for stagecoaches traveling to Seville, Málaga and Córdoba. Aguilar Montesinos [45] conducted a virtual reconstruction of this building, which is published on a Youtube channel. In 1900, through barter, the building was purchased by the Trillo-Figueroa family, who then sold it in 1907. After going through different owners, it was bought by the founder of the Cobos cellars, Mr. J. Cobos Ruiz, in 1907, who named it "Parador de San Francisco Solano", although it was exclusively used as a winery and cellar (Figure 9). Today it is a ruined building.

3.5.3. Case Study: Lagar de La Inglesa (Winery of the English Woman)

The "Lagar de La Inglesa" (winery of the English woman) is one of the most significant wineries of Montilla, for both its history and its architecture. It is located in the mountainous region of Montilla, in the neighbourhood of "La vereda del Cerro Macho", in one of the areas of greatest scenic and wine interest. Its name refers to English lady Luisa Rebeca de Ward, who was married to Mr. Diego de Alvear y Ponce de León. It was built in 1870, possibly by Mr. Francisco de Alvear y Ward (1817–1896), father of the Count of La Cortina. This building was renovated in successive interventions, until the Count of La Cortina, Francisco de Alvear y Gómez de la Cortina, gave it the look that is fully preserved today.

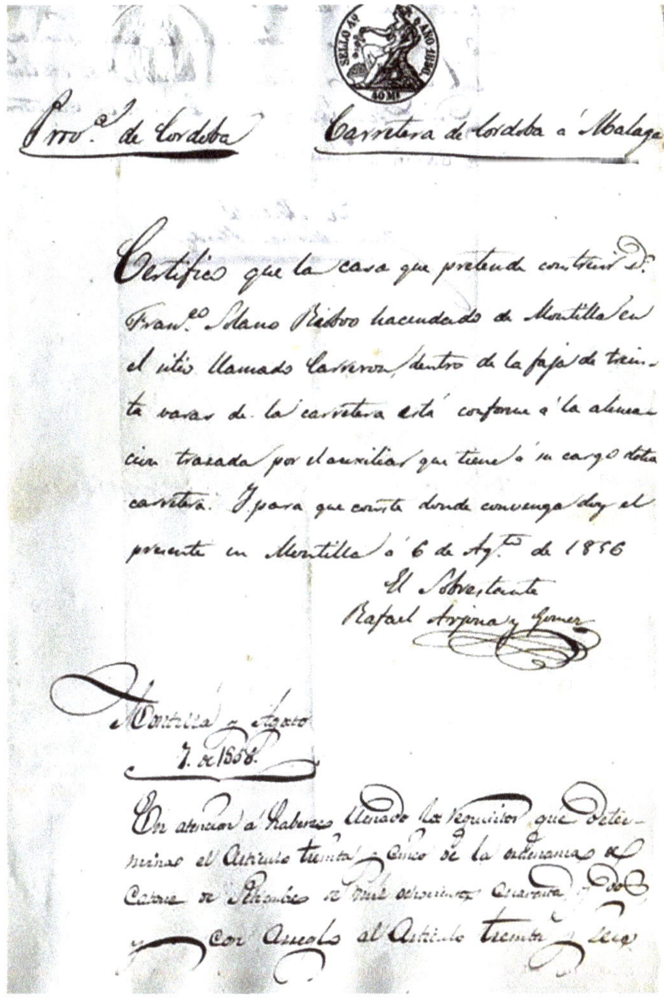

Prov.a de Córdoba — Carretera de Córdoba á Málaga

Certifico que la casa que pretende construir D. Fran.co Solano Pedrero hacendado de Montilla en el sitio llamado Caserón, dentro de la faja de treinta varas de la carretera está conforme á la alineacion trazada por el auxiliar que tiene á su cargo dicha carretera. Y para que conste donde convenga doy el presente en Montilla á 6 de Ag.to de 1856

El Sobrestante
Rafael Arjona y Gomez

Montilla 7 Agosto de 1856

Figure 8. Certification of August 6th, 1856 from overseer Mr. R. Arjona about the measurements for the construction of the Montilla Inn (Córdoba) [44].

The "Lagar de La Inglesa" is a mansion of pure English style, perfectly preserved by its current owner, wine producer Antonio Doblas (Moriles). The different rooms are distributed in a rectangular plan around a central yard (Figure 10). The building is entered through a tower-like central body, finished by an accessible terrace that was used as a lookout (Figure 11), and its interior includes: the hall, the lordly rooms and the staircase. The back of the building was crowned with the bell-gable of a chapel facing the interior of the yard [46]. Moreover, there is a large storage room under the roof along the entire building, which completes the service rooms. Initially, this house was a winery and a cellar; currently, grapes are not ground in it, since the fermentation cones were removed. However, wine is matured there in wooden barrels. Therefore, it is only considered a cellar, with over 600 casks of American oak.

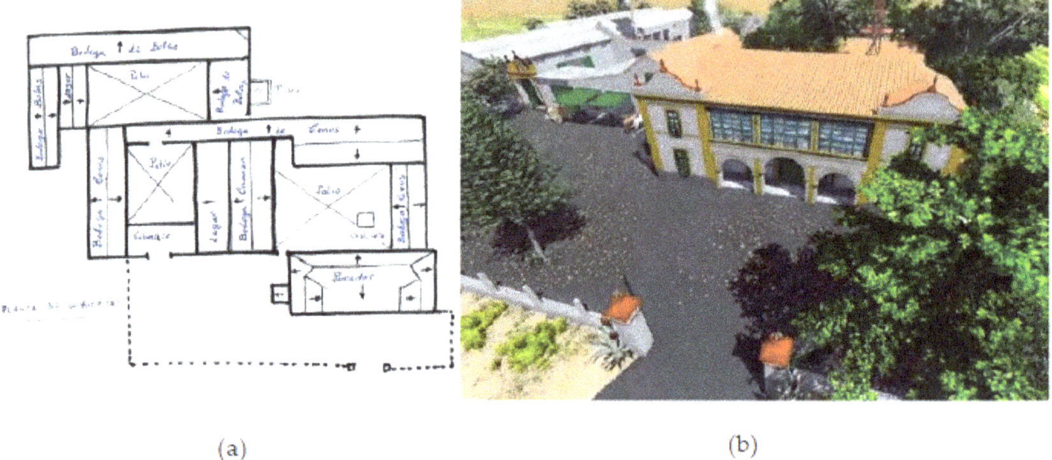

(a) (b)

Figure 9. (**a**) Roof plan of "Lagar el Parador" in Montilla (Córdoba) (Image provided by Francisco de Paula Montes Tubío); (**b**) panoramic view of the virtual reconstruction of "Lagar el Parador" [45].

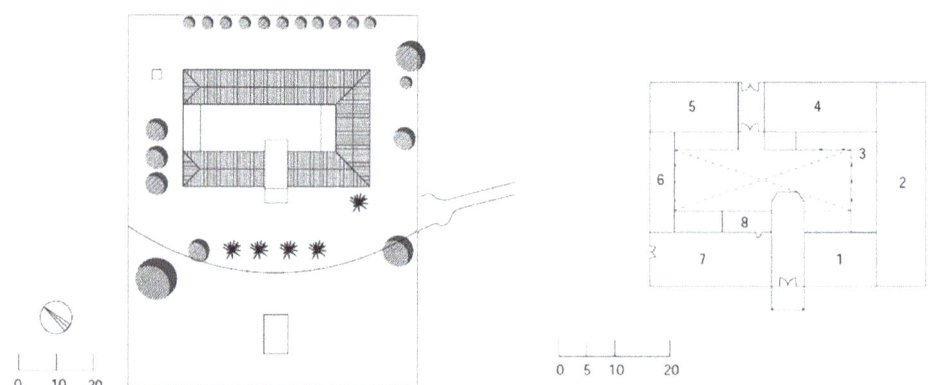

Figure 10. Ground plan of the "Lagar de La Inglesa" in Montilla (Córdoba): (1) lordly rooms; (2) maturation cellar in the ground floor and lordly rooms in the upper floor; (3) winepress; (4) winery; (5) kitchen in the ground floor and staff rooms in the upper floor; (6) old stable; (7) chapel; (8) vestry [18].

Brick is the main material used in the construction of the façades, with both building and ornamental purposes [47]. This is one of the multiple manifestations of clay, an autochthonous material that was widely used in the winery house [8]. This material was used to build, cover holes and outline horizontal bands that simulate the traditional mudéjar factory; in addition, worked in detail, it appears in the ledge and the high frieze of geometric decoration that surmounts the upper floor, thus gaining height [18].

Figure 11. Entrance façade to the "Lagar de La Inglesa" (photograph provided by Francisco de Paula Montes Tubío).

3.5.4. Case Study: Lagar de Las Monjas (Winery of the Nums)

The "Lagar de Las Monjas" (winery of the nuns) is located in the mountainous regions of Montilla, in the area of higher quality, 5 km from Montilla and 2.5 km from the "Lagar de La Inglesa". The first document about this building dates from the year 1722, with the following description: *"Old oil mill founded by the convent of Santa Ana in the year 1722. It preserves the press nave, divided by a central arcade and covered by a gable roof of ceramic tiles. It also has stables, storage rooms and other work rooms connected to adjacent yards. It preserves the front façade with an engraving of Santa Ana, ceiling stones and the spindle of the beam press integrated in the building itself. It was later transformed into an industrial oil mill. It is currently in poor state of preservation, with the roof partially torned down."* (Figure 12).

Figure 12. (**a**) State of the roof, partially torn down, of the "Lagar de Las Monjas" in Montilla (Córdoba) [43]; (**b**) lintelled stone signed by the implementer in the year 1769 in the mill of the "Lagar de las Monjas" [43].

In 1853, the "Lagar de Las Monjas" was turned into an industrial oil press, likely coinciding with the ecclesiastical confiscation or with its purchase by individuals, which is its current use. However, in 1974, the owner of the Las Monjas country house, Mrs. Elisa Valderrama Rioboó, ordered the construction of a cellar and a winery (Figure 13) for her own vineyard exploitation.

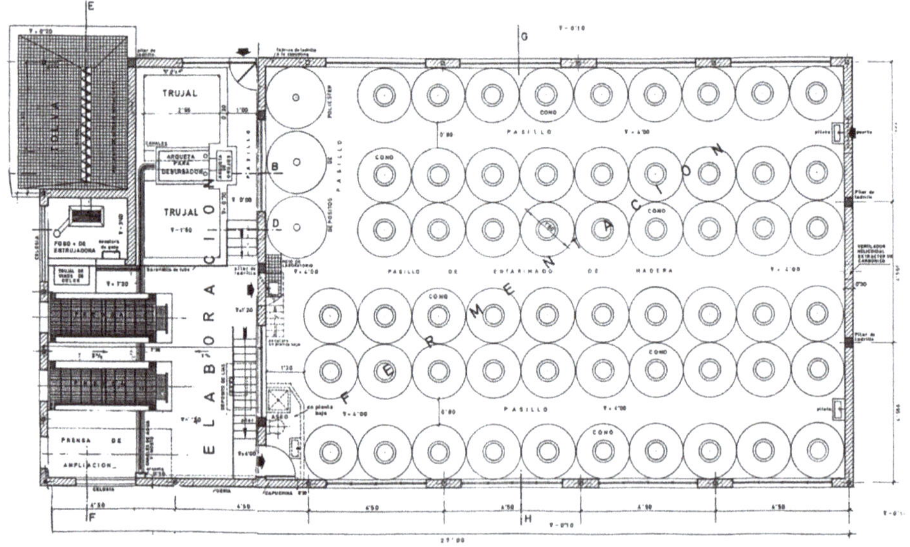

Figure 13. Ground plan of the production nave as of 1974 of the "Lagar de las Monjas" in Montilla (Córdoba) [48]

4. Conclusions

This paper describes the most prominent features of the historical wineries in Montilla-Moriles (PDO). The novelty of this paper is that it presents the description of these four representative wineries in the Montilla-Moriles area, from the data collected in the unpublished files that address these particular constructions, as part of the industrialisation of wine production, and valuable heritage insights that should be preserved for the future sustainable development.

The analysis conducted in this study indicates that the architecture of the historic wineries of Montilla is identified by the autochthonous materials used, such as stone, the masonry, stucco and brick factories and the roofs, generally gable roofs on wooden structures. Their design follows a traditional model, characterised by the main yard around which a set of rooms are distributed, including the accommodation area.

As a consequence of the decrease in the land area occupied by vineyards, and due to the greater reception capacity of wine-making cooperatives, numerous historic wineries have disappeared or nearly disappeared in Montilla, such as the "Lagar del Parador", which is in the red list of the heritage, near collapse and at imminent risk of falling down. None of the examples presented in this study have reached the present time with their original role of fruit transformation that was carried out in them; only the "Lagar de la Inglesa" maintains the maturation cellar.

To sum up, the physical abandonment ruin of numerous wineries has a negative effect on the Montillian landscape, where they stood out for the traditionality of the agrarian region in which they were located, being also part of our identity. With their disappearance, as is the case of "El Parador" and the "Lagar de La Capellanía", an important part of the industrial cultural heritage of Montilla-Moriles (PDO) is lost, and only other alternatives resources, such as virtual reconstructions [49,50], allow preserving the historical memory of these buildings. The identification of these historic wineries based on old literature documentation and local people's inner insights through interviews is indisputably important for cultural value conservation, fostering cultural sustainability. This manuscript helps to conserve and spread the word about the fundamental features of traditional old wineries, as industrial cultural assets.

To sum up, protecting and disseminating the industrial cultural heritage in Montilla-Moriles (PDO) would help policymakers in its adoption concerning a more sustainable cultural policy, ensuring the sustainability of rural world heritage sites.

Author Contributions: Conceptualization, A.M.-A. and F.d.P.M.-T.; methodology, A.M.-A. and F.d.P.M.-T.; validation, A.M.-A., F.d.P.M.-T. and I.L.C.-G.; investigation, P.T.-T. and A.V.-G.; writing—original draft preparation, A.M.-A., F.d.P.M.-T. and P.T.-T.; writing—review and editing, F.-J.M.-C. and P.T.-T.; supervision, I.L.C.-G., F.-J.M.-C. and P.T.-T. All authors have read and agreed to the published version of the manuscript.

Funding: This research received no external funding.

Conflicts of Interest: The authors declare no conflict of interest.

References

1. United Nations. Sustainable Development Goals. The Sustainable Development Agenda. 2019. Available online: https://www.un.org/sustainabledevelopment/development-agenda/ (accessed on 18 April 2021).
2. Ministerio de Cultura. El Plan Nacional de Patrimonio Industrial ("The National Plan for Industrial Heritage"). Madrid. 2011. Available online: http://www.culturaydeporte.gob.es/planes-nacionales/dam/jcr:b34f01e5-c3d9-497b-bc7a-ecba1ac65d22/04-industrial-eng.pdf (accessed on 16 April 2021).
3. Serrano, E.; González-Amuchastegui, M.J. Cultural Heritage, Landforms, and Integrated Territorial Heritage: The Close Relationship Between Tufas, Cultural Remains, and Landscape in the Upper Ebro Basin (Cantabrian Mountains, Spain). *Geoheritage* **2020**, *12*, 86. [CrossRef]
4. UNESCO. *Culture for the Agenda 2030*; United Nations Educational, Scientific and Cultural Organizations: Paris, France, 2018; Available online: http://www.unesco.org/culture/flipbook/culture-2030/en/Brochure-UNESCO-Culture-SDGs-EN2.pdf (accessed on 27 April 2021).
5. Calvo-Serrano, M.; Castillejo-González, I.; Montes-Tubío, F.; Mercader-Moyano, P. The Church Tower of Santiago Apóstol in Montilla: An Eco-Sustainable Rehabilitation Proposal. *Sustainability* **2020**, *12*, 7104. [CrossRef]
6. Gullino, P.; Beccaro, G.L.; Larcher, F. Assessing and Monitoring the Sustainability in Rural World Heritage Sites. *Sustainability* **2015**, *7*, 14186–14210. [CrossRef]
7. Pj, C.M.S. Preservation and Conservation of Rural Buildings as a Subject of Cultural Tourism: A Review Concerning the Application of New Technologies and Methodologies. *J. Tour. Hosp.* **2013**, *2*, 1000115. [CrossRef]

8. Naranjo Ramírez, J. El valor paisajístico de lo utilitario. La casa rural en el viñedo cordobés: "los lagares". In *En: Martínez De Pison, E. Y Ortega Cantero, N.*; del Paisaje, L.v., Ed.; Autonomous University of Madrid: Madrid, Spain, 2019; pp. 293–316.
9. Portero Laguna, J. Metodología Para El Conocimiento De Un Edificio Agroindustrial Integrando Distintas Disciplinas. Ph.D. Thesis, University of Cordoba, Córdoba, Spain, 2017.
10. Triviño-Tarradas, P.; Carranza-Cañadas, P.; Mesas-Carrascosa, F.-J.; Gonzalez-Sanchez, E.J. Evaluation of Agricultural Sustainability on a Mixed Vineyard and Olive-Grove Farm in Southern Spain through the INSPIA Model. *Sustainability* **2020**, *12*, 1090. [CrossRef]
11. Landi, S. Rural Landscapes of the 20th Century: From Knowledge to Preservation. *Arch. Civ. Eng. Environ.* **2019**, *12*, 47–56. [CrossRef]
12. Slámová, M.; Belčáková, I. The Role of Small Farm Activities for the Sustainable Management of Agricultural Landscapes: Case Studies from Europe. *Sustainability* **2019**, *11*, 5966. [CrossRef]
13. Prada Llorente, E.I.; Riesco Chueca, P.; Herrero Tejedor, T. Landscape and image: Heritage and form in the cultural construction of agrarian land. *Estudios Geograficos* **2013**, *74*, 557–583. [CrossRef]
14. Vlamia, V.; Kokkorisb, I.P.; Zogarisc, S.; Cartalisd, C.; Kehayiasa, G.; Dimopoulos, P. Cultural landscapes and attributes of "culturalness" in protected areas: An exploratory assessment in Greece. *Sci. Total Environ.* **2017**, *595*, 229–243. [CrossRef] [PubMed]
15. Quirós García, M. El Libro de Agricultura de Gabriel Alonso de Herrera: Un texto en busca de edición. *Criticón* **2015**, *123*, 105–131. [CrossRef]
16. Herrera, G.A. El Libro De Agricultura. In *Libro Segundo*; 1790; Volume CXXI, p. 71. Available online: http://bdh-rd.bne.es/viewer.vm?id=0000079905&page=1 (accessed on 20 May 2021).
17. MAPA. Ministerio de Agricultura, Pesca y Alimentación. Gobierno de España. Obra de Agricultura, 1513. Primera Edición. 2020. Available online: https://www.mapa.gob.es/es/ministerio/servicios/informacion/plataforma-de-conocimiento-para-el-medio-rural-y-pesquero/centenario/indice-1513.aspx (accessed on 12 April 2021).
18. Junta de Andalucía. Consejería de Fomento, Infraestructuras y Ordenación del Territorio. Cortijos, haciendas y lagares: Arquitectura de las grandes explotaciones agrarias de Andalucía. Provincia de Córdoba. Sevilla. 2006. Available online: https://ws147.juntadeandalucia.es/obraspublicasyvivienda/publicaciones/01%20ARQUITECTURA%20Y%20VIVIENDA/cortijos_haciendas_y_lagares_en_andalucia/cortijos_haciendas_cordoba/cordoba_tomo2.pdf (accessed on 28 April 2021).
19. Madoz, P. *Diccionario Geográfico-Estadístico-Histórico De España Y Sus Posesiones De Ultramar*; Establecimiento tipográfico de P. Madoz y L. Sagasti: Madrid, Spain, 1845; Volume 11.
20. Ramírez de las Casas Deza, L.M. *Corografía Histórico-Estadística De La Provincia Y Obispado De Córdoba (1840–1842)*; Monte de Piedad y Caja de Ahorros: Cordoba, Spain, 1840.
21. Vizetely, H. Facts about Aherry Gleaned in the Vineyards and Bodegas of Jerez, Seville, Moguer & Montilla Districts during the Autum of 1875. Londres, 1876. Ward, Lock, and Tyler, Warwick House. Available online: https://euvs-vintage-cocktail-books.cld.bz/1876-Facts-About-Sherry-by-Henry-Vizetelly/X/ (accessed on 16 April 2021).
22. IAPH (Instituto Andaluz del Patrimonio Histórico). *Paisaje Vitivinícola de Montilla*; Data Sheet; Consejería de Cultura: Sevilla, Spain, 2016.
23. Ramírez Pino, J. Montilla, 1950–1975. In *Entre La Historia Y La Memoria*; Montilla: Cordoba, Spain, 1994.
24. Triviño-Tarradas, P.M. Evolución y mejoras para el manejo sostenible de la explotación vitivinícola en la D.O.P. Montilla-Moriles. Evaluation and Improvements for the Sustainable Management of the Wine-Making Farm in the PDO Montilla-Moriles. Ph.D. Thesis, University of Cordoba, Cordoba, Spain, 2019.
25. Badia-Miró, M.; Tello, E.; Valls, F.; Garrabou, R. The grape phylloxera plague as a natural experiment: The upkeep of vineyards in catalonia (spain), 1858–1935. *Aust. Econ. Hist. Rev.* **2010**, *50*, 39–61. [CrossRef]
26. Duran, J.I.D.L. Wine cathedrals: Making the most of masonry. *Proc. Inst. Civ. Eng. Constr. Mater.* **2013**, *166*, 329–342. [CrossRef]
27. Instituto de Estadística y Cartografía de Andalucía. Consejería de Economía y Conocimiento. Atlas de Histórica Económica de Andalucía. SS XIX-XX. 2020. Available online: https://www.juntadeandalucia.es/institutodeestadisticaycartografia/atlashistoriaecon/atlas_cap_17.html (accessed on 19 April 2021).
28. Consejo Regulador de Montilla-Moriles. Estadísticas. 2020. Available online: https://www.montillamoriles.es/es/ (accessed on 28 April 2021).
29. Baraja Rodríguez, E.; Plaza Gutiérrez, J.I.; Prada Llorente, E.I. Naturaleza, Territorio y Ciudad en un Mundo Global. Actas del XXV Congreso de la Asociación de Geógrafos Españoles. In *Atributos y Valores Patrimoniales de los Viñedos Tradicionales en las Provincias de Zamora y Salamanca: El caso de los arribes del Duero*; A.G.E. (ASOCIACIÓN GEÓGRAFOS ESPAÑOLES): Madrid, Spain, 2017.
30. Valdés, M. *Montilla y Su Comarca: Enciclopedia-Guía Antológica*; Valdés M.: Sevilla, Spain, 1979.
31. González San José, M.L. *Enoturismo y Entornos Sostenibles.* 2017, Volume 193, p. a399. Available online: http://arbor.revistas.csic.es/index.php/arbor/article/view/2207 (accessed on 20 May 2021).
32. Pardo Abad, C.J. *El Patrimonio Industrial En España: Análisis Turístico Y Significado Territorial De Algunos Proyectos De Recuperación*; Boletín de la Asociación de Geógrafos Españoles: Seville, Spain, 2010; Volume 53, pp. 239–264.
33. García, N.T. La Memoria del Pasado Industrial. Conservación, Reutilización Y Creación De Nuevos Equipamientos. *RPH Rev. Electrónica Patrim. Histórico* **2016**, *19*, 72–99.

34. Mansilla Plaza, L. El Parque Minero de Almadén. Un Modelo De Recuperación Del Patrimonio Minero Industrial. *Herit. Museography* **2011**, *1*, 13–24.
35. Arnáiz, M.M.; Rodríguez, E.B.; Hernando, F.M. Criterios de la UNESCO para la declaración de regiones vitícolas como paisaje cultural: Su aplicación al caso español. *BAGE* **2019**, *80*, 2614. [CrossRef]
36. Martínez Carrión, J.M.; Ramon Muñoz, J.M. La vitivinicultura y la valorización de su patrimonio industrial, cultural y natural. In *Áreas. Revista Internacional De Ciencias Sociales*; University of Murcia: Murcia, Spain, 2010; Volume 29, pp. 148–155. Available online: https://revistas.um.es/areas/article/view/115631 (accessed on 18 May 2021).
37. Aladro-Prieto, J.M. Bodegas, Lagares y Casas de Viñas. AH Andalucía en la Historia, año XVII, 66, Octubre-Diciembre. 2019; pp. 14–17. Available online: https://www.centrodeestudiosandaluces.es/publicaciones/descargando/2356/documento (accessed on 18 May 2021).
38. Álvarez-Areces, M.A. Patrimonio industrial. Un futuro para el pasado desde la visión europea. In *APUNTES*; 2008; Volume 21, núm. 1; pp. 6–25. Available online: http://www.scielo.org.co/pdf/apun/v21n1/v21n1a02.pdf (accessed on 18 May 2021).
39. Luque Romero, F.; Agudo Torrico, J.; Cobos, J. *Cultura Material De Carácter Tradicional En La Provincia De Córdoba*; Gever S.L.: Cordoba, Spain, 1985, Volume IV, pp. 137–193.
40. Marine, N.; Arnaiz-Schmitz, C.; Herrero-Jáuregui, C.; Cabeza, M.R.O.; Escudero, D.; Schmitz, M.F. Protected Landscapes in Spain: Reasons for Sustainability of Conservation Management. *Sustainability* **2020**, *12*, 6913. [CrossRef]
41. Sobrino Simal, J. *Arquitectura de la industria en Andalucía*; Instituto de Fomento de Andalucía: Jaen, Spain, 1998.
42. Bellido Vela, I.D. Diego de Alvear y Ward. Un innovador de la agroindustria. La prensa hidráulica. Ph.D. Thesis, University of Cordoba, Cordoba, Spain, 2016.
43. Llamas Salas, M. El molino del duque de Montilla Y La Influencia Del Monopolio Señorial En La Arquitectura Oleícola. Ph.D. Thesis, University of Cordoba, Cordoba, Spain, 2016.
44. Archivo Histórico Municipal de Montilla (Cordoba-Spain). Legajo 980 B, Expediente 1. Available online: https://www.montilla.es/areas_municipales/cultura/archivo_historico (accessed on 20 May 2021).
45. Montesinos, F.J.A. Reconstrucción virtual del lagar El Parador. Unpublished Master's Degree Project, University of Cordoba, Córdoba, Spain. 2012. Available online: https://www.youtube.com/watch?v=rmjGV76s_n8 (accessed on 26 April 2021).
46. Junta de Andalucía. Consejería de Cultura y Patrimonio Histórico. Guía Digital del Patrimonio Cultural de Andalucía. 2020. Available online: https://guiadigital.iaph.es/inicio (accessed on 15 April 2021).
47. Muñoz, P.; Morales, M.; Mendívil, M.; Juárez, M.; Muñoz, L. Using of waste pomace from winery industry to improve thermal insulation of fired clay bricks. Eco-friendly way of building construction. *Constr. Build. Mater.* **2014**, *71*, 181–187. [CrossRef]
48. Cruz Marqués, M. *Evolución del diseño de proyectos de bodegas de vinos en la provincia de Córdoba*; University of Cordoba: Cordoba, Spain, 1998.
49. Torres, J.; Cano, P.; Melero, J.; España, M.; Moreno, J. Aplicaciones de la digitalización 3D del patrimonio. *Virtual Archaeol. Rev.* **2010**, *1*, 51–54. [CrossRef]
50. Bekele, M.K.; Pierdicca, R.; Frontoni, E.; Malinverni, E.S.; Gain, J. A Survey of Augmented, Virtual, and Mixed Reality for Cultural Heritage. *J. Comput. Cult. Herit.* **2018**, *11*, 1–36. [CrossRef]

Article

Intelligent Urban Planning and Ecological Urbanscape-Solutions for Sustainable Urban Development. Case Study of Wolfsburg

Joanna Dudek-Klimiuk [1,*] and Barbara Warzecha [2]

1 Department of Landscape Architecture, Institute of Environmental Engineering, Warsaw University of Life Sciences-SGGW, ul. Nowoursynowska, 159, 02-766 Warsaw, Poland
2 Independent Researcher, 00-201 Warsaw, Poland; barbara.maria.warzecha@gmail.com
* Correspondence: joanna_dudek_klimiuk@sggw.edu.pl

Abstract: Intelligent urban planning and ecological urbanism can be recognized as two of the key solutions to act against urban sprawl. This process is associated with suburbanization, blurring boundaries between the city and suburbs, and the undefined role of open and green spaces within new structures. It has been identified as the biggest and the most common problem worldwide. This non-central planning has a huge impact not only on economic aspects, but—most of all—on the ecological and landscaping balance within the urban area. This study covers not only the recognition of the outlined situation, but also a conceptual proposal to challenge the problems of urban sprawl. The city of Wolfsburg serves as a case study to which the tools of Ecological Urbanism and Intelligent Urbanism were applied. A corrective plan for the study area has been worked out, based on the main approaches in urban planning of the 21st century. The green transformation processes to achieve resiliency within urban areas are inevitable and will have to be conducted due to the rising number of the dwellers, steadily changing climate, and socio-economic conditions all over the world. The main solutions include mainly the system of green corridors, interconnectedness of open spaces, walkability with smart mobile options and social community as a nucleus of a local neighborhood.

Keywords: ecological urbanism; ecological urbanscape; green infrastructure; urban sprawl; Intelligent Urbanism; urban design

Citation: Dudek-Klimiuk, J.; Warzecha, B. Intelligent Urban Planning and Ecological Urbanscape-Solutions for Sustainable Urban Development. Case Study of Wolfsburg. *Sustainability* **2021**, *13*, 4903. https://doi.org/10.3390/su13094903

Academic Editor: Jacques Teller

Received: 7 March 2021
Accepted: 23 April 2021
Published: 27 April 2021

1. Introduction

The study focuses on a treatment proposal to create resilient urban areas influenced by the process of city sprawl. Core features of the sprawl are associated with suburbanization, blurring boundaries between the city and suburbs and the function of open and green spaces. It is understood as an unplanned, scattered, low density, automobile-dependent development at the urban periphery. It is caused by non-central planning and highly fragmented land-use governance. These actions result in a single land use, which is dominated and strengthened by car-oriented transportation. This common problem has a huge impact on the economic, ecological, and landscaping balance. Within the whole city–suburb complex, the sense of genius loci, and the esthetic and visual perception may be, therefore, significantly disrupted. To counteract the dispersed forms of the urban structure a holistic strategy is needed to be implemented so as to create a compact and resilient city.

This idea of sustainable growth dates back to the turn of the 19th and 20th centuries when it was an impulse for the creation of many corrective actions for constantly deteriorating living conditions of city dwellers [1,2]. Among them, the most important were the following: a Garden City [1,3,4], a Neighborhood Unit [5–7], and an Organic city [6]. Those three main planning strategies, even if aimed at repairing the living conditions, contributed to the spread of the cities. As a result, in the second half of the 20th century, environmental and climate issues became leading challenges for architects, landscape architects and planners, who looked for solutions to counteract these adverse changes and protect cities against threats. The search for corrective approaches for cities is not only based on the economic dimension [1]; it is necessary to respect the laws of nature, prevent, or minimize the negative impact of urbanization and build an intelligent, smart city characterized by sustainable development. Many strategic proposals have been worked out, the most important were Ecological and Intelligent Urbanism. The implementation of these theoretical assumptions (ideas) verified them, pointing to both their strengths and weaknesses.

The purpose of the research is to work out a corrective model (preconceptual proposal) for an urban structure of a previous car-oriented city of Wolfsburg (case study), located in the center of Germany, in the state of Lower Saxony, by using the tools of Ecological Urbanism and Intelligent Urbanism. A new design concept could be a solution to apply the tools of sustainable development in urban areas in a very early design stage. The implementation of systemic pro-ecological urban guidelines and rules of an Intelligent Urbanism can significantly improve the quality of residents' life, especially in large metropolises. The need to develop a new holistic design could be the biggest challenge to make our cities resilient and healthy. This approach should eventually focus not only on one district or precinct, but—most of all—on the urban area as a whole. The guidelines can serve as a basis for a modern and sustainable city design, starting with the genius loci in a local scale [1,8]. They should constitute an overall, comprehensive and holistic approach.

Having formulated the objective of the study which was described above, a number of research questions were defined:

- What is the origin of suburbanization? Which of its elements have influenced the current situation in Wolfsburg? What methods were implemented in order to counteract this process?
- Can the tools of Ecological and Intelligent Urbanism be used in the process of repairing the cities or their structures that are currently showing the features of the dispersed forms? What are the basic goals, values, and principles of Ecological and Intelligent Urban planning used in the process of urban repair?
- How can the needs of local communities be taken into account at the level of a preconceptional process? Can local communities be strengthened through such an urban design?

The answers to these questions have helped to describe the problem of the further coherence and sustainable development according to the Principles of Intelligent Urbanism and Ecological Urbanism, and—finally—to achieve the intended purpose of this study.

2. Literature Review

The era of rapid industrialization and early capitalism had an enormous impact on the transformation of cities in the 19th century [1]. A sudden increase in the urban population (which can be also observed nowadays) and the chaotic urban expansion related to that resulted in severe air pollution and poor sanitary and housing conditions. As a response to this development, countless urban planning proposals emerged to heal the ever-deteriorating living conditions in the cities [1,5] (pp. 51–54). Among many utopias, the one proposed by Ebenezer Howard—a Garden City—was based on a particular hierarchy and specific spatial order anchored in a central park or garden as an antithesis to a suburb [5] (pp. 59–62), [6] (pp. 58–60), [9,10]. According to some contemporary smart city researchers, this first smart strategic development appears nowadays to be a long-term vision of the future city rather than a reality [3,11–14]. Perry's Neighborhood Unit, which drew from the idea of a Garden City was organized around an elementary school as a self-sustaining, local-concentrated community [5] (pp. 62–66), [6] (pp. 401–403), [7] (pp. 256–270). This cellular unit developed further the notion of contemporary German cities, "of dividing a city into specialized zones" [7] (p. 263). Perry's new system of segregation of movements with blind access to the building complex (dead end), so called cul-de-sac, has become a prototype for new European housing solutions, for instance, "Organic city" by Hans Bernhard Reichow [5] (p. 65) (see chapter 4.1.). The organic city, with its proximity to nature (private garden, greenbelts, and parks) [6] (pp. 86–96), can be further described as an urban organism composed of interdependent neighborhoods—different parts of the city's body, based on balance between built and unbuilt areas.

As a consequence, the deliberating ideas of a garden city and neighborhood units have caused a spatial detachment of whole dwelling systems from the city center, which is described today as "urban sprawl". This tendency dates back to the 1920s [15–17], was described in the 1930s [18] (p. 177), [19] (pp. 52–53), and the 1940s [18] (p. 177), [20] (p. 15) of the last century and was first observed on an enormous scale in the United States after World War II [1,5] (p. 76). Litwinska mentions that there is no single definition of the phenomenon of urban sprawl [15] (p. 146) and there are many, sometimes mutually exclusive, characteristics [21] (p. 108). Consequently, the impulse for the formation process in Great Britain was rail transport [15] (p. 146), whereas in America—the mass produced cars [22], which forced inhabitants to use private vehicles as a means of transportation [5] (p. 65), [6] (pp. 400–404), [15] (pp. 140–141), [21] (pp. 107–118). This heritage of Anglo-American ideas for the corrective plans of healing for the urban tissues had been mutually influenced by the European ones and further developed. Further in Europe, the new planning systems were adapted. This process transformed planning strategies worldwide. The present times, urban planning as a discipline deals with the construction and expansion of cities, remaining in the mutual correlation with the town's policy, governance, and long-distance planning strategy. However, it is essential that the new approaches do not cause further uncontrolled sprawl of cities. The goal should not be to found the new, ideal colonial cities, but rather to cope with the existing imperfect urban tissues. Ecological concerns and urban objectives should rather complement each other, creating the planning coexistence. It is not the issue of spreading the built-up land into the landscape-on the contrary-the main corrective plan could be based on introducing the ecology standards into the city. According to Martha Schwartz, ecological urbanism, as a comprehensive approach, has the strength to combine these conflictual conditions. She is convinced that there is a great challenge for landscape architects to bring functional and rational planning of open spaces into the city, to include human systems as a part of ecology. This process is and should be taken into consideration to develop a sustainable urbanscape [23] (pp. 524–525); consequently, Ecological Urbanism must be focusing on green [24] and blue infrastructures as planning objectives of modern city planning.

Ecological aspects of urbanism are based nowadays principally on the sustainable development and resiliency of urban tissue. Ecological urbanism or an 'ecological definition of urbanism' focuses on the basic characteristics of urban planning, policies, and logic between the structure and form. It is also a new proposal in the relationship and perception between a man and the environment [17]. The values of ecological urbanism, such as public good and cultural ethics, are universal and differ from one continent to another due to geographical or weather conditions, orientation, or level of pollution. Resiliency to rapidly changing living conditions around the globe and an increasing number of the urban population (according to the UN two-thirds of the world population will live in cities by 2050, [25]) should be the main concern for this movement.

The determining criterion that advocates for the sense of ecological urbanism named by Mostafavi remains the density [23]. It is crucial to set a long-term plan towards compact cities with a close collaboration between public and private sectors. The city governors should take a huge responsibility to set directions for the future urban development. The responsibilities of city governance are, nowadays, far beyond only planning the city's maintenance [23]. Mostafavi stresses the fact that ecological urbanism should be based on such values as biodiversity, but not only on the level of quantity of species. The goal should be rather the 'biologically diverse urban landscapes' as a place for recreation and social interactions. It cannot remain only a basic green infrastructure [24]. The emphasis should also be put on the local production of food, even within the city tissue, strengthened by the cooperation with local farmers and markets [23].

Corrective actions undertaken by the political powers are inevitable to apply the ideas of Ecological Urbanism against uncontrolled suburbanization. Zuziak [17] emphasizes that political interaction in the strategy and the democratization of public life support the integration of spatial and strategic planning, which is not always implemented and often remains at the level of pure political ideology. Litwinska [15] (pp. 141–145), on the other hand, lists European metropolises like Bremen in Germany or Brussels in Belgium, whose municipal managers and governors approach strategically and deliberately the issues of the sustainable development of their cities as a whole. To prevent urban sprawl, algorithms and simulations have been prepared. They show the consequences of specific development directions based on the extension of public transportation. This emphasizes how complicated and complex sustainable urban planning is and how many disciplines it takes to organize the urban landscape that is the most beneficial to the health of the city and its inhabitants. This is a good example for modern governance and the importance of rational control of urban sprawl.

Intelligent city development, created in the 21st century [26], promises a much more holistic approach to contemporary urbanism, where digitality and reality come together. Therefore, the identification of an intelligent city is often replaced by a "smart city" or "digital city". Nevertheless, all these definitions have not been characterized consistently [3,27–30]. One thing seems to be certain—its goal is to create a coherent, compact, urban entity, unity of all components, an integrated urban ecosystem [3]—an antithesis to the sprawl. This long-term strategic development assumes a permanent improvement of living conditions and reduction of functional costs at the local and regional levels. This concept is aimed at effective management of the city's resources in the process of its development and during its operation. The Principles of Intelligent Urbanism are as follows: a balance with nature, a balance with tradition, conviviality, efficiency, human scale, opportunity matrix, regional integration, balanced movement, institutional integrity, and vision [26,31–33]. They are used to control and act against the urban sprawl and transform the city into a compact one.

Intelligent development integrates diverse urban planning solutions which are based not only on well-defined streets, open spaces, building layout [26,31–33], and cultural amenities [34] but also redefine the intelligent systems and automated means of transport [35,36]. In order to regain "the significant relationships between theory and practice, understanding and proposal, and between physical dimensions and economic and social dimensions of change towards intelligent territorial planning", an integrated approach must be adopted [37] (p. 120). The same applies to models of ecological connectivity [37]—a smart environment [38]. Sustainability objectives, which should serve as 'guidelines' for the European cities, can enhance intelligent and sustainable urban development. This development towards a 'green' city with all ecological features and logical connections provides cities with resiliency and tools to 'act against' climate change successfully [37,38]. According to the European Commissioner for the Environment, Kermenu Vella, the "[. . .] green cities can make more: they offer a better quality of life to their inhabitants and new business opportunities" [37] (p. 119), [39]. The key characteristics include a set of planning strategies based on innovative and optimized use of public services in the crucial areas, mobility, energy, and environmental efficiency. Their goal is to meet the needs of citizens and improve the quality of urban life [3,30,37–43], which "[. . .] should be expressed by a number of criteria that are individually formulated for each city" [38] (p. 9). It is established on cultural heritage, natural and human resources. Special attention should be, therefore, drawn to the historical–cultural resources of cities, which remain a great existing capital to shape the urban structures anchored around main landmarks. It seems that a spatial development based on such physical values enhances the urban atmosphere and creates an authentic urban lifestyle [43,44]. Consequently, the technology remains only a 'tool' of smart–sustainable development [35,36]; the human interests and activities are foregrounded [30,38,45].

Green infrastructure as a part of intelligent planning in terms of the ecological, environmental, economic, and social benefits, can be achieved through harmony with nature [24,37,43]. 'Nature' can be broadly represented by natural or designed green spaces as a part of the sustainable development of the urban fabric. Together with the blue infrastructure, it becomes a holistic feature of an ecosystem, able to mitigate the consequences of climate change [37,38]. "The green infrastructures are identified as 'eco-duts', ecological corridors, hedges, rows, green bridges, and all those linear entities that allow to reconnect natural or semi-natural areas (point area entities), which have been artificially fragmented by artifacts, buildings, roads, or railway lines" [37] (p. 123), [46]. The long-term perspective of their development should be regarded in such values as air and water quality, physical activity, aesthetic [38,47], social integration, inclusiveness, and sense of security [48]. Water features like rivers, ponds, wetlands, waterfronts complement green ecosystems, create the hydrographic networks and strengthen green corridors, which help to balance violent consequences of climate change. They also help with water retention and air cooling.

To sum up, the values of Ecological Urbanism are based primarily on sustainable development and biodiversity understood as the presence of a high number of species and a variety of open spaces. A compact city with balanced proportions of the urbanscape elements should be the goal of the public–private partnership. Intelligent Urbanism is based on the considerate approach towards nature, focus on the efficiency of resources and use of the appropriate technology in coping with environmental issues. The idea of implementing the human scale approach in any actions in the local governance helps to achieve regional integrity. The elements of mobility are balanced with the walkability within the pedestrian-friendly street life.

The active participation of residents and businesses together with good governance through the bottom-up approach is necessary to promote new solutions and political decision-making [3,37,38,42]; hence, Intelligent City should be a city for all citizens [3,42,48–51]. It should reflect the basic human values and needs as optimal conditions represented on the level of a place, urban space and urbanity. Smart electronic solutions can improve access to public services as well as information and increase the partaking at the level of design and operation of the city; consequently, they should contribute to reduction of social exclusion [36,48,50]. Social participation can be thus called "the main factor influencing the outcome of the changes, people must be willing to use and participate in smart solutions" [38] (p. 10) and can be given that chance through the use of electronic means [36,50]; therefore, it is next to the environmental and economic concerns one of core elements of sustainable development. It can be even observed that at the interregional level individual cities compete with each other for the title of a smart space of dynamic growth connected to economic city branding [52], and creative and physical capital and its mobility, which appears to be measurable in economic costs–taxes income [45]. However, "place attachment is the ability to use public space [48] (p. 82), [51]; consequently, the core of urban living lies in the human and social capital [29,42,48,51]. Therefore, the solutions of Intelligent Urbanism cannot be implemented only partially—all the objectives are required in all sectors, generations, and domains [3,48–51,53]. This is featured by the environmental layer in regard to the blue-green infrastructure and social one—as active local centers of human interactions, facing the cultural challenges of intelligent communities [45,51].

High quality of life remains one of the main ambitions of the European Commission, which is committed to supporting the development of green cities [30,37,54]. The awarded European cities are characterized by intelligent planning strategies adjusted to local conditions and their successful adaptation in order to create an environmentally friendly urban development, which can guarantee economic resiliency and qualitative sustainability. It often seems to be the case that the probability of designing and implementing objectives of Smart City is higher in cities that are already familiar with executing intelligent characteristics and policies [50,55,56]. Market pressure and weak governance can contribute to a failure to implement smart theory into practice [56]; hence, the performance of planning objectives can only be possible in the "presence of a strong political–technical leadership and an institutional framework with instruments of government of the territory able to make complex interventions possible" [37] (p. 122), [50,56,57].

The implementation of smart growth as a holistic strategy seems to be limited rather for the wealthiest and well-developed countries [55]. Since their impact on the environmental commonwealth has been the strongest—their actions towards sustainable growth are expected to be applied accordingly [3]. Even so, the natural limits of our planet and its natural resources required for such transformation remain the obstacle, which prevents sustainable development from happening. Additionally, there is the question of whose interest is represented here the most: marketing needs of corporates or social intelligence required for cities to be smart. Smart as one city might eventually become, it is essential to concentrate on a community-led transition which achieves resiliency to acute social, economic, and cultural challenges [58].

3. Materials and Methods

The Methodology is based on:

- a logical argumentation (literature review, analysis and description of the state of the research),
- interpretive–historical research with the case study of the city of Wolfsburg, Germany
- simulation and modeling research with a conceptual proposal for this case study.

The research was divided into several stages, in which different research methods were used depending on their specificity. In the first stage, the method of logical argumentation was applied and the literature on the subject was reviewed in order to compile basic definitions related to suburbanization and their critical interpretation. On this basis, the historical background and basic conditions of both suburbanization processes and attempts to counteract them (old and modern) were delineated. Materials and publications used at this stage of the research were quoted in the text of the article.

Based on the results (features characterizing the phenomenon of suburbanization), the object of further, deepened research was selected.

The main part of the presented research concerns Wolfsburg, Germany, a city facing strong suburbanization. This part of the research was conducted with the help of a case study method, which was accompanied by an historical–interpretative approach. The tools used in this part were as follows: literature research, analysis of logical argumentation, correlation of events, and facts. This part was mainly based on publications, including exhibition materials on the history of the city and archives kept at the city hall. The assessment of the land development condition and its use was based on the analysis of planning materials (Spatial Development Plan, digital ALKIS Stadtgrundkarte (2018)) as well as on the analyses of orthophotos and environmental diagnosis maps (noise, flood risk). On top of that, field observations constituted a vital portion of this evaluation. The demographic data was obtained from a socio-political analysis (2018) and publicly available statistical data.

The city of Wolfsburg was selected as the study area due to its size and the current state of the problematic urban structure. It seems to be an adequate object for further studies and considerations. The openness of the City Hall and the willingness of its administrators to cooperate had a significant impact on this study.

Thanks to the data collected, it was possible to describe the development of the city, its history, the correlation between historical and social facts as well as the city's spatial development. A diagnosis of the condition was developed and potential directions of corrective actions were indicated. The intended target, which—in this case—was developed in the form of a project—a corrective model, utilizes the tools of Ecological Urbanism (simulation and model studies).

The solution which was developed proves that it is possible not only to prevent further suburbanization but also to repair structures (cities) already affected by this problem.

4. Conceptual Proposal

The presented planning proposal for the city of Wolfsburg was taken from an earlier thesis entitled "Coherent city of Wolfsburg" by Warzecha, B. [59]. It aims to present the objectives of Intelligent Urbanism and Ecological Urbanism and is an important element of problem-solving possibilities presented in the article. The study area (in situ research) is situated in the northern part of the city in the vicinity of the Old Wolfsburg and Volkswagen factory (Figure 1). The project of coherent Wolfsburg described below as a case study has been entered for the student ideas competition of the city of Wolfsburg "WOLFSBURG AWARD for urban vision" in 2020 [60].

A number of project problems related to the case study were formulated:

- What types of intelligent transportation can make neighborhoods resilient? How to act against traffic congestion within the city core? How to create an intelligent and environmentally friendly transport concept?
- Can the quality of life of city dwellers be improved with special regard to open and green spaces? How to create an 'edible city' for humans and city animals?
- Can biodiversity and quality of green areas within the urbanscape and their interconnectedness be improved? What kind of plants are to be used in order to create biodiversity within specific green spaces?
- What should be the functional programme of single green spaces within an urban ecological corridor? How should they interact with the surrounding urban tissue, greenbelts, and landscape?

Figure 1. City of Wolfsburg nowadays. Housing estates seem to be scattered and incoherently planned. The VW plant is the most dominant figure of the city. The red area represents the Alt-Wolfsburg and Vorfsfelde, yellow—planning site/study area. Without a scale. Own elaboration based on the digital ALKIS Stadtgrundkarte (2018) of the city of Wolfsburg.

4.1. *Historical Background of the Urban Development of Wolfsburg. The Green City of Wolfsburg*

Wolfsburg is a city that has struggled with the urban planning and development processes just like any other city. However, what makes it unique (and is rarely the case in Germany), is the fact that the cityscape has been shaped by historical conditions and many urban planning models (only) since the 20th century. The most important planning strategies to be mentioned are the following: functional garden city (1938–1944); organic and automobile city (1947–1963); and spatially separated, sprawled city (1961–1998)—the characteristics of which are described above. This has resulted in a development of a patchwork family of the city districts, a specific small-scale structure and an island-like, inconsistent living space [61] (p. 8).

The first settlement of Alt-Wolfsburg was founded in the Renaissance era and was limited to the castle and accompanying buildings, which crowned over the swamp landscape. The first urban development was organized at the newly build Mittellandkanalin the middle of the former German Reich, in the immediate vicinity of the industrial center Braunschweig-Salzgitter-Magdeburg [61] (p. 14). The railway station and the production line for Volkswagen cars with the adjacent factory settlement, were established in May 1938—at Hitler's request. This first spatial concept (Figure 2) [62] was planned by Peter Koller as a model, functional, "industrial city as a garden city" (city center and Steimker Berg district, 1938–1944) [61] (p. 9). It was characterized by a strict distinction between factory (to the north) and city (to the south) [63] (pp. 31–33). The center of the new settlement with the Klievensberg Hill was connected to the Alt-Wolfsburg Castle via the southwest–northeast axis (further described as a "Cultural and Recreational Axis"), on which the new city was spatially organized. Most of the three-storey housing units were arranged in an open and green block perimeter development, with access roads leading into it from all sides. Due to the acute shortage of materials and the lost war, the first project of the city failed and the growing settlement structure took on the name of a nearby castle—Wolfsburg [61] (p. 20).

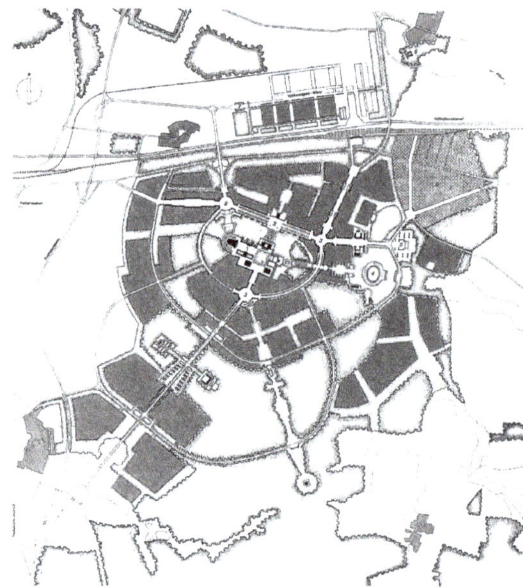

Figure 2. The city of the Volkswagen car, concept plan of an 'industrial city as a garden city' aimed at the north–south direction of the city's development, draft by Peter Koller, 1938, [62] (p. 43).

The economic boom between 1947 and 1963 enabled the city to grow again. The new expansion of a small precinct was planned by Hans Reichow according to a new axis to the west-east direction, which now covered only the southern part of the city. As depicted

(Figure 3), the already existing Garden City (in the middle with black building figures) was surrounded by new settlements, which seem to be chaotically submerged in greenery; there is no coherence between single housing estates. This new planning strategy of a "scattered and loose city", "organic and automobile city" sprawled into the juxtaposed landscape (e.g., the Rabenberg settlement) and changed the strategy of the future urban development enormously. From now on, the city was divided into functional zones, as new residential areas were planned and built as 'stepping stones' in green areas and only on the southern bank of the Mittellandkanal.

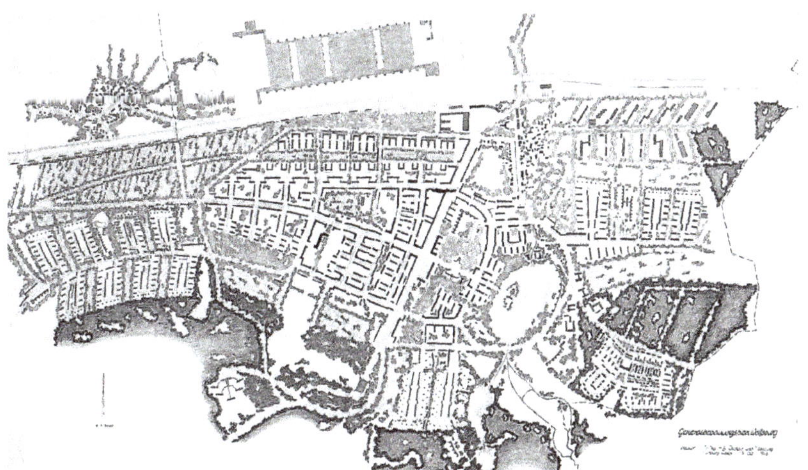

Figure 3. The organic and car-friendly city of Wolfsburg [62] (p. 41).

The most significant development of the city, which triggered the actual beginning of its sprawl character, took place between 1961 and 1998 (see also Figure 1). The further city growth intended to include a variety of building forms and social mix (but not a mixed use). This spatial and architectural experiment was designed as self-concentrated satellites, composed with no surrounding attachment. In this island-like urban structure the individual settlements were integrated into the urban landscape independently of one another over the course of time (eventually also in the northern part of the city). The patchwork-like building tissue seemed to have been created in an unplanned way and competed with the open space structures. The most significant characteristic of that design was the spatial separation of functions between "working", "living", and "leisure". An attempt was made to gradually reflect the proportional relationships between buildings and the green and traffic areas, also between the density and insolation in each district [61]. The northern part of the city, which had not been included in spatial strategies of the city for decades, has recently gained new island-like settlements. They have been organized around the historical parts of the city—the castle of Alt-Wolfsburg and the core of Vorsfelde district (further described as a "Historical Axis"). Due to the fact that this is the youngest and dynamically growing part of the city, it has an enormous potential for changing its sprawling character, starting with the study area and further on spreading on the city as a whole. Therefore, the preconceptual proposal will focus on the creation of a compact built-up area, using the tools of Intelligent and Ecological Urbanism.

Holistically speaking, Wolfsburg is nowadays divided into northern and southern parts by the Mittellandkanal, the railway line and the Aller valley which emphasize its spatial incoherence. Surrounded by a flat landscape with floodplains, filled with forests and farmlands, Wolfsburg offers a unique quality of living and working. The whole city area is as big as 20,452 ha with a huge proportion of the diverse structure of open, recreational and protected spaces (biotopes). Water features amount to over 3% of city land coverage (use); recreational areas—13.6%; forests—24%; and agricultural land—41.2%. All in all, around 60% of the urban tissue is filled now with open spaces, which makes Wolfsburg the greenest German city [64]. It is important for the residents to continue living in country-like surroundings and to be able to use the existing infrastructure. A strategic development should also address the issue of a growing residential population with a special regard to the balance between the green and built-up areas. It is essential to create such a development, which opens up to green spaces as an additional, functional part of the urban tissue. The tools of Intelligent and Ecological Urbanism can be used here as an overall sustainable strategy.

4.2. Analysis of the Current Situation

The edge of city structure (Figure 4) is easy to isolate from vast open spaces within the town's boundaries. The northern part of the city of Wolfsburg houses the Volkswagen plant and recreational areas with the floodplains of the Aller River. It also has a clear island-like structure of the built-up area. The traffic system has its hierarchy and streets are characterized by different widths. Spatially separated, individual settlements are embedded in the open space. They are linked to each other through the district's main streets, which function as traffic collectors within built-up neighborhoods. Organically formed residential streets depict the flow of the traffic system in new districts. The contemporary streets of Vorsfelde follow the old structure of field paths. They are based on a grid plan.

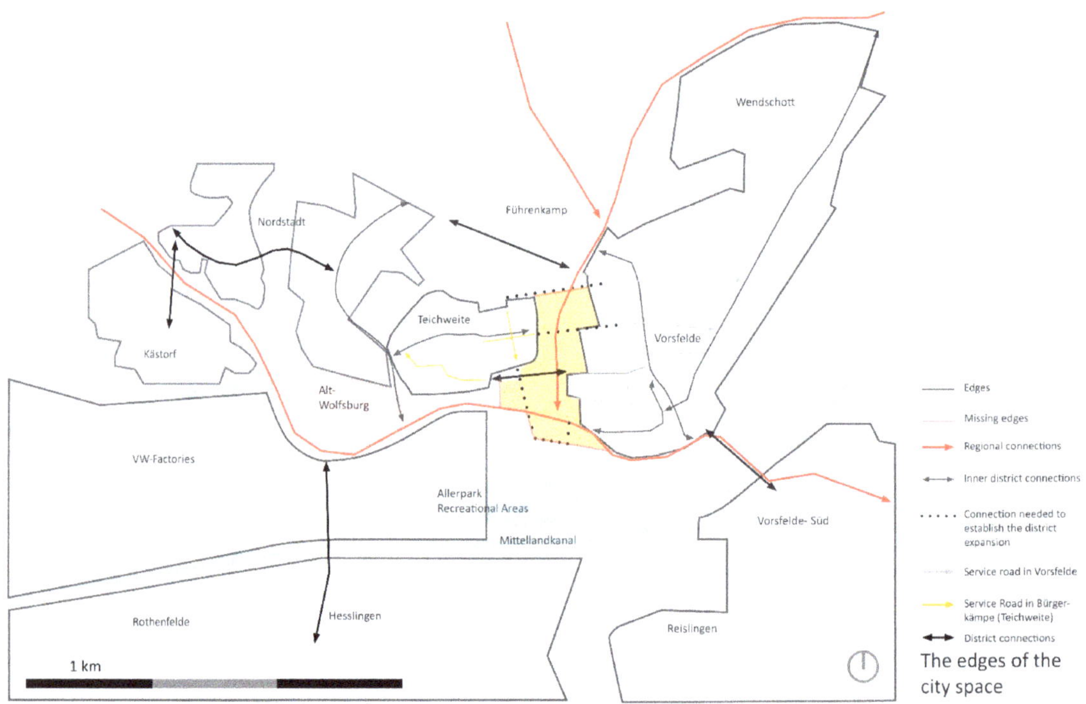

Figure 4. The edges of the city space. The scattered settlements are situated between the intercity roads. Own elaboration based on the digital ALKIS Stadtgrundkarte (2018) of the city of Wolfsburg.

A federal road, which does not impair settlement structures, runs through open spaces between districts in the north–west direction. The country road L188 runs similarly in a north–south direction and joins the federal road at the main junction.

The connection with the city center from the north is provided by a bridge east of the border of the VW plant. Another bridge connection is situated between Vorsfelde and Vorsfelde Süd and has no direct road connection with the city center. The northern part of the city appears to have neither a central point or street nor a landmark.

The city of Wolfsburg is very conveniently located on the Mittellandkanal, in the eastern part of the state of Lower Saxony, Germany. The highways A2 and A39 and the west–east railway lines run through the town. The A 39 motorway and the north–west federal highway 188 run through the city in a north-south direction. The city center and structures of Alt-Wolfsburg and Vorsfelde lie on the natural course of the Aller River. The train (InterCity Express ICE) connects the city with the capital of Germany in only one hour (up to two trains an hour), the regional train connects the entire region. The neighboring city of Braunschweig is only 20 min away. The city of Gifhorn can be reached by private transport within 30 min or by regional train in 15 min. Braunschweig-Wolfsburg Airport is located 80 min away if one travels by public transport or just 25 min away if one chooses a car.

The city of Wolfsburg is known as a commuter city due to a good network and high staffing requirements of the VW factory (76,771 commuters in 2017). The Volkswagen factory and associated facilities are classified as a special district of the town. The city is made up of a total of 13 districts.

According to Wolfsburg statistics for 2018, around 125,244 people lived in the city in 2017, 50.2% of whom were women. The city had a negative natural balance (−239) and a positive migration balance (+222). Around 61.5% of the population were in the productive age, in 2025 this will probably be 60.9%. The number of minors was around 16% in 2017 and will be 17.7% in 2025; seniors: 22% and 21.4% respectively [64].

In order to present the data for the city of Wolfsburg against the background of other cities in the state of Lower Saxony (German: Land Niedersachsen), two cities comparable in terms of total area from the closest region of Wolfsburg were taken into account. It was decided to omit such urban complexes as Bremen with Bremerhaven (a separate state) or Hanover (state capital) due to their administrative status. Hanover, although it has a city area similar to the one of Wolfsburg, has many times more inhabitants (20,430 ha, 535,061 inhabitants, 2,144,120 inhabitants in the agglomeration [65]). Even though Hanover and the Bremen complex are industrial centers, they will not be taken into account for the above-mentioned reasons in a further comparison. Another urban complex that will be excluded in this process is the city of Göttingen, which due to its function as a student and tourist city has a different development profile (it is not an industrial city).

In addition to Wolfsburg, the following cities can be, therefore, enumerated in the region: Braunschweig with an area of 19,270 ha with 250,361 inhabitants [66] and Salzgitter with an area of 22,392 ha with 107,014 inhabitants [67] (data as of 31 December 2017). Converting the area of the city to one inhabitant, we get 0.16 ha/inhabitant in Wolfsburg, 0.08 ha/inhabitant in Braunschweig, and 0.21 ha/inhabitant in Salzgitter.

The total number of green areas in Wolfsburg is 60% [64], in Braunschweig—53.2% [66], and in Salzgitter—71.5% [67] respectively. Converting the green area to one inhabitant, we get 0.10 ha/inhabitant in Wolfsburg, 0.08 ha/inhabitant in Braunschweig and 0.15 ha/inhabitant in Salzgitter.

All residents of the city declare their main residence in Wolfsburg, an additional 8901 as a secondary residence. The households consisted of an average of 1.9 people, 47.2% were only inhabited by one person. In Salzgitter and Braunschweig, this structure is similar.

The unemployment rate was 4.7% and the city had 3150 unemployed inhabitants registered. 72,844 people were employed in the manufacturing sector at the place of work, 76,771 people commuted to the city. Wolfsburg is a car-oriented city—134,756 cars were registered in 2017, which means that there are around 2 cars per household [64]. To sum up, there are more cars in this city than there are inhabitants.

The original urban axis from 1938 visually connects the Porsche Street (today a pedestrian zone) in the southern part of the city with the castle that crowns the city in the northern part. Unfortunately, this infrastructural and spatial linkage was interrupted, when the *Autostadt* (delivery center for the new cars produced at the VW plant, museum, and an amusement park) was rebuilt in the early 2000s. The pedestrian crossing to recreational areas at Allersee takes place through two existing traffic bridges upon the Mittellandkanal (one at the VW *Autostadt*, the second one in Vorsfelde) and the pedestrian bridge on the Allersee. An additional car-free connection between the new quarter and the city center as the southern part of the city would improve mobile networking (Figure 5).

Figure 5. As a typical automobile-dependent city, Wolfsburg is dominated by the vast roads, mostly intercity networks, which are very often jammed by the commuters (red area) from the Volkswagen factory. Own elaboration based on the digital ALKIS Stadtgrundkarte (2018) of the city of Wolfsburg.

The supraregional and local railway station is located directly by the VW plant, the line follows the Mittellandkanal. Moreover, the federal road runs through the northern side of the town in the north-western direction and absorbs a high level of traffic. During shift changes, it is severely affected by traffic jams at main intersections. A bus line with a network of stops runs parallelly on the main streets of the district. The bike and car routes are separated from the pedestrian traffic and embedded in the green spaces, which is typical of an automobile city.

The land use types of the northern part of the city shows a homogeneous structure of the built and unbuilt spaces. It functions mainly as a "bedroom city" with a significant proportion of residential areas. The schooling facilities with sports areas, as well as educational centers that are spread across the residential islands, are intended for the youngest residents. The main landmark in the northern urbanscape is the VW plant—the city's main area of commerce and industry, located in the western part. Other industrial areas are linked to the canal in the south. A belt of forests and dense parks, which seems to share the urban tissue with built-up areas, plays an equally important role. A single shopping street is found in the center of Vorsfelde (Figure 6).

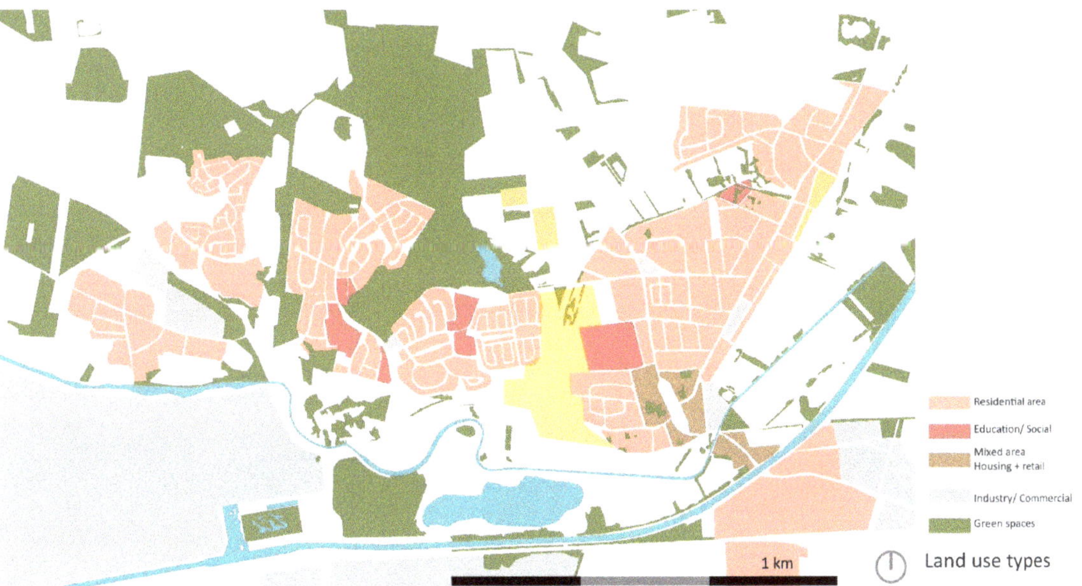

Figure 6. The settlements are situated at the urban periphery and represented mostly by residential areas submerged in the landscape. Own elaboration based on the digital ALKIS Stadtgrundkarte (2018) of the city of Wolfsburg.

The two core areas of Old Wolfsburg and Vorsfelde, which date back to the 17th century, are dominant structures of the northern part of Wolfsburg. Solitary building structures that were distributed throughout the area are represented by spacious educational facilities and sports infrastructure.

The western side of the district was created after 1950 and is typologically represented by attached and detached houses. This district is embedded in greenery and has a high proportion of open space structures (private gardens, community green). The settlement area of Vorsfelde is marked by mixed building types. The areas of single/multi-family houses, attached houses, as well as special typologies, e.g., knitted out by so-called carpet houses can be found here (Figure 7).

Figure 7. The urban area is characterized by a low density with a carpet made of blocks of flats (maximum 3–4 floors) and detached houses. The yellow area—the planning site/study area. Own elaboration based on the digital ALKIS Stadtgrundkarte (2018) of the city of Wolfsburg.

Several green and ecological corridors lead through the city. One of the most important green connections is formed by the floodplains and waters of the Aller, Allersee and Mittellandkanal rivers, which act as a distinctive link between the north and south of the city. Around 60% of Wolfsburg area consists of green and open spaces (Figure 8), and some of their main characteristics were often decisive when it comes to the location of new settlements (e.g., Klieversberg–Kliever Mountain or Waldsiedlung—"Forest settlement"—in the southwestern part of the city). Forests and floodplains play an enormous role in the urban landscape. The arable land, which is 'right in front of the door' of the residents, shapes the overall image of the area.

Figure 8. The city of Wolfsburg is rich with diverse green and open spaces and already has some existing ecological interconnectedness. Nevertheless, the north–south axis seems still to be undefined and requires a rational and programmed solution. Own elaboration based on the digital ALKIS Stadtgrundkarte (2018) of the city of Wolfsburg.

The quality of the existing urban landscape in the study area does not have a consistent sequence of rooms and clarified open space typology. Therefore, it remains unattractive for the potential users with regard to the facilities available. Currently, these areas exist as wasteland or replacement planting for the already built investments. It is essential to include these new green areas into the ecological corridor in the west–east direction. The new quarters should not completely close the development gap in the north–south axis but should offer a rich program by creating an additional connection to the southern part of the city. As a result, the affected open spaces are activated to strengthen the coherence of the city and bring both banks closer together.

The cityscape was strongly influenced by the cultural landscape (Figure 9). The historical development of the northern part still shows the old connection between Alt-Wolfsburg and Vorsfelde and results in the "Historical Axis", which is also linked to the course of the Aller. In order to create the coherence of the northern part of the city, it is necessary to bridge the development gap (design area) along this axis and to give this area a clear identity. This would result in getting a compact settlement structure in the overall picture.

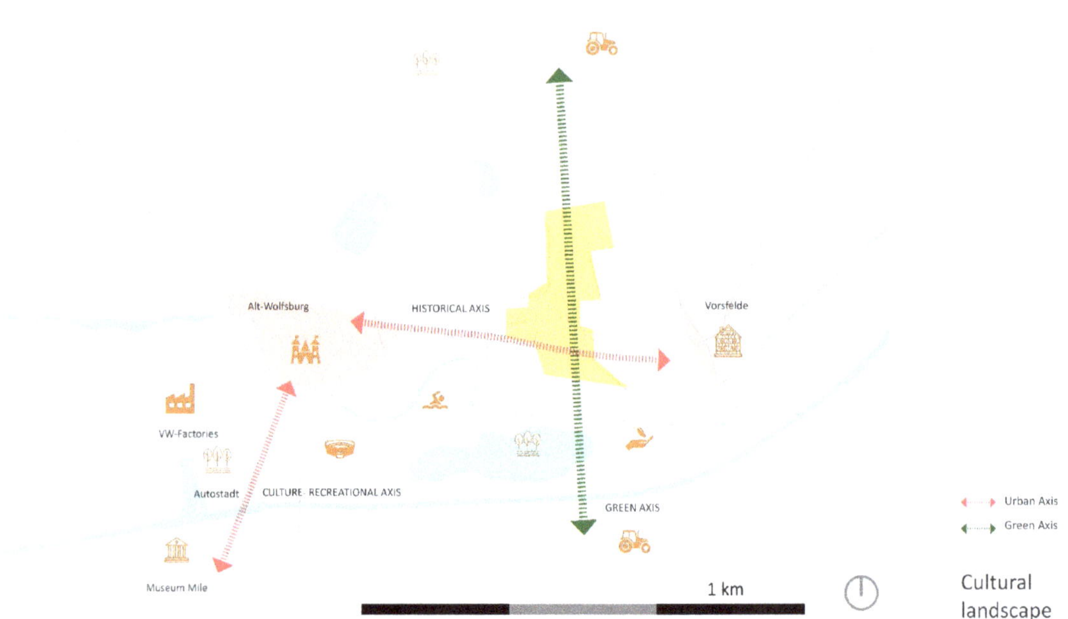

Figure 9. Cultural landscape of Wolfsburg. The "Historical Axis" connects the Alt-Wolfsburg with Vorsfelde, the "Culture-Recreational Axis" connects the Alt-Wolfsburg and Porsche Street, and the "Green Axis" is to be strengthened. Own elaboration based on the digital ALKIS Stadtgrundkarte (2018) of the city of Wolfsburg.

With the creation of the "Green Axis", an additional un-urban connection with the southern part of the city could be created, which is intended primarily for pedestrians or cyclists (no traffic connection) and should activate one of the most important open spaces in the city.

The "Culture–Recreational Axis", which was already designed in 1938, extends between the castle and Porsche Street (nowadays a pedestrian zone). Further south, it opens the "Museum and Culture Mile" with the city's landmarks: *phaeno* by Zaha Hadid, library by Alvaro Aalto or Scharoun's Theater am Klieversberg.

Furthermore, the cultural heritage should be strengthened in the north–south line, mostly by the tools of intelligent, Ecological Urbanism which allows the green infrastructure to create a sustainable solution within the urban tissue.

The federal road that runs through open spaces is currently a significant source of the noise. The development of the road space at the mouth of the country road will reduce traffic frequency and noise at the same time. The reduced speed on the city promenade will improve living conditions and minimize noise. The road permeability is additionally supported by two parking garages, which are intended to accommodate commuter traffic at the entrance to the city and support other mobile routes.

The ultimate goal should be the creation of a new, integrated urban area within the city boundaries based on long-term solutions. It ought to create sustainable open spaces as a part of an urban design with a cultural heritage background. Wolfsburg is rich with diverse green and open spaces and already has some existing ecological connections. Nevertheless, the north–south axis is still undefined and seems to require a rational and programmed solution.

5. Results

The task of this elaboration is to create coherence in the northern part of the city, to strengthen the edges and to create an uniform panorama image as well as attractive urban spaces. Strengthening of the precinct through infrastructural and green networking with innovative means of mobility and urban planning solutions supports the approach of the car-poor city, which should represent an antithesis to the previous car-friendly one.

5.1. Conceptual Guidelines

The main idea for the new district is to create coherence as an antithesis to the previous urban development (living separate from work, recreation and networking). Closing the edges of the space as a whole and emphasizing the building through a mix of typologies act as the foundation stone for coherent urban development. The functional mix: combining living with relaxation and work on a superordinate level as well as improving the infrastructural networking are guarantees for the all-round development of new quarters. Embedding in the landscape creates the backdrop and shapes the overall picture of the new district. Many forests and parks, moats, a very high proportion of agricultural land (relevant for self-sufficient food supply) and floodplains with their moat network shape the character of the district and are a contribution to Ecological Urbanism.

On the figure ground plan (Figure 10), it can be seen that the new precinct bridges a considerable gap between existing patchwork-like settlement islands and acts as an antithesis of Wolfsburg's urban development as a car-friendly and functional city. The open space plays a major role—the existing areas are reformulated and offer a mixture of buildings and valuable free spaces. Missing avenues are added in existing districts—Vorsfelde and Bürgerkämpe and further towards the castle—in order to emphasize the sequence of rooms and to guarantee the continuity of the ecological corridors. New building forms fit into the gap creating coherence and a clear settlement structure in the northern part of Wolfsburg.

Figure 10. Figure ground with the new neighborhood. Own elaboration based on the digital ALKIS Stadtgrundkarte (2018) of the city of Wolfsburg.

The overarching green corridor is reformulated with several open space types and offers a sequence of open spaces for the use of residents and guests. The area entrance is emphasized by selective high-rise buildings in the north and south and, thus, creates an opening to the landscape. The new district is orientated and linked to two differently formulated axes (Figure 11). The first, "Urban", is seen as the city promenade and is directly connected with the outside area of the daycare center in the south. The parallel "Green" leads through the open space with a wealth of leisure activities and ends in the balcony area in the south. The link with the Wolfsburg city center in the south is indicated and, at the same time, the edge of the new district on the northern side of the river Aller is marked.

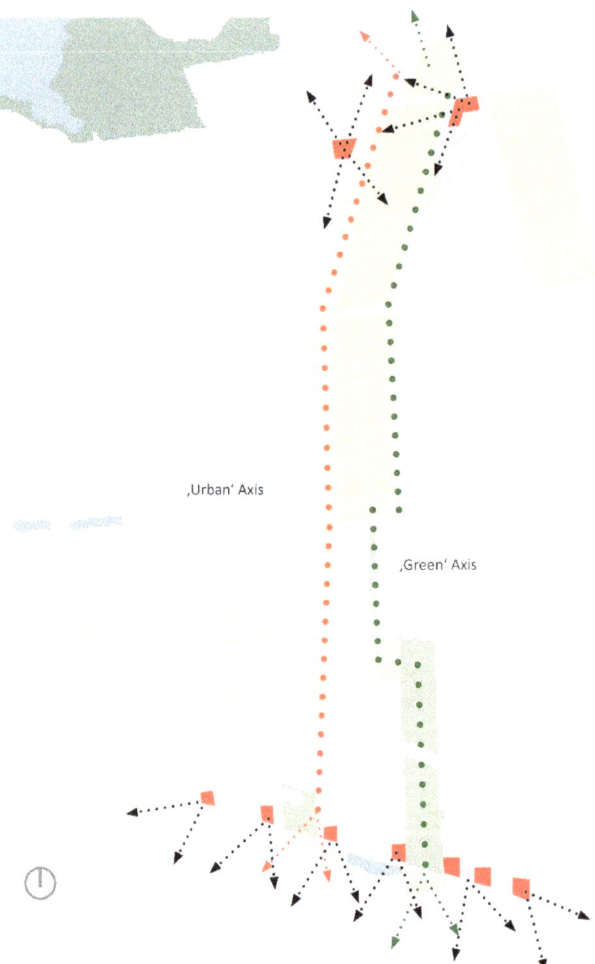

Figure 11. The two new Axes of the new neighborhood. The entrance area is emphasized by selective high-rise buildings in the north and south—opening into the landscape. The free space axes (1) the red one: "Urban Axis", as a city promenade, ends with the outside area of the day care center (2) the green one: "Green Axis", as a green corridor, as a parallel-ending with a park and a floodplain balcony. Diagram, without a scale. Own elaboration based on the digital ALKIS Stadtgrundkarte (2018) of the city of Wolfsburg.

The design of free spaces (Figure 12) is understood as a network of open spaces, where each area has a different use and function. The urban park serves as the main outdoor salon of the new district. The leisure offer is the richest here. Starting with the head-district square with multifunctional space, playrooms for children (water/playground, climbing equipment, etc.) and adults (boules, table tennis, etc.). The spacious park trees provide shade, the grassy area invites to linger. The pedestrian zone located on the eastern edge of the park gives the opportunity to spend a quiet time as well as to go cycling or roller skating. This area is well maintained and functions as the center of the district.

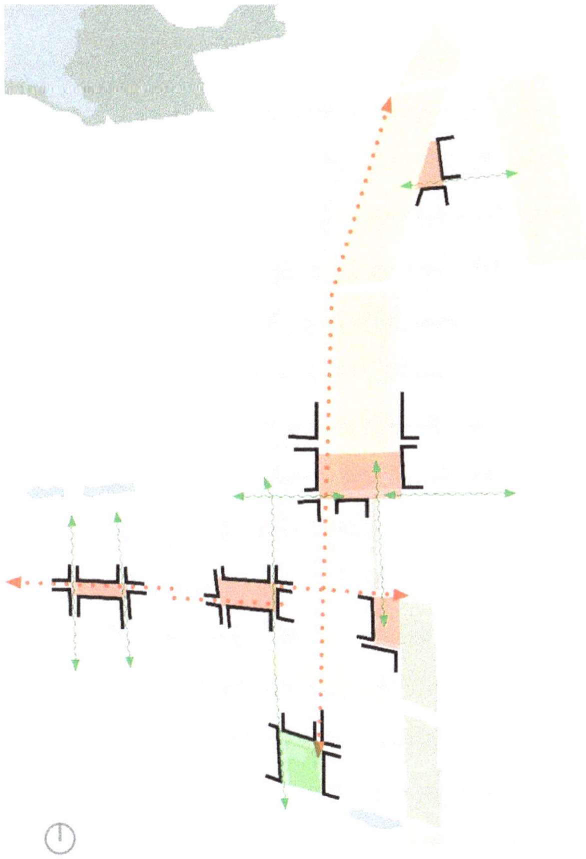

Figure 12. The concept of the new neighborhood squares is based on the networking of the neighborhood squares in the direction of N–S and W–E. A sequence of a Park-Plaza-Open space is supported by the networking of the open spaces (different types of program). Diagram, without a scale. Own elaboration based on the digital ALKIS Stadtgrundkarte (2018) of the city of Wolfsburg.

The orchard is mowed a maximum of two times a year, thus creating the best conditions for a flower meadow to grow. Fruit trees with edible fruits bloom here and provide shade on hot summer days. This community area is located directly next to the retirement home and can be maintained with the help of the residents. Everyone is invited to take part.

The square in the *Farmers District* can be made available for weekly markets with local and regional products. This strengthens togetherness and belonging to the region and the city.

The park at the *Water District* is laid on existing retention basins. The identity is emphasized by the new planting of white willows (*Salix alba* L.) and reinforced by water playgrounds for children and nature trails.

The network of the open spaces in the precinct has its accumulation with the balcony plaza in the *Field District*. This is not only a square with a small café and playground for children, but also a connection to the hiking and cycling path that goes over the Aller River and Mittellandkanal and is intended to connect the southern and northern parts of the city.

The new squares play an important role in the neighborhoods/units as meeting places and fit into the existing green spaces. Their networking in the north–south and west–east direction is supported by the sequence of different urban landscape types and strengthened by different attractions.

The edges of the new precinct are in line with neighboring open spaces (Figure 13). This highlights the densification of the district in the northern area through closed blocks of flats and spacious parking areas. The networking of the open space is accompanied and guided by the street space (city promenade and pedestrian zone also as a parking space). The *Field District* opens up to the Aller Park: towards the south with partially open inner courtyards, in the north the opening followed by the *Water District*.

Figure 13. The edges of the built-up space and relation to the landscape. A significant densification of the district in the northern area (closed blocks); interconnectedness between the green/open and the street spaces. The Field District and Water District open onto the open space to the south and north. The courtyards are partly opened. Diagram, without a scale. Own elaboration based on the digital ALKIS Stadtgrundkarte (2018) of the city of Wolfsburg.

5.2. Urban Structure and Quality of Open Spaces

The formulation of four differentiated *District* (Figure 14) with clear spatial delimitation makes up the overall picture of the precinct: in the north and south, there is a strong reference to the surrounding landscape, in the west and east existing structures are connected. It is important to interlink with very distinctive landscape structures such as agricultural fields, floodplains, as well as the belt of forests and parks, which play an important role for the residents as open and recreational areas. The superior ecological corridors are re-formulated for the greatest added value of the space. The centrally located urban park connects the areas of Vorsfelde and Bürgerkamp, which have been expanded accordingly in terms of urban planning, and connects them with a district square. A maximum density is intended to be reached on the park side and on the main street (city promenade)—this should be achieved with strong, closed and space-forming structures as well as many public and community areas. This appears to be the best location in the whole district. The prominent high points reinforce the key positions and provide orientation points for the sequence of urban spaces.

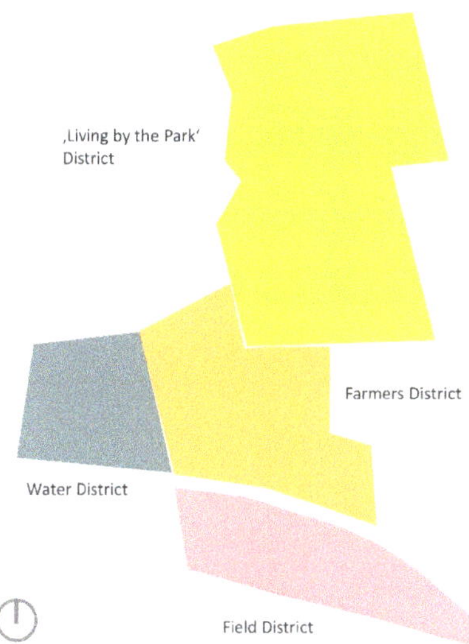

Figure 14. The New Districts of the study area, diagram, without a scale. Diagram, without a scale. Own elaboration based on the digital ALKIS Stadtgrundkarte (2018) of the city of Wolfsburg.

The southern districts offer looser and open development with a very strong relationship to the landscape and more private open spaces within the block.

In the *Water District* (Wasserquartier), residents will not only be able to enjoy the differentiated building types, but also a considerable proportion of private gardens. In the construction site, there is only one maintenance path for economic and maintenance reasons. The communal areas are located in surrounding parks and spacious quarters.

In the *Field District* (Auenquartier), residents are given the same share of private space. In addition, they have extensive floodplain landscapes as well as parking areas and playgrounds in front of their houses.

The *Living by the Park District* ("Wohnen am Park" Quartier) with its dense development offers communal areas within the block as inner courtyards. This is the place of rest, retreat and coexistence. In addition to the fairground with a high proportion of restaurants, the newly designed park also offers public play areas with a wide range of leisure activities. The new single-family home area is enriched with private gardens.

This is what defines the special character of each *District* and creates differentiated atmospheres and spatial types with correspondingly high density and building typologies that are as manifold as possible.

The ultimate goal is to develop a lively, sustainable and mixed development concept for the whole precinct (Figure 15).

Figure 15. Conceptual proposal for the study area. Own elaboration based on the digital ALKIS Stadtgrundkarte (2018) of the city of Wolfsburg.

The new precinct can be seen as a model plan for the future development of Wolfsburg. The goals of town planning concentrate on creating an anti-car city, unlike before, with the help of multifunctional means of transport and reducing traffic (dismantling a stretch of the L291 highway, residential street, residential path, and walkability). Creating an attractive place to live and work is an overarching theme. The new residents should be able to enjoy differentiated and attractive open spaces and their interconnectedness. The strengthening of self-sufficiency through the use of an "edible" city, the variety of native tree species and biological diversity answer the questions of sustainability and climate friendliness (resiliency). The concept of the walkable city of short distances is supported by an urban mix (vertical and horizontal) and intelligent urbanity.

The new intelligent and environmentally friendly transport concept (Figure 16) is intended to support the redevelopment of the precinct and represents the opportunity to focus on climate-friendly and adapted means of mobility. The goal is to create mixed mobility that also takes into account previous, traditional car models and partially car-poor areas. Alternative means of transport such as electromobility (E-Bike, and E-Car) is of particular importance for the Wolfsburg-Braunschweig region, which has been shaped by automotive history. Future issues such as energy generation, the expansion of the charging station network but also the use of electric vehicles should influence the further development of the city.

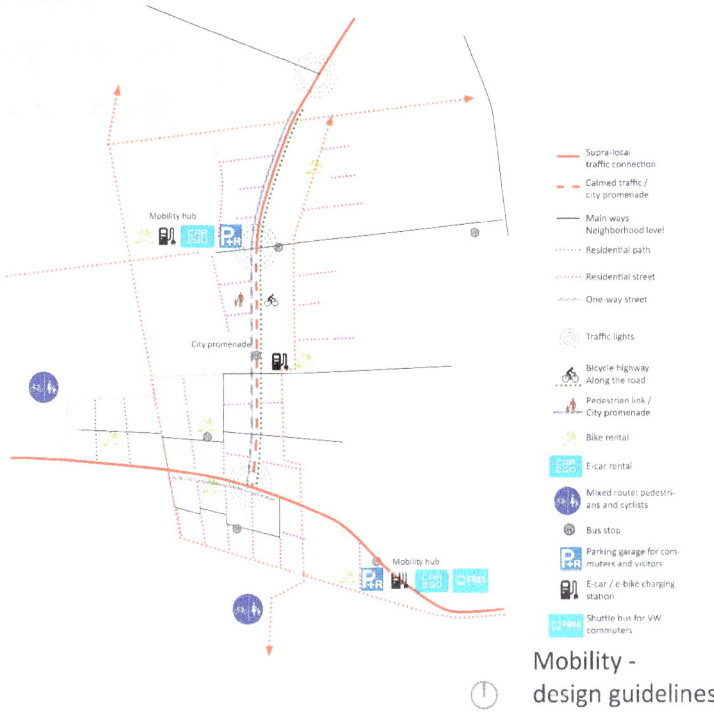

Figure 16. Mobility concept for the study area is based on the two mobility hubs located in the eastern and northern part of the study area. Their aim is to collect the commuters before entering the city area. The mobility should be strengthened with the pedestrian zones and cycling paths. Diagram, without a scale. Own elaboration based on the digital ALKIS Stadtgrundkarte (2018) of the city of Wolfsburg.

The new area should be planned as sustainably as possible. The main design idea to create the new urban landscape is based on the existing structures such as moats or oak avenues, which should be implemented in the further planning process. The planting of new districts should reflect the variety of local tree species and, thereby, support fauna and flora. Fruits can be partially edible (orchards with apples, fruit trees in the private garden). Specially selected local tree cultivars give the urban animals the opportunity to find food on site during the winter or early spring when there might be a shortage of food. Trees were selected according to what they are used for, resulting in the following categories: avenue trees (pedunculate/English oak—*Quercus robur* L., Norway maple—*Acer platanoides* L., small-leaved lime—*Tilia cordata* Mill., honey locust—*Gleditsia triacanthos* L., gray alder—*Alnus incana* (L.) Moench), trees with edible fruits for animals (serviceberry—*Amelanchier lamarckii*, rowan—*Sorbus aucuparia* L., horse chestnut—*Aesculus hippocastanum* L., locust—*Robinia pseudoacacia* L., pedunculate/English oak—*Quercus robur* L., rowan—*Sorbus aucuparia* L., small-leaved lime—*Tilia cordata* L., white willow—*Salix alba* L.), park trees (common beech—*Fagus sylvatica* L., horse chestnut—*Aesculus hippocastanum* L., locust—*Robinia pseudoacacia* L., small-leaved lime—*Tilia cordata* Mill.). The representative city promenade and other main avenues, also in the existing neighborhoods, are accentuated by pedunculate/English oak-*Quercus robur* L., which is very characteristic in this area. The *Water District* is characterized by the use of moisture-loving trees (white willow—*Salix alba* L., gray alder—*Alnus incana* (L.) *Moench*). Honey locust—*Gleditsia triacanthos* L. will be characteristic in the *Field District*. The heavily blooming rowan—*Sorbus aucuparia* L. and bird cherry—*Prunus avium* L. sprinkle the orchards and courtyards in the *Farmers District*; here the small-leaved lime—*Tilia cordata* L. exudes a strong aroma and provides food for the bees. Park trees characterize the *Living by the Park District*—the small-leaved lime dominates here as an avenue tree, other tree species such as common beech, horse chestnut and robinia reinforce the Urban Park. Juneberry and rowan berries bloom in the inner courtyards. The edible city is home for the young and the old.

Intelligent urban design integrates cohesively existing settlements and bonds the northern and the southern parts of the city. The sustainable, ecological interconnectedness has been strengthened through the mixed vertical and horizontal use, landmarks, diversified program of the green and recreational spaces, and alleys along the traffic routes. Biodiversity is emphasized with distinctive biologically active green spaces and a variety of species (trees and animals).

6. Discussion

The Principles of Intelligent Urbanism and Ecological Urbanism emphasize different aspects of sustainable urban development. They also agree on some points and set similar goals to be achieved. They both contribute to the prevention of urban sprawl and creation of a compact city. The Principles were taken into account into initial phase of the project (identification of design problems) and implemented in the conceptual proposal for the city of Wolfsburg.

The solutions to the problems of the overcrowded and polluted city of the end of the 19th century described by a number of authors as the construction of towns or settlements-gardens were also implemented in Wolfsburg at the turn of the 1930s and 1940s. It can be evidenced by the dispersed development of the central district or Steimker Berg. In the urban structure of Wolfsburg there is also a clear division into residential and industrial-production parts. It is rooted in the 1960s concept of an organic-scattered city. These three historical factors described by Schubert, Angelidou, Chmielewski, Lynch, Howard, Mumford, Nechyba, Black, Osborn, and others, contributed to the sprawl and inconsistency of today's city. It can be concluded that the period particularly responsible for this phenomenon has its origin in the second half of the 20th century, which seems to be parallel with the trends of European cities. What Wolfsburg can be distinguished for is the scale of this dispersion.

The proposal from Mostafavi of basing the ecological urban planning on biodiversity and introducing the idea of urban gardening and 'edible city' was considered possible and justified to implement in the design concept. Various tree species were selected based on the microclimate conditions and their utility for humans and the city animals. This strengthens biodiversity on the micro and macro-levels, creating an 'edible city'. It shows that the Principles of Ecological Urbanism mentioned in this study have been successfully implemented in the conceptual proposal. The infrastructural elements (blue infrastructure) like water retention on site and delayed water drainage are crucial and must be mentioned. The Principles of Intelligent Urbanism such as balance with nature, regional integration and vision have been, therefore, also accomplished.

Similarly, the conditioning of the sustainable growth of the urban areas was recognized as one of the elements of the individual cultural and landscape values along with the respect for the tradition of the place [26,31–33]. It was considered as the basic element indicating the direction of the development of the new spatial order. This principle was implemented by highlighting Wolfsburg's historical axes or the urban scale of the newly designed neighborhoods.

Intelligent Urbanism focuses on the sustainable use of the balanced movement: well-defined streets, provision of cycle paths as a contribution to public transport, hybrid transportation, and implementation of electromobility. The width of the new streets was hierarchized and provided the opportunity for walkability and sharing space. The imperfection of using e-mobility is based primarily on the introduction of the appropriate infrastructure (charging stations, subsidies for the solar panels, e-car, and e-bikes), which can basically be implemented only in the richest countries. This Principle can be, therefore, considered to be non-universal.

The main social goal was to develop a sense of community among residents. It was possible thanks to the application of Principles from both systems. In this matter, amenities and social facilities were located centrally in the new neighborhood unit or alongside main traffic arteries. Moreover, there are such elements as a vibrant building structure, private gardens, and access to green and open spaces. One can find areas for recreation and social interactions. This approach is in harmony with the existing conditions of the green urbanscape. Moreover, the central Allerpark has become a connecting element between both sides of the city. Its eastern zones have been, thanks to this design, activated and they offer new amenities. The biggest structural infrastructure is based on the interconnectedness of the system plaza-park, which, on a city level, helps sustain the north–south and west–east ecological corridors.

Ecological and Intelligent Urbanism share the idea that the local government should participate in the recovery process and coordinate it with a top-down approach. Alternatively, it can be mentioned that, at the design stage, such an assumption could have been made.

The city of Wolfsburg was lively interested in creating a new neighborhood for the commuters of the VW plant. However, it is difficult to discuss the social role of this project and whether there was a reaction of the local community to such changes. It would require the implementation of this case study and, above all, the further involvement of the local government and residents through social consultations. This conceptual proposal is a kind of an opportunity, a first step towards further development strategy. Any actions undertaken by the city would take into account the future residents. For objective reasons, the strengths of this study can be indicated as follows: it is possible that this proposal could be used in future works on city development strategy, so that it could be considered for the implementation as a platform for the future talks between local authorities, planners, and architects with the city residents about the further shape and functions of the city of Wolfsburg. The results and conclusions should correct the presented plan, thanks to which there is a chance for its full or partial implementation. The case study of Wolfsburg was not ordered for implementation purposes.

The goal of developing a repair model of Ecological and Intelligent Urbanism has been, at this planning stage, achieved. It offers a rational and sustainable development plan for urban areas.

7. Conclusions

The aim of this study was to create coherence for the northern part of the city and to suggest a new direction in the urban development for the city of Wolfsburg. The previous patchwork family of the settlement islands competed with landscape structures—the new quarter bridges the urban gap, strengthens the building structure of the city as a whole and creates clear urban edges. The northern part of the city was redefined and connected to the city center in the south through the spacious central park with Allersee and Allerpark. The new neighborhood is sustainably planned and can be integrated into the existing structures with the help of the infrastructure (streets and avenues) and open spaces (parks, orchards, playgrounds, plazas, and tree alleys) found in the transitional area. The new quarters will be sprinkled with local squares and communal areas and strengthen the local community. New residents and users will be given multifunctional mobility options as well as transport possibilities. The designed urban concept enables also the implementation of innovative means of electromobility. They are operated as environmentally friendly as possible thanks to the energy that is produced on site (solar panels, and wind energy). The green areas are planned to be inclusive and rich in programs to create a vibrant urban life and interaction with the landscape. Additional green spaces on roofs emphasize the relationship to open space and social life.

The new, integrated urban area is nothing like the existing suburban situation. This proposal rejects the idea of scattered, undefined urban and landscaping structures. The new infill with a denser architecture and green infrastructure offers a mixed use of open spaces and building forms. This coherent and meticulously planned urban development ensures intelligent growth and implements the tools of ecological urbanism as an offer for sustainable landscape architecture.

Author Contributions: Conceptualization, J.D.-K. and B.W.; methodology, J.D.-K.; validation, J.D.-K. and B.W.; formal analysis, J.D.-K. and B.W.; investigation,. J.D.-K. and B.W.; resources, B.W.; writing—original draft preparation, J.D.-K. and B.W.; writing—review and editing J.D.-K. and B.W.; visualization, B.W.; supervision, J.D.-K.; project administration, J.D.-K. and B.W. All authors have read and agreed to the published version of the manuscript.

Funding: This research received no external funding.

Institutional Review Board Statement: Not applicable.

Informed Consent Statement: Not applicable.

Data Availability Statement: Not applicable.

Acknowledgments: To the head Professor Heinz Nagler from the Chair of Urban Design at the Brandenburg University of Technology Cottbus-Senftenberg in Germany, for the inspiration for this article.

Conflicts of Interest: The authors declare no conflict of interest.

References

1. Schubert, D. Cities and plans—The past defines the future. *Plan. Perspect.* **2019**, *34*, 3–23. [CrossRef]
2. Angelidou, M. Smart cities: A conjuncture of four forces. *Cities* **2015**, *47*, 95–106. [CrossRef]
3. Basiri, M.; Zeynali Azim, A.; Farrokhi, M. Smart City Solution for Sustainable Urban Development. *Eur. J. Sustain. Dev.* **2017**, *6*, 71–84. [CrossRef]
4. Hall, P. *Cities of Tomorrow: An Intellectual History of Urban Planning and Design in the Twentieth Century*, 3rd ed.; Wiley-Blackwell: Hoboken, NJ, USA, 2002.
5. Chmielewski, J.M. *Teoria Urbanistyki w Projektowaniu i Planowaniu Miast;* Oficyna Wydawnicza Politechniki Warszawskiej: Warszawa, Poland, 2010.

6. Lynch, K. *A Theory of Good City Form*; MIT Press: Cambridge, MA, USA, 1981.
7. Mumford, L. The neighborhood and the neighborhood unit. *Town Plan. Rev.* **1954**, *24*, 256–270. Available online: https://www.jstor.org/stable/40101548 (accessed on 27 December 2020). [CrossRef]
8. Cretan, R. Who owns the name? Fandom, social inequalities and the contested renaming of a football club in Timisoara, Romania. *Urban Geogr.* **2019**, *40*, 805–825. [CrossRef]
9. Howard, E. *Garden Cities of To-Morrow*; S.Sonnenschein & Co., Ltd.: London, UK, 1902.
10. Garden Cities of Tomorrow. Available online: http://urbanplanning.library.cornell.edu/DOCS/howard.htm (accessed on 27 December 2020).
11. Komninos, N.; Pallot, M.; Schaffers, H. Open Innovation towards Smarter Cities. In *Open Innovation 2013*; European Commission, Directorate-General for Communications Networks Content and Technology: Luxembourg, 2013; pp. 34–41. [CrossRef]
12. Navigant Research. Smart Cities; Intelligent Information and Communications Technology Infrastructure in the Government, Buildings, Transport, and Utility Domains (Research Report). 2011. Available online: http://www.navigantresearch.com/research/smartcities (accessed on 9 February 2012).
13. Schaffers, H. Empowering Citizens to Realizing Smart Cities. Results from FIREBALL Smart City Case Studies (presentation). In Proceedings of the 2012 Future Internet Assembly (FIA), Aalborg, Denmark, 10–11 May 2012. Available online: http://www.fiaalborg.eu/downloads/Session_1.1_Hans_Schaffers.pdf (accessed on 15 December 2020).
14. Wolfram, M. Deconstructing Smart Cities: An Intertextual Reading of Concepts and Practices for Integrated Urban and ICT Development. In Proceedings of the REAL CORP 2012, Schwechat, Austria, 14–16 May 2012. Available online: http://www.corp.at/archive/CORP2012_192.pdf (accessed on 15 December 2020).
15. Litwińska, E. Modelling of metropolitan structure in aspect of urban sprawl. *Tech. Transl. Archit.* **2010**, *1-A*, 139–148. Available online: https://repozytorium.biblos.pk.edu.pl/redo/resources/32875/file/suwFiles/LitwinskaE_ModelowanieStruktur.pdf (accessed on 15 December 2020).
16. Słodczyk, J. Wielkość miast i ich struktura przestrzenna w świetle kryteriów rozwoju zrównoważonego. In *Przemiany Bazy Ekonomicznej i Bazy Przestrzennej Miast*; Wydawnictwo Uniwersytetu Opolskiego: Opole, Poland, 2002; pp. 323–331.
17. Zuziak, Z. Ecological definition of urban sciences. *Czas. Tech. Wydaw. Politech. Krak.* **2007**, *7-A*, 9–20. Available online: https://repozytorium.biblos.pk.edu.pl/redo/resources/34951/file/suwFiles/ZuziakZ_EkologiczneDefiniowanie.pdf (accessed on 15 December 2020).
18. Nechyba, T.J.; Walsh, R.P. Urban sprawl. *J. Econ. Perspect.* **2004**, *18*, 177–200. [CrossRef]
19. Black, J.T. The Economics of Sprawl. *Urban Land* **1996**, *55*, 52–53.
20. Osborn, F.J. Preface. In *Garden Cities of To-Morrow*; Faber and Faber: London, UK, 1946; p. 15.
21. Ewing, R. Is Los Angeles-Style sprawl desirable? *J. Am. Plan. Assoc.* **2007**, *63*, 107–126. [CrossRef]
22. Couch, C.; Leontidou, L.; Arnstberg, K.O. The origins of suburbia and urban sprawl in Europe and the USA. In *Urban Sprawl in Europe: Landscape, Land-Use Change and Policy*; Couch, C., Leontidou, L., Petschel-Held, G., Eds.; Blackwell Publishing Ltd.: Oxford, UK, 2007; pp. 6–14.
23. Mostafavi, M.; Doherty, G. *Ecological Urbanism*; Harvard University Graduate School of Design: Cambridge, MA, USA; Lars Müller Publishers: Baden, Switzerland, 2009.
24. Szulczewska, B. *Zielona Infrastruktura—Czy Koniec Historii?* Studia KPZK: Warsaw, Poland, 2019; Volume 189. Available online: http://scholar.googleusercontent.com/scholar?q=cache:Yk0_4ZKW-EQJ:scholar.google.com/+szulczewska+barbara&hl=pl&as_sdt=0,5 (accessed on 27 March 2021).
25. United Nations. Available online: https://www.un.org/development/desa/en/news/population/2018-revision-of-world-urbanization-prospects.html (accessed on 12 February 2021).
26. Benninger, C. Principles of Intelligent Urbanism: The case. *Ekistics* **2002**, *69*, 60–80. Available online: https://about.jstor.org/terms (accessed on 4 January 2021).
27. O'Grady, M.; O'Hare, G. How Smart Is Your City? *Science* **2012**, *335*, 1581–1582. [CrossRef] [PubMed]
28. Caragliu, A.; del Boro, C.; Nijkamp, P. Smart Cities in Europe. In Proceedings of the 3rd Central European Conference in Regional Science–CERS, Košice, Slovacka, 7–9 October 2009; pp. 45–59. Available online: https://inta-aivn.org/images/cc/Urbanism/background%20documents/01_03_Nijkamp.pdf (accessed on 22 March 2021).
29. Sikora-Fernandez, D. Smart city jako nowa koncepcja funkcjonowania i rozwoju miast w Polsce. *Res. Pap. Wrocław Univ. Econ.* **2014**, *339*, 175–181. [CrossRef]
30. Ibanescu, B.-C.; Banica, A.; Eva, M.; Cehan, A. The puzzling concept of smart city in central and eastern Europe. *Lit. Rev. Des. Policy Dev. Transylv. Rev. Adm. Sci.* **2020**, *61E*, 70–87. [CrossRef]
31. Serag el Din, H.; Shalaby, A.; Farouh, H.E.; Elariane, S.A. Principles of urban quality of life for a neighborhood. *HBRC J.* **2013**, *9*, 86–92. [CrossRef]
32. Principles of Intelligent Urbanism. Available online: https://missurasa.wordpress.com/2013/05/09/principles-of-intelligent-urbanism/ (accessed on 27 December 2020).
33. Bugadze, N. Theory and Practice of "Intelligent Urbanism". *Bull. Georgian Natl. Acad. Sci.* **2018**, *12*, 145–151. Available online: https://www.researchgate.net/publication/329908058_Theory_and_practice_of_intelligent_urbanism (accessed on 29 December 2020).
34. Rytelewska-Chilczuk, N. Rola architektury miejskiej w budowaniu tożsamości mieszkańców miast. *Ogrody Nauk Sztuk* **2018**, *8*, 224–233. [CrossRef]

35. Nikitas, A.; Michalakopoulou, K.; Njoya, E.T.; Karampatzakis, D. Artificial Intelligence, Transport and the Smart City: Definitions and Dimensions of a New Mobility Era. *Sustainability* **2020**, *12*, 2789. [CrossRef]
36. Santoso, S.; Kuehn, A. Intelligent Urbanism: Convivial Living in Smart Cities. In Proceedings of the iConference 2013, Fort Worth, TX, USA, 12–15 February 2013; pp. 566–570. [CrossRef]
37. Iacomoni, A. European green capitals, Best practices for sustainable urban development. *AGATHÓN Int. J. Archit. Art Des.* **2019**, *6*, 114–125. [CrossRef]
38. Jopek, D. Intelligent urban space as a factor in the development of smart cities. *Tech. Transl. Archit. Urban.* **2019**, *9*, 5–15. [CrossRef]
39. European Green Capitals Call for Global Action. Available online: ec.europa.eu/environment/efe/news/european-green-capitals-call-global-action-2019-01-24_en (accessed on 30 September 2019).
40. Caragliu, A.; Del Bo, C.; Nijkamp, P. Smart Cities in Europe. *J. Urban Technol.* **2011**, *18*, 65–82. [CrossRef]
41. Ghezzi, P.; Daole, F.; Ottaviani, G. *Pisa—Piantare Alberi per Mettere Radici—Città Resilienti, Infrastrutture Verdi*; Paysage: Milan, Italy, 2017.
42. Wahlstrom, M.H.; Kourtit, K.; Nijkamp, P. Planning Cities4People–A body and soul analysis of urban neighbourhoods. *Public Manag. Rev.* **2020**, *22*, 687–700. [CrossRef]
43. Kourtit, K.; Nijkamp, P.; Wahlstrom, M.H. How to make cities the home of people—A 'soul and body' analysis of urban attractiveness. *Land Use Policy* **2020**, *104734*, 1–9. [CrossRef]
44. Kourtit, K. Cultural heritage, smart cities and digital data analytics. *East. J. Eur. Stud.* **2019**, *10*, 151–159. Available online: https://ideas.repec.org/a/jes/journl/y2019v10p151-159.html (accessed on 12 April 2021).
45. Batabyal, A.A.; Nijkamp, P. Interregional Competition for Mobile Creative Capital with and Without Physical Capital Mobility. *Int. Reg. Sci. Rev.* **2021**. [CrossRef]
46. Andreucci, M.B. *Progettare Green Infrastructure—Tecnologie, Valori e Strumenti per la Resilienza Urbana*; Wolters Kluver Italia: Milan, Italy, 2017; Available online: http://hdl.handle.net/11573/1265073 (accessed on 22 March 2021).
47. Zuziak, Z. An introduction into the new philosophy of urbanism. *Urbanity Archit. Files* **2020**, *48*, 181–206. [CrossRef]
48. Malovics, G.; Cretan, R.; Berki, B.M.; Toth, J. Urban Roma, segregation and place attachment in Szeged, Hungary. *Royal Geogr. Soc. Area* **2019**, *51*, 72–83. [CrossRef]
49. Dernbach, J.C. Achieving sustainable development: The Centrality and multiple facets of integrated decision making. *Indiana J. Glob. Legal Stud.* **2003**, *10*, 247–285. Available online: https://www.repository.law.indiana.edu/ijgls/vol10/iss1/10 (accessed on 22 March 2021). [CrossRef]
50. Turner, J.; Holmes, L.; Hodgson, F.C. Intelligent Urban Development: An Introduction to a Participatory Approach. *Urban Stud.* **2000**, *37*, 1723–1734. [CrossRef]
51. Nassar, U. Principles of Green Urbanism: The Absent Value in Cairo, Egypt. *Int. J. Soc. Sci. Humanit.* **2013**, *3*, 339–343. [CrossRef]
52. Vesalon, L.; Cretan, R. "Little Vienna" or "European Avant-Garde city"? Branding narratives in a romanian city. *J. Urban Reg. Anal.* **2019**, *11*, 19–34. [CrossRef]
53. Stoddart, H.; Schneeberger, K.; Dodds, F.; Shaw, A.; Bottero, M.; Cornforth, J.; White, R. A Pocket Guide to Sustainable Development Governance. Stakeholder Forum. 2011. Available online: https://www.circleofblue.org/wp-content/uploads/2012/07/PocketGuidetoSDGEdition2webfinal%E2%80%93Stakeholder-Forum.pdf (accessed on 15 December 2020).
54. Commissione Europea, Relazione della Commissione al Parlamento Europeo, al Consiglio, al Comitato Economico e Sociale e al Comitato delle Regioni—Riesame dei Progressi Compiuti NELL'ATTUAZIONE della Strategia dell'UE per le Infrastrutture Verdi, 184 Final. 2019. Available online: ec.europa.eu/transparency/regdoc/rep/1/2019/IT/COM-2019-236-F1-IT-MAIN-PART-1.PDF (accessed on 22 October 2019).
55. Caragliu, A.; del Bo, C.F. Do Smart Cities Invest in Smarter Policies? Learning From the Past, Planning for the Future. *Soc. Sci. Comput. Rev.* **2015**, *34*, 657–672. [CrossRef]
56. Grant, J. Theory and Practice in Planning the Suburbs: Challenges to Implementing New Urbanism, Smart Growth, and Sustainability Principles. *Plan. Theory Pract.* **2009**, *10*, 11–33. [CrossRef]
57. Cappochin, G. The sense of a research project: Superurbano, EcoDistricts, European Green Capitals/EcoCities. In *European Green Capitals—Experiences of Sustainable Urban Regeneration*; Cappochin, G., Botti, M., Furlan, G., Lironi, S., Eds.; LetteraVentidue: Siracusa, Italy, 2017; pp. 7–10.
58. Deakin, M. From intelligent to smart cities. In *Smart Cities. Governing, Modelling and Analysing the Transition*, 1st ed.; Deakin, M., Ed.; Routledge: London, UK, 2013; pp. 15–32. Available online: https://www.taylorfrancis.com/books/edit/10.4324/9780203076224/smart-cities-mark-deakin?refId=b7692393-30a7-493b-b012-df4b388d4d12 (accessed on 13 April 2021).
59. Warzecha, B. Kohärente Stadt Wolfsburg. Master's Thesis, Brandenburg University of Technology Cottbus-Senftenberg, Cottbus, Germany, 18 September 2019.
60. Wolfsburg Award for Urban Vision_2020. Wolfsburg's Student Concept Competition on the Theme "Wolfsburg: City | Space | History". Available online: https://www.wolfsburg.de/wolfsburgaward (accessed on 15 April 2021).
61. Reichold, O. … *Erleben, wie eine Stadt Entsteht. Städtebau, Architektur und Wohnen in Wolfsburg 1938–1998*; Begleitband zur Ausstellung vom 26. Mai bis 28. August 1998 in der Bürgerhalle des Wolfsburger Rathauses; Meyer: Braunschweig, Germany, 1998.
62. Beier, R. *Aufbau West—Aufbau Ost. Die Planstädte Wolfsburg und Eisenhüttenstadt in der Nachkriegszeit*; Buch zur Ausstellung des Deutschen Historischen Museums vom 16. Mai bis 12. August 1997; Ostfildern-Ruit: Hatje, Germany, 1997.

63. Schneider, C. *Stadtgründungen im Dritten Reich: Wolfsburg und Salzgitter. Ideologie, Ressortpolitik, Repräsentation*; Moos: München, Germany, 1979.
64. Statisches Informationssystem Wolfsburg 2018. Available online: http://rs000455.fastrootserver.de/WolfsburgInteraktiv/JSP/main.jsp?mode=Detailansicht&area=Stadt&id=A&detailView=true, (accessed on 17 March 2019).
65. Statistische Berichte Niedersachsen, Landesamt für Statistik Niedersachsen, Bevölkerungsdichte der Kreisfreien Städte und Landkreise—Stand 31.12.2017. Available online: https://www.statistik.niedersachsen.de/startseite/themen/bevolkerung/themenbereich-bevoelkerung-statistische-berichte-172949.html (accessed on 19 February 2021).
66. Braunschweig in der Statistik. Available online: http://www.braunschweig.de/politik_verwaltung/statistik/jahrbuch.php (accessed on 19 February 2021).
67. Statistisches Jahrbuch 2017, Salzgitter. Available online: https://www.salzgitter.de/rathaus/downloads/jahrbuecher/Jahrbuch_2017.pdf (accessed on 19 February 2021).

Article

City of Waste—Importance of Scale

Bogusław Wowrzeczka

Department of Architecture and Visual Arts, Faculty of Architecture, Wroclaw University of Science and Technology, 50-317 Wroclaw, Poland; boguslaw.wowrzeczka@pwr.edu.pl

Abstract: By 2050, the world population is expected to reach 9.7 billion, almost 90% of which will live in urban areas. With such a fast growth in population and urbanization, it is anticipated that the annual waste generation will increase by 70% in comparison with current levels, and will reach 3.40 billion tons in 2050. A key question regarding the sustainability of the planet is the effect of city size on waste production. Are larger cities more efficient at generating waste than smaller cities? Do larger cities show economies of scale over waste? This article examines the allometric relationship between the amount of municipal waste (total and per capita) and the populations, city area, density, and wealth of city residents. The scope of the research concerned 930 Polish cities. Using the allometric equation, the waste scaling factors were calculated for selected parameters, and the Hellwig method was used to optimize their selection for cities with more than 50,000 inhabitants. The calculations show that the parameter population (1.059) and then the city area (0.934) are important elements influencing the scaling of the amount of municipal waste in cities of all sizes, but none came close to the value of the animal metabolism model (0.75). In response to the question of whether larger cities show benefits from economies of scale, it should be stated that, for the model of city size in Poland, such a regularity does not exist.

Keywords: municipal waste; city; scaling; population; population density; city area; GDP

Citation: Wowrzeczka, B. City of Waste—Importance of Scale. *Sustainability* **2021**, *13*, 3909. https://doi.org/10.3390/su13073909

Academic Editor: Åsa Gren

Received: 7 March 2021
Accepted: 29 March 2021
Published: 1 April 2021

1. Introduction

Urbanization is the hallmark of the 21st century, which is characterized by tremendous demographic changes and a rapid development of urban areas and the built environment on a large scale. Most of the future population growth in the remaining part of this century will occur in urban areas. The increase in global waste production due to population growth and wealth will have a significant impact on the sustainable development of cities.

The world produces 2.01 billion tons of municipal solid waste each year, with at least 33 percent not being managed in a safe way for the environment. All over the world, the amount of waste generated per person per day averages 0.74 kg but varies widely from 0.11 to 4.54 kg. Although they constitute only 16 percent of the world's population, high-income countries (high-income countries—78 countries with GDP above 12,000 $/year) generate about 34 percent, or 683 million tons, of the world's waste [1,2].

Considering the fact that urban populations will have increased by 2–3 billion by the end of the 21st century, understanding the way in which the size of cities affects the municipal waste volume can provide us with an insight into how city size can be part of a larger regional or national strategy for waste reduction [3]

1.1. *The Importance of Scale for the Production of Municipal Waste*

Galileo developed the idea of allometric growth in his treatise 'Discorsi e demonstrationi matematiche, intorno a due nuove scienze', which was published during his house arrest in 1638 [4]. He noticed that the bones of larger animals grew thicker at a faster rate than they grew in length compared to the same bones in smaller animals. Thus, the height-to-circumference ratio decreases along with the animals' growth.

Therefore, what is 'scaling'? In its most elementary form, it simply refers to the reaction of the system when its sizes change [5].

Scaling characterizes the way a given system quantity, y, depends on the size of the system. The scaling law is shown in the form of the following exponentiation relation:

$$y = ax^b \tag{1}$$

where x is the linear size of the system and y is its measure, whereas a is the proportionality coefficient, and b is the exponent specifying the exponentiation law.

The scaling laws apply to both natural phenomena and those resulting from human activity [5].

In particular, the scaling laws refer to models of spatial organization of cities and their growth—a well-known example of a scaling relation is 'Zipf's Law', which states that a city's population decreases inversely with its rank among other cities in the same city system [6] (Zipf's Law or Estoup–Zipf's Law—the law that describes a frequency principle of using individual words in any language. Zipf's law was mathematically expressed in Zipf's equation: r × f = constans, where: r is the rank of a word in a text or a group of texts, and f is its frequency of occurrence. [6]).

Cities offer benefits resulting from the economy of scale. Concentrations of people, large-scale infrastructure, and economic activities enable innovation and efficiency. Recent studies have shown that cities may exhibit different types of scaling in different urban phenomena or properties [7]. Nonlinear scaling (when exponents take a value less than 1) resembles the parallel allometric scaling laws observed in living organisms, and represents the benefits of the scale resulting from the increase in efficiency by sharing infrastructure; it is exposed, inter alia, in electric networks (by the length of electric cables) and road systems (length of roads or amount of road surfaces). Superlinear scaling (when the exponent b is greater than 1) seems to be unique to social systems, and is connected to the concept of network effects which lead to human ingenuity and creativity. Superlinear scaling has been identified in the number of new patents, inventors, research and development, employment, total salaries, etc. Linear scaling (when the exponent b is approximately equal to 1) means a proportional increase in urban phenomena/measures along with the size [8].

The size of a city's population, as well as its spatial organization and structure, can influence the amount of waste. Data from cities around the world suggest that climate, technology, density, and wealth are important determinants of waste generation.

The subject of the research is to establish allometric dependencies between the size of a city and the production of municipal waste in 930 Polish cities. The results show that this dependency varies across cities of different size, area, population density, and per capita income. In analogy to Kleiber's law [9] (Kleiber's Law, named after Max Kleiber because of his biology in the early 1930s, is based on the observation that, in most animals, the metabolic rate increases to $\frac{3}{4}$ of the strength of the animal's weight [9]), the amount of municipal waste, along with the increase in the city's population, should decrease due to the benefits resulting from the use of the understood service and network infrastructure of cities, which, in many cases, obeys the law of allometric growth. Are larger cities more economical in terms of waste production than smaller cities? Moreover, it is important to determine the importance of the city's basic spatial and economic indicators, i.e., area, population density, or GDP per capita for municipal waste produced. The knowledge of these relations can be fundamental to the optimization of the size of waste collection and processing facilities in cities.

1.2. Universal Quantifiable Features of Cities

Parysek [10] claims that, since the formulation by Ludwig von Bertalanffy [11] of the general theory of systems, it has been increasingly used in determining the subject of research in various fields of knowledge. The systemic approach to the subject comes from biology, where systems are living organisms. He further states that, by analyzing the spatial and functional structure of the city, we can conclude that the organism is an adequate model for the city system.

In a city, as in any living organism, there is a conversion of matter and energy. It is a specific form of metabolism, consisting not only of the consumption of various forms of energy of materials, but also of capital flow, knowledge, skills, information, etc. This form of metabolism can be identified with urban metabolism and, similarly to living organisms, can be studied using the scaling law [12]. Organisms, as metabolic engines, are characterized by indicators of energy consumption, growth rate, body size, and lifetimes, and therefore have a clear reference to urban systems [13,14]. Bettencourt states that cities manifest remarkably universal, quantifiable features. As to be expected, "size is the major determinant of most characteristics of a city; history, geography and design have secondary roles" [8,15,16].

The basic discoveries of allometric relationships describe the relationship between the total area of a city and its number of inhabitants [17–19] and the relationship between the city's area and the total length of its borders (fractal nature) [20,21]. Other relations concern the relationship between the city's surface and the total surface of its roads [22].

Kennedy [23] and others have quantified the energy and material flows across 27 megacities worldwide with a population of over 10 million people since 2010. It was confirmed that the flows of resources and waste generation across megacities largely follow the laws of scaling.

Few studies have investigated the scaling performance of solid waste disposal through statistical analysis with empirical data. Pan, Yu, and Yang [24] tested a sample of 651 cities in China using a correlation analysis and grouping model that determined the characteristics and overall trends of solid waste generation in five city groups of varying scales between 2007 and 2016.

Kleiber's law of allometry says that the metabolic rate is based on body weight with an exponent of 0.75. It should be assumed that the elimination of waste from the body is a system proportional to the metabolic process, that is: the more metabolic waste is produced, the more metabolic waste must be excreted [25]. Similarly, in an urban "organism", the amount of waste generated should be scaled to the number of inhabitants (or other city parameters) with an exponent of 0.75.

1.3. Municipal Waste Problem

Waste collected by or on behalf of municipalities includes household waste originating from households (i.e., waste generated by the domestic activity of households) and similar waste from small commercial activities, office buildings, institutions such as schools and government buildings, and small businesses that treat or dispose of waste at the same facilities used for municipally collected waste [26]

The amount of generated municipal waste depends on many factors, of which the most important are life standard, population rate, and goods consumption scale and intensity.

Municipal solid waste is remarkably diverse in terms of physical and chemical composition. It mainly depends on the equipment of buildings with technical and sanitary devices (mainly the heating method), the type of buildings, and the living standards of the inhabitants. Most often, municipal waste contains approx. 40–50% of organic substances, approx. 50–60% are mineral parts. Waste composition in the OECD region (The Organization for Economic Co-operation and Development (OECD) is an international organization that works to build better policies for better lives. It counts 37 member countries that span the globe, from North and South America to Europe and the Asia-Pacific.) is slightly different due to high income, and for organic waste it is below 30% [27].

The components contained in municipal waste, mainly organic, undergo biochemical changes, and affect the environment through decomposition products: carbon dioxide, ammonia, hydrogen sulfide, methane, nitrates, nitrites, sulphates, and others.

Municipal waste he poses a threat to the environment due to the possibility of the contamination of air, groundwater, and surface water with pathogenic microorganisms, for which it is a medium [28]. Solid muncipal waste varies greatly in terms of physical and chemical composition.

Based on the volume of waste generated, its composition, and how it is managed, it is estimated that 1.6 billion tons of carbon dioxide (CO_2) equivalent greenhouse gas emissions were generated from solid waste treatment and disposal in 2016, driven primarily by open dumping and disposal in landfills without a landfill gas capture system. This is about 5 percent of global emissions. Solid waste-related emissions are anticipated to increase to 2.6 billion tons of CO_2 equivalent per year by 2050 if no improvements are made in the sector [2].

In the European Union (EU), the amount of municipal waste generated per person in 2018 amounted to 492 kg, 5% less compared with its peak of 518 kg per person in 2008. In total, 220 million tons of municipal waste were generated in the EU in 2018, and this was slightly higher than in 2017 (218 million tons). With 766 kg, Denmark generated the most municipal waste per person among the EU Member States in 2018. At the other end of the scale, Romania generated 272 kg of municipal waste per person in 2018 and Poland generated 329 kg per person [29]. There was a perceptible change in trends in municipal waste management in the EU-28, with an apparent shift from disposal methods to prevention and recycling. Less waste is being landfilled because of reductions in the generation of some wastes and increases in recycling and energy recovery. Municipal waste landfilled decreased by almost 43% [30], but still corresponded to about 3% of total EU greenhouse gas emissions [31].

The correlation analysis of the population in the individual EU countries and the total volume of municipal waste and volume of waste per capita and GDP per capita in 2018 indicates that, in the first case, the Pearson coefficient (Rp) is 0.98, and 0.60 in the second (28 countries with the UK). The coefficients of determination (Rp^2) are 0.97 and 0.52, respectively, which proves that the correlation model in the first case is very good and works 97%, and in the second case it is worse and works 52%.

In the case of Poland, a detailed analysis of the relationship of GDP per capita to the volume of waste per capita indicates that Poland produces much less municipal waste in comparison with other EU countries with a similar GDP per capita.

The question about the reasons for such a phenomenon, in view of the insufficiently developed waste management infrastructure in Poland, remains unanswered and requires further research.

In 2018, the European Commission (CE) presented the amended content of the Waste Framework Directive, setting new targets for increasing the reuse and recycling of municipal waste to a minimum of 65% by 2035. [32].

In the face of the growing amount of municipal waste generated in European countries in recent years, it has become especially important to search for sustainable methods of municipal waste management. CE activities in waste management are recommended based on the ReSOLVE framework (regeneration, sharing, optimization, loop, virtualization, and replacement) [33] (The ReSOLVE framework was developed by the Ellen MacArthur Foundation and McKinsey, which are important bodies in the development of tools supporting the transformation process towards a circular economy in the EU. The circular economy has been described as a concept that mimics living systems. A helpful reframing, consistent with a circular economy approach, is to think of cities as living systems that rely on a healthy circulation of resources [33]).

It should be noted that in the European Union legislation on new urban waste management programs, there is no direct reference to urban planning issues, for example, regarding elements of city infrastructure, city area, or population density.

2. Materials and Methods

2.1. Database

The analyses used data on the volume of municipal waste, population, area, population density, and GDP of inhabitants from 930 cities and municipalities at the city level in Poland from 2018, ordered from the Central Statistical Office (Data from the Central

Statistical Office on 03/09/2019: GUS-DK02.601.1481.2019—municipal waste [34].) and obtained from the 2018 statistical yearbook of the Republic of Poland [34]. A city in Poland is defined as a settlement unit with a predominance of compact development and nonagricultural functions, having town privileges or the status of a city granted in accordance with the regulations [35] (the act of 29 August 2003 on the official names of settlements and physiographic objects (Journal of Laws of 2019, item 1443). Pursuant to the Act on Municipal Self-Government of 1990, the Council of Ministers decides by means of an ordinance on granting or abolishing the status of a city [35]). According to the act, for a settlement to receive town privileges, over 2000 people must be registered in it, it must have urban buildings (not farm buildings), and at least 2/3 of the inhabitants cannot be employed in agriculture. Not all cities in Poland meet this definition. Among 930 cities having over 2000 inhabitants, there are 863 cities, and below 2000, there are 67 cities. Both the total number of cities and, according to the definition, cities with more than 2000 inhabitants were examined. Among the 863 cities with over 2000 inhabitants, 4 groups of cities with the following number of inhabitants were specified (Table 1):

- small cities: from 2000 to 20,000 inhabitants—645 cities
- medium cities: from 20,000 to 50,000 inhabitants—134 cities
- big cities: from 50,000 to 100,000 inhabitants—46 cities
- large cities: over 100,000 inhabitants—38 cities

Table 1. The amount of waste and the number of inhabitants for the groups of cities in percentages.

Size of the City by Number of Inhabitants	Number of Cities	Amount of Waste T	% Share of Municipal Waste in a Group of Cities	Population for Categories of Cities	% Share of the Number of Inhabitants in a Group of Cities	Average Amount of Waste for Cities kg/inhab.
Large >100,000	38	4,356,281	49.33%	10,705,482	46.40%	389.0
Big from 50,000 to 100,000	46	1,120,595	12.69%	3,116,448	13.51%	357.0
Medium from 20,000 to 50,000	134	1,594,406	18.05%	4,246,564	18.41%	374.0
Small from 2000 to 20,000	645	1,723,621	19.52%	4,894,832	21.21%	342.0
Very small <2000	67	36,093	0.41%	108,648	0.47%	335.0
∑ values	930	8,830,996	100%	23,071,974	100%	349.0

In the research of the impact of GDP on the production of municipal waste, due to the scope of available data from the Central Statistical Office, the analysis was limited to the cities with more than 50,000 inhabitants.

In 38 large cities with over 100,000 inhabitants, the total amount of municipal waste accounted for about 50 percent of the waste from all cities in the research period (2018), and for 84 cities with over 50,000 inhabitants, it accounted for over 62 percent. This indicates that most of the municipal solid waste in Poland is produced by large- and medium-sized cities.

Total municipal waste is generated in large cities, although not directly proportional to the number of inhabitants—there is about 6% more of it than expected in relation to the number of inhabitants. In the remaining groups of cities, it is quite the opposite, namely, these differences are insignificant in favor of less waste by percentage in relation to the percentage of inhabitants. The data analysis shows a significant deviation in the average volume of waste per capita in the group of cities with 50,000 to 100,000 inhabitants.

2.2. Method

The subject of the analyses focused on the interrelationships between the amount of waste and the number of inhabitants, population density, city area, and GDP per capita. The exponentiation equation in the form of $y = ax^b$ (1) was applied. For exponentiation regression curves, a transformation to a linear model was performed, the exponentiation regression curve is computed according to equation $y = ax$ ^ b, which is converted to

$ln(y) = ln(a) + bln(x)$. For each case, the coefficient of determination, otherwise known as the coefficient of specificity, or R-squared, was calculated, which is a measure of what percentage of the variability of the dependent variable is explained by the independent variable.

In the analyzed problem of the urban scaling of waste, a cross-sectional analysis was used: "it produces exponents and residuals with greater temporal stability and the quality of fitting for scale-invariant relations is consistently better" [36].

Hellwig's method was chosen (for a set of cities >50,000) to determine a subset of predictors that are independent of each other, but strongly correlated with the dependent variable (section Discussion) [37,38].

All calculations were performed using Excel 2019.

3. Results of the Analysis

The results are presented in Table 2. The results were divided into groups depending on the size of the cities, but the first group applies to all cities. The exception here is the dependency between the number of inhabitants and GDP/capita due to the limited availability of data—this dependency applies only to cities with over 50,000 inhabitants.

Table 2. Waste scaling calculation results (by the author).

No.	Description of Dependencies	Exponent of Scaling b	Exponentiation Equation $y = ax^b$	Coefficient of Determination R^2
		All Cities (930 Cities)		
1.	Number of inhabitants/total waste	1.057	$y = 0.1978x^{1.057}$	0.935
2.	Number of inhabitants/waste per capita	0.061	$y = 191.18x^{0.061}$	0.051
3.	City area/total waste	0.934	$y = 3.5089x^{0.934}$	0.454
4.	City area/waste per capita	0.049	$y = 233.029x^{0.049}$	0.021
5.	Population density/total waste	0.924	$y = 7.5866x^{0.924}$	0.368
6.	Population density/waste per capita	0.058	$y = 227.19x^{0.058}$	0.025
		Cities >50,000 Inhabitants (84 Cities)		
7.	GDP per capita */total waste	1.527	$y = 0.0033x^{1.527}$	0.383
8.	GDP per capita */waste per capita	0.164	$y = 63.715x^{0.164}$	0.124
		Cities Over 2000 Inhabitants (863 Cities)		
9.	Number of inhabitants/total waste	1.055	$y = 0.2022x^{1.055}$	0.987
10.	City area/total waste	0.929	$y = 3.9643x^{0.929}$	0.475
11.	Population density/total waste	0.905	$y = 9.2308x^{0.905}$	0.341
		Cities From 2000 To 20,000 Inhabitants (645 Cities)		
12.	Number of inhabitants/total waste	1.069	$y = 0.1785x^{1.069}$	0.822
13.	City area/total waste	0.375	$y = 145.12x^{0.375}$	0.126
14.	Population density/total waste	0.464	$y = 119.53x^{0.464}$	0.192
		Cities From 20,000 To 50,000 Inhabitants (134 Cities)		
15.	Number of inhabitants/total waste	1.197	$y = 0.0469x^{1.197}$	0.475
16.	City area/total waste	0.209	$y = 2161.8x^{0.209}$	0.068
17.	Population density/total waste	0.057	$y = 7319.9x^{0.057}$	0.004
		Cities From 50,000 To 100,000 Inhabitants (46 Cities)		
18.	Number of inhabitants/total waste	1.227	$y = 0.0285x^{1.227}$	0.683
19.	Number of inhabitants/waste per capita	0.227	$y = 28.392x^{0.227}$	0.069
20.	City area/total waste	0.198	$y = 4484.4x^{0.198}$	0.195
21.	Population density/total waste	−0.107	$y = 51.759x^{-0.107}$	0.046
22.	GDP per capita/total waste	0.211	$y = 2.5174x^{0.211}$	0.054
		Cities >100,000 Inhabitants (38 Cities)		
22.	Number of inhabitants/total waste	1.059	$y = 0.1867x^{1.059}$	0.959
23.	Number of inhabitants/waste per capita	0.061	$y = 181.6x^{0.061}$	0.073
24.	City area/total waste	1.079	$y = 2.9992x^{1.079}$	0.707
25.	Population density/total waste	0.842	$y = 161.08x^{0.842}$	0.211
26.	GDP per capita/total waste	1.424	$y = 0.015.7x^{1.424}$	0.445

Source: the author. * GDP per capita data apply to cities > 50,000 inhabitants (no data for other cities).

3.1. Characteristics of the Dependency between the Size of Waste and the City Population

The dependency between the amount of municipal waste and the constant population for all cities is on the superlinear level, with a scaling factor of 1.06 and a determination factor of 0.935 (Table 2). This means that an increase in the number of inhabitants by one unit causes an increase in the amount of waste by an exponent of 1.06 in relation to the previous value, i.e., that the increase in waste in cities is faster than the increase in its population. Analyzing the dependence in the four groups of cities, it can be concluded that the scaling factor increases for cities with 50,000 to 100,000 inhabitants, where it is 1.23, then decreases to a level of 1.06 for cities with over 100,000 inhabitants. The lack of more cities with sizes exceeding 1 million inhabitants makes it impossible to study the problem of the further impact of increasing population on the amount of waste in Poland.

However, the scaling factor is different for waste per capita—the scaling factor is 0.06, with an exceptionally low level of determination of 0.051. This shows that there is no correlation between these values, while the adopted model is not sufficiently useful for determining dependencies. In this case, the values are spread on both sides of an almost horizontal trend line, which means that an increase in the city's population does not result in an increase in the amount of waste per capita. This means that the amount of waste per capita fluctuates around an average value, independent of the city's population size. The indicators in particular groups of cities are similar, where the highest scaling factor is for the group of cities with 50,000 to 100,000 inhabitants, and amounts to 0.23, with low determination amounting to 0.2 (Table 2).

3.2. Characteristics of the Relation between the City Area and the Size of Waste

Empirical analyses of the scaling relations for many urban systems suggest that there are quantitatively coherent agglomeration effects in different sizes of cities in many urban systems. Bettencourt and Lobo [19] carried out research on the properties of systems of European cities in terms of agglomeration and scaling effects for the systems, which showed, in most cases, a distribution in accordance with Zipf's law. For example, the analysis of the dependency between the number of inhabitants and the area of an urbanized area of cities shows that, for a city system, the European average scaling exponent is subline and amounts to 0.93. The developed city systems in Germany, France, Italy, Great Britain, and Spain were researched [19].

In the case of the dependency between the area of cities and the amount of municipal waste analyzed, here, we obtain the result of sublinear scaling with an exponentiation exponent of 0.93, with a determination coefficient of 0.45 (Table 2). This indicates a limited impact of city area increase on waste growth, at least lower than the previously presented result of the allometric dependency of the population and waste volume (which is 1.1). A more detailed analysis of this problem for various groups of city sizes proves the differentiation of scaling in a very wide range, from 0.2 in the group of cities from 50,000 to 100,000 inhabitants up to 1.1 for cities with over 100,000 inhabitants. In the first case, it means that there is a significant mismatch of the dependency between the city area and the amount of waste [15] (Systems reach final equilibrium when the allometric scaling exponent is close to 1. Otherwise, the system loses balance [15]).

The analysis of the allometric correlation of the city area and the amount of waste per capita (Table 2) leads to the conclusion that such a correlation is very weak (the scaling exponent is 0.05), with practically almost zero determination value (0.02). This correlation is even lower than the previously discussed correlation between the population and waste per capita, which indicates that there is no impact of urbanized area increase on the amount of waste per capita.

3.3. Characteristics of the Dependency between the City's Population Density and the Amount of Waste

Table 2 shows that the dependency between the amount of waste and the population density is sublinear (the exponent is 0.92), which proves that the weight of waste grows slower than proportionally to the increase in population density, revealing the benefits of the scale (Table 2).

The analysis of correlation in groups of cities shows a slight negative dependency of the amount of waste on the density in the group of cities with 50,000 to 100,000 inhabitants, with a scaling factor value of −0.107, but with an exceptionally low determination coefficient of 0.05 (Table 2). This shows a certain tendency in the correlation of the population density with the amount of total waste, namely, an increase in density means a decrease in waste mass.

In the group of cities with over 100,000 inhabitants, the scaling factor for the correlation of population density and waste mass is 0.842, which demonstrates the benefits of reducing the amount of waste in the case of population density increase in large cities (Table 2).

3.4. Characteristics of the Dependency between GDP Per Capita and the Amount of Waste

The analysis was based on data from 84 cities with over 50,000 inhabitants. For such limited data resources, the analysis showed a good correlation between the amount of GDP per capita and the total amount of waste generated in cities. The scaling factor is superlinear, and amounts to 1.53, which means that it is the highest of all the dependencies analyzed. In this model, the coefficient of determination (0.38) explains the waste amount variable at 38% (Table 2).

The scaling factor for the correlation between GDP per capita and the weight of waste per capita is 0.166 for a 12% determination, and indicates an insignificant impact of the GDP level on the generation of waste per capita.

3.5. Results Verification

To verify the obtained scaling results, an analysis of the significance of all variables was carried out in relation to the total amount of waste generated in cities with more than 50,000 inhabitants. The Hellwig [37,38] method was used to choose optimal predictors in the analysis of municipal waste production as the results of the influence of four variables (population, city area, population density, and GDP/capita) in 84 cities for which all data were available (>50,000 inhabitants).

As a result of the implementation of Hellwig's algorithm, a list of predictor combinations in descending order of the integral index of information capacity H was obtained. Results of 15 combinations of the values of this index are given in Table 3.

Table 3. Integral index of information capacity calculated using the Hellwig method.

Integral Index of Information Capacity (H)	Combination of Predictors	Combination of Predictors
0.987	C1	population (optimal value of a single variable)
0.937	C11	population + city area + population density (optimal value of many variables)
0.937	C5	population + city area
0.911	C15	population + city area + population density + GDP/capita
0.907	C13	population + city area + GDP/capita
0.885	C8	area + population density
0.884	C7	population + GDP/capita
0.832	C14	city area + population density + GDP/capita
0.820	C12	population + population density + GDP/capita
0.817	C6	population + population density
0.783	C9	city area + GDP/capita
0.767	C2	city area
0.539	C4	GDP/capita
0.520	C10	density + GDP/capita
0.088	C3	population density (lowest value)

It was identified that one-element combination C1—city population, provided the highest classification accuracy in the combination set. The best combination of more than one element turned out to be C11, i.e., a combination of three predictors: city population, city area, and population density. The combination of C5, two predictors: population and city area, had the same value of the integral index of information capacity as the previous C11.

The lowest value of the integral index was calculated for the one-element C3 combination, i.e., population density.

The result of the search for the optimal predictor confirms the previous findings from the scaling analysis about the dominant role of the population in generating municipal waste.

3.6. Analysis of Municipal Waste and Population Fluctuation over the Time

Bettencourt [38] described a method for analyzing urban scaling over changing time. He pointed to the problems appearing in the time analysis, which consist of the fact that "temporal exponents are sensitive to the intensive growth and circumstances when population growth is vanishes, leading to instabilities and infinite divergences". Bettencourt claims that this effect does not occur in cross-sectional scaling.

Time analysis requires a large amount of data, which is missing for the analyzed problem of scaling municipal waste in Polish cities.

The Polish Central Statistical Office (GUS) has complete data on the weight of waste in individual cities only for 2017, 2018, and 2019. Therefore, the comparative analysis of the waste scaling exponent in cities applies only to the period of these three years.

The waste scaling exponent calculated for 2017 in relation to the population was 1.072, and its determination was 0.935. In 2019, the waste scaling exponent was 1.057, and the determination was 0.947.

Comparing both exponents to the exponent of 1.057 tested for 2018, it can be concluded that it is identical to 2019, and slightly lower than in 2017. Additionally, the coefficient of determination for 2018, amounting to 0.935, is similar to the one obtained for 2017 and 2019. It can therefore be concluded that there is a coincidence of the scaling and determination exponents (Table 4).

Table 4. Waste scaling exponent results by year.

Year	Description of Dependencies: All Cities (930–946 cities)	Exponent of Scaling b	Exponentiation Equation $y = ax^b$	Coefficient of Determination R^2
2017	Number of inhabitants/total waste	1.072	$y = 0.1643x^{1.072}$	0.947
2018	Number of inhabitants/total waste	1.057	$y = 0.1978x^{1.057}$	0.935
2019	Number of inhabitants/total waste	1.057	$y = 0.2057x^{1.057}$	0.935

Source: the author.

Due to the lack of data on the amount of waste in cities over a longer period of time, the annual data available from 2005 was used to analyze changes in the amount of waste in subsequent years, and was correlated with the size of the population in cities in the corresponding years (Figure 1).

The Pearson linear correlation plot (Figure 1) is almost horizontal (0.055), so there is no correlation between the population and the amount of waste over the past 15 years (2005–2019). Slight fluctuations (increases and decreases) in the population of cities had practically no effect on the amount of municipal waste generated. The lack of population growth inhibits the growth of cities, and results in slight increases in the amount of waste caused by the increase in the living standard of the inhabitants and the demand for consumer goods (higher GDP) and the enlargement of the city area for the needs of low-intensity development. Based on the time analysis, we can conclude that the dynamics of changes in the population and in the production of municipal waste was small and consisted of frequent changes—increase and decrease—in individual values. On this basis, we can assume that the cross-sectional method on which the analyzes are based is correct, and its results indicate a constant trend in waste scaling independent of time.

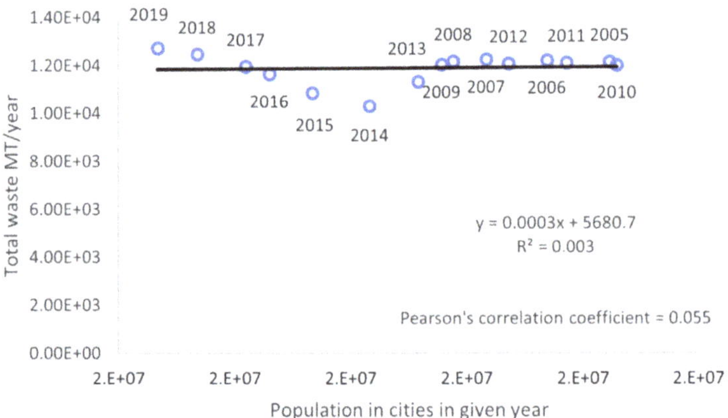

Figure 1. Correlation between municipal solid waste per year and the total population of cities in a given year (period of time 2005–2019).

Slight increases in the amount of municipal waste in the last dozen or so years in Poland go hand in hand with the better management of this waste. The amount of waste deposited in landfills decreased from 61% in 2010 to 43% in 2019. At the same time, recycling increased, which in 2019 was 26%, compared to 14% in 2010. The greatest increase was recorded in thermal waste treatment, with the use of waste for the production of electricity and heat, incineration increased from 1% to 23% (There are 9 thermal waste treatment installations operating in Poland (2021), producing electricity and heat from waste, 4 more are under construction and 10 new incinerators are planned.) The biological disposal of waste in composting plants has significantly decreased in recent years from 16% in 2016 to 9% in 2019.

4. Discussion

The hypothesis assumed at the outset that, along with the increase of the population, the amount of generated waste will decrease; however, this was not confirmed in Polish cities. The specificity of Polish cities consists of the fact that there are not enough big cities for this hypothesis to be either rejected or confirmed.

The choice of a region of the world to explore major cities does not matter, because cities manifest remarkably universal, quantifiable features: size is the major determinant of most characteristics of a city [14].

To check this additional hypothesis, large cities located in Central and South America, Asia, and the Middle East with populations of over 100,000 were researched [28]. The results of scaling the population of 508 cities in Central and South America, Asia, and the Middle East and municipal waste per day for each city are presented in the scatter plot below. The plot shows that in the sample of large cities with over 100,000 inhabitants, the scaling of the amount of waste occurs on a superlinear level. The calculations indicate that, along with the increase of the population, the amount of waste increases by 114% per unit of increase in the population, with the verifiability of such a model amounting to almost 89% (Figure 2). This shows that in large cities, the amount of waste is growing much faster than the population.

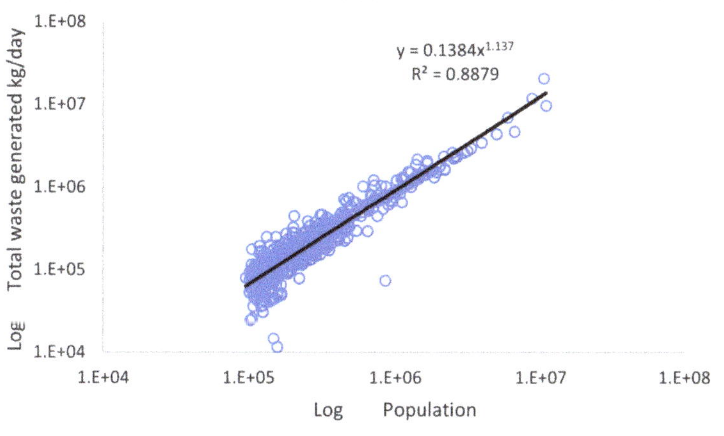

Figure 2. Scatter plot of exponentiation regression for 508 large cities located in Central and South America, Asia, and the Middle East with a population over 100,000, shows a superlinear waste growth in cities along with the population increase. The scaling exponent is 1.137 and the determination coefficient is 0.888.

Research conducted in China on a sample of 651 cities, most of which are cities over 500,000 inhabitants (491 cities), indicated that, similarly to Poland, waste is generated mainly in the group of very large cities [24]. Chinese research specified that the most influential factor in generating waste in megacities (>5 million inhabitants) is GDP per capita, similar to in Poland for the group of the largest cities (>100,000 inhabitants). In China, the city area is a common vital component that has an impact on the generation of waste in cities of all sizes. On the other hand, studies conducted for cities in Poland indicate that it is the population, and city area as second, are most influenced on the municipal waste generation (for all cities).

5. Conclusions

The main aim of the research was to define a model of municipal waste scaling depending on the population of Polish cities.

The most important conclusion about city waste scaling is that a systematic data analysis of 930 Polish cities has shown that a simple allometric scaling with a scaling exponent of 0.75 is not an appropriate method to predict "excreted" municipal waste.

For the basic model, the mean exponent of waste scaling in relation to the population shows superlinear values for all cities, and ranges from 1.06 to 1.07 (in years: 2017, 2018, 2019), with the highest value of 1.23 in the group of cities from 50 up to 100 thousand inhabitants ($R^2 = 0.683$).

In Poland, due to the tendency of population decrease in cities, demographic changes will not have a significant impact on changes in the amount of waste generated. However, the area of the city will be more and more important, followed by GDP (especially in the largest cities). Limiting the surface area of cities (compact cities), limiting consumption, and striving for a recirculation economy are the basic elements of the waste management strategy in cities in the near future in Poland.

5.1. Detailed Conclusions

The analysis of the scaling results of waste in individual groups of cities allows for the formulation of specific conclusions:

- the scaling exponent in the model with the city's population variable as the predictive one is the most stable in all groups of cities ($R^2 > 50\%$),

- for the population density of cities with more than 100,000 inhabitants, the indicator was the closest to the allometric exponent ($b = 0.84$),
- in the next model of the correlation of the city area with the total waste mass, the scaling exponent is sublinear ($b = 0.93$), and the coefficient of determination is almost 50%, and, at the same time, has the greatest impact on the amount of waste in large cities with more than 100,000 inhabitants ($b = 1.08$).

In spatial planning, it is crucial to understand whether this relationship applies to cities in countries at different stages of development, and whether the results are similar in other countries. The analysis of scaling municipal waste carried out for large cities in selected countries of South and Central America, Asia [24], and the Middle East confirms its superlinear character in a cross-sectional study ($b = 1.14$).

5.2. Limitations and Further Research

The limitations of the study are mainly due to the limited access to data: detailed data on municipal waste (in all cities) concerning only the last few years, the lack of complete information on GDP for small towns, and no data on the number of permanent city users (tourists, students, etc.). These data have not been analyzed. In addition, several other parameters and indicators that can affect the amount of waste in cities have not been studied, i.e., urban road networks that affect access to services, green spaces per capita that affect biomass waste, commercial and service space per capita that affect consumption, the sale of goods over the internet, and more.

The methodology and results of this study open the way for future research. First, studies should be carried out on representative collections of cities in other countries that confirm or deny the similarity of waste production to the Kleiber metabolism model. Additionally, in selected countries with representative types of urban metabolism: energy-consuming or material-consuming, the analysis of scaling the value of municipal waste and population as well as other independent variables identified as predictive attributes of municipal waste generation, i.e., energy-consuming or material-consuming, may answer the question of which type of metabolism is more beneficial due to the reduction of waste production [39].

Funding: This research received no external funding.

Institutional Review Board Statement: Not applicable.

Informed Consent Statement: Not applicable.

Data Availability Statement: Data Availability Statement: Data available in a publicly accessible repository (https://stat.gov.pl/) (accessed on 20 May 2019). The Figure 2 (scatter plot) was calculated on data available in [Hoornweg, D.; Bhada-Tata, P. What a Waste: A Global Review of Solid Waste Man-agement; Urban Development Series; knowledge papers no. 15; World Bank: Washington, DC. USA, 2012; Available online: https://openknowledge.worldbank.org/handle/10986/17388 License: CC BY 3.0 IGO] (accessed on 25 September 2020).

Conflicts of Interest: The authors declare no conflict of interest.

References

1. United Nations, Department of Economic and Social Affairs, Population Division (2019). World Population Prospects 2019: Highlights (ST/ESA/SER.A/423). Available online: https://population.un.org/wpp/Publications/Files/WPP2019_Highlights.pdf (accessed on 7 October 2020).
2. Kaza, S.; Yao, L.; Bhada-Tata, P.; Van Woerden, F. *What a Waste 2.0: A Global Snapshot of Solid Waste Management to 2050*; Urban Development; The World Bank: Washington, DC, USA, 2018; Available online: https://openknowledge.worldbank.org/handle/10986/30317 (accessed on 9 October 2019).
3. United Nations Environment Programme (UNEP). Global Waste Management Outlook United Nations. 2016. Available online: https://books.google.pl/books?id=uibGDwAAQBAJ (accessed on 18 June 2020).

4. Galilei, G. *Rozmowy i Dowodzenia Matematyczne w Zakresie Dwóch Nowych Umiejętności Dotyczących Mechaniki i Ruchów Miejscowych (r. 1638), Discorsi e Dimonstrazioni Matematiche Intorno a Due Nuove Scienze*; Mianowskiego Instytutu Popierania Nauki: Warsaw, Poland, 1930; Available online: https://rcin.org.pl/dlibra/show-content/publication/edition/1827?id=1827 (accessed on 20 August 2020). (In Polish)
5. West, G. Scaling: The surprising mathematics of life and civilization, Foundations & Frontiers of Complexity. *Santa Fe Inst. Bull.* **2014**, *28*. Available online: https://medium.com/sfi-30-foundations-frontiers/scaling-the-surprising-mathematics-of-life-and-civilization-49ee18640a8 (accessed on 3 March 2019).
6. Zipf, G.K. *Human Behavior and the Principle of Least Effort*; Addison-Wesley: Reading, MA, USA, 1949; Available online: https://archive.org/details/in.ernet.dli.2015.90211/page/n15/mode/2up (accessed on 3 August 2019).
7. Bettencourt, L.M.A.; Lobo, L.; Helbing, D.; Kühnert, C.; West, G.B. Growth, innovation, scaling, and the pace of life in cities. *Proc. Natl. Acad. Sci. USA* **2007**, *104*, 7301–7306. Available online: https://www.pnas.org/content/104/17/7301 (accessed on 9 May 2019). [CrossRef] [PubMed]
8. Bettencourt, L.M.A.; Samaniego, H.; Youn, H. Professional diversity and the productivity of cities. *Sci. Rep.* **2014**, *4*. [CrossRef] [PubMed]
9. Kleiber, M. Body size and metabolic rate. *Physiol. Rev.* **1947**, *27*, 511–541. [CrossRef] [PubMed]
10. Parysek, J. Miasto w ujęciu systemowym. *Ruch Praw. Ekon. Socjol.* **2015**, *1*, 27–53. Available online: http://cejsh.icm.edu.pl/cejsh/element/bwmeta1.element.ojs-doi-10_14746_rpeis_2015_77_1_3/c/931-1302.pdf (accessed on 20 September 2016). [CrossRef]
11. Von Bertalanffy, L. *General System Theory: Foundations, Development, Application*; G. Braziller: New York, NY, USA, 1968; Available online: https://monoskop.org/images/7/77/Von_Bertalanffy_Ludwig_General_System_Theory_1968.pdf (accessed on 7 November 2019).
12. Bettencourt, L.; West, G. A unified theory of urban living. *Nature* **2010**, *467*, 912–913. Available online: https://www.researchgate.net/publication/47510517_A_unified_theory_of_urban_living (accessed on 18 July 2019). [CrossRef] [PubMed]
13. Macionis, J.; Parrillo, V.N. *Cities and Urban Life. Upper Saddle River*; Pearson Education: Cranbury, NJ, USA, 2004.
14. Lobo, J.; Bettencourt, L.M.A.; Strumsky, D.; West, G.B. Urban Scaling and the Production Function for Cities. *PLoS ONE* **2013**, *8*, e58407. [CrossRef] [PubMed]
15. Batty, M. The Size, Scale, and Shape of Cities. *Science* **2008**, *319*, 769–771. Available online: https://www.researchgate.net/publication/5593135_The_Size_Scale_and_Shape_of_Cities (accessed on 25 February 2020). [CrossRef] [PubMed]
16. Calder, W.A., III. *Size, Function, and Life History*; Harvard University Press: Cambridge, MA, USA, 1984.
17. Batty, M.; Longley, P.A. *Fractal Cities: A Geometry of Form and Function*; Academic Press: London, UK, 1994; Available online: http://www.fractalcities.org/book/fractal%20cities%20low%20resolution.pdf (accessed on 23 February 2020).
18. Nordbeck, S. Urban Allometric Growth. *Geogr. Ann. Ser. B Hum. Geogr.* **1971**, *53*, 54–67. [CrossRef]
19. Bettencourt, L.M.A.; Lobo, J. Urban Scaling in Europe. *J. R. Soc.* **2015**, *13*, 20160005. Available online: https://www.researchgate.net/publication/282639661_Urban_Scaling_in_Europe (accessed on 6 June 2020). [CrossRef] [PubMed]
20. Makse, H.A.; Havlin, S.; Stanley, H.E. Modelling urban growth patterns. *Nature* **1995**, *377*, 608–612. Available online: https://www.researchgate.net/publication/252380094_Modelling_urban_growth_patterns (accessed on 6 April 2020). [CrossRef]
21. Zhang, Y.; Yu, J.; Fan, W. Fractal features of urban morphology and simulation of urban boundary. *Geo Spat. Inf. Sci.* **2008**, *11*, 121–126. [CrossRef]
22. Samaniego, H.; Moses, M.E. Cities as organisms: Allometric scaling of urban road networks. *J. Transp. Land Use* **2008**, *1*, 21–39. Available online: https://www.jtlu.org/index.php/jtlu/article/view/29 (accessed on 5 May 2020). [CrossRef]
23. Kennedy, C.A.; Stewart, I.; Facchini, A.; Cersosimo, I.; Mele, R.; Chen, B.; Uda, M.; Kansal, A.; Chiu, A.; Kim, K.-G.; et al. Energy and material flows of megacities. *Proc. Natl. Acad. Sci. USA* **2015**, *112*, 5985–5990. [CrossRef] [PubMed]
24. Pan, A.; Yu, L.; Yang, Q. Characteristics and Forecasting of Municipal Solid Waste Generation in China. *Sustainability* **2019**, *11*, 1433. [CrossRef]
25. Jansen, K.; Casellas, C.P.; Groenink, L.; Wever, K.E.; Masereeuw, R. Humans are animals, but are animals human enough? A systematic review and meta-analysis on interspecies differences in renal drug clearance. *Drug Discov. Today* **2020**, *25*, 706–717. Available online: http://www.sciencedirect.com/science/article/pii/S1359644620300441 (accessed on 11 June 2020). [CrossRef] [PubMed]
26. OECD. Municipal waste. In *Environment at a Glance 2015: OECD Indicators*; OECD Publishing: Paris, France, 2015. [CrossRef]
27. OECD. *Environment at a Glance Indicators*; OECD Publishing: Paris, France, 2020. [CrossRef]
28. Hoornweg, D.; Bhada-Tata, P. *What a Waste: A Global Review of Solid Waste Management*; Urban Development Series; World Bank: Washington, DC, USA, 2012; Available online: https://openknowledge.worldbank.org/handle/10986/17388 (accessed on 25 September 2020).
29. Rada Eurostat: European Statistics. Available online: http://ec.europa.eu/eurostat (accessed on 15 August 2019).
30. Eurostat (European Commission). Energy, Transport, and Environment Statistics—2020 Edition. DDN-20200318-1. 2020. Available online: https://op.europa.eu/en/publication-detail/-/publication/cabb084a-32bb-11eb-b27b-01aa75ed71a1/language-en (accessed on 30 November 2020).
31. EEA. Indicator Assessment: Diversion of Waste from Landfill, European Environment Agency, WST 006. 2019. Available online: https://www.eea.europa.eu/data-and-maps/indicators/diversion-from-landfill/assessment (accessed on 1 December 2019).

32. EEA Annual European Union Approximated Greenhouse Gas Inventory for the Year 2018. EEA Report No 16/2019; European Environment Agency. 2019. Available online: https://www.eea.europa.eu/publications/approximated-eughg-inventory-proxy-2018 (accessed on 13 December 2019).
33. European Commission. Directive (EU) 2018/851 of the European Parliament and of the Council of 30 May 2018. Available online: https://eur-lex.europa.eu/legal-content/EN/TXT/?uri=uriserv:OJ.L_.2018.150.01.0109.01.ENG (accessed on 13 December 2019).
34. *Statistical Yearbook of the Republic of Poland*; Central Statistical Office: Warsaw, Poland, 2018.
35. The Act of August 29, 2003 on Official Names of Places and Physiographic Objects (Journal of Laws of 2019, item 1443), Pursuant to the Act on the Municipal Self-Government of 1990, the Council of Ministers Decides by Way of a Regulation. Available online: http://isap.sejm.gov.pl/isap.nsf/DocDetails.xsp?id=wdu20031661612 (accessed on 20 May 2019).
36. Bettencourt, L.M.A.; Yang, V.C.; Lobo, J.; Kempes, C.P.; Rybski, D.; Hamilton, M.J. The interpretation of urban scaling analysis in time. *J. R. Soc. Interface* **2020**, *17*, 20190846. Available online: https://www.researchgate.net/publication/339045629_The_interpretation_of_urban_scaling_analysis_in_time/citation/download (accessed on 3 May 2020). [CrossRef] [PubMed]
37. Hellwig, Z. *Toward a System of Quantitative Indicators of Components of Human Resources Development*; UNESCO: Paris, France, 1968; Available online: https://unesdoc.unesco.org/ark:/48223/pf0000158559 (accessed on 10 October 2020).
38. Hellwig, Z. Wielowymiarowa analiza porównawcza i jej zastosowanie w badaniach wielocechowych obiektów gospodarczych. In *Metody i Modele Ekonomiczno-Matematyczne w Doskonaleniu Zarządzania Gospodarką Socjalistyczną*; Welfe, W., Ed.; PWE: Warsaw, Poland, 1981.
39. Ferrão, P.; Fernández, J. *Sustainable Urban Metabolism*; MIT: Cambridge, MA, USA, 2013.

Article

The Importance of Water and Climate-Related Aspects in the Quality of Urban Life Assessment

Agnieszka Ptak-Wojciechowska [1,*], Anna Januchta-Szostak [2], Agata Gawlak [3] and Magda Matuszewska [3]

1 Faculty of Architecture, Poznan University of Technology, 61-131 Poznan, Poland
2 Institute of Architecture and Physical Planning, Faculty of Architecture, Poznan University of Technology, 61-131 Poznan, Poland; anna.januchta-szostak@put.poznan.pl
3 Institute of Architecture, Urban Planning and Heritage Protection, Faculty of Architecture, Poznan University of Technology, 61-131 Poznan, Poland; agata.gawlak@put.poznan.pl (A.G.); magda.matuszewska@put.poznan.pl (M.M.)
* Correspondence: agnieszka.a.ptak@doctorate.put.poznan.pl; Tel.: +48-609-331-169

Abstract: Global challenges such as urbanization, aging societies, climate change, and environmental and water crises are becoming increasingly important in terms of the impact they might have on the quality of life (QoL) in cities. Appraisal instruments for QoL assessment, such as rankings and guides, should therefore include these aspects. The aim of this research was to verify the significance of water and climate-related aspects in assessment tools. A comparative analysis of 24 selected QoL assessment tools shows to what extent these aspects are included in the domains, criteria, and indicators proposed in the instruments. The method of verification is a comparison of the position of winning cities in QoL rankings and city resilience rankings. The results show that water and climate-related aspects are still underestimated in the QoL rankings and guides, and only a few cities with the highest quality of life ranked highly in sustainability and climate resiliency ratings. Our results suggest that the tools for the evaluation and comparison of cities need remodeling, taking into account the most important global risks and Sustainable Development Goals (SDGs), in order to create aging-friendly and climate-neutral cities.

Keywords: aging societies; appraisal instruments; blue-green infrastructure; quality of life (QoL); livability; climate resilience; water in cities; urban environments

Citation: Ptak-Wojciechowska, A.; Januchta-Szostak, A.; Gawlak, A.; Matuszewska, M. The Importance of Water and Climate-Related Aspects in the Quality of Urban Life Assessment. *Sustainability* **2021**, *13*, 6573. https://doi.org/10.3390/su13126573

Academic Editors: Jan K. Kazak, Katarzyna Hodor and Magdalena Wilkosz-Mamcarczyk

Received: 29 April 2021
Accepted: 5 June 2021
Published: 9 June 2021

1. Introduction

Cities are becoming the main place of human habitation on Earth and are areas with the greatest concentration of ecological and climate challenges [1]. Therefore, one of the most important indicators of the implementation of the Sustainable Development Goals (SDGs) [2] is the quality of life of citizens. Indicators used to assess the quality of life (QoL), including dynamic processes and growing threats, can support decision makers in developing policies and transforming urban structures, making them more user-friendly and climate neutral [3]. Research units and international bodies, as well as business entities, evaluate the quality of urban life using appraisal instruments based on selected domains and metrics, comparing cities and countries, and selecting the most livable locations. Some tools also take age-friendliness into account [4]. The rapid development of digital techniques and computational models has resulted in an increase in the number of tools and indicators for assessing quality of life in cities, but these tools are often inconsistent and incomparable [4,5]. Furthermore, some have mainly been designed for commercial and marketing purposes.

The global population is aging, which is one of the most important issues of this century. The share of the population aged 65 years or over within the total world population has been increasing. According to the United Nations *World Urbanization Prospects*, it has been estimated that the urbanization rate for the total population will reach 4.5 billion in

2021 [6]. In many European cities, the share of the population aged 65 and over is constantly increasing [7]. Moreover, the historical centers of European cities are usually inhabited by a large group of seniors [8–10]. In most appraisal instruments cities are assessed as homogenous entities, whereas living standards and quality can differ at the neighborhood level [5]. The WHO has defined an age-friendly city as a place that has an "inclusive and accessible urban environment that promotes active aging" [11]. Historic districts are mostly deprived of greenery and are vulnerable to the urban heat island phenomenon, as well as threats of heat waves, droughts, and floods, which are expected to increase with climate change [12,13]. This has a significant impact on the quality of life and health of this social group [14–16].

According to the WEF's *Global Risk Report 2021* [17], environmental risks dominate among the top risks for humanity, both in terms of likelihood and impact. The most serious include extreme weather conditions and failure of climate action, human environmental damage, and biodiversity loss, as well as natural resource crises. The risk of infectious diseases related to the COVID-19 pandemic, among other hazardous incidents, is the only societal risk at the top of the list by impact, while coming in fourth place on the list by likelihood. In compliance with this event, during the pandemic, urban water and green structures became even more meaningful for quality of life. All risks mentioned in the *Global Risk Report 2021* pose a threat to the quality of life in cities, particularly for the elderly.

Climate change is one of the most urgent and greatest long-term challenges that humanity faces [18–20]. The consequences of climate change include increased intensity, frequency, and duration of weather extremes, such as urban temperature variation, drought and water scarcity, coastal flooding, sea level rise related to storm surges, and heavy downpours [16]. The strong link between climate change and urbanization has been confirmed in the literature [21–24]. The key affected urban sectors are water supply, wastewater and sanitation, energy supply, transportation and telecommunication infrastructure, built environment (e.g., housing), and recreation, as well as natural and cultural heritage sites [13,16,25].

Valuing water [26] and the natural environment is fundamental to achieving the Sustainable Development Goals (SDGs) of the United Nations 2030 Agenda for Sustainable Development [27] and high quality of urban life. Water aspects are mainly considered in terms of resource availability, water supply and sanitation, and potential flood risk [16]. Meanwhile, proximity and access to water are essential for human culture and urban heritage, as well as for health, well-being, and disease prevention [28]. The well-being and safety of residents, as well as community involvement, are highly associated with water [29]. According to the Arcadis Sustainable cities water index, "Cities which carefully and creatively use their water assets for strategic urban advantage will ultimately be more livable, safe and competitive. Cities that are truly distinguished by a thriving relationship with their waterscape can make a huge contribution to the quality of life of their residents [..]" [30]. Moreover, urban developments have a consequential influence on the natural environment and climate. Sustainable urban water management and the protection of water sources can be used to mitigate climate change [29].

The impact of climate change on the quality of life had already been noted in the 1970s [31]. However, climate-related issues are still poorly considered in quality of life assessments [3,32,33]. Newton (2012) [33] compared the EIU's Quality of Life with the WWF's 2008 Living Planet Index of Ecological Footprints [34]. The results showed that the "liveability of cities is being achieved at the expense of environmental sustainability" [33]. Jones and Newsome (2015) [35] examined the relationships between livability indicators and environmental factors using the example of Perth, which scored poorly in the Economist Intelligence Unit global livability ranking and report of August 2013, concerning its accessibility to nature. However, the analysis was limited to just three rankings, and only indicators proposed by the Economic Intelligence Unit (EIU) were analyzed in detail, while Mercer's and Monocle's were described more generally. The authors of the article indicated the limitations of the EIU evaluation method and proposed environmental factors that should be included as livability indicators (green space, access to protected areas, remnant vegetation, biodiversity, air quality, and unpolluted water) [35]. Another example of analysis of the tools for QoL assessment concerns the verification of the relationship between the quality of life in cities and their environmental sustainability. Estoque et al. (2019) noted that "Climate change and variability affect QoL and human well-being in many ways, rendering it one of the most pressing and significant challenges of the present day" [32]. They reviewed quality of life (QoL) assessments and indicators, and found them to be poorly connected with climate-related issues. The authors proposed a new "QoL-Climate" assessment framework, designed to capture the socio-ecological impacts of climate change and variability [32].

As there is still little research on the connection between the quality of urban life and the environmental sustainability of cities, the authors of the present study conducted a comparative analysis of the instruments that assess urban quality of life and verified the inclusion of climatic, environmental, and water factors.

The recipients of the results from the available guides and rankings are often local government units and decision makers who create sustainable development policies and, thus, have an impact on improving living conditions and creating more sustainable and resilient cities. To develop more precise appraisal instruments to assess the sustainability and quality of life in cities, it is worth verifying the following: (i) Which water and climatic factors are considered to have the greatest impact on quality of life? (ii) Which tools assess water and climate aspects, and to what extent? (iii) Do international and Polish QoL assessment tools differ significantly and in which indicators? (iv) Which indicators are the most useful and reliable and which of them have been marginalized? Finally, which (v) are the cities with the highest QoL that are also sustainable and resilient?

2. Materials and Methods

The main goal of the research was to verify the integration and significance of water and climate-related aspects in the considered appraisal instruments. Analysis of the number and type of indicators, as well as the weight function of water and climate-related aspects, helped us determine which of the instruments most fully capture the impact of water and climate on QoL. Comparing the top ten cities in the QoL rankings with those in the sustainability and resilience rankings allowed us to identify which cities have most effectively combined strategies to improve the quality of life along with adaptation to climate change, as well as environmental and water crises.

The research subjects are the most influential, prestigious, and available international and Polish tools, which concern the features and indicators of livable cities for the assessment of quality of urban life and comparison of various countries, cities, and communities.

The first stage in the analysis of the importance of water and climate-related aspects in the assessment of quality of urban life in aging cities is based on the reports and a scientific literature review on topics relating to global changes (i.e., aging populations, urbanism, climate change, and water crises), as well as the analysis and selection of appraisal instruments (i.e., guides, rankings, reports from surveys) for the evaluation of livability of cities.

In the second stage, twenty-four tools were subjectively selected, including three international guides and one Polish one for age-friendly communities, thirteen international and two Polish rankings, and five other Polish appraisal instruments. The selected instruments were divided into two main groups: (i) Guides including core indicators for senior-friendly cities, and (ii) rankings measuring QoL and comparing the best countries and cities to live in. In both groups, international and Polish tools were selected to compare the indicators used. Additionally, several other Polish tools assessing the quality of life through a set of indicators were taken into account. The tools listed in Table 1 were assigned symbols (e.g., 1.1 for *Global Age-friendly Cities—A Guide*), which are used instead of their full names in the following figures.

Table 1. Selected tools of the quality of urban life assessment and institutions that prepared the documents.

Type	No.	Name	Institutions
International guides	1.1	*Global Age-friendly Cities—A Guide* [36]	World Health Organization
	1.2	*Measuring the age-friendliness of cities. A guide to using core indicators* [37]	WHO Kobe Centre
	1.3	*Age-friendly rural and remote communities: a guide* [38]	Healthy Aging and Wellness Working Group of the Federal/Provincial/Territorial (F/P/T) Committee of Officials (Seniors)
Polish guide	1.4	*System wsparcia osób starszych w środowisku zamieszkania—przegląd sytuacji, propozycja modelu. Synteza (System of support for the elderly in the living environment—overview of the situation, model proposal. Synthesis)* [39]	Komisja Ekspertów ds. Osób Starszych przy Rzeczniku Praw Obywatelskich
International rankings	2.1	*EIU's Global Liveability Ranking* [40]	The Economist Intelligence Unit (The EIU)
	2.2	*Mercer's Quality of Living Ranking* [41]	Mercer LLC
	2.3	*Monocle's Quality of Living Survey* [42]	Monocle
	2.4	*Deutsche Bank Liveability Survey* [43](Quality-of-Life Indices)	Deutsche Bank AG/London (data collected from Numbeo)
	2.5	*Euro Health Consumer Index* [44]	Health Consumer Powerhouse
	2.6	*IMD Smart City Index* [45]	IMD and SUTD (Singapore University for Technology and Design)
	2.7	*Best Cities for Successful Aging* [46]	Milken Institute Center for the Future of Aging
	2.8	*Human Development Report* [47]	Human Development Report Office
	2.9	*Quality of life (well-being of Europeans)* [48]	Directors of Social Statistics Quality of Life Expert Group (EG)
	2.10	*The European Quality of Life Survey* (EQLS) [49]	Eurofound
	2.11	*How's Life? 2020 Measuring Well-being* [50]	Household Statistics and Progress Measurement Division of the OECD Statistics and Data Directorate
	2.12	*Quality of life in cities. Perception survey in 79 European cities* [51]	European Commission, Directorate-General for Regional and Urban Policy
	2.13	*Active Ageing Index* [52]	IRCCS INRCA: National Institute of Health and Science on Ageing, Centre for Socio-Economic Research on Ageing, Ancona, Italy
Polish rankings	2.14	*Ranking jakości życia. Wymiary szczęścia (Quality of life ranking. Dimensions of happiness)* [53]	POLITYKA i Akademia Górniczo- Hutnicza
	2.15	*Uciekające metropolie. Ranking 100 polskich miast (Runaway metropolises. Ranking of 100 Polish cities)* [54]	Klub Jagielloński
Polish appraisal instruments	3.1	*Jakość życia w Polsce. Edycja 2017 (Quality of life in Poland. 2017 edition)* [55]	Główny Urząd Statystyczny Departament Badań Społecznych i Warunków Życia Departament Analiz i Opracowań Zbiorczych
	3.2	*Zadowolenie z życia–CBOS (Life satisfaction–Public Opinion Research Center)* [56]	CBOS
	3.3	*Diagnoza społeczna 2015 Warunki i jakość życia Polaków (Social Diagnosis 2015 Objective and Subjective Quality of Life in Poland)* [57]	Rada Monitoringu Społecznego
	3.4	*Jakość życia mieszkańców Łodzi i jej przestrzenne zróżnicowanie (Quality of life of Lodz inhabitants and city's spatial diversity)* [58]	Uniwersytet Łódzki, Wydział Ekonomiczno-Socjologiczny, Instytut Socjologii Katedra Socjologii Ogólnej
	3.5	*Jak się żyje osobom starszym w Polsce (How is life for the elderly in Poland)* [59]	Główny Urząd Statystyczny

A comparative analysis has been elaborated in a previous study by the authors [4]; however, it included eight additional Polish assessment instruments and, thus, the coverage matrix presented in Figure 1 has been extended. The analysis covered the evaluation and synthesis of different domains, criteria, and indicators used in appraisal instruments. Seven common areas (six main domains: Architecture and urbanism, Infrastructure, Nature, Health and well-being, Social environment, and Development, and one domain of Supplementary criteria) and 29 categories were developed, in order to allow for universal classification of the metrics used in the various appraisal instruments. The authors used the WHO report *Global Age-friendly Cities—A Guide* as the reference tool in view of the methodology for assessing criteria in the aspects of architecture and urban planning. This tool has been assessed as being the most extensive, in terms of metrics related to the spatial structure of the cities. In an additive way, the criteria pool was extended with the new metrics provided by other analyzed instruments. Using a method of expert assessment, the collected criteria were categorized. The matrix presents the structure of the adopted criteria and domains (Figure 1). Indicators from all the evaluated tools were interpreted and assigned to domains and categories recommended by the authors.

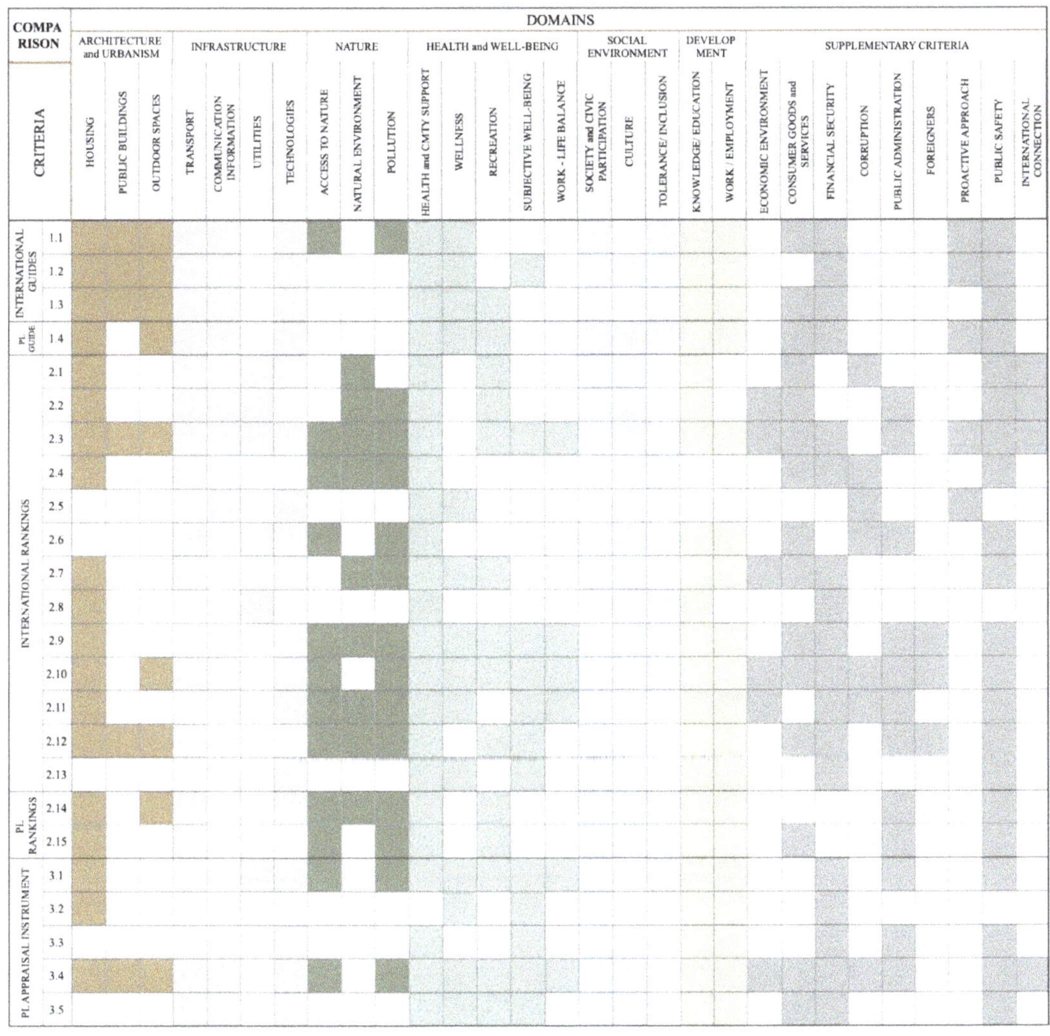

Figure 1. Share of respective criteria in the assessment of selected tools—coverage matrix.

The third step was to check the indicators and to determine the weight function of the water and climate aspects in all tools for the urban quality of life evaluation. First, the importance of domain-related (e.g., nature) indicators in particular tools (e.g., 2.12 presented in Table 2) was measured. Second, the average weight function for a domain (e.g., nature) in the set of 13 rankings was calculated. Third, analyses of the importance of water and climate aspects in all compared appraisal instruments (consisting of checking all single indicators) were conducted. Metrics assessing water or climate-related aspects were marked and the average weight functions were calculated, according to the following ratio: number of indicators concerning water/climate aspects linked to the proper domain to the number of all indicators in the analyzed tool. Finally, the average weight function of all climate-related indicators in the set of 13 rankings were calculated. To present our findings in figures, the results were subtracted from the average weight function of the domain in a set of indicators. Accordingly, the importance of climate-related aspects was measured in the considered guides. Numbers were rounded to one decimal place, to make rough calculations easier and to present clearer figures. The limitations mainly concerned a lack of availability and transparency in the definition of particular metrics. Therefore, the results were estimated.

Table 2. Exemplary analysis of one of the ranking's indicators (*Quality of life in cities. Perception survey in 79 European cities*), concerning water- and climate-related aspects highlighted in light grey.

Domains and Criteria	Indicators	AV. Weight
Environment	Green spaces, such as parks and gardens	
	The quality of the air	2.63%
	The noise level	
	Cleanliness	
	City is committed to fight against climate change (e.g., energy efficiency, green transport)	2.63%
The most important issues facing cities	Safety	
	Air pollution	2.63%
	Noise	
	Public transport	
	Health services	
	Social services	
	Education and training	
	Unemployment	
	Housing	
	Road infrastructure	
No. of indicators: 38	Importance of climate-related indicators in tool	7.89%

The fourth stage of the study concerned the comparison of the QoL assessment tools in terms of their content and type of water and climate-related indicators, considering the subject and scope of the assessment. International and Polish QoL assessment tools were compared. This analysis identified the most comprehensive rankings, taking into account water and climate aspects and the role of blue–green infrastructure.

In the fifth step, the top ten cities included in the most comprehensive international QoL rankings were compared with the results of the sustainability and resilience (S&R) rankings, in order to determine whether the cities with the highest QoL are also sustainable and resilient, or if the QoL improvement came at the expense of sustainability. The list of top cities was taken from the websites on which the results of the rankings selected for comparison were published.

3. Results

3.1. Integrating Water and Climate-Related Aspects in the QoL Assessment Tools

Comparison of the content and type of indicators related to water and climate aspects in selected QoL assessment tools allowed us to notice that, in the vast majority of instruments, these aspects were omitted or treated as a secondary issue. Additionally, the manner of defining the discussed aspects and their meanings differed significantly.

3.1.1. Importance of Water-Related Aspects in the Analyzed Tools

Our analysis revealed that the water topic has been omitted in many international and Polish guides, and most other Polish instruments (Figure 2).

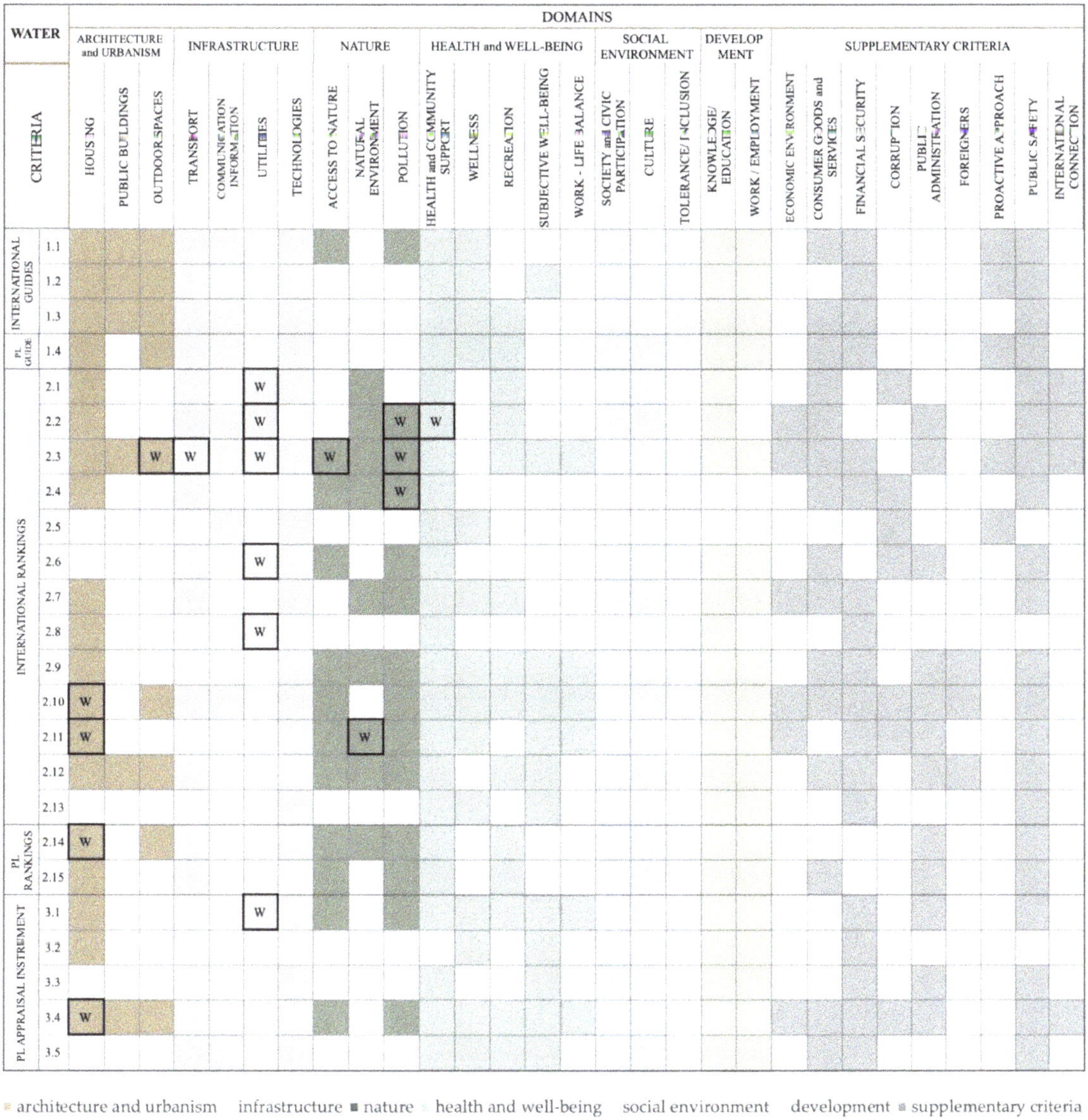

WATER		ARCHITECTURE and URBANISM			INFRASTRUCTURE				NATURE			HEALTH and WELL-BEING					SOCIAL ENVIRONMENT			DEVELOP MENT		SUPPLEMENTARY CRITERIA								
CRITERIA		HOUSING	PUBLIC BUILDINGS	OUTDOOR SPACES	TRANSPORT	COMMUNICATION INFORMATION	UTILITIES	TECHNOLOGIES	ACCESS TO NATURE	NATURAL ENVIRONMENT	POLLUTION	HEALTH and COMMUNITY SUPPORT	WELLNESS	RECREATION	SUBJECTIVE WELL-BEING	WORK - LIFE BALANCE	SOCIETY and CIVIC PARTICIPATION	CULTURE	TOLERANCE/ INCLUSION	KNOWLEDGE/ EDUCATION	WORK / EMPLOYMENT	ECONOMIC ENVIRONMENT	CONSUMER GOODS and SERVICES	FINANCIAL SECURITY	CORRUPTION	PUBLIC ADMINISTRATION	FOREIGNERS	PROACTIVE APPROACH	PUBLIC SAFETY	INTERNATIONAL CONNECTION
INTERNATIONAL GUIDES	1.1																													
	1.2																													
	1.3																													
PL GUIDE	1.4																													
INTERNATIONAL RANKINGS	2.1						W																							
	2.2						W				W	W																		
	2.3			W	W		W		W		W																			
	2.4										W																			
	2.5																													
	2.6						W																							
	2.7																													
	2.8						W																							
	2.9																													
	2.10	W																												
	2.11	W								W																				
	2.12																													
	2.13																													
PL RANKINGS	2.14	W																												
	2.15																													
PL APPRAISAL INSTREMENT	3.1						W																							
	3.2																													
	3.3																													
	3.4	W																												
	3.5																													

architecture and urbanism infrastructure nature health and well-being social environment development supplementary criteria

Figure 2. Coverage matrix of respective criteria in the assessment of selected tools, considering water aspects. The squares outlined with a bold black line indicate criteria where the indicators concern water aspects.

The aspects of water were included in eight of the international rankings: *EIU's Global Liveability Ranking* (2.1), *Mercer's Quality of Living Ranking* (2.2), *Monocle's Quality of*

Living Survey (2.3), *Deutsche Bank Liveability Survey* (2.4), *IMD Smart City Index* (2.6), *Human Development Report* (2.8), *The European Quality of Life Survey (EQLS)* (2.10), and *How's Life? 2020 Measuring Well-being* (2.11); two Polish tools (the instruments' names are marked with symbols proposed in Table 1): *Jakość życia w Polsce Edycja 2017* (3.1) and *Jakość życia mieszkańców Łodzi i jej przestrzenne zróżnicowanie* (3.4); and one Polish ranking: *Ranking jakości życia. Wymiary szczęścia* (2.14). The water aspects were considered in four domains and eight categories, such as housing, outdoor spaces, transport, utilities, access to nature, natural environment, pollution, health, and community support. The most diverse attitude towards water (up to five different QoL categories) was observed in the *Monocle's Quality of Life Survey (2.3)*.

The results indicated that water-related metrics have only been proposed in four domains: architecture and urbanism, infrastructure, nature, and health and well-being. However, the topic of water was not considered in the areas of social environment and development, nor in the supplementary criteria (Figures 2 and 3).

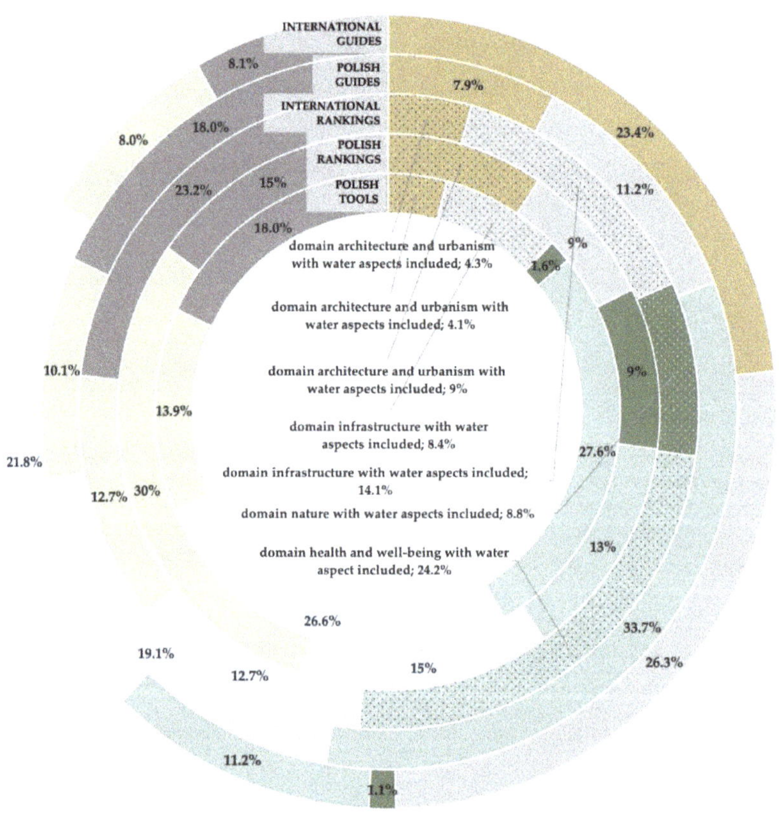

Figure 3. Share of the domains with water-related indicators used for the assessment of the quality of life (QoL), in reference to other domains, taken into account in the selected tools.

The average weight of the water aspects was very low. Concerning international rankings, for the domain of architecture and urbanism, it was only 0.3%; in the domain of nature it was 0.9% and, for the domain of health and well-being, it was 0.2%. The most significant average weight was for the domain of infrastructure (1.4%). For Polish rankings, the weight of water aspects for the domain of architecture and urbanism was 1%;

for Polish tools, the weights of water aspects for the domain of architecture and urbanism and infrastructure were equal to 0.3% (Figure 4).

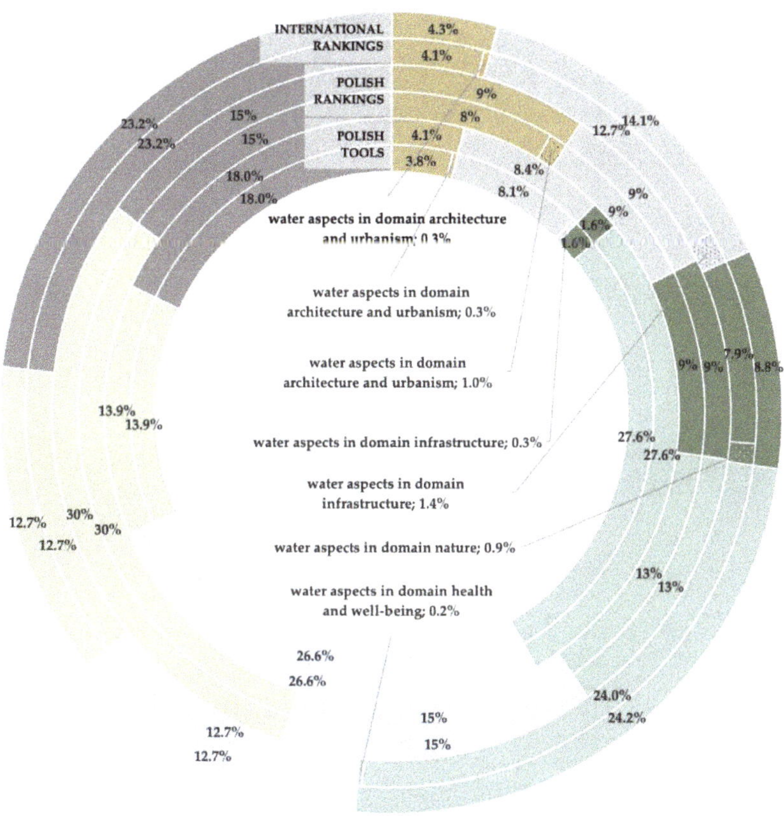

Figure 4. Share of water aspects in the domains for the quality of life assessment (in reference to other domains) taken into account in the Polish and international rankings, as well as other Polish tools.

3.1.2. Importance of Climate-Related Aspects in the Analyzed Tools

This analysis takes into account weather phenomena, temperature, and air conditions as short-term indicators of climate change. The outcomes are shown in Figure 5. Climate-related aspects were poorly covered in 10 of the international rankings: *EIU's Global Liveability Ranking* (2.1), *Mercer's Quality of Living Ranking* (2.2), *Monocle's Quality of Living Survey* (2.3), *Deutsche Bank Liveability Survey* (2.4), *IMD Smart City Index* (2.6), *Best Cities for Successful Aging* (2.7), *Quality of life (well-being of Europeans)* (2.9), *The European Quality of Life Survey (EQLS)* (2.10), *How's Life? 2020 Measuring Well-being* (2.11), and *Quality of life in cities. Perception survey in 79 European cities* (2.12); in two Polish rankings: *Ranking jakości życia. Wymiary szczęścia* (2.14) and *Uciekające metropolie. Ranking 100 polskich miast* (2.15); and in two Polish tools: *Jakość życia w Polsce Edycja 2017* (3.1) and *Jakość życia mieszkańców Łodzi i jej przestrzenne zróżnicowanie* (3.4). The climate-related aspects were considered in two domains and four categories, including transport, technologies, natural environment, and pollution (Figures 5 and 6).

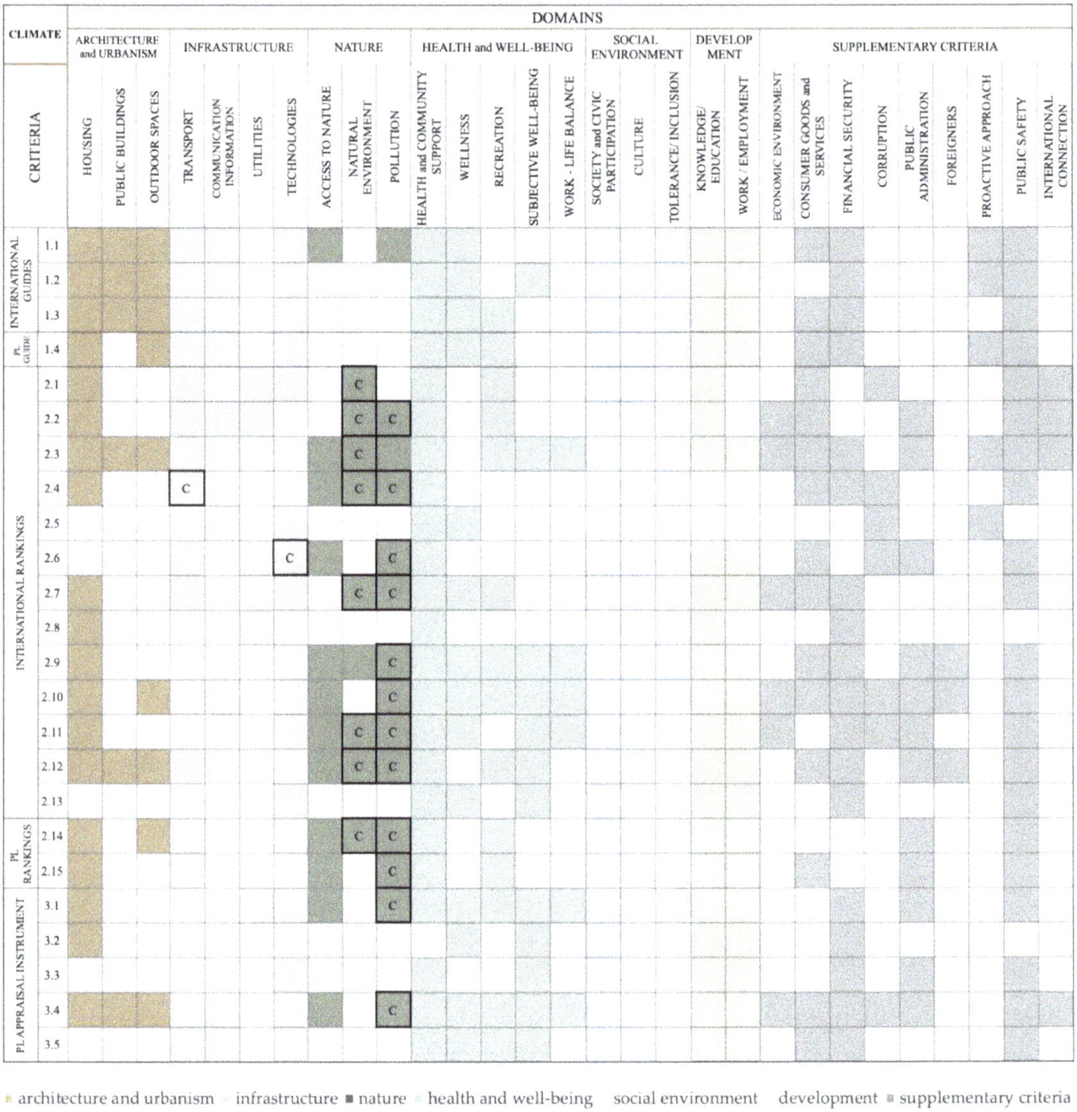

CLIMATE		DOMAINS																												
		ARCHITECTURE and URBANISM			INFRASTRUCTURE				NATURE			HEALTH and WELL-BEING					SOCIAL ENVIRONMENT			DEVELOP MENT		SUPPLEMENTARY CRITERIA								
CRITERIA		HOUSING	PUBLIC BUILDINGS	OUTDOOR SPACES	TRANSPORT	COMMUNICATION INFORMATION	UTILITIES	TECHNOLOGIES	ACCESS TO NATURE	NATURAL ENVIRONMENT	POLLUTION	HEALTH and COMMUNITY SUPPORT	WELLNESS	RECREATION	SUBJECTIVE WELL-BEING	WORK - LIFE BALANCE	SOCIETY and CIVIC PARTICIPATION	CULTURE	TOLERANCE/ INCLUSION	KNOWLEDGE/ EDUCATION	WORK / EMPLOYMENT	ECONOMIC ENVIRONMENT	CONSUMER GOODS and SERVICES	FINANCIAL SECURITY	CORRUPTION	PUBLIC ADMINISTRATION	FOREIGNERS	PROACTIVE APPROACH	PUBLIC SAFETY	INTERNATIONAL CONNECTION
INTERNATIONAL GUIDES	1.1																													
	1.2																													
	1.3																													
PL GUIDE	1.4																													
INTERNATIONAL RANKINGS	2.1									C																				
	2.2									C	C																			
	2.3									C																				
	2.4				C					C	C																			
	2.5																													
	2.6							C			C																			
	2.7									C	C																			
	2.8																													
	2.9										C																			
	2.10										C																			
	2.11									C	C																			
	2.12									C	C																			
	2.13																													
PL RANKINGS	2.14									C	C																			
	2.15										C																			
PL APPRAISAL INSTRUMENT	3.1										C																			
	3.2																													
	3.3																													
	3.4										C																			
	3.5																													

■ architecture and urbanism ■ infrastructure ■ nature ■ health and well-being ■ social environment ■ development ■ supplementary criteria

Figure 5. Coverage matrix of respective criteria in the assessment of selected tools, considering climate aspects. The squares outlined with a bold black line indicate criteria where the indicators concerned climate aspects.

The data showed that metrics related to weather and climate aspects were developed in only two areas: Infrastructure and nature (Figures 5 and 6) and they seemed to be of little significance.

In international rankings, climate factors were included in the infrastructure domain, with a weight of 0.4%; whereas for the domain of nature it was 4.1%. In Polish rankings, all indicators in the area of nature included climate-related aspects (5.6%). For Polish tools, the significance of climate factors for the domain of nature was equal to 0.5% (Figure 7).

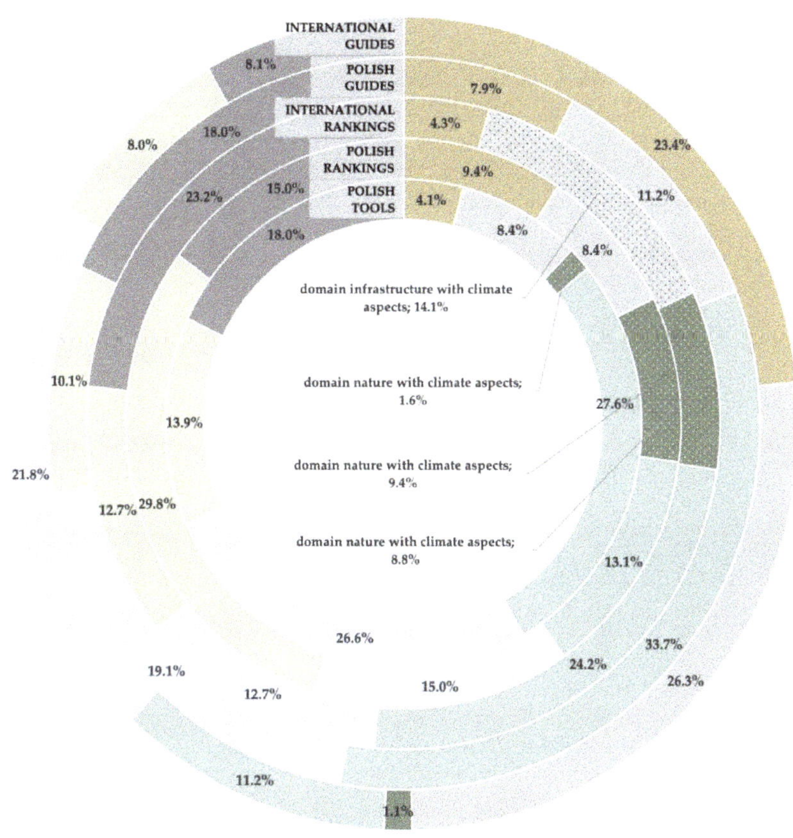

Figure 6. Share of the domains with climate-related indicators used for the assessment of quality of life (in reference to other domains) taken into account in the selected tools.

3.2. Comparison of the Indicators Related to Water and Climate Aspects

3.2.1. Water-Related Indicators in the Analyzed QoL Assessment Tools

Indicators related to water were totally omitted in 13 appraisal instruments: 1.1, 1.2, 1.3, 1.4, 2.5, 2.7, 2.9, 2.12, 2.13, 2.15, 3.2, 3.3, and 3.5. In all 11 other tools, the indicators were concerned with basic, physiological aspects, such as water quality (environment and water supply), availability, and sanitary aspects. Most of the tools assessed only 1–2 indicators. Social and transport aspects of water were proposed in one ranking (2.3), whereas aspects related to safety or a proactive approach were not used at all.

Water quality/pollution was measured only in two tools:

- *Water quality* (2.3); and
- *Water pollution* (2.4).

Tap water quality and supply was measured in four tools, all of which used different indicators:

- *Quality of water provision* (2.1);
- *Water potability* (2.2);
- *Quality of tap water* (2.3); and
- *Drinking water quality* (2.4).

In *Quality of life in Poland. 2017 edition* (3.1), an indicator concerning polluted water—*exposure to pollution or other environmental problems in the neighbourhood (% of households)*—was also included.

Water availability was also assessed in only four tools: 2.2, 2.4, 2.8, and 2.11. *How's Life? 2020 Measuring Well-being* (2.11) measured it using two (different from the rest) indicators: water stress (total) and *water stress (internal)*.

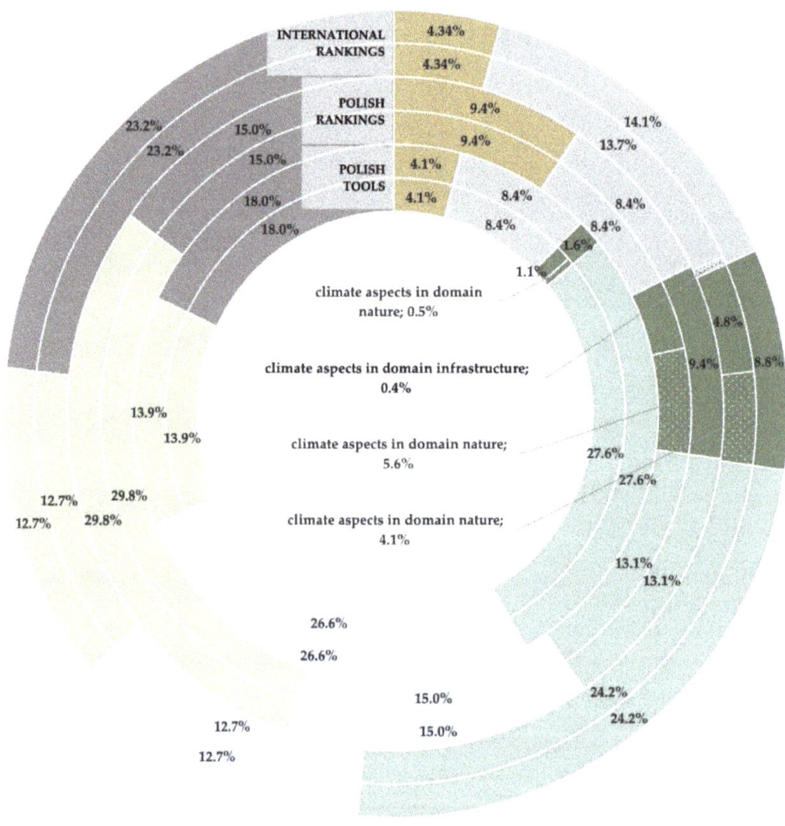

Figure 7. Share of the climate aspects concerned in the domains for the quality of life assessment (in reference to other domains) taken into account in the Polish and international rankings, as well as other Polish tools.

Sanitary aspects were verified in eight tools, with indicators measuring:

- Sewage (2.2, 2.14);
- Sanitary facilities (2.8, 2.10, 2.11); and
- Sanitary conditions (2.6, 3.1).

In the Polish tool *Jakość życia mieszkańców Łodzi i jej przestrzenne zróżnicowanie* (3.4), this area was the most extensive, with the following indicators: *warm running water, bathroom with bathtub and/or shower, flush toilet*, and *sewers*.

The most varied instrument, containing three indicators from all categories, including *water potability, water availability*, and *sewage*, was *Mercer's Quality of Living Ranking* (2.2).

The most extensive (with five indicators related to water) was *Monocle's Quality of Living Survey* (2.3), which considered social and transport aspects of water, and assessed the following indicators: *Access to water, waterfronts*, and *water as means of transport* (Figure 8).

	WATER-RELATED ASPECTS	INTERNATIONAL GUIDES			PL GUIDE	INTERNATIONAL RANKINGS													PL RANKINGS		PL APPRAISAL INSTRUMENT				
		1.1	1.2	1.3	1.4	2.1	2.2	2.3	2.4	2.5	2.6	2.7	2.8	2.9	2.10	2.11	2.12	2.13	2.14	2.15	3.1	3.2	3.3	3.4	3.5
BASIC, PHYSIOLOGICAL	1. Water quality/pollution (environment)							1	1												1				
	2. Tap water quality and supply					1	1	1	1																
	3. Water availability						1		1				1			2									
	4. Sanitary aspects						1				1		1		2	1			1		1			4	
SOCIAL	5. Access to water							1																	
	6. Waterfronts							1																	
EXTRA	7. Water as means of transport							1																	

Figure 8. Water-related aspects including thematically grouped indicators.

3.2.2. Climate-Related Indicators in the Analyzed QoL Assessment Tools

Indicators related to climate were not included in 10 of the analyzed tools: 1.1, 1.2, 1.3, 1.4, 2.5, 2.8, 2.13, 3.2, 3.3, and 3.5. In the other 14 appraisal instruments, the metrics assessed aspects that can be assigned into three categories: weather, climate, and safety/environmental initiatives. Most of the tools assessed only 1–3 indicators, while four instruments considered 4–6 metrics.

All aforementioned 14 tools verified weather aspects such as:

- *Air pollution/quality* (2.2, 2.4, 2.6, 2.7, 2.9, 2.10, 2.11, 2.12, 2.14, 2.15, 3.1, 3.4);
- *Humidity/temperature* (2.1, 2.3, 2.4);
- *Hours of sunshine* (2.3);
- *Weather (2.7); which was assessed as a composite score using heating degree days, cooling degree days, humidity, sunshine, and precipitation.*

The most extensive range of weather-related indicators, such as *dew point, temperature, average high humidex, and air quality,* was found in the *Deutsche Bank Liveability Survey (2.4).*

Climate aspects were underestimated in most of the appraisal instruments. They were varied and assessed in only five rankings. Four of them proposed only one indicator:

- *Discomfort of climate to travellers (2.1);*
- *Carbon neutral (2.3);*
- *CO_2 Emmission Index (2.4); and*
- *Climate (2.2);* which was explained using the example of characteristics of the Four Seasons.

The most extensive tool, *How's Life? 2020 Measuring Well-being* (2.11), evaluated three indicators: *Greenhouse gas emissions (domestic production), carbon footprint, and material footprint.*

Aspects related to safety or/and environmental initiatives were included in six tools. Environmental initiatives, friendliness, and problems were observed in five tools, including:

- *Environmental initiatives, walkable, and environmentally friendly in Monocle's Quality of Living Survey* (2.3);
- *Perception of pollution, grime, or other environmental problems in Quality of life (well-being of Europeans)* (2.9);
- *Recycling rate and renewable energy in How's Life? 2020 Measuring Well-being* (2.11);
- *City is committed to fight against climate change* (2.12);
- *City expenditure on air and climate protection* (2.14).

Aspects of safety were verified only through a *record of natural disasters* indicator in *Mercer's Quality of Living Ranking* (2.2).

The most differential tools, in terms of number and classification of indicators, were the rankings *How's Life? 2020 Measuring Well-being* (2.11), with six indicators assessing all three categories, and *Monocle's Quality of Living Survey* (2.3), with five metrics also evaluating all categories of weather and climate aspects (Figure 9).

	CLIMATE-RELATED ASPECTS	INTERNATIONAL GUIDES			PL GUIDE	INTERNATIONAL RANKINGS													PL RANKINGS		PL APPRAISAL INSTRUMENT				
		1.1	1.2	1.3	1.4	2.1	2.2	2.3	2.4	2.5	2.6	2.7	2.8	2.9	2.10	2.11	2.12	2.13	2.14	2.15	3.1	3.2	3.3	3.4	3.5
WEATHER	1. Air pollution/quality						1		1		2	1		1	1	1	1		3	1	1			2	
	2. Humidity/temperature					1		1	3																
	3. Hours of sunshine							1																	
	4. Weather											1													
CLIMATE	5. Discomfort of climate to travellers					1																			
	6. Carbon neutral							1																	
	7. CO2 Emmission								1							2									
	8. Climate						1																		
	9. Material footprint															1									
SAFETY, ENVIRONMENTAL INITIATIVES	10. Environmental initiatives							1																	
	11. Walkable and environmentally friendly							1																	
	12. Perception of environmental problems													1											
	13. Recycling rate															1									
	14. Renewable energy															1									
	15. City is committed to fight against climate change																1								
	16. City expenditure on air and climate protection																		1						
	17. Record of natural disasters						1																		

Figure 9. Climate and weather-related aspects including thematically grouped indicators.

3.2.3. Additional Indicators Related to Green Infrastructure in the Analyzed QoL Assessment Tools

Even if the rankings did not directly assess aspects of climate management, some of them appreciated the ecosystem services of urban greenery.

Six appraisal instruments included metrics related to urban greenery, with indices such as:

- *Green spaces* which were not specified (2.15), amount/number of green spaces, parks, gardens, and so on (2.3, 2.12, 2.14, 3.4), *and natural and semi-natural land cover* (2.11);
- *Intact forest landscape and protected areas* (terrestial and marine) (2.11);
- *Red List Index* (threatened species) (2.11); and
- *Soil nutrient balance* (2.11).

In three rankings (2.3, 2.10, 2.11), *access to nature* was also measured. Moreover, in *Quality of life of Lodz inhabitants and city's spatial diversity* (3.4), this aspect was measured through the indicator *"feeling a lack of green areas"*.

The quality of and satisfaction of citizens with green and recreation areas were assessed in six tools (1.1, 2.4, 2.6, 2.9, 3.1, 3.4). Moreover, *in Quality of life ranking. Dimensions of happiness* (2.14), there was one extra indicator—*expenditure on maintaining green areas in the years*—related to involvement of city authorities in designing green areas.

As BGI has a significant impact on climate and water management in cities [60–62], it is essential that it is also reflected in the rankings. The ranking *How's life going? 2020 Measuring well-being* (2.11) used as many as nine indicators (Figure 10).

URBAN GREENERY- RELATED ASPECTS		INTERNATIONAL GUIDES			PL GUIDE	INTERNATIONAL RANKINGS													PL RANKINGS		PL APPRAISAL INSTRUMENT				
		1.1	1.2	1.3	1.4	2.1	2.2	2.3	2.4	2.5	2.6	2.7	2.8	2.9	2.10	2.11	2.12	2.13	2.14	2.15	3.1	3.2	3.3	3.4	3.5
BLUE-GREEN INFRASTRUCTURE	1. Green spaces							2								3	1		1	1				1	
	2. Intact forest landscapes															1									
	3. Protected areas (terrestrial/marine)															2									
	4. Red List Index (threatened species)															1									
	5. Soil nutrient balance															1									
ACCESS	6. Access to nature, green areas							1							1	1									
	7. Feeling a lack of green spaces																							1	
QUALITY/ SATISFACTION	8. Green spaces quality	1							1															1	
	9. Satisfaction with recreational and green areas										1			1							1				
	10. Satisfaction with living environment (landscape)													1											
EXTRA	11. Expenditure on maintaining green areas																		1						

Figure 10. Urban greenery-related aspects including thematically grouped indicators.

The analysis of the types and scope of indicators in QoL assessment tools proves that they differ significantly (Figure 11); the reason for so many simplifications and imperfections is, inter alia, a shortage of commonly available and comparable data in various countries and cities (including Poland). One limitation in the research was also the lack of availability and transparency in methodology and metric descriptions in some of the tools.

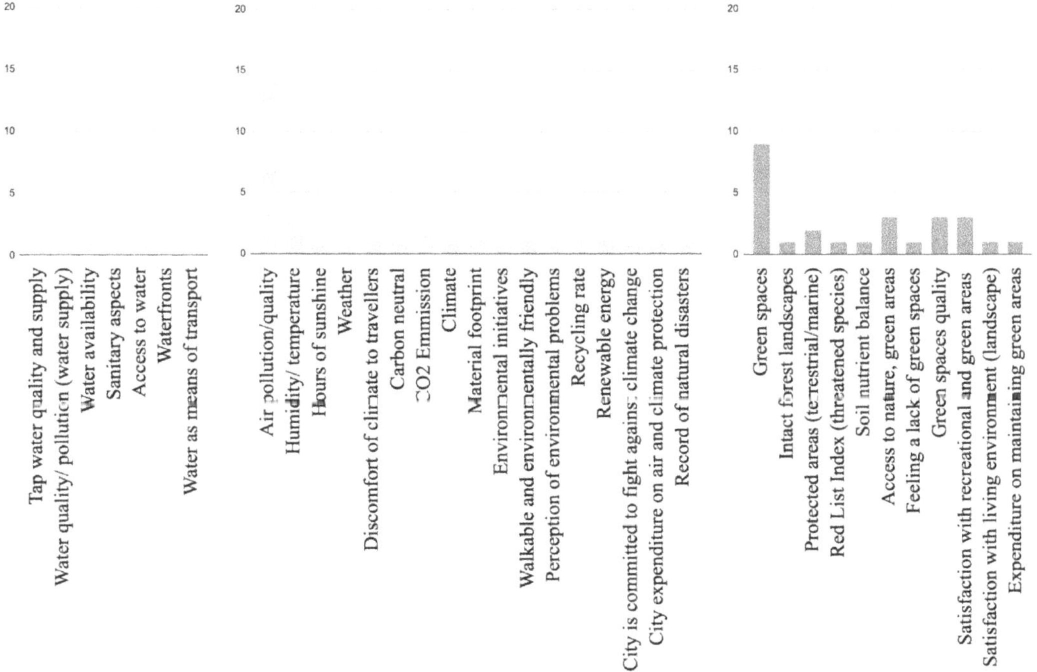

Figure 11. Number of water-, climate-, and urban greenery-related indicators grouped thematically.

3.3. Comparison of the Top Cities in the Most Comprehensive International QoL Rankings with the Results of the Sustainability and Resilience (S&R) Rankings

In order to determine whether the cities with the highest quality of life are also sustainable and resilient, the top ten cities in the most comprehensive international QoL rankings (*Mercer Ranking QoL* [41], *Monocle's QoL Survey* [42], and Deutsche Bank Liveability Survey [43]) were compared with the top ten cities in S&R rankings (Table 3). The ranking *How's life going? 2020 Measuring well-being* (2.11) has been excluded as it compares only countries.

Table 3. Comparative list of the winning cities (highlighted with color) in the QoL rankings and Sustainable and Resilient Cities rankings. The lists were obtained from the websites of selected rankings.

QOL Rankings						**Sustainable and Resilient Cities Rankings**							
Mercer's Quality of Living Ranking [41]		**Monocle's Quality of Life Survey [42]**		**Deustche Bank Liveability Survey [43]**		**Arcadis Sustainable Cities Water Index [30]**		**Arcadis Sustainable Cities Index [63]**		**Grosvenor Group Resilient Cities [64]**		**Global Ranking of Top 10 Resilient Cities [65]**	
	City		**City**		**City**		**City**		**City**		**City**		**City**
1	Vienna	1	Zurich	1	Zurich	1	Rotterdam	1	London	1	Toronto	1	Copenhagen
2	Zurich	2	Tokyo	2	Wellington	2	Copenhagen	2	Stockholm	2	Vancouver	2	Curitiba
3	Vancouver	3	Munich	3	Copenhagen	3	Amsterdam	3	Edinburgh	3	Calgary	3	Barcelona
3	Munich	4	Copenhagen	4	Edinburgh	4	Berlin	4	Singapore	4	Chicago	4	Stockholm
3	Auckland	5	Vienna	5	Vienna	5	Brussels	5	Vienna	5	Pittsburg	5	Vancouver
6	Dusseldorf	6	Helsinki	6	Helsinki	6	Toronto	6	Zurich	6	Stockholm	6	Paris
7	Frankfurt	7	Hamburg	7	Melbourne	7	Frankfurt	7	Munich	7	Boston	7	San Francisco
8	Copenhagen	8	Madrid	8	Boston	8	Sydney	8	Oslo	8	Zurich	8	New York
9	Geneva	9	Berlin	9	San Francisco	9	Birmingham	9	Hong Kong	9	Washington DC	9	London
10	Basel	10	Lisbon	10	Sydney	10	Manchester	10	Frankfurt	10	Atlanta	10	Tokyo

All three QoL rankings jointly listed certain cities as high-ranking, such as: Zurich, Vienna, and Copenhagen. Moreover, Monocle ranked Tokyo and Munich highly (second and third positions), and Mercer rated Vancouver highly (third position). The cities mentioned were also listed in the S&R rankings, but only Copenhagen ranked highly in two of the four S&R rankings. Some top cities in the S&R rankings (i.e., Rotterdam, London, and Toronto) were not on the QoL ranking lists.

4. Findings and Discussion

4.1. The Relationship between Water and Climate Aspects and Domains of the Quality of Urban Life

Cities' particular sensitivity to the effects of climate change should be taken into account in developing a vision of sustainable development for urban policy. Therefore, the cities' development perspective should be changed from the priority of economic development to ensuring a high quality of life for citizens. Nowadays, it is the QoL that states the competitiveness of cities [66].

Water and the climate have direct or indirect effects on all areas of human life, but remained poorly incorporated in the domains and criteria of the considered QoL assessments. Of the seven domains (including 29 categories) used in these studies, the water-related metrics were included in four: Architecture and urbanism (two categories: housing and outdoor spaces), Infrastructure (two categories: transport and utilities), Nature (three categories: access to nature, natural environment, and pollution), and Health and well-being (only one category: health and community support). Surprisingly, only one out of the 24 studied QoL tools took into account the effects of water on health and well-being, whereas the results of studies, such as the BlueHealth project [28], indicated the benefits of water bodies, in terms of public health and well-being.

The climate-related aspects were considered in two domains only: Infrastructure (two categories: transport and technologies) and Nature (two categories: natural environment and pollution). According to the literature [67–71], the types of land cover, buildings, and urban structures have a strong influence on the urban topoclimate (especially the radiant temperature and urban heat island); however, none of the rankings considered the impact of this aspect on the quality of life in the domains of Architecture and urbanism or Health and well-being.

To address the challenges of water management and adaptation to climate change in cities—especially in their historic centers—it is necessary to change the approach towards the perception and assessment of the roles of water and greenery [72–74]. Leading international organizations, such as the IWA, have provided solutions—for example, by developing principles for Water Wise Cities (2017) [75]—to encourage collaborative action "so that local governments, urban professionals, and individuals actively engage in addressing and finding solutions for managing all waters of the city" [29].

4.2. The Most Comprehensive QoL Assessment Tools Including Water and Climate Aspects

In the vast majority of instruments, the water and climate-related indicators were omitted or underestimated. The water-related indicators were included only in eight of the international rankings, two Polish tools, and one Polish ranking, whereas climate-related aspects were incorporated to a small degree in the 10 international and two Polish rankings, as well as in two Polish tools. The most diverse and comprehensive tools turned out to be *Monocle's Quality of Life Survey* (up to five different water-related indicators and up to five metrics and three different categories related to climate), *Mercer's Quality of Living Ranking Survey* (up to three indicators from all sub-categories of water, and all categories of climate-related aspects), *Deutsche Bank Liveability Survey* (up to three different water-related aspects, and up to five metrics and two different categories related to climate), and *How's life going? 2020 Measuring well-being* (up to three different water-related aspects, and up to six metrics from all categories of climate-related aspects.

QoL appraisal instruments can be used not only to assess the quality of life in entire cities, but also to compare living standards in individual districts, and especially downtown neighborhoods. However, the tools need to be remodeled in order to take into account the most important global challenges [32,33], the SDGs, and the growing vulnerability of an aging city population [76].

4.3. Comparison of the International and Polish QoL Assessment Tools

The International and Polish QoL assessment tools differed significantly, according to the water-related indicators. This may have resulted from an outdated approach to water management in Poland [77] and still low levels of climate and ecological awareness [78]. In international rankings, the indicators concerned a variety of aspects, such as *pollution, quality and availability of water, sanitation,* and *water stress;* whereas in Polish tools, only *pollution-* and *sanitation*-related indicators were included. International rankings were also more varied than Polish tools with regard to climate and weather-related indicators. However, the most common metrics (related to air pollution) were used in both groups of tools. In one Polish tool and several international rankings, indicators related to environmental initiatives were also included. The indicators that were omitted in Polish tools were related to *temperature, humidity, weather, climate, CO_2 emissions,* and *material footprint*.

4.4. Indicators Used in QoL Assessment Tools

The water aspects related to the domain of Architecture and urbanism were measured only by five appraisal instruments. Almost all metrics concerned the topic of *sanitation* in accommodation, consisting of *indoor flushing toilet and bath or shower* (2.10), *basic sanitary facilities* (2.11), *apartment connected to the sewage system* (2.14), and *warm running water, bathroom with bathtub and/or shower, flush toilet, sewers* (3.4). The most distinctive indicator was Monocle's *waterfronts* (2.3), which may be used to assess the image, aesthetics, form, and function of historic water districts. In terms of urban structures and architecture, there were no indicators linking land cover and types of buildings with the heat and humidity conditions in the city (i.e., green cover area, green roofs, share of low-carbon/ecologically certified buildings, rain water retention, and so on).

Metrics related to sustainable urban infrastructure and materials are missing. Based on the IWA Principles [29], we can propose some indicators related to the design of buildings and outdoor spaces, which allow for climate change adaptation and regenerative water services (e.g., *green roofs*), quick disaster recovery (e.g., *vital infrastructure),* the reduction of flood risks (e.g., *drainage solutions integrated with urban infrastructure design*), amplifying the livability with groundwater (i.e., *roadside green infrastructure*), and minimizing the impact of urban materials on water pollution (e.g., *roof and wall materials not emitting pollutants*). All aforementioned metrics should be considered in the context of urban heritage.

Almost half of the remaining water-related metrics corresponded to infrastructure, especially to the utilities category and for the evaluation of sanitation (2.6, 2.8, 3.1), as well as water quality (2.1, 2.3, 2.8) and availability (2.3, 2.4). Moreover, there was one water-related indicator in the transport category, *water as means of transport* (2.3), and one climate-related factor in the same category, *CO_2 Emissions Index,* which is *an estimation of CO_2 consumption due to traffic time* (2.4). One special climate-related indicator—*a website or App allows effective monitoring of air pollution* (2.6)—should be distinguished, as it can be classified as belonging to the technologies category. An essential indicator that was omitted in all compared tools was *wastewater reuse,* developed for the Cities Water Index [30].

The other water and climate-related indicators were correlated with the domain nature. There were aspects that may be included in the *pollution* category, such as *sewage* (2.2) *water pollution* (2.4), *clear water* (2.3), or *air pollution/quality* (2.2, 2.4, 2.6, 2.7, 2.9, 2.10, 2.11, 2.12, 2.14, 2.15, 3.1, 3.4); as well as in the natural environment category*: Water stress* (2.11), *humidity/ temperature* (2.1, 2.3, 2.4), *hours of sunshine* (2.3), *climate* (2.2), *weather* (2.7), *carbon neutral* (2.3), *material footprint* (2.11), *renewable energy* (2.11), *record of natural disasters* (2.2), and *environmental initiatives* (2.3, 2.12, 2.14).

Some of the most important aspects that were totally marginalized are those related to public safety and proactive approach. The key issue for citizens is their safety during extreme meteorological and hydrological phenomena. Extreme weather affects particularly vulnerable groups, such as the elderly [66].

For already-formed historic urban zones with specific surfaces, the indicators proposed by the IWA, such as *investing in coastal storm risks mitigation as well as flood and drought early warning systems* and *using an WRM framework and plan for drought mitigation strategies, seem to be highly reasonable* [29].

Metrics related to climate and water management may be developed on the basis of the scientific literature, recommendations (IWA Principles [29]), and appraisal instruments (e.g., Water City Index [79], Arcadis Sustainable Cities Water Index [30], Arcadis Sustainable Cities Index [63], Grosvenor Group Resilient Cities [64], and Global Ranking of Top 10 Resilient Cities [65]), and then adjusted to the domains and criteria proposed by the authors.

The results of our research confirm the conclusions of Estoque (2019) [32]: there is still a need to expand the range of indicators to incorporate climate-related issues into QoL assessment tools. It is crucial to include some essential climate variables, such as temperature and rainfall [80–82], as well as indicators such as exposure and sensitivity to climate hazards [83], thermal comfort, natural disaster exposure, and preparedness [32,84]. These effects of climate change, such as the urban heat island effect and rising temperatures, impact the well-being of citizens, especially the most vulnerable groups such as seniors [76,85].

4.5. Comparison of the Winning Cities in QoL and S&R Rankings

Only a few cities (Vienna, Zurich, Tokyo, Vancouver, Munich, Copenhagen, and Edinburgh) that ranked high in the QoL rankings were listed in the rankings evaluating the sustainability and climate resiliency (S&R) of cities. This is a serious cause for concern. Given the SDGs [2] and urban development agendas (e.g., ICOMOS_SDGs_Policy_Guidance_2021 [86], European New Bauhaus [87]), the significance and construction of global city rankings definitely necessitate a recalibration of the basics [88].

The top cities listed in both types of rankings were completely different, in terms of size, number of inhabitants, culture, and so on. The common issue was that each of these cities are among the richest places to live in the world. Environmental awareness and local politics are likely to be driven by financial resources [89–91], which are invested into eco-programs, and the existence of the world's leading universities, which shape public awareness [92]. This should be the subject of further research. Among them there are also old towns with historic districts that, despite the limitations of their spatial structures, face new challenges. They are all waterside cities, and all implement green spatial policies. A more detailed analysis of these policies and strategies—especially regarding natural and cultural heritage—can improve the management of urban environments and the associated quality of life.

The Paris Agreement (2015) [93] which recognises the role of cities, regions, and local authorities in addressing climate change, and invites them to "uphold and promote regional and international cooperation". The discrepancy between cities (rankings' leaders) emphasizes their role in looking for a common policy against climate threats.

5. Conclusions

Climate, environmental, and water crises affect the quality of life and well-being of residents in many ways, comprising one of the most urgent and important challenges facing modern cities. In the context of a broad debate on QoL assessment, sustainability, and livability indices [5,32,94], the socio-ecological paradigm emerges [83,88,94], which highlights the close links between QoL and climate and environmental variability [3,31–33]. Moreover, the aging of societies poses new challenges for the assessment of factors determining the quality of life in cities [4,36,46]. Our research identifies gaps in many QoL rankings based on a comparative analysis of 24 selected QoL assessment tools in terms of the scope of application of water and climate-related indicators. The authors' recommendation is that the rankings need to be refined if they are to help create climate-neutral and age-friendly cities.

Our review revealed that water and climate-related aspects are still underestimated in the appraisal instruments for QoL assessment. It has been analyzed that out of 24 QoL tools surveyed, only four rankings—Monocle's [42], Mercer's [41], Deutsche Bank's [43], and *How's life going? 2020 Measuring well-being* [50]—contain a broad spectrum of indicators related to water, climate, and environmental quality. Since international and Polish QoL assessment tools differ significantly, the tools should be remodeled to be more comprehensive and useful for decision-makers and urban planners. It has been found that water-related indicators most often focus on the efficiency of sanitation and water supply, disregarding broad ecosystem services, whereas climate-related metrics are based on current thermal and humidity conditions in cities, disregarding climate variability and increasing hydrometeorological extremes. Moreover, comparing the results of the QoL and S&R rankings showed that the top cities differed, whereas those that ranked highly in both types, such as Copenhagen, provide a valuable source of information on how to combine sustainability and resilience with high quality of life, by shaping the city structure as well as social and environmental policies.

Some of the QoL assessment tools take into account measures of the availability of urban green areas and the quality of the natural environment; however, ecosystem services—especially regulatory and cultural ones—remain underestimated. Moreover, the rankings do not take into account the dynamics of environmental and climate changes and their long-term consequences.

Additionally, the importance of the presented research for urban planning and cultural heritage is a result not only of the growing climatic and environmental threats and their impact on the quality of life. Concentration of water and climate-related problems also affects the possibility of modernizing the urban fabric in historic centers of large cities. Among them there are the urban heat island effect, heat waves, and coastal, fluvial, and pluvial floods.

The historical districts of many cities (including Polish ones) are typically inhabited by aging communities [8–10], particularly sensitive to extreme weather phenomena [85]. There has been little research addressing the relationship between quality of life in aging cities and sustainable urban development [95]. We have attempted to partially address this gap. The next step should be to remodel the rankings and use them to evaluate not only entire cities, but also individual districts [5], especially downtown. This can be helpful in detailing climate change adaptation plans and urban renewal programs.

However, we came across certain limitations, mainly concerning a lack of availability of materials and transparency in the definition of particular metrics used in appraisal instruments. Institutions often do not reveal the details of their research methodology [96]; thus, it is hard to deduce what the indicators consider. Appraisal instruments differ significantly, in terms of their purpose, sample size, contracting authorities, intended audience of the study, indicators and categories; thus, they are often incomparable [3–5]. That pose also challenges in adopting such indicators in urban planning [3].

Meeting global challenges and the SDGs requires the improvement of QoL assessment tools, in order to support municipalities in their decision-making process and to avoid choosing between quality of life and sustainable urban development and resilience.

Author Contributions: Conceptualization, A.P.-W. and A.J.-S.; methodology A.G., M.M. and A.P.-W.; formal analysis, A.P.-W.; investigation, A.P.-W.; resources, A.P.-W., A.G., M.M. and A.J.-S.; data curation, A.P.-W.; writing—original draft preparation, A.P.-W.; writing—review and editing, A.P.-W., A.G., M.M. and A.J.-S.; visualization, A.P.-W.; supervision, A.G., M.M. and A.J.-S.; project administration, A.P.-W.; funding acquisition, A.G. All authors have read and agreed to the published version of the manuscript.

Funding: This research was funded by Poznan University of Technology, grant number 012/SBAD/0181.

Institutional Review Board Statement: Not applicable.

Data Availability Statement: The data that support the findings of this study are available from the corresponding author, A.P.-W., upon reasonable request.

Conflicts of Interest: The authors declare no conflict of interest. The funders had no role in the design of the study; in the collection, analyses, or interpretation of data; in the writing of the manuscript, or in the decision to publish the results.

References

1. World Bank Group; Freire, M.; Lee, M.J.; Lee, M.J.; Bhada-Tata, P. *Cities and Climate Change: Responding to an Urgent Agenda*; World Bank Publications: Herndon, VA, USA, 2011; ISBN 978-0-8213-8667-5.
2. United Nations. *The Sustainable Development Goals Report 2019*; UN: New York, NY, USA, 2019. [CrossRef]
3. Lowe, M.; Whitzman, C.; Badland, H.; Davern, M.; Aye, L.; Hes, D.; Butterworth, I.; Giles-Corti, B. Planning Healthy, Liveable and Sustainable Cities: How Can Indicators Inform Policy? *Urban Policy Res.* **2015**, *33*, 131–144. [CrossRef]
4. Gawlak, A.; Matuszewska, M.; Ptak, A. Inclusiveness of Urban Space and Tools for the Assessment of the Quality of Urban Life—A Critical Approach. *Int. J. Environ. Res. Public Health* **2021**, *18*, 4519. [CrossRef]
5. Mittal, S.; Chadchan, J.; Mishra, S.K. Review of Concepts, Tools and Indices for the Assessment of Urban Quality of Life. *Soc. Indic. Res.* **2020**, *149*, 187–214. [CrossRef]
6. *World Urbanization Prospects: The 2018 Revision, Custom Data Acquired via Website*; United Nations. Department of Economic and Social Affairs, Population Division: New York, NY, USA, 2018; Available online: https://population.un.org/wup/DataQuery/ (accessed on 28 April 2021).
7. Statistics on European Cities; Eurostat: Luxembourg. 2019. Available online: https://ec.europa.eu/eurostat/statistics-explained/index.php/Statistics_on_European_cities#Population (accessed on 23 April 2021).
8. Temelová, J.; Dvořáková, N. Residential Satisfaction of Elderly in the City Centre: The Case of Revitalizing Neighbourhoods in Prague. *Cities* **2012**, *29*, 310–317. [CrossRef]
9. Chong, K.H.; To, K.; Fischer, M.M. Dense and Ageing: Social Sustainability of Public Places amidst High-Density Development. *Grow. Compact. Urban Density Sustain.* **2017**. [CrossRef]
10. Jancz, A.; Trojanek, R. Housing Preferences of Seniors and Pre-Senior Citizens in Poland—A Case Study. *Sustainability* **2020**, *12*, 4599. [CrossRef]
11. GHO | By Category | Life Expectancy and Healthy Life Expectancy-Data by Country. Available online: https://apps.who.int/gho/data/node.main.688 (accessed on 19 May 2021).

12. Gandini, A.; Garmendia, L.; San-Mateos, R. Towards sustainable historic cities: Adaptation to climate change risks. *Entrep. Sustain. Issues* **2017**, *4*, 319–327. [CrossRef]
13. Dastgerdi, A.S.; Sargolini, M.; Pierantoni, I. Climate Change Challenges to Existing Cultural Heritage Policy. *Sustainability* **2019**, *11*, 5227. [CrossRef]
14. Yang, H.; Lee, T.; Juhola, S. The Old and the Climate Adaptation: Climate Justice, Risks, and Urban Adaptation Plan. *Sustain. Cities Soc.* **2021**, *67*, 102755. [CrossRef]
15. Leyva, E.W.A.; Beaman, A.; Davidson, P.M. Health Impact of Climate Change in Older People: An Integrative Review and Implications for Nursing. *J. Nurs. Scholarsh.* **2017**, *49*, 670–678. [CrossRef] [PubMed]
16. Revi, A.; Satterthwaite, D.E.; Aragón-Durand, F.; Corfee-Morlot, J.; Kiunsi, R.B.R.; Pelling, M.; Roberts, D.C.; Solecki, W. Urban areas. In *Climate Change 2014: Impacts, Adaptation, and Vulnerability. Part A: Global and Sectoral Aspects. Contribution of Working Group II to the Fifth Assessment Report of the Intergovernmental Panel on Climate Change*; Field, C.B., Barros, V.R., Dokken, D.J., Mach, K.J., Mastrandrea, M.D., Bilir, T.E., Chatterjee, M., Ebi, K.L., Estrada, Y.O., Genova, R.C., et al., Eds.; Cambridge University Press: Cambridge, UK; New York, NY, USA, 2014; pp. 535–612. [CrossRef]
17. World Economic Forum. *The Global Risks Report 2021. Insight Report*; World Economic Forum: Cologny/Geneva, Switzerland, 2021.
18. *Global Warming of 1.5 °C. An IPCC Special Report on the Impacts of Global Warming of 1.5°C above Pre-Industrial Levels and Related Global Greenhouse Gas Emission Pathways, in the Context of Strengthening the Global Response to the Threat of Climate Change, Sustainable Development, and Efforts to Eradicate Poverty*; IPPC: Rome, Italy, 2019; Available online: https://www.ipcc.ch/sr15/ (accessed on 22 April 2021).
19. Rosenzweig, C.; Solecki, W.D.; Hammer, S.A.; Mehrotra, S. *Climate Change and Cities: First Assessment Report of the Urban Climate Change Research Network*; Cambridge University Press: Cambridge, UK, 2011; Available online: https://books.google.pl/books?hl=en&lr=&id=9GHW24KCuBYC&oi=fnd&pg=PR7&dq=climate+change+impact+on+seniors%27+quality+of+life+and+cities++&ots=_4rBX1v_MC&sig=ZrrYpvXBrrvRfMIRffRmnyKlpbg&redir_esc=y#v=onepage&q=climate%20change%20impact%20on%20seniors\T1\textquoteright%20quality%20of%20life%20and%20cities&f=false (accessed on 23 April 2021).
20. Benevolenza, M.A.; DeRigne, L. The impact of climate change and natural disasters on vulnerable populations: A systematic review of literature. *J. Hum. Behav. Soc. Environ.* **2019**, *29*, 266–281. [CrossRef]
21. Hoornweg, D.; Sugar, L.; Lorena Trejos Gomez, C. in press; Cities and Greenhouse Gas Emissions: Challenges and Opportunities. *Environ. Urban.* **2011**, *23*, 207–227. [CrossRef]
22. Mbow, H.-O.P.; Reisinger, A.; Canadell, J.; O'Brien, P. *Special Report on Climate Change, Desertification, Land Degradation, Sustainable Land Management, Food Security, and Greenhouse Gas Fluxes in Terrestrial Ecosystems (SR2)*; IPCC: Geneva, Switzerland, 2017; Available online: https://www.ipcc.ch/report/srccl/ (accessed on 10 August 2020).
23. Liang, L.; Deng, X.; Wang, P.; Wang, Z.; Wang, L. Assessment of the Impact of Climate Change on Cities Livability in China. *Sci. Total Environ.* **2020**, *726*, 138339. [CrossRef]
24. Hunt, A.; Watkiss, P. Climate Change Impacts and Adaptation in Cities: A Review of the Literature. *Clim. Chang.* **2011**, *104*, 13–49. [CrossRef]
25. Megarry, W. *Future of Our Pasts: Engaging Cultural Heritage in Climate Action: Report of the ICOMOS Workings Group on Climate Change and Cultural Heritage*; ICOMOS: Paris, France, 2019; Available online: https://adobeindd.com/view/publications/a9a551e3-3b23-4127-99fd-a7a80d91a29e/g18m/publication-web-resources/pdf/CCHWG_final_print.pdf (accessed on 19 May 2021).
26. *The United Nations World Water Development Report 2021: Valuing Water*; UNESCO: Paris, France, 2021. Available online: http://www.unesco.org/reports/wwdr/2021/en/download-the-report (accessed on 23 March 2021).
27. The 17 GOALS | Sustainable Development. Available online: https://sdgs.un.org/goals (accessed on 19 May 2021).
28. Grellier, J.; White, M.P.; Albin, M.; Bell, S.; Elliott, L.R.; Gascón, M.; Gualdi, S.; Mancini, L.; Nieuwenhuijsen, M.J.; Sarigiannis, D.A.; et al. BlueHealth: A Study Programme Protocol for Mapping and Quantifying the Potential Benefits to Public Health and Well-Being from Europe's Blue Spaces. *BMJ Open* **2017**, *7*, e016188. [CrossRef]
29. *The IWA Principles for Water Wise Cities*, 2nd ed.; IWA: London, UK, 2016; Available online: https://iwa-network.org/publications/the-iwa-principles-for-water-wise-cities/ (accessed on 19 May 2021).
30. Arcadis. *Sustainable Cities Water Index: Which Cities Are Best. Placed to Harness Water for Future Success*; Arcadis: Amsterdam, The Netherlands, 2016; Available online: https://www.arcadis.com/en/global/our-perspectives/which-cities-are-best-placed-to-harness-water-for-future-success-/ (accessed on 23 April 2021).
31. Roberts, W.O. Climate Change and the Quality of Life for the Earth's New Millions. *Proc. Am. Philos. Soc.* **1976**, *120*, 230–232. Available online: https://www.jstor.org/stable/986562 (accessed on 23 April 2021).
32. Estoque, R.C.; Togawa, T.; Ooba, M.; Gomi, K.; Nakamura, S.; Hijioka, Y.; Kameyama, Y. A Review of Quality of Life (QOL) Assessments and Indicators: Towards a "QOL-Climate" Assessment Framework. *Ambio* **2019**, *48*, 619–638. [CrossRef] [PubMed]
33. Newton, P.W. Liveable and Sustainable? Socio-Technical Challenges for Twenty-First-Century Cities. *J. Urban Technol.* **2012**, *19*, 81–102. [CrossRef]
34. China Council for International Cooperation on Environment and Development and WWF. *Report on Ecological Footprint in China*; China Council for International Cooperation on Environment and Development and WWF: Beijing, China, 2008.
35. Jones, C.; Newsome, D. Perth (Australia) as One of the World's Most Liveable Cities: A Perspective on Society, Sustainability and environment. *Int. J. Tour. Cities* **2015**, *1*, 18–35. [CrossRef]

36. *Global Age-Friendly Cities: A Guide*; World Health Organization: Geneva, Switzerland, 2007; Available online: https://www.who.int/ageing/publications/Global_age_friendly_cities_Guide_English.pdf (accessed on 23 April 2021).
37. Van der Weijst, L. *Measuring the Age-Friendliness of Cities: A Guide to Using Core Indicators*; WHO: Kobe, Japan, 2015; Available online: https://www.who.int/ageing/age_friendly_cities_guide/en/ (accessed on 23 April 2021).
38. Seniors, F.M.R. *Age-Friendly Rural and Remote Communities: A Guide*; Federal/Provincial/Territorial Ministers Responsible for Seniors: Toronto, ON, CA, 2007; Available online: https://www.phac-aspc.gc.ca/seniors-aines/alt-formats/pdf/publications/public/healthy-sante/age_friendly_rural/AFRRC_en.pdf (accessed on 23 April 2021).
39. Błędowski, P.; Szatur-Jaworska, B.; Bakalarczyk, R.; Łuczak, P.; Szweda-Lewandowska, Z.; Zrałek, M.; Plak, J.; Zubrzycka-Czarnecka, A. *System Wsparcia Osób Starszych w Środowisku Zamieszkania: Przegląd Sytuacji, Propozycja Modelu (System of Support for the Elderly in the Living Environment-Overview of the Situation, Model Proposal. Synthesis)*; Biuro Rzecznika Praw Obywatelskich: Warsaw, Poland, 2016; ISBN 978-83-65029-19-5.
40. *The Global Liveability Index 2019*; The Economist Intelligence Unit: London, UK, 2019; Available online: https://www.eiu.com/public/topical_report.aspx?campaignid=liveability2019 (accessed on 23 April 2021).
41. Mercer. Quality of Living-Location Reports. Available online: https://mobilityexchange.mercer.com/Insights/quality-of-living-rankings (accessed on 23 April 2021).
42. Monocle. Quality of Life Survey. *Monocle* **2019**, *13*, 41–65. Available online: https://monocle.com/magazine/issues/125/quality-of-life-survey/ (accessed on 23 April 2021).
43. Reid, J.; Nicol, C.; Allen, H. *Mapping the World's Prices 2019. Thematic Research*; Deutsche Bank AG: London, UK, 2019; Available online: https://www.dbresearch.com/PROD/RPS_EN-PROD/PROD0000000000494405/Mapping_the_world%27s_prices_2019.pdf?undefined&realload=F2~{|IVP2mCVDyfT99gzd7uKrqddmpspNMhF8BAP9Ave1vRpxfiznJaIQdrVH3WOK73x~{|GZGNgCWzIHs~{|wiDkeQw== (accessed on 23 April 2021).
44. Björnberg, A.; Phang, A.Y. *Euro Health Consumer Index 2018*; Health Consumer Powerhouse: Marseillan, France, 2019; Available online: https://healthpowerhouse.com/publications/ (accessed on 23 April 2021).
45. IMD World Competitiveness Center's Smart City Observatory, Singapore University of Technology and Design. *IMD Smart City Index*; The IMD World Competitiveness Center: Lausanne, France; Beach Road, Singapore, 2019; Available online: https://www.imd.org/research-knowledge/reports/imd-smart-city-index-2019/ (accessed on 23 April 2021).
46. Kubendran, S.; Soll, L.; Irving, P. Best Cities for Successful Aging 2017; Milken Institute, Santa Monica, California, US 2017. Available online: https://milkeninstitute.org/reports/best-cities-successful-aging-2017 (accessed on 23 April 2021).
47. Human Development Report 2019. *Beyond Income, beyond Averages, beyond Today: Inequalities in Human Development in the 21st Century*; UNDP: New York, NY, USA, 2019; Available online: http://hdr.undp.org/sites/default/files/hdr2019.pdf (accessed on 23 April 2021).
48. *Final Report of the Expert Group on Quality of Life Indicators*; Publications office of the European Union, Eurostat: Luxembourg, 2017. [CrossRef]
49. *European Quality of Life Survey 2016: Quality of life, Quality of Public Services, and Quality of Society*; Publications Office of the European Union: Luxembourg, 2017. [CrossRef]
50. *How's Life? 2020: Measuring Well-Being*; OECD: Paris, France, 2020; Available online: https://www.oecd-ilibrary.org/economics/how-s-life/volume-/issue-_9870c393-en (accessed on 19 May 2021).
51. *Quality of Life in Cities: Perception Survey in 79 European Cities*; Publications Office: Luxembourg, 2013. [CrossRef]
52. *2018 Active Ageing Index. Analytical Report*; UNECE/European Commission: Geneva, Switzerland, 2019; Available online: https://unece.org/population/active-ageing-index (accessed on 23 April 2021).
53. Polityka & Akademia Górniczo-Hutnicza. *Ranking Jakości Życia. Wymiary Szczęścia (Quality of Life Ranking. Dimensions of Happiness)*; Polityka & Akademia Górniczo-Hutnicza: Cracow; Warsaw, Poland, 2018.
54. Wałachowski, K.; Król, S. *Uciekające Metropolie. Ranking 100 Polskich Miast (Runaway Metropolises. Ranking of 100 Polish Cities)*; Klub Jagielloński: Kraków, Polska, 2019.
55. Bendowska, M.; Bieńkuńska, A.; Ciecieląg, P.; Luty, P.; Sobesjański, K.; Wójcik, J. *Jakość Życia w Polsce. Edycja 2017 (Quality of Life in Poland 2017 Edition)*; Central Statistical Office: Warsaw, Poland, 2017.
56. CBOS. *Zadowolenie z Życia (Life Satisfaction)*; CBOS: Warsaw, Poland, 2020.
57. Batorski, D.; Białowolski, P.; Czapiński, J.; Grabowska, I.; Kotowska, I.E.; Panek, T.; Pawlak, K.; Pytkowska, J.; Saczuk, K.; Strzelecki, P.; et al. *Diagnoza Społeczna 2015. Warunki i Jakość Życia Polaków. Raport (Social Diagnosis 2015. Objective and Subjective Quality of Life in Poland. Report)*; Czapiński, J., Panek, P., Eds.; Rada Monitoringu Społecznego: Warsaw, Poland, 2015.
58. Rokicka, E.; Petelewicz, M.; Woźniak, W.; Dytrych, J.; Przybylski, B.K.; Kruczkowska, P.; Drabowicz, T. *Jakość Życia Mieszkańców Łodzi i Jej Przestrzenne Zróżnicowanie. (Quality of Life of Lodz Inhabitants and City's Spatial Diversity)*; Rokicka, E., Ed.; Wydawnictwo Uniwersytetu Łódzkiego: Lodz, Poland, 2013.
59. *Jak Się Żyje Osobom Starszym w Polsce (How Is Life for the Elderly in Poland)*; Central Statistical Office: Warsaw, Poland, 2012. Available online: https://stat.gov.pl/obszary-tematyczne/warunki-zycia/dochody-wydatki-i-warunki-zycia-ludnosci/jak-sie-zyje-osobom-starszym-w-polsce-,8,1.html (accessed on 19 May 2021).
60. Száraz, L. The Impact of Urban Green Spaces on Climate and Air Quality in Cities. *Geogr. Locality Stud.* **2014**, *2*, 326–354.
61. Kozak, D.; Henderson, H.; de Castro Mazarro, A.; Rotbart, D.; Aradas, R. Blue-Green Infrastructure (BGI) in Dense Urban Watersheds. The Case of the Medrano Stream Basin (MSB) in Buenos Aires. *Sustainability* **2020**, *12*, 2163. [CrossRef]

62. Hamel, P.; Tan, L. Blue–Green Infrastructure for Flood and Water Quality Management in Southeast Asia: Evidence and Knowledge Gaps. *Environ. Manag.* **2021**, 1–20. [CrossRef]
63. Citizen Centric Cities. *The Sustainable Cities Index 2018*; Arcadis: Amsterdam, The Netherlands, 2018; Available online: https://www.arcadis.com/en/global/our-perspectives/sustainable-cities-index-2018/citizen-centric-cities/ (accessed on 21 February 2021).
64. *Resilient Cities: A Grosvenor Research Report*; Grosvenor: San Francisco, CA, USA, 2014; Available online: https://www.adaptationclearinghouse.org/resources/resilient-cities-a-grosvenor-research-report.html (accessed on 28 April 2021).
65. Cohen, B. *Global Ranking of Top 10 Resilient Cities*; TriplePundit: Northampton, MA, USA, 2011; Available online: https://www.triplepundit.com/story/2011/global-ranking-top-10-resilient-cities/76411 (accessed on 28 April 2021).
66. Wyzwania i Rekomendacje Dla Krajowej Polityki Miejskiej. *Raport Tematycznych Grup Eksperckich Kongresu Polityki Miejskiej 2019*; Kongres Polityki Miejskiej: Katowice, Poland, 2019; Available online: http://obserwatorium.miasta.pl/wp-content/uploads/2020/07/raport-wyzwania-i-rekomendacje-krajowa-polityka-miejska-Rajmund-Ry%C5%9B-Pawe%C5%82-G%C3%B3rny-Agnieszka-Sobol-Alina-Muzio%C5%82-Wec%C5%82awowicz.pdf (accessed on 19 May 2021).
67. Oke, T.R. City Size and the Urban Heat Island. *Atmos. Environ.* **1973**, *7*, 769–779. [CrossRef]
68. Oke, T.R.; Mills, G.; Christen, A.; Voogt, J.A. *Urban Climates*; Cambridge University Press: Cambridge, UK, 2017.
69. Voogt, J.A.; Oke, T.R. Thermal remote sensing of urban climates. *Remote Sens. Environ.* **2003**, *86*, 370–384. [CrossRef]
70. Davtalab, J.; Deyhimi, S.P.; Dessi, V.; Reza Hafezi, M.; Adib, M. The impact of green space structure on physiological equivalent temperature index in open space. *Urban Clim.* **2020**, *31*, 100574. [CrossRef]
71. Fortuniak, K. Numerical estimation of the effective albedo of an urban canyon. *Theor. Appl. Climatol.* **2008**, *91*, 245–258. [CrossRef]
72. Januchta-Szostak, A. *River-Friendly Cities*, 1st ed.; Peter Lang: Berlin, Germany; Bern, Switzerland; Bruxelles, Belgium; New York, NY, USA; Oxford, UK; Warszawa, Poland; Wien, Austria, 2020. [CrossRef]
73. Krauze, K.; Wagner, I. From Classical Water-Ecosystem Theories to Nature-Based Solutions—Contextualizing Nature-Based Solutions for Sustainable City. *Sci. Total Environ.* **2019**, *655*, 697–706. [CrossRef] [PubMed]
74. Kundzewicz, Z.W. Non-structural Flood Protection and Sustainability. *Water Int.* **2002**, *27*, 3–13. [CrossRef]
75. Koop, S.H.A. Towards Water-Wise Cities: Global Assessment of Water Management and Governance Capacities. Ph.D. Thesis, Utrecht University, Utrecht, The Netherlands, 2019. Available online: https://dspace.library.uu.nl/handle/1874/378386 (accessed on 23 April 2021).
76. Watts, N.; Amann, M.; Ayeb-Karlsson, S.; Belesova, K.; Bouley, T.; Boykoff, M.; Byass, P.; Cai, W.; Campbell-Lendrum, D.; Chambers, J. The Lancet Countdown on Health and Climate Change: From 25 Years of Inaction to a Global Transformation for Public Health. *Lancet* **2018**, *391*, 581–630. [CrossRef]
77. Januchta-Szostak, A.; Banasik, K.; Chudziński, P.; Drzewiecki, S.; Hausner, J.; Jania, J.; Kundzewicz, Z.; Kutek, K.; Konieczny, R.; Licznar, P.; et al. Alert Wodny Nr 3–Woda w Miastach. *Gospod. Wodna* **2020**, *8*, 4–6.
78. Sobol, A. *Raport Roboczy Grupy Eksperckiej Kongresu Polityki Miejskiej Ds. Środowiska i Adaptacji Do Zmian Klimatu, Obserwato-Rium Polityki Miejskiej*; IRMiR: Cracow, Poland, 2019.
79. Ćmielowski, M.; Głowacki, J.; Hausner, J.; Kudłacz, M.; Kutek, K. *Water City Index 2020. Ranking Efektywności Wykorzystania Zasobów Wody w Polskich Miastach*; Arcadis: Amsterdam, The Netherlands, 2020; Available online: https://oees.pl/water-city-index-2020/ (accessed on 23 April 2021).
80. Li, G.; Weng, Q. Measuring the Quality of Life in City of Indianapolis by Integration of Remote Sensing and Census Data. *Int. J. Remote Sens.* **2007**, *28*, 249–267. [CrossRef]
81. Rao, K.R.M.; Kant, Y.; Gahlaut, N.; Roy, P.S. Assessment of Quality of Life in Uttarakhand, India Using Geospatial Techniques. *Geocarto Int.* **2012**, *27*, 315–328. [CrossRef]
82. Morais, P.; Camanho, A.S. Evaluation of Performance of European Cities with the Aim to Promote Quality of Life Improvements. *Omega* **2011**, *39*, 398–409. [CrossRef]
83. Estoque, R.C.; Murayama, Y. A Worldwide Country-Based Assessment of Social-Ecological Status (c. 2010) Using the Social-Ecological Status Index. *Ecol. Indic.* **2017**, *72*, 605–614. [CrossRef]
84. PwC. *Cities of Opportunity*; PwC: London, UK, 2016.
85. Lee, J.Y.; Kim, H. Projection of Future Temperature-Related Mortality Due to Climate and Demographic Changes. *Environ. Int.* **2016**, *94*, 489–494. [CrossRef]
86. Labadi, S.; Giliberto, F.; Rosetti, I.; Shetabi, L.; Yildirim, E. *Heritage and the Sustainable Development Goals: Policy Guidance for Heritage and Development Actors*; ICOMOS: Warsaw, Poland, 2021; Volume 1, pp. 1–69.
87. New European Bauhaus: Beautiful, Sustainable, Together. Available online: https://europa.eu/new-european-bauhaus/index_en (accessed on 19 May 2021).
88. Bieri, D. Are Green Cities Nice Places to Live? Examining the Link between Urban Sustainability and Quality of Life. *Mich. J. Sustain.* **2013**, *1*, 44–73. [CrossRef]
89. *Promoting Investment for Sustainable Development in Cities*; UNCTAD: Geneva, Switzerland, 2019; Available online: https://unctad.org/system/files/official-document/diaepcbinf2019d1_en.pdf (accessed on 26 May 2021).
90. Marco, K.; Mihir, P.; Hannes, B. *Financing Sustainable Urbanization: Counting the Costs and Closing the Gap*; UN-Habitat: Nairobi, Kenya, 2020; Available online: https://unhabitat.org/sites/default/files/2020/02/financing_sustainable_urbanization_-_counting_the_costs_and_closing_the_gap_february_2020.pdf (accessed on 26 May 2021).

91. *2019 Forum of the Standing Commitee on Finance: Climate Finance and Sustainable Cities;* United Nations. Framework Convention on Climate Change: Bonn, Germany, 2019. Available online: https://unfccc.int/sites/default/files/resource/SCF%20Forum%202019%20report_final.pdf (accessed on 26 May 2021).
92. Mochizuki, Y.; Fadeeva, Z. Competences for Sustainable Development and Sustainability: Significance and Challenges for ESD. *Int. J. Sustain. High. Educ.* **2010**, *11*, 391–403. [CrossRef]
93. *Paris Agreement;* United Nations: Bonn, Germany, 2015. Available online: https://unfccc.int/sites/default/files/english_paris_agreement.pdf (accessed on 26 May 2021).
94. Leach, J.M.; Braithwaite, P.A.; Lee, S.E.; Bouch, C.J.; Hunt, D.V.; Rogers, C.D. Measuring Urban Sustainability and Liveability Performance: The City Analysis Methodology. *Int. J. Appl. Sci. Technol.* **2016**, *1*, 86–106. [CrossRef]
95. Qian, Q.K.; Ho, W.K.; Ochoa, J.J.; Chan, E.H. Does Aging-Friendly Enhance Sustainability? Evidence from Hong Kong. *Sustain. Dev.* **2019**, *27*, 657–668. [CrossRef]
96. Acuto, M.; Pejic, D.; Briggs, J. Taking City Rankings Seriously: Engaging with Benchmarking Practices in Global Urbanism. *Int. J. Urban Reg. Res.* **2021**, *45*, 363–377. [CrossRef]

Article

Multi-Attribute Analysis of Contemporary Cultural Buildings in the Historic Urban Fabric as Sustainable Spaces—Krakow Case Study

Ernestyna Szpakowska-Loranc

Faculty of Architecture, Cracow University of Technology, 31-155 Krakow, Poland; eszpakowska@pk.edu.pl

Abstract: This study concerns contemporary cultural buildings in the historic city centre of Krakow, Poland, and their assessment in terms of sustainability. The paper aims to bridge a research gap in previous studies on pluralistic values and the impact of cultural heritage on sustainability. The comparative case study conducted in Krakow aims to evaluate the functioning and potential of the space towards achieving the following five goals: accessibility, conservation, mix of functions, aesthetics, comfort and sociability. The perception of these buildings and the public space around them by the city residents, as well as their operation during unexpected circumstances, such as the COVID-19 pandemic, were also evaluated. The author combined an on-site analysis, behavioural mapping and a survey. The results correlate the liveability and aesthetics of public spaces along with the amount and quality of greenery found there with the comfort of users and the popularity of particular places. This paper highlights how important it is to create cultural spaces in a historic city to develop a range of their activities linked to the surrounding public spaces and green areas. Activating cultural spaces and connecting them to sustainability goals is especially important when faced with declining tourism.

Keywords: sustainable development; cultural heritage; building heritage; inclusive space; COVID-19 urban planning; post-pandemic urban planning; Krakow

Citation: Szpakowska-Loranc, E. Multi-Attribute Analysis of Contemporary Cultural Buildings in the Historic Urban Fabric as Sustainable Spaces—Krakow Case Study. *Sustainability* **2021**, *13*, 6126. https://doi.org/10.3390/su13116126

Academic Editor: Tomonobu Senjyu

Received: 1 May 2021
Accepted: 24 May 2021
Published: 28 May 2021

Publisher's Note: MDPI stays neutral with regard to jurisdictional claims in published maps and institutional affiliations.

1. Introduction

Implementing high-budget engineering solutions for sustainability is not always possible for economic reasons. The countries of lower economic power are lagging behind in this race [1], which is also evident in Central and Eastern European countries when compared to their wealthier western neighbours [2,3]. Central and Eastern European countries are among those that still rely heavily on fossil fuels [4–6] and struggle with an adequate percentage of recycled rubbish [7] and small-scale water retention in built-up areas [8]. This does not mean that the SDGs (Sustainable Development Goals) are to be abandoned with no effort to achieve them. A certain degree of sustainability should be achieved through other methods and then it should be extended to include the measures mentioned above. Such measures can be found in the three pillars of sustainability [9].

The problem is exacerbated in the areas with historic structures, representing a significant percentage of the built-up area in Europe. This is evidenced by the 26.4% share of dwellings built before 1945 [10]. Being of indisputable historical, social, aesthetic and economic value, these heritage areas are worth protecting, but they also generate environmental problems due to their typology of traditional towns. Characterised by dense development and limited greenery, historic centres of large cities become urban heat islands [11], contributing to increased air pollution and other environmental pathologies [12]. The conversion of protected historic heritage into NZEBs (nearly zero-energy buildings) presents many difficulties [13]. Given the need to protect cultural values in these areas, measures of a different kind must be planned and implemented. These measures must include urban layout, buildings and public interiors. It is easier to introduce eco-friendly solutions where urban investments are concerned, as they are generally less driven by the desire for investment profit. When planned properly, public buildings, including cultural facilities and adjacent urban spaces, can both improve the ecology of a given space and increase its social value [14]. The areas in question are strongly affected by the phenomenon of gentrification, which was particularly evident in 2020. The decline in tourism during the COVID-19 pandemic demonstrated the lack of resilience in districts that proved to be mono-functional desolate spaces when deprived of foreign visitors [15]. The cultural sector in Krakow is facing a crisis, halting its programme activities and moving from the urban space to the Internet [16].

The research examines contemporary public buildings serving a cultural function in the historic urban fabric in the context of sustainable development and spatial inclusion (Figure 1). Central and Eastern European cities constitute a good field to analyse the situation of developing metropolises. Krakow, chosen as the research site, is among them. It is the second-largest Polish city whose number of tourist visitors has steadily increased over the past 10 years to 14,050,000 in 2019, but dropping to 7,950,000 in 2020 due to the pandemic [17,18]. Although the city has SDG ideas embedded in its strategy [19], it does not introduce NZEB technologies in public buildings.

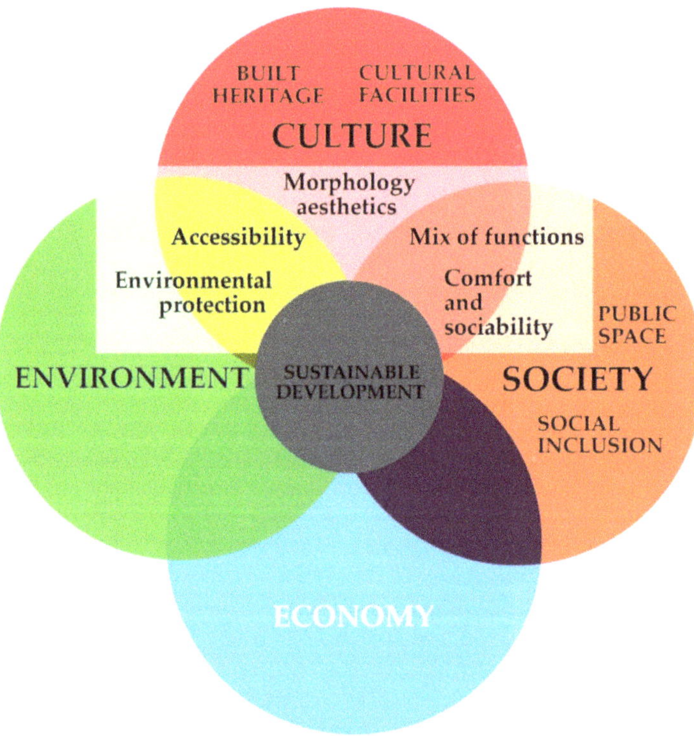

Figure 1. Scope of the research.

1.1. *Sustainability and Culture*

Despite a considerable amount of research on sustainability [20,21] and an emerging number of studies that link this issue to the functioning of cultural buildings in the urban fabric [22], a research gap in this topic is still present. This has been confirmed by Petti, Trillo and Tsube Macone who focus on gauging the implementation of the SDGs with reference to cultural heritage and recommend 'future research into the pluralistic values and impact of cultural heritage in achieving sustainable development' [23]. The lack of the studies on the functioning of buildings excluding their main cultural functions, but related to the surrounding public spaces, should be remedied by helping these facilities and areas to function more effectively for city residents. This is particularly relevant given the fact that the majority of the European metropolitan and smaller town centres are historic areas that should be subject to the process of achieving the SDGs alongside new developments. Dessein et al. [24], in the article entitled 'Culture in, for and as Sustainable Development', established three models of the relationship between culture and sustainability: (1) culture as an achievement in development, (2) culture as a resource and condition for development and (3) culture as a semiosis and development as a cultural process [25]. Of these three models, even the first one requires extensive measures in terms of urban sustainability. The development of cultural buildings in cities as public spaces can serve the purposes of walkable communities [14], or implement the six core design theories for health and well-being characterised by Cushing and Miller [26]. This model of linking culture and sustainability addresses the notion of culture and heritage most concisely (as the fourth pillar of sustainable development alongside the ecological, social and economic pillars). Nonetheless, it requires further research and implementation of measures for cultural institutions other than those directly related to the mission and programme activities of museums, as examined by Stylianou-Lambert et al. [27] and characterised by McGhie [28]. According to Dessein et al., the model risks being a limited approach, however, focused on protecting assets deemed cultural that are valued ('giving culture a voice of its own and an equal value'); it is sometimes too easily limited to a narrow definition of culture as the arts and creative-cultural sector [24].

Despite the concept of cultural sustainability and the 2015 ICOMOS (International Council on Monuments and Sites) declaration being introduced [29], there is published research that presents this issue with reference to the socio-economic post-communist reality conditioning the urban space in Central European countries. Moreover, apart from a few exceptions [22], the operation of the cultural buildings in Polish cities is commanded by museums. The research conducted by museums is mainly limited to the reception of their offers by visitors [30], whereas no evaluation takes into account their auxiliary functions, the layout of spaces enabling and customising their use and the public access to external space. This model of use or the problem of excluding these areas from the functioning urban fabric emerged during the COVID-19 pandemic, when the lockdown of cultural institutions in Poland (and elsewhere) resulted in the creation of deserted (ghost) spaces within the urban fabric. As the events of 2020 showed, the future is unknown [31]. According to a plausible hypothesis, Krakow, which has been packed with tourists until now [17], will have less tourist traffic and cultural institutions will be visited by fewer individual tourists and organised groups (also due to periods of distance learning). At the same time, due to periods of partial or total lockdown, the urban space is used for recreation and sport to a greater extent than before [32], and new methods of using it are emerging, such as the search for non-standard and small-group outdoor activities. Outdoor recreation and observing others are becoming even more desirable in public spaces than previously reported by Gehl [33], Carmona [34], London [14] and Cushing and Miller [26].

1.2. Research Aims

The following research aims to evaluate the spatial-functional potential of cultural buildings in the historic urban fabric for achieving sustainable development.

The following specific aims correspond to the general objective:

1. Examining which SDGs and to what extent are met by the buildings analysed in Krakow.
2. Examining how these buildings are perceived by the city residents. Do the citizens notice and respond well to the principles of urban ecology used in buildings?
3. Examining how these buildings operate during the pandemic. Are they likely to be flexible, inclusive spaces for other unforeseen events?

4. Finding a correlation between the formal (aesthetic) and functional shaping of space and the perception and use of space.

The research objective is the evaluation of contemporary cultural buildings in the historic central urban structure of Krakow (the area entered into the UNESCO World Heritage List and its buffer zone) in terms of urban ecology and inclusive, liveable and walkable public space. The findings are collated with municipal strategies.

The applicational relevance of the study for cultural and urban management professionals, as well as designers and environmental psychologists, consists of the search for a new design model in the historic urban tissue of Krakow. The identification of the shortcomings and correction of individual problems within cultural sustainability would enable the fulfilment of all six aspirations of the city set out in the strategy [19], as shown in Figure 2.

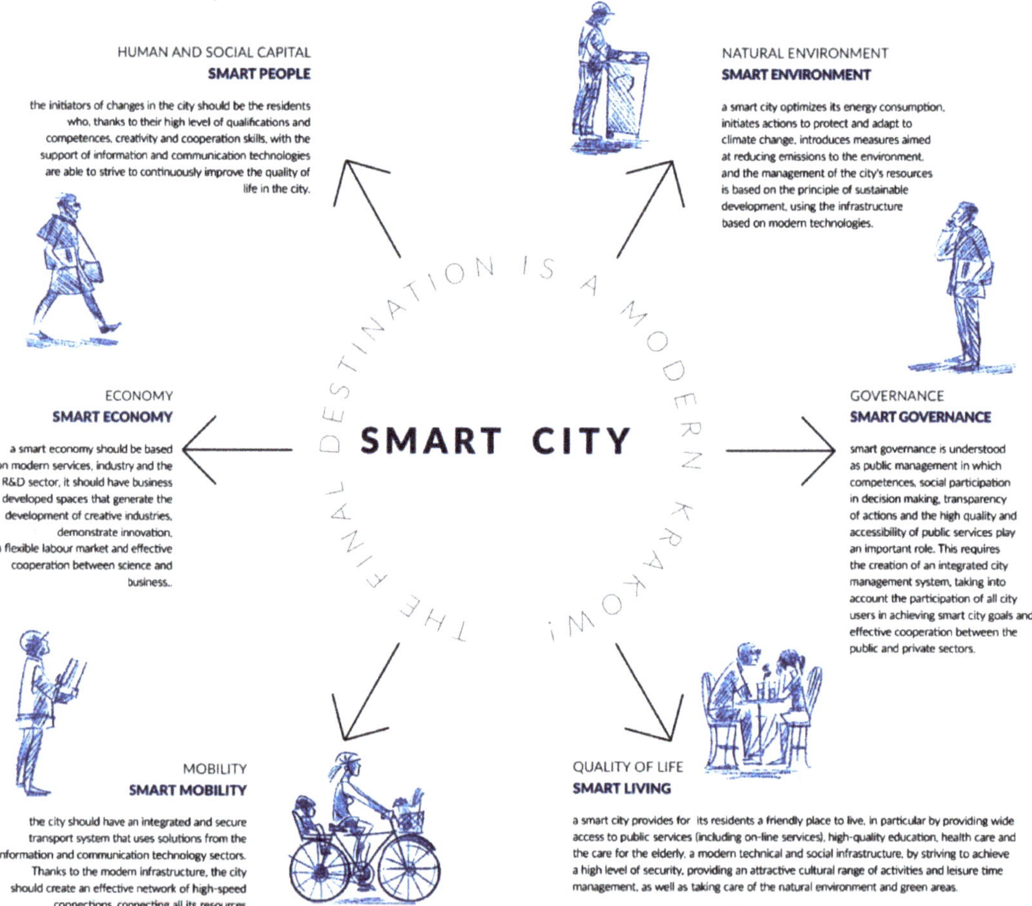

Figure 2. Smart city aspirations from Krakow's development strategy [19] (public domain). Publication developed on the basis of the contents of the Annex to Resolution No. XCIV/2449/18 of the Krakow City Council, multi-author text, editorial, graphic design E. Przybylska, graphic design M. Flis, translation IDEAGroup.

2. Materials and Methods

2.1. Selected Cases

The case study concerns the centre of Krakow, Poland's second-largest city with a historic fabric inscribed on the UNESCO list, located in Central Europe. This geopolitical situation is associated with limited economic resources in public institutions and problems resulting from the post-war development characteristics. In the second half of the 20th century and at the beginning of the 21st century, the historic fabric of Central European cities was first surrounded by modernist housing estates, then by low-rise detached and semi-detached sprawling development patterns and more recently by pathological dense multifamily housing developments devoid of greenery and public services. Central European cities are also overcoming deficiencies in transport infrastructure and public institutions as they are characterised by the underdevelopment of metropolitan functions [35]. Such conditions necessitate the search for sustainability in small-scale solutions with limited resources rather than high-budget engineering solutions. The study defines cultural buildings constructed after 1989 (Poland's liberation from Soviet domination and the transition from a centrally planned to a capitalist economy) as contemporary ones. Redeveloped or renovated sites are not examined. The spatial delimitation includes Krakow's UNESCO protection zone together with its buffer zone (Figure 3).

In terms of sustainability, Krakow was examined by Zachariasz, presenting its historic and contemporary green areas [36]; Telega, Telega and Bieda, measuring walkability with GIS [37]; Jarosińska and Gołda, opting for protecting the existing green areas [38]; Gyurkovich and Gyurkovich, analysing New Housing Complexes in Post-Industrial Areas [39]; and Dudzic-Gyurkovich, assessing urban development and compositional pressure in Mlynowka Park [40], among others. Porebska et al. [41] evaluated flood protection against the UNESCO world heritage site protection, while Kwartnik-Pruc and Trembecka [42] focused on the implementation of the Public Green Space Policy. In terms of culture, the city is the capital of the province, with the largest number of cultural centres in Poland [43]. Some of the city's contemporary cultural buildings have been researched by Jagodzińska in numerous publications ([44] among others) and Gyurkovich ([45] among others). Its cultural infrastructure is examined in terms of its offer and the residents and tourists who use it, as well as its reception [46]. The author has identified a research gap in the form of a functional-spatial analysis of contemporary cultural facilities in Krakow and their reception by city residents. Polish cultural institutions have been studied in terms of organised events and participants [43], but no clear link between new cultural buildings and sustainability has been shown.

At the same time, the observation of selected examples of cultural buildings in the historic fabric of Krakow and the analysis of relevant documents have revealed that the idea behind building these cultural facilities (coming from the municipal authorities most often) was to increase the offer of the city and districts, encouraging people to settle near the city centre and workplaces, which also contributes to the idea of the sustainable compact city [45,47]. The model of a cultural building in the space of Krakow after 1989 can be characterised as buildings on small plots in the city centre. Ten facilities were built in the UNESCO zone and buffer zone, and three outside them. Being a cultural heritage area, the research site is subject to a series of conservation acts affecting the form of buildings. The small size of the plots results in limited space for auxiliary functions, car parks and surrounding greenery.

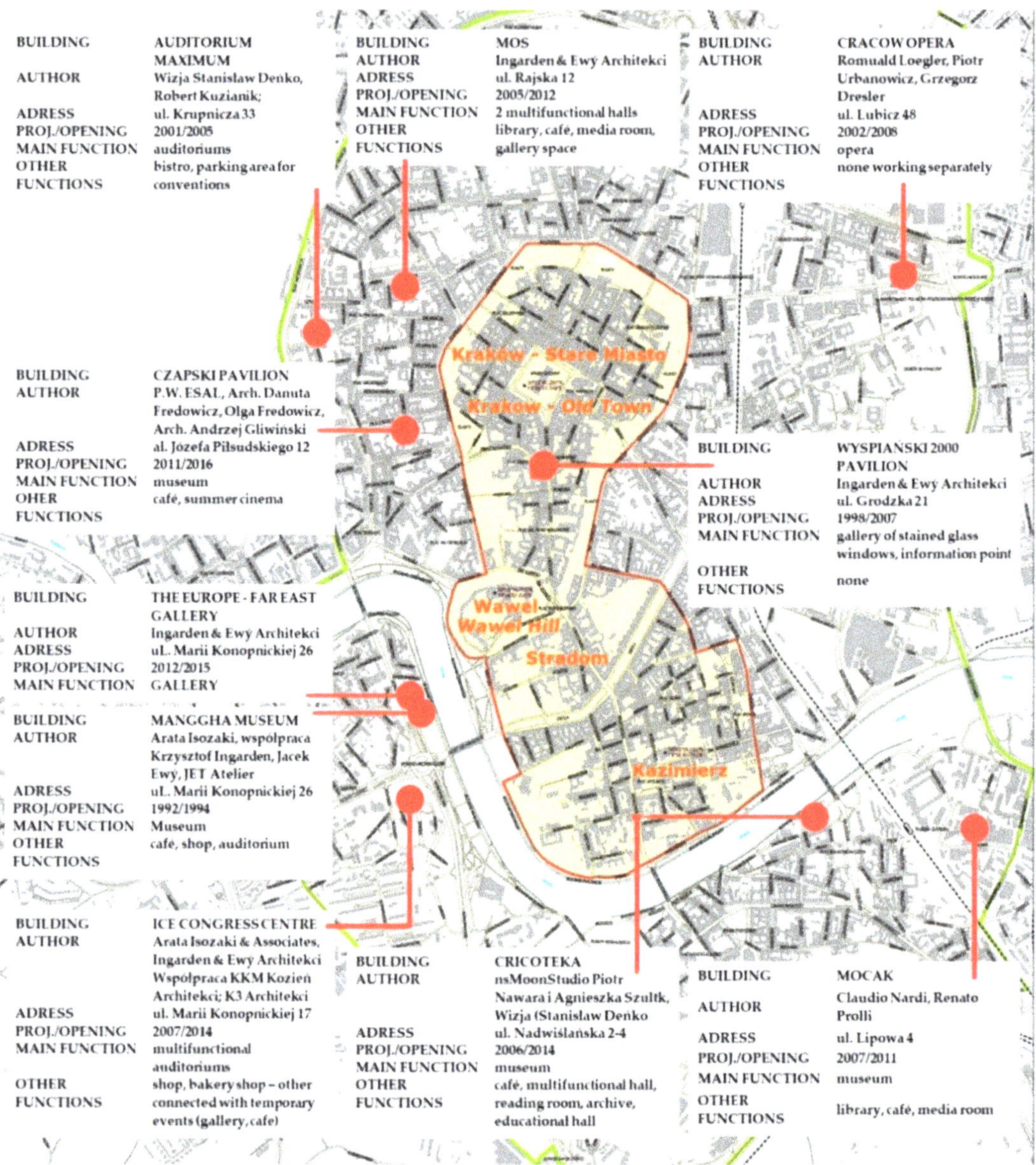

Figure 3. Map of cases shown in the map of inscribed World Heritage property. Red dots—location of cases. Red line—borders of the area entered into the UNESCO World Heritage List. Green line—buffer zone borders. Elaborated by the author, adapted from ref. [48] (public domain).

Figures 4 and 5, Supplementary Material S1 presents the analysed buildings.

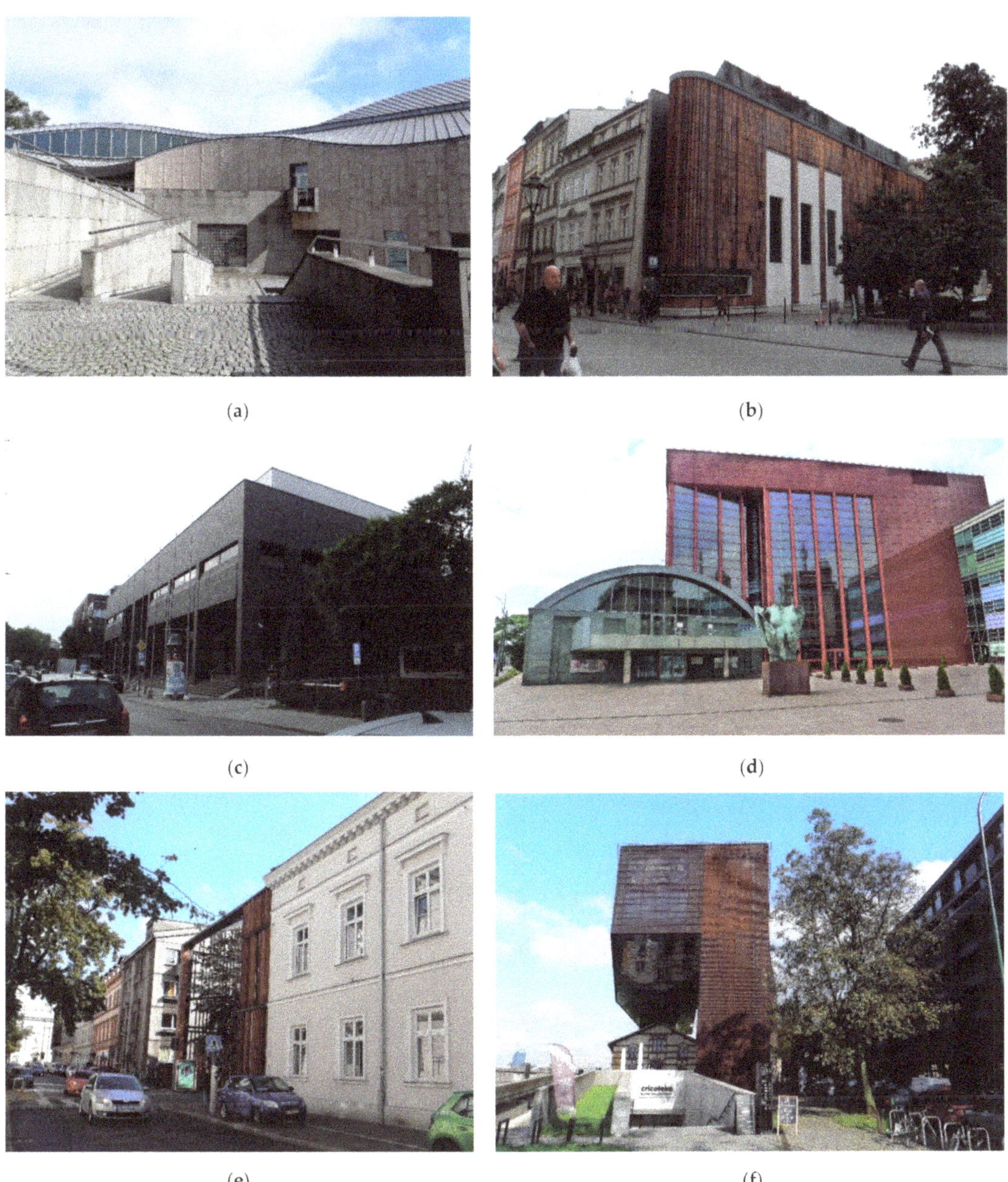

(a) (b) (c) (d) (e) (f)

Figure 4. (**a**) Manggha museum; (**b**) Wyspianski Pavilion; (**c**) Auditorium Maximum; (**d**) The Opera House; (**e**) MOS; (**f**) Cricoteka. Photos by the author.

Figure 5. (**a**) MOCAK; (**b**) ICE; (**c**) Czapski Pavilion; (**d**) Manggha complex: The Europe–Far East Gallery. Photos by the author.

2.2. Mixed Methods

The research employs Creswell's concurrent triangulation mixed methods approach which produces reliable results in a relatively short time [49]. The author collected qualitative and quantitative data separately by means of observation, survey and behavioural mapping and then compared them. The methods were chosen because the preliminary study, which was a qualitative evaluation of a selected case (the Krakow Opera), indicated the need to support the observational results with quantitative data. This approach is justified by Silverman, among others [50]. The author combined three methods, giving priority to method A due to limitations of methods B and C. The research methodology is presented in Figure 6. The research phases, methods and techniques are listed in Table 1.

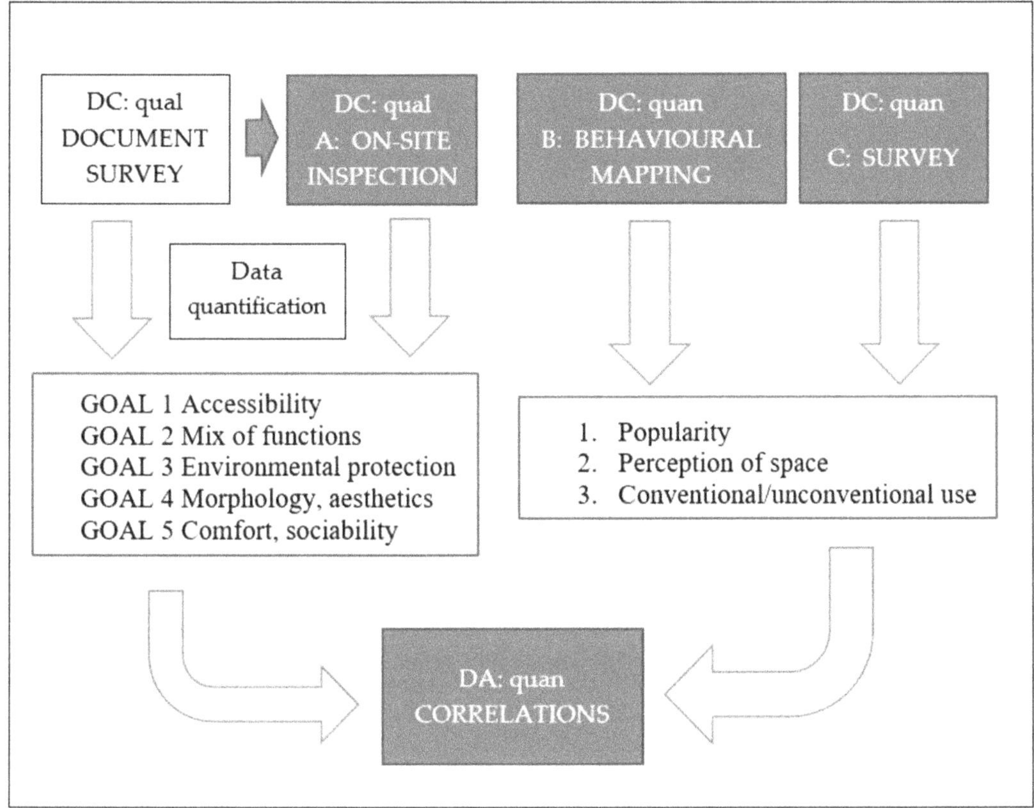

Figure 6. Research diagram. DC, data collection; DA, data analysis; qual, qualitative data; quan, quantitative data.

Table 1. Research methods and techniques.

	Method	Technique	Specific Purpose
A.	field research, literature and document review	review of authors' published descriptions, drawings of the buildings and magazine articles	indication of the degree of fulfilling urban ecology determinants
		review and critical studies of city strategies, statistical data	
		on-site inspection (query, critical studies, notes, photos, walkthrough)	
B.	behavioural mapping	behavioural mapping	analysis of the social perception of spaces and how the spaces function during the pandemics
C.	survey	a survey among inhabitants of Krakow	analysis of the social perception of spaces

2.2.1. Method A

The on-site observation and document review served as a qualitative and quantitative assessment of five groups of spatial factors that contribute to the realisation of the following five goals:

GOAL 1 Accessibility—the visual accessibility and infrastructure provision, as well as the quality of walking, cycling and public transport, have been assessed with respect to road transport and the needs of the disabled.

GOAL 2 Mix of functions—the number and variety of functions in and adjacent to the cases assessed, their arrangement, accessibility and presence in the external space have been assessed.

GOAL 3 Environmental protection—the quantity and quality of green infrastructure, pro-environmental solutions applied in buildings, including ways to reduce the urban heat island effect, have been assessed.

GOAL 4 Morphology, aesthetics—cultural heritage, urban morphology as well as image, forms and aesthetics of space were assessed.

GOAL 5 Comfort and sociability—the quality of space and social potential have been assessed.

Table 2 presents the source of the analysis criteria adopted based on the literature review. The criteria have been adopted drawing on expert knowledge in the field of architecture and urban planning. A five-point assessment scale is used for each factor.

Table 2. References of research tools in the method A.

Reference	Tool/Selection of Features Based on	Indicators
Gehl [51]	*'Life between buildings'*	goals 1, 5
Kesik [52]	*'Analysis of Pavements for Disabled Pedestrians in Metropolitan Cities'*	goal 1
Place Alliance [34]	The Ladder of Place Quality	goals 1, 4
NACTO [53]	*'Urban Bikeway Design Guide'*	goal 1
London [14]	*'Healthy placemaking'* principles assessment methodology	goal 1 goal 2
Gehl [33]	Quality criteria for the pedestrian environment, analysing ground floor of neighbouring buildings	goals 1, 2, 4
Cushing, Miller [26]	*'Creating great places'*, Salutogenic design	goal 2
Lehmann [54]	The 15 Principles of Green Urbanism: Energy and Materials, Water and Biodiversity	goal 3
Elliott, Eon, Breadsell [55]	GIDS to urban cooling	goal 3
US Environmental Protection Ag. [56]	*'EPA Heat Islands Cooling Strategies'*	goal 3
Center for Active Design [57]	*'Assembly: Civic Design Guidelines'*, Incorporating nature criteria	goal 3
Project for Public Spaces [58]	*'What makes a great place?'*	goals 4, 5
Lewicka [59]	characteristics of a phenomenological conservative concept of a place (as opposed to a non-place)	goal 4
Lofland [60]	elements contributing to the satisfaction of staying in a public space	goal 4
Gehl Institute [61]	*'The Inclusive Healthy Places Framework'*	goal 5

2.2.2. Method B

The fourth stage of the research involved behavioural mapping of the space around buildings using the snapshot method. The author observed the behaviour of pedestrians around the buildings, counted and classified them. The method aimed to see which venue generates the most traffic and what activities take place there, regardless of its cultural offer. This will allow one to discover the vitality of a place independent of the attractiveness of the programme offered by the cultural institution, and therefore lasting even during the pandemic.

The author visited all the places three times:

1. 23 September/2 October 2020—13:00–14:30;
2. 3 October 2020—16:40–17:50;
3. 6 October 2020—11:15–13:00.

Each visit took place in similar weather conditions for all the cases, i.e., approximately 20 °C. Such weather allowed the reflection of the environmental potential of the site as the aim was to compare behaviour in the spaces of the sites between them. During each 5-min visit at one location, the author counted the number of users and divided the people observed into three groups: users, passers-by, employees.

The behavioural mapping tool and the counting were based on the evaluation by the Project for Public Spaces [58] and Mantey's research that concerned the public spaces of Warsaw's suburbs [62]. Due to the need to move quickly between the buildings to obtain comparative results, the results were not plotted on a map but entered into a table.

The correlation between the results of the mapping and method A (on-site inspection with document survey) was examined. The author assumes that there should be a correlation between the outcomes of this method and the survey.

2.2.3. Method C

The author collected 173 questionnaire survey responses from Krakow's residents who were randomly assigned to the sample. The survey aimed to generalise the data from the sample and obtain conclusions about the perception of the buildings under study and the space around them. The survey method ensured quick collection of a large amount of data, quick orientation in the data set and access to the most diverse group of people providing information. The author aimed to collect the largest possible sample to obtain the most authoritative results.

The first phase of the research was conducted in September 2020, employing self-completion paper-based questionnaires, distributed by the author and her friends (20 questionnaires). To ensure the most reliable results, the author wanted to reach people who do not necessarily or rarely use the Internet. However, due to the need to increase the sample size in a shorter period and the threat of discretionary sampling and the presence of an interviewer skewing the results, the author decided to change the survey method to online self-completion questionnaires, ensuring a random selection of individuals. The author treated this phase as a pilot test, later included in the final results.

Subsequent surveys were conducted and collected in two rounds via Survio (free version) in October 2020 and March and April 2021 propagating information about them via Facebook in the following groups: Spotted Krakow, Krakowskie Mamy, Stowarzyszenie Architektów Oddział Kraków, Spotted Vilo. The survey was designed as a cross-sectional study. Due to the need to collect more data from people aged 60 and over, the author introduced a questionnaire distributed to the U3A (University of the Third Age) students in a further stage.

The survey included respondents' particulars, presented in the table below (Table 3). This was not used to compare data, but to check on an ongoing basis whether the over-representation of any group (e.g., age or artistic occupations) would not potentially skew the outcomes.

The core part of the questionnaire inquired, with regard to each facility under study, whether the respondent:

1. Knows the building;
2. Likes or dislikes each building;
3. Wants or does not want to enter the building when they are nearby;
4. Likes or dislikes its surroundings;
5. Wants or does not want to spend their free time around the building;
6. Feels the space around the building is suitably spacious (or too small/too big);
7. Feels that there is or there is not enough greenery there;
8. Will or will not come back to that place.

The full questionnaire is included in Supplementary Material S2.

Table 3. Respondents' particulars contained in the survey.

Question	Possible Answers
How old are you?	18–26/27–35/36–59/60 and over
What is your education?	Primary/lower secondary/vocational/upper secondary/higher
Do you have a profession related to art (painter, sculptor, architect, conservator, actor, musician, etc.)?	Yes/no
How many years have you lived in Krakow?	Less than one year/1–2 years/3–5 years/6–10 years/More than 10 years
How often do you use the city's cultural offer (exhibitions, cinema, theatre, etc.)?	Several times a month/About once a month/About once a quarter/1–2 times a year/Less often or not at all

The responses to questions 2–8 about each building could be given on a five-point semantic scale (−2, −1, 0, 1, 2). In the online survey, the slider was initially set to 0. The answers to these questions were not mandatory as the respondents might not have been familiar with all the buildings. As the slider remained in the 0 position when no response was given, the results did not reflect the specific number of responses from each group. However, the analysis of the number of options selected −2, −1, 1, 2 allowed for a case-by-case comparison. The author compared total scores being the products of ranks and frequencies of similar responses for each case. A stratified sample with equal allocation was applied. The data were presented on a semantic differential scale.

The data were used to compare buildings between individual cases in terms of perception of space in three groups of issues: knowledge of the building (I: question 1), positive or negative attitude towards space (II: questions 3, 4, 5, 6, 9) and perception of specific elements of space (III: questions 2, 7, 8). The questions in group II examined the distinctive and memorable appearance of the space, appreciation of the building and its surroundings, willingness to go inside or spend time next to it, as well as the willingness to return to this place. In group III, respondents answered additional questions concerning the accessibility by public transport, the amount of greenery and the feeling that the space around the building is the right size.

The analysis of correlations between individual questions verified the accuracy of the responses (e.g., between relatively similar questions 2, 3, 4, 5) or yielded the results on the complete perception of the space around the building, the form of the buildings and the perception of the presence of greenery or openness of the space. In turn, the analysis of the correlation between the answers to the questions and the results of the mapping and analysis of the space provided a full spectrum of the findings.

2.2.4. Correlations

Correlations between individual survey responses and the results of on-site analysis and behavioural mapping were examined using the Pearson correlation coefficient. The following correlation intervals were adopted: 0–0.3—weak correlation, 0.3–0.5—moderate correlation, 0.5–0.7—strong correlation, 0.7–1—very strong correlation.

3. Results

3.1. Method A

In general, in this phase of the study, MOS (Polish for The Małopolska Garden of Arts) and Cricoteka received the best scores, while the Krakow Opera and the Auditorium Maximum received the worst and lowest scores. The graph in Figure 7 presents a comparison of the results of the on-site analysis, spread over different groups of attributes (goals 1–5).

Figure 7. Summary of the results of method A.

The cultural facilities inscribed on the UNESCO World Heritage List in Krakow and in its buffer zone were built on degraded, disused plots of land (e.g., MOCAK, MOS, Cricoteka, Krakow Opera) but are now revitalising those neglected spaces by introducing new functions. MOCAK (the Museum of Contemporary Art in Krakow) has become one of the focal points of the now highly popular district of Krakow. In some cases, the effect of urban acupuncture is visible, e.g., the space around the ICE Congress Centre is gradually changing from neglected wasteland to new and renewed plots. In turn, the former working-class neighbourhoods around MOCAK and Cricoteka with their numerous cheap vacant lots left after the closure of factories have become a Mecca for new enterprises, such as art galleries, studios and workshops. The existing town centre has been redeveloped and revitalised. The buildings—as large as the local plans allow—have densified the surrounding areas. Through this densification process, Krakow seeks to be a compact city with sustainable and self-sufficient neighbourhoods. Most of those facilities form the nucleus of cultural and entertainment spaces.

3.1.1. Goal 1

Table S1 in Supplementary Material S1 presents the detailed results of the analysis in terms of sustainable transport and visual accessibility. The Wyspianski Pavilion, Cricoteka and ICE are rated best as far as the quality of walkability, cycling infrastructure and access to public transport are concerned. Auditorium Maximum scored worst—the building is located near the busy Trzech Wieszczow Avenues, poorly connected for cyclists and hardly visible from afar.

All of the analysed buildings are easily accessible on foot, although access to the Krakow Opera and Manggha is hampered by proximity to busy streets. Krakow is striving to improve the quality of the pedestrian environment by banning individual cars from the city centre, limiting maximum speed and extending paid parking zones. It is a walkable space with no barriers for pedestrians in most cases, occasionally disturbed by the multitude of cars parked in front of the buildings (Auditorium Maximum, Cricoteka) or even on their premises (the Krakow Opera, MOCAK).

During the pandemic, municipal authorities introduced several traffic changes, e.g., converting car lanes into cycle lanes. It should be mentioned, however, that although Krakow claims to be a cycle-friendly city, only two of the analysed venues are accessible by cycle lanes (ICE, the Krakow Opera). The other three can be reached by both pedestrian and cycle lanes; the rest by counterflow lanes, pavement or riding in city traffic. All nine of the buildings have bicycle stands, but their number varies.

Access for wheelchair users is usually provided from ground level or via a ramp but access to the Jozef Czapski Pavilion is hindered by the uneven surface of the cobbled pavement.

The location in the very centre of the city provides good access to all the facilities via public transport. The distance from the public transport stops varies between 50–390 m. MOCAK, Cricoteka and the Krakow Opera are also located within 350–900 m from the railway stations—a means of agglomeration transport.

The visual accessibility is most effective when the facilities are located by the Vistula River (Manggha Centre, Cricoteka) and near the less densely built-up Grunwaldzkie and Mogilskie Roundabouts (Manggha Centre, ICE and the Krakow Opera). The Wyspianski Pavilion is also well visible from various points in the Old Town, whereas the Jozef Czapski Pavilion, Auditorium Maximum and MOCAK are hidden within the dense fabric.

3.1.2. Goal 2

The MOS and ICE were rated best in terms of mixed-use programmes, contributing to the liveability and health of the community, while the Krakow Opera and Auditorium Maximum were rated worst as far as this aspect is concerned. Table S2 in Supplementary Material S1 shows the results.

Better results are related to more features available at different times of the day, including those that are free of charge and available during the hours when the main cultural area is closed. The worst results correspond directly to the limited programme of functions that are unavailable independently, resulting in the abandonment of a given space during the lockdown of cultural and educational institutions.

The analysis also included functions in the immediate surroundings of the buildings and adjacent public spaces. The results show maximum diversity of functions in six out of nine cases in the adjacent development, but poor filling of public spaces with functions allowing for their optional use. These are usually cafés with tables outside. There is also an outdoor cinema in the Jozef Czapski Pavilion, a museum lapidarium beside it, and a nursery school garden right next to the MOS. Apart from these elements, there are no other functions for children or seniors or health-promoting functions. The presence of NGOs in the immediate vicinity is not visible, either.

Hence, general deficiencies are evident in this group of factors: the buildings are connected to culture but lack new models for cultural development and fostering creativity (such as affordable and flexible studio spaces in historic buildings and warehouses).

3.1.3. Goal 3

Regarding climate change mitigation, green spaces are present in all cases except the Krakow Opera, including a large green area at the Manggha Centre with a bamboo garden and a tea pavilion. In contrast, there are no water facilities next to the buildings to provide comfortable cooling and physical recreation, nor are there renewable energy solutions, a zero-waste concept or attention to shorter supply chains. There are no green roofs or walls, other than a single wall with climbing plants at the Wyspianski Pavilion.

Among other solutions minimising the urban heat island employed in the analysed cases, one can mention the use of light colours (white at the Jozef Czapski Pavilion, ICE and Manggha Centre) and canopies (the pergola at the MOS and the Cricoteka building that creates a roofed space itself—although the idea was scenographic rather than ecological). Apparently, the UHI reduction was not considered by all architects in their designs. The dark building of Auditorium Maximum certainly does not counteract the urban heat island.

Located in a valley, Krakow has ventilation problems that are not alleviated by new development. On the contrary, it is getting worse as the municipal authorities are gradually allowing for ventilation corridors to be built over. The buildings in question may not be the most problematic, but they do not help in solving the smog problem either. Table S3 in Supplementary Material S1 shows detailed results.

Along with the expansion of green areas for natural retention, these measures are in line with the municipal strategy for counteracting the effects of the urban heat island, heavy rainfall and flooding, as well as the exceedance of air pollution standards. While the more publicly accessible, shaded seating areas along the transport and walking routes are one of them, only cafés provide them at MOCAK and the Manggha Centre—i.e., they are paid and therefore not entirely public. There are mixed ratings for weaving natural elements into parks and playgrounds, designing public spaces to reflect local geography and supporting direct interaction with nature through free and low-cost activities. The Manggha Centre and Cricoteka reflect the local geography with their location by the Vistula River. In turn, the MOS, ICE and the Jozef Czapski Pavilion offer free-of-charge activities in their green spaces: an open-air cinema at the Czapski Pavilion, a congress square at ICE and a variety of activities in the greenery under the roof at the MOS, which are very popular with Krakow residents.

3.1.4. Goal 4

As far as urban morphology and aesthetics are concerned, a compact, consistent pattern of development surrounds all the examples. It should be mentioned here that the city centre of Krakow has in general a pedestrian scale, a dense network of streets and natural surveillance. It is evident that the city is striving to find a balance between the preservation of historic heritage and new forms and functions of the buildings. In terms of the attributes listed, all examples were rated highly. Table S4 in Supplementary Material S1 shows the detailed results of the attributes in this group. The designers of the Krakow Opera and Cricoteka incorporated existing historic buildings into the new projects. The architects of the MOS and the Wyspianski Pavilion adhered to the historical alignment of the riding arena and the Lipka Tenement House, respectively. When designing MOCAK, Claudio Nardi Architects formally conformed to the characteristic development in the neighbourhood. The incorporation of the historic heritage into the new structures has added a heritage conservation value to the buildings. Furthermore, almost all the examples received total points as enclosed concentric authentic spaces characterised by genius loci and continuity of historic buildings in the place/non-place category following Lewicka's criteria [59]. Regarding the features of Lofland's sources of aesthetic pleasure [60], the places are characterised by perceptual innuendo, i.e., pleasure derived from glimpsing a small piece of the building (the canopy in front of the MOS from both directions of Rajska Street, the white façade of the Jozef Czapski Pavilion facing the courtyard of the Emeryk Hutten-Czapski Museum, MOCAK seen from the entrances of the neighbouring streets) and unexpectedness or whimsy (the aforementioned examples as well as the curved roof of the

Manggha Centre, the reflective underside of the overhanging in Cricoteka, the red glazed body of the building of the Krakow Opera). Historic Layering or Physical Juxtaposition between contemporary forms of the buildings and historic roofing is perceived in virtually all examples. Some of them offer the possibility of public solitude (Auditorium Maximum, the Jozef Czapski Pavilion, the Manggha Centre and the space at the rear of ICE), others stimulate diversity (the Wyspianski Pavilion, Cricoteka, the front of ICE).

The analysis of the ground floors of the buildings according to Gehl's criteria revealed that the most favourable situation—i.e., narrow fronts, numerous doors, diverse functions, façade relief and rich details—is present around the Wyspianski Pavilion, and the least favourable one occurs at the ICE Congress Centre.

3.1.5. Goal 5

Although the abovementioned examples have undeniable aesthetic qualities, acknowledged by numerous awards, the level of cleanliness of the space in some cases is low. The Wyspianski Pavilion is a neglected building (the renovation of its façade started in October 2020), polluted by pigeons, with homeless people present in the surroundings. The Manggha Centre is a dirty, crumbling building with bins and litter scattered at the back entrance. These facilities scored the worst, while the Jozef Czapski Pavilion was rated the best. Noise levels harmful to health, i.e., LDWN > 65 dB, occur in the vicinity of the Manggha Centre, Auditorium Maximum, the Krakow Opera and ICE.

The quantity and quality of seating (along with its adequate protection) was rated best at the Manggha and Cricoteka, and the least at the Krakow Opera. Regarding the other elements of street furniture (such as art in public space, furniture, facilities for children, seniors and the disabled), the results are fairly even since there are slightly different elements in each example. Overall, the MOS with its canopy, mural and bike stand as a spatial sculpture scored best, and Auditorium Maximum and ICE scored worst.

Detailed results for the group of five attributes can be found in Table S5 in Supplementary Material S1. The social capital assessment included in the table also encompasses the impact of a venue evident by the number of posts and followers on Instagram, the presence of collective actions and events in the surroundings and the number of people in the area that are not a crowd. Additional points were given if a restaurant, bistro or café was located in the building, which increases interest, also evident in behavioural mapping. Here, MOCAK and the ICE Congress Centre were rated highest, whereas the Manggha Centre was rated the lowest. For a description of the activities of the MOS, MOCAK and Cricoteka, see the final reports on the cultural sector survey [46,47].

3.2. Method B

Behavioural mapping revealed that there is little traffic at open venues, which, according to tourism statistics [18], is most likely a lockdown effect on the use of public space (especially among tourists and organised groups). This effect is visible even in the case of open museums. The research showed a variation in the number of users of the space (see Table 4). The highest number of users could be seen at the Manggha Centre and Cricoteka, which may be related to their location by the Vistula boulevards. There were practically no people engaged in 'optional activities', according to Gehl's notion [51] near ICE, the Krakow Opera and Auditorium Maximum. These places lack greenery and auxiliary functions, but above all, created an urban interior that would allow for 'public solitude' or a place where one can sit down and enjoy a cup of coffee or a meal (the patisserie in the ICE Congress Centre is located at the rear of the building, facing a busy street).

The majority of passers-by, however, can be seen around the MOCAK building, located inside the city block. This place has 'taken root' in the district, becoming a pleasant thoroughfare. People walk their dogs there. Passers-by are hardly to be seen around the buildings located away from the main public spaces of the city (e.g., Auditorium Maximum, the Josef Czapski Pavilion), but there are plenty of them when the building is situated along a pedestrian route, as is the case with the Wyspianski Pavilion. However, the location of the Wyspianski Pavilion in the centre did not result in pedestrians entering the building or stopping by it. Therefore, they were not counted and included in the mapping table. They move 'independently' from the building and thus, there is no spontaneous activity around it. The situation looks better at the Jozef Czapski Pavilion, where there are no passers-by, but people can be seen sitting at the outdoor café tables.

Table 4. Mapping results.

	Manggha Complex	Wyspian. Pavilion	Auditor. Maxim.	Opera	MOS	Cricoteka	MOCAK	ICE	Czapski Pavilion
employees	6	0	2	4	1	11	7	11	0
passers-by	4	5	0	3	15	2	18	4	0
other users (café, benches, etc.)	38	0	0	0	11	34	13	0	14
all	48	5	2	7	27	47	38	15	14
conclusions	UU	OU	OU, OP	OU	PP	UU	PP	OU	OP

UU, the most users; OU, no users; PP, the most passers-by; OP, no passers-by.

The centre of Krakow is a populous mixed-use neighbourhood with potential for social inclusion and functional diversity, except for the gentrified Old Town. The city centre is characterized by a high concentration of businesses and workplaces, as well as being home to universities and schools, which reduces the need for commuting between neighbourhoods. Although pedestrianisation is relatively high, the potential of some facilities has not been exploited at all. The MOS constitutes an exception, as apparently Krakow residents treat it as an iconic place. The mapping showed groups of young people coming here to take photos, organize photo-shoots and small pandemic gatherings, sitting and drinking both at café tables and on benches (also as part of hen and stag parties).

3.3. Method C

Figure 8 presents the results of the questionnaire survey. Overall, the MOS, Cricoteka and the Manggha Centre were rated highest, while the Krakow Opera was rated worst (except for the question about pedestrian access and the recollection of the appearance of the building), scoring below zero in four categories. Respondents commented negatively on the appearance of the surroundings of the building, their desire to spend time there, the amount of greenery around it and their perception of the size of the space. The results indicate that the choice of the site location by municipal authorities is misguided. According to the respondents, the most characteristic building is the ICE Congress Centre (they remember it best), while the Jozef Czapski Pavilion is the least recognizable. This building is one of the newer ones and is therefore the least known.

The respondents' answers to questions whether they like the building and its surroundings, whether they would like to go inside when passing by and spend time in its surroundings, form a coherent whole. The MOS was rated best in these categories, followed by Cricoteka and the Manggha Centre with the Far East Gallery, while the Krakow Opera and Auditorium Maximum were rated worst. The differential curves at these venues look interesting. In all cases, except for the Wyspianski Pavilion, respondents liked the building itself more than its surroundings. This difference is most pronounced in the case of ICE and the Krakow Opera, which may be related to the heavy traffic and high noise levels in the surroundings. The Wyspianski Pavilion, on the other hand, received significantly higher scores for its surroundings, which is most likely due to its location in the historic fabric of the city centre along the Royal Road. The respondents indicated less frequently than in the question about aesthetics that they felt like entering the building when passing by. They perceive the building and its surroundings more as an aesthetic site than as a place for any activity, which may be due to the functional programme or the rather closed form of the building.

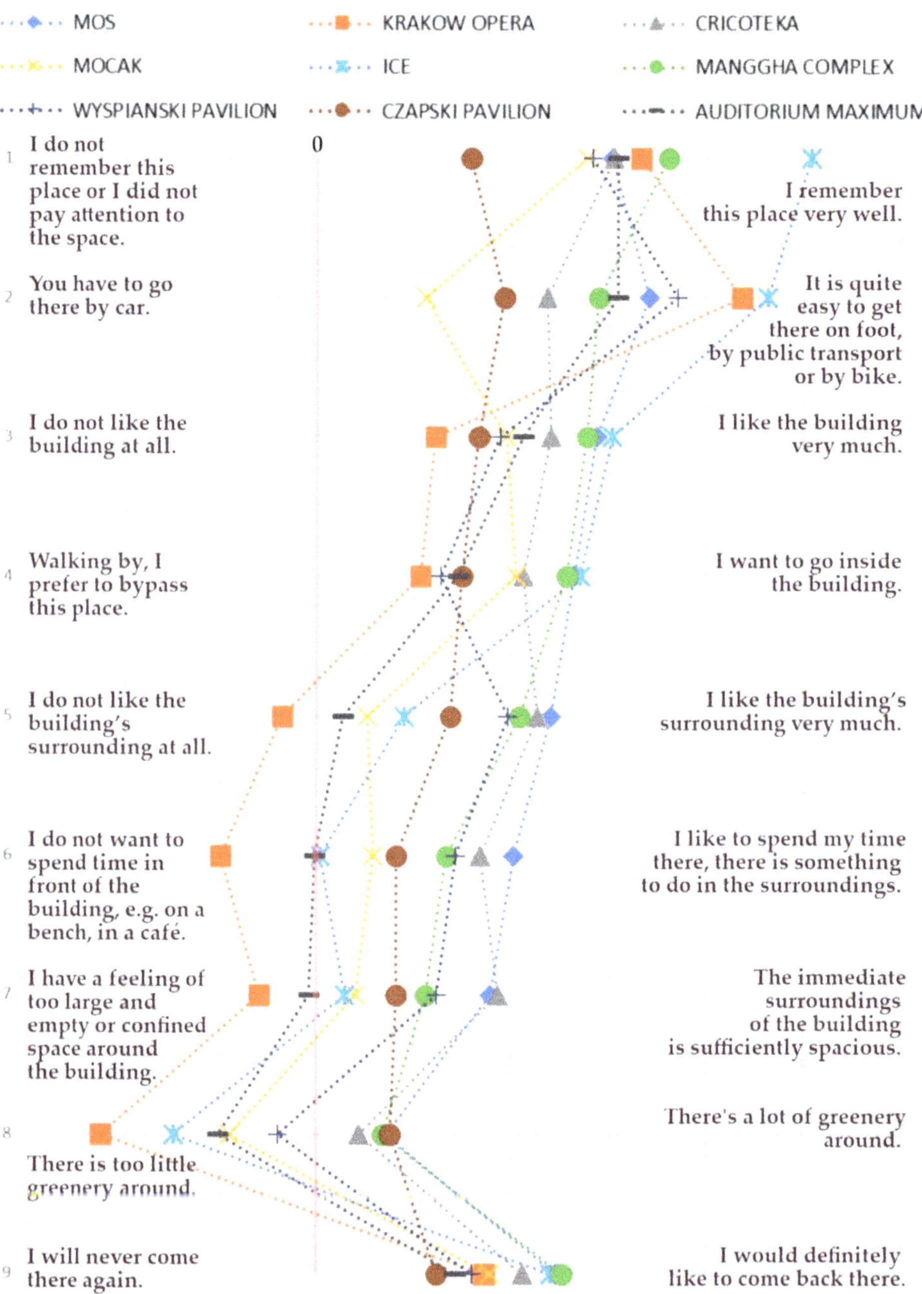

Figure 8. The survey results. The detailed survey results are available in Supplementary Material S2.

In all cases, except for MOCAK, respondents liked the building more than they wanted to go inside and liked the surroundings more than they thought that there was anything to do there. The answers to the last two questions overlap in the case of MOCAK. In turn, respondents indicated that they did not feel like spending time actively at the Krakow Opera, ICE and Auditorium Maximum, which is consistent with the results of Method B.

The results regarding access by public transport, on foot and by bicycle show different patterns—the respondents rank the vehicle and walking access to the Krakow Opera and ICE as the best, and to MOCAK and the Jozef Czapski Pavilion as the worst. For those venues, they most often choose the option of having to commute by car. The highest-rated facilities are actually located near bus and tram stops with a large number of lines. Access to green spaces was rated the worst of all issues by respondents, who stated that there were too few green spaces around the facilities studied. This is where the semantic differential curves slump, reaching the lowest function value in all cases. The highest scores here were given to four facilities located close to each other: the Manggha Centre, the MOS, Cricoteka and the Jozef Czapski Pavilion. The remaining venues received total scores below zero.

The perception of the size of the space around the building was rated slightly better than the amount of greenery. The results were worse than in the question on greenery only in the Jozef Czapski Pavilion. The respondents most probably regard this space as too cramped. In turn, the two worst scores regarding this question were given to the Krakow Opera and ICE which are actually located in a sparser urban fabric, with wide arterial roads. One can guess that the respondents were pointing to the lack of enclosed space delineating the venue, as seen in Lewicka's assessment [56].

When summarising the correlations between the responses, a very high correlation can be seen between the responses to questions on whether the respondents like the building and are willing to go inside (Q3 and 4: 0.93) and whether they like the surroundings and want to spend time there (Q5 and 6: 0.96). The willingness to return to a place correlates with the willingness to go inside a building (Q4 and 9: 0.87) and with the question on whether respondents like the building (Q3 and 9: 0.82). The above correlations can be considered as a validation of the reliability of completing the questionnaires. The questions were formulated in such a way that the author expected them to appear.

The answers to the question on how the size of the space around a building is perceived correlate with the answers to the questions on whether the surroundings of the building are liked (Q5 and 7: 0.97) and whether respondents feel like spending time there (Q6 and 7: 0.98). The answer to the question about whether there is enough greenery around the building correlates most strongly with opinions on whether the respondents enjoy spending time around the building (Q6 and 8: 0.88).

Table 5 shows all correlations between responses to the questions.

Table 5. Correlations between the survey answers.

Q NO.	1	2	3	4	5	6	7	8	9
1	1.00	0.67	0.59	0.52	−0.13	−0.27	−0.21	−0.47	0.67
2		1.00	0.17	−0.01	−0.16	−0.35	−0.27	−0.49	0.29
3			1.00	0.93	0.57	0.50	0.51	0.37	0.82
4				1.00	0.51	0.47	0.48	0.39	0.87
5					1.00	0.96	0.97	0.47	0.47
6						1.00	0.98	0.88	0.37
7							1.00	0.20	0.41
8								1.00	0.20
9									1.00

3.4. Correlations

Table 6 shows the correlation between the responses to the questionnaire survey and Method A. The strongest correlated factors were the responses to the question on the amount of greenery and the aesthetic perception of the building (0.93). The biophilia of the city residents is clearly evident here. A slightly lower correlation coefficient was obtained between aesthetics (GOAL 4) and the questions on the perception of the surroundings and the willingness to spend time there as well as the perception of the size of the space around the building (0.78–0.85) and between the answer to the question on the amount of greenery and environmental solutions (GOAL 3: 0.79).

Table 6. Correlations between the survey and on-site analysis results (methods A and C).

Q NO.	1	2	3	4	5	6	7	8	9
GOAL 1	0.46	0.36	0.56	0.46	0.61	0.52	0.60	0.17	0.55
GOAL 2	0.17	−0.04	0.75	0.73	0.76	0.75	0.75	0.56	0.53
GOAL 3	0.09	−0.21	0.70	0.71	0.85	0.75	0.80	0.79	0.56
GOAL 4	−0.64	−0.47	0.18	0.18	0.78	0.85	0.79	0.93	0.01
GOAL 5	−0.09	−0.45	0.38	0.43	0.35	0.41	0.50	0.39	0.13

The issue of walking, cycling and public transport access (GOAL 1) was not very strongly correlated with responses to any survey question (maximum 0.61), nor was social capacity (GOAL 5). In contrast, responses to questions on whether respondents liked the building and its surroundings, their willingness to go inside and spend time around the building as well as activities in and around the building (GOAL 2) and environmental issues (GOAL 3) were very strongly correlated (0.7–0.85). This confirms the potential of gathering users for both necessary and optional activities by filling the space with diverse functions and paying attention to environmental solutions.

The correlations between the answers to the survey questions were checked using method B (see Table 7). There is a correlation between the number of users of the space and the total number of users and pedestrians in all responses concerning the surroundings of the buildings (questions 5–8). The strongest correlations were the positive responses given by the respondents on whether there is enough space around the buildings and enough greenery in the surrounding area (0.6–0.69). Again, these correlations indicate biophilia.

Table 7. Correlations between the survey and behavioural mapping (methods B and C).

Question Number	Users	Users and Passers
1	−0.13	0.04
2	−0.50	−0.51
3	0.33	0.45
4	0.46	0.64
5	0.59	0.52
6	0.57	0.55
7	0.60	0.57
8	0.69	0.53
9	0.44	0.61

4. Discussion and Conclusions

The case study evaluation of the spatial-functional potential of cultural buildings in the historic urban fabric to achieve sustainable development, based on selected buildings in Krakow, demonstrated a relationship between the attributes of sustainable development and the 'popularity of buildings'—positive perception and the intensity in using the space around the buildings.

The most strongly correlated features relate to the presence of green areas and blue infrastructure and the service potential in terms of auxiliary functions connected to the furnishing of the space rather than the cultural mission of the buildings—places to eat and drink (café, bistro, restaurant, etc.), canopies, seating areas (including non-standard ones, i.e., walls, steps, railings) and the presence of art in the public space. These elements appear to fit into the individual genius loci of a place, conditioning its positive perception. The issues of activities available in the surroundings, denoting the surrounding buildings and access by public transport (meeting 11.2 SDG target quite well) turned out to be less important. This may be due to the comparably good accessibility of all sites, which paradoxically limited the study by the lack of contrasting cases and the comparably compact pattern of development, dense street network and natural surveillance during daytime. While affecting the walkability of the places, the less visible correlations between the morphology and aesthetics of the buildings and their perception indicate only the negatively perceived cases, without clearly differentiating the best ones.

The research shows that the residents notice and appreciate principles of urban ecology employed in buildings. If a building has more auxiliary functions and a better connection to green areas, it is more likely to remain functional in the city during the lockdown period, even when its main function is temporarily suspended and the number of tourists is significantly reduced. This affects the flexibility of the space in certain contingencies. Urban space becomes more resilient to both biological and human-induced disasters. It must be noted, however, that the environmental solutions implemented in Krakow are limited to the simplest ones, incorporating nature in the built environment, and thus reducing the UHI effect. No renewable energy and zero waste concepts are employed in the buildings. Hence, the implementation of 11.6 SDG target has been progressing very slowly, while that of 11.7 SDG target has been faster. That leads to the bitter conclusion that creating open active green spaces owned by a city or a province poses a problem in the centre of Europe that is striving to turn entire districts into zero-energy ones [63].

Specific site conditions (e.g., small plots, spatial chaos, no parking spaces, dense urban structure) combined with historic cultural heritage of Krakow have resulted in a new model of cultural buildings, in which designers either preserve the existing plots, historic buildings and their remnants, or use traces of the past, such as the geometry of roofs, materials, colours, details and proportions. Due to irregular shapes of the plots and various elements of the spatial context, adequate architectural forms had to be found for the new buildings. This balance of historic heritage with new forms and functions targets 11.4 SDG well.

The research showed that the city which is 'learning' to be an inclusive, citizen-friendly space—and Krakow seems to be doing so—introduces mixed-use buildings and public spaces connected with green areas. The architecture under study has been evolving from pursuing the 'wow' factor through more hybrid structures whose architects cared more for genius loci and inclusive space that can be used by the citizens for various purposes (sitting, talking, observing and participating in the social theatre, walking with children and dogs, cycling, etc.) to small lapidary forms inscribed in the context but without 'pretending' to be structures from the past, 'donning' a historical costume. Such contextualisation has the potential to link the spaces around buildings with the municipal system of inclusive public spaces, hitherto untapped in Krakow as shown by the mapping. According to research by the Center for Active Design, exploiting this potential of the so-called 'third places' in the city would promote social interaction and raise the level of civic trust [57].

The Krakow case study also revealed the low presence of culture in the public space. The cultural institutions do not use the outdoor space in the city for exhibitions, lectures, etc. This deficiency has been confirmed by the results of the reports commissioned by the city hall [46,64]. Their authors postulate the use of parks, green areas and the introduction of contemporary art into the public space. Such a measure would also be consistent with the Streets for Pandemic Response and Recovery strategy, in which NACTO envisaged a new purpose for streets during the COVID-19 pandemic, namely gatherings, events, play, school, etc [65]. Exhibitions, workshops, cinema screenings and other cultural events would fit well into the catalogue of activities, preventing the physical alienation of the community. At the same time, the closure of museums and other cultural venues in a forced semi-lockdown during the pandemic would not derail the continued use of the buildings regardless of cultural offer, thus lowering stress levels.

The postulate to incorporate cultural buildings into the public space system is also connected with the need to deagglomerate the municipal cultural offer and strengthen the culture-forming potential of districts located further from the central Old Town [46]. The sites analysed in the research contribute to this scheme, but they fail to expand the scope of their activities. The study commissioned by the Malopolska Voivodeship Office has illustrated the positive impact of the new facilities on the city space, as well as the prestige of the district and the surrounding economy [46]. At the same time, it is desirable to support the natural processes in which creative hubs emerge around such facilities as Cricoteka and MOCAK that are built in post-industrial districts [46]. The report from the study on Krakow's cultural sector includes a proposal to counteract the gentrification of Zabłocie and Podgórze districts as this process is stifling gatherings of Krakow's bohemians around these cultural venues. Unfortunately, the analysis of land prices and the functions of new investments shows that the protection of these areas is unsuccessful as more and more expensive housing developments fill the post-industrial areas, supplanting semi-formal artistic users [46].

By means of behavioural mapping, Mantey has shown that the greatest influence on the use value of a given space is its diversity [62]. The heterogeneity of users contributes to integration, and the more groups a given space is designed for, the more people in it. The study presented here found deficiencies (or rather lack of such elements) in the design of spaces for different age groups. Thus, one may speculate that the spaces developed with facilities for seniors and children would generate more traffic. There is also a correlation with pedestrian traffic—the more traffic around an area (busier point in the city), the higher the activity level of the space. The diversity of the space increases its use value.

The study had limitations, due to time limits, data collection during the pandemic and subjectivity of the observations. However, to counteract the arbitrariness of the sample, the author examined all contemporary cultural buildings located in the historic zones of Krakow as cases. They were built in the 1990s and later. Earlier buildings of this type were constructed up to the 1930s in Krakow [66].

The author supplemented the subjective partial observational results in method A with quantitative data and then verified them by means of quantitative data from methods B and C. These methods also had their limitations. The behavioural mapping conducted before the complete closure of museums during the second wave of the covid pandemic in Poland could not be repeated for time reasons. The respondents' particulars in the survey indicated a high interest in the questionnaire among people whose profession is related to art (45 out of 173 responses), people aged 18–26 (66 responses) and those with higher education (96 responses). Such a profile of respondents does not correspond to the demographics of Krakow's residents, yet it is similar to the profile of people who positively assess the impact of cultural institutions on the quality of life of the local community (the impact of infrastructure projects). It may therefore be assumed that those individuals were interested in completing the questionnaire.

The questionnaire survey yielded 173 responses, which, with a population of 771,069 people in Krakow in 2019 [67], a fraction size of 0.5 and a confidence level of 95%, resulted in a statistical error of 7%. Due to pandemic constraints, it is hardly possible to collect a larger survey group at the moment. Further research, e.g., using the PAPI (Pen-and-Paper Personal Interview) method, would potentially allow for more extensive results, but it would require direct contact with larger groups of Krakow residents.

Another limitation of the behavioural mapping method was the lack of comparison of the concentration of people in other places in Krakow—the less and more familiar public spaces with different functions and amenities. Comparing the results of the selected cases with other spaces would provide a background for determining their level of use and enable the identification of model sites with greater potential. Such a study requires more time and human resources and could be the subject matter of a separate paper, as well as serving as the monitoring of the places at different times of the year and the changing reality of the pandemic.

Thus, one must take into account the limitations of the research when analysing its results. Time constraints and lack of direct contact during the pandemic lockdown precluded continuation of the research. In turn, if continued after the return of increased tourist traffic, it would not show the potential of the buildings to function in a public space geared towards Krakow's residents.

Urban governance, leadership and best practice were not analysed in the study either. The Center for Active Design and the Gehl Institute point to the positive role of participatory design in public spaces [57,61]. Requiring separate research methods, these issues need further analysis that would answer the question of the extent to which the city authorities implement their professed care for environmental issues and social participation in governing. A preliminary examination of the topic suggests that they do so, to a certain extent. Some degree of social engagement of experts in the development of cultural buildings in Krakow is also noticeable. The majority of the designs discussed here were selected by means of a call for proposals, i.e., expert assessments.

The main message of the paper is to indicate how important it is for the creation of cultural spaces in the city to develop a range of their activities, linked to the surrounding public space and green areas. The relationship with greenery is central to this issue, especially when it replaces different technologies employed in a building to counteract climate change. Activating cultural spaces in the city and linking them to sustainability goals is especially important in a post-pandemic reality when the number of tourists is decreasing and the gentrified districts need to regain their former vibrancy.

Supplementary Materials: The following are available online at https://www.mdpi.com/article/10.3390/su13116126/s1.

Funding: This research received no external funding.

Institutional Review Board Statement: Not applicable.

Informed Consent Statement: Not applicable.

Data Availability Statement: The data presented in this study are available in [Supplementary material] and [repository name e.g., FigShare] at [doi], reference number [reference number]. Some Data are available in a publicly accessible repositories of the different branches of the Office of the City of Cracow and the General Statistics Office. They are all showed in References section.

Conflicts of Interest: The authors declare no conflict of interest.

References

1. Al-Saeed, Y.W.; Ahmed, A. Evaluating Design Strategies for Nearly Zero Energy Buildings in the Middle East and North Africa Regions. *Designs* **2018**, *2*, 35. [CrossRef]
2. D'Agostino, D.; Zangheri, P.; Castellazzi, L. Towards Nearly Zero Energy Buildings in Europe: A Focus on Retrofit in Non-Residential Buildings. *Energies* **2017**, *10*, 117. [CrossRef]
3. Strategies for a Nearly Zero-Energy Building Market Transition in the European Union. Available online: https://zebra2020.eu/publications/strategies-for-a-nearly-zero-energy-building-market-transition-in-the-european-union/ (accessed on 11 April 2021).
4. IEA. Coal Information: Overview, IEA, Paris. Available online: https://www.iea.org/reports/coal-information-overview (accessed on 11 April 2021).
5. Eurostat Coal Production and Consumption Statistics—Statistics Explained (europa.eu). Available online: https://ec.europa.eu/eurostat/statistics-explained/index.php?title=Coal_production_and_consumption_statistics. (accessed on 10 April 2021).
6. Coal 2019, IEA, Paris. Available online: https://www.iea.org/reports/coal-2019 (accessed on 10 April 2021).
7. Eurostat, Municipal Waste Statistics. Available online: https://ec.europa.eu/eurostat/statistics-explained/index.php/Municipal_waste_statistics-Municipal_waste_treatment (accessed on 12 April 2021).
8. Mioduszewski, W.; Okruszko, T.; Kardel, J.; Fehér, J.; Gáspár, J.; Tamas, J.; Mosný, V.; Muller, R.; Istenič, D.; Potokar, A. *Natural Small Water Retention Measures: Combining Drought Mitigation, Flood Protection, and Biodiversity Conservation—Guidelines*; Global Water Partnership Central and Eastern Europe: Bratislavia, Slovakia, 2015. Available online: https://www.gwp.org/en/gwp-cee/ (accessed on 25 May 2021).

9. Pope, P.; Annandale, D.; Morrison-Saunders, A. Conceptualising sustainability assessment. *Environ. Impact Assess. Rev.* **2004**, *6*, 595–616. [CrossRef]
10. Troi, A. Historic buildings and city centres—The potential impact of conservation compatible energy refurbishment on climate protection and living conditions. In Proceedings of the Conference Energy Management in Cultural Heritage, Dubrovnik, Croatia, 6–8 April 2011.
11. Oke, T. Towards Better Scientific Communication in Urban Climate. *Theor. Appl. Climatol.* **2006**, *84*, 179–190. [CrossRef]
12. Kumar, P.; Druckman, A.; Gallagher, J.; Gatersleben, B.; Allison, S.; Eisenman, T.S.; Hoang, U.; Hama, S.; Tiwari, A.; Sharma, A.; et al. The nexus between air pollution, green infrastructure and human health. *Environ. Int.* **2019**, *133*, 105181. [CrossRef]
13. Ramos, J.S.; Domínguez, S.Á.; Moreno, M.P.; Delgado, M.G.; Rodríguez, L.R.; Ríos, J.A.T. Design of the Refurbishment of Historic Buildings with a Cost-Optimal Methodology: A Case Study. *Appl. Sci.* **2019**, *9*, 3104. [CrossRef]
14. London, F. *Healthy Place Making*, 1st ed.; RIBA Publishing: London, UK, 2020; ISBN 978-1-85946-883-8.
15. Cristiano, S.; Gonella, F. 'Kill Venice': A systems thinking conceptualisation of urban life, economy, and resilience in tourist cities. *Hum. Soc. Sci. Commun.* **2020**, *7*, 143. [CrossRef]
16. Krakowska Kultura w Czasie Epidemii COVID-19. Available online: www.krakow.pl (accessed on 20 April 2021).
17. Borkowski, K.; Grabiński, T.; Seweryn, R.; Rotter, L.; Mazanek, L.; Grabińska, E. *Ruch Turystyczny w Krakowie w 2019 roku*, 1st ed.; Małopolska Organizacja Turystyczna: Kraków, Poland, 2020.
18. Ruch Turystyczny w Krakowie. UM Kraków. Available online: www.bip.krakow.pl (accessed on 15 March 2021).
19. Office of the City of Krakow (UMK). Strategia Rozwoju Krakowa "Tu Chcę żyć. Kraków 2030". Available online: https://www.bip.krakow.pl/?dok_id=94892 (accessed on 27 April 2021).
20. Roggema, R. The Future of Sustainable Urbanism: Society-Based, Complexity-Led, and Landscape-Driven. *Sustainability* **2017**, *9*, 1442. [CrossRef]
21. Heymans, A.; Breadsell, J.; Morrison, G.M.; Byrne, J.J.; Eon, C. Ecological Urban Planning and Design: A Systematic Literature Review. *Sustainability* **2019**, *11*, 3723. [CrossRef]
22. Jagodzińska, K.; Sanetra-Szeliga, J.; Purchla, J.; Van Balen, K.; Thys, C.; Vandesande, A.; Van der Auwera, S. Cultural Heritage Counts for Europe: Full Report; International Cultural Centre: Kraków, Poland. Available online: http://blogs.encatc.org/culturalheritagecountsforeurope/ (accessed on 18 May 2021).
23. Petti, L.; Trillo, C.; Makore, B.N. Cultural Heritage and Sustainable Development Targets: A Possible Harmonisation? Insights from the European Perspective. *Sustainability* **2020**, *12*, 926. [CrossRef]
24. Dessein, J.; Soini, K.; Fairclough, G.; Horlings, L. Culture in, for and as Sustainable Development; Conclusions from the COST ACTION IS1007 Investigating Cultural Sustainability 2015. Available online: https://www.cost.eu (accessed on 15 April 2021).
25. Soini, K.; Dessein, J. Culture-Sustainability Relation: Towards a Conceptual Framework. *Sustainability* **2016**, *8*, 167. [CrossRef]
26. Cushing, D.F.; Miller, E. *Creating Great Places: Evidence-Based Urban Design for Health and Wellbeing*, 1st ed.; Routledge: New York, NY, USA, 2020; ISBN 9780429289637.
27. Stylianou-Lambert, T.; Boukas, N.; Christodoulou-Yerali, M. Museums and cultural sustainability: Stakeholders, forces, and cultural policies. *Int. J. Cult. Policy* **2014**, *20*, 566–587. [CrossRef]
28. McGhie, H.A. *Museums and the Sustainable Development Goals: A How-To Guide for Museums, Galleries, the Cultural Sector and Their Partners*; Curating Tomorrow: Liverpool, UK, 2019. Available online: www.curatingtomorrow.co.uk (accessed on 13 April 2021).
29. ICOMOS 2016. Cultural Heritage, the UN Sustainable Development Goals, and the New Urban Agenda; ICOMOS Concept Note for the United Nations Agenda 2030 and the Third United Nations. Available online: www.icomos.org (accessed on 5 March 2021).
30. Baki Nalcioğlu, Ü.Z.S. The cultural aspect of sustainability in museums. *Milli Folk.* **2021**, *129*, 124–135.
31. Honey-Rosés, J.; Anguelovski, I.; Bohigas, J.; Chireh, V.; Daher, C.; Konijnendijk van den Bosch, C.; Litt, J.; Mawani, V.; McCall, M.; Orellana, A.; et al. The Impact of COVID-19 on Public Space: A Review of the Emerging Questions. *Cities Health* **2020**, 1–17. [CrossRef]
32. Streets for Pandemic Response & Recovery | 09/24/2020. Available online: https://nacto.org/publication/streets-for-pandemic-response-recovery/ (accessed on 7 March 2021).
33. Gehl, J. *Cities for People*, 1st ed.; Island Press: Washington, DC, USA, 2010; ISBN 9781597265737.
34. Matthew Carmona, for the Place Alliance, Place Value, and the Ladder of Place Quality. Available online: http://placealliance.org.uk/research/place-value/ (accessed on 2 March 2021).
35. Koncepcja Przestrzennego Zagospodarowania Kraju 2030. Available online: https://miir.bip.gov.pl/strategie-rozwoj-regionalny/17847_strategie.html (accessed on 15 March 2021).
36. Zachariasz, A. Development of the System of the Green Areas of Krakow from The Nineteenth Century to The Present, In The Context of Model Solutions. *IOP Conf. Ser. Mater. Sci. Eng.* **2019**, *471*, 112097. [CrossRef]
37. Telega, A.; Telega, I.; Bieda, A. Measuring Walkability with GIS—Methods Overview and New Approach Proposal. *Sustainability* **2021**, *13*, 1883. [CrossRef]
38. Jarosińska, E.; Gołda, K. Increasing Natural Retention—Remedy for Current Climate Change in Urban Area. *Urban Clim.* **2020**, *34*, 100695. [CrossRef]
39. Gyurkovich, M.; Gyurkovich, J. New Housing Complexes in Post-Industrial Areas in City Centres in Poland Versus Cultural and Natural Heritage Protection—With a Particular Focus on Cracow. *Sustainability* **2021**, *13*, 418. [CrossRef]

40. Dudzic-Gyurkovich, K. Urban Development and Population Pressure: The Case of Młynówka Królewska Park in Krakow, Poland. *Sustainability* **2021**, *13*, 1116. [CrossRef]
41. Porębska, A.; Godyń, I.; Radzicki, K.; Nachlik, E.; Rizzi, P. Built Heritage, Sustainable Development, and Natural Hazards: Flood Protection and UNESCO World Heritage Site Protection Strategies in Krakow, Poland. *Sustainability* **2019**, *11*, 4886. [CrossRef]
42. Kwartnik-Pruc, A.; Trembecka, A. Public Green Space Policy Implementation: A Case Study of Krakow, Poland. *Sustainability* **2021**, *13*, 538. [CrossRef]
43. Główny Urząd Statystyczny. Statistics Poland. Kultura w 2019 r. Culture in 2019. Available online: Stat.gov.pl (accessed on 18 March 2021).
44. Jagodzińska, K. *Czas Muzeów w Europie Środkowej. Muzea i Centra Sztuki Współczesnej (1989–2014)*, 1st ed.; Międzynarodowe Centrum Kultury: Kraków, Poland, 2014; ISBN 9788363463229.
45. Gyurkovich, M. *Polskie Przestrzenie Kultury. Wybrane Zagadnienia*, 1st ed.; Wydawnictwo Politechniki Krakowskiej: Kraków, Poland, 2019; ISBN 978-83-65991-33-1.
46. Rudolf, A.; Marciniak, M.; Pieniążek, W.; Chojecki, J.; Penszko, P.; Jońca, A. *Raport Końcowy z Badania Sektora Kultury KRAKOWSKA Kultura–Stan Obecny i Perspektywy Rozwoju*; Agrotec Polska: Warszawa, Poland, 2015. Available online: http://krakow.pl (accessed on 5 April 2021).
47. Biuro Badań Społecznych Question Mark. *Wpływ Projektów Infrastrukturalnych z Zakresu Kultury Zrealizowanych w Małopolsce na Otoczenie Społeczno-Gospodarcze. Raport z Badań*; Małopolskie Obserwatorium Rozwoju Regionalnego, Departament Polityki Regionalnej: Kraków, Poland, 2018. Available online: https://www.obserwatorium.malopolska.pl/ (accessed on 15 April 2021).
48. UNESCO Worls Heritage Centre. Historic Centre of Kraków. Available online: Whc.unesco.org (accessed on 5 October 2020).
49. Creswell, J. *Research Design: Qualitative, Quantitative and Mixed Methods Approaches*, 3rd ed.; SAGE Publications: New York, NY, USA, 2011; ISBN 10 1412965578.
50. Silverman, D. *Doing Qualitative Research. A Practical Handbook*, 2nd ed.; SAGE Publications: New York, NY, USA, 2005; ISBN 10 1412901960.
51. Gehl, J. *Life between Buildings: Using Public Space*, 6th ed.; Island Press: Washington, DC, USA, 2011; ISBN 978-1-59726-827-1.
52. Kesik, O.; Demirci, A.; Karaburun, A. *Analysis of Pavements for Disabled Pedestrians in Metropolitan Cities*, 1st ed.; Lambert Academic Publishing: Chisinau, Moldova, 2012; ISBN 9783846588888.
53. National Association of City Transportation Officials. *Urban Bikeway Design Guide*, 1st ed.; Island Press: Washington, DC, USA, 2013; ISBN 9781610914949.
54. Lehmann, S. Green Urbanism: Formulating a Series of Holistic Principles. *S.A.P.I.En.S* **2010**, *3*, 1–10. Available online: http://journals.openedition.org/sapiens/1057 (accessed on 6 October 2020).
55. Elliott, H.; Eon, C.; Breadsell, J.K. Improving City Vitality through Urban Heat Reduction with Green Infrastructure and Design Solutions: A Systematic Literature Review. *Buildings* **2020**, *10*, 219. [CrossRef]
56. United States Environmental Protection Agency. EPA Heat Islands Cooling Strategies. Available online: https://www.epa.gov/heatislands/heat-island-cooling-strategies (accessed on 2 April 2021).
57. Center for Active Design. Assembly: Civic Design Guidelines. Available online: https://centerforactivedesign.org/assembly (accessed on 14 March 2021).
58. Project for Public Space. What Makes a Successful Place? Available online: https://www.pps.org/article/grplacefeat (accessed on 14 March 2021).
59. Lewicka, M. *Psychologia Miejsca*, 1st ed.; Scholar: Warszawa, Polska, 2012; ISBN 978-83-7378-476-7.
60. Lofland, L. *The Public Realm. Exploring the City's Quintessential Social Territory*, 1st ed.; Aldine de Gruyter: New York, NY, USA, 1998; ISBN 0-202-30607-0.
61. Gehl Institute. Inclusive Healthy Places. A Guide to Inclusion & Health in Public Space: Learning Globally to Transform Locally. Available online: https://gehlinstitute.org (accessed on 23 February 2021).
62. Mantey. *Wzorzec Miejskiej Przestrzeni Publicznej w Konfrontacji z Podmiejską Rzeczywistością. Przykład Podwarszawskich Suburbiów*, 1st ed.; Wydawnictwo Uniwersytetu Warszawskiego: Warszawa, Poland, 2019; ISBN 978-83-235-3615-4.
63. Saheb, Y.; Shnapp, S.; Paci, D. *From Nearly-Zero Energy Buildings to Net-Zero Energy Districts*; Publications Office of the European Union: Luxembourg, 2019; ISBN 978-92-76-02915-1. [CrossRef]
64. Wydział Kultury i Dziedzictwa Narodowego Urzędu Miasta Krakowa. Program Rozwoju Kultury w Krakowie do roku 2030. Available online: www.krakow.pl (accessed on 5 March 2021).
65. National Association of City Transportation Officials. Available online: https://nacto.org/publication/streets-for-pandemic-response-recovery/ (accessed on 20 April 2021).
66. Purchla, J.; Fabiański, M. *Historia Architektury Krakowa w Zarysie*, 1st ed.; Wydawnictwo Literackie: Krakow, Poland, 2001; ISBN 9788308031667.
67. Urząd Statystyczny w Krakowie. Statistical Office in Kraków. Available online: www.stat.gov.pl (accessed on 16 April 2021).

Article

Towards an Understanding of the Pre-War Landscape Transformations in the Face of Contemporary Urban Challenges on the Example of Gajowice in Wrocław

Aleksandra Gierko

Department of Public Architecture, Basics of Design and Environmental Development, Faculty of Architecture, Wrocław University of Science and Technology, 50-317 Wrocław, Poland; aleksandra.gierko@pwr.edu.pl

Abstract: This paper discusses the results of desk and field studies conducted in the Gajowice estate in Wrocław. The aim of the paper is to identify the original assumptions of the development of areas around multifamily buildings and to examine the process of their transformation to the present day. The research hypothesis states that the used solutions would now be defined as green infrastructure or nature-based solutions. This was confirmed with the help of comparative cartographic studies. Research on the original land development of the interwar period allows for identifying the principles based not only on compositional aspects, but also the recognition of natural values in the variety of green forms used in a given area and the important role of trees with large target sizes, in addition to the principle of shaping the green system that permeates the urban tissue, creating ecological corridors and positively influencing the local climate. Thus, the historical development is in line with the contemporary postulates of climate resilient cities.

Keywords: greenery; land development; green areas; green infrastructure

Citation: Gierko, A. Towards an Understanding of the Pre-War Landscape Transformations in the Face of Contemporary Urban Challenges on the Example of Gajowice in Wrocław. *Sustainability* **2021**, *13*, 5962. https://doi.org/10.3390/su13115962

Academic Editors: Katarzyna Hodor, Magdalena Wilkosz-Mamcarczyk and Jan K. Kazak

Received: 30 April 2021
Accepted: 21 May 2021
Published: 25 May 2021

1. Introduction

The rapid growth of the European urban population since the nineteenth century has become a canvas for new urban concepts in which greenery has been playing a significant role. The regulatory plan of Berlin, developed under the supervision of James Hobrecht, became a guideline for other plans within the German Empire's cities, including Wrocław. Established in 1862, the so-called Hobrecht Plan assumed to solve the problems of overpopulation and the hygiene of the inhabitants' lives. A basic city block module was accompanied by a large courtyard that helped to ventilate the flats and allow sunlight to enter buildings. The planners' assumption was also to introduce green areas into the cityscape [1].

The first attempts to establish and implement zoning plans in Wrocław at the beginning of the twentieth century were slowed down by the First World War, which was followed by the Spanish flu epidemic, and an economic crisis only intensified the problems of the overpopulated city. The response to this instance in the form of well-thought-out urban solutions in the 1920s and 30s was the result of strategic planning by the city authorities in combination with the activities of construction societies and architects who understood the need to implement modern solutions.

Almost a hundred years after the housing estates' construction, the urban tissue was transformed, also as a result of the damage caused during the Second World War. Nevertheless, historical urban thought can still be read, especially with the use of comparative studies. The uniqueness of the architecture built in the interwar period was noticed by post-war specialists, and its building complexes were documented and recognized as a cultural heritage, mainly in terms of buildings, but not through the prism of landscape or greenery. Meanwhile, the study of the period's city planning rules allows one to see the holistic concept of shaping urban fabric with greenery. In this aspect, greenery shall be treated as a historical green infrastructure, influencing not only the local environment and microclimate, but also the wellbeing of city dwellers.

At present, increasingly frequent studies show that the availability of green areas, including allotment gardens, near places of residence affects the physical and mental wellbeing of residents [2,3], which can also be associated with the social properties of public spaces [3,4]. The principles of urban landscape planning introduced intuitively by urban designers over a hundred years ago are now reflected in ideas such as One Health. According to this approach, the health of humans, animals, and the entire ecosystem are interconnected [5]. Thus, the city is to be a biodiverse ecosystem, in which greenery plays not only an aesthetic role, but also provides ecological connections for animals and their habitats. Such a multi-layered attitude towards the role of the green system in the urban fabric is currently being promoted by the European Green Deal initiatives. The European Biodiversity Strategy was drafted under this plan's umbrella. It emphasises the role of areas created as a result of natural processes and the necessity of their conservation, but also draws one's attention to green urban spaces as places that provide a number of ecosystem services, both for humans and other organisms [6]. In the case of greenery, it is particularly important to introduce it into the urban space at different scales and so that it creates a network of connections that permeates the cityscape [7] (pp. 65–69). Land development adjacent to housing estates built in accordance with the state of knowledge at the time, in response to welfare and health problems, could be a reference for contemporary solutions combining design, ecology, and social and economic accessibility.

This paper discusses the results of desk and field studies conducted in the Gajowice estate in Wrocław and is a part of the broader research of the spatial development of pre-war housing estates. The aim of the paper is to identify the original assumptions of the development of areas adjacent to multifamily buildings, bearing in mind the wider urban planning context of the interwar period, and to examine the process of their transformation to the present day. Because of the numerous changes that land development has undergone, the original assumptions are now difficult to read using direct observation. The research hypothesis states that due to the adopted design principles, these estates were planned with great care for the buildings' surroundings; therefore, the used solutions would now be defined as green infrastructure or nature-based solutions. Thus, the historical development is in line with contemporary postulates of climate-resilient cities. Solutions examined with the help of comparative cartographic studies could become an introduction to a catalogue of local solutions, mitigating climate change, supporting biological diversity and human health that is also based on cultural heritage.

2. Materials and Methods—Scope and State of Research

As the study is a part of a broader research encompassing several housing estates in Wrocław, a spatial and thematic delimitation of the study was made for the purposes of this paper. The temporal scope of the issue is the same as for the entire study and concerns housing estates from the moment of their construction, i.e., from the interwar period to the present day. However, studies on urban tissue provenance date to the beginning of the nineteenth century. The spatial extent, shown in the figure below (Figure 1), was determined by the thematic scope. The subject of interest of the study was the development of areas adjacent to multifamily buildings, which jointly possessed the following features:

- Apart from general spatial dispositions, it was possible to recreate the details of land development on the basis of archival materials;
- The development showed the features of a broader urban concept;
- Post-war transformations did not completely erase the pre-war plan.

The spatial scope (Figure 1) encompasses a part of the Gajowice housing estate in Wrocław, whose border runs along present-day Krucza, Kwaśna, Grochowa, Jemiołowa, Krucza, Gajowicka streets, Generała Józefa Hallera Avenue, and a part of Wrocław's railway bypass. Contemporary Polish place names have been consistently used in this paper.

Figure 1. Spatial scope of the study with the coverage of the iconographic and cartographic sources. Original work based on historical cadastral maps [8–10] and contemporary aerial photography from 2018.

Among the used research methods, two groups shall be distinguished: chamber studies and field studies. The essence of chamber research was the collection and appropriate development of cartographic, iconographic, and literature materials, which could then be used to outline the study background and comparative cartographic analyses that were then supplemented by field studies. Comparative analyses, commonly used in landscape studies [11,12], were aimed at assessing the dynamics of landscape transformation and assessing the state of preservation of pre-war land development. The comparison of pre-war cartographic materials and architectural drawings with the current state was made in the GIS. Materials in the form of raster files were georectified: a geographical reference was added to them on the basis of control points. Map overlaying for tracking landscape changes is widely used in studies at different scales and for different purposes [13–15]. When analysing changes, attention was also paid to factors that determined the manner of land rearrangement related to period trends, thus referring to the method of landscape biography [16]. An important aspect of the field study was a detailed familiarisation with the examined objects and the confrontation of the actual state of preservation with the performed chamber studies. The field research was carried out using the method of direct observation combined with photographic documentation in August 2019 and February 2021.

The study background was drawn on the basis of a review of written sources from the pre-war period: a book publication [17], an article in a sector literature [18], and a local law act with its graphic appendix [19,20]. The first post-war publications from the 1950s concern general issues of Wrocław's urban development [21]. Embedding local tendencies in a broader context was possible thanks to a review of topical publications [22–29]. The current neighbourhood profile was outlined based on the city scale planning document [30], a consultation report [31], and an overview of participatory budget projects [32–34]. However, all these sources have their limitations, as mentioned in the subsection.

The contemporary research on architecture and its spatial arrangement during the interwar period in Wrocław was studied by, inter alia, Wanda Kononowicz [35]. Nevertheless, the spatial transformation of the estate's greenery has not been studied before. Single entries regarding the development of the areas covered by the article can be found in the Green Lexicon of Wrocław. Present access to cartographic and iconographic sources, including modern digital data, allows for in-depth comparative research and tracing the changes that green areas have been subjected to. The sources were obtained as a result of a query of archives in several institutions, which are summarised in Table 1. The list shows where a given document is stored, their typology, and dating, as well as an overview of their accuracy in terms of the land development study.

Table 1. A summary of iconographic, cartographic, and modern digital data sources used for the purposes of the study in the article.

Location	Source Type [1]	Dating	Description [2]
Wroclaw University Library	large-scale, manually drafted plans [20,36]	1865 and 1926	general urban layout; cadastral division, road system, layout of buildings, watercourses, and overall land cover record in the plan from 1865 [36]; division into building classes in the plan from 1926 [20]
Wrocław Construction Archive of Museum of Architecture	pre-war manually drafted plans and designs collected by City Magistrate and Construction Police [8–10,37–44]	1911–1943	depending on source; plans [8–10] with cadastral division, road system, layout of buildings, and overall land arrangement records such as water, trees, and terrain; plans [37–39] with general site purpose (urban disposition, future public areas); designs [40–44] with details of land development (playgrounds, benches, pavement materials, single trees, low planting types)
Municipal Water and Sewerage Company S.A. in Wrocław Archive	pre-war manually drafted sewage system plans [45,46]	1912 and 1932	design [45] and plan [46] of the sewage system with a site plan for buildings, road network, and land development
Herder Institute for Historical Research on East Central Europe	aerial black and white perspective photographs from the Hansa-Luftbild Archives [47–56]	1929–1934	land development details (elements of playgrounds, individual trees, low planting types), land use (school gardens, utility yards, sports grounds, etc.)
Military Historical Bureau in Warsaw	aerial black and white orthogonal photographs [57–60]	1947	land development details (greenery system) and the scale of post-war damage

Table 1. *Cont.*

Location	Source Type [1]	Dating	Description [2]
Main Office of Geodesy and Cartography in Warsaw	aerial black and white [61–64] and colour [65] orthogonal photographs	1974, 1985, and 1995	depending on the extent and season; photographs from 1974 and 1995: urban development on the scale of the greenery system (it is possible to identify individual trees); photographs from 1985: details of land development (playgrounds, etc.), also under tree canopies
Wrocław Spatial Information System	property map and orthoimagery [66]	2015–2021	cadastral division [66] and details of land development in 2015 and 2018 [66]
other online resources	pre-war iconography [67]	no exact date	details of land development: type of greenery and appearance of elements such as benches

[1] References in square brackets. [2] Including the degree of accuracy.

3. Results

3.1. Historical Background of the Study

After the First World War, Wrocław was one of the most populous cities of the Weimar Republic. Its population density reached 114 people per hectare of the total urban area, while the country's 46 major cities had an average number of 41.3 people per hectare [17]. This was mainly due to the small amount of land occupied by built-up area when compared to cities of an even smaller population, such as Frankfurt am Main. The post-war issue of housing shortages in the cities of Weimar Republic caused a critique of the free market and its role in shaping poor living conditions. It resulted in establishing affordable housing construction programmes on the national level. Low-profit enterprises were to solve the problems of not only accommodation, but also unemployment [23]. It was possible also due to rationalization of architectonic forms and building process. Technology was to improve living conditions, as perceived by architects [26,28].

The above-mentioned free market critique also involved issues of cityscape. In 1919, landscape architect Leberech Migge published 'Green Manifesto' in which he developed his idea of planning productive landscapes within the city. Migge was influenced by the Garden City Movement, but upscaled this idea to the whole country that was to become a garden. He perceived the connection of a city and land as a way to healthy society: not only in terms of physical health, but also economic self-sufficiency and freedom [24]. These ideas were transferred to the narratives of architects, such as Bruno Taut, who emphasized the essence of common access to land in his theoretical treatise and implementations [27]. In later years, Leberech Migge incorporated architectural thought into his considerations, writing about the important role of light in the animal and plant world, especially in temperate climate. In the concept, he proposed locating residential spaces on the southern side of the building [29].

The Garden City Movement itself, however, had influenced Berlin's planning policies before the First World War. The 'Greater Berlin' plan by Hermann Jansen was the first attempt to limit the spontaneous development of the city. The green belt around the built-up areas created an urban climate system and has given them a spatial order [22]. Jansen's plan remained a theoretical model until administrative reform and expansion of the city in the 1920s. In 1929, his postulate gained a real dimension through the provisions of the first urban development plan, prepared under the direction of Martin Wagner. Wagner saw in green spaces particular social and public health values. Under the influence of those ideas, the city authorities, through programmes implemented in the 1920s, increased the availability of recreational areas for a large number of residents and made green sites occupied about 20% of the urban area [25].

The municipal authorities of Wrocław decided to solve the overpopulation problem by implementing a wide-ranging programme of building affordable flats, combining public activities with those of purposively founded construction companies that dealt with not only building up individual plots of land near the pre-war tenement buildings, but also the construction of entire estates. The areas attached to the city in 1868 were largely developed. The beginning of the 1920s spurred the construction of vast housing estates, such as Sępolno or Popowice, with an extensive functional and spatial programme. Those ventures were the responsibility of the Siedlungsgesellschaft Breslau joint-stock company, whose activity stood out among those of other building societies of interwar-period Wrocław. Low-rise housing estates, mostly two-and three-storey, uniformly shaped, were built. Care was taken to provide adequate flat insolation, ventilation, and size. Two-and three-room flats constituted the majority, i.e., almost 90% of all units [17]. The activities of building societies were characterised by a planned organisation and economic approach, visible in the standardisation of apartments and building elements. The societies were also obligated to carry out certain projects in public areas, including the construction of underground utility grids, planting street trees, and wall climbing plants [35]. Moreover, famous Wrocław architects were invited to design, which had a significant impact on the appearance of the housing estates. They were built in the spirit of Modernism, using new technologies, following a holistic approach to shaping the urban fabric, including green areas. Framing with greenery was perceived by the architects of the time as an activity that influenced the harmonisation of the building with the surroundings [18]. A holistic urban planning approach, modern architectural forms, and technological solutions, such as central heating and a shared steam laundry—first introduced in 1926 in the block limited by what are now Krucza, Kwaśna, and Stalowa streets—made it so that the estates of the interwar period did not fall behind those built in other contemporaneous cities of the Weimar Republic, such as Schillerpark in Berlin designed by Bruno Taut.

The appropriate proportions of green sites in relation to built-up areas were strictly regulated by law. In the 1920s, a new zoning plan was developed for Wrocław, based on pre-war ordinances from 1912. According to the new building code, a part of the plot had to remain undeveloped—it could serve as an internal courtyard or garden, but it could not be used for utility functions, such as fuel depots or animal pens [19]. At present, this could be referenced to areas designated for parking lots or waste disposal sites. In those regulations, an emphasis was placed on the health-related aspects of living: unit insolation and ventilation. Buildings higher than two storeys had to have yards with an area of no less than 80 m^2 [19]. The entire city was divided into building classes, with a greater share of green areas in newly built housing estates compared to the older city centre. In the area of today's Gajowice, the standard was 6/10 of the undeveloped plot area [19,20].

The crisis of 1929 slowed down the pace and momentum of construction projects. Political changes and the necessity to deal with another war continued this trend. Despite this, post-war Wrocław, thanks to the actions of pre-war urban planners, could boast a high percentage of two- and three-room apartments compared to other cities within the borders of the new state [21].

3.2. General Pre-War Assumptions and Implementation of the Spatial Development of Gajowice Greenery

At the time of including today's Gajowice site within the boundaries of Wrocław, in the 1860s, the areas on which the current housing estate is located were arable lands belonging to the village of Gabitz, located on their western side. The map from 1865, an excerpt of which is presented in Figure 2, shows the road and water system as well as the division of arable land [36]. An important element of agricultural areas was the so-called Sour Source (Sauerbrunn in German)—a water reservoir created at the site of a water exudate, probably surrounded by greenery, which at the time of urban regulation became a canvas for arranging a central recreational space in the district. Part of the road system was respected when designing the urban plan [37]. It became, among other things, the basis for the delineation of the block limited by the present-day Krucza, Kwaśna, and Stalowa streets.

Figure 2. Indicative location of the development line of the quarter limited by the present streets of Krucza, Kwaśna, and Stalowa (white line) on the excerpt of the plan from 1865 [36]—original work. Field roads became the basis for marking out the street system.

The urban plans of the district were created at the turn of twentieth century. The site was developed naturally from the city centre, from the north-east, gradually towards the open urban fringe. The area of interest for this article developed mainly in the 1920s and 1930s, although regulatory plans were drawn up prior to the First World War. Then, the formal character of the intersection of Krucza, Stalowa, Mielecka, and Bernarda Pretficza streets was planned. The arrangement of buildings around the intersection was to create a square space. The urban layout was to be closed on three sides, while the corner at the intersection of Krucza and Bernarda Pretficza streets was to remain undeveloped and intended for a formal green space. Newly planted trees were to be an important element of street space, especially the double lane running in the middle of Mielecka Street [45]. After the First World War, the plans were verified: Mielecka Street was marked out at a different angle and in a smaller width, in the 1930s, buildings were introduced in the area that previously had been planned for a green square. The urban tissue that had been destroyed during the next war was later rebuilt in the same layout, as shown in Figure 3.

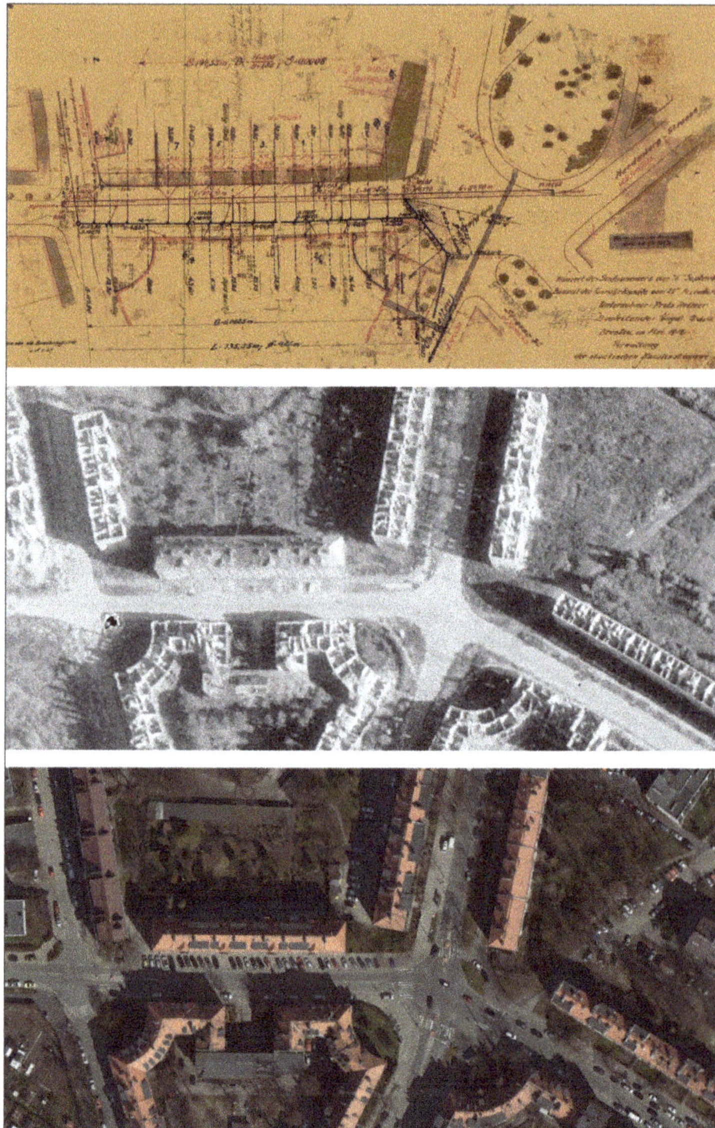

Figure 3. Development of the intersection of Stalowa, Krucza, Mielecka, and Bernarda Pretficza streets: a juxtaposition of a part of the project from 1912 [45] with aerial photographs from 1947 [60] and 2018 [66].

At the turn of the 1920s and 1930s, a tram line was built along Krucza Street (D.1 in Figure 1) and ended at the above-mentioned intersection [46,52]. Krucza Street was the main artery of the estate leading from east to west: from Gajowicka Street towards Grabiszynek. Its width was measured at 30 m between building lines. The cross-section at the height of Wróbla Street, going from the north towards the south, was as follows:

- sidewalk, 6.23 m wide;
- greenery with a row of trees, 0.50 m wide;
- bicycle road, 1.50 m wide;

- roadway, 8.00 m wide;
- bicycle road, 1.50 m wide;
- hedge, 0.50 m wide;
- tram track, 5.60 m wide;
- hedge, 0.80 m wide;
- sidewalk, 5.37 m wide [46].

It was a well-furnished street, with wide pavements and bicycle paths on both sides of the road. The tram track was separated from the carriageway using hedges. A row of trees, probably lime trees, was continued on the northern side of Krucza Street up to the railway embankment [9,60]. The southern part of the street from Mielecka Street to the embankment was probably treated as a reserve for the extension of the tram line. Street trees were an important element of land development. Noteworthy is the strip of trees located at Bernarda Pretficza Street (D.2 in Figure 1), at the height of Wróbla Street, which was established between two carriageways [9,57], currently with a mixed species composition and the last relic planting of catalpa trees.

Perhaps the most important public greenery site in the estate was the area covering a significant part of the block encircled by Krucza, Stalowa, Kwaśna, Grochowa, and Jemiołowa streets. The park at Sour Source (B.1/E.6 in Figure 1) was established at the turn of the century on the basis of a rebuilt pond that collected spring water. Surviving documentation from many archival sources would allow for a careful study of the area, almost no trace of which has survived to the present day, and could be subject to separate research, similarly to the playground at Gajowicka Street (B.2 in Figure 1). Temporary recreational areas were also depicted in pre-war aerial photographs. The photographs from 1932 and 1934 show a complex of tennis courts adjacent to the present Bernarda Pretficza Street (E.4 in Figures 1 and 4) [52,55,56], which in the 1940s was transformed into a built-up area adjacent to Wróbla Street.

Figure 4. The aerial photograph taken above Generała Józefa Hallera Avenue towards the north-west in 1932. Buildings along Wróbla Street in the foreground. In the upper right corner, a school on Krucza Street and school gardens on the other side of the street. Next to it, tennis courts (bright area on the right) [52].

The aerial photographs also show large areas designated for cultivated gardens adjacent to buildings, which provided part of the household food supply. Sites that were not yet built-up at the time were intended for temporary allotment gardens, as in the case of much of the land lying in the quarter of Stalowowolska, Połaniecka, Mielecka Streets, and Generała Józefa Hallera Avenue (C.1 in Figure 1) [38]. Before the war, the school at Jemiołowa Street had an extensive school garden stretched between Krucza and Bernarda Pretficza Streets (E.5 in Figures 1, 4 and 5), a small part of which was built up in the 1940s with a house perpendicular to the Bernarda Pretficza Street [57]. The garden was probably a place for students' education and the basis of their school meals. Tall trees were planted along the border of the lot. The alleys of trees marked the internal divisions of the garden, and the main axis of the composition was an avenue that ran across the green area from north to south [55]. The garden was liquidated after the war, and new residential buildings were erected in its place at the end of the 1960s, thus blurring the entire development, including trees. The only relic is probably a magnificent oak growing at the back of houses on Wróbla Street. Another example of the completely different needs of the post-war years are two plots limited by Generała Józefa Hallera Avenue, Buska, Sztabowa, and Stopnicka Streets (C.2 in Figures 1 and 5) that were private and mainly used as a horticultural farm. In the 1940s, they were designated for public green areas with a playground and a promenade [39], in line with the systemic approach to greenery and city policy pursued since the turn of the twentieth century, to buy private plots of land and transform them into parks and squares. The area was still used for the production of plants until the 1970s [61]; however, in the 1980s, it was built over by erecting an eleven-storey multifamily building. The current manner of development of the pre-war green areas is shown in Figure 5.

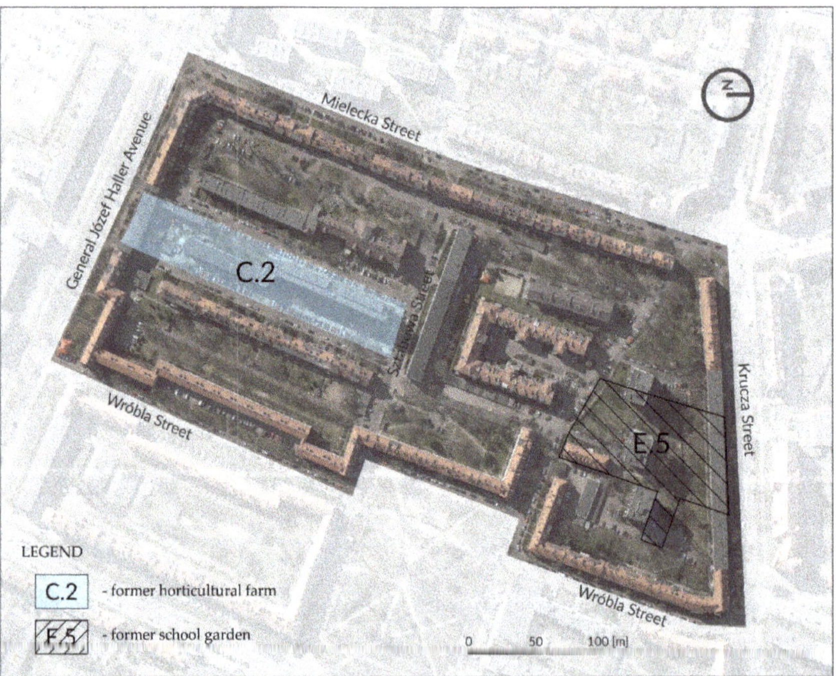

Figure 5. Pre-war green areas of the land designated for a public green space (C.2) and school garden (E.5) within the pre-war cadastral divisions [9,10] plotted on aerial photography from 2018 [66].

3.3. The Development of Areas Belonging to Multifamily Buildings

The main focus of the study is areas adjacent to multifamily buildings. These sites were intended for the everyday needs of residents. Nevertheless, in a broader planning perspective, the interwar designers noticed their significant impact on human health. In the preserved materials from Wrocław Construction Archive of the Museum of Architecture [40–44], compared to other cartographic and iconographic sources and the current state, a wider urban context taken into account in those projects can be seen. The modern systemic approach to greenery proves its societal and environmental potential.

The first case under study is the interior courtyard of a residential building in the triangle of Krucza, Kwaśna, and Stalowa streets (A.1/E.1 in Figure 1). As mentioned before, the plot was marked out according to the course of field roads (Figure 2). As seen on a map from 1865, the plot had been located between two roads, and a watercourse had run along the present Stalowa Street, bending to the east and joining Sour Spring [36]. The regulatory plan from 1911 still showed the canvas of field roads that marked the course of the present segments Kwaśna and Krucza Streets [37]. In fact, the course of the former dirt road leading to the Grabiszynek village had been visible even in 1929, before the actual construction of the road occurred and was immortalised on an aerial photograph, documenting the erection of the housing estate on the southern side of Krucza Street [48].

The estate on both sides of this part of Krucza Street (A.1/E.1 and E.2 in Figure 1) was given the name 'Am Sauerbrunn', which referred to its location and can be translated from German as 'At Sour Spring'. The contemporary name of Kwaśna Street (Sour in English) is a reference to this name. The buildings constructed in the triangle of Krucza, Kwaśna, and Stalowa streets were the first to be heated via central heating in interwar Wrocław. The entire complex, with a large, coherent garden in the interior surrounded by buildings predominantly with three-room apartments, was designed by a well-known and respected architect: Hermann Wahlich. The implementation was published in the trade magazine at the time [18], and it was also well documented in aerial photos of the interwar period [47–49,52–55]. A drawing showing the land development plan has been preserved in the Wrocław Construction Archive [40].

Aerial photographs allow the assessment of the actual pre-war use of the space between buildings and the verification the accuracy of archival drawings. The photographs were found to be consistent with the drawing deposited in the Wrocław Construction Archive and at the same time to indicate the use of presumably preliminary concepts as an illustration in a press article. A photograph of a mock-up of a building block published in a trade magazine shows the general site plan of the surroundings [18]. From the side of Krucza Street, front gardens bordered by low hedges were designed, while in the courtyard, an interior road running along the facades of buildings, a strip of low greenery separated by hedges and a line of trees surrounding the central square were planned. The layout of the courtyard's interior repeats the triangular outline of the facade. The published drawing shows perhaps the first concept of a site development plan, which differs from the mock-up and the actual state. The designed outlines of green patches are softer and more organic compared to the resulting simple geometric development drawing. The entrances to the centre of the yard are accentuated with pairs of trees, and in the centre, there is a double row of trees.

The completed project from 1927, shown in Figure 6, assumed hedging the front gardens and accentuating the entrances to the building with plantings on both sides of the entrance, located on the pavement side [40]. Based on individual trees preserved in these places, it can be assumed that they were hawthorns of a pink blooming variety. The aerial photograph from 1932 [53] shows spherically shaped small trees, which may confirm this hypothesis. Trees of the same size are visible at the entrances to the building on both sides of the gateway from the side of Stalowa Street. The photograph also shows a row of trees—probably lime trees—planted along the northern part of Krucza Street, which was a section of a wider concept that was described earlier. However, the archival drawing does not include the green strip between the bicycle path and the pavement. Based on comparison of the drawing with the iconographic sources and the current measurements, it can be assumed that the cross-section from the building wall to the road edge could be as follows:

- front garden, 5.0 m wide;
- sidewalk, 2.5 m wide;
- greenery with a row of trees, 1.5 m wide;
- bicycle road, 2.0 m wide.

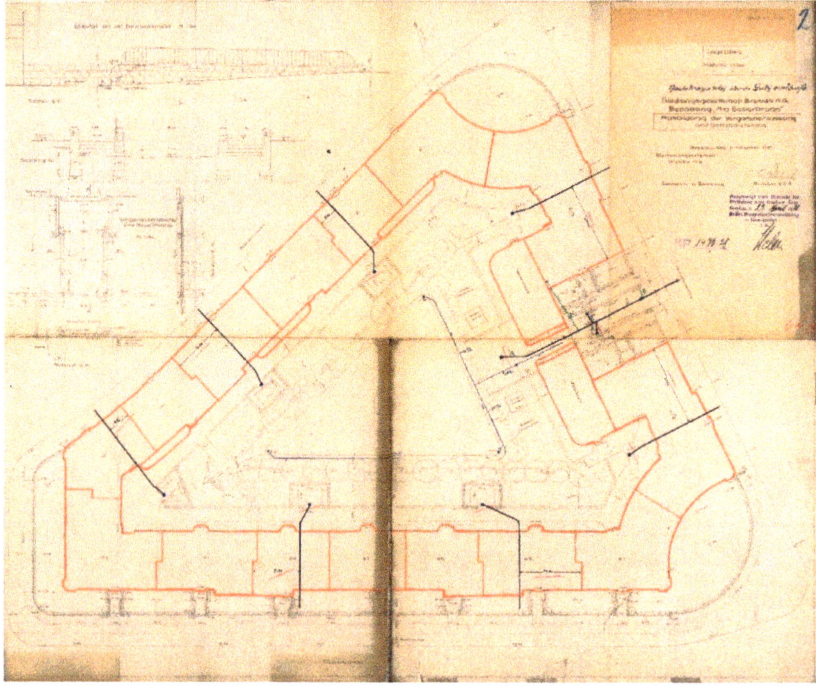

Figure 6. Implemented courtyard design from 1927 by Hermann Wahlich. The buildings outline marked with red colour. Drawing is north-east oriented [24].

The facades of the buildings shown in the photograph described above are mostly covered with climbing plants, both from the street and the courtyard sides (Figure 7). A comparison of this image with almost the same shot from three years before (1929) shows a significant growth of greenery, including climbers, hedges, and trees [48]. The corner in the form of an inverted arch, located at the intersection of Krucza, Stalowa, and Mielecka streets, overgrown with creepers and additionally planted shrubs of different heights in a well-thought-out composition emphasising the facade form and the formal character of the street's intersection, looks particularly impressive.

Figure 7. The aerial view at 'Am Sauerbrunn' estate along Kurcza Street towards the west in 1932. On the right is the triangle courtyard designed by Hermann Wahlich. On the left side of the street are wide lawns—a probable reserve for tram line extension [52].

The implemented design of the courtyard assumed 2.5 m wide pedestrian circulation spaces along the internal northern and southern elevations [40]. On the west side, there was a gateway leading to a yard with a square outline, closed on the sides with tall, probably hornbeam, hedges. There, symmetrically located on both sides of the square, were separate playgrounds with sandboxes and benches. Along the remaining two elevations, there was a strip of low greenery, 8 m wide, with hornbeam hedges encompassing utility boxes with carpet hangers. In several places, access was provided to the central triangular part that was underlined with a line of lime trees, along which there were bench niches. The inner square with a triangular outline, covered with grass, was separated from the surroundings by paths, and in its centre, a solitaire was planted, emphasising the composition. The interior design was simple, and it resulted from rectangular geometric forms. Greenery was treated as an urban material that complements the architecture.

The preserved aerial photographs show an additional, compared to the design, planting of bushes on the corners of the central square, which is also confirmed by an undated photo taken from the level of a pedestrian [67]. It was probably made in the 1930s and shows perennial beds emphasising the bypass along the buildings and simple forms of benches with concrete legs and wooden seats and back supports. On the basis of the photography, it could also be supposed that the internal pathways were constructed with gravel mixture, so that they had a partially permeable surface.

Post-war transformations in the development of the area under study can be traced via comparisons of aerial photographs taken in 1947 [60], 1974 [61], and 1985 [63] and present-day orthoimagery [66]. The photograph taken after the war shows the damage done to the buildings, while the composition of the tall greenery, both street and courtyard trees, was preserved. Bearing in mind not only the transformation of the site, but the broader context of the area under study, the Gajowice estate, it would be crucial to trace the changes by verifying materials from the 1950s and 1960s. However, there is substantial probability such that such materials do not exist. Nevertheless, it was at this time that much of the urban fabric of the estate began to be systematically supplemented and rebuilt. At that time, the Sour Source was presumably also covered.

A photograph from 1974 [61] recorded the state of the interior in the triangle of Krucza, Kwaśna, and Stalowa streets after rebuilding. The road system in the courtyard was reorganised—a 2.5–3 m wide circular road was introduced, which was paved with concrete blocks. The road was moved away from the facade of the building and separated from the wall with a strip of greenery and sidewalks. The permeable surface under tree canopies was narrowed to a width of about 3.5 m. This circulation system has remained to this day. In the 1970s, the tall greenery was still complete and compact. In the 1980s, only two trees remained from the street row, and the roadside lane began to be intensively used as a parking space for cars. The photograph from 1985 [63] also shows the functional layout that has been preserved to this day: the waste collection point on the axis of the entrance from the side of Stalowa Street and the playground in place of the central lawn.

Until today, only 23 limes remain out of the 41 trees planted before the war. The courtyard interior is a parking place for residents, which also has a significant impact on the high greenery condition (Figure 8). The greenery of the middle floor has grown at different times, and there are no intentionally composed shrubs or perennials. The custom of keeping creepers on the walls of buildings has not been continued. The composition of the space emphasises the utility functions of the interior, which had been supposed to serve the recreation of residents according to the original concept. Nevertheless, a large number of permeable surfaces and large trees constitute the natural and cultural value of this place.

Figure 8. The current state of land development: the obliteration of the historical layout and the use of the courtyard for utility purposes. Parking under trees adversely affects their condition.

The second case analysed in this study is a part of the estate established in the same period at the opposite side of Krucza Street. The entire urban layout marked E.2 in Figure 1 could be studied on the basis of aerial photographs [47–55]. The houses located in today's Kolbuszowska, Stalowowolska, Tarnobrzeska, and Połaniecka streets were built mostly in the mid-1920s. Due to the low development density—only a few families lived in each house—the adjacent areas were arranged as home gardens. The gardens had ornamental functions, their layout was often geometric, based on one or two axes, and they served a utility function. In accordance with the assumptions prevailing at the time, which were also used in the no-longer-existing housing estate in Popowice in Wrocław, most of the gardens had tool sheds shared by residents. The public front gardens had a simple geometric layout, and the trees were used only in a double lane planted in a small green area at the end of Kolbuszowska Street. Shared semi-public courtyards were laid out in a sequence of interiors adjacent to the buildings constructed along Krucza Street (Figure 9).

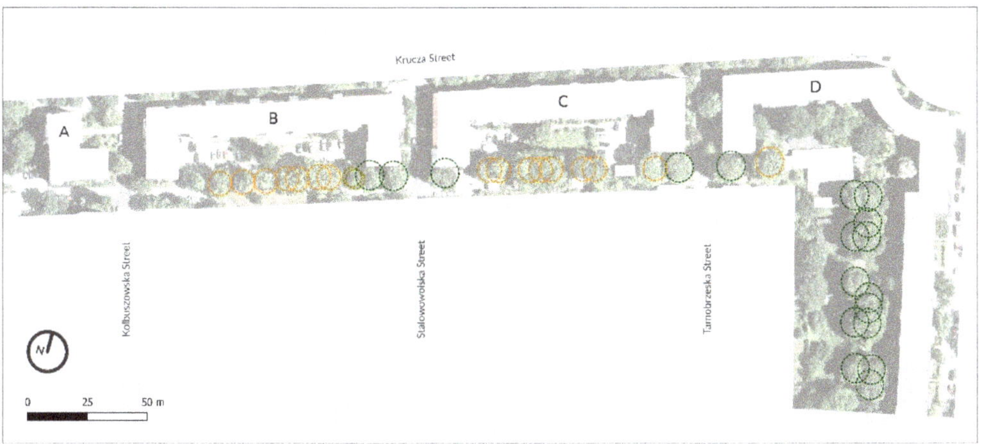

Figure 9. Courtyards with relics of historical greenery in the buildings' interior along Krucza Street. Chestnut trees marked in green, hornbeams in yellow, and maple trees in orange. Plotted on aerial photography from 2015 [66].

The buildings were erected in 1928–1929 as a project by Siedlungsgesellschaft Breslau. The buildings marked with the letters A and B in the figure were designed by Heinrich Rump, and buildings with the letters C–D by Hermann Wahlich. Concerning site plan drawings, only Heinrich Rump's drawing has been preserved in the Wrocław Construction Archive [41]. However, on the basis of comparative analyses of iconographic and cartographic resources with site studies, it is possible to reconstruct the idea of landscape development. Building A has an L-shaped plan. Such an arrangement of buildings determined that the remaining part of the plot was divided into two sections, each with a different character [41]. The front part was a formal entrance to the building, with a pavement running along its walls and a square-shaped greenery site in the middle. The green area was surrounded by a hedge, and the corners were planted with shrubs accentuating the composition. The part behind the building served recreational and utility functions. The courtyard was fenced, with two gates leading to it and an exit directly from the building. The western part, separated from the public promenade by a hedge, had a space with a sandbox and a U-shaped bench around it, and a separate spot with a carpet hanger (Figure 10).

The space behind building B, also designed by Heinrich Rump, was set on the north–south axis, perpendicular to the building and dividing the backyard into two, almost symmetrical parts. The front gardens were fenced with hedges, and the entrances were accentuated with bushes or hawthorns in stem form. The yard had the shape of an extended rectangle, about 20 m deep and about 80 m long. Directly next to the building, a 2.5 m wide green strip was marked, separated by access routes to the entrances. There was a path around the yard, probably permeable, 2.5 m wide. The middle part consisted of two large lawns, separated by a 2.0 m wide passage located on the north–south axis. From the side of the building, the passage was arranged in the form of a small square, 6.0 m wide, separated from the space by a hedge. The corners on both sides of the lawns near buildings were also separated by hedges and featured carpet hangers. At the end of the courtyard, in the southern part, there were two sandboxes and benches mirrored on both sides of axis. Six trees were planted on each side of the southern fence. On the basis of field studies and measurements, it can be concluded that maple trees were planted here, and the distance between them was 4.5 m. At the corners of the building, on the south side, bench niches were arranged on the projection of a part of an arch, planted with formed hedges. On the west side, the niche closed the yard with 3/4 of the arch, and there were two benches there, while on the eastern side, the niche was a half-arch with one bench. A passage was to be made to Stalowowolska Street, towards the east. In reality, however, as recorded in the aerial photographs [51,53,55], a westward exit was made instead. The same photographs also show that the southern facades were planted with climbing plants. On the basis of field studies of the preserved trees, it can be added that the hedges at the bench niches were planted with hornbeams. A pair of chestnut trees was planted at the side of the unrealised entry. Upon inspection in the field, it can be stated with certainty that this has become a spatial rule of this part of estate: the streets were flanked by plantings of chestnut trees that were closing the rows of maples planted inside the courtyards, as shown in Figures 9 and 11.

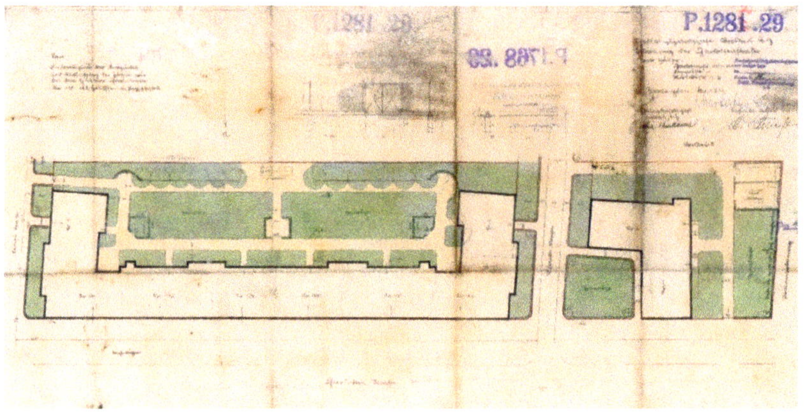

Figure 10. Implemented land development plan for buildings along Krucza Street from 1929 by Heinrich Rump. Drawing is south oriented [41].

The interior of the courtyard at building C was realised as a mirror image of the courtyard at building B, with an exit towards Tarnobrzeska Street, which is recorded in the aerial photographs [53–55]. Maple trees were planted at shorter distances from each other (4 m), so there were seven or eight trees on each side of the axis. In addition, the southern facades were planted with climbing plants. The series of interiors was closed by a fenced yard at building D. The character of land development refers to the general principle applied throughout all yards, which means that despite the divisions, both by street and fences, the sequence of backyards could be interpreted as a coherent assumption, both from a functional and environmental perspective. The yard closing line belonged to three communities in the part of building D [9]. It was fenced off from the longer, eastern part of the green space. The space was about 25 by 25 m. From the side of the building there was a green belt, in the middle there was a path around the lawn. In the corner of the building, there was a separate area with a carpet hanger, fenced with a hedge [53–55]. A row of probably three maple trees was planted at the southern border.

Figure 11. View along Stalowowolska Street towards the north. Chestnut trees closing the courtyard interiors. Summer 2019.

The last green space in this part of the estate is an elongated yard that is 115 m long and 40 m wide, located at building D, along Mielecka Street (Figure 12). From the side of the buildings, an escarpment ran along the entire length, moved away from the buildings by about 6 m [9]. It probably resulted from the necessity to level the area for construction. Thus, most of the green area was elevated in relation to the space next to the buildings by about 60 cm. It was possible to get to the higher placed area via stairs. In the northern part, the outline of the slope followed the arched outline of the facade, so that a small, irregular fragment with bushes was marked out from the rectangular interior [55]. The functional layout was realised in bands along the north–south axis: from the side of the buildings there was a recreational part, beyond a utility part with a lawn and places for drying linen and a zone of vegetable gardens adjacent to the back of gardens belonging to houses located on Tarnobrzeska Street [54].

The recreational part had a path running along the top of the slope and another one separating the site from the utility part. Along the path, three squares open to the west, with chestnut trees on the other sides were arranged. Two squares were 6 × 15 m in size; the last one was about 6 × 6 m. Six chestnut trees were planted on the larger squares, three were planted on the smaller square at its corners, perhaps emphasising the possibility of developing the composition towards the south. In addition, to access from the north–south path, the larger squares were accessed from the buildings through paths perpendicular to the facade.

Figure 12. The aerial view at 'Am Sauerbrunn' estate towards the north-east in 1932. The courtyard at the intersection of Krucza and Mielecka Streets with three small squares encompassed by trees [55].

As in the case of the quarter at Kwaśna Street, in the cases discussed above, only the trees are a remnant of the pre-war development. The changes, the effects of which remain to this day, and which are visible in aerial photographs from the 1970s and 1980s [61,63,64], were of a degrading nature. Although it was necessary to adapt the courtyards to modern functions, e.g., waste storage and parking, it could be done in a thoughtful way. A glaring example is the last discussed yard, with a waste disposal site arranged in the middle, at the height of the middle pre-war square with chestnuts. Due to such a location, it was necessary to change the entire path system. The road was improved between chestnut trees, completely erasing the historical layout. In other cases, new functions were implemented—parking lots were introduced in the courtyards, next to playgrounds and waste disposal sites. As shown in Figure 9, the historical arrangement of the trees is incomplete. This is especially noticeable in the lanes of maple trees that grow in a narrow green strip where parking is carried out (Figure 13).

The third discussed area is the site at Wróbla Street, in the section between Generała Józefa Hallera Avenue and Sztabowa Street: both the space in front of buildings (A.3 in Figure 1) and the backyards (E.3a and E.3b). Its pre-war development can be studied and documented using surviving maps and drawings from the Wrocław Construction Archive [10,42], as well as an aerial photograph from 1932 from the collections of the Herder Institute [52] in comparison with the post-war photograph from 1947 from the Military Historical Bureau in Warsaw [60]. The buildings in this section of the street have been shaped away from the street. Both the houses located on Generała Józefa Hallera Avenue are set back from the road by about 20 m, and the street interior of Wróbla Street has been shaped in such a way that, while at the entrances, the walls of the opposite three pairs of buildings are 25 m apart, the middle nine pairs of buildings are separated from each other by 50 m. This allowed for the arrangement of wide green areas separated only by pedestrian circulation. The cadastral map from 1941 [10] shows that three of five of the lawns at Generała Józefa Hallera Avenue were shaped as terrain depressions, which is also visible in the post-war aerial photograph. The map also shows a double row of trees on both sides of Wróbla Street, which is not confirmed by the photograph taken after the war. They were also not recorded on the land development drawing.

Figure 13. Backyard of building B designed by Heinrich Rump with a historical row of trees in summer 2019.

The development of the street space and courtyards is characterised by simplicity and functionality (Figure 4). Perhaps it is also a reflection of the crisis period at the turn of the 1920s and 1930s. Due to the fact that the building line was moved away, the yards belonging to the building were narrowed down to approximately 22 and 13 m or E.3a and E.3b, respectively. As in the case of the last described green area in the Am Sauerbrunn estate, also in this case, the terrain difference had to be solved with the help of a slope next to the buildings, which further reduced usable space. The slopes were planted with low shrubs, and the entrances were equipped with stairs [52]. Along the entire building, in the strip behind the escarpment, there were racks for laundry and carpet hangers. This part was edged with hedges. Wider spaces at the ends of the plot were also separated with hedges and arranged as neighbourhood spaces with sandboxes and benches. As the yard at the western side was wider, it was possible to distance the utilitarian part from the facade with the help of lawns. The paths heading to the back of the yard were accented with pairs of trees. They can be seen on the aerial photograph from 1947, as well as overgrown and not maintained hornbeam hedges, but it was not possible to determine the species, as these trees are no longer exist.

After the war, the wide spaces between the buildings at Wróbla Street were split in half to place parking spaces next to the buildings. Plantings along the street were implemented or restored, with ash trees on both sides, which had a positive effect on the aesthetic and ecological values of the street area. In the post-war period, the utility function of the courtyards changed—the laundry racks disappeared and waste collection areas were introduced instead. Interestingly, the recreational function of yard endings has been maintained for a long time. In the 1970s and 1980s, a parking lot was arranged only in the south-west corner [61,63]. The recreational areas have not survived [66]. Unmaintained hornbeam hedges have grown, thanks to which the spaces between the buildings today have a unique character (Figure 14).

Figure 14. Backyard of the building by Wróbla Street, western part with overgrown hornbeam hedge. Summer 2019.

The final cases analysed here are two courtyards placed at the beginning of Wróbla Street (A.4 and A.5 in Figure 1). The land development was created in 1937 and 1938, respectively, and, in the author's opinion, did not meet the criterion adopted in the method of 'showing the features of a broader urban concept'. Nevertheless, they contribute to understanding the changes that occurred in the interwar period land development of the Gajowice estate, and thus are an important element in constructing the landscape biography of the estate. In both cases, it is only possible to compare the state as designed [43,44] with the current one [66]. Both of them show the features of an intimate ornamental garden development with utilitarian sections. The land development at the eastern side of the street was a courtyard with five houses. The site could only be entered through the buildings. The plot had a triangular shape, which affected the circulation layout: next to the buildings there were paths made of granite blocks, along the fence, the path was made of decorative broken stone. Near the buildings, in the middle of the path, there are places with carpet hangers, and at their ends there are sandboxes and a place to sit. In the middle of the establishment, there was a round square planted with four lime trees and three benches below them. The fence was planted with clumps of shrubs, in the southern part there was one additional tree, and the corner of the building was emphasised with a composition of shrubs (Figure 15).

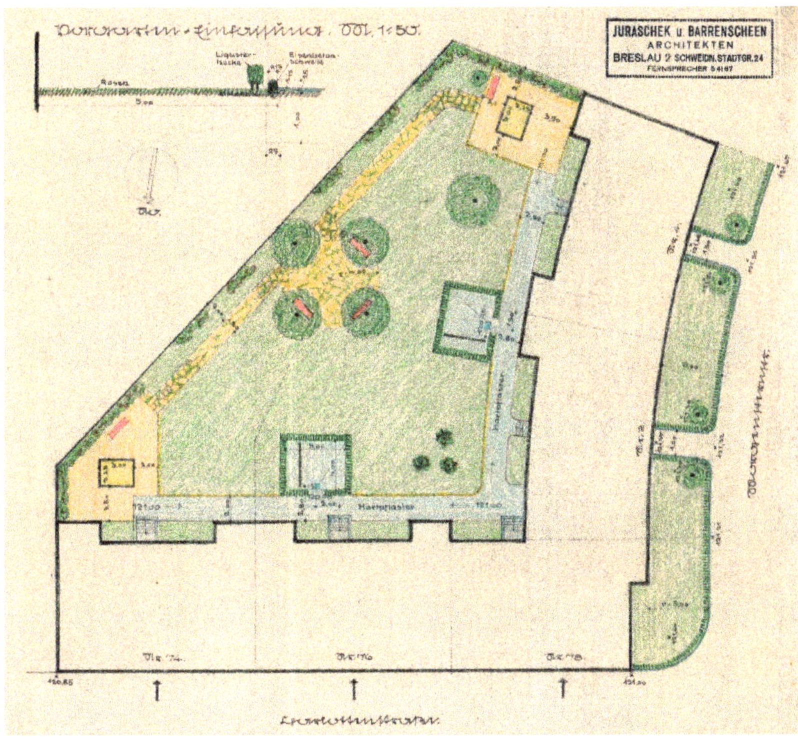

Figure 15. The design of the courtyard of the buildings at 2 Wróbla Street from 1937 by Juraschek u. Barrenscheen Architekten. Drawing south oriented [43].

On the opposite side of the street, the courtyard belonged to three houses. The plot was outlined in an irregular rectangle. In the corner, away from the buildings, an elevated part of a yard was made, emphasised by the corner planting of a bush or a tree. Below, behind a retaining wall, there are three benches on a path that surrounds the central lawn. The entire path was made of broken stone. The yard was fenced off with a formed hedge, and the entrance was only possible from buildings. At the facade of the building on the eastern side, there was a circular section with a diameter of 3 m with a bypass that was 2 m wide, planted with single shrubs, perhaps a sandbox. On both sides of Wróbla Street, the front gardens were 5 m wide. The entrances were accentuated with pairs of bushes, and the whole area was framed with privet hedges, as the description of the project shows.

The examples discussed here may indicate a change in the scale of urban planning assumptions, which reflects transformations of the interwar period economic situation. Wide ranging activity of construction companies was then past. Moreover, in this part of Wróbla Street, the whole urban concept was not completed before the war and was added to in the 1940s, but never fully completed. Regarding the examples of development under discussion, a composition of four lime trees encircling a round square is present to this day [66].

3.4. Current Neighbourhood Profile

The spatial development plan on a city scale, that is in force, defines the Gajowice as a part of a larger unit: Śródmieście Południowe [30]. The unit area is 378 ha, while the spatial scope of the study encompasses 53 ha. The land use data shows that residential areas predominate the unit: they occupy more than 30% of the total area. In 2016, the number of inhabitants was 52,700 people. Greenery sites occupy 33 ha, i.e., more than 8%

of the total area. The document states that the greenery system is rich and in different forms of arranged squares, semi-private greenery accompanying residential buildings, and street trees. Nevertheless, as judged by the authors, the unit faces many challenges concerning not only raising the standard of greenery in the areas of residence, but also creating systemic connections with green public sites. Challenges related to sustainable mobility are the development of rail transport and the creation of pedestrian and bicycle connections with neighbouring units.

Taking into account the needs expressed by the spatial development plan, in 2019, the city authorities established guidelines for the so-called complete estates, organizing participatory meetings, and working, among other things, on the case of the Śródmieście Południowe unit [31]. The aim of the consultation was to obtain feedback from the residents in four areas of interest:

- history and identity of the unit;
- character of public spaces and greenery;
- structure of local services;
- type of mobility within and outside the unit.

It should be noted that only 20 inhabitants of Śródmieście Południowe unit took part in the survey, which does not constitute a representative sample. Nevertheless, the response of the residents was analysed when determining the spatial dispositions for the complete estate.

The existence of local marketplaces and broadly understood post-German heritage were noted by responders in terms of the unit's identity. Regarding the second aspect, residents indicated that they are satisfied with the greenery in their surroundings, but noted numerous problems, such as damage due to illegal parking practices. They pointed to the lack of large parks or squares in the vicinity. The inhabitants were not able to clearly indicate the place of local integration and called for extending cultural, recreational, and gastronomic offers. During the consultations, the residents emphasized that their needs are met in terms of services. Regarding mobility, the discussed problem was the insufficient number of parking spaces and poor technical condition of sidewalks. At the same time, residents expected further development of the road network, including bicycle routes and public transport.

Another insight into the needs of the inhabitants gives the review of the participatory budget that has been established several years ago in Wrocław. This process enables residents to directly influence decisions to allocate part of the public budget to projects submitted annually by citizens. The review of the projects submitted in the last three years in the Gajowice [32–34] allows one to see the distribution of needs into different categories, which is summarised in Table 2.

Table 2. Participatory budget projects of Gajowice estate in the years 2019–2021, in the author's division into categories.

Year [1]	Project Categorisation			
	Restoration of Courtyards, Incl. Green Infrastructure	Sports and Recreation Facilities in Public Areas	Improvement of Road Safety	Restoration of Grey Infrastructure, Incl. Bikeways
2021 [2]	3	2	2	1
2020	1	2	1	0
2019	4	2	1	2

[1] A different unit division before 2019 does not allow for data comparison. [2] Projects before evaluation by the municipality units.

As seen above, in 2019 and 2021, restoration of areas neighbouring the residential buildings represented the main category of submitted projects. The restoration of greenery within those projects is articulated, but the accents are distributed differently among elements such as car parks, playgrounds, blue infrastructure, green infrastructure, and other components of the spatial arrangement of the courtyards. It may indicate a wide range of needs and different issues experienced by the residents. The courtyard renovation projects are followed by those that concern the issue of equipping public spaces, especially schools, with sports and recreational facilities. Projects of this type prevailed in 2020. Another important issue is the improvement of road safety around pedestrian crossings. The last category concerns ameliorating the quality of grey infrastructure, including cycling infrastructure. In fact, in 2021, one project concerned equipping bus stops with timetable displays, and in 2019, one concerned a bikeway and another car park in one of the courtyards.

4. Discussion

Urban concepts of healthy cities, taking into account the comfort of living and access to public spaces, derived from the Garden City Movement and written down in the form of planning regulations at that time, were verified by the experience of war, called by people then the Great War. The post-war period was characterised by large-scale development of urban planning in German cities, including Wrocław. The role of greenery as a factor shaping the space of human life was emphasised in the regulations [19,22,25], which was also reflected in architectural concepts, as these subjected to the study. In those assumptions, first attempts to treat greenery as a system that offers a number of benefits can be seen. The specialists of that time perceived green areas as an element shaping social equality and individual's freedom, postulating access to one's own piece of land for all [24]. However, in the case of areas intended for temporary gardens in the Gajowice, perhaps one should speak of necessity rather than economic self-sufficiency. The crisis of the late 1920s slowed down the pace of extensive changes; however, the idea of a green and accessible city, but also an affordable flat is still valid today.

The development of the southern suburbs of Wrocław, which started in the middle of the nineteenth century, is a history of the transformation of agricultural land into an urban area. Nevertheless, the spatial planners at the turn of the century paid attention to not only the functional layout of the unit, but also the place's context. An urban continuum is visible in practices such as respecting the previous road system or transforming a place with its own identity into the district's main recreation site, as on the example of Sour Source. In the spatial arrangement practices noted in the Gajowice estate before the Second World War, features of perspective planning could be seen. The streets had been designed with an appropriate width, so that after the construction of the buildings, trees were planted. An adequate reserve was left for the tram line extension. Various forms of urban mobility such as public transport and bicycles were taken into account. The surviving documents contain transformation's plans of a private plot into publicly accessible green areas. On the other hand, it shall be noted that some areas were built-up, although originally there, had been intended for greenery or temporarily for allotment gardens or recreational sites. Nevertheless, the Gajowice greenery system of the interwar period encompassed different forms of green infrastructure: street trees and hedges, urban park and playgrounds, allotment gardens, and courtyards rich in various forms of vegetation.

Research on the original land development showed that the designers of the interwar period adopted principles based on not only compositional or functional aspects, but also recognized the natural values in the variety of green forms used in one place, including wall climbing plants, and the important role of trees with large target sizes. It can be seen that the greenery system was planned as to permeates urban tissue. Looking from a contemporary perspective, those solutions could be identified as nature-based, designed to create ecological corridors and positively influence the local climate. Another important observation is that trees are a stable element of the cityscape; they are often only remnants of a historical layout. An overview of the land development adjacent to the buildings from the 1920s to 1930s in the Gajowice estate shows the transition of courtyards' functional and spatial programs: from complex ones to more simple utility forms and even small ornamental gardens. That probably reflects the overall economic situation of a country, the context that has already been recognized in other publications [25,35]. Nevertheless, in almost all studied cases, the recreational function of the courtyards played a major role.

The study shows that post-war changes resulted in the blurring of the green system and designers' original assumptions, as evidenced by the numerous transformations of large green areas of the Gajowice estate, such as the park at Sour Source, a school garden, or the former gardening school. It should be added, however, that sometimes the lack of maintenance—as in the case of numerous hornbeam hedges—influenced the development of large trees, that have higher value than shrubs when viewed through the prism of natural assets. Mostly, however, the depletion of green sites is visible: only some of the trees are left of various greenery forms in the courtyards. The courtyards are mainly used for utility needs related to waste storage and parking, which was not the original intention, neither the design nor the legal one. On the other hand, it partly reflects contemporary requirements.

The current needs of the unit's inhabitants, included in the spatial policies and expressed in the participation processes, although broader than only related to the quality of greenery, are referred to in many aspects. The issue is not only the poor standard of green sites in the areas of residence, but also the scarcity of systemic connections among them. Moreover, the inhabitants are not able to clearly indicate the place of local integration. They notice the lack of large parks or squares in the vicinity. Interestingly, post-German heritage is perceived by the residents in terms of the unit's identity. Nevertheless, it seems that the variety of needs requires a rational approach to the planning of 'green' policies.

5. Conclusions

This study of a landscape change of the Gajowice estate showed that many elements of the greenery system were planned with reflection and the long perspective of sustaining the dwellers areas of recreation. Various forms of greenery were used as coverings for building walls, partitions, and fences, and finally as elements determining the local climate, shading streets, and squares. Therefore, they should be seen, in addition to the buildings, as an essential element of cultural heritage, and in combination with their environmental properties, they shall be treated as green infrastructure. In the face of the effects of the climatic crisis, it is evident that it is not enough to develop based on the relationship between man and inanimate nature, but it is necessary to also take biological diversity into account. Looking from the perspective of adverse changes, the conservation of the historical system's remains seem more justified. Steps need to be taken to preserve its remnants and restore the system's continuity. The land development character shall be adapted to the contemporary requirements of cities that are resistant to climate change. Spatial solutions examined with the help of comparative cartographic studies could become an introduction to a catalogue of local solutions, supporting biological diversity and human health, that is based on cultural heritage.

Funding: This research received no external funding.

Data Availability Statement: Not applicable.

Acknowledgments: The author would particularly like to thank the Wrocław Construction Archive of the Museum of Architecture and Herder Institute for Historical Research on East Central Europe for consenting to the use of iconographic materials from their collections that helped to illustrate and greatly enriched the article.

Conflicts of Interest: The author declares no conflict of interest.

References

1. Gray, M.W. Urban sewage and green meadows: Berlin's expansion to the south 1870–1920. *Cent. Eur. Hist.* **2014**, *47*, 275–306. [CrossRef]
2. Brown, S.C.; Perrino, T.; Lombard, J.; Wang, K.; Toro, M.; Rundek, T.; Gutierrez, C.M.; Dong, C.; Plater-Zyberk, E.; Nardi, M.I.; et al. Health disparities in the relationship of neighborhood greenness to mental health outcomes in 249,405 U.S. Medicare beneficiaries. *Int. J. Environ. Res. Public Health* **2018**, *15*, 430. [CrossRef] [PubMed]
3. Robinson, J.M.; Brindley, P.; Cameron, R.; MacCarthy, D.; Jorgensen, A. Nature's role in supporting health during the COVID-19 pandemic: A geospatial and socioecological study. *Int. J. Environ. Res. Public Health* **2021**, *18*, 2227. [CrossRef] [PubMed]

4. Carter, M.; Horwitz, P. Beyond proximity: The importance of green space useability to self-reported health. *EcoHealth* **2014**, *11*, 322–332. [CrossRef] [PubMed]
5. American Veterinary Medical Association. One Health: A New Professional Imperative, AVMA, 2008. pp. 13–14. Available online: https://www.avma.org/sites/default/files/resources/onehealth_final.pdf (accessed on 21 April 2021).
6. European Commission. EU Biodiversity Strategy for 2030. Bringing Nature Back into Our Lives. 2020, pp. 12–13. Available online: https://eur-lex.europa.eu/resource.html?uri=cellar:a3c806a6-9ab3-11ea-9d2d-01aa75ed71a1.0001.02/DOC_1&format=PDF (accessed on 21 April 2021).
7. European Environment Agency. Nature-Based Solutions in Europe: Policy, Knowledge and Practice for Climate Change Adaptation and Disaster Risk Reduction. 2021, Volume 17, pp. 65–69. Available online: https://www.eea.europa.eu/publications/nature-based-solutions-in-europe (accessed on 16 May 2021).
8. Plan von Breslau. Blatt 331, Städtisches Vermessungsamt, Wrocław Construction Archive of the Museum of Architecture, 1939, No Inventory No., Paper Map in a Scale of 1:1000.
9. Plan von Breslau. Blatt 2311, Städtisches Vermessungsamt, Wrocław Construction Archive of the Museum of Architecture, 1940, Inventory No. 110376, Paper Map in a Scale of 1:2000.
10. Plan von Breslau. Blatt 2321, Städtisches Vermessungsamt, Wrocław Construction Archive of the Museum of Architecture, 1941, Inventory No. 110378, Paper Map in a Scale of 1:2000.
11. Chmielewski, T.J. *Systemy Krajobrazowe. Struktura-Funkcjonowanie-Planowanie*, 1st ed.; Wydawnictwo Naukowe PWN SA: Warszawa, Poland, 2012; pp. 247–269.
12. Stahlschmidt, P.; Swaffield, S.; Primdahl, J.; Nellemann, V. *Landscape Analysis. Investigating the Potentials of Space and Place*, 1st ed.; Routledge: London, UK; New York, NY, USA, 2017; pp. 36–39, 54–77.
13. Bender, O.; Boehmer, H.J.; Jens, D.; Schumacher, K.P. Using GIS to analyse long-term cultural landscape change in Southern Germany. *Landsc. Urban Plan.* **2005**, *70*, 111–125. [CrossRef]
14. Statuto, D.; Cillis, G.; Picuno, P. Using historical maps within a GIS to analyze two centuries of rural landscape changes in southern Italy. *Land* **2017**, *6*, 65. [CrossRef]
15. Van Eetvelde, V.; Antrop, M. Indicators for assessing changing landscape character of cultural landscapes in Flanders (Belgium). *Land Use Policy* **2009**, *26*, 901–910. [CrossRef]
16. Kolen, J.; Renes, H.; Bosma, K. Lanscape bibliography. In *Research in Landscape Architecture. Methods and Methodology*, 1st ed.; van den Brink, A., Bruns, D., Tobi, H., Bell, S., Eds.; Routledge: London, UK; New York, NY, USA, 2017; pp. 120–135.
17. Magistrat der Hauptstadt Breslau. *Siedlung und Stadtplanung in Schlesien. Heft 1: Breslau*, 1st ed.; Verlag Grass, Barth & Comp. W. Friedrich: Breslau, Poland, 1926.
18. Langer, K. Breslauer Siedlungsbauten. *Ostdeutsche Bau-Zeitung*, 4 May 1927, Volume 35, pp. 202–205, 209.
19. Berger, O. *Die Bauordnung vom 20. Mai 1926 und Andere Baurechtliche Bestimmungen für Breslau*, 1st ed.; Verlag Grass, Barth & Comp. W. Friedrich: Breslau, Poland, 1926.
20. Plan von Breslau, Städtisches Vermessungsamt, Wroclaw University Library, 1926 (Update), Inventory No. 5596-IV.C., Paper Map in a Scale of 1:15 000. Available online: https://www.bibliotekacyfrowa.pl/dlibra/publication/32794/edition/41391/content (accessed on 27 April 2021).
21. Maleczyński, K.; Morelowski, M.; Ptaszycka, A. *Wrocław. Rozwój Urbanistyczny*, 1st ed.; Budownictwo I Architektura: Warszawa, Poland, 1956; pp. 256–275.
22. Drapella-Hermansdorfer, A. Wielki Berlin. Poszukiwanie krajobrazowej tożsamości miasta. In *Kształtowanie Krajobrazu: Idee, Strategie, Realizacje. Cz. 1, Saksonia, Branderburgia, Berlin*, 1st ed.; Drapella-Hermansdorfer, A., Ed.; Oficyna Wydawnicza Politechniki Wrocławskiej: Wrocław, Poland, 2004; pp. 38–40.
23. Everett, P.D. The organic machine: City planning in the Weimar republic. In *Urban Transformations: From Liberalism to Corporatism in Greater Berlin, 1871–1933*; University of Toronto Press: Toronto, ON, Canada, 2019; pp. 198–234.
24. Haney, D.H. Leberecht migge's "green manifesto": Envisioning a revolution of gardens. *Landsc. J.* **2007**, *26*, 201–218. [CrossRef]
25. Jackisch, B.A. The nature of Berlin: Green space and visions of a new German capital, 1900-45. *Cent. Eur. Hist.* **2014**, *47*, 307–333. [CrossRef]
26. Seelow, A.M. The construction kit and the assembly line—Walter Gropius' concepts for rationalizing architecture. *Arts* **2018**, *7*, 95. [CrossRef]
27. Speidel, M. Bruno taut: Landscape as aesthetic substrate. In *Nature Modern: The Place of Landscape in the Modern Movement*, 1st ed.; Kirchengast, A., Ed.; Jovis Berlin: Berlin, Germany, 2018; pp. 87–103.
28. Struhařík, D. Bauhaus and the beginning of mass housing: How residential open building reacts to changes in society. *Archit. Artibus* **2019**, *11*, 54–60. [CrossRef]
29. Urbanik, J. Krajobraz nowoczesnych osiedli okresu międzywojennego w Niemczech w dziełach Leberechta Migge. In *Architektura Pierwszej Połowy XX w. i jej Ochrona w Gdyni i w Europie*, 1st ed.; Sołtysik, M., Hirsch, R., Eds.; Miasto Gdynia: Gdynia, Poland, 2011; pp. 35–42.
30. Biuro Rozwoju Wrocławia. Studium Uwarunkowań i Kierunków Zagospodarowania Przestrzennego. Wrocław, 2018. Available online: https://geoportal.wroclaw.pl/studium/ (accessed on 9 May 2021).

31. Fundacja na Rzecz Studiów Europejskich, Raport. Osiedla kompletne I i II. Konsultacje Społeczne Wytycznych Przestrzennych dla Obszaru Jagodna, Jerzmanowa-Jarnołtowa, Śródmieścia Południowego i Maślic Małych na Rzecz Osiedla Kompletnego. Wrocław, 2020. Available online: https://www.wroclaw.pl/rozmawia/files/dokumenty/27040/Osiedla-kompletne-I-II-RAPORT.pdf (accessed on 9 May 2021).
32. Lista Projektów WBO 2019. Available online: https://www.wroclaw.pl/budzet-obywatelski-wroclaw/projekty-2019/ (accessed on 9 May 2021).
33. Lista Projektów WBO 2020. Available online: https://www.wroclaw.pl/budzet-obywatelski-wroclaw/projekty-2020/ (accessed on 9 May 2021).
34. Lista Projektów WBO 2021. Available online: https://www.wroclaw.pl/budzet-obywatelski-wroclaw/projekty-2021/ (accessed on 9 May 2021).
35. Kononowicz, W. Rozwój urbanistyczny i polityka mieszkaniowa wrocławia w okresie republiki weimarskiej. In *Leksykon Architektury Wrocławia*, 1st ed.; Eysymontt, R., Ilkosz, J., Tomaszewicz, A., Urbanik, J., Eds.; Via Nova: Wrocław, Poland, 2011; pp. 93–98.
36. Plan von Breslau und den angrenzenden Ortschaften Gabitz; Höfchen; Neudorf; Lehmgruben; Huben; Morgenau; Zedlitz; Scheitnig; Hoffmann, A.; Sadebeck, M. Wroclaw University Library, 1865, Inventory No. 12874-IV.C., Paper Map in a Scale of 1:10 000. Available online: https://www.bibliotekacyfrowa.pl/dlibra/publication/35037/edition/41674/content (accessed on 27 April 2021).
37. Regulatory Plan of the Area Limited by Today's Streets Krucza, Jemiołowa, Grochowa, Bernarda Pretficza and not Existing Street, Städtisches Vermessungsamt, Wrocław Construction Archive of the Museum of Architecture, 1911, No Inventory No., Paper Map without Scale.
38. Ownership Map of the Areas Located at the Height of Property No. 6 at Połaniecka Street, Städtisches Vermessungsamt, Wrocław Construction Archive of the Museum of Architecture, 1936, Inventory No. 84158, Paper Map in a Scale of 1:1000.
39. Lageplan des Grundstueckes von Herrn Ernst Schirdewan, Construction Police, Wrocław Construction Archive of the Museum of Architecture, 1943, Inventory No. 64857, Paper Drawing without Scale.
40. Land Development Plan for the Quarter Limited by Today's Streets Krucza, Kwaśna, and Stalowa, Wahlich, H., Wrocław Construction Archive of the Museum of Architecture, 1927, Inventory No. 86531, Paper Drawing in a Scale of 1:200.
41. Land Development Plan for Courtyards Located at the Back of Krucza Street between Numbers 130 and 142, Rump, H., Wrocław Construction Archive of the Museum of Architecture, 1929, Inventory No. 87192, Paper Drawing in a Scale of 1:200.
42. Land Development Plan for Front Gardens along Wróbla Street between Numbers 36 and 70, Baugeschaeft Simon & Halfpaap, Wrocław Construction Archive of the Museum of Architecture, 1932, Inventory No. 88089, Paper Drawing in a Scale of 1:500.
43. Land Development Plan for Backyards at 2 Wróbla Street, Juraschek u. Barrenscheen Architekten, Wrocław Construction Archive of the Museum of Architecture, 1937, No Inventory No., Paper Drawing in a Scale of 1:200.
44. Land Development Plan for Backyards at 1 Wróbla Street, Juraschek u. Barrenscheen Architekten, Wrocław Construction Archive of the Museum of Architecture, 1938, No Inventory No., Paper Drawing in a Scale of 1:200.
45. Sewage System Design Plan along Stalowa Street between Krucza and Kwaśna Streets, Verwaltung der Staedtischen Kanalisationswerke Breslau, Municipal Water and Sewerage Company S.A. in Wrocław Archive, 1912, Inventory No. 805, Paper Drawing, Scale Illegible. Available online: https://gis.mpwik.wroc.pl/jarc-gui/views/main.xhtml (accessed on 27 April 2021).
46. Sewage System Design Plan along Krucza Street between Stalowa and Gajowicka Streets, Verwaltung der Staedtischen Kanalisationswerke Breslau, Municipal Water and Sewerage Company S.A. in Wrocław Archive, 1932, Inventory No. 621, Paper Drawing in a Scale of 1:500. Available online: https://gis.mpwik.wroc.pl/jarc-gui/views/main.xhtml (accessed on 27 April 2021).
47. Eichborn-Siedlung in Breslau-Gräbschen, Hansa-Luftbild, Herder Institute for Historical Research on East Central Europe, 1929, Inventory No. 60062, Aerial Photography. Available online: https://www.herder-institut.de/bildkatalog/iv/60062 (accessed on 17 April 2021).
48. Eichborn-Siedlung in Breslau-Gräbschen, Hansa-Luftbild, Herder Institute for Historical Research on East Central Europe, 1929, Inventory no. 60493, Aerial Photography. Available online: https://www.herder-institut.de/bildkatalog/iv/60493 (accessed on 17 April 2021).
49. Siedlung Sauerbrunn in Breslau von N, Hansa-Luftbild, Herder Institute for Historical Research on East Central Europe, 1930, Inventory No. 60063, Aerial Photography. Available online: https://www.herder-institut.de/bildkatalog/iv/60063 (accessed on 17 April 2021).
50. Hardenberghügel Breslau, Hansa-Luftbild, Herder Institute for Historical Research on East Central Europe, 1932, Inventory No. 60320, Aerial Photography. Available online: https://www.herder-institut.de/bildkatalog/iv/60320 (accessed on 17 April 2021).
51. Hardenberghügel Breslau, Hansa-Luftbild, Herder Institute for Historical Research on East Central Europe, 1932, Inventory No. 60321, Aerial Photography. Available online: https://www.herder-institut.de/bildkatalog/iv/60321 (accessed on 17 April 2021).
52. Siedlung an der Kürassierstraße Breslau, Hansa-Luftbild, Herder Institute for Historical Research on East Central Europe, 1932, Inventory No. 59988, Aerial Photography. Available online: https://www.herder-institut.de/bildkatalog/iv/59988 (accessed on 17 April 2021).
53. Siedlung Sauerbrunn in Breslau von O, Hansa-Luftbild, Herder Institute for Historical Research on East Central Europe, 1932, Inventory No. 60637, Aerial Photography. Available online: https://www.herder-institut.de/bildkatalog/iv/60637 (accessed on 17 April 2021).

54. Siedlung Sauerbrunn in Breslau von SO, Hansa-Luftbild, Herder Institute for Historical Research on East Central Europe, 1932, Inventory No. 60636, Aerial Photography. Available online: https://www.herder-institut.de/bildkatalog/iv/60636 (accessed on 17 April 2021).
55. Siedlung Sauerbrunn in Breslau von SW, Hansa-Luftbild, Herder Institute for Historical Research on East Central Europe, 1932, Inventory No. 60638, Aerial Photography. Available online: https://www.herder-institut.de/bildkatalog/iv/60638 (accessed on 17 April 2021).
56. Caroluskirche Breslau von O, Hansa-Luftbild, Herder Institute for Historical Research on East Central Europe, 1934, Inventory No. 60042, Aerial Photography. Available online: https://www.herder-institut.de/bildkatalog/iv/60042 (accessed on 17 April 2021).
57. Borek and Południe Estates in Wrocław, Military Historical Bureau in Warsaw, 1947, Inventory No. M33035-1947-014-6973-10k, Aerial Photography.
58. The North-Eastern Part of the Grabiszyn Estate in Wrocław, Military Historical Bureau in Warsaw, 1947, Inventory No. M33035-1947-013-6917-10k, Aerial Photography.
59. The Northern Part of the Grabiszyn Estate in Wrocław, Military Historical Bureau in Warsaw, 1947, Inventory No. M33034-1947-013-6918-10k, Aerial Photography.
60. The Southern Part of the Grabiszyn Estate in Wrocław, Military Historical Bureau in Warsaw, 1947, Inventory No. M33034-1947-014-6975-10k, Aerial Photography.
61. Grabiszyn, Południe, Borek, and Gaj Estates in Wrocław, Main Office of Geodesy and Cartography in Warsaw, 1974, Inventory No. 1_7388, Aerial Photography. Available online: https://opendata.geoportal.gov.pl/ZdjeciaLotnicze/miniatury/72227/72227_2524871_1_7388.jpg (accessed on 27 April 2021).
62. The Central Part of the Grabiszyn Estate in Wrocław, Stalowa Street in the Centre of Photography, Main Office of Geodesy and Cartography in Warsaw, 1985, Inventory No. 7_7045, Aerial Photography. Available online: https://opendata.geoportal.gov.pl/ZdjeciaLotnicze/miniatury/70893/70893_2325020_7_7045.jpg (accessed on 27 April 2021).
63. The Central Part of the Grabiszyn Estate in Wrocław, Mielecka Street in the Centre of Photography, Main Office of Geodesy and Cartography in Warsaw, 1985, Inventory no. 8_7073, Aerial Photography. Available online: https://opendata.geoportal.gov.pl/ZdjeciaLotnicze/miniatury/70893/70893_2324819_8_7073.jpg (accessed on 27 April 2021).
64. The Central Part of the Grabiszyn Estate in Wrocław, Stalowowolska Street in the Centre of Photography, Main Office of Geodesy and Cartography in Warsaw, 1985, Inventory No. 8_7074, Aerial Photography. Available online: https://opendata.geoportal.gov.pl/ZdjeciaLotnicze/miniatury/70893/70893_2325043_8_7074.jpg (accessed on 27 April 2021).
65. Grabiszyn, Południe, Borek, Gaj, and Tarnogaj Estates in Wrocław, Main Office of Geodesy and Cartography in Warsaw, 1995, Inventory No. 1_2963, Aerial Photography. Available online: https://opendata.geoportal.gov.pl/ZdjeciaLotnicze/miniatury/62736/62736_364007_1_2963.jpg (accessed on 27 April 2021).
66. Property Map of Wrocław and Orthoimagery of Wrocław from 2015 and 2018, Wrocław Spatial Information System. Available online: https://www.geoportal.wroclaw.pl/en/ (accessed on 27 April 2021)
67. Photo from the Sanding Man Point of View Taken in the Interior of the Courtyard at Kwaśna Street. View towards the South-East. Available online: https://polska-org.pl/756328,foto.html (accessed on 27 April 2021).

Article

Using Composition to Assess and Enhance Visual Values in Landscapes

Magdalena Gyurkovich and Marta Pieczara *

Faculty of Architecture, Poznań University of Technology, 61-131 Poznań, Poland; magdalena.gyurkovich@put.poznan.pl
* Correspondence: marta.pieczara@put.poznan.pl

Abstract: (1) The research presented in this paper aims to study the value attributed to a landscape composition's visual elements and their overall influence on how they are perceived. The historical and contemporary visual approaches to a landscape constitute its background, for example, geographical, aesthetic, iconographic, phenomenological. (2) The visual assessment method elaborated by the Polish school of landscape architecture is used in the first part of this study. It is built of three steps with corresponding tools: landscape inventory, composition analysis, and evaluation. Moreover, an expert survey is used to complete the study. The work's novelty is completing the visual approach with an expert inquiry, which aims to solve the subjectivity issue, an inherent visual evaluation controversy. The study area comprises urban and suburban locations from the agglomeration of Poznań, Poland. (3) The research results indicate the significant contribution of three visual elements to the positive assessment of landscape values: greenery, built heritage, and water. The importance of the composition is also demonstrated. (4) The main research findings show that visual evaluation tools should be implemented as part of sustainable spatial planning. Their implementation permits identifying the essential positive value in the existing landscape and creating guidelines for its preservation or enhancement. The article's significance is the effect of proposing real and possible guidelines to improve the spatial planning policy, making landscape management more sustainable.

Keywords: landscape; panorama; composition; visual values; landscape evaluation; landscape management; sustainable spatial planning

Citation: Gyurkovich, M.; Pieczara, M. Using Composition to Assess and Enhance Visual Values in Landscapes. *Sustainability* **2021**, *13*, 4185. https://doi.org/10.3390/su13084185

Academic Editors: Jan K. Kazak, Katarzyna Hodor and Magdalena Wilkosz-Mamcarczyk

Received: 8 March 2021
Accepted: 3 April 2021
Published: 9 April 2021

1. Introduction

This work aims to examine the role of the selected types of visual elements in a landscape's composition as perceived by humans. The research applies visual assessment methods to identify the landscape elements that can be claimed to create its essential positive values. The scope of the study remains limited to the area of the Poznań agglomeration. However, it covers different types of the cultural landscape, both urban and suburban. While the assumed research results may indicate the need for new landscape protection strategies, their essential background includes discussing the landscape's definition in its complexity and presenting both historically and contemporarily applied analytical methods and approaches towards landscapes.

The term "landscape" can have different meanings depending on the field of study. For a geomorphologist, for example, the landscape represents the Earth's surface and is considered as the result of the formational physical processes. Meanwhile, a landscape ecologist would consider a landscape in the light of interactions that once took place or now take place within it. The focus on the interaction is equally felt behind the definition of the landscape provided by the European Landscape Convention (ELC), which considers a landscape as "an area, as perceived by people, whose character is the result of the action and interaction of natural and/or human factors" [1] (art. 1). This definition combines three significant aspects of the landscape: its geographical origins, anthropogenic modifications, and human perception.

While the natural processes involved in landscape formation are the subject of geographical research with a long-established tradition "evolving from naturalists such as Alexander von Humboldt and Darwin" [2] (p. 2), the recognition of human participation in shaping the landscape came subsequently [2,3]. This was a starting point of the holistic approach to the landscape. By combining research approaches and methods typical for the natural (physical) and social (human) sciences, landscape studies have become an interdisciplinary field that contributes to overcoming the disciplinary division between the two scientific branches [4,5]. The multidisciplinary nature of landscape studies and the need to analyze landscapes as part of a holistic approach has already been noticed by von Humboldt, who is credited with defining landscape as "the total character of a region of the Earth" [6] (p. 27). By using the word "total", this definition describes the landscape "as a holistic entity perceived by humans and having a distinct character or identity" [7] (p. 188). As can be deduced from the cited explanation, the holistic approach to the landscape incorporates human aspects in terms of the anthropogenic influence on its shaping process and perception. Besides the fact that landscape as a concept includes the material reality resulting from "a continuous dynamic interaction between natural processes and human activity" [7] (p. 188), it also refers to "the immaterial existential values and symbols of which the landscape is the signifier" [7] (p. 188). The mutual relationships between the social culture and the landscape can thus be represented as processes occurring between two endpoints—the first being land molding by human labor and the other being the landscape's symbolic expression of a culture. In other words, the landscape is shaped by society members so as to materialize the values of their immaterial culture, and, in a feedback loop, its final appearance "shapes the citizens' attitudes and behavior" [8] (p. 11).

Every cultural landscape shaped by human labor is characterized by a set of culture-related attributes, some of which are perceived visually—for example, aesthetic, expressive, symbolic—or aimed at identification [8]. The importance of cultural features in the interpretation of landscapes was demonstrated by Kobayashi [9], who emphasized that the communication of meanings within the landscape is subject to cultural limitations. The effectiveness of linguistic expression in conveying understandable messages depends on the clarity of a landscape's structure, with a high formality acting in its favor [9] (p. 180). Hence, the landscape is implied as a structured semiotic system, built of elements that play the role of signifiers. The relevance of semiotic theory in the study of visual design representations, and specifically in landscape design, was demonstrated by Raaphorst et al. [10]. Among the basic semiotic systems, the visual one seems to be most suitable for the image-based analysis of the landscape.

The importance of a landscape's visual aspect has been approached from different perspectives. Cosgrove highlighted the importance of visual perception in both forming and understanding landscapes by stating that "the landscape idea represents the way of seeing" [11] (p. 1). This statement also means that the perception of a single landscape can change depending on the viewer's background. "Semiotics and iconography teach us that there are as many meanings as there are stakeholders" [10] (p. 130). Iconography's approach to landscape treats its representations as "consistent images of its meaning or meanings" [12] (p. 1), making an image equal to the reality it represents. Iconographic research perceives landscape as an image or symbol, being at the same time based on the study of the symbolic imaginary [12].

By defining the role of symbols as objects representing, or denoting, something else [13,14], the image-based approach to landscape refers to the semiotics. Derived from linguistics, the theory of semiotics views language as "a system of signs where there is nothing essential except the union of meaning and the acoustic image" [15] (p. 32). Like the verbal semiotic system, the visual one also implies the unambiguous connection of a signifier (a sign) with its denotation (a meaning). At the same time, the differences in comprehension depend on the stakeholder's background [10,16]. As applied to landscape studies, the image-based approach thus aims to "identify the symbolic meanings and messages contained in the landscape" [14] (p. 212). Hence, the landscape is considered an organized system of symbolically represented values that are perceived visually. "Landscape carries meaning as well as minerals and agricultural wherewithal" [14] (p. 245). Using a linguistic metaphor, a signifier within a landscape can be presented as a visual element (e.g., tower), whereas the signified, or its meaning, refers to the relevant idea (e.g., the source of power).

The decoding process of a landscape image can link one sign with additional secondary meanings, just like one architectural object can communicate different secondary functions [17]. Backed with the semiotic theory of logic developed by American philosopher Charles Sanders Pierce, the triadic understanding of semiosis is a key to decode sign-systems other than language, including visual ones [10,16]. According to the triadic model, each sign has an equivalent referent [10], or non-coded message [16]. In addition, it can connote diverse coded messages, or connotations [16], which are interpretations of the sign [10]. Taking for example a tower in a landscape, its denoted meaning (the referent) would be a source of dominative power (the rule), while the connoted interpretation can be a king's castle, a sacral building, or a bank headquarters. What decides the appearance of different interpretations is the context. Decoded meanings tend to depend on the viewer's background and experience, as well as on his knowledge [10,16]. Going further with the words of Muir, "viewers will tend to evaluate landscapes according to their perceived merits, which will include aesthetic and ecological considerations as well as others, like cultural characteristics" [14] (p. 182).

The iconography of landscape, backed up with the theory of semiotics, forms the aesthetic approach to the landscape. It aims to explain what features of a landscape make people like it and the reasons behind this. In the words of Appleton, "what is it that we like about landscape, and why do we like it?" [13] (p. xv) and [14] (p. 244).

The perception of the landscape, which gains core importance in the aesthetic approach, relies considerably on its characteristic visual features. The definition of landscape in the Oxford Dictionary indicates this, describing this interdisciplinary concept as "all the visible features of an area of land, often considered in terms of their aesthetic appeal". However, the importance of visually perceived landscape characteristics is not limited to the aesthetic approach and is also used in different analysis scales. For example, the renowned patch-corridor-matrix model [18] also applies a visual assessment method, to an extent, to analyze the land mosaic.

Landscape analysis methods based on distinguishing visual and non-visual elements form the basis of several significant contributions to the theory of landscape perception. First of all, the phenomenological approach must be mentioned. The concept of a phenomenon at its core is usually defined as something observable, manifesting itself. The idea of the phenomenon was derived from ancient Greek philosophy and was later re-defined by Immanuel Kant [19]. Kant placed it in opposition to the noumenon concept, which he described as representing the essence of things—such as, for example, truths and values, which cannot be observed and therefore are recognizable uniquely through reason. Revived in modern times by Kant, the concept of this phenomenon formed the beginning of the philosophical movement of phenomenology, which is described by Edmund Husserl [20] as focused on consciousness structures. The phenomenology trend continued in Martin Heidegger's [21] concept of Fourfold (das Geviert), which inspired further distinguishing and classifying phenomena that can be identified in a landscape.

The philosophical movement of phenomenology inspired an analytical method for studying landscape that was developed and applied by Christian Norberg-Schulz, a Norwegian architect, theorist, and historian. According to Norberg-Schulz, phenomena are tangible things that build the world surrounding us [22]. They are interconnected in a complex and sometimes even contradictory way. They can be classified according to their nature (i.e., natural or artificial), location (i.e., Earth or sky), or adoption (i.e., inside or outside) [22]. The phenomenological approach decomposes the landscape into elements, or entities, that have specific meanings and connotations in the range of landscape studies [23].

The phenomenological approach towards landscape has numerous distinguished contemporary successors. Breaking the landscape down into diverse visual and non-visual elements is their common denominator. The components identified are subsequently examined in terms of their impact on the overall perception of a landscape. Such an approach has been adopted, among others, by Górka, who distinguished creative and passive images of the landscape [24]. Particularly important for architectural and urban studies, the landscape's creative image is built prevailingly from visual elements, including forms and patterns [24]. Such an explicit image of the landscape finds its counterpart in imaginary values that refer to the collective consciousness. The social awareness of "the values attaching to landscapes and the issues raised by their protection, management and planning" [1] (art. 6 B) is at present considered crucial. This fact emphasizes the significance of landscape studies and visually oriented research, contributing to increased knowledge about the landscape. It is vital to shape citizens' expectations regarding a landscape's quality and improve their sustainable development responsibility [24]. Human perception is hence considered an indispensable factor of landscape integration in terms of sustainable development [25]. Particularly, a balance between landscape protection, enhancement, and sustainability issues needs to consider its perception by humans as well as the values they attach to it [25]. As shown by the work of Serraino and Lucchi [25], sustainable development must use an interdisciplinary approach to landscape, integrating multiple diverse aspects: technical (e.g., energy efficiency), ecological (e.g., preventing pollution, conforming with Green Deal policy), cultural (e.g., heritage protection) and humanistic (e.g., perception of the values).

The recognition of a landscape's visual value as one of the necessary conditions for any appropriate sustainable development strategy [1,24,26] contributes to the appreciation of visual landscape research as particularly important. Contemporarily applied visual methods include the Landscape Physiognomy Assessment (LPA) and Landscape Visual Capacity Assessment (LVCA) [26]. The latter approach, modeled over the landscape capacity assessment analysis carried out across the UK as part of preparing a local plan [27], defines a landscape's visual capacity as its resilience to changes resulting from the absorption of new investments [26]. Applied as a part of the integrated landscape management strategy, the method contributes to recognizing visual values in landscape and minimizing visual hazards due to the appearance of new investments (e.g., residential settlements).

Despite the difficulty of avoiding natural subjectivity [10,16], the visual assessment of landscape quality has entered both discussion and practice in landscape-related studies. Its usefulness for land management policies has been proven theoretically [28–30] and through practical examples of using visual analysis to design a landscape protection strategy [28] (pp. 117–118, 136). Largely inspired by the development of perception studies, the idea of using landscape evaluation for planning purposes profoundly influenced the approach adopted towards landscape by British geographers in the 1970s [30] (p. 46). This approach, which considered the preservation of the visual qualities in the landscape as an integral part of any consistent planning strategy, was originally short-lived [30]. More recently, its essential elements have received new attention in the form of contemporarily applied methods of landscape assessment—for example, Landscape Character Assessment (LCA) [27], Landscape Physiognomy Assessment (LPA), and Landscape Visual Capacity Assessment (LVCA) [26].

The approach to landscape adopted in this research is one of many possible ways to assess the landscape. Other active approaches derive from different landscape conceptualizations that are adopted in different disciplines, for example, geography, ecology, landscape ecology, history, historical ecology, archaeology, environmental psychology, or landscape architecture [31]. The actively used visual landscape assessment approaches mainly derive from landscape conceptualization in its physical aspect. The approach presented in this article refers to the existing visual assessment method that uses human sight to define its physical elements. However, it combines this approach with the semantic view, which is derived from iconography and aesthetics, aiming to assess their value as perceived by humans. The approach also uses selected elements of the Delphi technique, which are applied to a quantitative study of research results in order to check their compliance (see also Section 2: Materials and Methods).

As remarked by Daniel [32] the contemporary environmental management practice mainly uses an expert approach to landscape, while contemporary research is dominated by the perception-based approach. The two approaches differ in terms of landscape conceptualizations and "the relative importance of the landscape and human viewer components" [32] (p. 267). While landscape perception studies draw from the Gestalt holistic approach, considering landscape images through the prism of its conceptualizations, the environmental approach develops towards rigorous scientific studies. They aim to collect relevant data and apply analytical tools to build models with which to explain specific relationships between the condition of the environment and the viewer's impression. Both approaches seem incomplete if separated, hence this study will combine two stages: a visual study of landscape composition and a survey used as the basis of quantitative research. Recently, a need to create a more integrated approach to landscape has been identified [33].

From the point of view of architectural studies, which belong to visually oriented disciplines, both aesthetic and phenomenological approaches constitute the essential background of any research analyzing a landscape's composition and humans' perception of it. Treating the landscape as a structured system that can convey semantic messages irrevocably refers to visual elements' significance. If specific types of such elements could be assigned a positive or negative value, the questions of what we like about a landscape and why this is so could be answered. This is precisely the goal of the research presented in this paper, which uses visual assessment methods to test such a possibility for a few exemplary locations from the Poznań agglomeration. The study's expected results can form a starting point for a new landscape management strategy, integrating landscape visual quality with the traditional geographical view.

2. Materials and Methods

The study area is the agglomeration of Poznań, which forms a major urban area and is a capital of the Greater Poland region. Instead of providing a map of the study area in this section, which would not be fully legible, most viewpoints discussed in this article will be provided with their geographical coordinates, whenever it was possible to accurately retrace. This information allows the reader to locate them in any GPS or GIS system.

The research is organized in two stages. The first stage is empirical and uses the composition analysis method, which is consequently applied across the Polish school of landscape architecture for both study and design purposes. The method derives from the analytical procedure formulated by Bogdanowski [34,35] and aims at understanding the landscape as the human perceives it. The basic assumption of the method is that when trying to orientate in an outdoor space, the same means are used as in an interior one. Namely, a human intuitively finds reference points such as walls and free-standing objects. According to Bogdanowski, each landscape interior can be analyzed within four categories: horizontal plane, walls, vaults (ceilings), and free-standing objects [34,35]. At the same time, their mutual relations determine the perception of the whole.

Most importantly, the horizontal plane's proportions define whether an interior is wide, elongated, or circular. Secondly, the proportions of the openings as compared to the surface of the walls define it as a concrete interior (clearly perceived with less than 30% of openings), an objective one (with openings covering between 30 and 60% of wall surfaces), or subjective one (with walls perceived subjectively due to their porosity above 60%). Moreover, a landscape's interior can be open or closed in the sense of its termination. Then, the number and layout of free-standing objects define it as simple or complex (Figure 1a). Secondly, vertical articulation and lines leading the viewer's sight are analyzed, defining the landscape's rhythm. The composition analysis permits listing all visual elements observed in the landscape and, more importantly, assess the role they play in the whole image (e.g., creating rhythm and symmetry, dividing long intervals, disturbing harmony) (Figure 1b).

ceiling walls horizontal plane openings

(**a**)

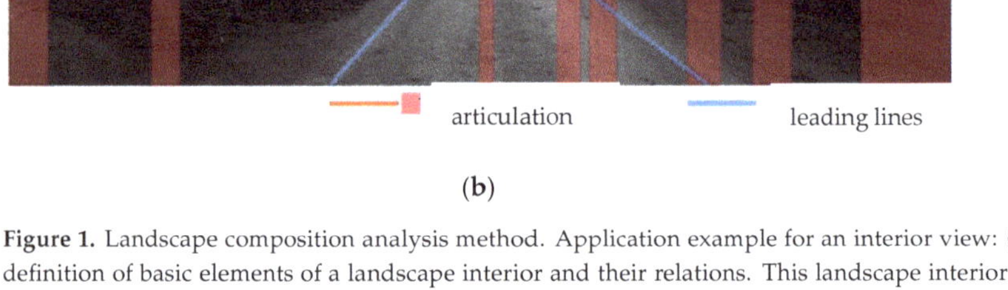

(**b**)

Figure 1. Landscape composition analysis method. Application example for an interior view: (**a**) definition of basic elements of a landscape interior and their relations. This landscape interior is defined as elongated, concrete, open, and simple; (**b**) Further analysis of rhythm, articulation, and lines leading the viewer's sight. Illustration author: Kacprzyk, M., PUT (Poznań University of Technology) 2021.

As a result, the elements and principles of the composition, which are visual, become the foundation for the systematic analysis of any type of landscape [34–40]. For open spaces, an equivalent procedure can be performed based on a panoramic view, which can be equally broken down into a set of elements whose interrelationships determine the meaning of a given object and its role in the overall image [34,37,40,41]. The essential elements of a panorama are dominant (strong spatial form), subdominant, accent (a form that distinguishes itself), main content, frames, background (uniform plane), and foreground (horizontal plane) [34] (Figure 2a).

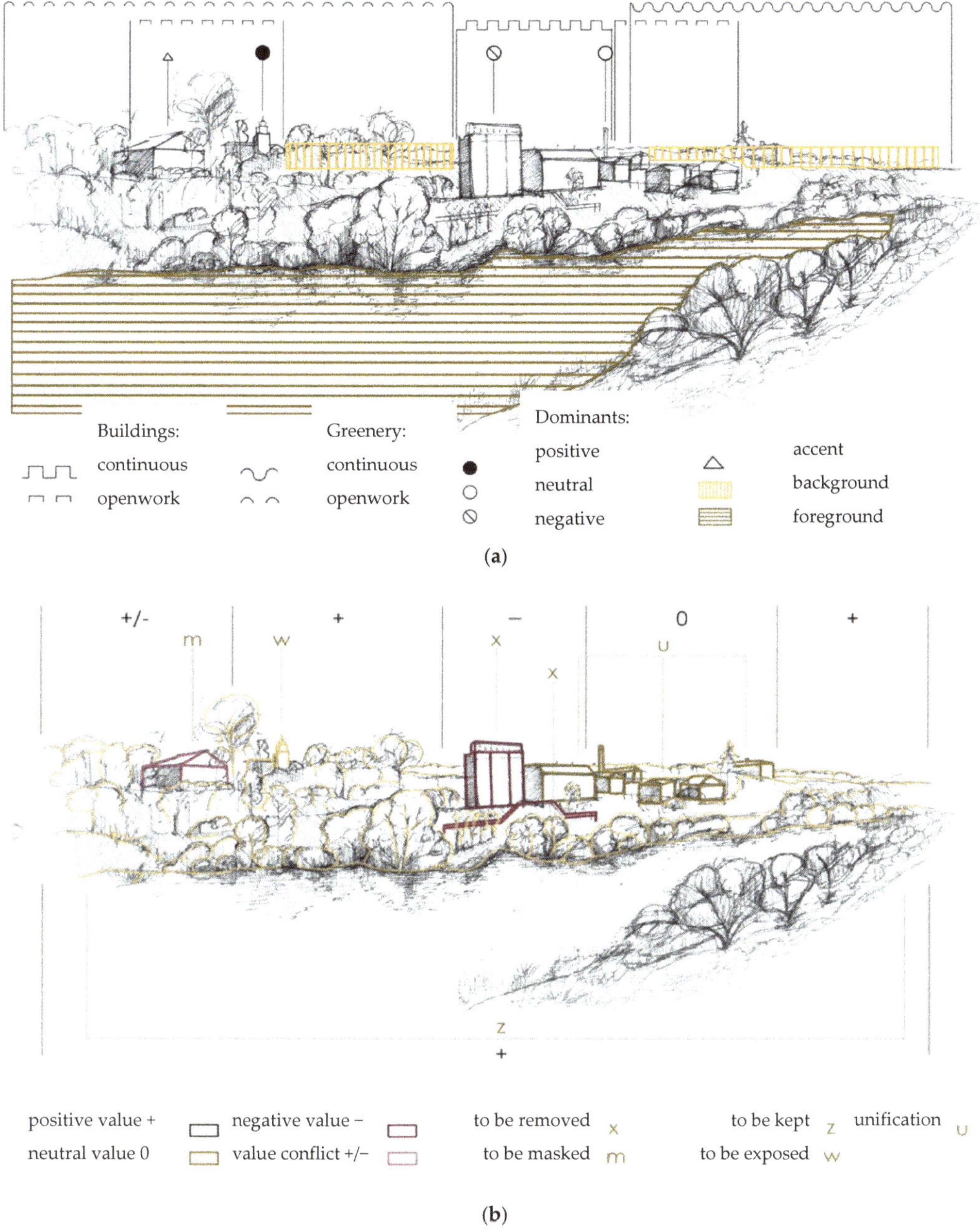

Figure 2. Landscape composition analysis applied to a panorama: (**a**) A belltower is identified as the main positive dominant, while the old warehouse structure (a ruin) is attributed to a negative value. An accent is a brick building, which distinguishes itself against the background due to its color. The surface of the Warta river forms the foreground. (**b**) Evaluation of the panorama is a consequence of the analysis step. The positive dominant is pointed as an element to be exposed ("w"), the negative dominant to be removed ("x"). It is also suggested to mask the accent ("m") and to unify the background ("u"). Greenery and water foreground are marked as positive content to be maintained. Illustration authors: Janczak, Sz.; Guzicka, N.; Plota, A. PUT, 2020.

A systematic inventory of panoramas and interiors is a fundamental step in identifying the main visual elements. Subsequently, the evaluation step attributes specific qualities, positive or negative, to the elements identified, as perceived by the observer (Figure 2b, top line). The evaluation of a landscape finally serves as a starting point for developing a policy to protect or improve its visual quality. It contains guidelines for modifications, which may consist of (1) removing or masking selected elements (e.g., using greenery), (2) preserving or protecting the existing state of the landscape, and even highlighting its selected features (e.g., unifying its background), (3) adding a new dominant feature to integrate the composition, or (4) standardizing (unifying) the main content (e.g., by using greenery or a color code) [23] (Figure 2b).

The material used in the first stage of this research consists of similar complete panorama analysis examples performed by architecture students at the PUT (Poznań University of Technology) for selected cultural landscapes located within the range of the Poznań agglomeration, in both urban and suburban areas. The first stage of the research aims to identify the visual elements that are the most recurrent and significant in the context of the study area. This part of the research will be described in Section 3.1.

The second stage of the research will try to confirm the results brought forth by the composition analysis study on a broader scale. It consists of an inquiry undertaken among architecture students after their fifth semester who had completed courses on urban composition, greenery design, and landscape architecture, making them competent experts in assessing a landscape's visual value. The inquiry respondents (33 persons) were provided with 20 images representing selected examples of cultural landscape panoramas from the study area, including urban and suburban locations. The panoramas are contained in Section 3.2.1, along with the presentation of survey results. The respondents were shown the images and asked three questions. In the first question, they were required to define whether the image they see contains primarily positive or negative values, a conflicting combination of the two, or neither positive nor negative values. In the second question, the respondents were asked to name the visually positive elements they could see, if such existed, and in the third question, they were asked to point out negative elements in the same way. What is important, in this stage of research the survey respondents were asked to provide their first impressions, without performing an analysis procedure. This feature makes the survey imitate the usual landscape perception, as it takes place in everyday conditions. In this way, the survey seeks to verify the results of the first research stage in a more realistic and ordinary aspect (Figure 3).

The second stage of the research is to ascertain the positive or negative value that the landscapes' visual elements represent. Therefore, our work on the survey results is a quantitative study, combining aspects of the Delphi approach to identify respondents' agreement on specific aspects [42,43].

The Delphi technique belongs to the group of heuristic methods and is commonly used for forecasting, but also for policymaking and reporting guidelines. It relies on a panel of experts and usually strives to achieve their agreement on a specific issue, which is the focus of the study [42]. In this research, two principal elements of this approach will be used: relying on a group of experts and checking their agreement rate regarding the perception of landscape values. The survey is limited to only one round, which is a major difference compared to normal applications of the Delphi technique. The reason is that, as shown in the introduction, landscape perception naturally differs depending on the viewer's personal background. Full agreement would not be possible to achieve. Hence, the goal of using selected elements of the Delphi approach is to study the realistic agreement rate among experts, without striving to achieve their complete consent.

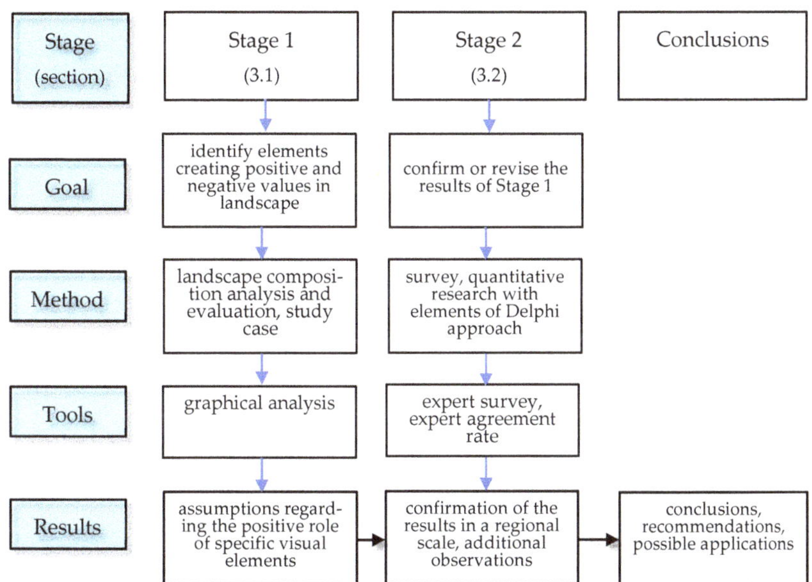

Figure 3. A graphic scheme explaining the research organization and the link between its two stages. Own elaboration.

3. Results

The presentation of the results is divided into two sections, following the above-described division of the research into two organizational stages. The first part discusses the significance of landscape composition, citing examples of a panorama-based analytical procedure to examine the landscape's visual aspect structure. The second part discusses the results of the inquiry regarding the values attributed by the respondents to specific visual elements of the landscape.

3.1. Landscape Composition Analysis

With regard to the landscape, the composition analysis procedure treats it as a visual semiotic system. It hence aims to examine its structure, focusing on the mutual relationships between its different visual elements. These relations can be perceived in both static [34,35] and dynamic [39] ways. In this study, the first approach is adopted, focusing on the landscape's structure as it can be perceived through an image, usually a hand drawing.

The landscape analysis procedure examples used for this part of the study came from multiple locations within the agglomeration of Poznań. They were realized by the architecture students of the PUT within the framework of Architectural Design in the landscape studio. Figure 4 presents an example of the landscape analysis performed for a typical urban panorama in the center of Poznań.

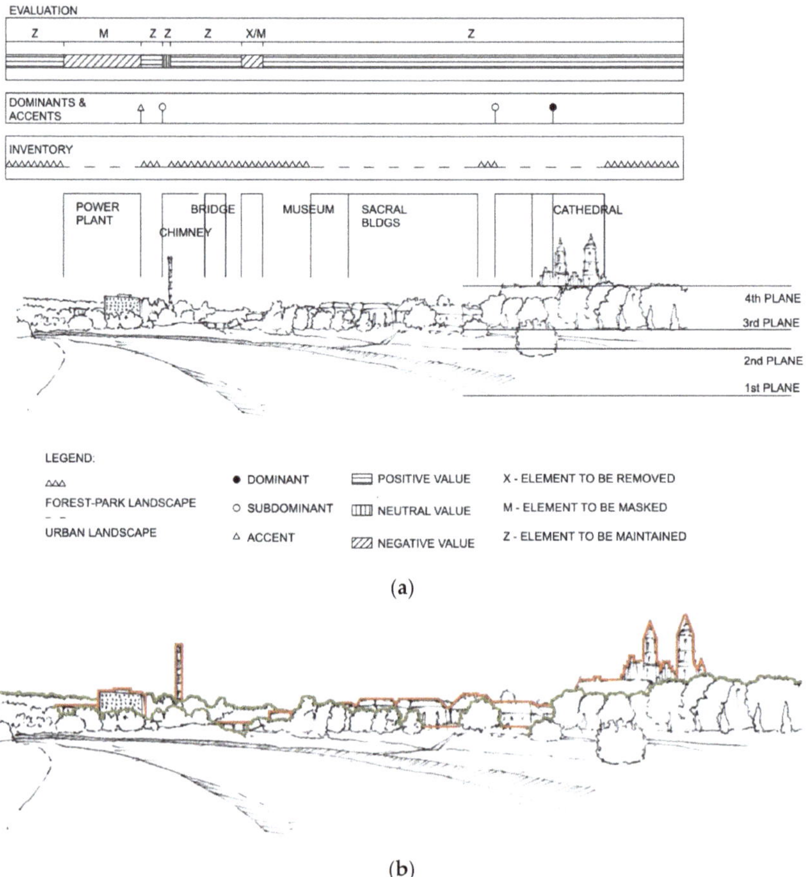

(**a**)

(**b**)

Figure 4. Example of complete landscape analysis with (**a**) inventory, composition analysis, and evaluation stages; (**b**) softness (**green line**) and hardness (**red line**) assessment. Illustration authors: Marciniak, W.; Łoniewska H. PUT, 2020.

The analysis procedure started with the inventory of the selected panorama. Students were encouraged to draw the chosen panorama by hand rather than use photographs to deepen their understanding of the landscape's structure. While drawing the panorama, they were asked to identify the types of landscape they observe (e.g., urban, park, forest, rural) and inventory its actual contents (e.g., specific buildings) (Figure 4a, inventory line). The panorama inventory was not limited to the view itself. It was supplemented by an analysis of the surroundings based on the plan, which helped to break the view down into first, second, and subsequent planes (Figure 4a, right side).

In the second step, the students proceeded to analyze the landscape composition. The composition's main elements (i.e., dominants, subdominants, accents, foreground, and background plans) were usually distinguished in the first step (Figure 4a, middle line). In the example shown above, the cathedral was identified as a main positive dominant of the composition. A modern utility building was identified as an accent. Then, other characteristic features were recorded, including the definition of the soft (e.g., vegetation) and rigid (e.g., buildings) content (Figure 4b). Other examples also analyzed the articulation and rhythm or involved the creation of a color chart. Less common landscape features, such as the mirror effect produced by the presence of surface water, were also noted at the composition analysis stage.

The phases of inventory and composition analysis prepared the ground for the next stage of the procedure: landscape evaluation. This stage aimed to recognize the primary values by assessing the landscape's visual quality and indicating vital problematic or conflict areas. This knowledge subsequently allowed determining actions to either protect or improve the landscape, as noted in the presented study's top line (Figure 4a). In this example, the right half of the panorama, containing mainly heritage landmarks and greenery, was indicated to be maintained in its present state. The same applies to a short section on the extreme left side of the panorama, which primarily consists of greenery. The two panorama sections marked with letters "M" and "X" contain elements indicated to be masked (power plant) or removed (old utility building). In this way, the analysis's conclusions were translated into both design guidelines and land management policies.

Another example of panorama analysis from an urban location (Figure 5) shows that positive values are not necessarily attributed to a dominant. In this case, the dominant (a modernist block of flats) was considered neutral. Positive values were attributed to the foreground of the composition, which is formed by the waterfront and a landmark bridge. The improvement guidelines are a natural consequence of the evaluation stage, suggesting maintenance of the green waterfront and unification (or calming down) of the large residential unit. What is interesting, the possibility of improving the panorama with a new dominant is suggested (Figure 5c, letter "D").

Urban locations prevailed among the panoramas selected for the study (12 out of 14 selected examples). As shown in the example of a complete analysis realized for an urban panorama (Figure 4), the most positive visual values are regularly attributed to historical landmarks. This situation is the most recurrent for urban environments. In such panoramas, the greenery's role is usually secondary, forming the background or foreground in the landscape's composition. The visual value attributed to the vegetation in such panoramas ranges from positive to negative, and its maintenance remains a critical factor in locating its exact weight on this scale. Next, the flat surface of water or a lawn is usually identified as a suitable foreground, enhancing the perception of other panorama elements, which are, therefore, attributed a positive or neutral value. Finally, the panorama elements being regularly attributed negative values include industrial buildings and infrastructure (e.g., heat and power plants, exhaust towers) and numerous modernist and international-style buildings.

Less frequently selected by the students (2 out of 14 examples chosen for this study), suburban locations deliver additional information on how specific elements of a landscape's composition are given a positive, neutral, or negative value in the individual's perception. Like urban sites, historical landmark buildings are associated with assigning these elements a positive value. In cases where such objects do not exist, a positive dominant role is usually played by greenery, primarily solitary or rhythmically aligned trees (Figure 6a,b). Particular importance was attributed to the articulation of large trees, which create a rhythm in the panorama (Figure 6c). On the other hand, negative values in suburban locations are frequently given to buildings with low aesthetic quality or poorly maintained buildings (Figure 6d). The students usually identified agricultural land or meadows as a neutral foreground for a composition.

Composition analysis

● dominant
△ accent
continuous buildings
openwork buildings
continuous greenery
openwork greenery

(**a**)

Evaluation

(**b**)

Design guidelines

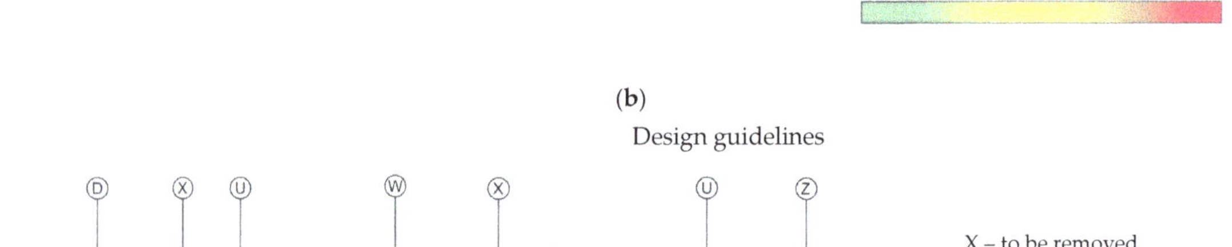

X – to be removed
Z – to be maintained
W – to de exposed
D – new dominant proposed
U – unification

(**c**)

Figure 5. Example of a landscape analysis with (**a**) inventory and composition analysis; (**b**) evaluation; (**c**) improvement guidelines. Illustration authors: Kulikowski, T.; Tabert, D.; Strymer, K. PUT, 2020.

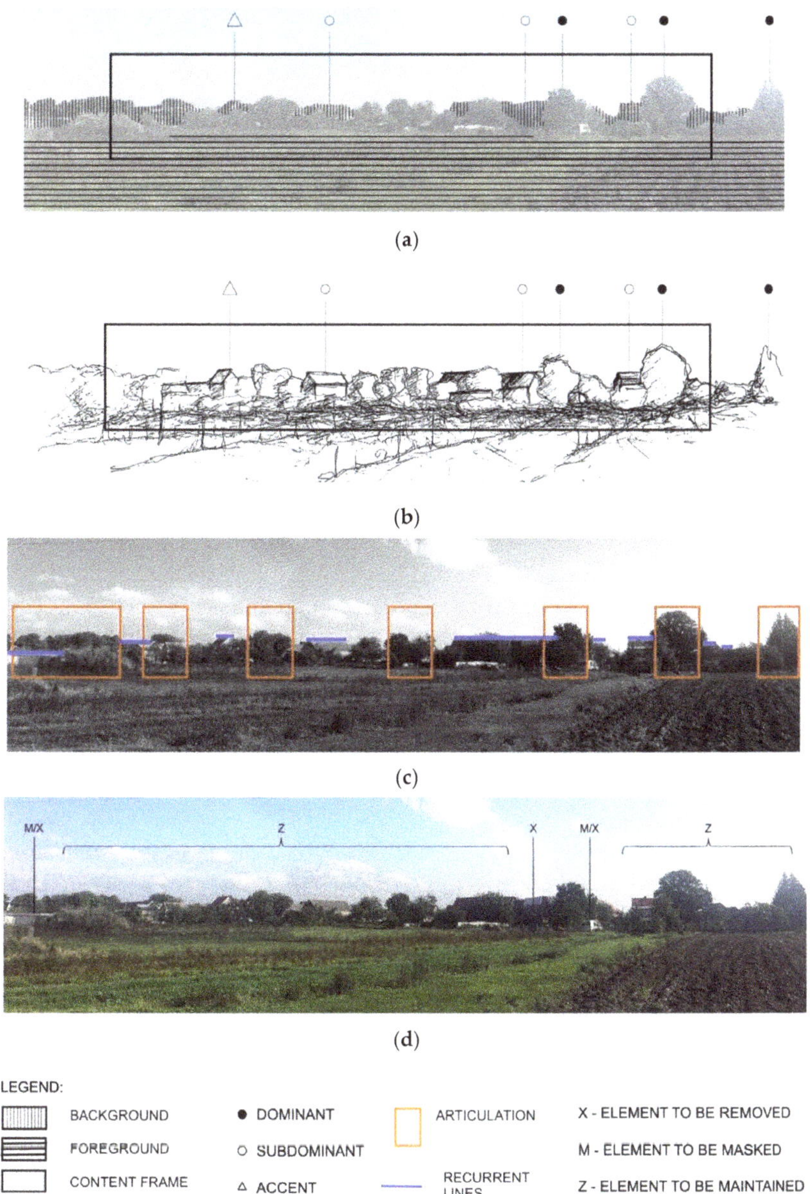

Figure 6. Example of suburban panorama analysis: (**a**) composition analysis; (**b**) composition analysis in freehand drawing version; (**c**) articulation; (**d**) improvement guidelines. Illustration author: Kacprzyk, M. PUT 2020.

As a result of applying the landscape composition analysis and evaluation procedure to selected panoramas from various Poznań agglomeration locations, it could be observed that positive visual values are consistently attributed to heritage landmark buildings and frequently also to greenery. Since the study case method does not generalize such observations made for the specific examples analyzed, the second part of this research will seek to confirm or revise these results. It will be based on a survey answered by a broader group of experts (see also Section 2: Materials and Methods).

3.2. Landscape Evaluation Survey

The landscape evaluation survey used as a part of this research aimed to study, in a broader way, the mechanism for assigning visual values (positive, neutral, or negative) to specific elements of a landscape. The inquiry results were expected to confirm the observations made in the first part of the research or provide new assumptions. This step was necessary in order to assess whether the mechanisms observed in the preceding part of the study represent general rules of attributing visual values to specific elements of landscape composition.

The landscape evaluation survey consisted of 20 sections. Each of them contained one panorama and three questions. The images were shot in different seasons. In the first question, the respondents expressed their general impression of the landscape they saw (see also Section 2: Materials and Methods). In the next two questions, they were asked to name the elements they considered the landscape's most positive assets (question 2) and most negative ones (question 3). The respondents were allowed complete freedom to define these elements in their own words, so as not to influence the survey results.

The survey was completed by 33 persons considered competent experts based on having completed courses on urban composition, greenery design, and landscape architecture. During these courses, they acquired both the knowledge and skills necessary to analyze landscape composition, distinguish different elements, and evaluate their impact on landscape perception. The number of responses is sufficient to perform a quantitative study, which was the target form of research for this part.

The survey results' presentation follows the Delphi approach's essential feature that verifies the respondents' agreement on specific questions. The first issue to be discussed is their agreement on the overall visual quality of the selected examples of panoramas.

3.2.1. Experts' Agreement Rate

Of the 20 panoramas surveyed, a compliance rate of above 50% was obtained for images with a high blue-green content (Figures 7–17). The green-blue content was not calculated in percentage of the image surface, because the images were shot with different cameras and at different resolutions. Moreover, they were shot at different times of the year. As a result, the greenery is not always green, and the water surface is not always blue. For these reasons, the content of the panorama images was defined in a visual manner.

Figure 7. Panorama of Malta lake, Poznań. Panorama No. 4 in the survey. Viewpoint location: N 52.40266464803816, E 16.961630012717553. Photo: Bossy, A. PUT, 2019.

Figure 8. Panorama of Wiry. Panorama No. 17 in the survey. Viewpoint location: N 52.32180136588434, E 16.852350540368725. Photo: Michałowska, W. PUT, 2020.

Figure 9. Panorama over Kórnik lake near Poznań. Panorama No. 20 in the survey. Viewpoint location: N 52.24728286088842, E 17.08538305507003. Photo: Pieczara, M. 2018.

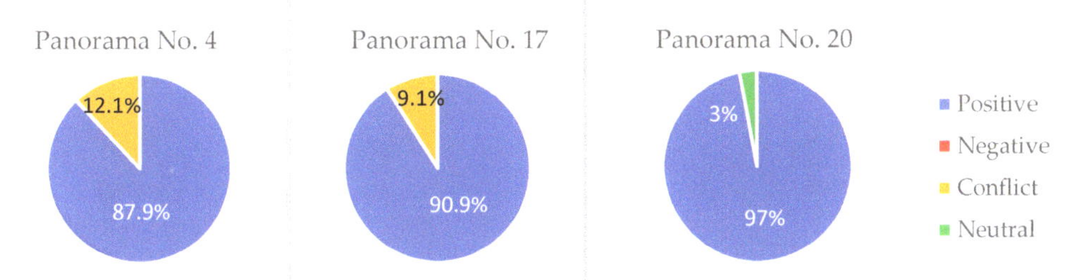

Figure 10. Experts' agreement rate graphs for the three most positively (>75%) evaluated panoramas. Own elaboration.

Figure 11. Panorama of Poznań. Panorama No. 1 in the survey. Viewpoint location: N 52.41060889253612, E 16.945173460522557. Photo: Pieczara, M. 2021.

Figure 12. Panorama of Poznań. Panorama No. 2 in the survey. Viewpoint location: N 52.413583890398286, E 16.94125026984746. Photo: Loos, E., Maćkowska, N. PUT, 2020.

Figure 13. Panorama of Poznań. Panorama No. 8 in the survey. Viewpoint location: N 52.4853225265442, E 16.90479324172883. Photo: Gyurkovich, M. 2021.

Figure 14. Panorama of Poznań over Malta lake. Panorama No. 18 in the survey. Viewpoint location: N 52.40107259667871, E 16.98750308423139. Photo: Sznura, S. PUT, 2019.

Figure 15. Example of a suburban panorama. Panorama No. 15 in the survey. Photo: Kacprzyk, M. PUT, 2020.

Figure 16. Example of a suburban panorama in Skórzewo. Panorama No. 16 in the survey. Viewpoint location: N 52.39104071303244, E 16.776569078162602. Photo: Kasińska-Andruszkiewicz, A. 2021. For this location, the survey used a Google Street View image that was shot in the spring season: https://goo.gl/maps/wSWYhaqz1mhySWFj8, accessed on 27 March 2021.

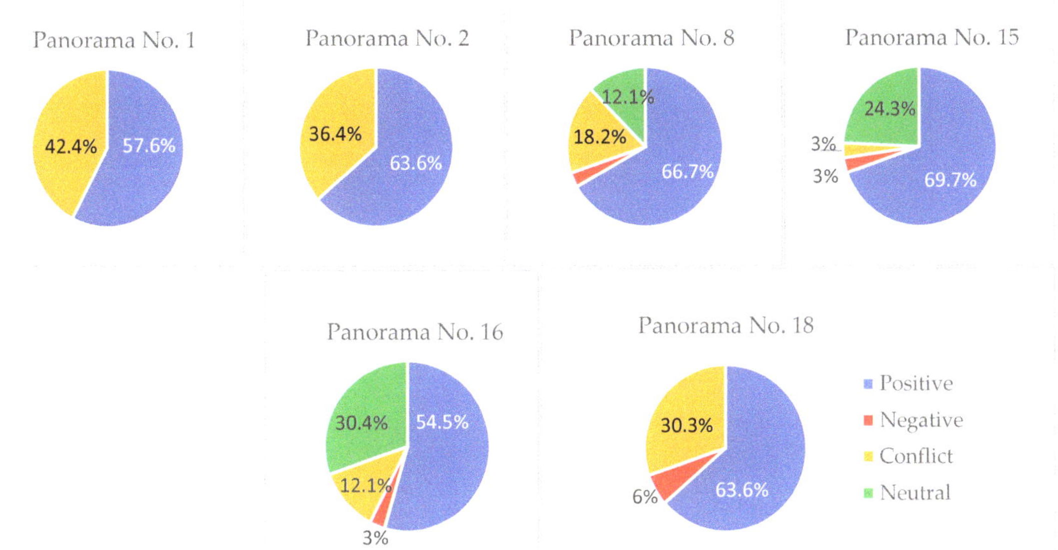

Figure 17. Experts' agreement rate graphs for panoramas evaluated positively by 50–75% of respondents. Own elaboration.

More specifically, all 9 out of the 20 panoramas rated by more than 50% of respondents as dominated by positive values contained a large amount of greenery in their composition, most recurrently in the form of a foreground (lawn) or a background (trees). They often had a water surface as their foreground. At the same time, responses other than "positive" for those detailed panoramas were more balanced. Namely, prevailing different reactions in these cases are either "conflict situation of positive and negative values" (5 of 9 cases) or "neither positive nor negative (neutral)" (2/9 cases). Hardly ever were "negative" values (opposite value) selected, but sometimes a mix of all possible alternative responses was observed (2/9 cases).

An agreement rate of more than 75% for overall positive perception was achieved for a mere three panoramas out of 20. One came from the urban area of Poznań city (Figure 7) and two others from its suburban zones (Figures 8 and 9). The first example represented the metropolitan type of recreational greenery landscape from Poznań (Figure 7). It was dominated by surface water and greenery (mainly large trees), to which the respondents also assigned the most significant positive values. The two other examples were suburban landscapes containing few elements (Figures 8 and 9). They consisted of greenery; landmark buildings (a church in the first and a palace in the second panorama); and, uniquely in the second case, surface water. Those three panoramas did not obtain any "negative" grades in the survey (Figure 10).

An agreement level for positive evaluation ranging between 50% and 75% was achieved mostly for cityscape or suburban panoramas that combine visible historical landmarks with greenery and other attributes—for example, surface water and residential and office buildings (Figure 11, Figure 12, and Figure 14). An interesting exception is Panorama No. 8 (Figure 13), which consists of a green foreground and a continuous mass of residential units.

In the open question section, respondents frequently mentioned historic buildings, greenery, and water as the essential positive assets. On the other hand, a negative role was usually assigned to communication and utility infrastructure, abandoned greenery fragments, and modernist architecture.

The suburban scenes that achieved a similar agreement result usually consisted of agricultural land, houses, and vegetation (e.g., Figures 15–17). The respondents most frequently pointed to the natural content (fields, trees) as their central positive values. On the downside, they mentioned a lack of maintenance of some greenery parts and low-quality structures (e.g., garages, sheds).

The highest agreement rate for negative evaluation was slightly above 50%, and it applied to only one panorama (Figures 18 and 19). The concerned panorama mainly consisted of modernist residential units, forming multiple planes in the perspective. It also had a dominant feature in the form of an exhaust chimney. The historic landmark buildings enclosed within this panorama, indicated by most of the respondents as positive values, were deprived of their dominant role. This situation resulted in an overall negative assessment of the panorama.

Figure 18. Panorama of Poznań from Winogrady. Panorama No. 10 in the survey. Image curtesy: fot. Radosław Żyto, www.fotokedar.pl. Available online: https://epoznan.pl/news-news-54545-panorama_poznania_z_winograd_zobacz_zdjecia (accessed on 8 April 2021)

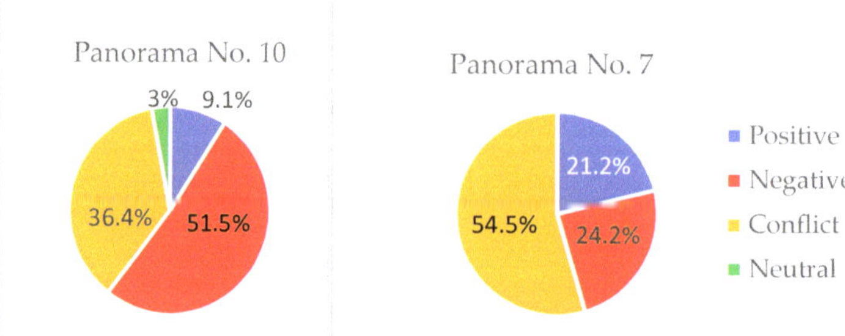

Figure 19. Experts' agreement rate graphs for panoramas evaluated negatively (**left**) and conflictingly (**right**) by more than 50% of respondents. Own elaboration.

A similar compliance rate of slightly above 50% was achieved for another panorama (one example), indicated as having a conflict situation of positive and negative values (Figures 19 and 20). The concerned panorama contained greenery (park) in the front and modernist housing blocks in the background. Greenery was unanimously indicated by the respondents as its positive attribute, while different housing block features (e.g., scale, materials, colors) were mentioned as creating negative values.

Figure 20. Piastowskie District in Poznan. Panorama No. 7 in the survey. Image curtesy: MOs810 CC BY-SA 4.0. Available online: https://upload.wikimedia.org/wikipedia/commons/b/bd/Piastowskie_Distr._Poznan.jpg (accessed on 8 April 2021)

Finally, the lowest rate of expert agreement was achieved for the remaining nine panoramas (Figures 21–29). All those images represent urban views with multiple types of visual elements—for example, historic and modern buildings of different forms and functions, historical and contemporary landmarks, infrastructure, exhaust chimneys, and advertisements. Their composition was complicated, while various elements were built into multiple planes. All but two panoramas (Figures 21 and 29) in this group had relatively little greenery.

Figure 21. Panorama from Dębiński Bridge in Poznań. Panorama No. 3 in the survey. Viewpoint location: N 52.36817652005035, E 16.928234475123023. Photo: Pieczara, M. 2021.

Figure 22. Panorama from Dworcowy Bridge, Poznań. Panorama 5 in the survey. Viewpoint location: N 52.40354002563637, E 16.911933353709713. "Bałtyk" office building designed by "MVRDV" architects on the extreme right. Photo: Pieczara, M. 2021.

Figure 23. Panorama from Queen Jadwiga Bridge, Poznań. Panorama No. 6 in the survey. Viewpoint location: N 52.398662696245225, E 16.940504605550498. Photo: Pieczara, M. 2021.

Figure 24. Panorama of Poznań downtown. Panorama No. 9 in the survey. Image curtesy: fot. Radosław Żyto, www.fotokedar.pl. Available online: https://epoznan.pl/news-news-54545-panorama_poznania_z_winograd_zobacz_zdjecia (accessed on 8 April 2021)

Figure 25. Panorama of Poznań, western part of the city. Panorama No. 11 in the survey. Image curtesy: fot. Radosław Żyto, www.fotokedar.pl. Available online: https://epoznan.pl/news-news-54545-panorama_poznania_z_winograd_zobacz_zdjecia (accessed on 8 April 2021)

Figure 26. Panorama of Poznań, western side of the city. Panorama No. 12 in the survey. Image curtesy: fot. Radosław Żyto, www.fotokedar.pl. Available online: https://epoznan.pl/news-news-54545-panorama_poznania_z_winograd_zobacz_zdjecia (accessed on 8 April 2021)

Figure 27. Panorama of Poznań, residential districts. Panorama No. 13 in the survey. Image curtesy: fot. Radosław Żyto, www.fotokedar.pl. Available online: https://epoznan.pl/news-news-54545-panorama_poznania_z_winograd_zobacz_zdjecia (accessed on 8 April 2021)

Figure 28. Panorama of Poznań city center seen from Winogrady. Panorama No. 14 in the survey. Image curtesy: fot. Radosław Żyto, www.fotokedar.pl. Available online: https://epoznan.pl/news-news-54545-panorama_poznania_z_winograd_zobacz_zdjecia (accessed on 8 April 2021)

Figure 29. Panorama of PUT campus, seen over the Warta river. Panorama No. 19 in the survey. Viewpoint location: N 52.405428087025406, E 16.945422813277442. Photo: Pieczara, M. 2021.

Taking a more complicated case, for example, Panorama No. 19 achieved a low expert agreement rate. This situation requires an in-depth analysis of experts' answers to the questions. They indicate greenery, groups of trees, and new buildings as positive assets. At the same time, the negative value is given to badly maintained greenery in the foreground (weeds and bushes) and modernist buildings in the background. These results demonstrate how the seemingly same visual elements (greenery vs. greenery, buildings vs. buildings) can be perceived in the opposite ways. For this reason, the geographical approach to landscape visual assessment cannot be fully effective if not completed with the iconographic view.

The distribution of the ratings in this group covered all possible responses, without the apparent domination of one category (Figure 30). The most frequently quoted downsides were visual chaos, too much architectural diversity, a lack of order, improper maintenance, and monotony. Even though the panoramas in this group had less greenery in their overall image, green elements (e.g., trees, groups of trees) were frequently indicated by the respondents to create positive visual values (8 out of 9 cases). Secondly, positive visual values were identified in the historic buildings (3 out of 9) and modern architectural dominants (2 out of 9 cases).

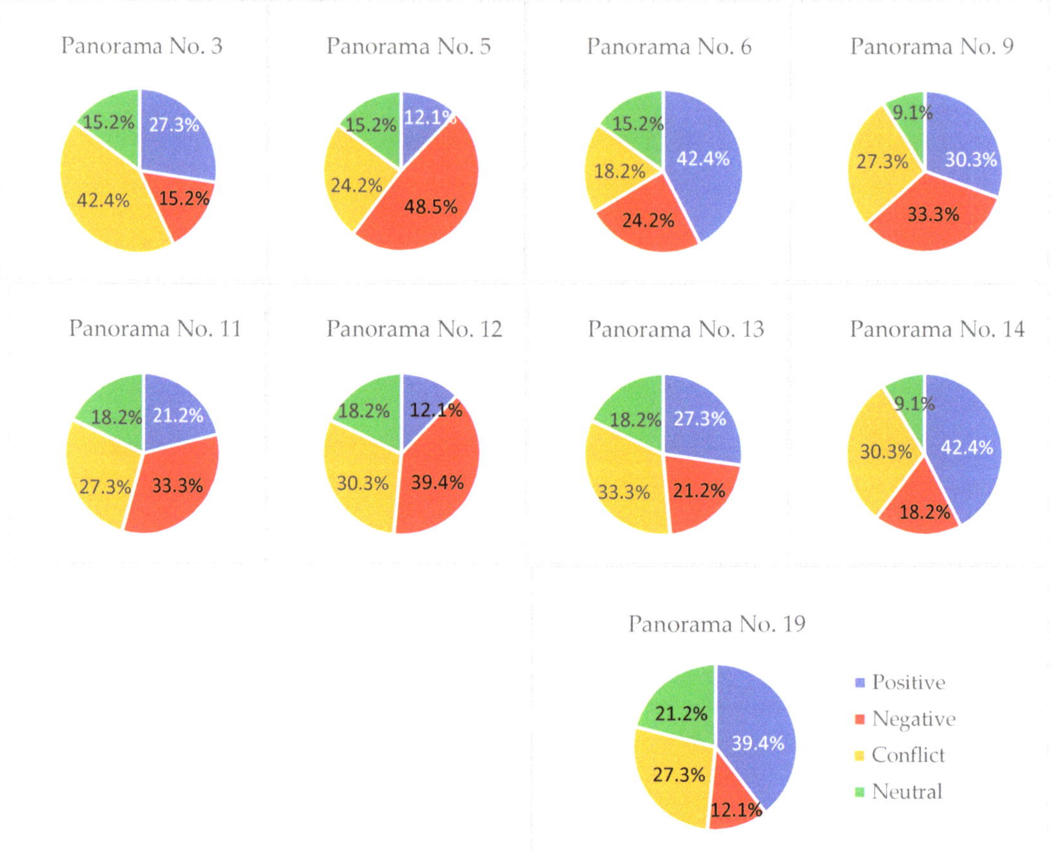

Figure 30. Expert agreement rate graphs for the nine least compliant panoramas. Own elaboration.

The assumptions that can be drawn from this part of the survey include the following:

- The simplicity of the composition has a positive effect on landscape visual evaluation based on its image (panorama);
- A lower number of elements create panoramas that are simpler to perceive, making the landscape informative content more legible;
- Cameral panorama views give less ambiguity of assessment than bird-eye views, which comprise more visual elements in the range of sight;
- Green elements are of crucial importance for creating positive visual values in the cultural landscape of the study area (Poznań agglomeration), even in the winter season;
- Historic landmarks create essential positive values;
- The type of element alone does not determine positive or negative perception. Its state of maintenance is of equal importance;
- Visual chaos or monotony bring positive perception down, as well as improper maintenance;
- Clarity of composition is decisive in terms of assessing landscape values.

3.2.2. Recurrent Positive Values

The survey, which was designated to analyze the relationships between specific visual elements and the viewers' positive or negative impressions, contained two open questions. The respondents directly named the most positive and negative assets. This form of open question was intended to minimize the risk of influencing the results. On the other end, however, this made it more complicated to present the results because the respondents often used different words to describe one concept. Therefore, this section presents a selection of the visual elements that were most recurrently indicated, often under different synonymous terms, by the respondents as the landscape's essential positive assets.

The most recurrent elements credited with positive values fit under the common term of greenery. The word "greenery" itself appeared 419 times in the answers about the panoramas' positive assets. The word "trees" appeared separately five times in the answers that did not indicate the general term "greenery". In percentage terms, 64% of the responses mentioned "greenery" or "trees" as essential positive values.

The second most recurrent group of visual elements recognized by the respondents as positive was the built heritage. The word "church" appeared 63 times in the answers, "cathedral" was indicated 26 times, "palace" appeared 6 times, and the word "historic" appeared 58 times concerning the buildings. These add up to 153 mentions altogether, representing 23% of all responses. The lower representation of this category of elements in the overall survey results does not mean that it was less significant. While greenery was present, to a different extent, in all the panoramas, cultural heritage was less frequently present. Built heritage legibly appeared in 12 out of 20 panoramas. The total of 153 mentions compared to the number of responses for the 12 panoramas represents 39%.

Another frequently mentioned element was water. The words "water" and "water surface" appeared 110 times in the results regarding positive values. The term "river" appeared 39 times, also in combination with the river name Warta. The word "lake" appeared 5 times, also combined with the lake name "Malta". Altogether, water was mentioned 154 times under different names. This represents 23% of all responses. Similar to the built heritage, water did not appear in every panorama. It was only present in nine of them. The total of 154 mentions related to the number of answers for the nine panoramas represents 52% of responses. Thus, for the panoramas which included water, 52% of the respondents mentioned it as being a positive asset in the landscape.

The survey results also show some less frequent but nonetheless interesting relationships. For example, one cityscape panorama (Figure 22) showed a recognizable area from downtown Poznań, with Poznań International Fair buildings and the Bałtyk office building designed by "MVRDV" architects. The results for this particular panorama mentioned the Bałtyk office building as its key positive element 15 times (45% of responses for this specific image). The Poznań International Fair complex's architecture was indicated as representing positive value 12 times in 36% of responses. This situation suggests that new buildings of modern architectural expression can become positive spatial landmarks.

In summary, the survey results confirm that greenery and cultural built heritage create essential positive value in the landscape of Poznań agglomeration. The third attribute that appeared to impact the landscape's positive evaluation significantly was water (Table 1). Knowledge about the positive impact of the attributes mentioned above on the landscape evaluation can be used both for visual protection purposes and to formulate design guidelines. It is also helpful for forecasting the visual perception of a specific landscape example. However, it must be highlighted that the clarity of composition, the legibility of its structure, and the state of maintenance also have a decisive influence on evaluation.

Table 1. Visual elements creating positive values in the landscape of Poznań agglomeration. Own elaboration.

Type of Elements	Element Name	Mentions No.	Percentage of All Responses
Greenery	greenery, trees, green foreground, green background	424	64
Built heritage	historical buildings, church, cathedral, palace	153	23 [1]
Water	water, water surface, river (also: Warta river), lake (also: Malta lake)	154	23 [2]

[1] Or 39% of the responses for 12 panoramas where built heritage was present. [2] Or 52% of the responses for nine panoramas with surface water.

3.2.3. Recurrent Negative Values

The most recurrent negative elements, on the other hand, comprise two groups of visual elements relative to buildings and infrastructure and one group of visual properties representing the condition of some elements or entire panorama (Table 2). The group of infrastructure elements assigned a negative influence on the perception mainly contained chimneys or exhaust towers (84 mentions), cranes (17 mentions), and railway (14 mentions). The latter case is worth attention, as railway tracks were present in one panorama (Figure 22) and they were indicated as being negative by 42% of respondents (14 out of 33). Among different building types, which altogether received 168 negative mentions (25%), housing blocks can be distinguished. They were mentioned 85 times, which represents 13% of all responses. The group of properties can be subdivided into two parts. The words referring to the state of an overwhelming multitude, like, for example, "chaos", "disorder", or "colors" (i.e., multiple colors), appear 48 times in the results. In percentage terms, this represents 7% of all responses. The opposite condition, referred to as "monotony" or "grey color", gained 12 responses, nearly 2%. It can be therefore observed that, in the study area, visual chaos is four times more frequently mentioned as a reason for negative assessment than monotony (Table 2).

Table 2. Visual elements creating negative values in the landscape of Poznań agglomeration. Own elaboration.

Type of Elements	Element Name	Mentions No.	Percentage of All Responses
Infrastructure	chimneys; railway; poles; fencing; cranes	127	19
Buildings	buildings; offices; housing blocks	168	25 [1]
Properties or qualities	chaos; disorder; colors; grey and monotony	48 12	7 2

[1] 13% for housing blocks only.

4. Discussion

The survey results confirmed the main assumptions adopted in the process of applying the composition analysis procedure to selected panoramas, helping to minimize the subjectivity inherent in landscape evaluation. According to semiotic theory, the perception of any message depends as much on its actual content as it does on the receiver's background [10,15,16]. The same also applies to the meaning contained within landscape structures. Therefore, landscape perception is filtered through an observer's aesthetic appreciation [28], which usually retains individual or cultural features. The sight itself, as a sense, is in reality not treated with the risks of subjectivity, but the interpretations of what is seen are. Hence, subjectivity is a natural feature of landscape assessment. It is also inherent in the empiricism linked to survey and drawing. In this research, it is minimized due to the application of two methodologically diverse research stages. The work's novelty is the improvement of the visual assessment method and its regionalization. By completing the visual approach with an expert inquiry, the research tries to solve the subjectivity issue, which is an inherent visual evaluation controversy.

The methodology presented in this article refers to the existing visual assessment methods deriving from the geographical paradigm (e.g., LCA, LPA, LVCA) by the principles it adopts in using human sight to assess physical aspects of the environment. However, it combines it with the semantic approach typical for iconography and aesthetics, placing attention on the values that we humans attach to landscapes. While most geographical approaches to landscape assessment are interested in describing its physiognomy, mapping different landscape types (e.g., LCA), or assessing the impact of the proposed development on the existing landscape (LVCA), this research identifies types of elements that are recurrently assigned positive visual value. This knowledge is essential in order to delineate areas to which particular attention should be given. Poland's existing planning tactics protect landscape fragments considered particularly valuable (e.g., centers of old towns, UNESCO heritage sites, natural monuments), building a city's recognizable image and bringing measurable benefits (e.g., in touristic competition). On the contrary, the ordinary cultural landscapes of suburbs and small towns are generally not protected. Considering how vast the areas covered with this landscape typology are, it is necessary to implement criteria-based selection principles. The landscape value assessment method presented in this paper can solve this problem. Instead of covering the entire scope of administrative units with landscape evaluation analysis, study areas can be delineated around specific objects recognized as potential positive qualities. According to the research results presented in this paper, such items include historical landmark buildings, built heritage objects, ordered complexes of high greenery, and surface water.

Consistent with these research results, the multidisciplinary interest in the problematics related to the built heritage as a part of the landscape identifies the need for its conservation [25,44] and points it out as an essential element of spatial identity [45–47]. Recently, its potential to bring significant image benefits in a comparative evaluation of urban public spaces has been noticed, distinguishing the historical urban fabric of a city as a value that guarantees "a distinctive atmosphere that attracts both residents and visitors" [45] (p. 17). The study results also confirm a relationship between the positive visual value and both the organizing and informative roles of heritage landmarks in the cityscape [46,47]. Another observation confirmed by the research results is that sacral landmarks, which refer more specifically to the cultural context of the study area by "highlighting the identity and memory of a place" [47] (p. 21), tend to have a particular bearing on the positive visual assessment of the landscape.

The presented study results also confirm the crucial role of architectural dominants, both sacral and secular, in landscape assessment [46,47] and evaluation. Characterized mainly by its scale and height and the fact that it usually interferes with the city's silhouette by surpassing its "skyline" [48], a dominant is frequently embodied by a strong form in terms of composition [49]. Such specific elements (strong forms) are significant for both the perception and identification of places. As proved by Gyurkovich and supported with the Gestalt theory, strong architectural forms that help to integrate or crystalize the sequences of spaces in how they are perceived, remembered, and recognized by the users [49] (p. 173) are not necessarily historical. This statement finds its reflection in the results obtained for a cityscape panorama comprising the Poznań International Fair and the "Bałtyk" office building, which were identified as positive values (Figure 22). These observations indicate the necessity to broaden the scope of visual protection policies to comprise newer structures that are contemporarily becoming essential landmarks and will become cultural heritage for future generations.

The latter insight has bearing on what is presently considered the sustainable development approach. Currently, spatial development policies in the study region more frequently include guidelines regarding the visual protection of historical urban centers, as Graczyk strongly postulated [46]. However, the present policies tend to emphasize the "inside views" on the historical urban center, at the same time frequently forgetting about the "outside views", which are equally significant in the perception of local identity [46] (pp. 183/184). "The places which are essential from the point of view of city identification have to be protected" [46] (p. 184). With this statement, Graczyk opens the discussion on what those places are in reality. This research, in particular the survey part, shows that such places include cultural heritage sites and valuable greenery complexes or units. Still, some contemporary architectural dominants are equally essential. A revision of the currently applied landscape management policies can hence be suggested to include modern landmarks.

In his assumptions made about the study area, which is also enclosed within the borders of the Poznań agglomeration, Graczyk stated that a "protective viewing zone for dominants should be created" [46] (p. 184). He also argued that architectural dominants' substantial presence within the cityscape is a vital sign of sustainable landscape management. Therefore, it is necessary to protect the architectural dominants' visibility also as part of sustainable development. However, contemporarily applied methods for designating visual protection zones tend to be reduced to studying "the visibility of objects in the field described by means of visibility diagrams and maps" [26] (p. 17) [46,50]. The research presented in this paper can add to the existing visual assessment studies by proposing a survey-based expert method to examine which specific local landscape type elements create essential positive values.

Another contribution that the presented research makes to the currently applied visual assessment methods is its regionalization scheme. As observers' background and individually defined priorities affect how they perceive a landscape [10,28], visual assessment methods that are successful in one region may not bring awaited results in another one. This emphasizes the importance of a landscape evaluation survey carried out among local experts, which facilitates the identification of cultural priorities regarding the visual qualities in landscape specifically for a defined area of study. Inviting local experts as respondents is vital in order to avoid the deviations observed between residents' and external experts' responses regarding their value perceptions and preservation attitudes [51]. As observed by Yang, Qiu, and Fu, residents and professionals differ considerably in the effect of "value orientation, place attachment, and its relationship with landscape preservation" [51] (p. 11). The method presented in this paper aims to minimize this paradox by addressing the survey to local professionals. On the other hand, this consists of a limitation of the approach adopted. In the light of participatory planning, involving daily users in the evaluation of the environment they inhabit would increase the effectiveness of the process [52]. However, as shown by Yang, Qiu, and Fu, combining evaluations of experts together with daily users creates a considerable ambiguity in the results [51]. Therefore, the results of this study, delivered in the effect of using experts only, should constitute the starting point for the development of new tactics of involving local users in the planning practice. This responds to the recently identified need for developing new strategies and tactics for participatory planning in the contemporary public realm [53].

Another source of variation in responses regarding landscape character judgment can be linked to aesthetic preference [54]. As demonstrated by Wang, Zhao, and Liu, "landscape types have a significant influence on judgment consensus. We conclude that a clean environment with a high degree of vegetation normally implies high judgment consensus among observers" [54] (p. 216). This observation is also confirmed by the results of the presented study, in which the panoramas with a high vegetation content achieved a higher agreement rate among the survey respondents.

The problem of landscape evaluation objectivity, which reveals itself in all attempts to assess a landscape's quality visually, inevitably lead the discussion towards Laurie's definition in 1975 [28]. Namely, Laurie presented landscape evaluation as a process of assessing its visual quality by making comparisons between one or more landscapes [14,28,29]. The comparative approach is revealed in several studies sacrificed to landscape assessment. For example, an integrated comparative synthesis method was used by Graczyk [46] to assess the identifying role of architectural dominants in the context of small towns situated within the Poznań agglomeration. The idea of evaluating a landscape through a study of comparative relationships impacted the research presented in this article. Namely, making comparisons is enhanced by the number of panoramas included in the survey stage and their differentiation, despite representing one agglomeration area.

The comparative analysis was equally important in ascertaining that landscape evaluation is context-dependent. Similar to ecological ones, visual landscape indicators were demonstrated to be interpreted differently depending on the context [55]. In terms of landscape imageability, currently applied visual indicators include the presence of iconic elements or historic landmarks, particular viewpoints, and long views (panoramas), often with the significant presence of water bodies [55] (p. 941). The presence of surface water, reported as located at the intersection of ecological and visual indicators [55], was distinguished in the present survey as an element positively influencing landscape evaluation. This fact can be considered as proof that ecology is both perceived and appreciated as a part of landscape perception [55,56]. Because it has been suggested that knowledge and education enhance the appreciation of ecological aspects in landscape perception [55,57,58], it was also recommended that "further empirical studies, e.g., landscape preference studies should address the aesthetic appreciation linked to ecological function" [55] (p. 944). It was subsequently stated that "such empirical studies should test indicators across different landscape types and with different groups of observers" [55] (p. 944). The research presented in this paper answers this call by addressing the role of vital visual indicators within a delineated area of the Poznań agglomeration and is realized by a specified group of observers (architecture students). As previously suggested by existing studies [46,55], the presented research results confirm a correlation between the landscape's visual attributes and the level of implementing sustainable development policies in terms of land management. For example, the exposition of historical landmarks and ordered greenery identified in the study as essential positive visual indicators of the landscape type under examination are both direct effects of a conscious and sustainable space management strategy. Tools enabling the ongoing control of spatial planning effects should constitute a substantial element of such an approach. Poland's current spatial planning policy is not burdened with all the consequences of the decisions made. The failure of responsibility for the landscape's visual quality does not act in its favor. It results in the lack of spatial cohesion, which is a major contemporary problem in the study area [59]. More importantly, the destruction of the landscape composition takes place systematically in the areas covered by active local plans, which demonstrates their low efficiency in the actual state.

The fact that the problem of landscape visual quality intersects with the concept of sustainability points to the importance of harmonious composition for the perception of a landscape's ecological dimensions. The maintenance of specific relationships between visual elements of a landscape can be considered as a sign of sustainability. This emphasizes the importance of landscape visual analysis and survey study that is performed for a specific landscape type and regionally delineated, leading to the recognition of its most significant assets. This step is essential in identifying the composition elements and their place in the whole image. The informative content of particular elements, for example, dominants, cannot be recognized without analyzing their relationships with other scene components [37,46,48]. A visual balance must be achieved to maintain an appropriate level of relationships that define the landscape's meaningful content. It can, therefore, be assumed that maintaining visual balance in the landscape should be part of any sustainable development strategy. Considering that the study results point to the built heritage and greenery as essential visual assets, such elements should gain more attention within this framework.

Elements of such a strategy resonate in the discussion on Poland's contemporary housing environment that is currently underway. Gyurkovich recently observed that "elements of natural and cultural heritage that could aid in the correct shaping of the housing environment have only recently gained significance in Poland" [60] (p. 31). Further on, they demonstrate that these two groups of elements are fundamental in creating a public space with its own "spirit" or genius loci. The presented study brings this idea further to the agglomeration scale, aiming to distinguish types of landscape elements that create visually positive values at the entire public domain level. Including them in the local plans should be considered a role model in treating the landscape composition within any sustainable development strategy framework. Because sustainable development assumes integration of economic, environmental, and social aspects, solving problems requires various actions taken simultaneously in different fields [61]. Related mainly to the environmental part, the visual balance of a landscape may trigger reactions also on the social (e.g., wellbeing) and economic (e.g., tourist competition) levels.

Poland's current spatial planning documents and tools still require improvement in order to guarantee sustainable development [62]. As recommended by Badach and Raszeja [63], the inclusion of selected landscape and greenspace indicators into the existing spatial planning instruments is necessary in order to develop a sustainable urban planning approach. This paper proposes including the visual evaluation tools it described and tested for the Poznań agglomeration area. Their implementation in the planning practice allows to (1) identify essential visual values characteristic of a specific landscape type (survey + landscape inventory and composition analysis) and (2) propose analytical tools to designate guidelines for their protection or enhancement (visual evaluation step).

The research results create precedence for elaborating a new landscape management strategy. It should follow the two points indicated above, creating two separate steps. In the first step, the landscape composition analysis method is applied to selected examples of the local landscape types in order to identify the most recurrent visual elements and their relations. This reproduces Section 3.1 of this research. In the effect, it helps to choose relevant panoramas that the survey should include. Then, the survey is set to confirm or revise the preliminary results. Reproducing Section 3.2, collecting and analyzing the survey results completes the first stage of the proposed landscape management policy. Once the first step is completed, the knowledge about the experts' agreement rate and their responses to questions allow localizing places that need visual protection or enhancement.

Taking the results of this study as an example, their direct application in the study area can be simulated. For instance, Panorama No. 20 (Figure 9) was evaluated positively in its existing state, which predestinates it directly for protection. Next, the results obtained for Panorama No. 17 (Figure 8), another example, indicate a conflict situation besides its generally positive reception. The landscape inventory, composition analysis, and evaluation tools should be used to analyze the situation and identify the source of conflict. The evaluation provides planning guidelines on how to fix the situation. A generalized strategy of incorporating the methodology tested in this study in the planning practice will be presented in the conclusions.

5. Conclusions

In this study, the key visually positive attributes of cultural landscapes were distinguished within the Poznań agglomeration area. They include four essential element types: cultural (built) heritage, landmarks, ordered greenery, and surface water. The results also showed that the clarity of composition and the limitation of elements' number have a positive influence on landscape evaluation. Considered a sign of sustainable development, maintaining the visual balance at a proper level or enhancing it to achieve such a level should become a standard component of spatial development policies. Depending on the situation, visual balance protection and perfection scenarios should be included in regional development studies. This task can be accomplished through the following steps:

1. Include expert surveys in spatial development studies to identify key visual values of the local landscape type(s);
2. Include landscape analysis and evaluation within the framework of spatial development studies, which will allow us to identify specific views to be protected or revised (perfected) depending on the current situation;
3. Systematically involve participatory planning tools to socially validate the results of landscape analysis and evaluation done by experts;
4. Set policies of visual balance protection or perfection—for example, by banning development in certain places or setting specific requirements regarding architecture and finishing materials;
5. Require local plans to realize the policies mentioned above;
6. Introduce a system for the ongoing monitoring of the effects of spatial policy and responsibility for decisions made.

From these six steps, the step 1 and 2 are direct applications of the methodology tested in this study.

We should also address greenery's influence on a landscape scene's positive assessment. Considering that the landscape panoramas characterized by a high vegetation content achieved higher levels of expert agreement in this survey study, the potential of greenery to improve visual balance should be rethought. Greenery can be considered as a fabric to create an exceptionally efficient background that will help the viewer focus on the informative landscape content (landmarks). Many visual conflict situations can be fixed with the proper use of greenery. At the same time, the study showed that greenery maintenance is vital for positive landscape perception.

Finally, the notion of the landmark within the landscape needs to be broadened. As demonstrated by this survey, positive values within the cultural landscapes in the Poznań agglomeration were in large part created by historic buildings (built heritage), but not exclusively. The experts recognized some new architectural dominants as creating positive values as well. The definition of the cultural (built) heritage in landscape studies should thus also comprise more recent structures that will presumably become landmarks for future generations. Expert surveys with open questions, like the one used in this study, provide the possibility to identify such elements. Following their identification as landmarks, such elements should then be placed under the same visual protection or enhancement policy procedures.

Author Contributions: Conceptualization, M.G. and M.P.; methodology, M.G. and M.P.; formal analysis, M.P.; investigation, M.G. and M.P.; resources, M.G. and M.P.; data curation, M.P.; writing—original draft preparation, M.P.; writing—review and editing, M.G. and M.P.; visualization, M.P.; supervision, M.G.; funding acquisition, M.G. and M.P. All authors have read and agreed to the published version of the manuscript.

Funding: This research was funded from statutory funds of the Poznań University of Technology, Faculty of Architecture. Funds numbers: 0111/SBAD/0406 and 0111/SBAD/2116.

Institutional Review Board Statement: Not applicable.

Informed Consent Statement: Not applicable.

Data Availability Statement: Data sharing not applicable.

Acknowledgments: The authors express their gratitude to the students for participating in the study by (1) taking part in the survey, (2) realizing landscape composition analyses, and (3) giving permission for the use of their graphic materials to illustrate the paper. The authors also thank the photographer Radosław Żyto for his consent to use his images in the study and publication. The authors express their gratitude to the reviewers, whose remarks and suggestions have helped to improve the work.

Conflicts of Interest: The authors declare no conflict of interest.

References

1. European Landscape Convention, European Treaty Series—No. 176, Council of Europe, Florence (2000). Available online: https://rm.coe.int/1680080621 (accessed on 26 June 2020).
2. Antrop, M. Geography and landscape science. *Belg. Rev. Belge Géographie* **2000**, *30*, 9–36. [CrossRef]
3. Vidal de La Blache, P. *Principes de Géographie Humaine: Publiés d'après les Manuscrits de l'auteur par Emmanuel de Martonne*; ENS Édition: Lyon, France, 2015.
4. Castree, N.; Demeritt, D.; Liverman, D. Introduction: Making sense of environmental geography. In *A Companion to Environmental Geography*; Castree, N., Demeritt, D., Liverman, D., Rhoads, B., Eds.; Wiley-Blackwell: West Sussex, UK, 2009; pp. 1–15. [CrossRef]
5. Franch-Pardo, I.; Napoletano, B.M.; Bocco, G.; Barrasa, S.; Cancer-Pomar, L. The Role of Geographical Landscape Studies for Sustainable Territorial Planning. *Sustainability* **2017**, *9*, 2123. [CrossRef]
6. Zonneveld, I.S. *Land Ecology: An Introduction to Landscape Ecology as a Base for Land Evaluation, Land Management and Conservation*; SPB Academic Publishing: Amsterdam, The Netherlands, 1995; p. 199.
7. Antrop, M. Sustainable landscapes: Contradiction, fiction or utopia? *Landsc. Urban Plan.* **2006**, *75*, 187–197. [CrossRef]
8. Bonenberg, W.; Zierke, P. *Dobra Kultury Współczesnej Jako Element Krajobrazu Powiatu Poznańskiego (en. Contemporary Cultural goods as an Element of the Landscape of the Poznań County)*; Poznań University of Technology: Poznań, Poland, 2014; p. 245.
9. Kobayashi, A. A Critique of Dialectical Landscape. In *Remaking Human Geography (RLE Social & Cultural Geography)*, 1st ed.; Kobayashi, A., Mackenzie, S., Eds.; Routledge: London, UK, 1989; pp. 164–185. [CrossRef]
10. Raaphorst, K.; Duchhart, I.; van der Knaap, W.; Roeleveld, G.; van der Brink, A. The semiotics of landscape design communication: Towards a critical visual research approach in landscape architecture. *Landsc. Res.* **2017**, *42*, 120–133. [CrossRef]
11. Cosgrove, D.F. *Social Formation and Symbolic Landscape*, 2nd ed.; The University of Wisconsin Press: Madison, WI, USA, 1998; p. 332.
12. Daniels, S.; Cosgrove, D. Introduction: Iconography and landscape. In *The Iconography of Landscape: Essays on the Symbolic Representation, Design and Use of Past Environments*, 1st ed.; Cosgrove, D., Daniels, S., Eds.; Cambridge University Press: Cambridge, UK, 1988; pp. 1–10.
13. Appleton, J. *The Experience of Landscape*, 2nd ed.; John Wiley and Sons Ltd.: Chichester, UK, 1996; p. 296.
14. Muir, R. *Approaches to Landscape*; Palgrave: London, UK, 1999; p. 336. [CrossRef]
15. de Saussure, F. *Cours de Linguistique Générale*; Payot: Paris, France, 1983; p. 509.
16. Barthes, R. Rhetoric of the Image. In *Image, Music, Text*; Heath, S., Ed.; Hill and Wang: New York, NY, USA, 1977; pp. 32–51.
17. Eco, U. Function and Sign: Semiotics of Architecture. In *The City and the Sign. An Introduction to Urban Semiotics*, 1st ed.; Gottdiener, M., Lagopoulos, A., Eds.; Columbia University Press: New York, NY, USA, 1986; pp. 55–86. [CrossRef]
18. Forman, R.T. *Land Mosaics: The Ecology of Landscapes and Regions*, 1st ed.; Cambridge University Press: Cambridge, UK, 1995; p. 656.
19. Kant, I. *The Critique of Pure Reason*; Guyer, P., Wood, A.W., Eds.; Cambridge University Press: Cambridge, UK, 1998; p. 785.
20. Husserl, E. *The Idea of Phenomenology*, 1st ed.; Hardy, L., Ed.; Husserliana: Edmund Husserl—Collected Works; Springer: Dordrecht, The Netherlands, 1999; Volume 8, p. VI 72.
21. Heidegger, M. *Basic Writings, Routledge*, 1st ed.; Routledge: Abingdon, UK, 1978; p. 392.
22. Norberg-Schulz, C.H. *Genius Loci. Paysage, Ambiance, Architecture*, 3rd ed.; Seyler, O., Ed.; Mardaga éditions: Brussels, Belgium, 1997; p. 216.
23. Pieczara, M. An architecture course to teach respect for the landscape. *World Trans. Eng. Technol. Educ.* **2020**, *18*, 450–455.
24. Górka, A. Landscape perception and the teaching of it in Poland. *World Trans. Eng. Technol. Educ.* **2020**, *18*, 124–128.
25. Serraino, M.; Lucchi, E. Energy Efficiency, Heritage Conservation, and Landscape Integration: The Case Study of the San Martino Castle in Parella (Turin, Italy). *Energy Procedia* **2017**, *133*, 424–434. [CrossRef]
26. Górka, A. Visual Capacity Assessment of the Open Landscape in Terms of Protection and Shaping: Case Study of a Village in Poland. *Sustainability* **2020**, *12*, 6319. [CrossRef]
27. Swanwick, C.; Fairclough, G. Landscape character: Experience from Britain. In *Routledge Handbook of Landscape Character Assessment. Current Approaches to Characterization and Assessment*, 1st ed.; Fairclough, G., Sarlöv, H.J., Swanwick, C., Eds.; Routledge: London, UK, 2018; pp. 21–36. Available online: https://www.taylorfrancis.com/books/e/9781315753423 (accessed on 2 February 2021).
28. Laurie, M. *An Introduction to Landscape Architecture*, 1st ed.; American Elsevier Publishing Company: New York, NY, USA, 1975; p. 214.
29. Zube, E.H.; Brush, R.O. *Landscape Assessment: Values, Perceptions and Resources*, 1st ed.; Zube, E.H., Brush, R.O., Fabos, J.G., Eds.; Dowden, Hutchinson & Ross: Stroudsburg, PA, USA, 1975; p. 367.
30. Cosgrove, D. Prospect, perspective and the evolution of the landscape idea. *Trans. Inst. Br. Geogr.* **1985**, *10*, 45–62. [CrossRef]
31. Antrop, M.; Van Eetvelde, V. Approaches in Landscape Research. In *Landscape Perspectives. Landscape Series*; Springer: Dordrecht, The Netherlands, 2017; Volume 23, pp. 61–80. [CrossRef]
32. Daniel, T.C. Whither scenic beauty? Visual landscape quality assessment in the 21st century. *Landsc. Urban Plan.* **2001**, *54*, 267–281. [CrossRef]

33. Bürgi, M.; Ali, P.; Chowdhury, A.; Heinimann, A.; Hett, C.; Kienast, F.; Mondal, M.K.; Upreti, B.R.; Verburg, P.H. Integrated Landscape Approach: Closing the Gap between Theory and Application. *Sustainability* **2017**, *9*, 1371. [CrossRef]
34. Bogdanowski, J. *Kompozycja i Planowanie w Architekturze Krajobrazu (en. Composition and Planning in Landscape Architecture)*, 1st ed.; Zakład Narodowy im. Ossolińskich: Wrocław, Poland, 1976.
35. Bogdanowski, J. *Metoda Jednostek i Wnętrz Architektoniczno-Krajobrazowych (JARK-WAK) w Studiach i Projektowaniu (en. The Units and Rooms Method (JARK-WAK) in Studies and Design)*, 2nd ed.; Cracow University of Technology: Cracow, Poland, 1990; p. 36.
36. Lynch, K. *The Image of the City*, 1st ed.; The MIT Press: Cambridge, MA, USA, 1960; p. 194.
37. Bell, S. *Elements of Visual Design in the Landscape*, 2nd ed.; Spon Press: London, UK, 2004; p. 230.
38. Smardon, R.C.; Palmer, J.F.; Felleman, J.P. (Eds.) *Foundations for Visual Analysis*, 1st ed.; Wiley-Interscience: New York, NY, USA, 1986; p. 374.
39. Wejchert, K. *Elementy Kompozycji Urbanistycznej (en. Elements of Urban Composition)*; Arkady: Warsaw, Poland, 1984; p. 280.
40. Zachariasz, A. Landscape architecture, landscape composition and specialist vocabulary—Preliminary considerations. *Diss. Cult. Landsc. Comm.* **2016**, *32*, 11–29.
41. Böhm, A. *Planowanie Przestrzenne dla Architektów Krajobrazu: O Czynniku Kompozycji: Podręcznik dla Studentów Wyższych Szkół technicznych (en. Spatial Planning for Landscape Architects: About the Composition Factor: A Textbook for Students of Technical Universities)*, 1st ed.; Cracow University of Technology: Cracow, Poland, 2006; p. 324.
42. Garson, G.D. *The Delphi method in Quantitative Research*; Statistical Associates Publishers: Asheboro, NC, USA, 2014; p. 38.
43. Wang, V.C.X. *Handbook of Research on Scholarly Publishing and Research Methods*, 1st ed.; Wang, V.C.X., Ed.; Information Science Reference IGI Global: Hershey, PA, USA, 2015; p. 589.
44. European Committee for Standardization (CEN). *Conservation of cultural heritage. Guidelines for Improving the Energy Performance of Historic Buildings, Standard FprEN 16883*; CEN: Brussels, Belgium, 2017.
45. Zagroba, M.; Szczepańska, A.; Senetra, A. Analysis and Evaluation of Historical Public Spaces in Small Towns in the Polish Region of Warmia. *Sustainability* **2020**, *12*, 8356. [CrossRef]
46. Graczyk, R. *Identyfikacyjna Funkcja Dominanty Architektonicznej w Strukturze Małego Miasta (en. Identifying Function of an Architectural Dominant in the Structure of a Small Town)*, 1st ed.; Poznań University of Technology: Poznań, Poland, 2015; p. 202.
47. Hodor, K.; Fekete, A. The sacred in the landscape of the city. *Tech. Trans.* **2019**, *116*, 15–22. [CrossRef]
48. Ozimek, A. Landscape dominant ELEMENT—An attempt to parametrize the concept. *Tech. Trans.* **2019**, *116*, 35–62. [CrossRef]
49. Gyurkovich, J. *Znaczenie form Charakterystycznych dla Kształtowania i Percepcji Przestrzeni: Wybrane Zagadnienia Kompozycji w Architekturze i Urbanistyce (en. The importance of Characteristic Forms for the Shaping and Perception of Space: Selected Issues of Composition in Architecture and Urban Planning)*, 1st ed.; Cracow University of Technology: Cracow, Poland, 1999; p. 216.
50. Ozimek, A. *Landscape Measure. Objectification of Views and Panoramas Assessment Supported by Computer Tools*, 1st ed.; Cracow University of Technology: Cracow, Poland, 2019.
51. Yang, H.; Qiu, L.; Fu, X. Toward Cultural Heritage Sustainability through Participatory Planning Based on Investigation of the Value Perceptions and Preservation Attitudes: Qing Mu Chuan, China. *Sustainability* **2021**, *13*, 1171. [CrossRef]
52. Smith, R.W. A theoretical basis for participatory planning. *Policy Sci.* **1973**, *4*, 275–295. [CrossRef]
53. Lodato, T.; DiSalvo, C. Institutional constraints: The forms and limits of participatory design in the public realm. In Proceedings of the 15th Participatory Design Conference: Full Papers—Volume 1 (PDC '18); Association for Computing Machinery: New York, NY, USA, 2018; pp. 1–12. [CrossRef]
54. Wang, R.; Zhao, J.; Liu, Z. Consensus in visual preferences: The effects of aesthetic quality and landscape types. *Urban For. Urban Green.* **2016**, *20*, 210–217. [CrossRef]
55. Fry, G.; Tveit, M.S.; Ode, Å.; Velarde, M.D. The ecology of visual landscapes: Exploring the conceptual common ground of visual and ecological landscape indicators. *Ecol. Indic.* **2009**, *9*, 933–947. [CrossRef]
56. Gobster, P.H.; Nassauer, J.I.; Daniel, T.C.; Fry, G. The shared landscape: What does aesthetics have to do with ecology? *Landsc. Ecol.* **2007**, *22*, 959–972. [CrossRef]
57. Fudge, R.S. Imagination and the science-based aesthetic appreciation of unscenic nature. *J. Aesthet. Art Crit.* **2001**, *59*, 275–285. [CrossRef]
58. Matthews, P. Scientific knowledge and the aesthetic appreciation of nature. *J. Aesthet. Art Crit.* **2002**, *60*, 37–48. [CrossRef]
59. Kołata, J.; Zierke, P. Assessment of spatial cohesion in suburban areas based on physical characteristics of buildings. *Ann. Warsaw Univ. Life Sci. SGGW Horticult. Landsc. Architect.* **2020**, *41*, 37–49. [CrossRef]
60. Gyurkovich, M.; Gyurkovich, J. New Housing Complexes in Post-Industrial Areas in City Centres in Poland Versus Cultural and Natural Heritage Protection—With a Particular Focus on Cracow. *Sustainability* **2021**, *13*, 418. [CrossRef]
61. Chruscinski, J.; Kazak, J.K.; Tokarczyk-Dorociak, K.; Szewranski, S.Z.; Swiader, M. How to Support Better Decision Making for Sustainable Development? *IOP Conf. Ser. Mater. Sci. Eng.* **2019**, *471*, 112008. [CrossRef]
62. Broża, N.; Birnbaum, K.; Garcia Castro, D.; Kazak, J.K. Spatial Absorbency Assessment for Sustainable Land Development. *Geomat. Environ. Eng.* **2020**, *14*, 5. [CrossRef]
63. Badach, J.; Raszeja, E. Developing a Framework for the Implementation of Landscape and Greenspace Indicators in Sustainable Urban Planning. Waterfront Landscape Management: Case Studies in Gdańsk, Poznań and Bristol. *Sustainability* **2019**, *11*, 2291. [CrossRef]

Article

Revitalization of Public Spaces in Cittaslow Towns: Recent Urban Redevelopment in Central Europe

Agnieszka Jaszczak [1,*], Katarina Kristianova [2], Ewelina Pochodyła [3], Jan K. Kazak [4] and Krzysztof Młynarczyk [1]

1 Department of Landscape Architecture, University of Warmia and Mazury in Olsztyn, Prawochenskiego St. 17, 10-719 Olsztyn, Poland; kfm@uwm.edu.pl
2 Institute of Urban Design and Planning, Faculty of Architecture and Design, Slovak University of Technology in Bratislava, Námestie slobody 19, 812 45 Bratislava, Slovakia; katarina.kristianova@stuba.sk
3 Department of Water Management and Climatology, University of Warmia and Mazury in Olsztyn, Plac Łódzki 2, 10-719 Olsztyn, Poland; ewelina.pochodyla@uwm.edu.pl
4 Institute of Spatial Management, Wrocław University of Environmental and Life Sciences, ul. Grunwaldzka 55, 50-357 Wrocław, Poland; jan.kazak@upwr.edu.pl
* Correspondence: agnieszka.jaszczak@uwm.edu.pl

Abstract: Revitalization of cities varies depending on the scale of a city, type of challenges, and the socio-environmental context in each case. While revitalization projects carried out in globally known cities are well described, there is still a gap in characterizing revitalization processes that aim to improve quality of life in smaller units like medium-sized towns. This paper fills this gap by the insight from 82 revitalization projects implemented in 14 towns of Warmia and Mazury region (Poland) which are associated in the Cittaslow movement. The study combines a quantitative assessment of statistical data describing these projects with their qualitative evaluation based on interviews with local experts. The results of conducted analyses show that socio-economic development plays a major role as, despite projects which directly refer to the social domain, social elements were found also in projects initially categorized as those targeted to architectural and spatial domains. On the other hand, the authors observed that environmental and ecological as well as cultural issues are treated unevenly or marginally in projects compared to social ones. Interviews with experts show that the least importance was assigned to cultural and historical domain. The obtained results might constitute important knowledge to understand the background of current revitalization processes outside of global metropolises to improve future mechanisms supporting urban renewal.

Keywords: public spaces; revitalization; Cittaslow; quality of life; liveability

Citation: Jaszczak, A.; Kristianova, K.; Pochodyła, E.; Kazak, J.K.; Młynarczyk, K. Revitalization of Public Spaces in Cittaslow Towns: Recent Urban Redevelopment in Central Europe. *Sustainability* **2021**, *13*, 2564. https://doi.org/10.3390/su13052564

Academic Editor: Carmela Cucuzzella

Received: 30 January 2021
Accepted: 23 February 2021
Published: 27 February 2021

1. Introduction

The contemporary urban development process should focus not on "developing more", but on "developing better". However, to complex problems related to urban development planning, it is difficult to find solutions, which are correct or false, or answers on how the idealized planning should function [1]. Experience of recent decades in urbanization processes [2] shows that ignoring the issue of quality of urban development may influence many aspects of life like poverty [3], crime [4], public health [5], socialization [6], and many others. That leads to demographic changes which are caused by migrations [7] and as a result influence the development of unsustainable urban forms like low-density urban sprawl in suburban zones [8]. Of course, the problem of urban sprawl mainly affects large cities. However, the effects of this process may affect small towns located in areas near large agglomerations [9]. On the other hand, in small towns in the provinces, one can observe the phenomenon of migration of people to large cities. In order to reverse these ineffective processes, urban renewal became a direction of development policies in many municipalities [10]. In its original meaning, renewal is a process that is intended to improve the condition of the urban environment by introducing direct changes to the housing structure of cities and poor neighborhoods. Improvement of living conditions and better aesthetics of the surroundings are to change the way residents think about space and care for the place where they live [11,12].

Urban regeneration is a process of transforming the economic and social conditions of a place. It requires action at the level of introducing a coordinated small town development policy and cooperation between the public and private sectors, as well as involving the local community in these activities. Regeneration refers to interventions and changes to maximize the outcomes of increasing social, cultural, environmental, and economic outcomes [13].

City revitalization in this context is defined as a response to the stimuli arising from the forces of disinvestment and deterioration (i.e., decline), including interracial demographic shifts, metropolitan suburbanization, intraregional economic competition, and economic globalization [14]. The revitalization of small towns is a process aimed at "revitalizing" them, which takes into account changes in space and the environment concerning the improvement of socio-economic conditions. Revitalization affects especially poor, marginalized places and places in need of intervention [15].

In recent years there is an increasing interest in the role of communities in urban revitalization. The concept of community action and its spatial dimension is understood as the action of a collective of individuals toward a common goal of improvement of the living conditions within their residence or environments [16]. However, there is no one general approach to revitalize a city or town, as every community has got its characteristics and different factors should be taken into account, so the practices of community actions are difficult to use as replicable models [16].

Revitalization projects all around the world differ as other issues have to be solved among cities, they vary according to natural conditions, available financial sources, and kinds of human activity that should be stimulated [15,17–24]. Due to their scale, some mega-projects like Big Dig from Boston (MA, USA) [25], Green Carpet from Maastricht (Netherlands) [26], waterfront regeneration in Malmö (Sweden) [27], Huangpu river revitalization in Shanghai (China) [28] or High Line in New York (NY, USA) are well known [29]. However, not every revitalization project is so spectacular, which is in strong relationship with the investment activity of each city [30]. Smaller towns for example may implement different strategies that can be connected also with smaller public spaces in order to create an opportunity for a cozy atmosphere, friendly environment to slow down and step out from very dynamic daily routine. One of the best-known organizations gathering such towns is the Cittaslow network.

The Cittaslow movement was created in 1999 and its goal is to resist globalization and homogenization of towns by promoting cultural diversity, protecting the environment, promoting traditional local products, and striving to improve quality of life. Towns that want to become a member of the network need to have a population under 50,000 citizens (according to European standards classified as medium-sized town [31]) as well as support and implement the goals of Cittaslow [32]. Local policy assessments rely on six main criteria: environmental policies, the safeguarding of autochthonous production, infrastructure, technologies and facilities for urban quality, hospitality, and awareness of the aims, procedures, and programs of the Cittaslow initiative [33]. In order to guarantee high quality of life, public spaces in towns associated in Cittaslow are also being revitalized by local authorities, however, they require different approaches in revitalization projects to maintain their calm character [34–40].

According to information supported by the Cittaslow network, an example coordinated revitalization of Cittaslow towns was implemented in Poland in the mid-2010s [41]. As with every public investment, the revitalization of public spaces in Cittaslow towns depends on financial feasibility. The Regional Operational Programme of the Voivodeship of Warmia and Mazury 2014–2020 [42], supported by the European Union, among others, from the European Regional Development Fund (ERDF) and European Social Fund (ESF), creates such financial framework for local and regional authorities. The region of

Warmia and Mazury plays a key role in the Polish Cittaslow network [43–47], as 23 out of 32 associated towns are located in this voivodeship. To manage public investments in that area in an integrated approach, the Supra-Local Revitalization Program of the Network of Cittaslow Cities for 2014–2020 (SLRP) was adopted in 2015 [48,49]. Currently, once the time frame is almost finished it is possible to evaluate how Cittaslow municipalities used this opportunity and what are the findings for future revitalization of public spaces in this specific group of municipalities.

The research aimed to define the role of the revitalization program in transforming public spaces in Cittaslow towns in the region of Warmia and Mazury (Poland). The assessment of the sustainable development of the analyzed towns was of great importance in the research and took into account not only the economic and social aspects but also (which is very important in the 21st century) the ecological and environmental aspects. The ratio of the number of all revitalization projects to projects related to public spaces and architecture (including spaces and architecture of historical importance) was determined. The ratio of the revitalized area was also determined in relation to the groups of revitalization projects (S-space and SA-space and architecture) adopted for the research purposes. The results of an expert interview on the purposefulness and perception of the revitalized space were analyzed (also in terms of addressing the issue of environmental protection and ecology).

The conducted analyzes were to answer the main questions:

- Did the membership in the Cittaslow town network and the inclusion of towns in the Supra-Local Revitalization Program (SLRP) increase the attractiveness of towns?
- To what extent are the projects implemented under the SLRP directly related to the main goal of Cittaslow towns, i.e., sustainable development and respect for identity?
- What was the main motivation for the revitalization of public spaces?
- What is the perception of changes in urban space after revitalization?

Additionally, the authors tried to find an answer to the supplementary question:

- Is the participation of towns in SLRP related to the increase in the number of projects based on pro-environmental, ecological, and cultural development, as well as their quality and effectiveness?

The paper is structured as follows: Section 2 includes literature review; Section 3 describes methods that were applied in the research and the materials that were used to perform the analyses; Section 4 contains results of the research presented in diagrams; discussion and conclusions of the obtained results are presented in Sections 5 and 6.

2. Literature Review

2.1. Revitalization of Small Towns and the Livability Concept

Revitalization involves the process of repairing public and private spaces that are in some way neglected in terms of spatial and functional, but also aesthetic and social. Physical space where people can assemble—libraries, schools, playgrounds, parks, public spaces, as well as commercial establishments create social infrastructure and facilitate sociality [50,51]. Examples can be found in which public art intervention has attempted to generate inclusion [52]. A revitalization is an important tool for the socio-economic development of urban structures [53,54], in this case, small towns. It manifests itself in undertaking initiatives in the field of renovation of facilities or areas that require immediate intervention or immediate repair tasks. A comprehensive approach is important in this case. Often, as a result of revitalization, the function and purpose of such objects changes, which often allows for economic, social, or cultural effects, while maintaining the identity of the place to a significant extent, sometimes by signaling the original meaning, or the historical sense.

Historic centers of small towns are often objects of revitalization. In comparison to other parts of an urban area, they require transformations leading to a better adjustment of historical structure to new functional uses [55,56]. Revitalization projects in small towns often focus also on improvements in mobility [57–59], or opportunities for cycling [60–63].

Terms livability and sustainability are used as guiding principles of revitalization processes [64,65], as they represent concepts expressing the quality of life and quality of the urban environment [66,67]. To improve the environmental quality and human well-being, policies of revitalization projects give attention to the revitalization of urban green infrastructure, and public green spaces [68,69].

Recent trends in revitalization policies take into account the need to apply nature-based solutions [70–78], as critical components of sustainability transitions [71], climate change mitigation, and adaptation strategies [79–84], important also for small municipalities [85–90]. The benefits of nature-based solutions are still not recognized enough in urban planning policies of cities [74]. There are still gaps and barriers in the impact assessments of nature-based solutions, however, it would be an essential part of promoting this concept and approach [70]. For the implementation of climate change adaptation strategies, multilevel coordination between cities and higher levels of government is required and insufficient responses erode the basis of sustainable development [80]. In the case of small municipalities, the local economic benefits are usually on top of the agenda and climate adaptation provisions become secondary [85]. Lack of priority, limited policy direction, and perceptions of climate change are barriers in implementing adaptation into local plans in small municipalities [87].

As a response to the current social and environmental issues, principles of circular economies are advocated, conceptualizing applications for cities [91–94] as in transitioning to a circular economy the role of the built, and particularly the urban environment, is crucial [93]. Core aspects include issues of energy, recycling resources, but at a local level, the key strategy promoted is waste management [95,96]. Recycling urban space and transformation of unused and dysfunctional built-up structures belong to the concepts of circularity [97,98].

Many efforts of revitalization projects addressing the development of cities from economic, social, and cultural points of view are oriented towards tourism, cultural and social activities [99–103]. Small towns need to adopt substantively different cultural strategies [102] and culture and leisure are often focal points both to local entrepreneurship and to local governments [103]. Small towns may attract tourism and experience economy, due to the beauty of their natural or constructed environment, or their history [102]. The role of citizen participation and activism in community development and revitalization and creative urban revitalization is essential [104]. Non-governmental organizations create often new stimuli for cultural development [105] and in revitalization planning and financing policies, participation becomes important [106].

2.2. The Role of the Revitalization Process in the Renewal of Small Towns in Warmia and Mazury

In the Revitalization Act of 9th of October 2015 [107], the issue of "revitalization" covers activities, both comprehensive and integrated, and aimed at leading cities and towns out of the state of crisis. These are also construction and technology investments on a small scale, such as minor repairs. On the other hand, from the administrative point of view, revitalization includes interventions initiated by the local government in neglected, unwanted, forgotten areas, but suitable for broadly understood renewal. With regard to small towns in the region of Warmia and Mazury, it should be noted that the renewal applies to areas with high unemployment, as well as marginalized areas, i.e., those located outside a large agglomeration. In the latter reference, one can speak of the so-called hinterland, i.e., in the spatial and social sense, located more than 50 km from large agglomerations [108]. Revitalization in small towns should also take into account the degree of diversification in the income of residents, by preparing places (e.g., flats) also for the less wealthy part of the the society. Another issue is the process of depopulation (especially the flight of young people to agglomerations) and the aging of the society. It is especially noticed in the case of towns in Warmia and Mazury, therefore all activities and treatments in the field of revitalization should take into account these demographic conditions. In terms of space, a sufficient number of places and facilities for health and care should be created. The issues of social development in relation to revitalization in Warmia and Mazury were emphasized in the Regional Operational Program of the Voivodeship of Warmia and Mazury 2014–2020 [42], as well as in other EU documents related to the 2014–2020 financial perspective. As we read in the Supra-Local Revitalization Program 2015 (SLRP) [48], "the revitalization process contributes to the protection of cultural heritage and the increase of local and regional awareness of the inhabitants, which is particularly important in the context of European integration. Therefore, the involvement of the local community, non-governmental organizations, entrepreneurs, and other stakeholders of social dialogue in the creation and implementation of the revitalization program has a chance to positively affect the increase in civic activity, engaging residents in the social life of towns, as well as in cooperation with public administration and business".

3. Materials and Methods

3.1. Assumptions of the Supralocal Revitalization Program

The SLRP for the towns of Cittaslow in the Warmian-Masurian Voivodeship was created to respond to economic, social, cultural, and demographic transformations and changes.

The program emphasizes more effective use of the existing possibilities related to financing individual regeneration activities. Combining investments and projects into integrated undertakings is intended to improve the situation in small towns of Warmia and Mazury selected for the projects. The complexity of activities in the selected area should lead to permanent changes in the quality and functionality of the area, which should contribute to sustainable development on various scales, from local to regional [48].

The SLRP is a long-term operational program, subordinate to the town's development strategy, adopted and coordinated by a local government unit. It is implemented according to a specific schedule and financed from available sources. The planned undertakings will be financed from various sources (ERDF, ESF, own funds). The scope of projects is also possible to be financed under the Regional Operational Program of the Voivodeship of Warmia and Mazury 2014–2020 [42,48]. Revitalization of selected small towns belonging to the Cittaslow association in the region of Warmia and Mazury is planned as activities within Integrated Investment Projects. An Integrated Investment includes at least two projects related to a selected topic, which constitute a common goal. The selection of projects, their assumptions, goals, and the manner of implementation are the tasks of the regeneration working groups established in each town of the supralocal program. The organization implementing the SLRP of Cittaslow towns is the Association of "Polish Cittaslow Cities".

3.2. Research Assumptions

SLRP for selected towns of Cittaslow from the Warmian-Masurian Voivodeship was established in 2015 and fourteen towns from the region joined it. A revitalization plan was developed for these towns, including objectives of individual projects, and then their implementation began in the following years. After 5 years, the effects of some activities can be seen, some projects have been abandoned, and others are still being implemented. In 2019, a new SLRP was developed, this time for 19 towns, including five new ones. This program has introduced changes compared to the 2015 program and has selected key priority projects. In this study, the authors included only 14 towns that participated in the first version of the revitalization program, due to the implementation of their projects. The projects in the new program and for new towns are at the planning stage.

3.3. The Study Area

Fourteen of all towns operating in the Polish Cittaslow Cities Association were selected for the study, which joined the SLRP in 2015. Fourteen towns were intentionally selected for the research, as these towns are included in the SLRP and implement regeneration projects under the program.

These are: Barczewo, Biskupiec, Bisztynek, Dobre Miasto, Gołdap, Górowo Iławeckie, Lidzbark Warmiński, Lubawa, Nidzica, Nowe Miasto Lubawskie, Olsztynek, Pasym, Reszel, Ryn. They are located in north-eastern Poland in the Warmian-Masurian Voivodeship (Figure 1).

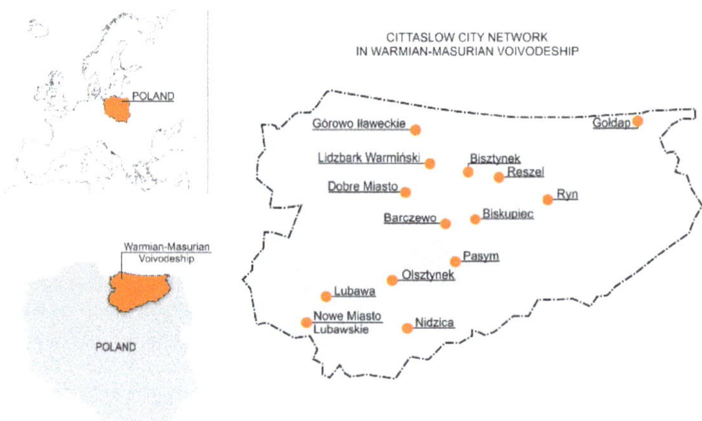

Figure 1. Location of Poland, Warmia and Mazury Voivodeship and selected Cittaslow Towns.

The smallest of the fourteen analyzed towns in terms of a number of inhabitants is Bisztynek (2437 inhabitants), while the largest is Lidzbark Warmiński (15,877), Table 1.

Table 1. The number of inhabitants in the analyzed 14 Cittaslow towns.

Town	Number of Inhabitants [1]
Barczewo	7290
Biskupiec	10,582
Bisztynek	2437
Dobre Miasto	10,741
Gołdap	13,593
Górowo Iławeckie	4021
Lidzbark Warmiński	15,877
Lubawa	10,269
Nidzica	14,166
Nowe Miasto Lubawskie	10,977
Olsztynek	7669
Pasym	2561
Reszel	4667
Ryn	2899

[1] Own elaboration based on data of the Central Statistical Office of Poland as of 2015.

3.4. Research Stages and Methods

The research was divided into five stages. In the first, the literature on the subject as well as monographic and cartographic documents, as well as strategic documents concerning the towns of Cittaslow, were studied. In the second stage, the degree of urban space development in relation to the revitalized space was analyzed. In the third stage, projects were selected according to the adopted criteria. In the fourth, the degree of development of public space was assessed concerning the projects related to space, and to architecture and space. The fifth stage concerned the evaluation of projects by a group of experts. The conducted research covers the period from the beginning of 2015 to December 2020. An environmental interview with experts was conducted in the second half of 2020.

3.4.1. Stage I. Analysis of Materials and Documents Related to Selected Towns

The analysis concerned strategic and planning materials and documents, including Strategies for the Development of Towns and Communes, Study of the Conditions and Directions of Spatial Development, Local Spatial Development Plans, current and previous (before 2015), Local Development Plans of Cities and Communes, Local Urban Revitalization Plans (before 2015) as well as Social and Economic Strategy of the Warmińsko-Mazurskie Voivodeship until 2015 [109], and Regional Operational Program of the Voivodeship of Warmia and Mazury 2014–2020 [42].

3.4.2. Stage II. Analysis of the Revitalized Area

At this stage, the degree of revitalization was determined in relation to the total area of towns and the revitalized area, as well as the number of town residents and the number of residents from the revitalized area. As it results from the SLRP, towns were guided by the following regulations, establishing that the area of revitalization is "all or part of a degraded area, characterized by a particular concentration of negative phenomena, ... is an area of significant importance for local development, ... cannot cover areas larger than 20% of the commune and may not be inhabited by more than 30% of the commune inhabitants" [48].

Due to the fact that the research covered only towns, without the entire commune area, the authors developed a revitalization unit (RU), which refers to the town area and the number of town inhabitants. It was also created to compare the degree of spatial development in the analyzed towns.

RU refers to a number that express the impact of the program on the town space, according to:

$$RU_i = \left(\frac{RA}{TA}\right)_i \times \left(\frac{IRA}{ITA}\right)_i \tag{1}$$

Then, the average of revitalization units was calculated according to the formula:

$$\overline{RU} = \frac{1}{n}\sum_{i=1}^{n}\left(\frac{RA}{TA}\right)_i \times \frac{1}{n}\sum_{i=1}^{n}\left(\frac{IRA}{ITA}\right)_i \tag{2}$$

where

RU—revitalization unit;
RA—revitalization area;
TA—total area;
IRA—number of inhabitants of the revitalization area;
ITA—number of inhabitants of the total area;
n—number of towns;
$\overline{RU}$—average revitalization unit.

3.4.3. Stage III. Selection and Analysis of Projects Taking into Account Goals and Functions

At this stage, the projects were grouped according to their goals and functions, divided into groups. The first group consisted of projects of a social nature (SOC), the second group of projects related to architecture (A), the third group of projects related to space (S), and the fourth group of combined projects related to the development of space with the renovation of architectural objects (SA). Due to the previously adopted research assumptions, projects from the last two groups were selected for further analysis.

3.4.4. Stage IV. Assessment of the Degree of Development of Public Space in Relation to Projects Related to Space (S), and Architecture and Space (SA)

As already mentioned, as a result of the selection, projects from groups S and SA were selected for further research. At this stage, the number of projects was compared by function and location, and the projects were collected into the following subgroups:

- S1—parks and green squares;

- S2—recreation and sport areas;
- S3—main squares, streets, parking spaces;
- SA1—buildings with a social and administrative (contemporary or with historical value) and surrounding areas;
- SA2—buildings with social/recreational/sports function and surrounding areas;
- SA3—buildings with cultural function (with historical values);
- SA4—buildings with transportation function and surrounded road infrastructure (railways stations and parking spaces).

Then, based on the data available in the SLRP, the area of the implementation of projects was calculated. The next step was to calculate the revitalized areas divided into individual groups (S and SA).

3.4.5. Stage V. Expert Assessment Regarding the Implementation of Projects

The expert evaluation aimed to find out the opinions of 20 experts (17 women, 3 men), who professionally represent various fields related to spatial planning and management, administration of small towns, education, and social animation. When selecting experts, the authors made use of the following criteria:

- Knowledge of the revitalization process in the small towns of the Warmian-Masurian Voivodeship;
- Expert's ability to assess changes in the public space of the fourteen sites selected for analysis, and thus knowledge of the spatial and functional structure of those sites;
- Experts are not authors or directly involved in the implementation of the projects.

Experts assessed the projects in the region where they live, so they know all the analyzed 14 towns. After familiarizing themselves with all projects in the revitalization program, the experts assessed the implemented projects in towns. Experts expressed their opinion on the usefulness of regeneration projects and their impact on changing the space of the analyzed towns. The interview protocol included 10 issues grouped into two blocks: a descriptive opinion and questions regarding the assessment of changes in space in relation to the types of design and functions of space.

Issues in the interview concerned the degree of increasing the attractiveness of the town after revitalization, the main motives for participation in the program, and thus the implementation of projects, detailed assumptions of implemented projects, quality and efficiency of projects, the impact of revitalization on improving the quality of life of residents, visible changes in public space, practical recommendations for the preparation of future projects for the analyzed towns and towns outside the revitalization program

Respondents were acquainted with the purpose of the interview and were informed about the research assumptions. Due to the pandemic, most of the interviews were conducted as online interview.

4. Results

4.1. The Results of the Analysis of Source Materials Regarding the Priority Objectives Included in the Supra-Local Revitalization Program

The analysis of source materials shows that the studied towns have an attractive geographic location and unique natural and cultural values, which creates opportunities for their sustainable development. At the same time, they have the development potential of areas intended for revitalization. Despite this, both the natural and cultural potential (including protected areas, monuments) is not used. The analyzed source materials show that the economic development of towns should be in line with environmental protection. Due to the specificity of towns and the above-mentioned natural and cultural values, a significant role is assigned to redefining the concept of recreational spaces for both residents and tourists. On the other hand, the poor condition of built-up areas, including historic architecture, technical infrastructure, and the transport network, is noticeable. There is a lack of integrated space management and ideas to stop environmental degradation. The use of alternative energy sources is very low. Another obstacle in the proper implementation of the revitalization process of public spaces is the unregulated situation in land ownership.

Therefore, among the priority objectives contained in the urban renewal plans and other analyzed documents, taking into account the development of towns, the following can be distinguished:

- With regard to the protection of cultural heritage—protection and improvement of cultural heritage objects, restoration or giving them a new function;
- In terms of space renewal—revitalization and modernization of public space, improvement of accessibility of space for disabled or excluded people, improvement of the quality of public space for residents and tourists, revitalization of publicly accessible recreational areas;
- In terms of improvement of infrastructure and residential areas—modernization of technical and road infrastructure, improvement of the accessibility of residential buildings and the environment, thermal modernization of residential buildings;
- In the field of social, educational, and promotional activities—improvement of ecological awareness and pro-ecological attitudes of residents, shaping a positive image of the town, improving conditions for rest and leisure by supporting initiatives for children, youth, and the elderly.

4.2. Results of the Assessment of the Degree of Spatial Development in Relation to the Revitalized Area in Towns

When comparing the revitalization areas in all towns, it can be seen that the largest area was designated for revitalization in Dobre Miasto, then in Biskupiec, Bisztynek, and Lidzbark Warmiński. The smallest area is in Nidzica and Pasym. However, taking into account the ratio of the revitalized area to the total area of the town, the largest area designated for changes was in Bisztynek, then in Dobre Miasto and Biskupiec, and then in Górowo Iławecki, Olsztynek and Lidzbark Warmiński. The smallest area was designated in Pasym and Nowe Miasto Lubawskie (Figure 2).

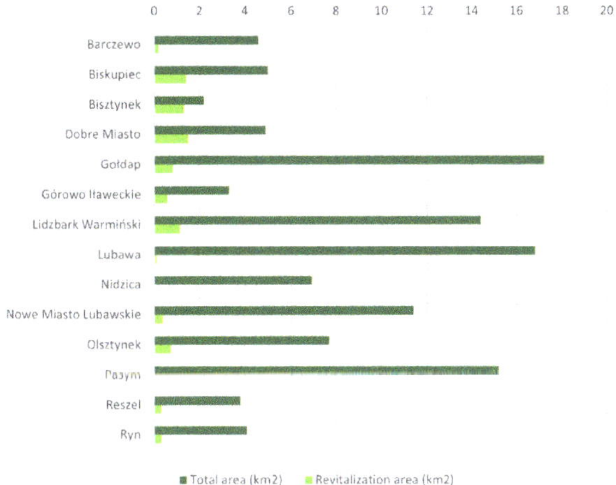

Figure 2. Comparison of towns in relation to the town area and the revitalized area in these towns.

Analyzing the number of inhabitants of the revitalized areas in all towns, the largest number of inhabitants lives in this area in Biskupiec, then in Lidzbark Warmiński, Dobre Miasto, and Gołdap, and the least in Ryn. However, when referring to the proportion to a total number of inhabitants in an individual town, the largest number of inhabitants living in the revitalized area is found in Bisztynek, then Biskupiec and Dobre Miasto, while the least in Lubawa and Nowe Miasto Lubawskie (Figure 3).

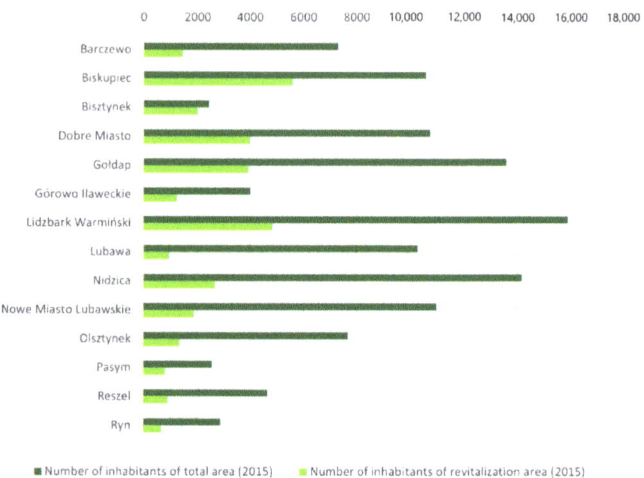

Figure 3. Comparison of towns in relation to the number of their inhabitants (in total) and the number of inhabitants of the revitalized area in 2015.

Comparing the revitalization units (RU) of towns with the average revitalization unit ($\overline{RU}$), it can be seen that Bisztynek stands out. The revitalization unit of Bisztynek is several times higher than the average value calculated based on data from all 14 towns, followed then Biskupiec and Dobre Miasto. In Górowo Iławeckie, the value approximate to the revitalization unit can be noticed (Figure 4).

4.3. Results of Analysis of the Projects Taking into Account Goals and Functions

In the Supra-Local Revitalization Program (SLRP), towns have designated a total of 82 projects for implementation. The largest group, 64%, are projects related to spatial development, renovation of architectural objects, and renovation of space around architectural objects (S, SA, A), while social projects (SOC) make up 36% of all activities planned in the program (Figure 5).

Social projects (SOC) most often concerned the activation of residents, especially the unemployed, socially excluded, and the poor. The main goal of these projects is broadly understood education, including improving professional qualifications and increasing the competitiveness of people in the labor market, as well as easier access to activities with a cultural, integrative, and health-promoting function. Most projects of this type have been or are being implemented in Gołdap and Olsztynek, while the least in Dobre Miasto, Nidzica, Nowe Miasto Lubawskie, Reszel, and Ryn (Table 2). In the case of Gołdap, the projects related to integrated services for families, therapeutic activities, and ways of spending free time together, as well as activities improving technical and information technology skills. Of course, social projects are important from the point of view of increasing the quality of life of residents, however, they are often not integrated with other projects, e.g., those related to the revitalization of public space.

Among urban development projects, revitalization projects of space (S) have the largest share, followed by projects aimed at renovating buildings with different functions (A), while there are slightly fewer projects identified by the authors as a combination (SA). Their goal is to renew the conditions of buildings along with the development of their surroundings (Figure 6).

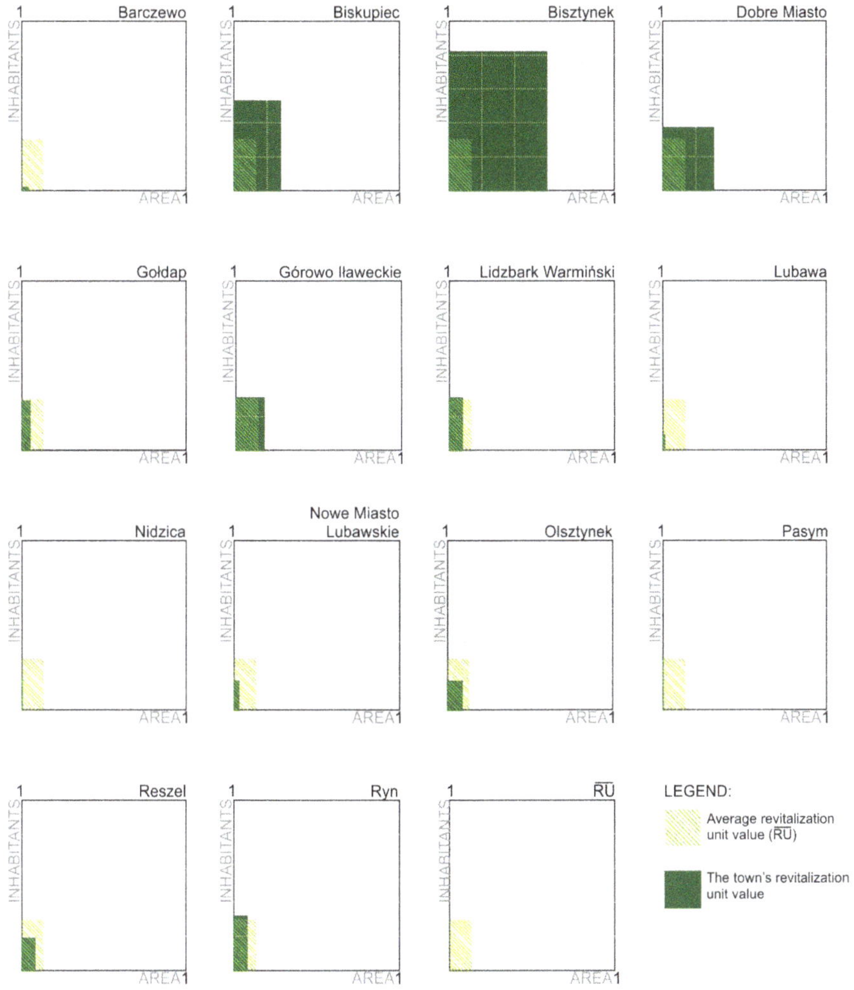

Figure 4. Comparison of the revitalization units of analyzed towns (RU) in relation to revitalization unit $(\overline{RU})$.

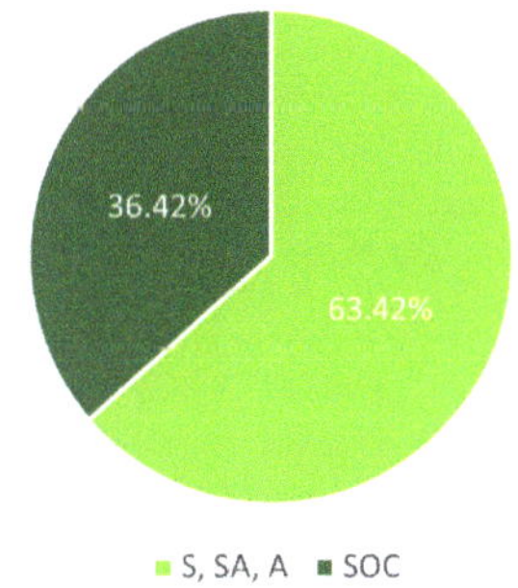

Figure 5. The number of projects (%) related to spatial development (S, SA, A) and social projects (SOC).

Table 2. The number of projects in the analyzed towns.

No.	Town	Together (with Social) *	** Without Social	Architecture (A)	Space (S)	Architecture and Space (SA)
1	Barczewo *	5	3	1	1	1
2	Biskupiec	5	3	-	1	2
3	Bisztynek	4	2	1	-	1
4	Dobre Miasto	3	2	-	1	1
5	Gołdap	13	5	4	1	-
6	Górowo Iław.	6	4	-	2	2
7	Lidzbark Warm.	10	8	2	3	3
8	Lubawa	4	2	1	1	-
9	Nidzica	6	5	1	3	1
10	Nowe Miasto Lub.	6	5	2	3	-
11	Olsztynek	6	3	1	1	1
12	Pasym	5	3	2	1	-
13	Reszel	3	2	-	1	1
14	Ryn	6	5	2	1	2
	Total	82	52	17	20	15

Own source based on Supralocal Revitalization Program (2015). * Number of projects in the local revitalization program (including those of a social nature). ** Number of projects related to space and architecture (excluding those of a social nature).

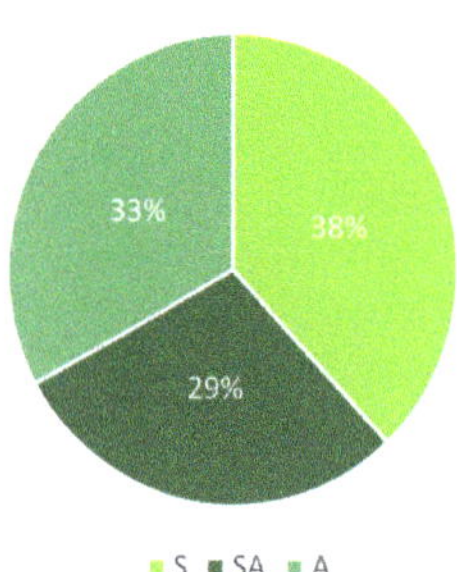

Figure 6. The number of projects (%) by groups S, SA, A.

The analysis of the number of projects focused on the revitalization of space and the renovation of architectural objects showed that most projects were implemented or are under implementation in Lidzbark Warmiński and then in Gołdap, Nidzica, Nowe Miasto Lubawskie and Ryn and least in Bisztynek, Dobre Miasto, Lubawa and Reszel (Table 2). As far as the development of the space itself (without architecture) is concerned, most of the projects were carried out in Lidzbak Warmiński, Nidzica and Nowe Miasto Lubawskie, while in the case of the revitalization of architectural structures and their surroundings, most of them were realized in Lidzbark Warmiński, and then in Górowo Iławeckie, Biskupiec and Ryn. Only in one town, Bisztynek, no space development has been planned as part of the revitalization program, while in Gołdap, Lubawa, Nowe Miasto Lubawskie and Pasym there are no joint projects SA (Table 2).

4.4. The Results of the Analysis of Spatial Development Divided into Functions in Groups S and SA

Based on the analysis, 20 projects were assigned to the S group and 16 projects to the SA group in all towns. In group S, the largest number are S2 (10) development projects, then S1 (6), and the least are S3 projects (4). Of the recreational and sports space projects, the largest number concerns sidewalks, cycling, and walking routes, and recreational areas by the water (rivers, lakes). Such areas have been or are under construction in Barczewo, Biskupiec, Dobre Miasto, Gołdap, Górowo Iławeckie, Lidzbark Warmiński, Olsztynek,

Ryn, and the most in Lidzbark (3). Regarding the revitalization of parks, it should be noted that 2 parks are being renewed in Nowe Miasto Lubawskie and Nidzica, and 1 in Reszel and Górowo Iławeckie. On the other hand, squares, street space, and car parks are being revitalized in Lubawa, Nidzica, Nowe Miasto Lubawskie and Pasym (Figure 7). As mentioned above, most revitalization projects in group S concerned recreational space. This space, unlike designed town parks, is most often connected with areas of great natural and ecological value. The results of the revitalization carried out largely took into account the reorganization of these places for use by residents. However, the projects lacked the environmental context relating to the care and protection of these unique values.

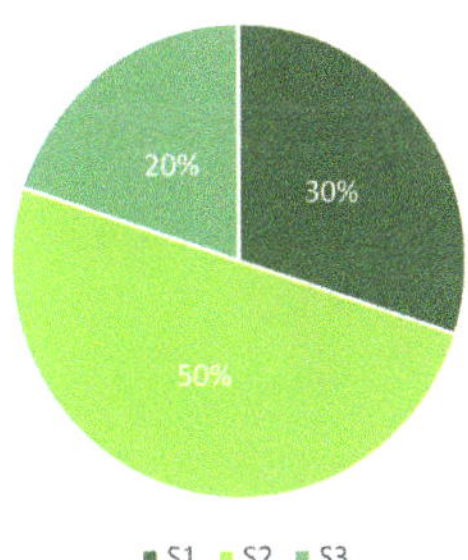

Figure 7. The number of projects (%) in group S (S1, S2, S3).

Most of the projects in the SA group are SA3 projects in 8 towns. They are implemented in Biskupiec, where the project covers up to 10 locations, as well as in Lidzbark (2 projects), Bisztynek, Dobre Miasto, Górowo Iławeckie, Lubawa, Olsztynek and Ryn. The projects from SA3 group usually concern the development of facilities with a small area outside the building. Most often these are areas with a historic water tower (for example in Ryn and Olsztynek), areas with fragments of historic defensive walls (Lidzbark Warmiński), or areas near cultural centers (Biskupiec, Górowo Iławeckie). Fewer projects are in the subgroup SA1, five projects include Górowo Iławeckie, Lidzbark, Nidzica, Reszl and Ryn. The least represented are the projects in SA2 subgroup (l project in Barczewo) and SA4 subgroup (one project in Biskupiec) (Figure 8). When it comes to the function and quality of the development of the space around the buildings, most of the projects show that it has been treated marginally in relation to the renovation of the buildings themselves. In most towns, projects of this type included resurfacing and infrastructure preparation, while they often lacked even a minimal amount of greenery or elements referring to the history and genius loci of the place. The focus was on the renovation of the facilities and their use by residents (social factor).

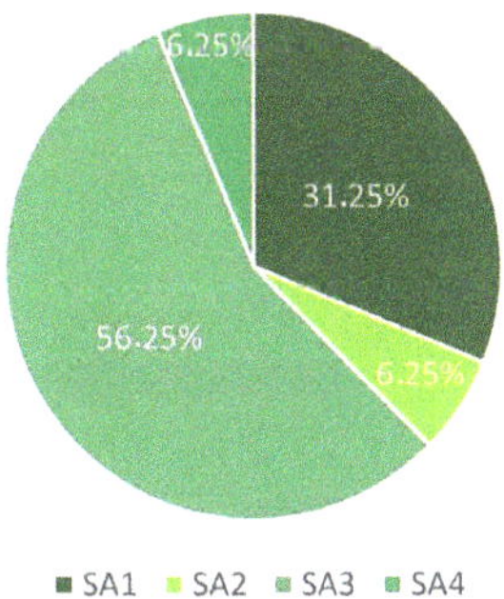

Figure 8. The number of projects in group SA (SA1, SA2, SA3, SA4).

The results of the analysis of the area included in the completed projects in group S in all towns indicate that the largest area was revitalized in the subgroup S1, then S3 and slightly less in the subgroup S2 (Figure 9). The largest area under revitalization in terms of subgroup S1 represents two parks and squares in Nidzica (11.5 ha), a slightly smaller area the park in Reszel (9.5 ha), and the smallest area is reported in Górowo Iławeckie (0.8 ha). In the subgroup S2, the largest revitalized area is a recreational space in Ryn (0.89 ha), and the smallest in Biskupiec (0.2 ha). The revitalized area of the S3 group is the largest in Lubawa (1.37 ha), and the smallest in Nowe Miasto Lubawskie (0.5 ha).

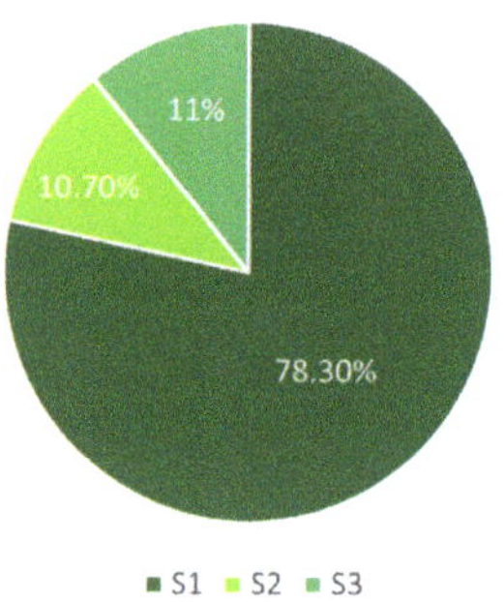

Figure 9. Revitalized area (%) in group S (S1, S2, S3).

In SA projects, the largest area was revitalized in the case of SA3, followed by SA2 areas and SA4 (Figure 10). Among the projects implemented in the SA group, the largest area in the SA1 subgroup is represented by the project in Ryn (0.5 ha), and the smallest area in Reszel (0.086 ha). In the SA2 subgroup, there is only one project in Barczewo (area 1.5 ha), while in SA3 the largest area was recorded for the area in Biskupiec (6.58 ha), it should be noted that this is the total area of revitalization sites located in different parts of the town. The smallest area of objects, together with the surrounding revitalized areas in this subgroup, is in Bisztynek (0.363 ha). The last group SA4 includes only one project in Biskupiec for the facility and adjacent areas with an area of 1.04 ha.

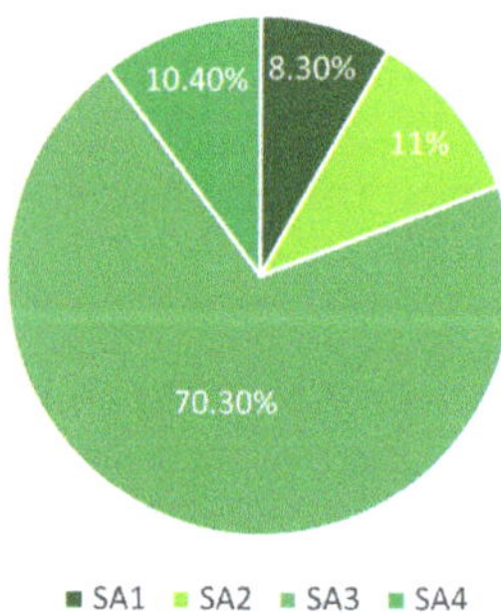

Figure 10. Revitalized area (%) in the SA group (SA1, SA2, SA3, SA4).

4.5. The Results of Interview with Experts

Experts about the relationship between towns belonging to the Cittaslow network and the implementation of the SLRP assumptions, and increasing the attractiveness of towns, indicate the following elements: making the image of towns more attractive, improving the quality of life of residents, improving aesthetics and spatial order (Figure 11, point I.):

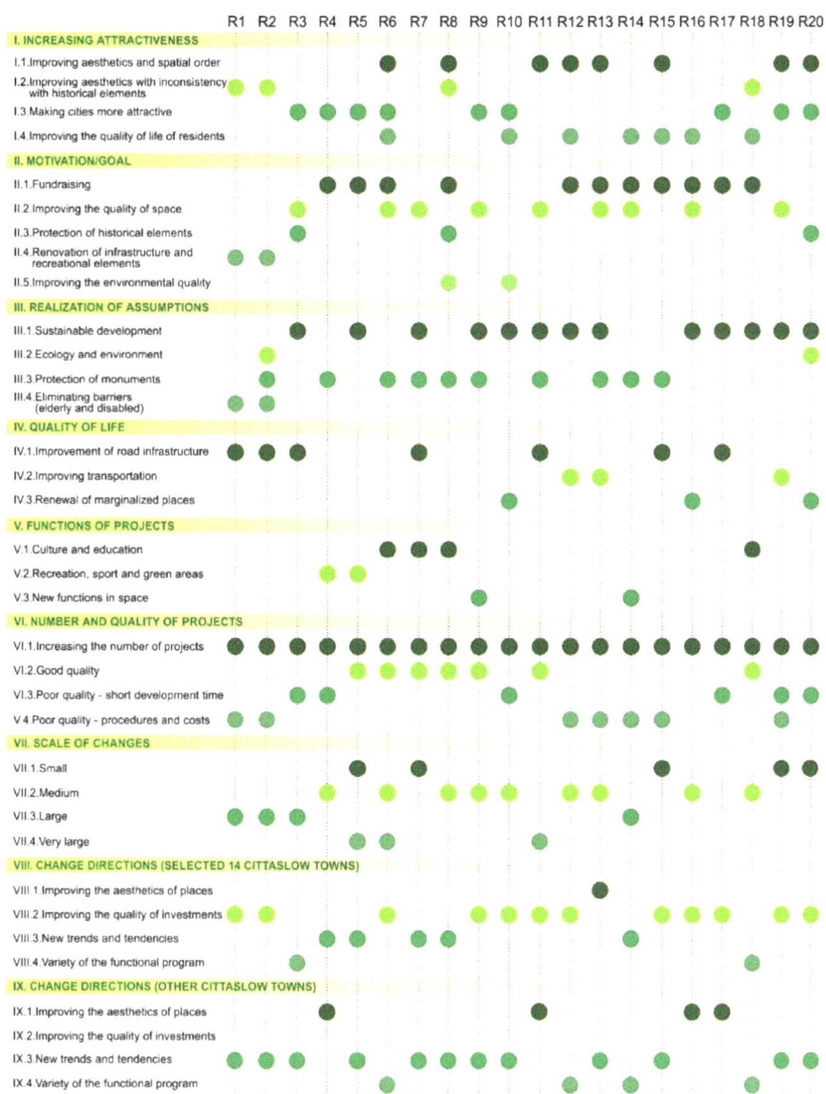

Figure 11. Matrix showing the range of experts' responses.

> "... *definitely belonging to the Cittaslow network and SLRP made it easier for Cittaslow towns to increase their attractiveness by implementing projects in line with the idea of words ... indicated in the program. SLRP Cittaslow allowed promoting towns that share similar issues* ... "

> "... *the towns become more beautiful, they become a tourist attraction; ... the inhabitants live calmer and better* ... "

At the same time, they point out the lack of coherence with the historical tissue in the revitalized spaces (Figure 11):

> "... *joining the SLRP has significantly increased the possibilities of improving the aesthetics of public spaces in towns, but there is no care for historical elements* ... "

" ... there is a noticeable dissonance between the revitalized areas and the existing tissue, historical buildings or the surrounding landscape; ... the point is not to introduce artificially historicizing elements, "pretending that they are from the past", but to perform revitalization with respect for the existing elements and genius loci ... "

With regard to the main motivations of Cittaslow towns in the implementation of revitalization projects and accession to the SLRP, the experts emphasized the possibility of financing projects, improving the quality of public space, and to a lesser extent, the protection of historic areas and buildings, renovation of transport and recreational infrastructure, or improvement of the quality of the environment (Figure 11, point II.):

" ... An important motivation was the pressure of the inhabitants to replace post-communist public spaces with more modern ones; ... the motivation was to a small extent to improve the condition of the environment, take care of monuments or protect natural resources ... "

" ... the most important motivation was the possibility of obtaining funds and increasing the quality of life of the inhabitants and changing the image of the revitalized area ... "

Regarding the detailed assumptions in the revitalization projects of the Cittaslow towns, experts believe that they should be based on sustainable development, respect for historical elements, shaping ecological attitudes among the population and the possibility of using space for all (accessibility for elderly and disabled) (Figure 11, point III.).

Regarding the selection of directions of revitalization related to the improvement of the quality of life of residents in space revitalization projects, experts point out that the projects most often concerned the renovation of road infrastructure (access roads, paths, parking) and technical infrastructure (Figure 11, point IV.1. and IV.4.):

" ... definitely important for the people managing the revitalization process was the improvement of the quality of infrastructure and the renewal of space; ... the aesthetics of architectural objects was secondary, if the remaining needs of the residents were not met ... "

" ... especially in the perspective of towns with a low level of investment activity, it was important in the first place to meet the infrastructural needs and adapt the infrastructure to the needs of the resident, meeting the expectations of a town user in the 21st century ... "

They also point to the need to focus on the following issue—the renovation of areas and facilities with a cultural, cultural and educational function, with high environmental potential, including green areas for recreational and health-promoting purposes. The changes in the functions of space are also of great importance for the respondents. At the same time, they point out that in the implemented projects, these issues were often not resolved correctly or were treated rather marginally (Figure 11, point V.).

" ... the sphere of culture and the preservation of its tangible and intangible heritage are also very important issues in the perspective of space revitalization in Cittaslow towns ... "

" ... regardless of the age group, it is important to arrange places for various activities, including active (sports, recreational) or passive (walking in silence, relaxing) ... "

Experts firmly stated that the participation of towns in the Cittaslow revitalization program increased the number of projects. Some of the respondents claim that the number of implemented projects does not refer to their quality. This is mainly due to the short time for pre-project analyses and the short time for preparing documentation. Another important issue here is the issue of tender procedures, which assume the selection of the company and contractor for the projects at the lowest price (Figure 11, point VI.).

" ... often the possibility of applying for subsidies is very limited in time, hence the need to prepare project documentation in a short time ... imposing the will and vision of officials on designers in the absence of public consultations may also be a problem ... "

"... *a disadvantage in the development of the SLRP was a short time that towns had to select the projects and prepare documentation* ... *it could translate into the quality and effectiveness of the projects* ... "

"... *the main criterion for selecting contractors is the price, which is not related to quality* ... *often poor quality materials are used* ... *work is carried out quickly, inaccurately, which causes manufacturing defects and the need for repairs* ... "

In terms of noticeable changes in the public space after the implementation of the projects, experts mostly believe that they are visible to a medium and low degree (Figure 11, point VII). Referring to the directions of revitalization in terms of functions of spaces and the preparation of future projects for the analyzed 14 towns, the experts pointed to the improvement of the quality of the projects and, consequently, the revitalized spaces, taking into account new trends and diversifying the utility program. One of the experts also noted the need to improve the aesthetics of the space (Figure 11, point VIII). On the other hand, when it comes to the directions of revitalization in terms of the functions of space and the preparation of future projects for the remaining Cittaslow towns in the Warmia and Mazury region, which were not taken into account in the SLRP, the interviewees proposed to refer to new trends and to the same degree to reduce attention to the aesthetics of places, as well as the diversification of the utility program adapted to all users (Figure 11, point IX).

The experts found that the most visible changes in relation to the revitalized spaces are seen in the S group, respectively in the case of S2, then S1, and finally S3. As for the SA group, they notice the greatest changes concerning the area in the case of SA1, then SA2 and SA3, and finally SA4 (Figure 12).

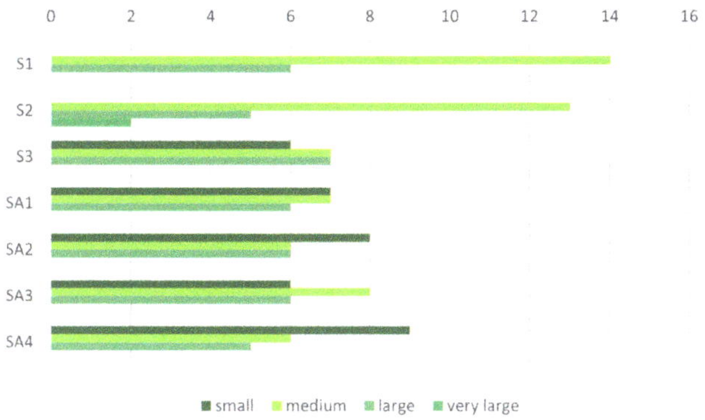

Figure 12. Assessment of the degree of changes in public space in relation to revitalization as part of the projects from groups S and SA (after 2015).

5. Discussion

As it results from the provisions of the revitalization programs of the analyzed towns and the SLRP, the main goal of revitalization is to carry out social and spatial changes in order to improve the living conditions of residents and the quality of space [48]. Improving quality, as understood by project managers, should have a direct impact on economic development.

It should also be emphasized that towns associate the effects of the revitalization process with the continued strengthening of the local economy through improved conditions for small businesses, the service industry and the socio-cultural offer. This vision assumed an increase in the attractiveness of towns as places to live, as well as the development of tourism. These issues are highlighted by [110–114]. As a result of revitalization, the image of revitalized areas is to be significantly improved, which will stimulate changes in neighboring areas.

The results of interviews with experts indicate to a large extent a positive perception of the SLRP itself, as well as the main goal and motives of the activities undertaken in the field of revitalization. Experts emphasize the improvement of the quality of life through, for example, the renovation of infrastructure and the possibility of obtaining funds for revitalization. However, when analyzing the detailed answers, it can be noticed that they also pay attention to the shortcomings of the project preparation process itself, and often their quality, as well as the quality of implementation (execution). This translates directly into changes of the space that are visible after implementation and during use. Research carried out by [115–118] also refers to problems related to the preparation process and the quality of projects.

Based on the analysis of source materials, it can be concluded that in the spatial and functional sphere, towns refer to the vision of socio-economic development, but also to improving the condition of the environment, or caring for space and objects of great cultural and historical importance. In contrast, research has shown that they have made a significant step in terms of social development and improving their image and promotion. This is indicated by the results of the analysis of the number of projects in individual groups. The SOC group accounts for 1/3 of all activities, while the projects in the S, SA and A groups also relate to social development. As can be seen from the interview, the experts also emphasized that the projects had largely social and economic significance. In other studies, the authors also pay attention to the social and economic aspects of revitalization [99,100,104,119].

On the other hand, when it comes to caring for the same features (including historical ones) and environmental protection and ecology, the situation is a bit different. Experts' statements indicate a poor reference in projects (and, consequently, in implementations) to historical or cultural characteristics. The results of the analysis of public spaces from the S group indicate that the number of marketplaces or central squares (S3) projects is the smallest. It should be emphasized that it is the center of small towns that have a historical character. Taking into account the developed area, the centers (markets) are comparable to the S2 group, but still, the area is smaller than the projects from the S1 group. However, when referring to the space around objects of cultural importance (SA3) and partially with a social function (not all) SA1, although the number of projects is quite large, according to experts, they often lack reference to the context of the place.

As it results from our analysis of the revitalization programs of individual town as well as the SLRP, special attention was paid to the pro-ecological approach to revitalization. The provisions of the SLRP deal with the need to care for the environment and broadly educate the public on ecology. While the revitalized public spaces referred to the willingness (not always the need) of pro-social changes, the issue of ecology and the use of pro-environmental solutions in these projects and later implementations is treated marginally or is neglected. Experts pay attention to this. Although according to our research, the projects of parks and lawns (S1) and recreational space (S2) are dominant in terms of numbers and areas, the pro-ecological program and environmental protection ideas were practically not taken into account in them. The evidence is often felling trees prior to land development, very little differentiation in the selection of species in the case of park greenery and squares, no reference to contemporary environmental problems (climate change, environmental pollution), no reference to contemporary pro-ecological design (NBS application). There is a need for a different approach to designing public space in towns [70–78,84,120,121].

Regarding the changes after revitalization in public spaces in relation to the area of towns, one can notice a differentiation depending on the function of space. In the perception of experts, the scale of changes generally ranges from small to medium. The results of the research show that the biggest changes refer to S1 and SA3, while experts say slightly different ideas and select groups S2 (then S1) and SA1 (then SA3). Other studies also show that the most visible space after revitalization are green squares and parks [122,123].

In view of the general need for a change in the direction of thinking about the design of public spaces in small towns, experts suggest the need to introduce new trends, pro-ecological trends and paying attention to the quality and content of projects by adapting them to the conditions and nature of towns (genius loci). This suggestion concerns significant future projects according to analyzed 14 towns, as well as other towns from the Cittaslow network, which were not included in the SLRP. In addition, for this other group, they propose the consideration of introducing a diversified utility program.

6. Conclusions

The objective of this study was to define the role of the revitalization program in transforming public spaces in Cittaslow towns in the region of Warmia and Mazury (Poland). The main conclusions are as follows:

- The participation of towns in SLRP undoubtedly brought the intended effects of drawing attention to the need for changes in public spaces and thus taking steps towards their revitalization. This, in turn, increased their attractiveness and influenced the perception of Cittaslow towns as standing out from the rest of the region and the country. This was confirmed, inter alia, by the results of the expert interview.
- The main motivation for the revitalization of public spaces was the possibility of obtaining EU funds for the implementation of projects and the preparation of common places for residents. Most of the projects concerned social function. In contrast, the participation of towns in SLRP did not increase the number of projects based on pro-environmental, ecological, and cultural development.
- Our research shows that the goal of social development and making the revitalized public spaces available to residents has been achieved. However, the implemented projects often did not take into account the objectives of improving ecological conditions or respect for identity on the same level as social objectives. This is important due to the declaration of sustainable development clearly defined in the strategic goals of Cittaslow towns, in the SLRP, and in the revitalization plans of selected towns.
- According to the research, changes in public spaces proceeded quickly, and the prepared projects were not always well thought out and implemented, which resulted in their lower quality and lower efficiency.

Therefore, in order to improve the revitalization process and the quality of revitalized public spaces in the future in the fourteen analyzed towns, other towns belonging to the Cittaslow network, as well as other towns, we recommend:

- Even distribution of the goals of revitalization, and thus paying attention not only to the economic and social aspect, but also, which is significant nowadays, to the issues of ecology and environmental protection. In future projects, attention should be paid to the specificity of the place and the reference to historical and cultural conditions.
- It is necessary to develop detailed pre-design analyzes, as well as pay attention to the need for a longer design process, allowing for the elimination of errors and the introduction of optimal solutions.
- It is important to conduct reliable interviews with experts, but also with residents in order to develop common goals regarding the use of space.

The presented research contributes to the knowledge of public space revitalization planning, through the use of a simple, multi-criteria analysis process aimed at the need for wide-ranging multi-directional activities for the sustainable development of small towns in Central and Eastern Europe regions. While many studies on the topic of the revitalization of public spaces in small towns are available globally, research from Central and Eastern Europe is still scarce, particularly that employing a multidimensional approach towards revitalization. The majority of studies focus only on the social role of revitalization while there is a lack of studies taking into account the assessment of the balance between social and economic, cultural, and environmental aspects in the implemented revitalization projects. Moreover, different geographic, spatial, and economic conditions of these regions should be emphasized to permit comparison with other regions around the globe in the future. The presented method can be used for the analysis of small towns in other regions of Europe, or the world. However, the condition for its application is the selection of towns with a similar structure, spatial and economic, and social conditions.

Author Contributions: Conceptualization, A.J. and K.K.; methodology, A.J.; software, E.P.; formal analysis, A.J., K.K.; investigation, A.J., K.K., E.P.; resources, A.J., E.P.; writing—original draft preparation, A.J., K.K.; writing—review and editing, A.J., K.K., J.K.K., K.M.; visualization, E.P.; supervision, A.J.; project administration, A.J. All authors have read and agreed to the published version of the manuscript.

Funding: This research received no external funding.

Institutional Review Board Statement: Ethical review and approval were waived for this study, due to the fact that respondents and their privacy were not violated. Responses were recorded without any personal data.

Informed Consent Statement: Informed consent was obtained from all subjects involved in the study.

Data Availability Statement: The analyzes were made on the basis of the information contained in the Supra-Local Revitalization Program of the Network of Cittaslow Towns for 2014–2020 (SLRP) (2015) available on www.cittaslowpolska.pl (accessed on 10 January 2021).

Acknowledgments: The authors would like to thank the Cost Action CA17133—Implementing nature-based solutions for creating a resourceful circular city and the project APVV SK-PL-18-0022 LIVA—The Concept of livability in the context of small towns funded by NAWA—Polish National Agency for Academic Exchange and SRDA—Slovak Research and Development Agency for enabling the work and interaction between the researchers required for the article. The research was also supported by scientific activity conducted within the Leading Research Group: Sustainable Cities and Regions at the Wrocław University of Environmental and Life Science.

Conflicts of Interest: The authors declare no conflict of interest.

References

1. Rittel, H.W.J.; Webber, M.M. Dilemmas in a general theory of planning. *Policy Sci.* **1973**, *4*, 155–169. [CrossRef]
2. Szymańska, D.; Matczak, A. Urbanization in Poland: Tendencies and transformation. *Eur. Urban Reg. Stud.* **2002**, *9*, 39–46. [CrossRef]
3. Świąder, M.; Szewrański, S.; Kazak, J. Spatial-temporal diversification of poverty in Wroclaw. *Procedia Eng.* **2016**, *161*, 1596–1600. [CrossRef]
4. Foryś, I.; Putek-Szeląg, E. The impact of crime on residential property value—on the example of Szczecin. *Real Estate Manag. Valuat.* **2017**, *25*, 51–61. [CrossRef]
5. Van Hoof, J.; Bennetts, H.; Hansen, A.; Kazak, J.K.; Soebarto, V. The Living Environment and Thermal Behaviours of Older South Australians: A Multi-Focus Group Study. *Int. J. Environ. Res. Public Health* **2019**, *16*, 935. [CrossRef] [PubMed]
6. Kajdanek, K. Newcomers vs. Old-Timers? Community, Cooperation and Conflict in the Post-Socialist Suburbs of Wrocław, Poland. In *Mobilities and Neighbourhood Belonging in Cities and Suburbs*; Watt, P., Smets, P., Eds.; Palgrave Macmillan: London, UK, 2014. [CrossRef]
7. Trojanek, M.; Tanaś, J.; Trojanek, R. *Changes in the Demographic Structure of the Central City in the Light of the Suburbanization Process (The Study of Poznań)*; SMARTCITY360; Springer: Berlin, Germany, 2016. [CrossRef]
8. Brezdeń, P.; Szmytkie, R. Current Changes in the Location of Industry in the Suburban Zone of a Post-Socialist City. Case Study of Wrocław (Poland). *Tijds. voor Econ. en Soc. Geog.* **2019**, *110*, 102–122. [CrossRef]
9. Karwińska, A.; Böhm, A.; Kudłacz, M. The phenomenon of urban sprawl in modern Poland: Causes, effects and remedies. *Zarządzanie Publiczne* **2018**, *45*, 26–43. [CrossRef]
10. Bieda, A. Urban renewal and the value of real properties. *Studia Reg. Lokal.* **2017**, *3*, 5–28.
11. Mehdipanah, R.; Marra, G.; Melis, G.; Gelormino, E. Urban renewal, gentrification and health equity: A realist perspective. *Eur. J. Public Health* **2018**, *28*, 243–248. [CrossRef] [PubMed]
12. Jackson, C. The effect of urban renewal on fragmented social and political engagement in urban environments. *J. Urban Aff.* **2019**, *41*, 503–517. [CrossRef]
13. Glasson, J.; Wood, G. Urban regeneration and impact assessment for social sustainability. *Impact Assess. Proj. Apprais.* **2009**, *27*, 283–290. [CrossRef]
14. Gale, D.E. Urban Planning: Central City Revitalization. In *International Encyclopedia of the Social & Behavioral Sciences*; Neil, J., Smelser, N.J., Baltes, P.B., Eds.; Imprint Pergamon, Elsevier: Amsterdam, The Netherlands, 2001; pp. 16040–16044. [CrossRef]

15. Ramlee, M.; Omar, D.; Yunus, R.M.; Samadi, Z. Revitalization of Urban Public Spaces: An Overview. *Procedia Soc. Behav. Sci.* **2015**, *201*, 360–367. [CrossRef]
16. Tricarico, L. Community action: Value or instrument? An ethics and planning critical review. *J. Archit. Urban* **2017**, *41*, 221–233. [CrossRef]
17. Grodach, C.; Ehrenfeucht, R. *Urban. Revitalization: Remaking Cities in a Changing World*; Routledge: New York, NY, USA, 2016.
18. Zielenbach, S. The Art of revitalization. In *Improving Condition of Distressed Inner-City Neighborhoods*; Garland Publishing Inc.: London, UK, 2000.
19. Zheng, H.W.; Shen, G.Q.; Wang, H. A review of recent studies on sustainable urban renewal. *Habitat Int.* **2014**, *41*, 272–279. [CrossRef]
20. Spaans, M. The implementation of urban regeneration projects in Europe: Global ambitions, local matters. *J. Urban. Des.* **2004**, *9*, 335–349. [CrossRef]
21. Temelová, J. Urban revitalization in central and inner parts of (post-socialist) cities: Conditions and consequences. In *Regenerating Urban Core*; Ilmavirta, T., Ed.; Center for Urban and Regional Studies C72, Helsinki University of Technology: Espoo, Finland, 2009; pp. 12–25.
22. Massey, R. Urban Renewal in South African Cities. In *Urban Geography in South Africa*; Massey, R., Gunter, A., Eds.; GeoJournal Library; Springer: Cham, Switzerland, 2020. [CrossRef]
23. Yi, Z.; Liu, G.; Lang, W.; Shrestha, A.; Martek, I. Strategic approaches to sustainable urban renewal in developing countries: A case study of Shenzhen, China. *Sustainability* **2017**, *9*, 1460. [CrossRef]
24. Nzimande, N.P.; Fabula, S. Socially sustainable urban renewal in emerging economies: A comparison of Magdolna Quarter, Budapest, Hungary and Albert Park, Durban, South Africa. *Hung. Geogr. Bull.* **2020**, *69*, 383–400. [CrossRef]
25. Purvis, P. Boston's "Big Dig" History and Reflection of a 15 Year Highway Construction Project. *PSU GEOG* **2019**, *1*, 482. Available online: https://storymaps.arcgis.com/stories/c3c18274fcb348af872822e9f2a1887a (accessed on 10 January 2021).
26. Struct on Green Carpet: Multipurpose Use of Space. Available online: https://strukton.com/en/projects/2019/03/avenue_2 (accessed on 10 January 2021).
27. Bo01, Malmö. Available online: https://www.urbangreenbluegrids.com/projects/bo01-city-of-tomorrow-malmo-sweden/ (accessed on 10 January 2021).
28. Jing, Y. Refining and Reappearance: Shanghai South Bund Revitalization. In Proceedings of the 47th ISOCARP Congress, Wuhan, China, 24–28 October 2011; Available online: http://www.isocarp.net/Data/case_studies/1963.pdf (accessed on 10 January 2021).
29. High Line. Available online: https://www.thehighline.org/ (accessed on 12 January 2021).
30. Przybyła, K.; Kachniarz, M.; Ramsey, D. The investment activity of cities in the context of their administrative status: A case study from Poland. *Cities* **2020**, *97*, 102505. [CrossRef]
31. Dijkstra, L.; Poelman, H. Cities in Europe the New OECD-EC Definition Regional Focus. A Series of Short Papers on Regional Research and Indicators Produced by the Directorate-General for Regional and Urban Policy RF 01/2012. Available online: https://ec.europa.eu/regional_policy/sources/docgener/focus/2012_01_city.pdf (accessed on 12 January 2021).
32. International Network of Cities Where Living is Good. Available online: https://www.cittaslow.org/ (accessed on 12 January 2021).
33. Ekinci, M.B. The Cittaslow philosophy in the context of sustainable tourism development; the case of Turkey. *Tour. Manag.* **2014**, *41*, 178–189. [CrossRef]
34. Mayer, H.; Knox, P.L. Slow Cities: Sustainable Places in a Fast World. *J. Urban. Aff.* **2006**, *28*, 321–334. [CrossRef]
35. Perano, M.; Abbate, T.; La Rocca, E.T.; Casali, G.L. Cittaslow & fast-growing SMEs: Evidence from Europe. *Land Use Policy* **2019**, *82*, 195–203. [CrossRef]
36. Zawadzka, A.K. Making Small Towns Visible in Europe: The Case of Cittaslow Network—The Strategy Based on Sustainable Development. *Transylv. Rev. Adm. Sci.* **2017**, *SI*, 90–106. [CrossRef]
37. Turner, J.; Morrison, A. Designing Slow Cities for More than Human Enrichment: Dog Tales—Using Narrative Methods to Understand Co-Performative Place-Making. *Multimodal. Technol. Interact.* **2021**, *5*, 1. [CrossRef]
38. Zagroba, M.; Szczepańska, A.; Senetra, A. Analysis and Evaluation of Historical Public Spaces in Small Towns in the Polish Region of Warmia. *Sustainability* **2020**, *12*, 8356. [CrossRef]
39. Ozmen, A.; Can, M.C. The urban conservation approach of Cittaslow Yalvaç. *Megaron* **2018**, *13*, 13–23. [CrossRef]
40. Semmens, J.; Freeman, C. The value of cittaslow as an approach to local sustainable development: A New Zealand perspective. *Int. Plan. Stud.* **2012**, *17*, 353–375. [CrossRef]
41. Revitalization of Cittaslow Town Launched. Available online: https://www.cittaslow.org/news/polish-national-network-revitalization-cittaslow-town-launched (accessed on 12 January 2021).
42. Regional Operational Programme of the Voivodeship of Warmia and Mazury 2014–2020. Available online: https://ec.europa.eu/growth/tools-databases/regional-innovation-monitor/policy-document/regional-operational-programme-voivodeship-warmia-and-mazury-2014-2020 (accessed on 10 January 2020).
43. Senetra, A.; Szarek-Iwaniuk, P. Socio-economic development of small towns in the Polish Cittaslow Network—A case study. *Cities* **2020**, *103*, 102758. [CrossRef]
44. Wierzbicka, W.; Farelnik, E.; Stanowicka, A. The development of the Polish National Cittaslow Network. *Olszt. Econ. J.* **2019**, *14*, 113–125. [CrossRef]

45. Wierzbicka, W. Socio-economic potential of cities belonging to the Polish National Cittaslow Network. *Oecon. Copernic.* **2020**, *11*, 203–224. [CrossRef]
46. Farelnik, E. Cooperation of slow cities as an opportunity for the development: An example of Polish National Cittaslow Network. *Oecon. Copernic.* **2020**, *11*, 267–287. [CrossRef]
47. Maćkiewicz, B.; Konecka-Szydłowska, B. Green tourism: Attractions and initiatives of polish Cittaslow cities. In *Tourism in the city*; Bellini, N., Pasquinelli, C., Eds.; Springer: Cham, Switzerland, 2017; pp. 297–309.
48. Supralocal Program of Revitalisation of Cittaslow Towns. Ponadlokalny Program Rewitalizacji Miast Cittaslow. 2019. Available online: https://cittaslowpolska.pl/images/PDF/PPR_08_2019.pdf (accessed on 12 January 2021).
49. Strzelecka, E. Network model of revitalization in the Cittaslow cities of the Warmińsko-Mazurskie Voivodship. *Barometr. Reg.* **2018**, *16*, 53–62.
50. Latham, A.; Layton, J. Social infrastructure and the public life of cities: Studying urban sociality and public spaces. *Geogr. Compass* **2019**, *13*, e12444. [CrossRef]
51. Klinenberg, E. *Palaces for the People: How Social Infrastructure Can Help Fight Inequality, Polarization, and the Decline of Civic Life*; Penguin: London, UK, 2018.
52. Sharp, J.; Pollock, V.; Paddison, R. Just Art for a Just City: Public Art and Social Inclusion in Urban Regeneration. *Urban Studies* **2005**, *42*, 1001–1023. [CrossRef]
53. Bachour, D.W. Socioeconomic impact of urban redevelopment in inner city of Ningbo. *J. Zhejiang Univ. Sci. A* **2006**, *7*, 1386–1395. [CrossRef]
54. Alpopi, C.; Manole, C. Integrated Urban Regeneration—Solution for Cities Revitalize. *Procedia Econ.* **2013**, *6*, 178–185. [CrossRef]
55. Zagroba, M. Issues of the Revitalization of Historic Centres in Small Towns in Warmia. *Procedia Eng.* **2016**, *161*, 221–225. [CrossRef]
56. Kristianova, K.; Jaszczak, A. Historical Centers of Small Cities in Slovakia—Problems and Potentials of Creating Livable Public Spaces. *IOP Conf. Ser. Mater. Sci. Eng.* **2020**, *960*, 022012. [CrossRef]
57. Kramarova, Z.; Kankovsky, A. Mobility in Public Spaces of Small Towns in the Czech Republic. *IOP Conf. Ser. Mater. Sci. Eng.* **2020**, *960*, 042090. [CrossRef]
58. Evans-Cowley, J. Sidewalk Planning and Policies in Small Cities. *J. Urban Plan. Dev.* **2006**, *132*, 71–75. [CrossRef]
59. Hu, H.; Xu, J.; Shen, Q.; Shi, F.; Chen, Y. Travel mode choices in small cities of China: A case study of Changting. *Transp. Res. D Transp. Environ.* **2018**, *59*, 361–374. [CrossRef]
60. Handy, S.L.; Heinen, E.; Krizek, K. Cycling in small cities. In *Cycling for Sustainable Transport: International Trends and Policies*; Pucher, J., Buehler, R., Eds.; MIT Press: Cambridge, MA, USA, 2012.
61. Xing, Y.; Handy, S.; Mokhtarian, P. Factors associated with proportions and miles of bicycling for transportation and recreation in six small us cities. *Transp. Res. D* **2010**, *15*, 73–81. [CrossRef]
62. Audikana, A.; Ravalet, E.; Baranger, V.; Kaufmann, V. Implementing bikesharing systems in small cities: Evidence from the Swiss experience. *Transp Policy (Oxf)* **2017**, *55*, 18–28. [CrossRef]
63. Jaszczak, A.; Morawiak, A.; Zukowska, J. Cycling as a sustainable transport alternative in polish cittaslow towns. *Sustainability* **2020**, *12*, 5049. [CrossRef]
64. Gough, M.Z. Reconciling Livability and Sustainability: Conceptual and Practical Implications for Planning. *J. Plan. Educ. Res.* **2015**, *35*, 145–160. [CrossRef]
65. Ruth, M.; Franklin, R.S. Livability for all? Conceptual limits and practical implications. *Appl Geogr* **2014**, *49*, 18–23. [CrossRef] [PubMed]
66. Kovacs-Györi, A.; Cabrera-Barona, P.; Resch, B.; Mehaffy, M.; Blaschke, T. Assessing and Representing Livability through the Analysis of Residential Preference. *Sustainability* **2019**, *11*, 4934. [CrossRef]
67. Van Kamp, I.; Leidelmeijer, K.; Marsman, G. Urban environmental quality and human well-being: Towards a conceptual framework and demarcation of concepts; a literature study. *Landsc. Urban Plan.* **2003**, *65*, 5–18. [CrossRef]
68. Kothencz, G.; Kolcsár, R.; Cabrera-Barona, P.; Szilassi, P. Urban green space perception and its contribution to well-being. *Int. J. Environ. Res. Public Health* **2017**, *14*, 766. [CrossRef]
69. Jančová, N. New approaches for research public spaces and urban green infrastructure in the context of livable urban environment. In Proceedings of the INTED 2019 Conference Proceedings, Valencia, Spain, 11–13 March 2019; pp. 3829–3832.
70. Dushkova, D.; Haase, D. Not Simply Green: Nature-Based Solutions as a Concept and Practical Approach for Sustainability Studies and Planning Agendas in Cities. *Land* **2020**, *9*, 19. [CrossRef]
71. Van der Jagt, A.P.N.; Raven, R.; Dorst, H.; Runhaar, H. Nature-based innovation systems. *Environ. Innov. Soc. Transit.* **2020**, *35*, 202–216. [CrossRef]
72. Shchur, A.; Lobikava, N.; Lobikava, V. Revitalization of (Post-) Soviet Neighbourhood with Nature-Based Solutions. *AHR* **2020**, *23*, 76–80. [CrossRef]
73. Frantzeskaki, N. Seven lessons for planning nature-based solutions in cities. *Environ. Sci. Policy* **2019**, *93*, 101–111. [CrossRef]
74. Zwierzchowska, I.; Fagiewicz, K.; Poniży, L.; Lupa, P.; Mizgajski, A. Introducing nature-based solutions into urban policy—facts and gaps. Case study of Poznań. *Land Use Policy* **2019**, *85*, 161–175. [CrossRef]
75. Ordóñez, C.; Grant, A.; Millward, A.A.; Steenberg, J.; Sabetski, V. Developing Performance Indicators for Nature-Based Solution Projects in Urban Areas: The Case of Trees in Revitalized Commercial Spaces. *CATE* **2019**, *12*, 1.
76. Hughes, S.; Chu, E.K.; Mason, S.G. *Climate Change in Cities. Innovations in Multi-Level Governance*; Springer: Cham, Switzerland, 2018.

77. Krauze, K.; Wagner, I. From classical water-ecosystem theories to nature-based solutions—Contextualizing nature-based solutions for sustainable city. *Sci. Total Environ.* **2019**, *655*, 697–706. [CrossRef]
78. Osei, G.; Pooley, A.; Pascale, F. A Community-Driven Nature-Based Design Framework for the Regeneration of Neglected Urban Public Spaces. In *Sustainable Ecological Engineering Design*; Scott, L., Dastbaz, M., Gorse, C., Eds.; Springer: Cham, Switzerland, 2020. [CrossRef]
79. Kabisch, N.; Korn, H.; Stadler, J.; Bonn, A. *Nature-Based Solutions to Climate Change Adaptation in Urban. Areas—Linkages between Science, Policy and Practice*; Springer Open: Cham, Switzerland, 2017.
80. Sanchez Rodriguez, R.; Ürge-Vorsatz, D.; Barau, A.S. Sustainable Development Goals and climate change adaptation in cities. *Nat. Clim. Chang.* **2018**, *8*, 181–183. [CrossRef]
81. Landauer, M.; Juhola, S.; Klein, J. The role of scale in integrating climate change adaptation and mitigation in cities. *J. Environ. Plan. Manag.* **2019**, *62*, 741–765. [CrossRef]
82. Shade, C.; Kremer, P.; Rockwell, J.S.; Henderson, K.G. The effects of urban development and current green infrastructure policy on future climate change resilience. *Ecol. Soc.* **2020**, *25*, 37. [CrossRef]
83. Stagrum, A.E.; Andenæs, E.; Kvande, T.; Lohne, J. Climate Change Adaptation Measures for Buildings—A Scoping Review. *Sustainability* **2020**, *12*, 1721. [CrossRef]
84. Roggema, R. *Adaptation to Climate Change: A Spatial Challenge*; Springer: Dordrecht The Netherlands, 2009; pp. 211–251.
85. Chowdhury, M.B.H. Coupling of Economic Benefits and Climate Change Adaptation: Reflection from a Small Municipality Development Plan. *WIT Trans. Ecol. Environ.* **2020**, *241*, 431–441. [CrossRef]
86. Dannenberg, A.L.; Frumkin, H.; Hess, J.J.; Ebi, K.L. Managed retreat as a strategy for climate change adaptation in small communities: Public health implications. *Clim. Chang.* **2019**, *153*, 1–14. [CrossRef]
87. Picketts, I.M.; Déry, S.J.; Curry, J.A. Incorporating climate change adaptation into local plans. *J. Environ. Plan. Manag.* **2014**, *57*, 984–1002. [CrossRef]
88. Major, D.C.; Juhola, S. Guidance for climate change adaptation in small coastal towns and cities: A new challenge. *J. Urban. Plan. Dev.* **2016**, *142*, 2516001. [CrossRef]
89. Bausch, T.; Koziol, K. New Policy Approaches for Increasing Response to Climate Change in Small Rural Municipalities. *Sustainability* **2020**, *12*, 1894. [CrossRef]
90. Kalbarczyk, E.; Kalbarczyk, R. Typology of Climate Change Adaptation Measures in Polish Cities up to 2030. *Land* **2020**, *9*, 351. [CrossRef]
91. Paiho, S.; Mäki, E.; Wessberg, N.; Paavola, M.; Tuominen, P.; Antikainen, M.; Heikkilä, J.; Rozado, C.A.; Jung, N. Towards circular cities—Conceptualizing core aspects. *Sustain. Cities Soc.* **2020**, *59*, 102143. [CrossRef]
92. Williams, J. Circular cities. *Urban Stud.* **2019**, *56*, 2746–2762. [CrossRef]
93. Pomponi, F.; Moncaster, A. A Theoretical Framework for Circular Economy Research in the Built Environment. In *Building Information Modelling, Building Performance, Design and Smart Construction*; Dastbaz, M., Gorse, C., Moncaster, A., Eds.; Springer: Cham, Switzerland, 2017. [CrossRef]
94. Bassens, D.; Kębłowski, W.; Lambert, D. Placing cities in the circular economy: Neoliberal urbanism or spaces of socio-ecological transition? *Urban. Geogr.* **2020**, *41*, 893–897. [CrossRef]
95. Petit-Boix, A.; Leipold, S. Circular economy in cities: Reviewing how environmental research aligns with local practices. *J. Clean. Prod.* **2018**, *195*, 1270–1281. [CrossRef]
96. Zeller, V.; Towa, E.; Degrez, M.; Achten, W.M.J. Urban waste flows and their potential for a circular economy model at city-region level. *Waste Manag.* **2019**, *83*, 83–94. [CrossRef]
97. Sołtysik, M.; Mazur-Belzyt, K. City Space Recycling: The Example of Brownfield Redevelopment. *IOP Conf. Ser. Mater. Sci. Eng.* **2020**, *960*, 042016. [CrossRef]
98. Basova, S.; Sopirova, A.; Kristianova, K. Potential of Recycling Urban Territories. *IOP Conf. Ser. Mater. Sci. Eng.* **2019**, *471*, 092053. [CrossRef]
99. Szaja, M. Social Aspects of Revitalization of Urban Public Spaces. *EJSM* **2018**, *28*, 463–469.
100. Polatoğlu, Ç.; Çıtak, C.; Kararmaz, Ö. Enhancing Socio-cultural and Socio-economic Values with Architectural Design—Revitalization of Place. In Proceedings of the 14th International Conference Design Principles and Practices, Advocacy in Design: Engagement, Commitment and Action, New York, NY, USA, 11–13 November 2020; pp. 1–10.
101. Jayne, M.; Gibson, C.; Waitt, G.; Bell, D. The Cultural Economy of Small Cities. *Geogr. Compass* **2010**, *4*, 1408–1417. [CrossRef]
102. Lorentzen, A. Sustaining small cities through leisure, culture and the experience economy. In *Cultural Political Economy of Small Cities*; Lorentzen, A., van Heur, B., Eds.; Routledge: Oxon, UK, 2013; pp. 65–79.
103. Środa-Murawska, S.; Biegańska, J.; Dąbrowski, L. Perception of the role of culture in the development of small cities by local governments in the context of strategic documents—A case study of Poland. *Bull. Geogr. Socio-Econ. Ser.* **2017**, *38*, 119–130. [CrossRef]
104. Sepe, M. Urban transformation, socio-economic regeneration and participation: Two cases of creative urban regeneration. *Int. J. Urban. Sustain. Dev.* **2014**, *6*, 20–41. [CrossRef]
105. Środa-Murawska, S.; Szczepańska, A.; Biegańska, J.; Senetra, A. The phenomenon of non-governmental organizations—new stimuli for cultural development in rural areas in Poland. In Proceedings of the Annual 20th ISC Research for Rural Development, Jelgava, Latvia, 21–23 May 2014; pp. 229–236.
106. Tylman, A. Participatory Revitalisation as the Determinant of Changes in Urban Policy Financing. *Entrep. Manag.* **2017**, *8*, 345–355.

107. Revitalization Act of 9th of October 2015. Available online: https://www.infor.pl/akt-prawny/DZU.2015.215.0001777,ustawa-o-rewitalizacji.html (accessed on 15 December 2020).
108. Jaszczak, A. Przyszłość miast Cittaslow. *Archit. Kraj.* **2015**, *1*, 70–81.
109. Social and Economic Strategy of the Warmińsko-Mazurskie Voivodeship until 2015. Available online: https://bip.warmia.mazury.pl/3/strategia-rozwoju-spoleczno-gospodarczego-wojewodztwa-warminsko-mazurskiego.html (accessed on 10 December 2020).
110. Lordkipanidze, M.; Brezet, H.; Backman, M. The entrepreneurship factor in sustainable tourism development. *J. Clean. Prod.* **2005**, *13*, 787–798. [CrossRef]
111. Wise, N. Outlining triple bottom line contexts in urban tourism regeneration. *Cities* **2016**, *53*, 30–34. [CrossRef]
112. Roundy, P.T. Back from the brink: The revitalization of inactive entrepreneurial ecosystems. *J. Bus. Ventur. Insights* **2019**, *12*, e00140. [CrossRef]
113. Sinkienė, J.; Kromalcas, S. Concept, directions and practice of city attractiveness improvement. *Viešoji Politika ir Administravimas (Public Policy and Administration)* **2010**, *31*, 147–154.
114. Alvarez, M.D. Creative cities and cultural spaces: New perspectives for city tourism. *Int. J. Cult. Tour. Hosp. Res.* **2010**, *4*, 171–175. [CrossRef]
115. Kusiak, J. Revitalizing urban revitalization in Poland: Towards a new agenda for research and practice. *UDI* **2019**, *63*, 17–23. [CrossRef]
116. Parysek, J.J. Urban Revitalization in Poland: Problems, Dilemmas, Challenges and Hopes. *Studia Reg.* **2017**, *50*, 103–121. [CrossRef]
117. Szulc, T. Organisational aspects of execution of the local revitalisation programmes in the largest cities in Poland—indication of directions of further research. In *Systems Supporting Production Engineering*; Review of problems and solutions; Kaźmierczak, J., Ed.; Wydawnictwo Pa Nova: Gliwice, Poland, 2014; pp. 91–100.
118. Belniak, S. A partnership of public and private sectors as a model for the implementation of urban revitalization projects. *J. Eur. Real Estate Res.* **2008**, *1*, 139–150. [CrossRef]
119. Babis, H.; Janowski, M. Revitalisation of problem areas as an instrument for social and economic activity in the Polish municipalities. *EJSM* **2018**, *28*, 17–25. [CrossRef]
120. Lafortezza, R.; Chen, J.; Konijnendijk van den Bosch, C.; Randrup, T.B. Nature-based solutions for resilient landscapes and cities. *Environ. Res.* **2018**, *165*, 431–441. [CrossRef]
121. Farelnik, E. Determinants of the development of slow cities in Poland. *Prace Naukowe Uniwersytetu Ekonomicznego we Wrocławiu* **2020**, *64*, 18–36. [CrossRef]
122. Jaszczak, A.; Kristianova, K. Social and Cultural Role of Greenery in Development of Cittaslow Towns. *IOP Conf. Ser. Mater. Sci. Eng.* **2019**, *603*, 032028. [CrossRef]
123. Jaszczak, A.; Kristianova, K.; Sopirova, A. Revitalization of public space in small towns: Examples from Slovakia and Poland. *Zarządzanie Publiczne* **2019**, *45*, 35–46. [CrossRef]

Article

Urban Development and Population Pressure: The Case of Młynówka Królewska Park in Krakow, Poland

Karolina Dudzic-Gyurkovich

Faculty of Architecture, Cracow University of Technology, 31-155 Kraków, Poland; kdudzic-gyurkovich@pk.edu.pl

Abstract: Green areas are necessary components of contemporary cities. They have a positive impact on the climate, ecological balance and resilience of the city structure, and provide numerous benefits to inhabitants. However, progressing urbanisation and a rise in urban population leads to increasing pressure on existing green spaces. Since the beginning of the twenty-first century, but especially over the past decade, the number of new developments in Krakow has been successively increasing. New multi- and single-family housing complexes are built not only in peripheral areas, but also as infilling and development of areas closer to the historical city centre. Simultaneously, the number of urban green spaces has increased only insignificantly. This paper analyses the example of Młynówka Królewska Park, located in the western part of Krakow, Poland. It focuses on city expansion processes and their relationship with public greenery. Furthermore, his study examines spatial and demographic issues that may have an impact on accessibility of the park area. The study was based on statistical data, analyses of recent planning documents, procedures and practices, as well as research performed in situ. The results indicate that the population pressure in the area under analysis changed significantly. According to the results, a correlation can be found between the latest urban development and population dynamics.

Keywords: Krakow; urban green areas; green infrastructure; population dynamics; urban planning; urbanisation

Citation: Dudzic-Gyurkovich, K. Urban Development and Population Pressure: The Case of Młynówka Królewska Park in Krakow, Poland. *Sustainability* **2021**, *13*, 1116. https://doi.org/10.3390/su13031116

Academic Editors: Jan K. Kazak, Katarzyna Hodor and Magdalena Wilkosz-Mamcarczyk
Received: 17 December 2020
Accepted: 16 January 2021
Published: 21 January 2021

1. Introduction

Since the mid-twentieth century, urbanisation processes have shown an unprecedented dynamic. International reports indicate a global shift between urban and rural areas, which, combined with demographic growth, can increase the share of urban communities in the global populations up to 68% in 2050. In Europe, 75% of the population is urban and it is estimated that this value is to increase [1]. Relatively speaking, the least urbanised of the world's regions is Africa, yet its urban population makes up over 40% of the total [2]. The global rise in the significance of cities implicates economic, social and spatial shift, while also contributing to climate change, natural resource consumption and the possibility of implementing sustainable development, which is the main concern of global development policy [3,4].

Living in the city can lead to increased exposure to adverse factors such as air, water and noise pollution, as well as extensive traffic. They all decrease quality of life and negatively impact human health. It is possible to address these tendencies, for instance by enabling contact with natural elements of the environment. The link between physical and psychological health and the use of green areas has been studied at length [5–8]. Attractive green areas were found to encourage physical activity [9–12], contribute to lowering stress levels [13] and also lead to an evening out of social inequities by creating places for meetings and social interactions [14]. Green areas were also found to contribute to general wellbeing in society by providing a reconnection to nature [7,13,14] and aesthetic satisfaction [15].

Urban green and blue spaces are currently a major element that conditions sustainable urban development. They were stated to be necessary to restoring and maintaining ecological balance [16], improve air quality [17] or improve city resilience to climate change [18,19]. Parks, gardens, greens and green corridors, as well as other forms of greenery like household gardens, green roofs or sports areas are a part of larger greenery systems described as green infrastructure. Green infrastructure is perceived to be equal to grey infrastructure in terms of acting as strategic components of the structure of cities and regions [20].

Progressive urbanisation is a challenge to green area management. In developing countries, open areas are being developed and cities continue to expand. Population growth is often highly dynamic. Processes that take place in urbanised areas, based on the densification of existing urban structures, also contribute to an increase in the urban population that makes use of green areas [21,22]. Many cities are under pressure to balance the potential conflicts of land use between green development and the growing housing demands. As a result, access to green areas is not equal and equitable to all city residents [8,23,24]. It can also be hindered or limited due to an insufficient amount of such areas, their low quality or uneven distribution within the city structure [25].

1.1. Related Works

Some scholars indicate that the number and size of urban green areas is a product of their geographical location [26,27]. In the conducted research, the cities of the European continent were taken into account. General conclusions point that cities located in the north and centre of the continent typically have more green areas than southern cities. European Union statistics that display the percentage share of green areas in the areas of cities in Member States confirm this conclusion only to a certain degree. Tallin in Estonia has a share of green urban areas of 36.7%, while Ljubljana in Slovenia, which is much further south, has a share of 72.5% of such areas. EEA data state that most Polish cities have a share of green spaces 50–65% [21].

A study conducted by Fuller and Gaston indicated that the amount of greenery is also affected by urban form, and cities with compact, dense fabric have a lower amount of green area per capita [28]. After studying 386 European cities, they demonstrated that this indicator ranges from ca. 3 to 4 m^2 per capita in cities in countries such as Spain or Italy, while it reached much higher values—i.e., 300 m^2 per capita—in certain cities in Belgium and Finland. Kabisch et al. showed that the share of the European urban population that lives within 500 m and 300 m of a green area of a size at least 2 ha is varied and ranges between 11% and 98% [29].

In recent years, the problem of green urban space accessibility has been discussed in a considerable amount of global, European and Polish academic publications. The literature review performed by Rigolon shows that three basic types of variables to evaluate access to urban parks are most commonly used:

1. Proximity refers to physical geographical distance in metres or minutes of walk;
2. Quantity tells how many parks, or hectares of parks, are within reach; and it can be measured in park area per capita or total park area ratio;
3. Quality refers to design issues, such as composition, furniture, facilities, as well as maintenance level [27].

Describing accessibility issues can be approached through network analyses such as Space-Syntax, where accessibility is associated with connections of lines or streets and is highly dependent on city layout [30,31]. Other related network-based approaches include centrality, which explains, among others, that 'some places are more important than others because they are more central' [32] (p.3). Centrality measures are applied mostly in the urban context. However, interesting findings concerning a mixed urban–rural environment were provided by a study conducted by Pérez-Campaña et al. [33], where several abandoned and unused sites were identified as those of high centrality values.

A study of green area accessibility was performed for voivodeship capital cities in Poland, and revealed that most residents do not have access to a minimum 2 ha green spaces within 300 m walking distance from their homes. The same study discusses the matter of spatial barriers such as rivers or railway lines that require additional effort and distance to cross [34]. Detailed studies of park accessibility in Gdańsk, Poland, conducted by Korwel-Lejkowska and Topa indicate that almost 78% of the city's area is outside of the accessibility zone of city parks as outlined by a travel distance lower than 20 min [35].

The abovementioned observations contribute to the development of spatial policy goals associated with the amount and accessibility of green areas in cities. In Europe, ensuring the proper amount and quality of urban green space has become a major development policy goal [2,36]. For instance, the European Environment Agency recommends a distance from green areas that is no greater than a fifteen minutes' walk [37,38]. The EEA's recommendations are nevertheless not obligatory, which results in adopting different indicators and values depending on local conditions and applicable strategies.

Insofar as all components that form green infrastructure are essential to sustainable development of cities and societies, urban parks are one of the most valuable and attractive [39]. Parks fulfil a broad range of roles: those of urban recreational areas serve as places of physical activity of varying degrees of intensity, such as walking, dog walking, jogging, cycling, hiking or even outdoor fitness [40], as well as allow people to come into direct contact with natural elements of the environment [41]. They occupy an important place in the network of public spaces and induce social contact among visitors. Previous studies have suggested that the features, attractiveness and composition of park space encourage its use and increase visitor count [12,42–44].

1.2. Research Question

Numerous studies have explored the problems of the quantity and accessibility of urban green areas from a territorial perspective. They largely focus on the distribution of such areas in a specific spatial and physical structure and focus on large scales: those of the city, region or country [27,28,34,43,45]. In reference to the local scale, studies from a human-based perspective predominate and refer to age or social groups [46], as well as the subjective perception of green spaces [7,47,48]. The vast majority of studies on urban green space accessibility and forms of use primarily focus on the problem of their insufficient quantity or difficulties in access to them, which can limit the number of persons willing to visit them. However, there is also an opposite problem—that of an excessive number of users which leads to the overcrowding of the most popular recreational spaces. As argued by A. Arnberger, in case of more sensitive persons, this can lead to discomfort and, as consequence, to less frequent visits [49,50].

Against this background, insufficient knowledge can be identified concerning local spatial and societal determinants that can affect accessibility and use of specific urban green areas. There is also a lack of such reports on Polish conditions. This study addresses the existing gap by studying the case of the Młynówka Królewska linear park in Krakow, Poland in terms of transformation its surrounding urban space over the past ten years. The aim of the study is to answer the question of whether there exists a link between tendencies in urban development, population dynamics, and the use of urban parks. The background for this study is provided by a review of municipal policy and planning frameworks towards urban greenery.

The research objectives are as follows:

1. Determine whether the built-up area in the proximity of the park has changed since 2009;
2. Examine whether a correlation exists between those changes and the general accessibility of the park;
3. Estimate the population pressure on the park area and its dynamics in the last decade.

2. Materials and Methods

2.1. Study Area Selection

2.1.1. Krakow, Poland

The study was conducted in Krakow, which is the second-largest Polish city, capital of the Lesser Poland Voivodeship and an important cultural centre. It is located in the southern part of the country, in the Vistula River valley (Figure 1). The city's area within its administrative limits is 337 km^2 and its population is close to 800 thousand. Its urban core was established in the High Middle Ages, and is currently one of the most valuable historical urban and architectural complexes, inscribed on the World Heritage List. Many of the more distant districts of the city developed around previously existing settlements that were incorporated into the city. As a result, since the beginning of the twentieth century, Krakow's area has increased by 320 km^2 [51]. Processes of Krakow's urbanisation were associated with shifts in geopolitical and economic situation, which resulted in complex and non-uniform city structure [52]. In order to get a broader perspective of the spatial context of the area in question, a brief overview of the development of Krakow will be presented in the following section.

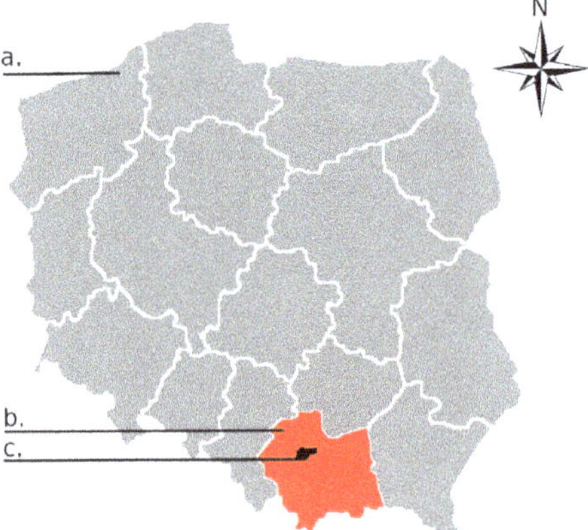

Figure 1. Map showing location of Krakow. (**a**) map of Poland with major administrative borders; (**b**) Lesser Poland Voivodeship; (**c**) Location of Krakow.

2.1.2. Development of the City Structure

Documented urban development of Krakow dates back to the tenth century, although archaeological excavations indicate that some forms of settlement were present in the river bend of the River Vistula centuries prior [51,53]. In the year 1257, the urban layout of the city was defined as a consequence of a town charter. Its centrally located market square and a regular grid of diverging streets are still present and visible in the city's structure. In the thirteenth century, a new system of fortifications was erected, with defensive walls, towers and fortified entrance gates gradually developed and modified in the following centuries. The first ring of fortifications determined and restricted the spatial development in subsequent centuries. The administrative borders of the city had not changed substantially up to the eighteenth century, which can also be associated with the geopolitical situation and partition of Poland between neighbouring countries.

The extension of the city limits in 1910 was an important step towards Krakow becoming a modern city. The so-called Greater Krakow was created by incorporating the surrounding villages and outskirts. The area of the city in the beginning of the twentieth century was near 50,165.3 km^2. However, the inhabitants were concentrated mostly in the historical centre. Krakow existed in this form until the Second World War, when it became the capital of the administrative-territorial unit called the General Governorate under Nazi German occupation. This 'capital city' on the one hand and as the perception of Krakow as 'an ancient, Germanic city' by the occupying forces on the other led to a strategy oriented towards development rather than exploitation [53]. In 1941, a decision was made to significantly extend the area of the city. The borders were expanded the and the city area increased more than threefold reaching 165.3 km^2, yet the population did not exceed 285 thousand [51,54].

The incorporation of the surrounding villages in 1951 had a substantial influence on the post-war development of Krakow. In their place, an industrial district was created with a metalworking complex and the housing estate. Nowa Huta, built in the Socialist Realist style, was a separate city for some time and has since become a district of Krakow [51]. During the last decades of the twentieth century, the area of the city was slowly reaching its contemporary size. However, the amount of built-up area was substantially smaller (Figure 2). Breakthrough moments in the modern city's development can be associated with the processes of extending its administrative borders by incorporating and urbanising surrounding territorial units. Consequently, the population increased, although the demographic changes were also driven by several other factors such as: fertility and mortality, migration patterns and economy (Table 1) [55].

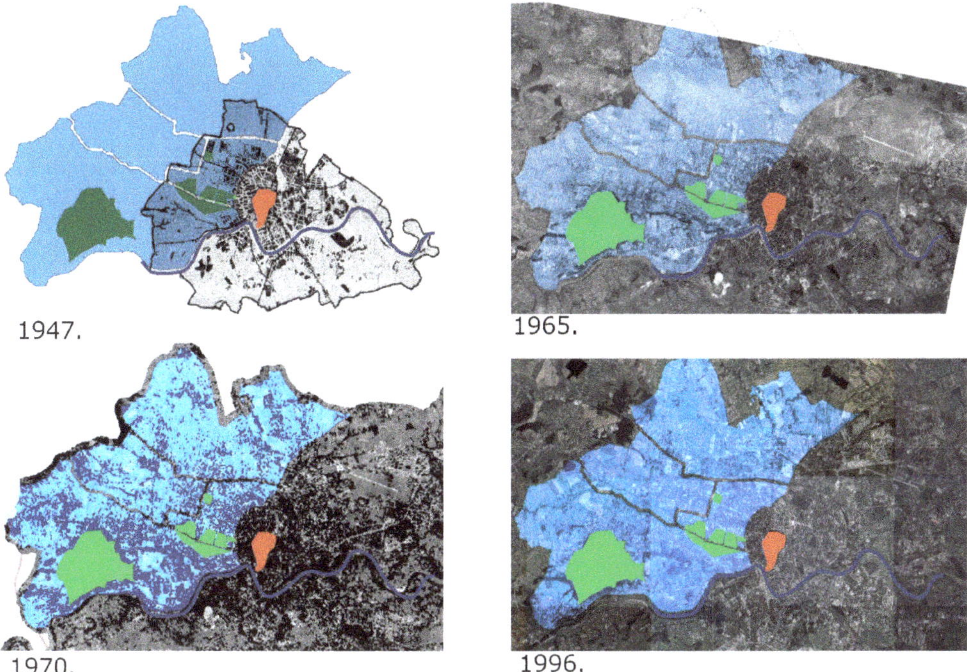

Figure 2. Map showing the western districts of the city in several stages of urbanization in the twentieth century. Figure based on cartographic materials available at the Office of the City of Krakow (UMK) webpage.

The end of the turn of the twenty-first century has been a period of dynamic urbanisation both of the city and its vicinity. Suburbanisation processes and urban sprawl are particularly visible, yet the city itself is also a subject to major development growth [56,57]. According to data published yearly by the Office of the City of Krakow (UMK) [58], the built-up area of the city increased significantly during the last decade, reaching the value of 9767 ha (Table 2).

Table 1. Table illustrating the main steps of changes in the area and population of Krakow 1915–2013. Table modified from UMK [54].

Year	Area of the City after Expansion/ha	Population of the City
1915	4690	183,000
1941	16,530	285,000
1951	22,990	355,000
1973	32,230	657,276
1986	32,680	743,652
2013	32,685	758,334

Table 2. Data on built-up area of Krakow 2009–2018 as presented in annual publications of UMK [58].

Year	Residential, Industrial and Other Built-Up Area/ha	Percentage of the City Area
2009	5964	18.24%
2010	6106	18.67%
2011	6246	19.10%
2012	8940	27.34%
2013	9641	29.48%
2014	9798	29.96%
2015	9406	28.76%
2016	9535	29.16%
2017	9535	29.16%
2018	9767	29.87%
increase	3803	

Krakow possesses several dozen public parks. They are located on municipal grounds and administered by the city. Apart from inner-city greenery, the presence of natural reserves and landscape parks in the surroundings of the city must be noted. The western part of the city of Krakow lies within a wedge of the Bielańsko-Tyniecki Landscape Park, which stretches almost to the city centre and covers several natural areas and forest parks. To the north is Tyniecki Landscape Park (Figure 3). The landscape parks are managed by the Complex of Landscape Parks at Voivodship level (ZPKWM) [59].

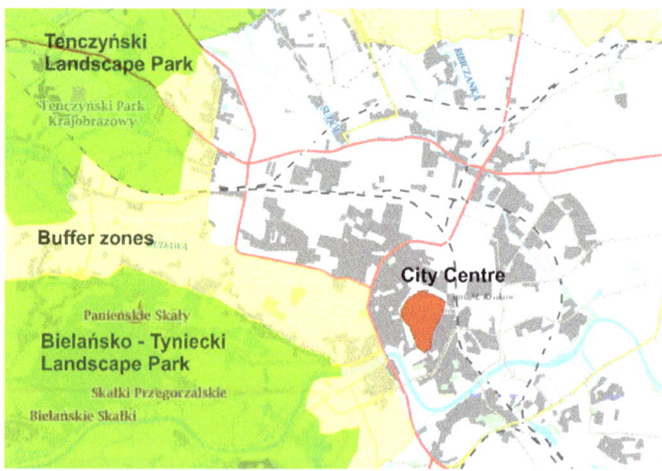

Figure 3. Map showing the landscape parks within the borders of the city in its western districts. Based on cartographic materials available at the Complex of Landscape Parks at Voivodship (ZPKWM) webpage [59].

2.1.3. Młynówka Królewska Park

The Młynówka Królewska Park is a unique example of a linear park, built along a canalized waterway of the same name. It runs through Krakow's western districts: Krowodrza and Bronowice, along a distance of over 8 km—stretching from the city centre to its western border. The transverse dimension of the park varies depending on the section and often is constrained by surrounding urban tissue. The Młynówka River was created in the thirteenth century as a canal that transported water from the distant Rudawa River to Krakow. The canal and the engineering structures, was of strategic significance, as it was the main source of utility water that filled moats and fish ponds, flowed in municipal fountains, powered water mills and was used in leatherworking shops, malt houses and slaughterhouses. The medieval water supply system operated up to the nineteenth century, yet it lost its significance along with the development of industry. In the following years, its successive sections were gradually filled in. Fragments of the riverbed were present in the landscape of the western part of the city up to as late as the middle of the twentieth century [51].

In 1995, a decision was made to establish a city park along the site of the Młynówka, which was to act a protective measure for an area of high cultural and landscape value. At present, Młynówka Królewska Park is an attractive pedestrian and cyclist destination equipped with recreational infrastructure, including seating, deck chairs, playgrounds, gyms and dog parks. It is a popular place for walking, cycling, jogging and social interactions. Młynówka Królewska Park is an attractive place of recreation and a component of the urban heritage of the city (Figure 4).

Figure 4. The park is attractive, well maintained and provides a range of recreational services.

It is also a part of the existing and planned system of green areas. Urban river parks are receiving more attention worldwide, as they reduce the heat island effect and flood risk, support biodiversity and provide city dwellers with places for recreation [60]. Młynówka Królewska, although partially covered and paved, is classified as a river park, of which there are several in Krakow. The most prominent is organised along the main river of the city—the Vistula. According to municipal policy, they are meant to act as the structural axes of the greenery system and form natural ecological corridors that connect smaller green areas or lead to green complexes of regional significance [61].

Along its length of 8 km, the park's character gradually changes, as does its surrounding development. The following urban patterns can be identified: compact nineteenth- and twentieth-century urban blocks, urban villas, mass housing from the 1970s and 80s, new dense multi-family housing, detached and single-family houses, as well as office buildings and complexes (Figure 5).

2.2. Revision of City of Krakow Policy towards Urban Green Space

In light of currently applicable regulations, Poland does not have cohesive legislation concerning public green space area per capita or per territorial unit. There is also a lack of codified and widely applicable urban planning standards that would define such matters [62]. The only binding regulations concerning green space area are listed in the Ordinance of the Minister of Infrastructure of 12 April 2002 on the technical conditions to be met by buildings and their placement [63]. However, the ordinance only refers to buildable plots and determines the minimum amount of biologically active surface area that must be provided when carrying out real estate development projects, and as such does not apply to urban public green spaces [61,64].

The last document to be binding in this respect was the standard passed in 1974 which, among other things, regulated the amount of green areas per urban unit, type of development and population count. The minimum values for green and recreational spaces were at 25–30 m^2 per capita, at a distance no greater than 800 m from a place of residence [64]. After political transformation towards democracy and the adoption of a market-based economy that Poland underwent in 1989, most ordinances were repealed. The previous centralised planning structure was dismantled and currently most competencies concerning planning are in the hands of municipal governments that make decisions concerning land use and development. While doing so, municipal authorities determine dependencies between the number of developed areas and urban public green spaces. At the same time, the restoration of private property has drastically altered the conditions of carrying out real estate development projects, and the necessity of intensive land use to ensure economic profitability of a project has led to the phenomenon of development pressure on, among others, urban green spaces [65,66].

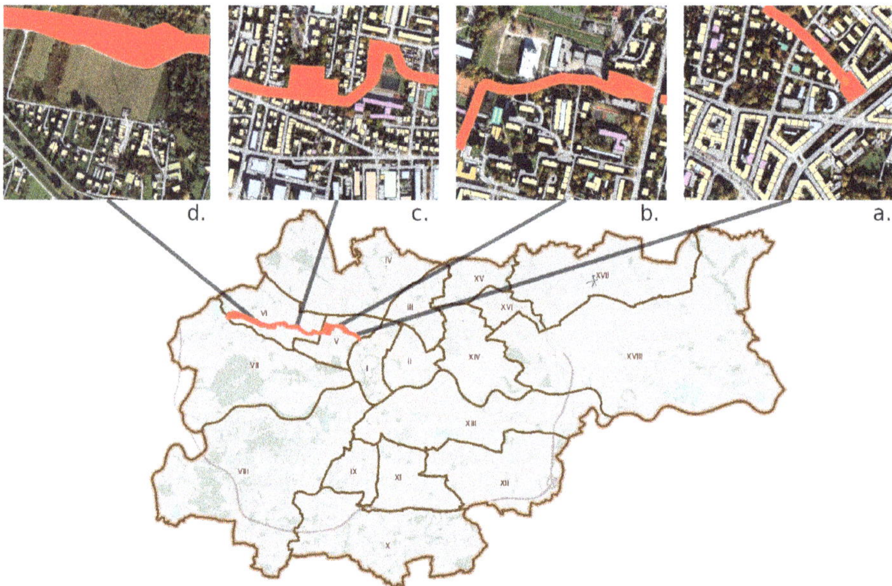

Figure 5. The location of Młynówka Królewska Park in Krakow and spatial patterns of the neighbourhood: (**a**) historical city blocks and urban villas; (**b**) mass housing estates; (**c**) low density multi-family housing, office complexes; (**d**) single-family housing. The location of the park has been marked in red.

In the present planning model, the main document that a municipality is obligated to draft is the Study of Conditions and Directions of Spatial Development (SUiKZP), which includes provisions concerning land use [67]. The SUiKZP, which is drafted for the entirety of a municipality's territory, assigns areas for housing and other uses, green space systems, the course of transport routes and strategic infrastructural paths. The study is not an act of local law and is merely an instrument of spatial policy implemented at the municipal level [68,69]. The document that directly affects land use and development is the local spatial development plan (MPZP), which is drafted for selected fragments of municipalities. Local plans cannot infringe on the provisions of the SUiKZP. However, the preparing of local plans is not obligatory and municipal governments are given significant freedom in the choice of areas plans are to be drafted for. In Krakow, in the second half of 2020, there were 384 valid local plans, and another 97 plans were being prepared. This situation is not stable as plans are often repealed in part or in whole, and new ones are prepared in their place.

In a situation when a given area does not have a local plan in place, the carrying out of any project can be based on an administrative decision that enacts development conditions (WZ). Such a decision is issued at the request of an interested party. It is a parallel path, but suffers from serious faults, such as its fragmentality (it applies only to a specific building plot), susceptibility to influence by the particular interests of parties or its inability to shape spatial order [70]. The legal basis for issuing WZ decisions does not obligate institutions to make them compliant with the SUiKZP, which leads to the development of areas that are not assigned for development and the uncontrolled spread and dispersion of urban structures. This procedure was intended to apply to the short period prior to the enactment of a local plan. However, it has grown to become a fully fledged tool for shaping new development. In 2019, 1097 WZ decisions were issued in Krakow, of which 350 pertained to the construction of multi-family residential buildings or complexes.

Under these conditions, the only effective means of protecting green spaces from development is by appropriate local plan provisions. In this procedure, the maintenance and protection of existing parks can be ensured, as well as the establishment of new urban public green areas. In 2016, the drafting of MPZPs for 215 of Krakow's wildlife areas began (over 10% of the city's territory). Green areas indicated in the SUiKZP and previously not covered by local plans have had such plans enacted. The objective of this endeavour was to protect green spaces and put a stop to growing development pressure that had been identified as a major threat to sustainable greenery management in the city. Other stages of the plan are being gradually prepared and implemented.

Another, indirect form of controlling development processes should be mentioned here. Areas within the borders of a landscape park are under legal protection due to their natural, historical and cultural value, although the level of protection is significantly lower than in the case of national parks or national preserves [71]. Some forms of protection also apply to buffer zones delimited in protection plans. Such plans are drafted for each park individually and are oriented towards the sustainable use of the area in terms of protecting rare species and conservation the natural elements of the environment. Thus, the placement and form of new development are regulated to a certain degree, as the protection plan provisions are included in the MPZP, or serve as an act of local law [71].

A vision of further development of green spaces is presented in the Directions of development and management of green areas in Krakow 2019–2030 (KRiZTZ). It is a proposal intended to determine the city's long-term policy towards maintaining existing public green spaces with the use of their ecological, societal and cultural potential. It also constitutes a basis for creating new areas that supplement the system of green infrastructure. The overall concept for further development of green areas is based on the mixed network model with interconnecting wedges formed by river parks and larger forested areas. The document's provisions emphasise the continuity of ecological and recreational corridors and connections between the city and the region [61].

In the document, it was assumed that by 2030, at least 86% of residents will have access to green areas within walking distance from their homes (300 m). A target value of 10 m^2 of recreational green area per capita was adopted based on a review of similar European documents and strategies [61,72]. In addition, the KRiZTZ provides several analyses of the actual state of green areas in each district. It reveals that most of them are currently below average in terms of the amount of public green areas, which can be explained by their uneven distribution in the city structure.

Analysis of the documents confirmed that there is no design standard that defines the distance to green space or its population capacity. Consistent guidelines were found in the Directions of development and management of green areas in Krakow 2019–2030 (KZiRTZ). As mentioned, this document is not legally binding, and it serves as a base for preparing local plans. A review of the document found that, despite progress in planning during the last ten years, a sizable portion of the city's area does not have a local plan in force (MPZP), which leads to an overuse of the procedure of administrative *decisions* on development conditions (WZ). However, this procedure does not provide tools to manage overall land use patterns.

2.3. Data Sources

The primary sources of reliable data about the city used in the study were publications and websites by the General Statistics Office (GUS), the different branches of the Office of the City of Krakow (UMK) and the Municipal Greenery Authority (ZZM). The most up-to-date population data on the City of Krakow were found in GUS publications, while the more detailed demographic maps were available through the UMK cartographic services. The demographic information and maps available at UMK, as well as GUS publications are based on permanent and temporary residency declarations, which are obligatory official forms filled by each resident of the city. In Poland, it is the primary form of managing the population register and keeping track of population count.

Since a 1991 administrative reform, Krakow has been divided into eighteen districts, whose functioning is subjected to the Office of the City of Krakow, and their competencies for local governance are highly limited [73]. However, the districts form the most important ancillary territorial units of the city. The north-western part of Krakow is within four neighbouring districts: IV Prądnik Biały, V Krowodrza, VI Bronowice and VII Zwierzyniec. Up to 1991, they formed a single administrative unit: the Krowodrza district. This historical division is still used by the General Statistics Office (GUS), which distinguishes four units for Krakow: Śródmieście, Krowodrza, Podgórze and Nowa Huta [74]. As such, the data obtained from the GUS refers either to the entirety of the city or its largest fragment (statistical unit) and they could not be used directly. Such generalised data includes, for example, the number of apartments handed over for use in each year, the average size of apartment and average number of residents or the number of building permits issued. Therefore, some data for the selected area had to be counted directly from cartographic sources (for instance: the number of housing complexes or population in the buffer zones under analyses). One valuable source of information was the Municipal Spatial Information System (MSIP), where a real-time population register is shown, which allows for extraction of data on plot sizes, building sizes and number of storeys. It was the primary source of data on newly built buildings.

To perform urban analyses concerning urban tissue, barriers and accessibility, maps and orthophotomaps supplied by the UMK were used. Available planning documents, such as local spatial development plans (MPZP) were helpful mostly in delimiting the border and the area of the park, as well as in determine the state of the art in spatial planning. The study of the city's development utilises information as well as historical plans and maps in possession of the UMK. All the above-mentioned maps were obtained in the form of graphic raster images (jpg, pdf, png).

The study is focused on the period of the last ten years. It has been a period of particularly intense new real estate development and an increase in development pressure in Krakow. This period also saw the start of many municipal projects focused on structuring and revitalising existing green areas. The possibility of obtaining data and maps necessary for the study was another criterion. Prior to 2009, the amount and quality of statistical and cartographic reports available in digital form was significantly lower.

2.4. Methods

At present, data that would directly illustrate the growth of development for a given city fragment and within a set timeframe, is not directly available. To assess changes in spatial development of the fragment of Krakow that occurred between 2009 and 2019, a number of methods was used. First, the areas of the park's pedestrian accessibility were outlined. To visualise the geographical accessibility of Młynówka Park, distances measured in a straight line were used. Based on the recent literature on the pedestrian mobility and accessibility, the following values were adopted: 750 m, 500 m and 300 m [23,38,75–77] (Figure 6). The adopted distances are in line with half-mile (ca.0.8 km) and quarter-mile distances (ca.0.4 km) found in the classic writings on urban design by C. Perry or C.S. Stein. The concept of the Radburn neighbourhood by Stein and Wright introduced the superblock of the single-family housing enclosed by arterial roads or natural features. The superblocks were arranged into neighbourhoods, where the maximum walking distance on pedestrian routes was half a mile [78]. In the neighbourhood unit by Perry, the radius of around a quarter of a mile or 5 min walking delimits the distance from residential to non-residential components of the unit [79]. These principles influenced Anglo-Saxon urbanism to a great extent and are still used, e.g., in determining the accessibility zones around public transport stations or distance to public spaces [80].

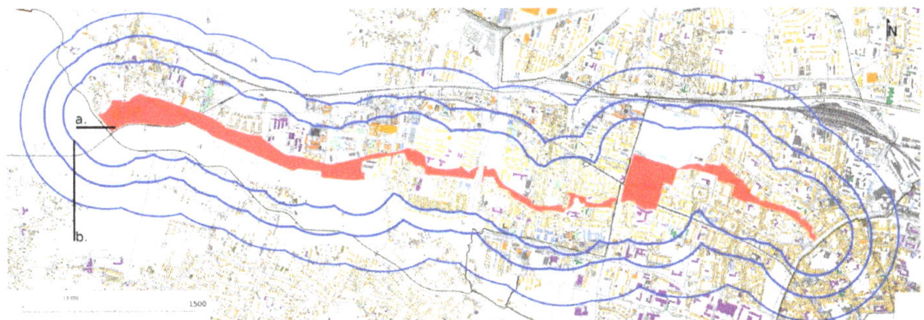

Figure 6. Młynówka Park accessibility zones; (**a**) territory of the park; (**b**) 300 m, 500 m, 750 m buffer zone outlines.

The outlines of buffer zones were created by setting series of rings with radii of adopted length in the sections of ca.150 m along the park boundary. Additional rings were placed at vertexes. The procedure was performed for each distance separately. The buffer zone of 750 m covers an extensive territory (ca. 31.76 km^2), yet this can be explained by the size and significance of the park on the scale of the city. The distances of 500 m and 300 m, especially, delimit the boundaries of the immediate neighbourhood.

Afterwards, a graphical comparative analysis of topographic maps and orthophotomaps was performed in order to identify the areas of new developments. Multi-family housing was placed at the centre of the study, as it increases the population number more than single-family development. This phenomenon can be associated with the land use intensity index. According to city planning documents, in case of high–intensity multi-family development the intensity of land use can reach the value of 2.0. This means that the total gross floor area of the building can be two times greater than its plot area. In the case of low intensity and single-family housing the value varies from 0.4 to 0.6, which is much lower [67]. Higher values of the index produce a greater amount of usable floor area, thus more inhabitants contribute to population growth. In addition, the demand for recreational greenery typically cannot be satisfied within the borders of a project's site, due to land use intensity or the presence of underground car parks, which hinders natural vegetation.

The areas of multi-family complexes and zones as existed in 2009 were identified on an archival ortophotomap. The next step included the identification of multi-family housing zones as indicated in the general planning study and the local planning documents. Afterwards, either completed multi-family housing complexes or those that are still under construction were marked on the map.

To assess greenery saturation in the city structure and estimate the potential of its use by residents, the k/n indicator is the most commonly used: k denotes the surface area of green spaces while n is the number of residents or users in a given area. However, Wen et al. note that this container approach has several shortcomings. They found that 'the recreational service that a green place can offer to a catchment area depends not only on its capacity but also on how many people in that catchment area must share the resources' [81]. Furthermore, this parameter also assumes the summation of the surface area of green spaces in a given area and the distribution of the total amount between all people. Therefore, for this study it cannot be fully utilised, as the focus is given to identifying whether there existed a population pressure on the selected park area and to determining its actual demographic capacity. Other existing green areas in the Krowodrza unit and in delimited buffer zones were not taken into account, although they contribute to the total amount of greenery. In order to determine the dependency between the size of the park and the size of its potential user population, the formula f = p/k was used, where p is the population count within the buffer zone and k is the size of the selected green area.

Changes in the number of inhabitants in the buffer zones were determined based on a series of estimates. First, the data published for each district by UMK were reviewed. These data include the area of the district and its population. In this case, the number of inhabitants was calculated from official residency declarations, which show the population count for the selected year. Afterwards, the population counts in each pedestrian accessibility zones for 2009 and 2019 (p0.1, p0.2, p0.3, p1.1, p1.2, p1.3) were calculated. The calculation was performed based on the real-time population register map supplied by the UMK, which shows residency declarations in aggregated form of regular hexagons with a side of 250 m (Figure 7). The values from clusters were counted for each buffer zone separately, and permanent and temporary residency declarations were summarised. The fragments of the hexagons in the buffer zones were included proportionally.

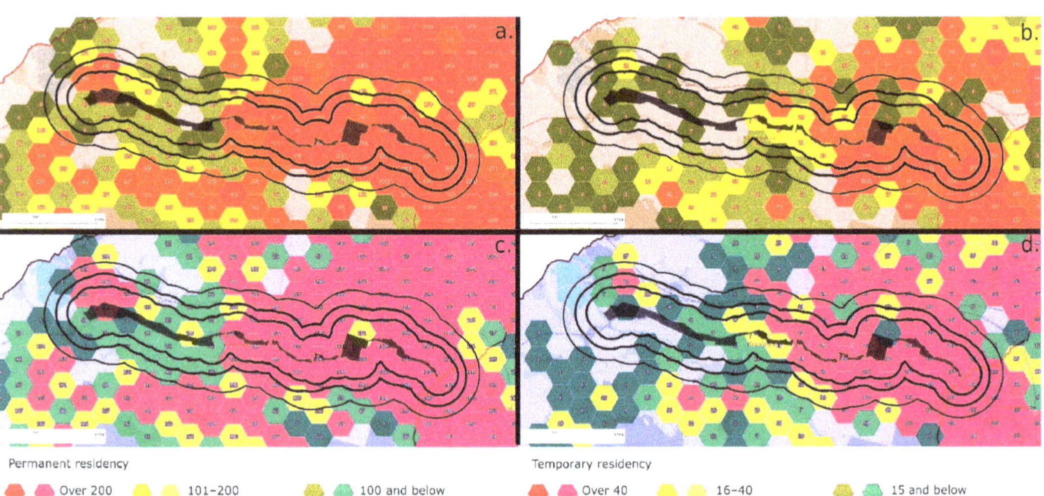

Figure 7. Map showing the population register; (**a**) permanent residency in 2009; (**b**) temporary residency in 2009; (**c**) permanent residency in 2019; (**d**) temporary residency in 2019.

The second estimation relies on analyses of spatial development, in particular the identification of new housing developments in the park's buffer zones (p2.1 ...) (Table 3). As there are no direct data on population capacity of the newly built complexes, it was counted based on the spatial data derived from cartographic materials (MSIP). These data included plot size, built-up area and number of storeys for each building. Hence, it was possible to calculate the usable floor area and number of apartments. To estimate the population of the selected building and housing complex it was necessary to adopt indices published by GUS: the average size of the apartment in multi-family building in Krakow (57 m^2) and the average number of people per apartment (2.63). These values were published for the city of Krakow in 2017.

The last estimation is based on available data on the number of new apartments built and handed over for use (p3.1., p3.2., p3.3) (Table 2). These data are published annually by GUS and provide a valuable source of information concerning the development of the housing sector. However as mentioned in the previous section, they refer to the Krowodrza unit in general. Therefore, to assess the population number the same indices of average apartment size and average amount of inhabitants were utilised. In addition, a rough approximation was carried out based on proportions between the population of the Krowodrza unit and the population of the buffer zones as calculated in the first estimation.

As a result of the calculation, the factor f illustrating the amount of people on the surface unit of the park shall take different values depending on the population count.

Table 3. The matrix of population count as used in further research.

Buffer Zone	Population 2009 [1]	Population 2019 [1]	Population 2019 [2]	Population 2019 [3]
750 m	p.0.1	p.1.1	p.2.1	p.3.1
500 m	p.0.2	p.1.2	p.2.2	p.3.2
300 m	p.0.3	p.1.3	p.2.3	p.3.3

[1] Population count counted from the real-time population register map from MSIP resources. [2] Population count assessed in population estimation performed for selected newly built housing complexes. [3] Population count as a result of approximate estimation based on data on housing stock published by the General Statistics Office (GUS).

3. Results

3.1. General Background

According to available data, the situation of Krakow and the Krowodrza unit in terms of demographics, built-up area and green spaces was found to be as follows: green areas of different types cover 62.3% of the city; urban parks, along with other wildlife areas described as valuable, comprise ca. 15% of the area, which is around 49.05 ha. In 2014, the index of recreational green public space per capita was 8.3 m^2 [67], while the population density was 2359 persons per km^2. The average population density in the Krowodrza unit is almost two times higher than in the whole city (Table 4). It was found that the green area per capita index takes different values depending on the administrative district. The lowest value of 4.50 m^2 is represented by district IV (Prądnik Biały), while the highest amount of green area—27.54 m^2 was found in district VII (Zwierzyniec), which is characterised by large share of low-density housing, as well as the presence of the biggest park in Krakow (Błonia Krakowskie).

Table 4. Data on green area and population of the Krowodrza unit in comparison to Krakow.

	Krakow	Krowodrza Unit
area	327 km^2 [1]	67.32 km^2 [2]
population 2019	779,115 [1]	146,422 [2]
population prognosis in 2030	779,104 [3]	x
population density	2359 p/km^2 [4]	4597 p/km^2 [4]
green area cover	62.30% [5]	x
green area per capita	8.3 m^2 [5]	4.50–27.54 m^2 [5]
total area of public parks	4.57 km^2 [6]	1.29 km^2 [6]

[1] Data on the city of Krakow published by the UMK in 2019. [2] Data on the districts published by the UMK in 2019. [3] Prognosis published by GUS in 2019. [4] Data published by GUS in 2018. [5] Data published in the KRiZTZ. [6] Data derived from Municipal Greenery Authority (ZZM) resources.

Current data show that since 2009, the built-up area of the city has increased by almost 4 thousand ha, while recreational areas, which include parks, have remained almost unchanged (Table 5).

Table 5. Data on built-up area and recreational area change in Krakow as presented in annual publications of the UMK for 2009 and 2019.

Year	Built-Up Area/ha	Percentage of City Area	Recreational and Leisure Area/ha	Percentage of City Area
2009	5964	18.24%	889	2.72%
2018	9767	29.87%	930	2.84%
increase	3803		41	

3.2. Planning and Urban Development

At present, the area of the park itself and its immediate surroundings are covered by four neighbouring MPZP plans. Land use calculations indicate that a total of 29.49 ha was assigned to public green space (parks). The area under the municipal administration, managed by the ZZM is 18.4 ha (Table 6), and this value is used in further calculations. In 2009, none of the local plans were in force, so the park area was formally not protected. Other valid MPZP cover only part of the selected area and comprise of separate, unconnected fragments (Figure 8).

Table 6. Area and planning status of the Młynówka Królewska park.

No.	MPZP Name	Area of MPZP/ha	Area of Public Greenery/ha	Area of the Park Managed by ZZM/ha
1.	Młynówka Królewska—Grottgera II	22.1	8.06	
2.	Młynówka Królewska—Zarzecze	5.95	3.85	
3.	Młynówka Królewska—Filtrowa	11.24	4.01	
4.	Młynówka Królewska—Zygmunta Starego	30	4.57	
	Total area	69.29	20.49	18.41

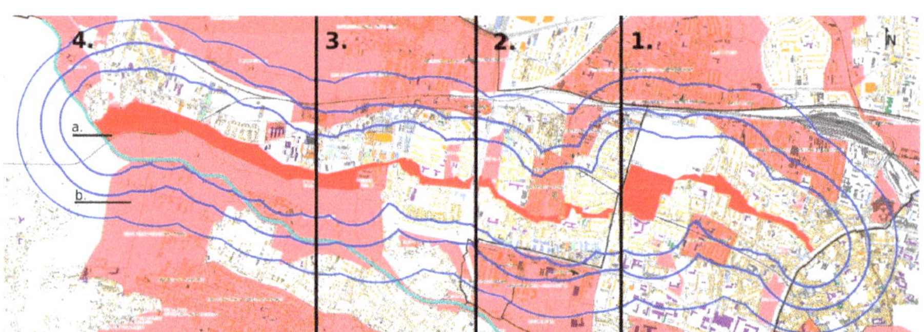

Figure 8. Map of the existing local plans in 2019; (**a**) territory of the park as delimited in the local plans 1–4 (MPZP); (**b**) other existing, valid local plans (MPZP).

Changes in the development structure have been identified through comparison of SUiKZP maps, local plans and available ortophotomaps that covered the selected timeframe. The most of the multi-family complexes existing in 2009 was concentrated in the eastern part of the area, while the western part was characterized by low-density, single-family urban patterns. Areas that are designated for multi-family housing purposes cover more land towards the west, but still a clear division between eastern and western part is visible. It was also observed that some completed multi-family housing projects are located outside of areas assigned in the SUiKZP for this type of development, which means that they were built on the basis of WZ procedure (Figure 9d). It leads to a situation where the land use balance cannot be accurately planned and controlled, as this type of procedure is restricted to a single building plot.

3.3. Demography

According to the UMK data on districts, within the last 10 years, the total population of the Krowodrza unit increased only by 2.08%, in addition, the district located closest to the city centre (V) recorded a clear decrease in the number of inhabitants (Table 7). This provides an overview of the demographic tendencies and population shift from central urban areas to more distant districts.

The population that declared permanent and temporary residency within the adopted buffer zones was calculated using the population register map, as described in the previous chapter. The 750 m isochrone was found to be inhabited by ca. 67,808, which was over 46% of the overall population in the Krowodrza unit in 2019. Within the 500 m isochrone the number was 50,183 (34.32%), and within 300 m—36,333 (24.86%). Only several percent increase was noted in this area as well. Table 8 illustrates the result of the performed population count.

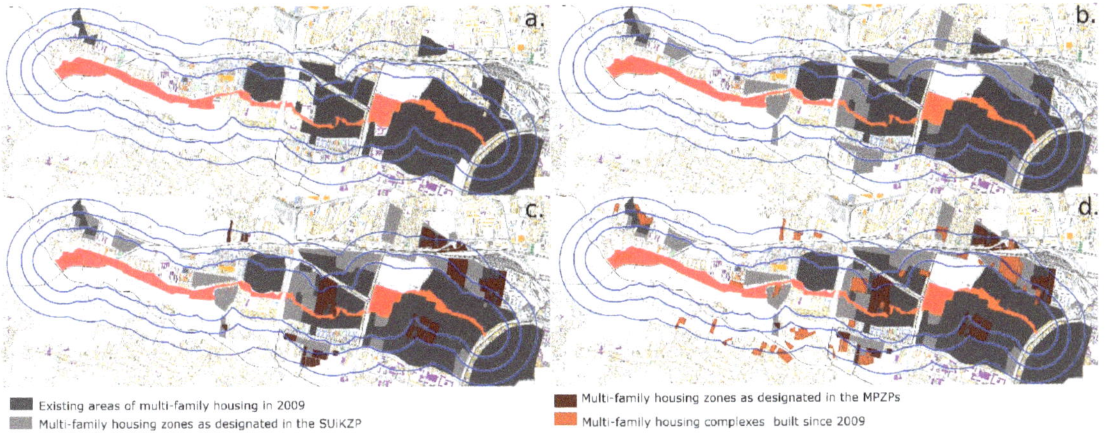

Figure 9. Map of spatial development of the area; (**a**) existing areas of multi-family housing in 2009; (**b**) multi-family housing zones as designated in the SUiKZP; (**c**) multi-family housing zones as designated in the local spatial development plan (MPZP); (**d**) multi-family housing complexes built since 2009.

Table 7. UMK Data on population change in administrative districts as presented in annual publications of the UMK for 2009 and 2019.

District's Number	District's Name	Area/ha	Population 2009	Population 2019	Population Change
IV	Prądnik Biały	2342	66,472	71,752	107.94%
V	Krowodrza	562	34,288	30,184	88.03%
VI	Bronowice	956	22,519	23,931	106.27%
VII	Zwierzyniec	2873	20,154	20,555	101.99%
	Total	6733	143,433	146,422	102.08%

However, these data do not fully illustrate the actual state, and the main reason for that assumption is the dynamic development of the housing sector. In the past several years, it has been marked by continuous growth, with new buildings and complexes built. According to GUS, the pool of existing dwellings has not decreased [74]. This tendency is not reflected in residency declaration increase (Table 7). In light of these observations, the demographic capacity of the area under study can be considerably greater.

Accounting for the significant increase in built-up area of the city, it can be assumed that this process also took place within the limits of Krowodrza unit within the specified timeframe. Since 2009, the area within 750 m of the park has become the site of several dozen new multi-family housing complexes, some of which are in close proximity to the park. The estimated number of new developments is 45, based on a comparison of the current and previous state of land cover as well as in-situ observations. This group includes both singular buildings and complexes of varying size. Three of the largest housing complexes, which comprise around a dozen multi-family buildings (m1, m2 and m3) were chosen to more detailed study (Figure 10). It was assumed that, due to their built-up area, number and size, the buildings can affect an increase in the overall number of residents and thus the park's users.

Table 8. Result of the population count performed for the selected buffer zones.

Buffer zone	Population 2009	Percentage of Unit Population 2009	Population 2019	Percentage of Unit Population 2019
750 m	65,912	45.95%	67,808	46.37%
500 m	49,716	34.66%	50,183	34.32%
300 m	36,006	25.10%	36,333	24.85%

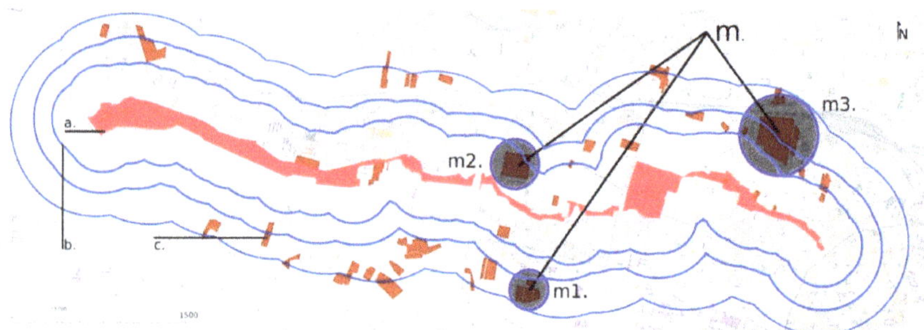

Figure 10. Location of the selected multi-family housing complexes (m1, m2, m3); (**a**) territory of the park; (**b**) buffers of 300 m, 500 m, 750 m; (**c**) multi-family housing complexes built since 2009.

To assess the scale of this phenomenon, an estimation was performed for the three selected development complexes (m1, m2, m3). The number of dwellings was counted based on the total built-up area, number of storeys in each building and average apartment size for Krakow—in 2017, this was 57 m^2. The average number of residents per apartment was assumed after the GUS publications to be at a level of 2.63 [74]. Therefore, the total number of residents in these complexes can be as high as 11 thousand (Table 9). As the three complexes are all within the 750 m zone, this value was used to assess population growth only in this zone.

In the last step of the population estimation, the statistical data on the housing development for the Krowodrza unit were taken into account. They indicate that over the past ten years, over 20 thousand dwellings were handed over for use in newly built edifices (Table 10).

The housing stock in the western part of the city has substantially increased in size, which is expected to have impact on the final number of park users. The number of residents of the new dwellings can contribute to the overall population of the Krowodrza unit. In this calculation, the same coefficient of 2.63 residents per apartment was used. As a result, the estimated number of people inhabiting the entire Krowodrza unit could be higher by over 53 thousand. This value can be added to the population of the unit derived from official UMK publications (Table 7). As a result, the total population of the unit can amount to 199,887 people (Table 11).

Table 9. Data on the size of the selected housing complexes (m1, m2, m3). The population is a result of estimation based on size of the development.

	m1	m2	m3	Total
Land area/m^2 [1]	35,650	5992	88,472	130,114
Built-up area/m^2 [1]	10,661	14,417	24,273	49,351
Number of storeys [1]	4–8	4–17	6–10	x
Number of dwellings [2]	677	1487	2265	4429
Population (m) [3]	1779	3911	5956	11,646

[1] Data derived from cartographic services MSIP. [2] The number of dwellings was calculated using average apartment size of 57 m^2 as published by GUS. [3] Population number was calculated using average number of residents per apartment (2.63) published by GUS.

Table 10. The number of new dwellings built in Krowodrza unit in each year according to data published by GUS.

Year	Number of New Dwellings	Number of Residents (Estimated) [1]
2009	3213	8450
2010	1330	3498
2011	1354	3561
2012	1449	3811
2013	1741	4579
2014	1355	3564
2015	1080	2840
2016	1459	3837
2017	2364	6217
2018	2222	5844
2019	2762	7264
Total	20,329	53,465

[1] Population count was calculated using average number of residents per apartment (2.63) published by GUS.

Table 11. Result of the population estimation for Krowodrza unit.

Population of the Unit 2019 [1]	Number of Residents in Newly Built Dwellings [2]	Total Number of Residents in Krowodrza Unit
146,422	53,465	199,887

[1] According to publications of UMK for 2019. [2] Population count is a result of calculation presented in Table 10.

However, this result refers to the statistical unit of Krowodrza, not to the specified buffer zones of the Młynówka Park. As mentioned, no data exclusively covering the selected part of the city were found, therefore another rough approximation was made. It was based on the supposition that the ratio between the number of persons in buffer zone and the population of the unit remains constant. This is a conceptual simplification, thus it is expected that the result of the last estimation can be the least precise. The ratio could have been changed by the new development, as it is not distributed evenly throughout the unit. Yet, it was found reasonable to perform the calculations and compare the results obtained throughout the study to gain a broader perspective on possible population change.

It was found in the previous stages of the research, that in 2019 the biggest (750 m) buffer zone is inhabited by 46.37% of the population of the unit. The buffer zones of 500 m and 300 m were inhabited by 34.32% and 24.85% respectively (Table 8). By changing the initial population number (from 146,422 to 199,887), it was possible to obtain values of the estimated population number in each of the buffer zones. Table 12 illustrates the results of the performed calculation.

3.4. Population Pressure

The final result of the calculations performed at previous stages of the research is a set of values of parameter *p*, which illustrates population changes in each of the three buffer zones (Table 13), and the resultant *f* values, which describe the relation between the size of the green space and the size of the population (Table 14). The outcomes vary between 0.196 and 0.503 persons per m^2 of the park area. The lowest expressions applied to the data from 2009 (f0.3), while the highest expression is associated with the size of the current housing stock for the Krowodrza unit (f3.1).

The city's goal for the amount of urban recreational greenery is at least 10 m^2 per inhabitant. Thus, the desired *f* value for every urban green area should be no bigger than 0.1. It is supposed to provide comfort and enough space for users. During the study, it was found that the actual values of *f* factor for Młynówka Królewska Park exceeds this number at least twice. The desired value was exceeded even in the case of the lowest population count in 2009 in the 300 m buffer zone.

Table 12. Result of the population estimation for the selected buffer zones.

Population of the Unit 2019 [1]	Buffer Zone	Percentage of Unit Population Inhabiting the Buffer Zone [2]	Population of the Buffer Zone (Estimated) [3]
199,887	750 m	46.37%	92,688
	500 m	34.32%	68,601
	300 m	24.85%	49,672

[1] The population number was counted in Table 11. [2] The percentage is a result of the calculations presented in Table 8. [3] Population count was calculated using the ratio between the number of persons in buffer zone and the population of the unit.

Table 13. The population in the buffer zones calculated in 2009 and 2019.

Buffer Zone	p0	p1	p2	p3
750 m	p0.1 = 65,912	p1.1 = 67,808	p2.1 = 79,454	p3.1 = 92,688
500 m	P0.2 = 49,716	p1.2 = 50,183	x	p3.2 = 68,601
300 m	P0.3 = 36,006	p1.3 = 36,333	x	p3.3 = 49,673

Table 14. The population pressure calculated in 2009 and 2019.

Buffer Zone	f0	f1	f2	f3
750 m	f0.1 = 0.358	f1.1 = 0.368	f2.1 = 0.432	f3.1 = 0.503
500 m	F0.2 = 0.270	f1.2 = 0.273	x	f3.2 = 0.373
300 m	F0.3 = 0.196	f1.3 = 0.197	x	f3.3 = 0.270

3.5. Urban Analyse

The urban analyses of the selected area reveal that, within the walking distance, the urban tissue varies in form, density and function. Urban blocks and parts of regular grid can be visible in the eastern part, close to the city centre. Towards the west, the development structure becomes observably less dense (Figure 11), and the last section of the park is adjacent to farmland and meadows located within the borders of the landscape park buffer zone.

The important factor that has impact on accessibility is the property of the land and plot divisions. It was found that towards the west plots became visibly bigger, as well as the share of private grounds increased (Figure 12).

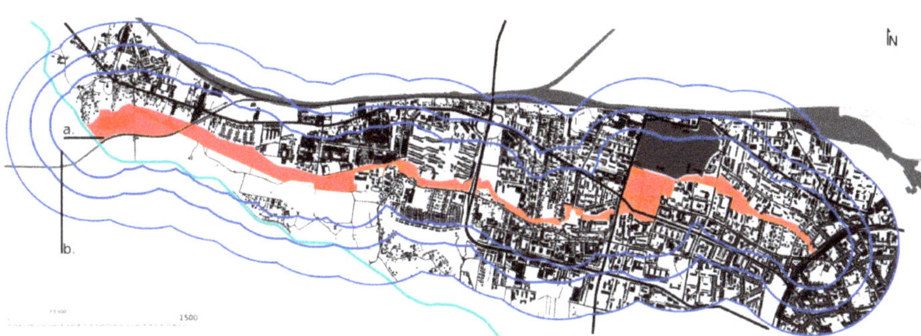

Figure 11. Urban tissue within walking distance from the park area; (**a**) territory of the park; (**b**) buffers of 300 m, 500 m, 750 m.

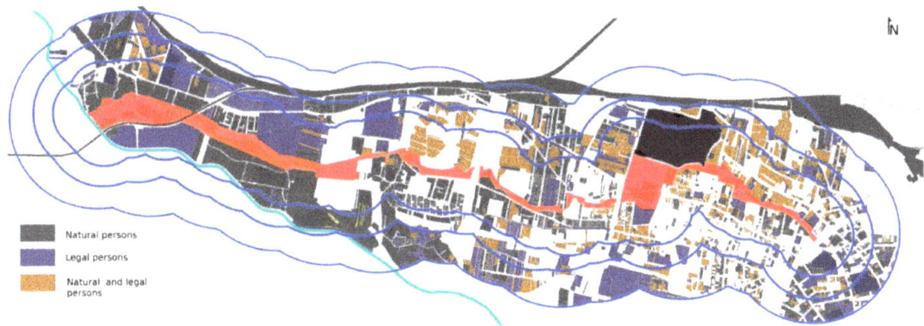

Figure 12. Map indicating ownership of the land.

Railway tracks, river, busy roads, and large closed areas were identified as fundamental structural barriers (Figure 13). They allow traversal passage only in a limited number of places, thereby extending pedestrian routes sometimes even twofold. The best access is observed in the eastern and central section of the park, where the urban structure is regular.

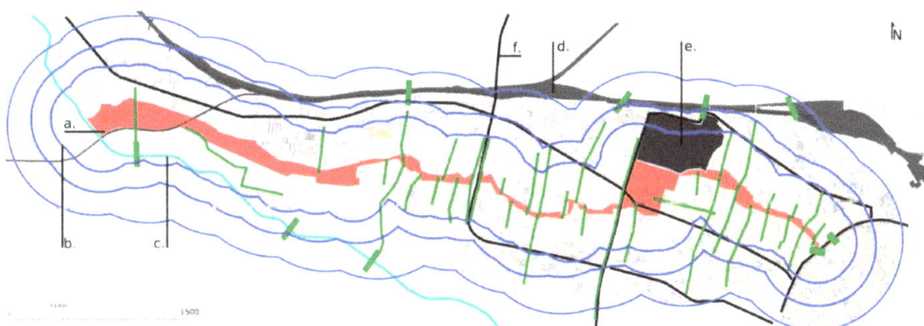

Figure 13. Młynówka Park pedestrian access possibilities and limitations; (**a**) territory of the park; (**b**) buffers of 300 m, 500 m, 750 m; (**c**) river Rudawa; (**d**) railway; (**e**) military area; (**f**) main roads. The major pedestrian passages that allow access are marked in green.

The ability to reach the park on foot varies depending on the section of the park. Towards the west, it decreases considerably due to a looser road network, the lack of pavements or the higher share of private, fenced areas.

4. Discussion

One of the important reasons for undertaking the work was to draw attention to the problem of increasing pressure on existing, attractive urban green areas, which is the result of the city's spatial development, particularly the increase in the amount of multi-family housing. Population pressure on Młynówka Królewska Park, as calculated in the study, is significant. It can lead to overcrowding in the space, a lack of comfort, and consequently to a lowering of its value as a recreational space. It can be discussed, to what extent the number people derived from secondary sources and approximations reflects the actual state. Certainly, the performed calculations give a general overview of the population changes, which occurred to be much bigger than initially expected. It was shown that, on the local scale, the processes of city growth are visible, and they can be associated with housing development.

In light of available data and previous research, it was not possible to determine the exact degree to which the number of residents carries over to the number of park users. In majority of the research the overall number of inhabitants was used, as the most reliable population data. In the case of Młynówka Królewska, it can be assumed that the landscape value, location and size of the park place it high in the hierarchy of attractiveness of urban green spaces in the western districts of the city. The residents could be more willing to undertake longer journeys to an attractive green area. Previous studies confirm that more attractive, easy to access and well-equipped green spaces are visited more often than others [40,42,43]. Accordingly, the analysis of cycling preferences performed by Campos-Sánchez et al. provides evidence, that green areas, together with other environmental conditions (e.g., proximity of university buildings, which occurs in this case) can play an attractive role in route choices [82].

Numerous studies point to an insufficient amount of green space in city structures or the associated problems with access to them by residents. Initial research revealed that existing empirical studies that analysed green space provision mostly conceptualised access as a geographic construct, and used GIS methods to get specific results [45,77,83]. However, this study proves that to measure accessibility, in-depth urban analyses are necessary. The problem of the urban green area accessibility is complex. Parks are situated within the urban fabric, which determines possibilities as well as limitations of the pedestrian travels. As shown in previous studies on centrality, location within the street network can play a vital role in the patterns of use of the public space. The areas of high centrality measures (e.g., reach, closeness, betweenness or straightness centrality) are better connected and experience more intense pedestrian flows [32,33].

It has been found that accessibility is related not only to geographical proximity, but also to urban form, walkability, lack of urban barriers and number of access points on the perimeter of the area [84]. Railway tracks, busy roads, large closed areas, or gated communities that make it necessary to take longer routes are fundamental structural barriers for pedestrians, making the distance travelled extend beyond the limit of subjective justification [34,85,86]. In particular, studies on pedestrian travel preferences can provide valuable information about preferences in mobility [87–90]. The general findings are also in line with Jan Gehl's approach, that some neighbourhoods have better quality in terms of walking conditions, safety and convenience [91].

During the study, the main obstacle was a shortage of precise local data and knowledge related to spatial, statistical and demographic issues pertaining to development. The demographic forecasts featured in the SUiKZP and yearly reports on the state of the city do not reflect the actual population count in Krakow for a number of reasons. First: the obligation to file residency declarations is not obeyed, which means that only some residents declare permanent or temporary residency. Second, dwellings are purchased as a form of capital investment. Such dwellings are often rented, and in the case of Krakow a large number of persons who rent apartments are university students, of whom there are 135 thousand in the city [68]. They are not obliged to file residency declarations in Krakow, thus they are not considered as residents, although they significantly contribute to overall population count. Finally, users of urban public spaces also include residents of neighbouring municipalities who work and study in Krakow.

In light of the above, it is clear that the official demographic data published by the UMK or GUS are far from reflecting the actual state. However the more precise data or coherent datasets are not yet available. The lack of coherent statistical data was also pointed out by other green space accessibility studies of Polish cities [34,92]. These conditions can be described as difficult not only form a research point of view, but also that of city policy towards providing green spaces.

5. Conclusions

The study was aimed to answer the question whether and if so, to what degree, does the spatial development model that has been implemented in Krakow for the past ten years can affect green spaces. The focus was put on an analysis of a specific case of a city park and its surroundings located within walking distance. The study demonstrated certain characteristics of the urban tissue in the delimited buffer zones of the Młynówka Królewska in Krakow. It was found that, in the selected timeframe, urban growth increased, and urbanisation processes manifested themselves in the form of, among others, the number of new multi-family housing complexes. It was shown that this phenomenon has an impact on use of the park area. It contributes to increasing the population in the surroundings of the park, and thus the number of potential park users, which if left uncontrolled can lead to the area becoming overcrowded. Therefore, a correlation between tendencies in urban development, population and the use of urban parks can be established.

By adopting a park-based perspective, this study fills a gap in the existing research by defining the tendency to experience the increased number of visitors of the green area. The transformations that were observed include especially the growth of built-up area, specifically dense multi-family housing. The study also demonstrated that the estimated number of the park's users could have increased by as much as 70% over the past decade. This is critical observation, which casts a new light on green spaces management on the local and city scale. The index of park users per surface area unit of the park makes it possible to determine the capacity of recreational areas and can be used to better understand the relationships between green spaces and the surrounding development.

It was found that the present legal framework and city policy show certain shortcomings. First, it utilises official statistical data, which are not sufficient, as proved during research. This may lead to an underestimation of green areas planned on the city scale. Second, the lack of local plans (MPZP) can contribute to dispersing areas of high density, as the alternative procedure (WZ) is being used. Based on that procedure, several housing complexes in the analysed area were built in the zones other than multi-family. According to this development model, it is difficult to manage and even predict the city growth processes. In consequence, the goal of 10 m^2 of recreational green area per capita, as mentioned in the KZiRTZ may be difficult to achieve.

The key limitation of the study derives mainly from its preliminary character. The selected type of analysis can provide only some of the answers to the complex range of problems connected with green areas provision, transformations of urban tissue and demographic processes. The available, yet far from sufficient spatial and statistical data were used in a series of estimations, which leads to approximate and not precise results on the actual population pressure on the park area. Nonetheless, the study provided interesting insight on the issue of housing development in context of green area provision. Additionally, it should be noted that in spatial planning and urban planning tools such as indicators, indices, factors or estimates are frequently used. The selected case is also a certain limitation, as it describes a unique situation. Młynówka Królewska Park is an attractive space, located in a big city with rich urban history, and, as a linear park, it is also well connected with surrounding urban tissue. Furthermore, managing urbanisation processes, and hence the provision of green areas, is highly dependent on applicable legal regulations and existing urban standards or lack thereof. The above considered, further research of the problem of population pressure and overcrowding of urban green areas should aim at the comparison of case studies from different locations. The experiences of other countries with various spatial and legal backgrounds may allow us to generalize the research findings.

Funding: This research received no external funding.

Institutional Review Board Statement: Not applicable.

Informed Consent Statement: Not applicable.

Data Availability Statement: Data available in a publicly accessible repositories of the different branches of the Office of the City of Cracow and the General Statistics Office. They are all showed in "References" section.

Conflicts of Interest: The author declare no conflict of interest.

References

1. Carneiro Freire, S.; Corban, C.; Ehrlich, D.; Florczyk, A.; Kemper, T.; Maffenini, L.; Melchiorri, M.; Pesaresi, M.; Schiavina, M.; Tommasi, P. *Atlas of the Human Planet 2019*; Publications Office of the European Union: Luxembourg, 2019.
2. OECD/European Commission. *Cities in the World: A New Perspective on Urbanisation*; OECD Urban Studies; OECD Publishing: Paris, France, 2020.
3. United Nations. Transforming Our World: The 2030 Agenda for Sustainable Development. A/RES/70/1. 2015, Volume 16301. Available online: https://sdgs.un.org/2030agenda (accessed on 4 October 2020).
4. Klopp, J.M.; Petretta, D.L. The urban sustainable development goal: Indicators, complexity and the politics of measuring cities. *Cities* **2017**, *63*, 92–97. [CrossRef]
5. Cohen, D.A.; McKenzie, T.L.; Sehgal, A.; Williamson, S.; Golinelli, D.; Lurie, N. Contribution of public parks to physical activity. *Am. J. Public Health* **2007**, *97*, 509–514. [CrossRef] [PubMed]
6. Rydin, Y.; Bleahu, A.; Davies, M.; Dávila, J.D.; Friel, S.; De Grandis, G.; Groce, N.; Hallal, P.C.; Hamilton, I.; Howden-Chapman, P.; et al. Shaping cities for health: Complexity and the planning of urban environments in the 21st century. *Lancet* **2012**, *379*, 2079–2108. [CrossRef]
7. Kothencz, G.; Kolcsár, R.; Cabrera-Barona, P.; Szilassi, P. Urban Green Space Perception and Its Contribution to Well-Being. *Int. J. Environ. Res. Public Health* **2017**, *14*, 766. [CrossRef]
8. Wolch, J.R.; Byrne, J.; Newell, J.P. Urban green space, public health, and environmental justice: The challenge of making cities "just green enough". *Landsc. Urban Plan.* **2014**, *125*, 234–244. [CrossRef]
9. Maas, J.; Verheij, R.A.; De Vries, S.; Spreeuwenberg, P.; Schellevis, F.G.; Groenewegen, P.P. Morbidity is related to a green living environment. *J. Epidemiol. Community Health* **2009**, *63*, 967–973. [CrossRef]
10. Koohsari, M.J.; Mavoa, S.; Villianueva, K.; Sugiyama, T.; Badland, H.; Kaczynski, A.T.; Owen, N.; Giles-Corti, B. Public open space, physical activity, urban design and public health: Concepts, methods and research agenda. *Health Place* **2015**, *33*, 75–82. [CrossRef]
11. Kaczynski, A.T.; Potwarka, L.R.; Saelens, P.B.E. Association of park size, distance, and features with physical activity in neighborhood parks. *Am. J. Public Health* **2008**, *98*, 1451–1456. [CrossRef]
12. Schipperijn, J.; Cerin, E.; Adams, M.A.; Reis, R.; Smith, G.; Cain, K.; Christiansen, L.B.; Dyck, D.; van Gidlow, C.; Frank, L.D.; et al. Access to parks and physical activity: An eight country comparison. *Urban For. Urban Green.* **2017**, *27*, 253–263. [CrossRef]
13. Wang, R.; Helbich, M.; Yao, Y.; Zhang, J.; Liu, P.; Yuan, Y.; Liu, Y. Urban greenery and mental wellbeing in adults: Cross-sectional mediation analyses on multiple pathways across different greenery measures. *Environ. Res.* **2019**, *176*, 108535. [CrossRef]
14. Jennings, V.; Bamkole, O. The relationship between social cohesion and urban green space: An avenue for health promotion. *Int. J. Environ. Res. Public Health* **2019**, *16*, 452. [CrossRef] [PubMed]
15. Wang, R.; Zhao, J.; Meitner, M.J.; Hu, Y.; Xu, X. Characteristics of urban green spaces in relation to aesthetic preference and stress recovery. *Urban For. Urban Green.* **2019**, *41*, 6–13. [CrossRef]
16. Xie, G.; Chen, W.; Cao, S.; Lu, C.; Xiao, Y.; Zhang, C.; Li, N.; Wang, S. The Outward Extension of an Ecological Footprint in City Expansion: The Case of Beijing. *Sustainability* **2014**, *6*, 9371–9386. [CrossRef]
17. Badach, J.; Dymnicka, M.; Baranowski, A. Urban vegetation in air quality management: A review and policy framework. *Sustainability* **2020**, *12*, 1258. [CrossRef]
18. Lafortezza, R.; Chen, J.; van den Bosch, C.K.; Randrup, T.B. Nature-based solutions for resilient landscapes and cities. *Environ. Res.* **2018**, *165*, 431–441. [CrossRef] [PubMed]
19. Russo, A.; Cirella, G.T. Modern compact cities: How much greenery do we need? *Int. J. Environ. Res. Public Health* **2018**, *15*, 2180. [CrossRef] [PubMed]
20. John, H.; Neubert, M.; Marrs, C.; Alberico, S.; Bovo, G.; Ciadamidaro, S.; Danzinger, F.; Erlebach, M.; Freudl, D.; Grasso, S.; et al. *Green Infrastructure Handbook—Conceptual & Theoretical Background, Terms and Definitions*; Interreg Central Europe Project MaGICLandscapes: Dresden, Germany, 2019.

21. European Environmental Agency Urban Green Infrastructure—Interactive Map. Available online: https://eea.maps.arcgis.com/apps/MapSeries/index.html?appid=42bf8cc04ebd49908534efde04c4eec8 (accessed on 4 October 2020).
22. Kabisch, N.; Qureshi, S.; Haase, D. Human-environment interactions in urban green spaces—A systematic review of contemporary issues and prospects for future research. *Environ. Impact Assess. Rev.* **2015**, *50*, 25–34. [CrossRef]
23. Wüstemann, H.; Kalisch, D.; Kolbe, J. Access to urban green space and environmental inequalities in Germany. *Landsc. Urban Plan.* **2017**, *164*, 124–131. [CrossRef]
24. Biernacka, M.; Kronenberg, J. Classification of institutional barriers affecting the availability, accessibility and attractiveness of urban green spaces. *Urban For. Urban Green.* **2018**, *36*, 22–33. [CrossRef]
25. Zhu, Z.; Lang, W.; Tao, X.; Feng, J.; Liu, K. Exploring the quality of urban green spaces based on urban neighborhood green index-a case study of Guangzhou city. *Sustainability* **2019**, *11*, 5507. [CrossRef]
26. Kabisch, N.; Haase, D. Green spaces of European cities revisited for 1990–2006. *Landsc. Urban Plan.* **2013**, *110*, 113–122. [CrossRef]
27. Rigolon, A.; Browning, M.; Lee, K.; Shin, S. Access to Urban Green Space in Cities of the Global South: A Systematic Literature Review. *Urban Sci.* **2018**, *2*, 67. [CrossRef]
28. Fuller, R.A.; Gaston, K.J. The scaling of green space coverage in European cities. *Biol. Lett.* **2009**, *5*, 352–355. [CrossRef] [PubMed]
29. Kabisch, N.; Strohbach, M.; Haase, D.; Kronenberg, J. Urban green space availability in European cities. *Ecol. Indic.* **2016**, *70*, 586–596. [CrossRef]
30. Karimi, K. A configurational approach to analytical urban design: 'Space syntax' methodology. *Urban Des. Int.* **2012**, *17*, 297–316. [CrossRef]
31. Batty, M. A New Theory of Space Syntax. *Casa Work. Pap.* **2004**, *44*, 1–36.
32. Crucitti, P.; Latora, V.; Porta, S. Centrality in Network of Urban Streets. *Chaos* **2006**, *16*, 015113. [CrossRef]
33. Pérez-Campaña, R.; Abarca-Alvarez, F.J.; Talavera-García, R. Centralities in the city border: A method to identify strategic urban-rural interventions. *Ri-Vista* **2016**, *14*, 38–53. [CrossRef]
34. Wysmułek, J.; Hełdak, M.; Kucher, A. The Analysis of Green Areas' Accessibility in Comparison with Statistical Data in Poland. *Int. J. Environ. Res. Public Health* **2020**, *17*, 4492. [CrossRef]
35. Korwel-Lejkowska, B.; Topa, E. Dostępność parków miejskich jako elementów zielonej infrastruktury w Gdańsku. *Rozw. Reg. Polityka Reg.* **2017**, *37*, 63–75. [CrossRef]
36. European Environment Agency (EEA). *Urban Sustainability Issues—What Is a Resource-Efficient City?* Publications Office of the European Union: Luxembourg, 2015; Available online: https://www.eea.europa.eu/publications/resource-efficient-cities (accessed on 14 October 2020).
37. Wüstemann, H.; Kalisch, D.; Kolbe, J. Towards a national indicator provision and environmental inequalities in Germany: Method and findings. *Landsc. Urban Plan.* **2016**, *164*, 124–131. [CrossRef]
38. Le Texier, M.; Schiel, K.; Caruso, G. The provision of urban green space and its accessibility: Spatial data effects in Brussels. *PLoS ONE* **2018**, *13*, e0204684. [CrossRef]
39. Chiesura, A. The role of urban parks for the sustainable city. *Landsc. Urban Plan.* **2004**, *68*, 129–138. [CrossRef]
40. Nam, J.; Dempsey, N. Place-Keeping for Health? Charting the Challenges for Urban Park Management in Practice. *Sustainability* **2019**, *11*, 4383. [CrossRef]
41. Cortinovis, C.; Zulian, G.; Geneletti, D. Assessing Nature-Based Recreation to Support Urban Green Infrastructure Planning in Trento (Italy). *Land* **2018**, *7*, 112. [CrossRef]
42. Lindberg, M.; Schipperijn, J. Active use of urban park facilities—Expectations versus reality. *Urban For. Urban Green.* **2015**, *14*, 909–918. [CrossRef]
43. Bahriny, F.; Bell, S. Patterns of Urban Park Use and Their Relationship to Factors of Quality: A Case Study of Tehran, Iran. *Sustainability* **2020**, *12*, 1560. [CrossRef]
44. Voigt, A.; Kabisch, N.; Wurster, D.; Haase, D.; Breuste, J. Structural diversity: A multi-Dimensional approach to assess recreational services in urban parks. *Ambio* **2014**, *43*, 480–491. [CrossRef]
45. Gupta, K.; Roy, A.; Luthra, K.; Maithani, S. GIS based analysis for assessing the accessibility at hierarchical levels of urban green spaces. *Urban For. Urban Green.* **2016**, *18*, 198–211. [CrossRef]
46. Onose, D.A.; Iojă, I.C.; Niță, M.R.; Vânău, G.O.; Popa, A.M. Too Old for Recreation? How Friendly Are Urban Parks for Elderly People? *Sustainability* **2020**, *12*, 790. [CrossRef]
47. Kothencz, G.; Blaschke, T. Urban parks: Visitors' perceptions versus spatial indicators. *Land Use Policy* **2017**, *64*, 233–244. [CrossRef]
48. Wright Wendel, H.E.; Zarger, R.K.; Mihelcic, J.R. Accessibility and usability: Green space preferences, perceptions, and barriers in a rapidly urbanizing city in Latin America. *Landsc. Urban Plan.* **2012**, *107*, 272–282. [CrossRef]
49. Arnberger, A. Urban densification and recreational quality of public Urban green spaces-A viennese case study. *Sustainability* **2012**, *4*, 703–720. [CrossRef]
50. Arnberger, A.; Haider, W. Social effects on crowding preferences of urban forest visitors. *Urban For. Urban Green.* **2005**, *3*, 125–136. [CrossRef]
51. Encyklopedia Krakowa. *Encyklopedia Krakowa*; Stachowski, A.H., Adamczyk, E., Eds.; PWN: Warszawa/Kraków, Poland, 2000.
52. Romańczyk, K.M. Krakow—The city profile revisited. *Cities* **2018**, *73*, 138–150. [CrossRef]
53. Purchla, J. *Kraków w Europie Środka*; BOSZ: Lesko, Poland, 2007.

54. Office of the City of Krakow (UMK). Poczet Krakowski. Available online: https://www.poczetkrakowski.pl/ (accessed on 1 January 2021). (In Polish).
55. Kurek, S.; Wójtowicz, M.; Gałka, J. The changing role of migration and natural increase in suburban population growth: The case of a non-capital post-socialist city (The Krakow Metropolitan Area, Poland). *Morav. Geogr. Rep.* **2015**, *23*, 53–70. [CrossRef]
56. Musiał-Malagó, M. Procesy suburbanizacji obszarów podmiejskich Krakowa. *Zesz. Nauk. Uniw. Ekon. Krakowie* **2014**, *936*, 63–77. [CrossRef]
57. Harańczyk, A. Procesy Suburbanizacji w Krakowskim Obszarze Funkcjonalnym. *Stud. Miej.* **2015**, *18*, 85–102.
58. Office of the City of Krakow (UMK). Raport o Stanie Miasta 1991–2019. Available online: https://www.bip.krakow.pl/?id=509 (accessed on 12 September 2020). (In Polish).
59. Zespół Parków Krajobrazowych Województwa Małopolskiego. Available online: https://zpkwm.pl/ (accessed on 29 December 2020).
60. Giannakis, E.; Bruggeman, A.; Poulou, D.; Zoumides, C.; Eliades, M. Linear Parks along Urban Rivers: Perceptions of Thermal Comfort and Climate Change Adaptation in Cyprus. *Sustainability* **2016**, *8*, 1023. [CrossRef]
61. Office of the City of Krakow (UMK). Kierunki Rozwoju i Zarządzania Terenami Zieleni w Krakowie na Lata 2017–2030 (In Polish); Kraków. 2017. Available online: https://zzm.krakow.pl/dla-mieszkancow/kriztz/tresc.html (accessed on 22 October 2020).
62. Dąbrowska-Milewska, G. Czy w Polsce potrzebne są krajowe standardy urbanistyczne dla terenów mieszkaniowych? *Arch. Artibus* **2010**, *2*, 12–16.
63. Ministerstwo Infrastruktury. Rozporządzenie Ministra Infrastruktury z dnia 12 kwietnia 2002 r. w Sprawie Warunków Technicznych, Jakim Powinny Odpowiadać Budynki i ich Usytuowanie, Wraz z Późniejszymi Zmianami. Dz.U. 2002 nr 75 poz. 690 z późn. zm. (In Polish). Available online: https://isap.sejm.gov.pl/isap.nsf/download.xsp/WDU20020750690/O/D20020690.pdf (accessed on 21 October 2020).
64. Dąbrowska-Milewska, G. Standardy urbanistyczne dla terenów mieszkaniowych—wybrane zagadnienia. *Arch. Artibus* **2010**, *2*, 17–31.
65. Węcławowicz, G. Urban Development in Poland, from the Socialist City to the Post-Socialist and Neoliberal City. In *Social Polarisation in the New Town Regions of East-Central Europe*; Institute for Sociology, Centre for Social Sciences Hungarian Academy of Sciences: Budapest, Hungary, 2016; pp. 65–82.
66. Taubenböck, H.; Gerten, C.; Rusche, K.; Siedentop, S.; Wurm, M. Patterns of Eastern European urbanisation in the mirror of Western trends—Convergent, unique or hybrid? *Environ. Plan. B Urban Anal. City Sci.* **2019**, *46*, 1206–1225. [CrossRef]
67. Office of the City of Krakow (UMK). STUDIUM Uwarunkowań i Kierunków Zagospodarowania Przestrzennego Miasta Krakowa—Dokument Ujednolicony; Kraków. 2014. Available online: https://www.bip.krakow.pl/?id=48 (accessed on 12 November 2020). (In Polish).
68. Hołuj, A.; Zawilińska, B. Planning Documents Issued in Poland at the Municipal Level. *Example of the Krakow Metropolitan Area. J. Settl. Spat. Plan.* **2013**, *4*, 122–124.
69. Musiał-Malagó, M.; Hołuj, A. Znaczenie dokumentów planistycznych w zagospodarowaniu przestrzeni Krakowa. *Zesz. Nauk. Uniw. Ekon. Krakowie* **2014**, *1*, 105–126. [CrossRef]
70. Wagner, M. Evading spatial planning law-Case study of Poland. *Land Use Policy* **2016**, *57*, 396–404. [CrossRef]
71. Badora, K. O potrzebie i możliwościach przebudowy Krajowego Systemu Ochrony Krajobrazu About need and possibilities of reconstruction of the National Landscape Conservation System. *Landscape* **2009**, *23*, 29–35.
72. Zachariasz, A. Development of the System of the Green Areas of Krakow from the Nineteenth Century to the Present, in the Context of Model Solutions. In *IOP Conference Series: Materials Science and Engineering*; Institute of Physics Publishing: Prague, Czech Republic, 2019; Volume 471, p. 110297.
73. Semczuk, M.; Serafin, P.; Zawilińska, B. Dzielnice samorządowe w świadomości mieszkańców miasta na przykładzie wybranych dzielnic Krakowa. *Biul. Kom. Przestrz. Zagospod. Kraj. Pan* **2018**, *270*, 159–173.
74. General Statistics Office (GUS). Available online: https://krakow.stat.gov.pl/ (accessed on 20 November 2020).
75. De Sousa Silva, C.; Viegas, I.; Panagopoulos, T.; Bell, S. Environmental Justice in Accessibility to Green Infrastructure in Two European Cities. *Land* **2018**, *7*, 134. [CrossRef]
76. Fan, P.; Xu, L.; Yue, W.; Chen, J. Accessibility of public urban green space in an urban periphery: The case of Shanghai. *Landsc. Urban Plan.* **2017**, *165*, 177–192. [CrossRef]
77. Mears, M.; Brindley, P.; Jorgensen, A.; Maheswaran, R. Population-level linkages between urban greenspace and health inequality: The case for using multiple indicators of neighbourhood greenspace. *Health Place* **2020**, *62*, 102284. [CrossRef]
78. Stein, C.S. The Radburn Plan. Notes on the new town planned for the city Housing corporation. In *The Writings of Clarence S. Stein: Architect of the Planned Community*; Parsons, K.C., Ed.; John Hopkins University Press: Baltimore, MD, USA, 1998; pp. 150–152.
79. Perry, C. The Neighborhood Unit: A Scheme of Arrangement for the Family Life Community. *Reg. Plan N. Y. Its Envion.* **1929**, *8*, 30.
80. El-Geneidy, A.; Grimsrud, M.; Wasfi, R.; Tétreault, P.; Surprenant-Legault, J. New evidence on walking distances to transit stops: Identifying redundancies and gaps using variable service areas. *Transportation* **2014**, *41*, 193–210. [CrossRef]
81. Wen, C.; Albert, C.; Von Haaren, C. Equality in access to urban green spaces: A case study in Hannover, Germany, with a focus on the elderly population. *Urban For. Urban Green.* **2020**, *55*, 126820. [CrossRef]

82. Campos-Sánchez, F.S.; Valenzuela-Montes, L.M.; Abarca-Álvarez, F.J. Evidence of Green Areas, Cycle Infrastructure and Attractive Destinations Working Together in Development on Urban Cycling. *Sustainability* **2019**, *11*, 4730. [CrossRef]
83. Chang, Q.; Li, X.; Huang, X.; Wu, J. A GIS-based Green Infrastructure Planning for Sustainable Urban Land Use and Spatial Development. *Procedia Environ. Sci.* **2012**, *12*, 491–498. [CrossRef]
84. Campos-Sánchez, F.S.; Abarca-Álvarez, F.J.; Reinoso-Bellido, R. Assessment of open spaces in inland medium-sized cities of eastern Andalusia (Spain) through complementary approaches: Spatial-configurational analysis and decision support. *Eur. Plan. Stud.* **2019**, *27*, 1270–1290. [CrossRef]
85. Dudzic-Gyurkovich, K. *Pokonywanie Barier Urbanistycznych Związanych z Układami Transportu na Obszarze Metropolii Barcelońskiej—Wybrane Problemy*; Wydawnictwo PK: Kraków, Poland, 2019.
86. Strohmeier, F. Barriers and their Influence on the Mobility Behavior of Elder Pedestrians in Urban Areas: Challenges and Best Practice for Walkability in the City of Vienna. *Transp. Res. Procedia* **2016**, *14*, 1134–1143. [CrossRef]
87. Talavera-Garcia, R.; Soria-Lara, J.A. Q-PLOS, developing an alternative walking index. A method based on urban design quality. *Cities* **2015**, *45*, 7–17. [CrossRef]
88. Badland, H.M.; Keam, R.; Witten, K.; Kearns, R. Examining public open spaces by neighborhood-level walkability and deprivation. *J. Phys. Act. Health* **2010**, *7*, 818–824. [CrossRef]
89. Lo, R.H. Walkability: What is it? *J. Urban.* **2009**, *2*, 145–166. [CrossRef]
90. Duncan, D.T.; Aldstadt, J.; Whalen, J.; Melly, S.J.; Gortmaker, S.L. Validation of Walk Score ® for Estimating Neighborhood Walkability: An Analysis of Four US Metropolitan Areas. *Int. J. Environ. Res. Public Health* **2011**, *8*, 4160–4179. [CrossRef] [PubMed]
91. Gehl, J. *Life between the Buildings: Using Public Space*; Island Press: Washington, DC, USA, 2011.
92. Feltynowski, M.; Kronenberg, J.; Bergier, T.; Kabisch, N.; Łaszkiewicz, E.; Strohbach, M.W. Challenges of urban green space management in the face of using inadequate data. *Urban For. Urban Green.* **2018**, *31*, 56–66. [CrossRef]

Article

Correlation between Land Use and the Transformation of Rural Housing Model in the Coastal Region of Syria

Nebras Khadour, Nawarah Al Basha, Máté Sárospataki and Albert Fekete *

Institute of Landscape Architecture, Urban Planning and Garden Art Budapest, Hungarian University of Agriculture and Life Sciences—MATE, 1118 Budapest, Hungary; Nebras.Khadour@hallgato.uni-szie.hu (N.K.); Albasha.Nawarah@hallgato.uni-szie.hu (N.A.B.); Sarospataki.Mate@uni-mate.hu (M.S.)

* Correspondence: Fekete.Albert@uni-mate.hu

Abstract: The phenomenon of urban sprawl has caused radical changes in the spatial structure of cities and rural areas all around the world. Syria is among the developing countries that have experienced this phenomenon. Some of the resulted processes of urban sprawl like urbanization and counter-urbanization have had a clear impact on the land use and lifestyle in both cities and the countryside of different regions in Syria. This research focuses on the coastal region and the spatial changes that affected the nature of social life, such as the rapid growth of the population, the expansion of cities, and the new developments, which in turn have led to considerable changes in the relationship and scale of the house, garden, and landscape. The research studies the development of the rural housing model in the coastal region and its relation to the surrounding landscape. It tracks three phases of the housing unit's development and conducts a comparative study on four villages using a questionnaire to evaluate the performance of those units. The results of this research show significant change in the relationship between rural and urban areas resulting from the new residential developments, as well as the relationship of land use and the historic plot structure and that of the garden and the house into the overall character of the landscape.

Keywords: landscape character; rural housing; urbanization; counter-urbanization; heritage protection; rural social life

Citation: Khadour, N.; Basha, N.A.; Sárospataki, M.; Fekete, A. Correlation between Land Use and the Transformation of Rural Housing Model in the Coastal Region of Syria. *Sustainability* **2021**, *13*, 4357. https://doi.org/10.3390/su13084357

Academic Editor: Jan K. Kazak, Katarzyna Hodor and Magdalena Wilkosz-Mamcarczyk

Received: 27 February 2021
Accepted: 9 April 2021
Published: 14 April 2021

Publisher's Note: MDPI stays neutral with regard to jurisdictional claims in published maps and institutional affiliations.

1. Introduction

The phenomena of urbanization and counter-urbanization involve demographic, social, and economic processes, which can explain some significant changes (both spatial and intangible) in rural communities, especially changes related to lifestyle and land use. During the course of the last century, Syria has experienced similar processes as a result of fundamental changes in political and economic fields. The rural landscape has been radically transformed because of the rapid growth of population and the expansion of cities, to the detriment of the adjacent rural areas, the development of infrastructure, and new highways, as well as many other important developments.

Agriculture used to be the primary function of rural areas in the past, and the dominance of farmstead buildings at varying densities and in different spatial arrangements was characteristic for the architecture in the villages [1]. However, as the urbanization process was intensified with time, the rural areas became better equipped with social and technical infrastructure, and interest in an urban way of life increased; thus, new functions started to prevail over the agricultural. Those new activities, accompanied by great progress in construction to accommodate the growing population of the region, have made many farmers to turn to non-farming activities. This has led to a certain diversification of land uses, as farmers put their land to a variety of uses, such as commercial activities [1–3].

This wide range of changes has been the generator of new morphogenetic structures of rural settlements and landscapes, with the housing units taking over the farmland. Our analysis aims at exploring the changes transforming the landscape of the coastal region, as well as identifying the positive and negative aspects of this transformation and how the relationship between man and nature is affected during the process.

2. Historical Review

In accordance with the major changes (in politics, economy, transport, society etc.) affecting the rural development of the region [4,5], three main phases of transformation of the coastal landscape are possible to identify. As a consequence, three main types of relationships between the residential building and the surrounding land and of indoor and outdoor functional layouts are possible to distinguish:

2.1. Until 1970: The Traditional Residential Unit in the First Phase

The traditional residential unit is a reflection of the feudal system, from the time when most of the people lived in poverty and had very limited capabilities, while the power and large estates were in the hands of a few feudal lords [4]. These traditional houses were built before the 1950s; the design was simple and functional, reflecting the rural lifestyle of the period. The houses had a related strongly to the surrounding land, where most of the inhabitants practiced agricultural activities as the main source of their income. The unit consisted of two parts: the inner space of the house itself that served mainly for sleeping, including a separate space to shelter and protect the domestic animals at night, and the outer space adjacent to the house, where most of the daily activities of the inhabitants occurred. This outer area resembled the living room of the house; people were very accustomed to spending their time and practicing their daily activities, such as social encounters, cooking, eating, and relaxing, in the outer space, with the exception of extreme weather situations, therefore, this area was the core of the traditional unit and allowed a strong relationship between the inhabitants and the land. The people used the surrounding land intensively as a source of food and a space for living without increasing the built area in the village (Figure 1).

Figure 1. A traditional residential unit in the first phase, Jableh countryside. (**a**) The house; (**b**) the layout (source: photo and figure by N. Khadour).

The residential units were built by the inhabitants themselves, with traditional methods and using only materials that existed in the surrounding nature (stone, wood, and clay), which gives these units a great ecological value, as they are able to provide a comfortable atmosphere during extreme weather conditions [5].

The layout of the village developed as a combination of large properties of agricultural land that belonged to the feudal lords and much smaller properties of those traditional residential units, where locals lived and owned only the space that they live in while working for the feudal lords on the bigger plots [6]. The distribution of these traditional units was also controlled by the feudal lords. The village lacked public services and had a poor connection to the city. Its economic system relied mainly on agriculture and animal products.

The feudal and bourgeois system continued in Syria until the new law of agrarian reform was announced in 1958 [7]. The law mandated the formation of a farmer's cooperative association to support farmers who reclaimed their lands. From a social point of view, the law aimed to reduce inequality in the distribution of land among social groups in rural areas, and to reduce the severity of social injustice and the poor distribution of income caused by poor distribution of agricultural property. From an economic point of view, the law aimed to direct rich groups to invest in other non-agricultural service sectors in order to create additional job opportunities and strengthen the national economy in general.

However, no significant change happened in the few years following the announcement of the law. It took around ten years for the results of the law start appearing clearly in the countryside in terms of transforming the social life and economic situation for the rural population. With this new development of the economic and social conditions in the countryside, a new change of the housing units and the village layout started to appear.

2.2. 1970–1990: The Traditional Residential Unit in the Second Phase

This period can be described as a transitional period between the feudal system and the communist system. It lasted for about thirty years and witnessed some simple but very essential changes regarding the rural lifestyle. After 1970, as a reflection of the improvement of the economic situation [8], the rural housing unit developed. The buildings became larger, and the animal space was separated from the inner space of the house, which reflected an improvement of the economic state of the community and their lifestyle. The outer area in front of each house remained an important part of the unit, as it still accommodates most daily activities, emphasizing the strong connection between the inhabitants and their land. This type of building has replaced the first type and dominated the rural housing scenery until the third type appears at the beginning of the 2000s.

New building materials started to appear, so while the walls of houses were still constructed of natural stone, concrete started to appear in the construction of the roofs, which led to the opportunity to create an additional floor, especially since the number of family members was increasing by marriage of the youth [9] (Figure 2).

Figure 2. A traditional residential unit in the second phase, Jableh countryside. (**a**) The house; (**b**) the layout (source: photo and figure by N. Khadour).

Although the improvement of the economic situation of the villagers had its effects on the rural lifestyle, the social ties, and in particular the kinship ties, remained very strong. In particular, all the inhabitants shared agricultural work as a common source of income, which provided them with a sense of unity and affinity.

The layout of the village has simply changed in the coastal region in this period, according to several factors, including

- Economic factors: the abolition of the feudal system and the distribution of large properties to the villagers, providing farmers the right to own the land on which they worked. As a result, the countryside is composed of agricultural land and rural houses, which were built randomly on the farms;
- Social factors: the increase of public services, such as schools and hospitals, as a result of the government's cultural and social service development plan for the countryside;
- Transportation/connectivity: the connectivity between the countryside and the cities has increased dramatically because of new developments in the road system, and this

led to the migration of some of the rural population to the cities, in order to access new job opportunities other than agriculture.

The general changes that started in Syria after 1970 at the national level, accelerating even more after 1990. With the stability of the political situation in the country, the economic revival was clearly apparent in all Syrian cities. Many factories and commercial companies were established, in addition to the recovery of tourism and import and export operations.

This atmosphere initiated the urbanization process of Syrian cities, because most of the development projects were located in the outskirts, encouraging people to leave the countryside and move closer to those new job opportunities, leading to urban expansion without compatible urban development plans for the long term.

In this sense, the urbanization process in the coastal region was also initiated. Commercial activities were closely linked to the presence of seaports in the cities of Tartus and Latakia [10], which were considered Syria's gateways to the world, and most of the goods that entered Syria were arriving through these ports. This led to the development of a new transportation network to connect the coast with the inner part of the country, and to facilitate the process of distributing the goods and materials from the ports to the rest of the country.

The new roads constructed had an impact on the character of the areas which they passed through, and led to significant changes of the land uses alongside. In general, most of these roads crossed rural areas, and this led to the creation of new activities and new land uses that did not exist before. Consequently, commercial and industrial activities increased at the expense of the agricultural activity in the countryside, and more urban activities started appearing in rural areas.

2.3. *After 2000: The Contemporary Residential Unit*

In addition to all the above-mentioned changes and their accelerated effects during this period, population growth increased rapidly (with a population growth estimate of 2.7% between 2000 and 2007 alone [9]), which resulted in raised demands on housing units. The most important change in this sense was the change in the type of housing, where traditional rural houses were disappearing and being replaced with modern concrete buildings, which brought also great changes in the social and rural lifestyle. Consequently, the landscape has changed radically.

The new housing unit was transformed into a multi-story building (mostly 3–4 storys), constructed mainly from concrete, while the use of traditional materials has disappeared. With a more urbanized lifestyle, and less dependence on cattle breeding, there has been no more need for spaces assigned for animals within or adjacent to the house, and the agricultural land that surrounded each unit was neglected as the new housing model weakened the relationship between the inhabitants and the surrounding landscape (Figure 3).

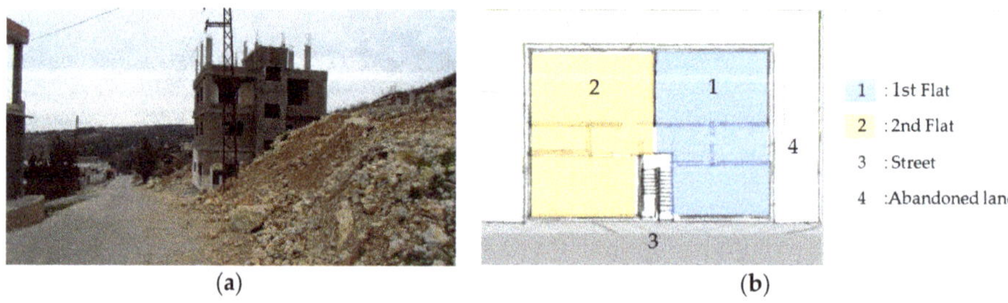

Figure 3. A temporary residential unit Jableh countryside. (**a**) The unit; (**b**) the layout (source: photo and figure by N. Khadour).

The structure of the village has changed drastically (especially after 2011, as the Syrian crisis resulted in families migrating from conflict areas and settling in the coastal region, which was considered as a safe area, thus inducing additional housing demands). The multi-story, concrete buildings dominated the views and the scenery of the village, introducing an urban way of living into the countryside, which was reflected in many aspects. The social ties and relationships of the inhabitants started to weaken as the new model of houses lost the outer open area, which was the core of the rural social life. The more enclosed type of the housing unit encourages independent life, creates boundaries, and decreases the chances of daily encounters and communication, which was a main and celebrated attribute of rural living. Furthermore, as the economic system continued developing, people started finding new investment opportunities to answer the needs of a new lifestyle, and new commercial and service activities started to replace the agricultural ones as they provided easier and faster income. As a result, the cultivated land around the villages became fragmented and abandoned. The Regional Planning Commission in 2005 [11] established that the percentage of workers in the agricultural sector in the coastal region decreased from 35% in 1990 to 11% in 2005, while 60% of the workers are in the services sector and 20% work in commercial activities. This indicates a very serious transformation in the overall status of the coastal region, as it is shifting from being an agriculturally productive region to a consumer region. As a consequence, the balance of the region in social, ecological, and economical fields is being compromised.

As shown in Figure 4, there is a clear relationship between the layout of the residential units and their distribution and the character of the landscape. In the first phase, the landscape is the dominant feature of the rural landscape, where housing units are small in number and allocated randomly, giving priority to preserving the agricultural land. In the second phase, we can observe the appearance of a more organized pattern, with an increasing number of houses around a center, which makes the agricultural land relatively peripheral but still retaining its important role in rural life. As the influence of urbanization grows, the third phase brings an increasing number of roads and a higher density of the built area on account of the agricultural land, which becomes fragmented, losing its dominance over the built area.

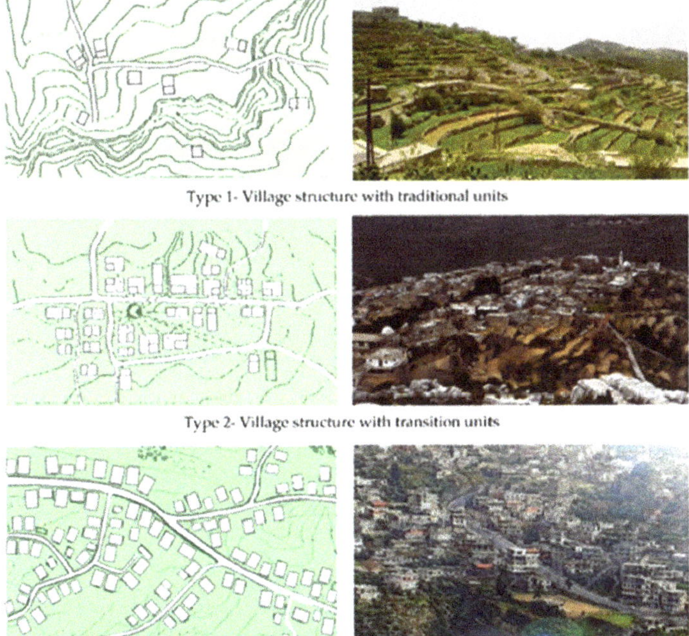

Figure 4. Evolution of the general layouts of the village types (source: photos and drawings by N. Khadour).

3. Materials and Methods

This research deals with rural development in the coastal region of Syria. The Syrian coast is located on the western coastal front of the Syrian state, between the Mediterranean Sea (as the western boundary) and the coastal mountain range that stretches along the coast from the eastern side. The coast extends to a length of 330 km and a width of 30–50 km [12], and includes two governorates: Latakia in the north and Tartous in the south.

The landscape of the Syrian coast is diverse, from the plains that extend from the sea in the west to the mountain areas in the east, with a width ranging between 15 and 20 km. There we can find urban and semirural settlements with fertile agricultural land. The highest point of the Coastal Mountain Range running parallel to the coast is Nabi Yunis (1575 m) near Latakia [13].

The western slopes catch moisture-laden winds from the Mediterranean Sea, and are thus more fertile and more heavily populated than the eastern slopes. In general, the coastal region has a mild Mediterranean climate with a dry summer, and the average temperature is 25.8 °C in the summertime (average June–August) and 12.8 °C in the wintertime (average December–February) [14].

The general analysis focuses on the mountain area of the coastal region; however, a deeper comparative analysis was conducted on four villages (Bet Yashot, Hellet Ara, Helbako, and Almnaizlah) that are located along a main highway axis connecting the coastal region with the middle region, each with a different height and distance from the city (Jableh) shown in Figure 5. The reason why those villages were chosen is related to their location in the mountain area, within the western part of the ridge, because the changes in the landscape were stronger and more spectacular here than the changes in the plains. The latter are further away from the urbanized areas, and thus have been less developed and have kept their vernacular traditions and system for a longer period of time.

Figure 5. Illustrative diagram of the study area (source: Google earth).

The four villages are located sequentially on an axis related to the city center of Jableh, which gives them particular attributes regarding their relationship to a highway connecting the coastal region to the middle region, which was constructed between 1985–1989.

This paper is based on an extensive international overview regarding urban and rural heritage protection and planning guidelines [15–20], comprehensive studies in the field of urban and rural planning regeneration and development [21–35], and case studies in good practice of urban and rural development [36–47]. Using the conclusions of international experiences in the field, the research proceeds with an analytical study of the housing models in the study area, in order to have a clear understanding of the morphological change of each model and its impacts on the social life and the interaction between man and nature. Furthermore, the comparative study on the four villages in the study area is conducted by the use of a questionnaire, in order to get feedback from the inhabitants of those villages and be able to evaluate the performance of each model in the end. Depending on the results, the research will define the correlation between the housing model and the land use as a main goal of the research and provide recommendations.

The analytical study of the four villages will rely on the data collected by the questionnaire, which was a "random sample" distributed to 200 people from the village inhabitants (The portion was close to 1.5% of the population in 2004 per village [48]). For the distribution of the questionnaires, an important consideration was that half of the participants were living in traditional houses and the other half were living in contemporary houses.

The questionnaire was distributed in different proportions to suit the population numbers of each village, as shown in Table 1.

Table 1. The questionnaire distribution numbers in the surveyed villages.

Village Name	Population [48]	Total Number of Questionnaires Distributed in the Village	Number of Questionnaires for Traditional Houses Group	Number of Questionnaires for Contemporary Houses Group
Bet Yashot	6115	90	45	45
Hellet Ara	3155	50	25	25
Helbako	1949	30	15	15
Almnaizlah	1633	30	15	15

The results of the questionnaire were divided into four parts:

- Part 1: General characteristics of the participating groups and their lifestyles;
- Part 2: Transportation system;
- Part 3: Social life;
- Part 4: Ecological aspect.

Each part includes a group of factors that are related and provide specific indicators for the assessment of the questionnaire results.

4. Results and Discussion

4.1. General Characteristics of the Participating Groups and Their Lifestyles

By analyzing the results of the first part of the questionnaire displayed in the summary table, we notice that the average age of the inhabitants living in traditional houses is higher than the average of those living in contemporary houses, and this comes as a logical result after the rapid growth of population during the last three decades, which has increased the demands on constructing more houses in a short time scale. Therefore, the younger generations moved to live in these newly constructed houses, as they are more capable of adapting to a new lifestyle than the older-aged groups, who can be much more attached to their inherited customs, traditions, and lifestyle (Table 2).

By analyzing the results of the first part of the questionnaire displayed in the summary.

Most of the residents of traditional houses own agricultural land, and most of the family members are engaged in agricultural activities. A significant difference can be observed in the second group, where the majority of the residents of new houses are employees in the service sector, and most of them do not own agricultural land (only 35% do). However, most of those who own agricultural land do not engage in any agricultural activity.

These data reflect the change of the concept of agricultural work among residents. For the residents of traditional housing, farmland is still considered as the main pillar of their lives, and agricultural income still counts as essential in their economic system. Therefore, they have stronger connection to their land, and they pay more attention to agricultural activities, because the design and the structure of their houses facilitate this connection to the farms and the surrounding landscape. In contrast, the lifestyle of the residents of new houses has changed. Agricultural work is no longer a priority, as it is no longer considered the primary source of income and a basis of their economic system. Instead, they started heading to government and private jobs that provide a stable income, unlike the agricultural income, which is very much dependent on the weather.

Table 2. Questionnaire results related to the characteristics of the participating groups and their lifestyles.

	Traditional Houses 100 Participants, 100%				New Houses 100 Participants, 100%			
Age average (σ = stander deviation)	50 Years (σ = 15.2)				39 Years (σ = 11.5)			
Professions	Employees 15%	Farmers 64%	Craftsman 11%	Other 10%	Employees 72%	Farmers 12%	Craftsman 7%	Other 9%
Where do you prefer to live?	In the village 78%		In the city 22%		In the village 51%		In the city 49%	
Which type of houses do you prefer to live in?	In a traditional house 69%		In a modern house 31%		In a traditional house 48%		In a modern house 52%	
Do you feel that your house fits the surrounding landscape ?	Yes 98%		No 2%		Yes 66%		No 34%	
Do you own agricultural land in the village?	98%		2%		35%		65%	
If yes, do you do any agricultural activities on this land?	78%		22%		42%		58%	
Any of your family members help with these activities?	55%		45%		25%		75%	
Do you feel that you belong to the place that you live in?	94%		6%		68%		32%	

Figure 6 represent the land use changes in the coastal region in general between 2010–2020 [49]. We can see the increase in the bare and uncultivated lands, especially in the mountainous area of the Syrian coast region, with an increase in the proportion of urban areas, while the plains region is still witnessing agricultural activities, due to the availability of favorable conditions for agriculture, such as fertile land, availability of equipment, and ease of transportation.

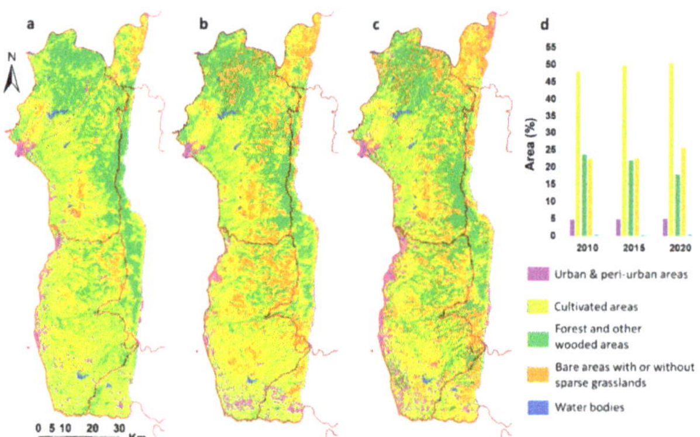

Figure 6. Land use and land cover in the Syrian coastal region between 2010–2020 (source: [49]).

Another factor that has fostered changes in the way of living is the increasing interest in education in rural areas. For educated people, the pursuit of specialized work suited to their education is more common than going into agricultural work that depends on acquired experience and physical effort. That is why we can see that the design of contemporary houses does not focus on the connection with surrounding nature; hence, the residents of these houses lost that feature, and have been relatively distanced from practicing the agricultural activities specific to this area, and even from small-scale manifestations like gardening. They also lost a certain form of social interaction and traditional gatherings, which were essential for achieving a sense of unity and closeness between the rural residents.

On the other hand, the fact that the residents of traditional houses have participated in the process of building their homes with their hands, and that the building materials are natural materials, provides this group with feelings of belonging to the rural environment, which cannot be said about the other group, whose houses came as a response to the increased demand on housing, without respecting the historical legacy of their environment.

To sum up, the relationship that links the residents of traditional houses with their environment and homeland is much more significant than for those living in contemporary houses. This can be explained by the different reality and conditions they are facing, as members of the latter group are in a daily contact with the city, since most of them work there. They will have different needs and develop a different mental image of their life routine related to the practical and artificial lifestyle of the city.

4.2. Transportation System

One of the statements on the questionnaire was "Please give an evaluation of the public transportation system in your village".

Through the results of the questionnaire were related to the effectiveness of the public transportation system in the studied villages, as shown in Figure 7, we can see that the evaluation of the transportation means by the inhabitants of the villages seems to be more positive and rated as effective in the villages closer to the city.

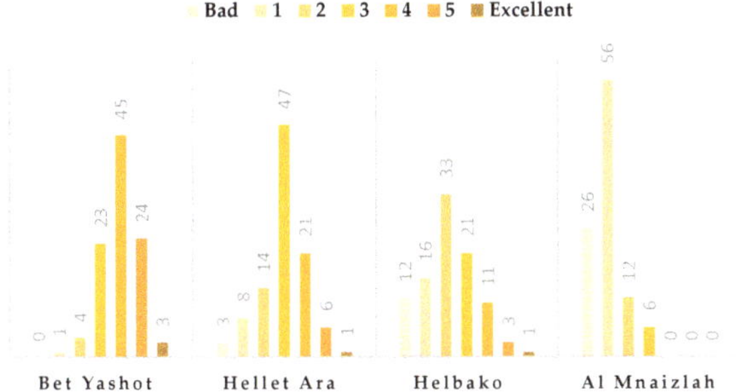

Figure 7. Questionnaire results related to the transportation system.

Comparing these results to the statistics issued by the Regional Planning Commission in Latakia in 2008, related to the proportions of workers in the agricultural sector shown in Table 3 and the proportion of contemporary and traditional housing in the study villages shown in Figure 8, we note that the ratio of workers in the agricultural industry decrease gradually in the villages that are closer to the urban center. At the same time, the amount of contemporary housing is increasing progressively by getting closer to this urban center, while the numbers of traditional houses are greater in the villages that are farther from the center of Jableh.

Table 3. Percentage of workers in the agriculture sector and other sectors in the studied villages, from the Regional Planning Commission in Latakia (2008).

Village Name	Distance from the City	Agriculture	Industry	Real Estates	Constructions	Restaurants Hotels	Transportation	Service Sector
Bet Yashot	20 Km	4.5%	3.0%	0.5%	4.7%	2.5%	2.7%	81.5%
Hellet Ara	25 Km	12.0%	2.5%	3.5%	5.0%	5.0%	4.5%	67.5%
Helbako	30 Km	22.0%	1.5%	3.0%	4.5%	4.0%	4.0%	61.0%
Al Mnaizlah	35 Km	35.0%	3.0%	2.5%	3.0%	2.5%	3.0%	51.0%

Similarly, we can see a clear relationship between the effectiveness of the transportation system and the increasing impact of urban areas on the surrounding rural environment, and the decisive factor appears to be the proximity of the new housing to the urban center.

This is largely related to the overlap of economic systems between cities and the surrounding rural centers, which has increased by facilitating connectivity through the transportation system. At the same time, rural areas farther from the urban center, with less effective transportation systems, seem to have preserved the traditional way of living—

that is, the villages that are difficult to reach have more independent economic systems, and their built and natural environments kept the traditional character.

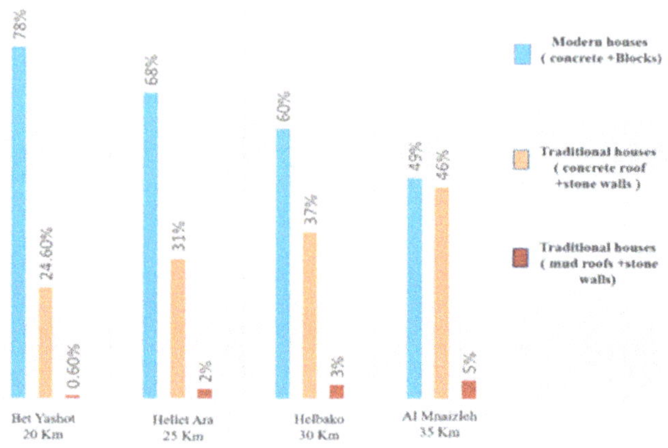

Figure 8. The ratio of modern and traditional houses in the four villages. Source: The Regional Planning Commission in Latakia, 2008.

4.3. Social Life

The third section of the questionnaire includes questions related to social life in the Syrian coastal countryside. Based on the results, it is clear that social relationships and ties have changed for the residents of the new and traditional houses, which has affected social life in the countryside in general. (Table 4)

Table 4. Questionnaire results related to social life.

	Traditional Houses: 100 Participants, 100%					New Houses: 100 Participants, 100%				
Do you do any activities in the nature around you?	No 8%	Picnic 14%	Sports 61%	Farming 84%	Other 22%	No 24%	Picnic 34%	Sports 53%	Farming 22%	Other 24%
Do you have good relations with the neighbors?	Yes 76%		Medium 16%		No 8%	Yes 33%		Medium 49%		No 18%
Do you share your house with another family (are the different generations live together)?	Yes 68%		No 32%			Yes 19%		No 81%		

The social relationships and ties that link the residents of traditional houses together are stronger and more durable, and this is due to several reasons; most important among them is the layout of the traditional residential units, which were open towards the outside, allowing a physical connection to the surrounding environment. This allowed visual connection to other houses, encouraging more social interaction and weakening boundaries. The rural society with traditional houses was considered to be an integral unit, where people coexisted in very harsh conditions in terms of political and economic organizational aspects. In particular, when the feudal system was dominant over the entire countryside of the coast, it was very important for the inhabitants to stand beside each other and strengthen their relationships.

On the other hand, a significant characteristic of the traditional residential unit was that it occupied a relatively large area and accommodated more than one family, as a result of difficult economic conditions that the Syrian coastal countryside witnessed in recent decades, so it was useful to increase the number of family members for a better life, as it was seen as an increase in the workforce in agriculture, and thus an increase in the financial income of the family. Due to this state of family unity and economic difficulties, when a family member wanted to get married and form his or her own family, a new room was constructed attached to the original house as a solution to avoid wasting agricultural land and money.

The changes in the rural economic system during the past three decades, along with increased connections to the nearby urban areas by the transportation system and the move of large number of rural residents to live or work in the city, have led to clear changes in social life in the countryside. The influence of urbanization has started to appear in every aspect of the rural inhabitant's daily life, and a tendency toward increased privacy and emphasis on territorial boundaries has started to develop. This process was reflected in the spatial structure of the village with the increase of multi-story buildings similar to those in the city, due to the profit they provide to the owner, since they can accommodate multiple families, with each having an independent home and life.

4.4. Ecological Aspect

According to the results displayed in Table 5, there has been a clear shift in the perception toward the land and how it can be used. Instead of the old traditional gardening and agricultural activities that were meant to provide the daily needs of inhabitants, people have started to allocate all of the land to grow more profitable crops that provide more and stable income, such as olive trees and tobacco (see Figure 9).

Figure 9. Tobacco cultivation in Bet Yashout (source: photo by N. Khadour).

It is clear that both new and traditional houses suffer from some functional problems, but the nature and quality of these problems differ between the two types.

The traditional houses have challenges in adopting innovations, such as the ability to allow electrical installations, since most of them need electrical connections to be installed inside the walls built of stone. In contrast, the contemporary houses were built taking into account requirements like electrical appliances and their extensions, as well as advanced sewage systems. On the other hand, most of the problems that the residents of contemporary houses suffer from are related to the special environmental and climatic conditions in this rural mountain area, which is characterized by cold, rainy winters and mild summer weather. Most of the residents suffer from moisture inside their homes and difficulties with thermal insulation, especially in the winter. The traditional houses are more adaptable to the environment in that sense; they are more efficient from an ecological aspect, because they are built with local natural materials and inherited traditional methods, which give these type of houses more ecological value. For example, the natural stone used in construction constitutes significant thermal insulation, and maintains a moderate temperature inside the house in summer and winter; in addition, the rainwater from the winter was stored in tanks built from stone adjacent to the house, to be used later in the summer. The ecological superiority of the traditional housing units is very clear compared to the incompetence of the contemporary ones, which were designed and built with no other consideration than a quick solution to accommodate the increasing population.

Table 5. Questionnaire results related to the ecological aspect for the houses in the studied villages.

	Traditional Houses: 100 Participants, 100%					New Houses: 100 Participants, 100%				
If you live in a new house: Do you have any connection with the surrounding landscape?						Yes, I have a garden and I use it 21%		Yes, I have a garden, but I do not use it 23%		No, I do not have any connection 56%
Do you have any problems with your house?	Yes 53%		No 47%				Yes 62%		No 38%	
If yes, which kind of problems do you have?	Thermal insulation 7%	Moisture problems 14%	Sanitation problems 43%	Electrical Problems 62%	Water problems 14%	Thermal insulation 63%	Moisture problems 71%	Sanitation problems 5%	Electrical Problems 11%	Water problems 54%

5. Conclusions

The major changes that occurred in the Syrian coastal region during the last century in political, social, and economic terms have led to fundamental changes in the planning of the coastal cities and fostered their economic growth, which has transformed them into polarization centers [50]. This has encouraged migration from rural areas towards the urban centers, because they provided more work opportunities. This has initiated the process of urbanization and counter-urbanization in the region.

The effects of urbanization have reached some remote villages of the Syrian coast, and this was facilitated by the development of the transportation system, which led to a significant change in the land use in these villages, especially in those with direct connection to the main transportation routes. The income provided by commercial and industrial functions developed along these routes has presented a more stable and reliable source of living than agricultural income, which is subject to weather. However, the construction of regional roads, such as the highway connecting the coastal region with the middle region in the study area, brings radical changes and development to the surrounding areas, and in some cases, it can be a threat to losing heritage values. That is why a comprehensive plan should address these aspects, and make sure that the impact of this construction is controlled and preplanned.

Over the last three decades, the impact of these changes has increased to include most rural areas on the Syrian coast, but in varying degrees, depending on its position relative to the urban centers.

Changes in land use in rural areas have been accompanied by changes in housing models. The recent residential development model represents a new style, with a greater resemblance to the urban modern residential units, and thus has abandoned some attributes and values to learn from architectural ingenuity in adapting to the close environment and the tangible evidence reflecting the inhabitants' traditions and strong connection with nature. Furthermore, the old traditional house was constructed of natural materials, and presents an ecological housing model that is compatible with the environment of the region, while the recent model has lost this advantage due to the use of concrete and artificial materials. The diversification and change of activities in the areas adjacent to the rural houses has had a great impact on the nature of interaction between the village residents and nature. The horticultural activities have deteriorated, and residents no longer rely on the land to provide their daily needs of food and supplements, as these were easy to access from newly found commercial businesses. The agricultural activities are now limited to producing crops for commercial purposes, especially tobacco and olive fruits.

The case study reflects the variety of impacts of urbanization and counter-urbanization processes, which appear clearly in every detail of the lifestyle of the community; however, the developments that are related to the change in the use of lands, which have a direct impact on houses and the new way of interaction between the dwellers and the surroundings of their houses, is very crucial. The questionnaire results indicate a loss of connection and change in the recognition of nature for the dwellers of the new housing model, which can create difficulties for future development plans, especially ones that aim to restore the values and characteristics of the coastal region, as it can be harder to motivate people to engage in the process without the appreciation of their values.

New technologies and materials in housing result in ecological disadvantages, as well as negative consequences for living conditions (poor heat isolation, dampness problems, connection to nature lost). The answers to questionnaires clearly reflected that ties to the local landscape are much stronger for those living in rural areas than for those residing in crowded contemporary urban housing. The economic dependency in agriculture and direct connection to nature result in a greater awareness towards the landscape among the inhabitants living in rural places, with lifestyle contributing to the preservation of the landscape character and identity.

Although the development projects for the region were planned to improve the quality of life of the rural areas and provide more services to these areas, they had also an impact on the identity and the cultural heritage of the countryside in the region, and made it take an undesirable direction. It gave the villages the characteristics of the city with no regards to the important assets of the natural environment and its contribution to the social life and spatial identity of the countryside.

Housing and rural development policies must focus on the importance of the cultural and social dimensions of rural society, which preserve the essence of the relationship between man and nature and prevent the transformation of rural housing into a consumer unit that forms a barrier between the inhabitant and the natural environment.

Author Contributions: Conceptualization, N.K. and N.A.B.; methodology, N.K. and N.A.B.; validation, M.S. and A.F.; investigation, N.K.; resources, N.K. and N.A.B.; data curation, N.K. and N.A.B.; writing—original draft preparation, N.K.; writing—review and editing, N.K., N.A.B., M.S. and A.F.; supervision, M.S. and A.F.; project administration, A.F. and M.S.; funding acquisition A.F. and M.S. All authors have read and agreed to the published version of the manuscript.

Funding: This research received no external funding. The publication was supported by Hungarian University of Life Sciences—MATE.

Institutional Review Board Statement: Not applicable.

Informed Consent Statement: Not applicable.

Data Availability Statement: Not applicable.

Conflicts of Interest: The authors declare no conflict of interest.

References

1. Banski, J.; Weslowsk, M. Transformations in housing construction in rural areas of Poland's Lublin region—Influence on the spatial settlement structure and landscape aesthetics. *Landsc. Urban Plan.* **2010**, *94*, 116–126. [CrossRef]
2. Sevenant, M.; Antrop, M. Settlement models, land use and visibility in rural landscapes: Two case studies in Greece. *Landsc. Urban Plan.* **2007**, *80*, 362–374. [CrossRef]
3. Unvin, T.; Nash, B. Township boundaries: Theoretical considerations and analytical implications. In *The Transformation of the European Rural Landscape: Methodological Issues and Agrarian Change 1770–1914*; Verhoeve, A., Vervoloet, J., Eds.; Société Belge d'Etudes Géographiques: Bruxelles, Belgium, 1992; Volume 61, pp. 116–127.
4. Weulersse, J. *Le pays des alaouites tours Tours, Arrault & cie, maiˆtres imprimeurs*; Uniberstè de Paris: Paris, France, 1940.
5. Zakkar, S.; Alwaraa, G. *The Countryside of Latakia between Architecture and Heritage from settling until 1925*(زكار. س، الورعة، غ. ريف اللاذقية بين العمارة والتراث من بداية الاستقرار وحتى العام 1925، دار التكوين); Dar Altaqueen: Dmascus, Syria, 2013.
6. Yaseen, A. *The Story of the Land and the Farmer* (ياسين. أ، حكاية الأرض والفلاحدار الحقائق, النسخة الأولى), 1st ed.; Dar Alhaqaeq: Beirut, Lebanon, 1979.
7. Act 161 of Land Reform, 1958—Syrian Parliament Official Website. Available online: http://www.parliament.gov.sy/arabic/index.php?node=201&nid=10641&ref=tree& (accessed on 3 November 2020).
8. Deep, K. *Contemporary History of Syria* (ديب. ك، تاريخ سورية المعاصردار النهار, النسخة الأولى), 1st ed.; Dar An-Nahar: Beirut, Lebanon, 2011.
9. Barut, J.M. *The First Basic National Advisory Report "Syria 2025"*; State Planning Commission and UNDP: Damascus, Syria, August 2007.
10. Sofi, A. Latakia seaport. Historical and economical study (مجلة العمران، نسخة خاصة عن الساحل السوري، نسخة). *Urban J. Special Issue Syr. Coast. Minist. Munic.* **1968**, *25–26*, 34–49.
11. Official Website of the Syrian Regional Planning Commission. Available online: http://www.rpc.gov.sy (accessed on 10 October 2020).
12. Fadel, I. The Demographic Characteristics and its Role in The Development of The Coastal Territory of Syria. *Tishreen Univ. J. Res. Sci. Stud. Arts Hum. Ser.* **2017**, 143.
13. Collelo, T. *Syria: A Country Study*; Federal Research Division, Library of Congress: Washington, DC, USA, 1988. Available online: https://www.loc.gov/item/87600488/ (accessed on 27 February 2020).
14. Latakia Climate. Available online: https://www.weather-atlas.com/en/syria/latakia-climate (accessed on 27 February 2020).
15. Buchanan, C.; Shanahan, M. *Cork Rural Design Guide: Building a New House in the Countryside*; Cork County Council: Cork, Ireland, 2003.
16. Müller, N.; Werne, P.; Kelcey, J.G. (Eds.) *Urban Biodiversity and Design*; Blackwell Publishing Ltd.: Oxford, UK, 2010.
17. Carmona, M. *Companion to Urban Design. Decoding Design Guidance*; Rutledge: New York, NY, USA, 2011; pp. 288–303.
18. Tsakiri, E. Polis Painting: The creation of a City Image based on Narrative, Urban Analysis and Fine Art principles. In *Changing Cities II: Spatial, Design, Landscape & Socio-Economic Dimensions*; Gospodini, A., Ed.; Grafima Publications: Thessaloniki, Greece, 2015; pp. 2331–2342. ISBN 978-960-6865-88-6.

19. Council of Europe Conferences of Ministers responsible for Spatial/Regional Planning. *European Rural Heritage Observation Guide*; 13 CEMAT (2003)4; Council of Europe: Brussels, Belgium, 2003. Available online: https://rm.coe.int/16806f7cc2 (accessed on 2 December 2020).
20. ICOMOS. ICOMOS-IFLA Principles Concerning Rural Landscapes as Heritage. Available online: https://www.icomos.org/images/DOCUMENTS/General_Assemblies/19th_Delhi_2017/Working_Documents-First_Batch-August_2017/GA2017_6-3-1_RuralLandscapesPrinciples_EN_final20170730.pdf (accessed on 30 July 2018).
21. Ellin, N. *Postmodern Urbanism*; Princeton Architectural Press: New York, NY, USA, 1996; pp. 67, 192.
22. Bengston, D.N.; Fletcher, J.O.; Nelson, K.C. Public policies for managing urban growth and protecting open space: Policy instruments and lessons learned in the United States. *Landsc. Urban Plan.* **2004**, *69*, 271–286. [CrossRef]
23. Koolhas, R. Whatever happened to urbanism. *Des. Q.* **1995**, *164*, 28–31. [CrossRef]
24. Bruegmann, R. *Sprawl, a Compact History*; University of Chicago Press: Chicago, IL, USA, 2005.
25. Riganti, P.; Nijkamp, P. The Value of Urban Cultural Heritage: An Intelligent Environment Approach. *Reg. Sci.* **2006**, *36*, 451–469. [CrossRef]
26. Gehl, J. *Cities for People*; Island Press: Washington, DC, USA, 2010; p. 93.
27. Offenhuber, D.; Ratti, C. *Decoding the City: Urbanism in the Age of Big Data*; Birkhäuser: Berlin, Germany; Boston, MA, USA, 2014.
28. Gehl, J.; Svarre, B. *How to Study Public Life*; Island Press: Washington, DC, USA, 2013.
29. Abdullah, J. City Competitiveness and Urban Sprawl: Their Implications to Socio-Economic and Cultural Life in Malaysian Cities. *Procedia Soc. Behav. Sci.* **2012**, *50*, 20–29. [CrossRef]
30. Ehrlich, M.V.; Hilber, C.A.L.; Schöni, O. Institutional settings and urban sprawl: Evidence from Europe. *J. Hous. Econ.* **2018**, *42*, 4–18. [CrossRef]
31. Lu, D.; Li, L.; Li, G.; Fan, P.; Ouyang, Z.; Moran, E. Examining Spatial Patterns of Urban Distribution and Impacts of Physical Conditions on Urbanization in Coastal and Inland Metropoles. *Remote Sens.* **2018**, *10*, 1101. [CrossRef]
32. Clark, K. Power of Place - Heritage Policy at the Start of the New Millennium. *Hist. Environ. Policy Pract.* **2019**, *10*, 255–281. [CrossRef]
33. Lima, E.G.; Chinelli, C.K.; Guedes, A.L.A.; Vazquez, E.G.; Hammad, A.W.A.; Haddad, A.N.; Soares, C.A.P. Smart and Sustainable Cities: The Main Guidelines of City Statute for Increasing the Intelligence of Brazilian Cities. *Sustainability* **2020**, *12*, 1025. [CrossRef]
34. Davis, P. *Ecomuseums: A Sense of Place*; Leicester University Press: London, UK, 1999.
35. Steiner, A.; Shenggen, F. Rural Areas are in Crisis. Revitalization is the Solution. Thomson Reuters Foundation News. 2019. Available online: http://news.trust.org/item/20190326150827-0ryhf/ (accessed on 25 March 2019).
36. García-Palomares, J.C. Urban sprawl and travel to work: The case of the metropolitan area of Madrid. *J. Transp. Geogr.* **2010**, *18*, 197–213. [CrossRef]
37. Zhao, P. Sustainable urban expansion and transportation in a growing megacity: Consequences of urban sprawl for mobility on the urban fringe of Beijing. *Habitat Int.* **2010**, *34*, 236–243. [CrossRef]
38. Rahnam, M.R.; Wyatt, R.; Heydari, A. What happened from 2001 to 2011 in Melbourne? Compactness versus sprawl. *Sustain. Cities Soc.* **2015**, *19*, 109–120. [CrossRef]
39. Masoumi, H.E.; Hosseini, M.; Gouda, A.A. Drivers of urban sprawl in two large Middle Eastern countries: Literature on Iran and Egypt. *Hum. Geogr.* **2018**, *12*, 55–79. [CrossRef]
40. Persson, Å.; Eriksson, C.; Lõhmus, M. Inverse associations betwwen neighborhood socioeconomic factorsand green structure in urban and suburban municipalities of Stockholm county. *Landsc. Urban Plan.* **2018**, *179*, 103–106. [CrossRef]
41. Taleshi, M.; Zianoushin, M.M. The physical transformations due to rural sprawl settlements of Hamedan periphery. *J. Res. Rural Plan.* **2018**, *70*, 141–160.
42. Li, G.; Li, F. Urban sprawl in China: Differences and socioeconomic drivers. *Sci. Total Environ.* **2019**, *673*, 367–377. [CrossRef]
43. Mehriar, M.; Masoumi, H.; Mohino, I. Urban Sprawl, Socioeconomic Features, and Travel Patterns in Middle East Countries: A Case Study in Iran. *Sustainability* **2020**, *12*, 9620. [CrossRef]
44. Oakes, T. Heritage as Improvement: Cultural Display and Contested Governance in Rural China. *Mod. China* **2013**, *39*, 380–407. [CrossRef]
45. Kuai, Y. Research on Traditional Villages' Protection System in China. *Mod. Urban Res.* **2016**, *31*, 2–9.
46. Andersen, J.E. Housing and The shape of the City Challenging the concept of 'informal' in Sub-Saharan African Cities The case of Maxaquene A, Maputo, Mozambique. In *Cities in Transformation: Housing and the Shape of the City*; Politecnico di Milano: Milano, Italy, 2014.
47. Di Fazio, S.; Modica, G. Historic Rural Landscapes: Sustainable Planning Strategies and Action Criteria. The Italian Experience in the Global and European Context. *Sustainability* **2018**, *10*, 3834. [CrossRef]
48. *Annual report 2004: General Statistic of Latakia Governorate*; Regional planning commission: Latakia, Syria, 2004. Available online: http://cbssyr.sy/ (accessed on 15 October 2013).
49. Mohamed, M. An Assessment of Forest Cover Change and Its Driving Forces in the Syrian Coastal Region during a Period of Conflict, 2010 to 2020. *Land* **2021**, *10*, 191. [CrossRef]
50. Al Zaied, I. Geographical distribution of the population in Latakia Governorate (أ.الزايد. مجلة جامعة دمشق المجلد 32 - التوزع الجغرافي للسكان في محافظة اللاذقية). *Damascus Univ. J.* **2016**, *32*, 391–413.

Article

Do We Need a New Florence Charter? The Importance of Authenticity for the Maintenance of Historic Gardens and Other Historic Greenery Layouts in the Context of Source Research (Past) and Taking into Account the Implementation of the Sustainable Development Idea (Future)

Marzanna Jagiełło

Department of Architecture Conservation and Restoration of Cultural Landscape, Faculty of Architecture Wrocław, University of Science and Technology, 50-317 Wrocław, Poland; marzanna.jagiello@pwr.edu.pl

Abstract: This year, 40 years have passed since the adoption of the basic document for the protection of historic gardens, i.e., the Florence Charter. During this time, its recommendations have been verified by both conservation and researchers' actions, who in various environments discussed its meaning as well as its essential shortcomings. Some of the provisions of the Charter were criticized in the context of the effects of their use, especially those relating to the issue of historic gardens fundamental protection, namely to authenticity in its various scopes with particular emphasis on the use of source research which raises many reservations for conservation actions. Moreover, their excessively superficial interpretation, which was demonstrated by the example of the most popular plant used in regular gardens, namely boxwood. This article presents and analyzes the most important theses of these discussions and the main axes of the dispute, dividing them into two parts, i.e., the first relating to authenticity and the other to the use of sources. On this basis, it was necessary to extend these considerations to all kinds of historic greenery. Attention was also paid to the meaning and scope of authenticity which changed along with the expansion of the semantic field in relation to heritage. Furthermore, the fact that since the adoption of the Florence Charter, some of the aspects of authenticity indicated in the article have been included in other documents developed under the auspices of ICOMOS, but usually relating to the heritage as a whole, sometimes considered regionally. Appendices contain the most important doctrinal documents referring separately to authenticity, meaning, as well as types and the scope of the usage of sources. This article presents new contexts in which authenticity connected with climate change and the postulates of the development doctrine should be considered. Additionally, the article indicates the need to extend the conditions in which to start considering the historical greenery areas, which should be treated as an element of green infrastructure. The article also points to the use of new techniques and tools in research on authenticity. The summary indicated the necessity to continue the discussion on aspects of authenticity in relation to historic greenery layouts. Taking into account all the above aspects and at the same time meeting the ICOMOS "Journeys for Authenticity" initiative, As part of the conclusions from the analyzes carried out in the article, a model of procedure was proposed. It aims to bring us closer to the preparation of a new document recommending the protection of historic greenery, addressing both critical comments about the present Florence Charta, as well as new challenges and opportunities. This model is shown by means of a diagram. Part of it is a set of themes around which around which the debate on the new Florence Charter could be launched. They were assigned to four panels, i.e., I. Historic greenery as an element of heritage: II. Authenticity of historic greenery complexes in research and conservation strategies; III. Authenticity of historic greenery complexes and sustainable development; IV. New techniques and tools in research on the authenticity of historic greenery layouts.

Citation: Jagiełło, M. Do We Need a New Florence Charter? The Importance of Authenticity for the Maintenance of Historic Gardens and Other Historic Greenery Layouts in the Context of Source Research (Past) and Taking into Account the Implementation of the Sustainable Development Idea (Future). *Sustainability* **2021**, *13*, 4900. https://doi.org/10.3390/su13094900

Academic Editor: Jan K. Kazak

Received: 28 February 2021
Accepted: 15 April 2021
Published: 27 April 2021

Keywords: Florence Charter; world heritage convention; cultural heritage; historic gardens; authenticity; criteria; documentation; preservation; restoration; landscape architecture; green infrastructure of the city; sustainable development; interdisciplinary approaches

1. Introduction

The subject of the analyses undertaken in this article covers historic gardens, defined by the fundamental and updated official document concerning their protection, which is the Florence Charter, as "an architectural and horticultural composition of interest to the public from the historical or artistic point of view [. . .] must be preserved in accordance with the spirit of the Venice Charter" [1] (Art. 3), to which also other forms of historic greenery were added such as cemeteries, urban green areas (promenades, parks, boulevards and avenues) and public green spaces characterized in the ICOMOS-IFLA Document on historic urban parks as "an essential and inalienable part of traditions and plans of many towns and settlements" [2] (preamble).

It should be recalled here what was meant when writing about "architectural and horticultural composition". Namely, it is about the layout, plant complexes, structural and decorative elements as well as water with the sky reflected in it [1] (Art. 4).

Most of the preserved gardens and green areas under protection are now much more than 100 years old. Some of their elements show significant durability, however plants, except for some species of trees, do not live that long. The Florence Charter recommends their cyclical replacement, which, however, raises a lot of controversy among researchers, even if care is taken (in historical terms) in the selection of plant species. Questions also appear when it comes to restoration combined with partial restitution of green complexes (full reconstruction as an action the result of which "could not be considered a historic garden" [1] (Art. 17) will in principle not constitute the subject of these considerations). On the one hand, criticism is given to the overly literal reading of the Florence Charter's records regarding the use of sources, as well as incomplete or even incorrect interpretation of source materials. On the other hand, there are methodological errors which seem to be missing a deeper interdisciplinary reflection on various aspects of authenticity in relation to historic gardens. All this resulted in the fact that in many countries (although not in all of them) recommendations for the protection of historic gardens contained in the Florence Charter, as well as in other documents of this rank, are contested or even ignored.

Relatively new aspects have also contributed to this, in the context of which historic green areas are considered, also in the context of authenticity, namely the management of historic gardens, and more broadly of historic green areas, which will take into account sustainable development programs, especially in the ecological context, i.e., maintaining biodiversity and managing ecosystems [3].

It all seems to raise a justifiable question about the need to discuss a new document on historic greenery layouts, taking into account both the current experience as well as new challenges.

2. Materials and Methods

This article is part of the series of research carried out earlier by the author (together with W. Brzezowski) on the problem of authenticity in historic gardens, which was analyzed in various scopes referring to the following: reconstruction [4], vegetation [5] and the importance of exotic plants [6]. These analyses were connected with the research on Silesian gardens (today the region of Poland, previously belonging successively to Poland, Bohemian Kingdom, the Habsburg Monarchy, Prussia and then to Germany until 1945) from the Middle Ages to the Baroque period, the results of which were included in a two-volume study [7,8]. At that time, significant discrepancies in the interpretations of various aspects, including source materials connected with historic gardens were noticed by different authors, the consequence of which were inconsistent research results, and then controversial in terms of authenticity conservation activities. The consistent expansion of the observation field allowed the author to go beyond the borders of Silesia and after a deeper analysis, to make an attempt at formulating comments of a more general character. They were made on the basis of two objectives which were identified for the purposes of this article in the following way:

1. firstly, the discussion of authenticity in relation to historic gardens and other forms of historic greenery were analyzed, taking into account various scopes and aspects in which this issue were addressed; in this context, both official documents (Appendix A) as well as contemporary theories and opinions of researchers were taken into consideration, including an interdisciplinary approach to the issue of authenticity; at the same time, attention was paid to the ecological aspects of protection of historic green complexes in the context of sustainable development;
2. secondly, types of sources and methods of analyzing them and then using the results obtained in this way in research on historic gardens in the context of their authenticity, and then in activities of a conservation character; at the same time, attention was drawn to the risks for authenticity resulting from the lack of a critical approach to the content of some sources; recommendations on the use of source materials contained in official documents were also compiled (Appendix B).

The basic methods, which were adopted in this article, are derived from the methodological canon adopted in the humanities, i.e., the queries of written sources, including old treatises on garden art, as well as historical descriptions and iconographic materials with particular emphasis on old engravings, paintings, designs, plans and maps. This made it possible to draw attention to the dangers resulting from their insufficiently critical use in many cases as well as to formulate suggestions of issues which could constitute the basis for further discussions on the character of the new document devoted to the historic greenery layouts.

In the theoretical considerations on authenticity, the documents relating to historic buildings (including gardens) were analyzed, starting with the Venice Charter [9], through the already mentioned basic document regulating protection of historic greenery, the Florence Charter (1981) which complements the Venice Charter with points regarding historic gardens, thanks to which their protection received a lot of support for the first time, allowing the emergence of a specialized discipline, after the Nara Document on Authenticity [10] in which attention was paid to cultural conditions in assessing the value and authenticity of monuments. Reference was also made to the 2017 ICOMOS-IFLA Document, in which, for the first time, official recommendations highlighted the protection of public green areas [2]. More general regulations were also applied, i.e., the London Charter which codified the principles of computer methods for visualizing cultural heritage [11] as well as the Xi'an Declaration on the Conservation of the Setting of Heritage Structures, Sites and Areas [12]. The latter emphasizes a multidisciplinary approach in research and documentation of the environment and the importance of various sources of information for full "understanding" and "interpretation" of the value and importance of the surroundings of buildings, sites, and areas which constitute heritage. Local regulations were also studied, especially those with a broader international range of influence, such as the Burra Charter [13] and the Declaration of San Antonio which was organized by ICOMOS and constituted a summary of InterAmerican Symposium on Authenticity in the Conservation and Management of the Cultural Heritage to discuss the meaning of authenticity in preservation in the Americas [14].

The above-mentioned documents were confronted with the previous conservation practice and related problems as well as with views on the issue of authenticity in the context of gardens [15–31]. In this context, the discussion on the Florence Charter in recent years in the environment of German and Swiss researchers turned out to be very interesting [19,31,32].

The publications devoted to historic green areas and their management in the context of sustainable development were also applied, including, among other things, those published in the last month of 2020 on the pages of "Sustainability" in Special Issue "Challenges for Historic Gardens Sustainability between Restoration and Management" [33,34]. At the same time, attention was paid to the increasing number of voices raising the issue of the importance of historic gardens in the biodiversity protection strategy [35] as well as to the dangers to maintaining authenticity of historic greenery layouts, especially those located in cities and connected with climate change and its consequences [36]. In this context, attention is drawn to the more and more frequent treatment of urban parks and gardens as green infrastructure of cities [33].

New possibilities of researching historic greenery layouts were also indicated, which made it possible to verify some source messages. They are enabled by modern technologies, including laser scanning (TLS and LIDAR), which are one of the most modern techniques of obtaining data for the numerical model of the area (including the garden). Geophysical methods used in archaeological research are also very useful, such as: Electromagnetic conductivity (EM) and ground-penetraiting radar (GPR) [37,38]; see Figure 1. Orthophoto is also used for research purposes; see Figure 2. The use of these modern methods has been sanctioned in the London Charter (2009), which also defined the "Principles for computer methods of visualizing cultural heritage". We can read in it, inter alia, that "In order to ensure the intellectual integrity of computer-based visualization methods and outcomes, relevant research sources should be identified and evaluated in a structured and documented way" [11] (Principle 3). This makes it possible to apply computer visualizations to present historic greenery layouts in situations when it is not possible to reconstruct their original state in terms of authenticity.

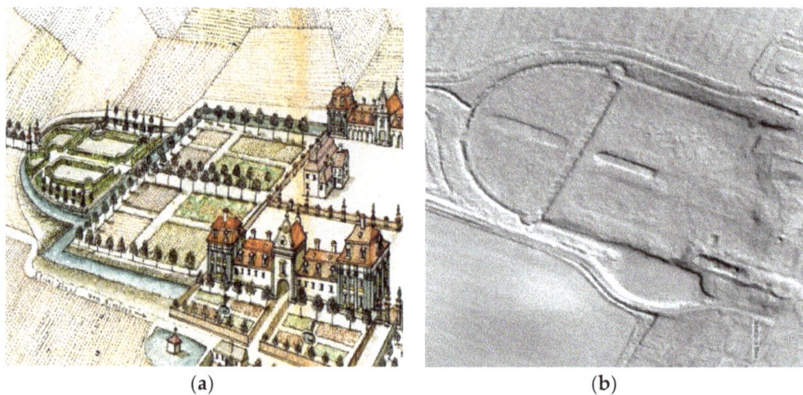

(a) (b)

Figure 1. Establishment of a manor and garden in Jezierzyce Małe (Silesia, Poland): (**a**) F.B. Werner drawing from around the middle of 18th century. In collection of University Library in Wrocław II, 497; and (**b**) model of the terrain in the LIDAR system showing the actual course of the canal surrounding the garden in Jezierzyce other than in Werner's drawing; from (public domain) http://www.geoportal.gov.pl/ (accessed on 8 September 2015).

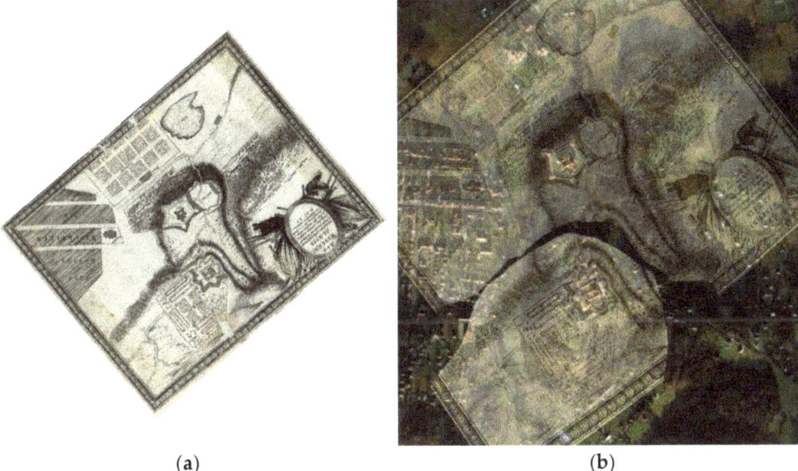

(a) (b)

Figure 2. Fortress in Wiśnicz (Poland): (**a**) on the engraving of Dahlberg, 1655. Author's collection. (**b**) on an orthophotomap with a "cut" Dahlberg plan, comp. M. Myczkowsk–Szałankiewicz, 2017.

3. Results

3.1. The Issue of Authenticity

For the first time in the official document, in the context of cultural heritage protection, the term "authenticity" appeared in the Venice Charter in relation to restoration and with the following recommendation "respect for original material and authentic documents" [9] (Art. 9). Just a bit more extensive was the reference to authenticity in the Florence Chapter, where we read that "The authenticity of a historic garden depends as much on the design and scale of its various parts as on its decorative features and on the choice of plant or inorganic materials adopted for each of its parts " [1] (Art. 9). Only in the Nara Document on Authenticity records, was authenticity recognized as the most important quality factor. For it says: "Our ability to understand these values depends, in part, on the degree to which information sources about these values may be understood as credible or truthful. Knowledge and understanding of these sources of information, in relation to original and subsequent characteristics of the cultural heritage, and their meaning, is a requisite basis for assessing all aspects of authenticity" [10] (Art. 9). At the same time, the criteria for assessing authenticity were relativized by saying: "It is thus not possible to base judgements of values and authenticity within fixed criteria" [10] (Art. 9). This was pointed out by David Lowenthal who noticed that "Authenticity is never absolute in practice, always relative" [28] (p. 4).

The role, meaning and aspects of authenticity defined in the Nara Document were commented on and developed in the Declaration of San Antonio. We refer to it here because of the findings contained therein which extend and at the same time organize the issue of authenticity, relating to "1. Authenticity and identity (The authenticity of our cultural heritage is directly related to our cultural identity); 2. Authenticity and history (An understanding of the history and significance of a site over time are crucial elements in the identification of its authenticity); 3. Authenticity and materials (The material fabric of a cultural site can be a principal component of its authenticity)". This Declaration also drew attention to other aspects of authenticity which were absent in earlier documents, namely "authenticity and social value", "authenticity in dynamic and static sites", "authenticity and stewardship" and "authenticity and economics" [24], referring them to all categories of objects and places which make up the heritage. This document also includes recommendations suggesting "That further consideration be given to the proofs of authenticity so that indicators may be identified for such a determination in a way that all significant values in the site may be set forth. The following are some examples of indicators: Reflection of the true value. That is, whether the resource remains in the condition of its creation and reflects all its significant history. Integrity. That is, whether the site is fragmented; how much is missing, and what are the recent additions. Context. That is, whether the context and/or the environment correspond to the original or other periods of significance; and whether they enhance or diminish the significance. Identity. That is, whether the local population identify themselves with the site, and whose identity the site reflects. Use and function. That is, the traditional patterns of use that have characterized the site" [8]. These recommendations, apart from the general ones mentioned above, refer to three groups of objects and places: 1. architecture and urbanism; 2. archeological sites group and 3. cultural landscape group. Although the last group is so capacious that some actions related to the landscape can also be related to the historic greenery layouts, it is noteworthy that this element of heritage is not mentioned directly (American specificity).

It should be noticed here that the authenticity of the broadly understood heritage was also discussed outside official documents, also in the context of other values which monuments carry with them, namely historical, emotional, cultural, aesthetic, artistic, the value of antiquity, identity, quotation, sociological and psychological and finally ecological values, to which various researchers of the issue added authenticity and tradition of a place–genius loci, including authenticity of a specific place–situs [39]. Let us also recall that the Declaration on the Preservation of the Spirit of Place was adopted in Québec by the General Assembly of ICOMOS only in 2008. It drew attention to the protection of intangible elements of cultural heritage.

Recognizing the importance of the issue of authenticity for the protection of cultural heritage, in 2010, on the initiative of ICOMOS, all international and local documents (of various ranks) as well as the findings of conferences and scientific symposia, in which the authenticity criterion was taken into account, were collected and put together. The whole comparison was divided into two parts (theoretical aspects and practical aspects) and included 282 titles, although only nine (including some which are connected with the Nara Document) were directly related to gardens [40]. This is an important observation because in the case of all other monuments and objects, in the unanimous opinion of researchers, theorists and in the provisions included in legal acts, "only an authentic monument is a carrier of historic and scientific values [...]. Therefore, preserving the full value of a monument is possible only when we record and conserve its original matter" [41]. In relation to historic gardens, we are forced to particularly modify the authenticity defined in this way. For in this special case, we are dealing (in its natural part because the remaining elements are essentially subject to the provisions of the Venice Charter) with the matter which, while alive, is subject to natural processes involving the life of individual plants. This became the basis for thinking about the garden as a dynamic composition, in which the durability of the whole depends on the durability of its components, mainly plant components, and the average lifetime of one generation of plants was estimated for about 100 years. In this case—in accordance with the recommendation of the Florence Charter—an extension of the spatial form in order to maintain a harmonious spatial continuity of all plant forms should take place by means of "prompt replacements when required and a long-term program of periodic renewal clear felling and replanting with mature specimens" [Florence Charter, Art. 11]. Such a procedure was accepted and has been practiced in many countries to this day, also in Poland [17] (pp. 16–17).

Let us make it clear that the recommendation for periodic replacement of plant forms was supplemented in the Florence Charter with the recommendation that the species selection of plants used in this process should be based on "selection with regard for established and recognized practice in each botanical and horticultural region, and with the aim to determine the species initially grown and to preserve them" [1].

Directly to authenticity, the Florence Charter refers—as already mentioned—to this very generally, thus opening the door to many far-reaching activities which will be presented later in this article. Its records mean that authenticity in the context of historical gardens should be considered in two aspects:

1. "design and scale of its various parts as on its decorative features"
2. "plant or inorganic materials adopted for each of its parts" [1] (Art. 9.)

The wording in the Charter on determining the appropriate vegetation "with the aim to determine the species initially grown and to preserve them" requires some commentary here [1] Art. 12. The question is how to relate it to cases so stylistically complex and at the same time rich in historical, artistic and social meanings and roles, as the Villa d'Este gardens in Tivoli, which was inscribed on the World Heritage List in 2001; see Figure 3.

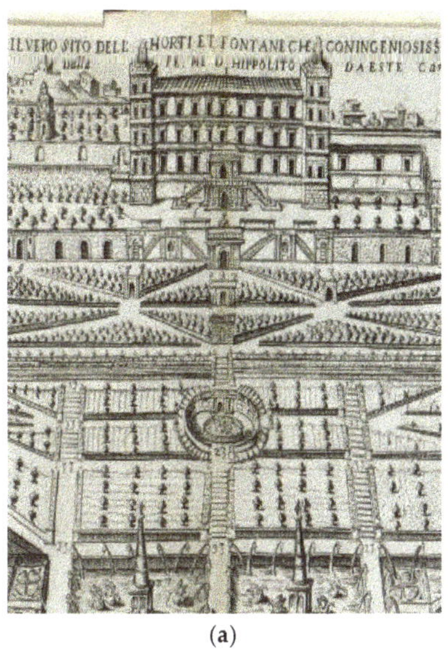

(**a**)

(**b**)

Figure 3. The main compositional axis of the Villa d'Este garden in Tivoli: (**a**) fragment from an etching by G.A. Brambilla; from (Public domain); https://i.pinimg.com/originals/1d/9d/fd/1d9dfd4177025bef3e43763884f041f6.jpg?epik=dj0yJnU9WlhPdzFPUlI4MGVMSjVrUENWdndpUDhKQTlZU3c2WXEmcD0wJm49UVZIbjFlbjBmaGREQXlVSF9GLVdCdyZ0PUFBQUFBR0FWWnp3 (accessed on 8 December 2020). (**b**) in contemporary photography. Photo W. Brzezowski, 2014.

We will come back to Tivoli Gardens, however, at this stage, we will turn to situations which are clearer, that is, to gardens that maintained their stylistic unity, or where such unity is sought through partially regenerative activities. In such cases, a botanical aspect considered in the context of authenticity deserves special attention. We noticed that failure to exercise due diligence in maintaining authenticity of gardens, understood as the initial nature appropriate for various historic periods, too often leads to the introduction of plants which had not been originally cultivated there, e.g., boxwood in many gardens with a medieval and Renaissance origin and located to the north of the Alps, and sometimes even in the historic past of many unknown gardens, such as plantain lilies (in Europe only since 1830) and hydrangeas (the first in Kew Gardens in 1789), which we can find in some old cloisters, for example in the territory of today's France and Belgium [42] as well as Poland [7] (pp. 47, 77–80).

Many researchers also criticized the point exchange, which was recommended by the Florence Charter and used in such cases, which aimed at restoring gardens to their original "youthful" condition. From the beginning of the 1990s, they began to treat this type of plant exchange as a reconstructive action consisting in the removal of evidence important for the diagnosis of the history of the garden. As already mentioned, an interesting discussion on this subject took place in the environment of German and Swiss researchers. The Swiss art historians, Brigitt Sigel encouraged to change the approach to such practices, which were recommended by the Florence Charter, i.e., "It is the duty of a conservator-restorer to preserve material traces of history, not to recreate a textbook model". In her opinion, it is impossible to "destroy the authentic traces of history" and then recreate them because "history cannot be recreated" [16] (p. 274). This topic was taken up by German landscape architect and researcher Erika Schmidt who paid attention to the importance of plants as historic sources [28] (p. 270) and postulated the extension of life of not only permanent elements, but also plants by means of gardening procedures which respected the limitation of intervention. Ingo Kowalik, an ecologist, went even more broadly in similar considerations, i.e., he connected protection of garden monuments with nature conservation, also in terms of ecology [18]. This debate can be considered as a summary expressed in 2006 in the publication published under the meaningful title "Der Garten–ein Ort des Wandels." [19], in which, after Italian conservator of gardens Lionella Scazossi, the authors used the term "opera aperta", treating the garden as an open work, constantly and dynamically changing, which clearly indicated the need to "preserve the traces of a complex past without interrupting this process". At the same time, it was acknowledged that it was "possible if the historical analysis did not serve as a starting point for reconstruction" [19]. Such an attitude brings us significantly closer to taking into account the goals set in the documents on sustainable landscape management as regards gardens. They were included in the European Landscape Convention (Florence Convention), where we read, among other things, "Landscape management" means action, from a perspective of sustainable development, to ensure the regular upkeep of a landscape, so as to guide and harmonize changes which are brought about by social, economic and environmental processes" [43] (Art. 1.e). Attention is drawn to the fact that, so far, no other doctrinal documents concerning the protection of heritage in the context of historic greenery have addressed this issue. Perhaps the results of research on various ecosystems in the context of ecology and biodiversity should be used for this purpose.

A specific voice in the above-mentioned debate could be the view which was formulated many years ago (1961) by Polish art historian and conservator-restorer Józef Dutkiewicz who used the "myth about overcoming death" in relation to monuments. In the garden context, it can be read as "eternal life" thanks to the rebirth of nature resulting from the vegetative process and from the repeated cycles of plant reproduction, the way of their genetic self-copying and in this biological sense, preserving authenticity and even "immortality" [44] (p. 6).

The discussion about treating historic gardens as an "opera aperta" was undertaken by Austrian historian of gardens Geza Hajos, who contrasted the garden defined as "a place of changes" [19] with "a historic garden–as a place of remembrance" because—as he particularly emphasized—"the historical garden is not a work of nature but a work of man" [22] (p. 23). Hajos also formulated ten "theses for discussion" which were connected with the protection of garden heritage. In terms of authenticity, he referred to the Nara Document, stating at the same time that some reconstruction activities (such as restoring the course of paths, flowering specific to a given period) should not arouse any controversies. He also warned that the polarization of two extreme methods, both long-term conservation (up to the disappearance of some plant forms and, consequently, the destruction of the garden structure) and the "artistic continuation" standing on the other extreme, is equally dangerous for historic gardens [22] (p. 220).

With a similar vein joined this exchange of views Stefan Rhoter, retired director of Bayerischen Verwaltung der staatlichen Schlösser, Gärten und Seen (The Bavarian Administration of State-Owned Palaces, Gardens and Lakes). By analyzing both of the above positions, he came to the conclusion that it was impossible to establish uniform and universal rules. He advises on an individual approach which should be adjusted to the circumstances. He recommends the use of all methods, even the reconstruction of missing elements, provided that their "original shape has been documented and the existence of which is essential for understanding the monument, which should be preceded by thorough research" [25] (p. 63). Rhotert also calls for "honest information to visitors" about any reconstructions and also draws attention to the costs of not only restoration and reconstruction activities, but also the subsequent maintenance of gardens, in the case of private layouts which often "overwhelm financially". He also writes about social aesthetic pressure, especially in the context of tourist facilities which are available for tourists, emphasizing that these "less restored, although not less historically valuable complexes often have a dark existence" [25] (p. 64).

Dorota Sikora, a Polish landscape architect and author of publications devoted to the specific nature of conservation actions in regular gardens, also referred to the issue of plant replacement [20]. She justified her decisions regarding the reconstruction of Branicki Garden in Białystok (see Figure 4) with source texts, referring to the recommendations contained in the treatise Dezalier d'Argenvillee [45], which recommended that for the formation of avenue trees and divine plantings, specimens up to nine meters high, so that they would be adapted to the rest of the garden composition. Consequently, it was supposed to mean periodic replacement of trees, although it is not known whether it was actually carried out in the past. Sikora also referred to some examples of contemporary activities in historic gardens based on the replacement of tree stands (Schwetzingen, Herrenhausen, Hampton Court, and also Versailles) [20] (pp. 32, 47).

Figure 4. The Branicki garden in Białystok, reconstructed on the basis of source materials and research. Fot. J. Dyr., 2013; from (Public domain) https://upload.wikimedia.org/wikipedia/commons/thumb/c/c1/150913Garden_of_the_Branicki_Palace_in_Bia%C5%82ystok-02.jpg (public domain) (accessed on 8 January 2021).

Let us return, however, to the consideration of authenticity. What else, in terms of the authenticity of historic gardens, does the Florence Charter not refer to? It does not take into account, for example, the loss of this value in cases connected with the movement—even justified for the purposes of protection—of monuments constituting the "equipment" of gardens, which in general is made possible—under certain conditions—by the provisions of the Venice Charter [9], (Art. 7, 8). The Polish example is provided by, for example, a complex of baroque sculptures currently decorating the upper and lower terrace of the former royal Wilanów Palace in Warsaw, coming from Brzezinka near Wrocław (until 1945, the German Briese); see Figure 5. A change in the original composition consisting of 22 statues in Wilanów led to a departure from the original, we would say "authentic" symbolic and content program, which probably was not in line with the iconographic message which was conveyed by the sculptures originally standing in Wilanów Garden and were then stolen by the Russian Army in 1804 [8] (pp. 346–352).

(a)

(b)

Figure 5. Baroque sculptures: (**a**) in the garden by the palace in Brzezinka in Silesia (Poland). Photo unknown author, circa 1930; from (Public domain) https://fotoposka.eu/Brzezinka/b22278,Palac_Kospothow.html?f=1276070-foto (accessed on 8 January 2021). (**b**) the same sculptures after they were moved to the garden of the residence in Wilanów (Warsaw, Poland). Photo M. Jagiełło, 2013.

At this point, we should pay attention to the fact that the symbolic and content-related significance played this time a key role in the case of a group of gardens known as Masonic (for example, Parc Monceau—France, Louisenlund—Germany and Nieborów—Poland). The awareness of this fact and its consequences for the composition of such layouts, in particular the form and role which were originally given to its individual elements and their mutual relations, should be of fundamental importance when seeking solutions to protect both their authentic substance and the original message [46,47].

The Florence Charter does not comment exhaustively on another important aspect of authenticity, namely the "garden existence" which was formulated by Dmitry Likhachov [48] (p. 78) and understood as the original purpose of the garden. This very often determined, for example, the width of the avenue, the height of hedges, the length of the canals, the extent of water mirrors and the area of meadows. It also referred to the use of certain elements of garden staffage some time ago, for example in the form of shepherds surrounded by a herd of cows or sheep, which was to transport observers to idyllic Arcadia. Today their reconstruction would not meet—as one can assume—with the proper reception in the context of authenticity.

As above-mentioned in this article, Erika Schmidt also drew attention to the aspect of authenticity connected with the "garden existence", referring it to the Florence Charter, where it was written that it was essentially the "place of enjoyment suited to meditation or repose" and was associated with various problems which result from the modern use of many historic gardens far beyond those defined in the Charter [29] (p. 85).

It is also worth adding that the quoted "garden existence" is also connected with the issue of authenticity relating to former users, who, being equipped with a kind of thesaurus constituting an attribute of every "well-mannered" and sometimes even well-educated person, easily read a symbolic code of the garden (we must add that it was different in subsequent epochs). It is usually unavailable to a modern recipient. However, it should not be overlooked in research on historic gardens.

When looking at fountains, cascades and other water arts gushing out in many iconographic representations, we should also think with reserve in assessing authenticity in the context of "garden existence". In most of the gardens (except for the Italian parts using natural springs and favorable topography) they were activated only for a moment when the owner was watching them. This was not always because of saving water (in the modern sense), but rather because of the difficulties which had to be overcome in order to deliver it in the days before the steam engine and electricity appeared, often using complicated devices for this purpose, such as the famous machine from Marly (see Figure 6), which worked in Versailles from 1684 or other devices and systems for obtaining and effective use of water, designs of which were included in the pages of many old treatises of works, among others [49–52]. Today, water, which is supplied by waterworks and the inflow of which we will probably have to limit again, gushes from them.

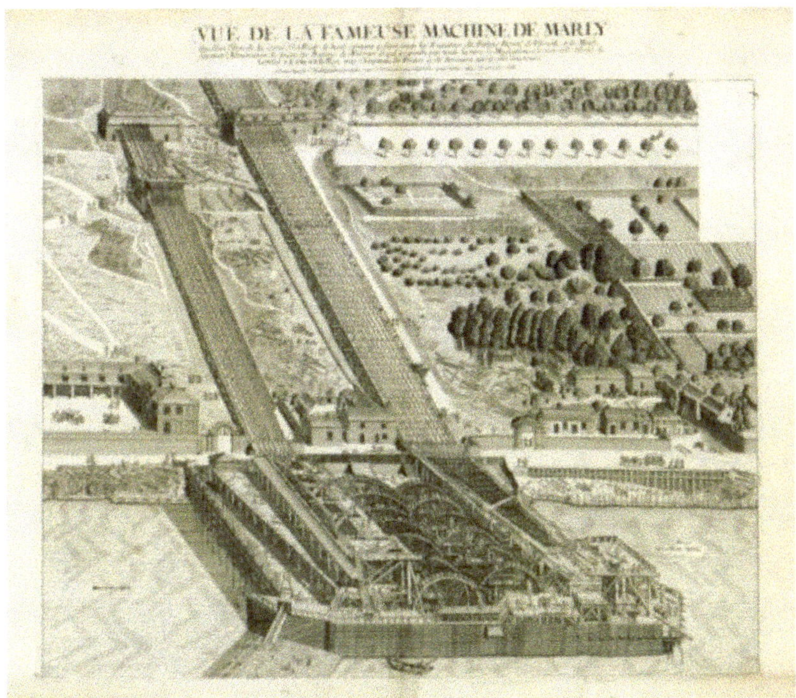

Figure 6. A machine from Marly that once fed the gardens of Versailles. A drawing from the beginning of the 18th century. Collections gallica.bnf.fr/Bibliothèque nationale de France; from (Public domain) https://gallica.bnf.fr/ark:/12148/btv1b6945480x/f1.item.r=machine%20marly.zoom (accessed on 8 January 2021).

The Florence Charter does not connect stylistic analyses with aesthetic criteria appropriate for the particular epochs, which are illegible or out of date today. This document was limited to a general statement that the garden is "a testimony to a culture, a style, an age, and often to the originality of a creative artist" [1] (Art. 5). As such, it happens that these aesthetic criteria acquired new different meanings like, for example, sublimity—once referring to the sphere of impressions and dynamically changing emotional experiences as well as picturesqueness—defined in the past as a sudden changeability, austerity and uniqueness, and nowadays it characterizes attractiveness. Let us recall here that picturesqueness created the picturesque style, in which the main role was played by the painting experience (replacing the gardening, architectural and engineering) of a garden designer, his knowledge of compositions, the ability to use chiaroscuro and color and to seek harmony [8] (p. 77). Perhaps in the case of this type of gardens, they should be included in the competences of conservators or people cooperating with them.

In the Florence Charter there is also no mention about the importance of cultivating old agrotechnical techniques and methods, which to a large extent determined the shape and character of gardens. A record of such a value of them did not appear until the Nara Document. This aspect of authenticity was noticed during, inter alia, the reconstruction of the famous Cornish Lost Gardens of Heligan, not only in terms of the layout and fidelity to the plant species once cultivated here, or the original irrigation system, but also in a meticulous approach to the old methods of their cultivation, which were recreated with reverence (e.g., pineapples) [53]. When thinking about techniques, we should not forget about old engineering and technical possibilities of setting up gardens. Gardens, as engineering undertakings in historical terms, are still waiting for thorough examination, also in the context of authenticity [54].

Another issue which raises controversy is the Florence Charter's record which recommends an analysis of broadly understood "documentation" and the possibility of obtaining knowledge on this basis, which could become the foundation for the restitution of certain parts of gardens "on the basis of the traces that survive or of unimpeachable documentary evidence" [1] (Art. 16). The rationale for such actions would be, for example, the reconstruction of the historical relationship between the garden and buildings so that it would become clear again ("reconstruction work might be undertaken more particularly on the parts of the garden nearest to the building it contains in order to bring out their significance in the design") [1] (Art. 16). Erika Schmidt raised an objection to this provision, appealing for "respect for the internal value of the garden and its individual historical development. It often took place independently of the development of the complex" [29] (p. 85).

We are able to quote numerous examples of actions taken in accordance with the recommendations of Florence Charter in order to reconstruct stylistic relationships between the garden and the building in recent years also in Poland in, among other things, Branicki Garden in Białystok (2009–2011), in the case of bishop's gardens in Wrocław (2002) and in Wilanów, where in the years 2010–2012 the immediate surroundings of the palace were reconstructed, labeling this action with the term "conservation creation", translated as "a creative continuation of the spirit of the composition" [55]. The restoration of the gardens on the basis of the "dialogue with historic sources" also took place at the gothic wing of the Royal Castle in Warsaw. As it seems, unfortunately, there is environmental acceptance for this type of actions, considered permissible because they are in accordance with recommendations of the Florence Charter. It was also included in the basic coursebook on the protection and conservation of garden monuments published in Poland [17] and reissued two years ago.

A certain kind of justification for such an attitude in Poland may be connected with destructions of the entire historic substance during the Second World War and then the post-war nationalization which also included garden complexes, which, apart from conservation measures, also required adaptation to completely new purposes subordinated to new social requirements and even political ones. At the same time, the lack of sufficient funds limited most of these actions to "a group of more valuable layouts, which, for didactic purposes, would make it possible to reconstruct some kind of model solutions from each style epoch" [56] (p. 15).

It seems, however, that the permissibility of restitution in the Florence Charter defined in this way, also became a kind of passport for other practices, namely, the so-called "re-creation" [57] (pp. 101–130) and "creative conservation" [58] (p. 126). It should be remembered here that these are not new practices. They correspond to the former activities of brothers Henri and Achille Duchêne in France and Great Britain, as well as to creative German reconstructions from the 1920s and 1930s (Brühl, Herrenhausen, Sanssouci). They also refer to the concept of "creative monument conservation" (Schöpferische Denkmalpflege) which was introduced in 1929 by Rudolf Esterer. Years later, this attitude was characterized by Dieter Hennebo as the one which permitted "both in the care and restoration of historic gardens for personal creative interference and changes in order to meet the current utility requirements" [15] (p. 72).

Concerned about the "limitless creativity", which was accepted by links in the records of both the Venice Charter and the Florence Charter, German garden researcher Clemens Alexander Wimmer claims that although conservation of historic gardens cannot be done without creativity, it cannot be unlimited, because then it becomes harmful. At the same time, he proposed three criteria for the valorization and activities carried out in historic gardens. "Ethical evaluation" seems to be a special one among them. In the opinion of this researcher, it should be included in the Florence Charter which was also revised in this respect [32].

The discussion around this issue in relation to all monuments took place also in Poland and focused on two attitudes, i.e., conservative, represented in the mainstream of the discourse between Ksawery Piwocki who convinced others that "the authenticity of a work of art is connected not only with the form of a work, but also with its substance" [59] (p. 27) and Jan Zachwatowicz who noticed the need for reconstruction and believed that "it is necessary that the monument should have the fullest and really suggestive form" [60] (p. 49).

Nowadays, it is not without significance for this approach to think in terms of activating tourism to which nice-looking complete gardens contribute, despite the fact that tourism is becoming an increasing burden for them. Let us also not forget about the frequent "entanglement of heritage in contemporary purposes", or the created (sometimes for political reasons) "celebration of feelings and creation emotional images of the past, which in practice may mean opposition to the historic truth" [61], (p. 3).

Regardless of the assessment of the views and recommendations presented above as well as the research and activities carried out on their basis, the foundation of all of them should be a properly conducted "documentary evidence" analysis, which is mentioned by the Florence Chart and Nara Document on Authenticity. However, how to effectively use them in research on the history of gardens and in conservation practice to avoid undermining the obtained scientific results and activities undertaken? The next part of the article focuses on drawing attention to the most important issues which are connected with this topic in the context of authenticity.

3.2. Analysis of Sources

The role and importance of source materials in research and actions which are connected with monuments was noticed in the Venice Charter in the chapter on restoration, where we read: "The restoration in any case must be preceded and followed by an archaeological and historic study of the monument" [9] (Art. 9). The Florence Charter also says—as already mentioned—about "unimpeachable documentary evidence", but only in the context of justifying the restoration [1], Art. 16. An extended comment can be found in the Nara Document which highlights the properties of sources which may differ depending on the culture and even within the same culture. The Document also emphasizes their diversity which may concern the "form and design, materials and substance, use and function, traditions and techniques, location and setting, and spirit and feeling, and other internal and external factors. The use of these sources permits elaboration of the specific artistic, historic, social, and scientific dimensions of the cultural heritage being examined" [10] (Art. 11, 13). References to the importance and scope of the use of sources for research and activities in the protection of the broadly understood heritage can also be found in the Burra Document and the XI'AN Declaration. Their summary, including all the main official documents, can be found in Appendix A.

However, it was the entry from the Florence Charter that provoked the reaction of many researchers who, like Hartmut Troll, rated the widely used in practical conservation of garden monuments and often thoughtless approach to sources as "naïve" and at the same time they expressed concern that the Florence Charter's message "may even support misuse of historic source material" [31] (p. 87). Stefan Schweizer went even further by saying that "in such a delicate matter as garden, iconographic sources often function as substitutes for the object. A large part of garden historiography is based on views and prints, which, due to insufficient consideration of conventions [...] and various functions which were performed in the epoch, could be mistakenly taken as almost authentic evidence of historic reality" [27] (p. 2) [31] (p. 86).

3.2.1. Written Sources

Referring to the above doubts about the meaning of sources and at the same time trying to answer the question posed in this article at the end of the section devoted to the issue of authenticity, we will start with the characteristics of written sources, including the group of herbaria, botanical and garden treatises, horticultural and agrotechnical guides as well as florilegia and plant catalogs. Firstly, it must be borne in mind that many of them had just a local influence and only some of them turned out to be, we would say, pan-European. Furthermore, the fact that some of them, especially the oldest ones, constituted indeed a summary of the contemporary knowledge of nature, but it was often based on ancient and medieval Arab works. Not everyone remembers that it is to the works of Columella and Pliny that we owe the so-called raised ridges and perches limited by wooden palisades or wicker fences. Additionally, the scientific basis for the experiments of a hydraulic character, thanks to which fountains and other water arts were powered, were provided by Arabic engineer al-Jazari, who, moreover, used the treatise of Heron of Alexandria [62]. Additionally, we ought to recall the fact that in the 16th century the famous work of Pietro de Crescenzi [63] in the version (as a manuscript in French in 1373, printed for the first time in German in 1471) which was translated into many European languages, including the Polish language, but without any comments, sometimes contained pieces of advice which were completely impractical due to climatic conditions and sun exposure in gardens located to the north of Europe, and, moreover, did not take into account the current knowledge of horticulture at the time of printing.

In written sources, we can also find a lot of information about plants, for example the aforementioned boxwood, which, as an effective evergreen and now easy to grow plant, is now planted in almost all regular gardens in Europe. However, when reading source materials carefully, we become aware of its late use in Europe to the north of the Alps (except for the British Isles, where it appeared thanks to the Romans) due to, inter alia, long-term dislike of boxwood (unpleasant smell, bitter honey and intoxicating properties) and even problems with its cultivation, for example in the 17th and 18th century Sweden, where it was frozen and in order to create decorative ground floors, local cowberries had to be used [5].

The absence of boxwood as a hedge plant or even non-existence of any hedges in the late-medieval gardens is also confirmed by a query which was conducted among the 15th century representations of gardens (both secular and symbolically religious) and which can be found in miniatures, drawings and paintings. We can see garden quarters which are framed by low walls, woven fences or wooden boxes, but never by hedges! When they appeared, in the case of border forms and decorative motifs, they were rarely made of boxwood (with the exception of Italy, Spain and England), which was confirmed by the work of French gardener of the three French kings Claude Mollet (1564–1649), i.e., "very few notable people wanted to have this plant in their gardens" [64] (p. 201). Additionally, although the knowledge of boxwoods in France, which however did not confirm their more widespread use, was revealed in Charles Estienne's treatise of 1554, which briefly mentioned that "boxwood hedges surround the garden and create opere topiaria" [65] (p. 29), nevertheless, a breakthrough for their later career in European gardens—according to Mollet—was to be the consent of Henry IV to plant boxwood in Saint-Germain-en-Laye in 1595 and then in Fontainebleau [64] (p. 202).

In the light of what was said about the introduction of boxwood into French gardens, there are reasonable doubts about the large-scale use of these plants for the re-creation of medieval gardens (see Figure 7), especially monastic ones, in almost all of Europe; see Figures 8 and 9. Even during the so-called reconstruction of Renaissance gardens on the example of Villandry, which were originally established before the mid-16th century. After the period of transformation into a landscape complex, they were re-established as "renaissance" in 1906 and then inscribed on the UNESCO World Heritage List as part of the Loire Valley complex in 2000; see Figure 10. We can find the effects of similar actions also in Poland, for example in the "renaissance" giardino segreto in Pieskowa Skała (see Figure 11), the "medieval" cloister of the Museum of Architecture in Wrocław (see Figure 12) as well as in the Wawel Royal Castle, where in recent years "royal gardens" with the use of boxwood were established in the 16th-century stylistics; see Figure 13.

Figure 7. Medieval garden reconstructed with the use of box trees in front of the Rodemack city walls (Lothringen, Germany). Photo Feles, 2005; from (Public domain) https://upload.wikimedia.org/wikipedia/de/thumb/4/43/Mittelalterlicher_Garten.JPG/766px-Mittelalterlicher_Garten.JPG (accessed on 20 January 2021).

Figure 8. Pandhof, the monks' garden located at the bell tower of the Gothic St. Martin in Utrecht. Photo Moblio, 2011; from (Public domain) https://upload.wikimedia.org/wikipedia/commons/thumb/9/95/Kloostergang_Domkerk.JPG/800px-Kloostergang_Domkerk.JPG (public domain) (accessed on 20 January 2021).

Figure 9. The courtyard of the medieval sanctuary of Mont-Saint-Michel (France). Fot. Libriothecaire; from (Public domain) https://upload.wikimedia.org/wikipedia/commons/8/8b/Cloitre_abbaye_mont_saint_michel.jpg (accessed on 20 January 2021).

Figure 10. Fragment of the "Renaissance" gardens of Villandry (France), an example of the so-called creative conservation. Photo W. Brzezowski, 2016.

Figure 11. Giardino segredo in Piaskowa Skala (Polish), translation of the re-creation with the use of box trees. Fot. Piotr Górski, 2005; from (Public domain) https://commons.wikmedia.org/wiki/File:Pieskowa_Ska%C5%82a_Ogr%C3%B3d_01.jpg (accessed on 10 January 2021).

Figure 12. The courtyard of the Museum of Architecture in Wrocław (Poland), once a medieval Bernardine monastery. In the foreground, boxwood and funky. Photo M. Jagiełło 2016.

Figure 13. The "Renaissance" gardens of King Sigismund at Wawel (Krakow, Poland) reconstructed mainly with the use of boxwood. Photo M. Jagiełło, 2019.

Reading some old treatises allows us to find out which plants were once used to shape hedge forms, for example in the German language zone. So, in a book by doctor and astronomer Peter Lauremberg, which was entitled "Horticultura" and published in 1631, the author recommended the following plants for hedges: Liguster, rosemary, and southern mosquito shrub, and for higher forms—hawthorn, juniper, barberry, laurel and lilac [66] (p. 61). In a work of gardener Heinrich Hesse entitled "Neue Garten-Lust", published in Leipzig in 1696, we find a chapter on how to "gracefully secure the garden and make the best use of hedges" [67] (pp. 7–9, 389–390). We can also read there that the French think a lot about a hedge in the garden, so we in Germany must also arrange it." Later, the author claimed that "hedges and trees are the highest decoration of the garden" and recommended dogwood (e.g., for avenues planted on both sides with trees or creepers, whose shoots intertwine over the road creating a vault), ligustra, holly, and also (although rather for the kitchen garden) gooseberries, currants, barberry and roses [67] (pp. 7–9. It is worth keeping this information in mind when thinking about plants which will replace boxwood that is ruthlessly exterminated by Cydalima perspectalis.

Continuing the considerations on the authenticity of the botanical world, we should also pay attention to catalogs of plants which grow in the particular gardens (florilegia constituted such a subgroup for flowers). The practice of compiling this kind of lists, most often richly ornamented, was initiated by German physician and botanist Joachim Camerarius in "Florilegium", published in 1589 [68]. He established a sort of canon which contained a short catalog and descriptive part as well as the rich illustrative one. Unfortunately, some of these studies lost their illustrations, which is even more painful as the old names assigned to individual plants differ significantly from today's names and only by comparing these names with catalogs created in a similar period, is it possible to determine a species list of plants planted in European gardens at that time in individual regions or language zones. For Silesia, such a comparative analysis was carried out for two of the most famous Wrocław gardens, i.e., the mannerist one which belonged to doctor Laurentius Scholz [7] (pp. 113–138) and the baroque one whose owner was merchant Caspar Wilhelm Scultetus [69] (pp. 471–482) (see Figure 14). We only have lists of plants growing there along with the names, partly regional, which were adopted at that time. For Scholz's garden the aforementioned florilegia were such a reference, for Scultetus and his collection of citruses—Johann Christoph Volkamer's "Nürnbergische Hesperides" from 1708 [70].

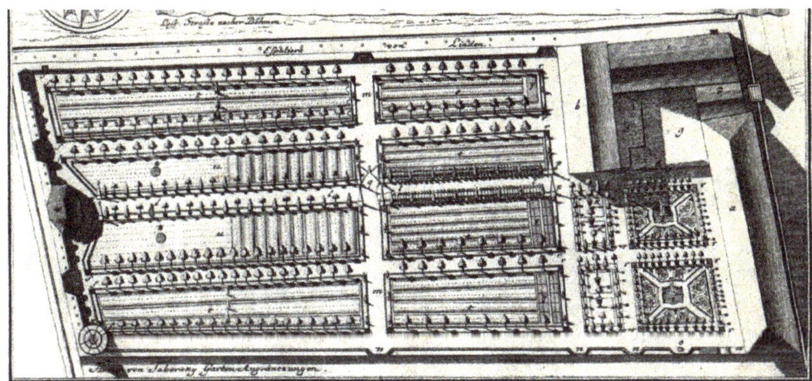

Figure 14. W. C. Scultetus' garden plan in Wrocław (Poland), 1731; from (Public domain) https://fotopolska.eu/765163,foto.html?s=1&cx=904&cy=649 (accessed on 2 February 2021).

When analyzing sources, we should also pay attention to the fact that today the extremely rich collections in state or university libraries, including, among other things, historic studies on gardens and horticulture, moreover often available online, say nothing about the actual access to these publications in the past. These expensive books, which were once printed in a limited number of publications, were available to owners and designers in a small number of titles. Moreover, some of those designers created and published their thoughts on gardens only on the "local market", dedicating them to wealthy investors. Only a careful analysis of these collections in terms of prior belonging to private or monastery libraries, as evidenced by ownership marks or other information proving the origin of these books and the way they traveled (antiquarian marks, marginal notes and donation entries), can provide indications about the character and the scope of their influence on the shape of the studied gardens. This kind of preliminary query, concerning the collections of the present Wrocław University Library, consisting of books and other prints which were previously owned by private collectors (nobility, aristocracy, but also rich burghers) as well as by monastic owners, was carried out by Wojciech Brzezowski [71].

When analyzing the content of such works, we should also take into account a completely different attitude of the people of that time to what we now call copyright. The studies devoted to gardening issues show the great ease with which their authors, very often without providing sources, included extensive fragments of books by their colleagues "in the trade" in their own publications (this is, of course, a topic for a separate discussion on the "authenticity" of these works in the source context).

The group of written sources important for the study of the history of gardens also includes economic inventories, often with lists of plants ordered for purchase or descriptions of investment activities (their character and costs) connected with gardens as well as chronicles, guides, letters, occasional prints (e.g., catalogs gardening exhibitions, advertisements for companies producing garden equipment) and press articles, essays, epic poems and poetry.

Instructive information is also provided by guidebooks, although not often used as a source. As an example, we can quote a fragment of one of them, which was written by August Zemplin in 1816, a health resort doctor in Szczawno (German: Bad Salzbrunn) in Silesia, who encouraged his patients to walk towards the castle and park complex in nearby Książ, namely "Who would paint this wonderful, a one-of-a-kind route, where brick arches meet the rocks and a hiker seems to hover above the stream rippling in the depths among the rocks. Steep, massive rock reefs, once standing above each other, once again climb up steep walls, alternate with lush clumps of forest and tell us about some ancient devastation, as a result of which this valley was formed" [72] (pp. 134–135). Reading about "mighty reefs", "wild streams" and "ancient desolation", it is hard to resist the impression that both the author of this description and the others who wandered these routes were familiar with the feelings of sublimity experienced by the first explorers of Alpine trails, including Joseph Addison who recorded his impressions in the following way: "The Alps filled my mind with a pleasantly felt kind of horror" [73] (p. 261). Apart from the sublimity, which was treated in aesthetic categories from the 18th century, separating it from beauty and entering into a philosophical discussion (let us recall that it was Kant who believed that the sublimity should be sought in wild nature) [74] (pp. 121–122), to which we have a different attitude today (marked in the colloquial sense by exaltation and naive pathos), we have one more problem with the Książ complex (but not only this) in the context of authenticity. It is the growth of vegetation, especially high vegetation, which covered most of the earlier view connections important for understanding the garden and changed the character of many places; see Figures 15 and 16. It seems that in this case, they can only be "reconstructed" by means of virtual reality, as shown by actions such initiatives taken for gardens, the physical reconstruction of which is impossible or unjustified; see Figures 17–19.

Figure 15. Książ (Silesia, Poland), view of the castle from the Pełcznica valley according to A. Duncker, c. 1860; from (Public domain) https://pl.wikipedia.org/wiki/Plik:Schloss_Fuerstenstein_Sammlung_Duncker.jpg (accessed on 20 January 2021).

Figure 16. Książ (Silesia, Poland), view of the castle and the Pełcznica valley, the present state. Photo W. Brzezowski, 2013.

Figure 17. King's Visualisation Lab, King 's College London; from http://www.kvl.cch.kcl.ac.uk/kew.html (accessed on 20 January 2021).

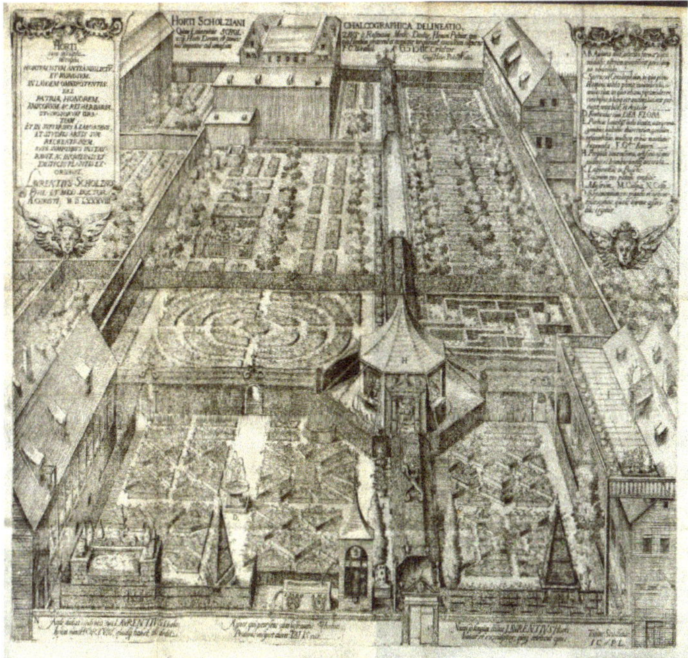

Figure 18. Laurentius Scholz's garden in Wrocław. An engraving by G. Hayer, 1598. Collections of the University Library in Wrocław, OSD, ref. 011185.

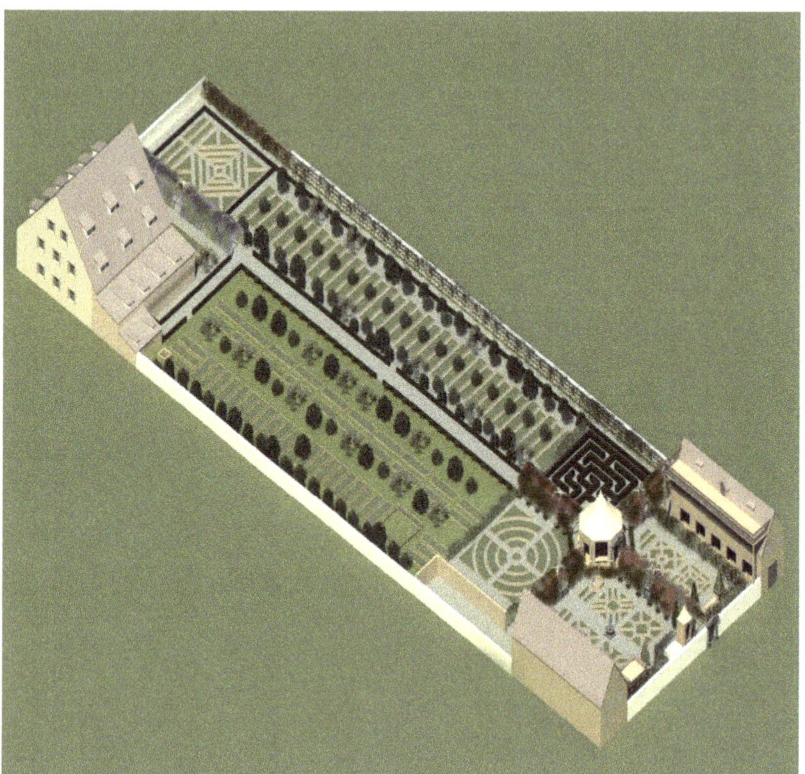

Figure 19. Virtual reconstruction of the axonometric view of the garden L. Scholz in Wrocław (Poland) based on an engraving by G. Hayer from 1588, analyzes of city plans and archaeological research, comp. by M. Jagiełło, K. Pietras, K. Lontkowska, 2013.

3.2.2. Iconographic Sources

Let us now focus on the second group of sources described as iconographic. We will start with comments about plans, vedute (understood as images of cities) and other views. Let us recall, then, that at the beginning, representations of gardens were most often connected with presentations of cities, towns, villages, castles, etc. The majority of them and the most precise in their iconographic message were made in the years 1629–1650 in the studio of Frankfurt engraver Matthäus Merian, first in the work entitled "Theatrum Europaeum" [75], and then in the 16-volume work entitled "Topographia Germaniae" which was continued after his death (until 1688), [76]. Among 1486 engravings placed there, we can also find those which show gardens in detail. This monumental study had its followers. At the beginning of the 18th century, the works by Georg Matthias Vischer (1628–1696), which covered the then Austria and then Styria in more detail, were published [77,78]. For Bavaria, such a collection containing over a thousand engravings made for 850 cities, monasteries, manors, castles and often with extremely detailed gardens, was prepared by Michael Wening (1645–1718) and his descendants in the years 1701–1726 [79], documenting their condition at the turn of the 18th century, i.e., the greatest baroque development. The above-mentioned examples were followed by, inter alia, Erik Dahlberg (1625–1703), a soldier in the service of the Swedish king Charles X Gustav and above all an excellent draftsman. On the basis of his sketches many etchings were made. They were included in two monumental studies glorifying the power of Sweden at that time, also through unusual and rich residential premises shown on numerous plans and views [80]. Dahlberg's inventory etchings were also used in the work of Samuel von Puffendorf, in which we can find several gardens from Poland [81]. We have similar documentation also for Silesia. It was developed in the years 1730–1760 in the form of several thousand colorful small but very detailed drawings by Friedrich Bernhard Werner (1690–1776) [82,83]. Similar in terms of subject matter, but graphical, was also made for baroque Vienna and its surroundings. It was prepared (in the years 1726–1731) by Austrian artist Salomon Kleiner [84] who specialized in vedute. We should also mention a group of works in which only gardens were presented collectively, such as "Ville e giardini di Roma" by Giovanni Battista Falda (1643–1678) [85] or the work of similar title "Li giardini di Roma con le loro piant" by Giacomo Giovanni Rossi (1627–1691) [86].

We cannot fail to mention the original design cards for the particular layouts as an invaluable iconographic source, however, they must always be verified by examining the scope of their actual implementation (also by means of archaeological research and by scanning the area). Studies which are also important are studies in the form of templates with visualizations of solutions relating to gardens of various functions and sizes as well as to various objects and forms of greenery filling them, which inspired both designers and investors. Such a series was developed for baroque gardens by, inter alia, Matthias Diesel [87], Antoine–Joseph Dezalier d'Argenville [45] and Jacques–François Blondel [88], and for the English–Chinese gardens, for example, Georges–Luis Le Rouge [89]. In Poland, Izabela Czartoryska and her "Myśli różne" [90] played an important role in this respect, whereas for the German gardens horticulture theorist Christian Cay Lorenz Hirschfeld [91].

It is also impossible to ignore the influence which individual complexes had, especially if they had monographic illustrated studies by Salomon de Caus (1576–1626) [92], for example for the gardens in Heidelberg Castle or Bad Muskau of Prince von Pückler–Muskau [93], or when as pioneering they set generally accepted trends, often on a pan-European scale, like Versailles. The knowledge about it was gained from numerous graphics prepared by, inter alia, Pierre Aveline (1656–1722), to whom we also owe an insight into other French baroque garden layouts.

The historical and cognitive benefits resulting from the analysis of this type of sources for the history of individual objects and regions seem obvious. They can also serve as a collection of information useful as a comparative material, if we take into consideration areas similar to each other in many ways, such as Bavaria and Silesia at the turn of the 18th century. Or when, on the basis of historic and environmental analyses, the zones of direct or indirect influence (of countries, artists, directions of their migration, and, finally, family or social ties of investors) are correctly identified. In this context, the importance of image memory and memories (from various journeys) is also significant, both in relation to garden designers and owners; in some cases, they are the same people, the so-called landscape amateurs. Here we have to take into account directions of journeys which change at different times and include them in diaries, journals, and correspondence. Personal contacts also turned out to be important, for example when passing on plants which were rare or new in Europe.

However, in-depth studies of the contents of various sources showed that we must treat them very carefully. For example, Swedish gardens from Dahlberg's work turn out to be in some cases an exaggerated projection behind which the then superpower policy of that country operated. Moreover, Werner's drawings, once burdened with many perspective errors [94] (pp. 63–70) and with strange abbreviations, sometimes constituted a kind of "visualizations" of plans which were not realized later.

Analyses of painting achievements also require a justified distance. Let us take the Polish example of Zygmunt Vogel, who, as commissioned by King Stanisław August Poniatowski, did a series of watercolor paintings documenting the earliest landscape gardens in Poland. Krystyna Sroczyńska, who dealt with the painting of this period, defined the properties of the artistic convention adopted by him, namely a decorative treatment of natural elements as frames for the presented architectural objects and the tendency to enlarge water mirrors, and thus artificially create the depth of the painting [95] (pp. 64–71). This shows that when considering the usefulness of artistic creativity for research on gardens, even the one which seems to have documentary features, we should refer to the evaluation of the overall oeuvre of the artist, preferably carried out from the perspective of research specific to art history.

Let us add that some of these inaccuracies can be verified, especially in terms of the plan, not only by means of archaeological research, but more and more often by using TLS and LIDAR scans, and first of all thanks to the analysis of maps. In the scope of our interest, the most accurate will be military topographic maps, although—which is obvious—all available cartographic materials should be analyzed. A rich collection of them concerning a significant part of Europe can be found on the Austrian portal mapire.eu [96]. Their precision and reliability results from their special purpose. They performed a role similar to that for which Erik Dahlberg prepared his inventory drawings for Poland, bearing in mind their military usefulness; see Figure 2.

For a slightly later period, such a group consists of military maps, the so-called Messtischblatten, which are very precise and compiled for German countries in the years 1876–1944. Of course, the cadastral plans are even more precise and detailed. They served tax purposes, hence the extraordinary detail of the plans connected with them, which was explained in extensive legends [97]. They were made in France in 1808 and in 1817 for Austria–Hungary, whereas the cadastre was also in force in the Kingdom of Prussia in 1819.

At the end of these considerations about sources, it should be emphasized that the usefulness of most of them should be limited to the benefits and research findings (registration and classification), whereas in the field of conservation, only to activities which do not raise suspicions of reconstruction actions.

4. Discussion

Forty years have passed since the adoption of the Florence Charter. Although it does not constitute a legal act but only a set of recommendations, it has left a significant mark on the actions carried out in historic gardens. Not always positive, but certainly inspiring for discussions, as evidenced by the exchange of views quoted in this article. They show that the recommendations of the Florence Charter were an expression of the then, not always up-to-date, state of knowledge and opinions on the issues of protecting garden heritage and another environmental and social awareness. Particular criticism was given to those provisions of the Charter that were related to the problem of maintaining authenticity. At the same time, the greatest controversy arises from allowing the restitution of historical gardens on the basis of source materials. The recommendations in the Florence Charter regarding the point exchange of plants (sometimes entire groups of plants) in order to maintain form stability are also opposed in many scientific and conservative communities. Following the spirit of the times, but also because of the critical attitude towards some parts of the records of the Florence Charter, national recommendations began to be developed independently of it. In 2007 such recommendations were prepared by the United Kingdom which in the document entitled Management and Maintenance—instead of making references to the Florence Charter—referred to the "Burra Charter" which was developed in 1979, but repeatedly updated (last time in 2013) and which consisted of a set of good practices (standards) in the scope of protection and management of Australian cultural heritage. It was accompanied by the "Code on the Ethics of Co-existence in Conserving Significant Places", which indicated the importance of the ethical attitude in cultural heritage management. When defining the value of cultural heritage, the "Burra Charter" also emphasizes "variability" by writing "Cultural significance may change as a result of the continuing history of the place" and adding "Understanding of cultural significance" may change as a result of new information" [13], Art.1, Explanatory notes. It also contains an important provision in the context of authenticity, i.e., "Changes to a place should not distort the physical or other evidence it provides, nor be based on conjecture" [13] (Art 3.2).

Italy also followed its own path. Still in 1981, the local ICOMOS proposed its version of the charter entitled "Proposta per una carta del restauro dei giardini storici". The entry of this document, which constitutes the key to our considerations on authenticity, is quoted by Erik Schmidt, after Maria Pozzana, i.e., "the conservation intervention must take into account the entire process which is connected with the history of the garden because its structure and appearance materialized in this process" [98] (p. 236) after [29] (p. 87). This provision results in the fact that the garden should be treated as a palimpsest. The essence of this procedure was very aptly described by Italian art historian Isa Bella Barsali, referring it to Tivoli, i.e., "In these gardens, little was done in the 19th century. Negligence and lack of care during different periods caused uncontrolled growth of vegetation. Today, Tivoli is no longer a place which praises its creators–the Cardinals of the House of Este. It is entirely a different garden. It is a romantic forest in which we stumble over architectural fragments scattered like islands around the garden. [. . .] Today, Tivoli Gardens have a special beauty which was formed by many layers that have overlapped over the years. [. . .] The problem with our inheriting gardens along with the complex Renaissance, Baroque and 19th century shape lies precisely in preservation of these features, not their restoration" [99] (p. 528); see Figure 20.

Figure 20. Villa d'Este gardens in Tivoli. Photo Palickap, 2017; from (Public domain) https://upload.wikimedia.org/wikipedia/commons/a/a1/Tivoli%2C_Villa_d%27Este%2C_giardino.jpg (accessed on 10 January 20210).

This approach to historic gardens is also supported by the current discussion on the definition of heritage, which is more and more often treated as "a process and not a type of resource" [Ashworth 2007, p. 32]. A change in heritage thinking was also made in the Burra Charter, where "places of cultural significance" were replaced by "sites and monuments" [13]. "This switched the emphasis from "stones and bones", material culture, towards the meanings of places, the significance that humans attribute to material culture [100] (p. 4).

Since the adoption of the Florence Charter, the semantic scope of the monument was also extended to new dimensions. As Polish historian Krzysztof Kowalski put it, referring to the findings of French researcher Nathalie Heinich, i.e., "the process described by Heinich led to a clear extension of the concept of a monument and the formation of a modern definition of the concept of heritage and new conservation strategies" [61] (p. 5) after [101] (pp. 17–21) as well as to the clear postmodern "dispersion of control over the past and its interpretations" [101] (p. 8.) "the world has changed" continues Kowalski "and along with postmodernity and the replacement of great stories with their local and often individual counterparts, the role of historic truth and authenticity of material substance has been weakened. Heritage has put emphasis differently, allowing for a multitude of interpretations and departing from the exclusivity which was lost by academic experts working on the matter of the past" [61] (p. 9).

Also important for the discussion about authenticity is the fact that more and more often considerations of gardens and other historic greenery layouts are included in the scope of regulations connected with sustainable management, which in the context of the observed changes (social, economic, and climate) seems to be a very urgent need [34,102]. For example, attention was paid to the detrimental influence of heat islands effect on entire groups of plants in historic gardens [36]. Additionally, on the activity of new pests and diseases which damage the historic stand as well as other plants, such as the aforementioned boxwood. This resulted in research on a new sustainable approach to the maintenance and restoration of historic gardens [33,34] and their role in the biodiversity protection strategy [35] as well as in maintaining the ecosystem structure of urban complexes [103]. Thus, to include the above-mentioned aspects in the considerations on the authenticity of historic greenery layouts, although so far most often on the basis of case studies.

With regard to the broadly understood heritage, the discussion on various aspects of authenticity has already started, initiating the project called "Journeys for Authenticity" on the initiative of ICOMOS, conceived as "platform for open discourses sharing ideas, solutions, and discussion about the concept of authenticity. Formats of communication might range from posters, live feeds, blog posts, reports, PowerPoints, video interviews or presentations, for example [104]. The ICOMOS Emerging Professionals Working Group (EPWG) hopes that the Journeys to Authenticity could be an opportunity for collaboration and mentorship between established ICOMOS members and emerging professionals, enabling intergenerational dialogue about these concepts and encouraging the use of diverse communication media and strategies" [105]. It seems that at the beginning it would be worth considering, analyzing, and perhaps using the existing provisions on authenticity contained in various official documents. The list of them is included in Annex B.

Let us also mention the fact that ICOMOS addressed the issues of heritage protection in the context of sustainable development, declaring the development of Sustainability Policy in 2019. In this context, it seems important to prepare recommendations regarding historic greenery layouts by researchers and conservators dealing with historic greenery, perhaps as part of ICOMOS-IFLA.

5. Summary

In the light of all that is presented in this article, do we need a new Florence Charter? It seems that there is no clear answer to this [29]. If so, it should, in our opinion, meet one essential condition. Its construction should be carefully considered so that, like the Burra Charter, it would be logical and transparent, and above all, flexible enough, and thus would enable periodic updating. Perhaps by analogy to the aforementioned "opera aperta", we should think of "carta aperta"?

Certainly, one should refer to all the most important controversies that arise from some of Florence Charter's recommendations. Then, take into account the experiences identified in this text resulting from the impact on heritage protection of other official documents. Especially in those aspects that relate to the issue of authenticity, as highlighted in this article.

Its establishment should be preceded by a further, thematically ordered discussion related to the protection of historic greenery assumptions, updated with the spirit of the transformations that we observe, in order to take into account the progress in understanding the theory and management of cultural heritage. Furthermore, by a discussion conducted as widely interdisciplinary as possible, including the field of sustainable development in relation to maintaining biological diversity and ecology [35]. The need for such a discussion is forced not only by changes in the approach to heritage, social changes or technological development, but also–which is becoming more and more important and urgent–those related to climate change. A model of such a procedure is presented in the diagram; Figure 21.

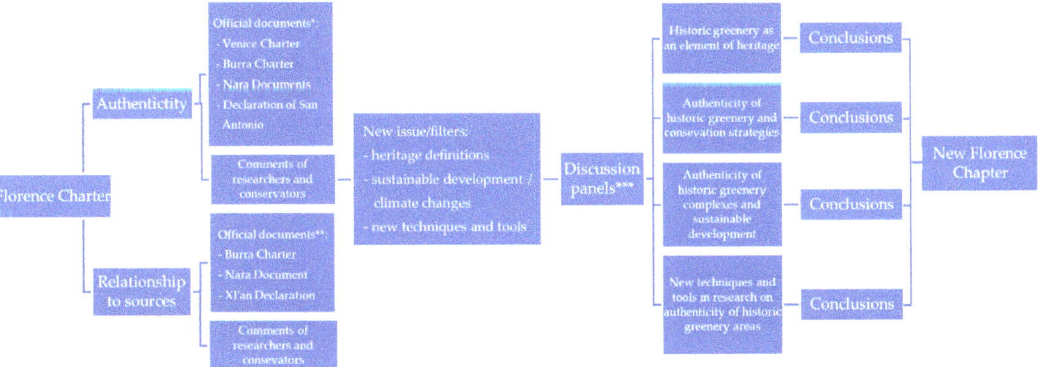

* Appendix A
** Appendix B
*** Detailed topics divided into four panels are presented in Appendix C

Figure 21. A diagram showing the results of the research and the method of further proceeding to prepare a New Florence Charter.

As part of this model, we propose four panels in which discussions about the new Florence Charter could be held:

1. Historic greenery as an element of heritage;
2. Authenticity in historic green complexes in research and conservation strategies;
3. Authenticity of historic green complexes and sustainable development;
4. New techniques and tools in research on the authenticity of historic greenery layouts.

A detailed range of topics for each panel is presented in Appendix C. Some of these discussions could be held under the auspices of UNESCO, for example within the "Journays for Authenticity" program. The place of considerations covering panel III. it could be for example the special issues of "Sustainability".

Funding: This research received no external funding.

Institutional Review Board Statement: Not applicable.

Informed Consent Statement: Not applicable.

Data Availability Statement: Not applicable.

Conflicts of Interest: The author declares no conflict of interest.

Appendix A

Official Doctrinal Documents	
Authenticity	
NO.	DOCUMENT
1.	**Venice Charter, 1964···** **(International Charter for the Conservation and Restoration of Monuments and Sites)** Art. 9. *The process of restoration is a highly specialized operation. Its aim is to preserve and reveal the aesthetic and historic value of the monument and is based on respect for original material and* ***authentic*** *documents. It must stop at the point where conjecture begins . . .*
2.	**The Burra Charter 1979/1981/1988/1999/2013** Art. 3.2. *Changes to a place should not distort the physical or other evidence it provides, not be based on conjecture.* *Explanatory Notes to Art. 5.2. A cautions approach is needed, as understanding of cultural significance may change.*
3.	**Florence Charter, 1981** Art. 9. *The authenticity of a historic garden depends as much on the design and scale of its various parts as on its decorative features and on the choice of plant or inorganic materials adopted for each of its parts.* Art. 11. . . . *the preservation of the garden in an unchanged condition requires both prompt replacements when required and a long-term programme of periodic renewal . . .* Art. 12. . . . *with the aim to determine the species initially grown and to preserve them.* Art. 21. *The work of maintenance and conservation, the timing of which is determined by season and brief operations which serve to restore the garden's authenticity, must always take precedence over the requirements of public use.*
4.	**Nara Document on Autenticity, 1994** Art. 9. *Knowledge and understanding of these sources of information, in relation to original and subsequent characteristics of the cultural heritage, and their meaning, is a requisite basis for assessing all aspects of authenticity.* Art.10. *Authenticity, considered in this way and affirmed in the Charter of Venice, appears as the essential qualifying factor concerning values. The understanding of authenticity plays a fundamental role in all scientific studies of the cultural heritage, in conservation and restoration planning, as well as within the inscription procedures used for the World Heritage Convention and other cultural heritage inventories.* Art. 13. *Depending on the nature of the cultural heritage, its cultural context, and its evolution through time, authenticity judgements may be linked to the worth of a great variety of sources of information. Aspects of the sources may include form and design, materials and substance, use and function, traditions and techniques, location and setting, and spirit and feeling, and other internal and external factors.* *Appendix A. Efforts to determine authenticity in a manner respectful of cultures and heritage diversity requires approaches which encourage cultures to develop analytical processes and tools specific to their nature and needs. Such approaches may have several aspects in common: • efforts to ensure assessment of authenticity involve multidisciplinary collaboration and the appropriate utilization of all available expertise and knowledge; • efforts to ensure attributed values are truly representative of a culture and the diversity of its interests, in particular monuments and sites; • efforts to document clearly the particular nature of authenticity for monuments and sites as a practical guide to future treatment and monitoring; • efforts to update authenticity assessments in light of changing values and circumstances.* Art. 5. *Particularly important are efforts to ensure that attributed values are respected, and that their determination includes efforts to build, as far as possible, a multidisciplinary and community consensus concerning these values.*
5.	**The Declaration of San Antonio (1996)** Art. B. 1. *Authenticity and identity (The authenticity of our cultural heritage is directly related to our cultural identity); 2. Authenticity and history (An understanding of the history and significance of a site over time are crucial elements in the identification of its authenticity); 3. Authenticity and materials (The material fabric of a cultural site can be a principal component of its authenticity); 4. Authenticity and social value; 5. Authenticity in dynamic and static sites; 6. Authenticity and stewardship; 7. Authenticity and economics.* Art. C. 1.c. *That further consideration be given to the proofs of authenticity so that indicators may be identified for such a determination in a way that all significant values in the site may be set forth. The following are some examples of indicators: i. Reflection of the true value. That is, whether the resource remains in the condition of its creation and reflects all its significant history. i. Integrity. That is, whether the site is fragmented; how much is missing, and what are the recent additions. iii. Context. That is, whether the context and/or the environment correspond to the original or other periods of significance; and whether they enhance or diminish the significance. iv. Identity. That is, whether the local population identify themselves with the site, and whose identity the site reflects. v. Use and function. That is, the traditional patterns of use that have characterized the site.* Art. C. 4.g. *Since in conserving the authenticity of cultural landscapes the overall character and traditions, such as patterns, forms, land use and spiritual value of the site may take precedence over material and design aspects, that a clear relationship between values and the proof of authenticity be established.* 4.h. *That expert multi-disciplinary assessments become a requirement for the determination of authenticity in cultural landscapes, and that such expert groups include social scientists who can accurately articu-late the values of the local communities.* 4.i. *That the authenticity of cultural landscapes be protected prior to major changes in land use and to the construction of large public and private projects, by requiring responsible authorities and financing organizations to undertake environmental impact studies that will lead to the mitigation of negative impacts upon the landscape and the traditional values associated with these sites.*

Appendix B

Official Doctrinal Documents	
Relationship to Sources	
NO.	DOCUMENT
1.	**Burra Charter 1979/1981/ 1988/1999/2013** Art. 6.1. *The cultural significance of a place and other issues affecting its future are best understood by a sequence of collecting and analysing information before making decisions.* Art. 19. *Restoration is appropriate only if there is sufficient evidence of an earlier state of the fabric.*
2.	**Florence Charter, 1981** Art. 15. *No restoration work and, above all, no reconstruction work on a historic garden shall be undertaken without thorough prior research to ensure that such work is scientifically executed and which will involve everything from excavation to the assembling of records relating to the garden in question and to similar gardens.* Art. 16. *Restoration work must respect the successive stages of evolution of the garden concerned. In principle, no one period should be given precedence over any other, except in exceptional cases where the degree of damage or destruction affecting certain parts of a garden may be such that it is decided to reconstruct it on the basis of the traces that survive or of unimpeachable documentary evidence.*
3.	**Nara Document on Autenticity, 1994** Art.11. *All judgements about values attributed to cultural properties as well as the credibility of related information sources may differ from culture to culture, and even within the same culture. It is thus not possible to base judgements of values and authenticity within fixed criteria.*
4.	**XI'AN Declaration on the conservation of the setting of heritage structures, sites and areas, 2005** Art. 3. *Understanding, documenting and interpreting the setting is essential to defining and appreciating the heritage significance of any structure, site or area.* Art. 4. *Understanding the setting in an inclusive way requires a multi-disciplinary approach and the use of diverse information sources. Sources include formal records and archives, artistic and scientific descriptions, oral history and traditional knowledge, the perspectives of local and associated communities as well as the analysis of views and vistas. Cultural traditions, rituals, spiritual practices and concepts as well as history, topography, natural environment values, use and other factors contribute to create the full range of a setting's tangible and intangible values and dimensions. The definition of settings should carefully articulate the character and values of the setting and its relationship to the heritage resource.*

Appendix C

Topics for Discussion	
I. Historic greenery as an element of heritage	1. "Garden as a place of change" (opera aperta, palimpsest) versus "garden as a place of remembrance".
	2. "Garden existence" in research on the authenticity of gardens versus adaptation to contemporary needs.
	3. Garden as a scientific and engineering project in the context of authenticity.
	4. Old agrotechnical methods and techniques as a research subject and the conservation issue in the context of authenticity.
	5. Historic green areas as a subject of interdisciplinary research.
	6. Spirit of Place in research on the authenticity of historic greenery areas.
	7. Historic greenery areas as an element of heritage in the postmodern narrative.
	8. Historic greenery areas in the discussion about heritage and new conservation strategies/models.
	9. The importance of ethical attitudes in the management of cultural heritage in the context of historical green areas.
	10. Cultural conditions in the assessment of the value and authenticity of historical green areas.
	11. Authenticity of historical green areas in the light of social identification.
II. Authenticity of historic green areas in research and conservation strategies	1. Conservation creation/creative conservation in the context of authenticity considerations.
	2. The use of sources in research on the authenticity of historic greenery areas and in conservation practice (written sources).
	3. The use of sources in research on the authenticity of historic greenery areas and in conservation practice (iconographic sources).
	4. The use of sources in research on the authenticity of historic greenery areas and in conservation practice (cartographic sources).
	5. Authenticity of the natural world of historic greenery areas in the context of source research.
	6. Authenticity of the natural world of historic greenery areas as a subject of research and conservation practice in the context of climate change and new diseases and parasites threats
III. Authenticity of historic greenery areas and sustainable development	1. A sustainable approach to maintaining and restoring historic green areas, taking into account aspects related to authenticity.
	2. The role of historic greenery in the strategy of protecting biological diversity and its authenticity.
	3. Historic greenery as an element of green infrastructure of cities.
	4. The influence of tourism on maintaining authenticity of gardens.
IV. New techniques and tools in research on authenticity of historic greenery areas	1. Possibilities and importance of research on historic greenery areas conducted with the use of modern techniques and tools (LIDAR, TSL), electromagnetic conductivity (EM), ground-penetraiting radar (GPR)
	2. Historic greenery areas and virtual reality, opportunities and threats to authenticity.

References

1. ICOMOS-IFLA. Historic Gardens—The Florence Charter. 1981. Available online: https://www.icomos.org/images/DOCUMENTS/Charters/gardens_e.pdf (accessed on 7 September 2020).
2. ICOMOS-IFLA. Document on Historic Urban Parks. 2017. Available online: https://www.icomos.org/images/DOCUMENTS/General_Assemblies/19th_Delhi_2017/Working_Documents-First_Batch-August_2017/GA2017_6-3-2_HistoricUrbanPublicParks_EN_final20170730.pdf (accessed on 9 September 2020).
3. Agenda 21. Available online: https://sustainabledevelopment.un.org/content/documents/Agenda21.pdf (accessed on 8 September 2020).
4. Jagiełło, M.; Brzezowski, W. Autentyzm a Rekonstrukcja Zabytkowych Zespołów Ogrodowych. In *Dziedzictwo Architektoniczne: Rekonstrukcje i Badania Obiektów*; Łużyniecka, E., Ed.; Oficyna Wydawnicza Politechniki Wrocławskiej: Wrocław, Poland, 2017; pp. 74–84. Available online: https://www.dbc.wroc.pl/publication/41252 (accessed on 10 January 2021).
5. Jagiełło, M., Brzezowski, W. Nie Tylko Bukszpany. O Autentyzmie Ogrodowej Szaty Roślinnej w Formach Żywopłotowych i Topiarycznych. In *Dziedzictwo Architektoniczne: Restauracje i Adaptacje Zabytków*; Łużyniecka, E., Ed.; Oficyna Wydawnicza Politechniki Wrocławskiej: Wrocław, Poland, 2018; pp. 23–40. Available online: https://www.dbc.wroc.pl/publication/86547 (accessed on 10 January 2021).
6. Jagiełło, M.; Brzezowski, W. Autentyzm Świata Przyrodniczego w Ogrodach Europejskich: Rośliny Egzotyczne. In *Dziedzictwo Architektoniczne: W Kręgu Świata Przyrodniczego i Budowli Miejskich*; Łużyniecka, E., Ed.; Oficyna Wydawnicza Politechniki Wrocławskiej: Wrocław, Poland, 2019; pp. 5–16. Available online: https://www.dbc.wroc.pl/publication/142344 (accessed on 10 January 2021).
7. Jagiełło, M.; Brzezowski, W. *Ogrody na Śląsku. T. 1, Od Średniowiecza do XVII Wieku*; Oficyna Wydawnicza Politechniki Wrocławskiej: Wrocław, Poland, 2014.
8. Brzezowski, W.; Jagiełło, M. *Ogrody na Śląsku. T. 2, Barok*; Oficyna Wydawnicza Politechniki Wrocławskiej: Wrocław, Poland, 2017.
9. ICOMOS. International Charter for the Conservation and Restoration of Monuments and Sites (the Venice Charter 1964). Available online: https://www.icomos.org/charters/venice_e.pdf (accessed on 8 September 2020).
10. ICOMOS. The Nara Document on Authenticity. 1994. Available online: https://www.icomos.org/charters/nara-e.pdf (accessed on 8 September 2020).
11. The London Charter for the Computer-Based Visualisation of Cultural Heritage. 2008. Available online: https://www.londoncharter.org/index.html (accessed on 8 September 2020).
12. ICOMOS, Xi'an Declaration on the Conservation of the Setting of Heritage Structures, Sites and Areas. 2005. Available online: https://www.icomos.org/charters/xian-declaration.pdf (accessed on 8 September 2020).
13. Burra Charter. The Australia ICOMOS Charter for Places of Cultural Significance Australia ICOMOS Incorporated International Council on Monuments and Sites. 2013. Available online: https://australia.icomos.org/wp-content/uploads/The-Burra-Charter-2013-Adopted-31.10.2013.pdf (accessed on 8 September 2020).
14. ICOMOS. The Declaration of San Antonio. 1996. Available online: https://www.icomos.org/en/resources/charters-and-texts/179-articles-en-francais/ressources/charters-and-standards/188-the-declaration-of-san-antonio (accessed on 8 September 2020).
15. Hennebo, D. *Gartengeschichte–Gartendenkmalpflege. Versuch einer Bilanz. Das Gartenamt*; Patzer Verlag: Hannover, Germany, 1985; pp. 168–182.
16. Sigel, B. Alles Erhaltene wird zum redenden Zeugnis. Das Gartendenkmal mit der Elle des Baudendenkmalpflegers gemessen. In *Die Gartenkunst*; Wernersche Verlags GmbH: Biblis Germany, 1993; pp. 273–282.
17. Majdecki, L. *Ochrona i Konserwacja Zabytkowych Założeń Ogrodowych*; Wydawnictwo Naukowe PWN: Warsaw, Poland, 1993.
18. Kovarik, I. Historische Gärten und Parkanlagen als Gegenstand eines Denkmal-orientirten Naturschutzes. In *Naturschutz und Denkmalpflege*; Kowarik, I., Schmidt, E., Sigel, B., Eds.; vdh Hochschulverlag an der ETH: Zürich, Switzerland, 1998. Available online: File:///C:/Users/mj/Downloads/Sigeletal.1998WegezueinemDialogimGarten%20(1).pdf (accessed on 10 January 2021).
19. De Jong, E.; Schmidt, E.; Sigel, B. ... sie Hätten Nicht Unter einer Glasglocke Gestandenm Sondern im Lebendigen Strom der Geschichte Eine Einführung in das Thema. In *Der Garten—Ein Ort des Wandels. Perpectiven für Die Denkmalpflege*; De Jong, E., Schmidt, E., Sigel, B., Eds.; Hochschil-Verlag an der ETH: Zürich, Switzerlad, 2006.
20. Sikora, D. Specyfika Działań Konserwatorskich w Ogrodach Regularnych. *Kur. Konserw.* **2010**, *7*, 34–42. Available online: http://bazhum.muzhp.pl/media//files/Kurier_Konserwatorski/Kurier_Konserwatorski-r2010-t-n7/Kurier_Konserwatorski-r2010-t-n7-s32-42/Kurier_Konserwatorski-r2010-t-n7-s32-42.pdf (accessed on 10 January 2021).
21. Sikora, D. *Konserwacja ogrodów regularnych*; SGGW: Warsaw, Poland, 2011.
22. Hajós, G. Der historische Garten—Ein Ort des Wandels ode rein Ort der Erinnerungen? *Die Gart.* **2006**, *2*, 385–394.
23. Hajós, G. Rekonstruktion in der Gartendenkmalpflege, eine Problemstellung. In *Rekonstruktion in der Gartendenkmalpflege*; Hajos, G., Wolschke-Bulmahn, J., Eds.; Zentrum für Gartenkunst und Landschaftsarchitektur (CGL) Leibniz Universität Hannover: Hannover, Germany, 2007; pp. 18–26.
24. Hajós, G. Notes on the problem of authenticity in historical gardens and parks in the light of the Austrian experience. *J. Gard. Hist.* **2012**, *15*, 67–83. Available online: https://www.researchgate.net/publication/254237272_Notes_on_the_problem_of_authenticity_in_historic_gardens_and_parks_in_the_light_of_the_Austrian_experience (accessed on 20 February 2021).

25. Rhotert, S. Fragen an die Gartendenkmalpflege. In *Rekonstruktion in der Gartendenkmalpflege*; Hajos, G., Ed.; Zentrum für Gartenkunst und Landschaftsarchitektur (CGL) Leibniz Universität Hannover: Hannover, Germany, 2007; pp. 60–62. Available online: https://www.cgl.uni-hanno-ver.de/fileadmin/cgl/Forschung/Publikationen/Broschueren/Broschu__re__Rekonstruktion_in_der_Gartendenkmalpflege_.pdf (accessed on 12 January 2021).
26. Lowenthal, D. Passage du Temps dans le Paysage. Le Globe. *Rev. Genev. Géographie* **2009**, *149*, 177–179. Available online: https://www.persee.fr/doc/globe_0398-3412_2009_num_149_1_1564 (accessed on 20 February 2021).
27. Scheitzer, S. *Möglichkeiten und Grenzen artefaktischen Wissens in dem Bild- und Textquellen einer Fragilen Gattung*; Manuscript: Würzburg, Germany, 2011.
28. Schmidt, E. Erhaltung historischer Pflanzenbestände. Möglichkeiten und Grenzen. *Die Grtenkunst* **1997**, *2*, 270–273.
29. Schmidt, E. Die Charta von Florenz nach dreissig Jahren kritisch betrachtet. In *Denkmalpflege in Bremen*; Das Landesamt für Denkmalpflege Bremen: Bremen, Germany, 2012; pp. 83–91.
30. Troll, H. Rekonstruktion in der Gartendenkmalpflege. Annäherungen. In *Rekonstruktion in der Gartendenkmalpflege*; Hajos, G., Ed.; Zentrum für Gartenkunst und Landschaftsarchitektur (CGL) Leibniz Universität Hannover: Hannover, Germany, 2007; pp. 60–62. Available online: https://www.cgl.uni-hanno-ver.de/fileadmin/cgl/Forschung/Publikationen/Broschueren/Broschu__re__Rekonstruktion_in_der_Gartendenkmalpflege_.pdf (accessed on 18 January 2021).
31. Troll, H. *Długi cień Karty Florenckiej—Ewolucja Standardów Konserwacji Zabytków Niemczech. Ochrona Zabytków*; Narodowy Instytut Dziedzictwa: Warszawa, Poland, 2016; Volume 1, pp. 73–92. Available online: http://cejsh.icm.edu.pl/cejsh/element/bwmeta1.element.desklight-fc8d27f3-c8cc-4b4d-988e-01a5ad2de541 (accessed on 12 January 2021).
32. Wimmer, C.A. Drei Hauptkomponenten in der Behandlung von historischen Gärten. In *Rekonstruktion on der Gartendenkmal-pflege*; Hajos, G., Ed.; Zentrum für Gartenkunst und Landschaftsarchitektur (CGL) Leibniz Universität Hannover: Hannover, Germany, 2007; pp. 38–40. Available online: https://www.cgl.uni-hanno-ver.de/fileadmin/cgl/Forschung/Publikationen/Broschueren/Broschu__re__Rekonstruktion_in_der_Gartendenkmalpflege_.pdf (accessed on 10 January 2021).
33. Gulliano, P.; Pomatto, E.; Gaino, W.; Devecchi, M.; Larcher, F. New Challenges for Historic Gardens' Restoration: A Holistic Approach for the Royal Park of Moncalieri Castle (Turin Metropolitan Area, Italy). *Sustainability* **2020**, *12*, 10067. [CrossRef]
34. Funsten, C.; Borsellino, V.; Schimmenti, E. A Systematic Literature Review of Historic Garden Management and Its Economic Aspects. *Sustainability* **2020**, *12*, 10679. [CrossRef]
35. Prigioniero, A.; Sciarrillo, R.; Spuria, L.; Zuzolo, D.; Marziano, M.; Scarano, P.; Tartaglia, M.; Guarino, C. Role of historic gardens in biodiversity-conservation strategy: The example of the Giardino Inglese of Reggia di Caserta (UNESCO) (Italy). *Plant Biosyst. Int. J. Deal. All Asp. Plant Biol.* **2020**, *155*, 1–11. Available online: https://www.tandfonline.com/doi/full/10.1080/11263504.2020.1810812 (accessed on 10 January 2021). [CrossRef]
36. Oishi, Y. Urban Heat Island Effects on Moss Gardens in Kyoto, Japan. *Landsc. Ecol. Eng.* **2019**, *15*, 177–184. Available online: https://link.springer.com/article/10.1007/s11355-018-0356-z (accessed on 10 December 2020). [CrossRef]
37. Counsell, J. An evolutionary Approach to Digital Recording and Information about Heritage Sites. In Proceedings of the 2001 Conference on Virtual Reality, Archeology, and Cultural Heritage—VAST'01, Glyfada, Greece, 28–30 November 2001; ACM Press: Glyfada, Greece, 2001; pp. 33–41, 362. Available online: https://dl.acm.org/doi/10.1145/584993.584999 (accessed on 10 January 2021).
38. Zapłata, R. Autentyzm Zabytkowej Architektury i Palimpsest w Przestrzeni Historycznej—Nowe Media a Prezentacja Dziedzictwa Kulturowego. *Architectus* **2016**, *1*, 97–114. Available online: http://www.architectus.arch.pwr.wroc.pl/45/45_09.pdf (accessed on 12 January 2021).
39. Buchaniec, A.J. Autentyzm—Podstawowa Wartość w Konserwacji Zabytków Architektury. Ph.D. Thesis, Wydział Architektury Politechniki, Cracow, Poland, 1999. Available online: http://fbc.pionier.net.pl/details/nntsbsn (accessed on 10 November 2020).
40. ICOMOS, Authenticity. A Bibliography. 2010. Available online: https://www.icomos.org/centre_documentation/bib/Biblio_authenticity_2010.pdf (accessed on 10 January 2021).
41. Rouba, B.J. Autentyczność i Integralność Zabytków. *Ochr. Zabyt.* **2008**, *4*, 37–57. Available online: https://www.nid.pl/upload/iblock/19c/19cf1f3005c3c71f2e0052ee43e22ffe.pdf (accessed on 18 January 2021).
42. Valéry, M.-F.; Toquin, A. *Jardins du Moyen Âge*; La Renaissance du Livre: Tournai, France, 2002.
43. European Landscape Convention, Florence. 20 December 2000. Available online: https://rm.coe.int/1680080621 (accessed on 12 January 2021).
44. Dutkiewicz, J.F. *Sentymentalizm, Autentyzm, Automatyzm*; Ochrona Zabytków Narodowy Instytut Dziedzictwa: Warszawa, Poland, 1961; pp. 3–16. Available online: http://yadda.icm.edu.pl/yadda/element/bwmeta1.element.desklight-b6b63f8c-ebf2-452c-9dfa-ed248eb20f57 (accessed on 12 January 2021).
45. Dezalier d'Argenville, A.-J. La Théorie et la Pratique du Jardinage; Paris 1709. Available online: https://gallica.bnf.fr/ark:/12148/btv1b8626274z/f7.image (accessed on 14 January 2021).
46. Swirida, I. W Poszukiwaniu Ukrytych Znaczeń. Park naturalny XVIII Stulecia a Wolnomularstwo. *Ars Regia* **1993**, *2*, 7–40.
47. Wegner, F. *Der Freimaureregarten*; Kulturförderverein Ruhrgebiet E.V. KFVR: Gladback, Germany, 2014.
48. Lichaczow, D. *Poezja Ogrodów*; Zakład Narodowy im. Ossolińskich: Wrocław, Poland; Warsaw, Poland; Kraków, Poland, 1991.
49. de Caus, S. Les Raisons des Forces Mouvantes; Paris 1615. Available online: https://library.si.edu/digital-library/book/raisonsdesforce00caus (accessed on 10 January 2021).

50. Schickhart, H. Beschreibung einer Reiß, welche ... Friderich Hertzog zu Würtemberg vnnd Teck, ... im Jahr 1599 selb neundt, auß dem Landt zu Würtemberg, in Italiam Gethan; Mömpelgard, 1602. Available online: https://reader.digitale-sammlungen.de/de/fs1/object/display/bsb11212296_00005.html (accessed on 10 January 2021).
51. Ramelli, A. Le Diverse et Artificiose Machine del Capitano Agostino Ramelli; Firenze 1588. Available online: https://digital.sciencehistory.org/works/4b29b614k (accessed on 12 January 2021).
52. Böckler, G.A. Architectura Curiosa Nova; Nürnberg ca 1664. Available online: https://blogs.ethz.ch/digital-collections/2012/08/17/georg-andreas-bockler-architectura-curiosa-nova-nurnberg-1664/ (accessed on 18 January 2021).
53. Smit, T. *The Lost Gardens of Heligan*; Orion Publishing: London, UK, 2000.
54. Thoday, P. Science and Craft in Understanding Historic Gardens and Their Management. In *Gardens & L in Historic Building Conservation*; Harney, M., Ed.; John Wiley & Sons, LTD.: Oxford, UK, 2014; pp. 141–148.
55. Myczkowski, Z.; Wowczak, J. Kreacja konserwatorska na przykładzie ogrodów wilanowskich. *Aura* **2010**, *9*, 10–14.
56. Ciołek, G.; Chrabelski, K. Uwagi o Potrzebie i Metodzie Odbudowy Zabytkowych Ogrodów. *Ochrona Zabytków* **1949**, *1*, 15–18. Available online: http://yadda.icm.edu.pl/yadda/element/bwmeta1.element.desklight-538a588b-75ed-4682-b19f-a46dbc2eadc0 (accessed on 10 January 2021).
57. Landsberg, L. *The Medieval Garden*; University of Toronto Press: Toronto, ON, Canada, 2003.
58. Jellicoe, G.; Jellicoe, S.; Goode, P.; Lancaster, M. *The Oxford Companion to Gardens*; Oxford University Press: Oxford, NY, USA, 2001; pp. 96–112.
59. Piwocki, K. Zabytek rezerwat, park kultury: Muzeum w plenerze. *Ochr. Zabyt.* **1961**, *14*, 54–55.
60. Zachwatowicz, J. *Ochrona Zabytków w Polsce*; Polonia: Warsaw, Poland, 1965.
61. Kowalski, K. Od zabytku do dyskursu. O kilku Źródłach Współczesnej Definicji Dziedzictwa. Zeszyty Naukowe Uniwersytetu Jagiellońskiego. *Pr. Etnogr.* **2017**, *45*, 1–14. Available online: https://www.ejournals.eu/Prace-Etnograficzne/2017/45-1-2017/art/10504/ (accessed on 15 January 2021).
62. van Buren, A.H. Reality and Literary Romance in the Parc of Hesdin. In *Medieval Gardens*; Macdougall, E.B., Ed.; Dumbarton Oaks: Washington, WA, USA, 1983; pp. 115–134.
63. De' Crescenzi, P. Ruralia commode; (manuscript) 1304. Ruralium Commodorum Libri Duodecima; Schüssler, Augsburg 1471. 1st edition.
64. Mollet, C. Théâtre des plans et jardinages; Paris 1652. Available online: https://gallica.bnf.fr/ark:/12148/bpt6k1042641w.image (accessed on 10 January 2018).
65. Estienne, C. Praedium rusticum [...]; Lutetia 1554. Available online: https://gallica.bnf.fr/ark:/12148/btv1b8608310f (accessed on 10 February 2018).
66. Lauremberg, P. Horticultura; Francofurti ad Moenum 1631. Available online: http://diglib.hab.de/wdb.php?dir=drucke/6-4-oec-1 (accessed on 10 February 2018).
67. Hesse, H. Neue Garten-Lust; Leipzig 1696. Available online: https://wellcomecollection.org/works/umwxfdfj/items?canvas=1&langCode=ger&sierraId=b30513285 (accessed on 22 January 2021).
68. Camerarius, J. *Camerarius Flirilegium, ca 1589, Complete Facsimile ed.*; Harald Fischer Verlag: Erlangen, Germany, 2004.
69. Jagiełło, M. Ogród Caspara Wilhelma Scultetusa (Scholza) na Przedmieściu Świdnickim we Wrocławiu i Jego Powiązania Artystyczne. In *Nie Tylko Zamki*; Chorowska, M., Ed.; Oficyna Wydawnicza Politechniki Wrocławskiej: Wrocław, Poland, 2005; pp. 471–482.
70. Volkamer, J.C. Nürnbergische Hesperides; Nürnberg 1708. Available online: http://digital.bibliothek.uni-halle.de/hd/content/pageview/242577 (accessed on 10 September 2020).
71. Brzezowski, W. The Silesians and Theories of Garden Art from the 16th to the 18th Century with New Research from the Wrocław University Library collection. *Architectus* **2013**, *1*, 3–10. Available online: http://architectus.pwr.edu.pl/online_33_01en.html (accessed on 10 September 2020).
72. Zemplin, A. Salzbrunn und Seine Mineralquellen. Im Anhange: Fürstenstein in der Gegenwart und Vergangenheit, Breslau 1822. Available online: https://books.google.pl/books?id=tnJeAAAAcAAJ&printsec=frontcover&hl=pl#v=onepage&q&f=false (accessed on 10 September 2020).
73. Addison, J. Remarks on Several Parts of Italy etc. in The Years 1701, 1702, 1703, London 1745. Available online: https://quod.lib.umich.edu/e/ecco/004846589.0001.000?view=toc (accessed on 10 January 2021).
74. Kościerza, I. Sublime (w) Historii. Wzniosłość Jako Narzędzie i Przedmiot Badań. *Pamięć Spraw.* **2012**, *11*, 115–137. Available online: http://cejsh.icm.edu.pl/cejsh/element/bwmeta1.element.desklight-7b96b0f2-399e-43fb-92a1-b2c2fa739a27 (accessed on 10 January 2021).
75. Merian, M. Theatrum Europaeum; Bd. 1–21, Franckfurt 1633–1738. Available online: https://en.wikipedia.org/wiki/Theatrum_Europaeum (accessed on 11 December 2020).
76. Merian, M.; Zeiler, M. Topographia Germaniae; Bd. 1–38, (1642–1660s). Available online: https://en.wikipedia.org/wiki/Topographia_Germaniae (accessed on 10 January 2021).
77. Vischer, G.M. Topographia Archiducatus Austriae Superiors Modernae, Augsburg 1674. Available online: https://fedora.phaidra.univie.ac.at/fedora/objects/o:25730/methods/bdef:Book/view (accessed on 10 January 2021).
78. Vischer, G.M. *Topographia Ducatus Styriae*; AGRIS: Graz, Austria, 1681.

79. Wening, M. Historico-Topographica Descriptio; Bd. 1–4, München 1701–1726. Available online: https://bildsuche.digitale-sammlungen.de/index.html?c=viewer&bandnummer=bsb00063022&pimage=00001&v=100&nav=&l=de (accessed on 12 January 2021).
80. Dahlberg, E. Suecia Antiqua et Hodierna; Holmiae 1698–1720. Available online: https://suecia.kb.se/F/?func=find-b&local_base=sah (accessed on 15 January 2021).
81. Puffendorf von, S. De Rebus a Carolo Gustavo Sveciae Rege Gestis; Norymb 1696. Available online: https://polona.pl/item/samuelis-liberi-baronis-de-pufendorf-de-rebus-a-carolo-gustavo-sveciae-rege-gestis,NzkyMzU3NQ/3/#info:metadata (accessed on 20 January 2021).
82. Werner, F.B. *Topographia Silesiae*; Geheims Staatsarchiv Preussischer Kulturbesitz Berlin-Dahlem: Berlin, Germany, 2018; Volume 1–5.
83. Werner, F.B. Topographia oder Prodromus Delineati Silesiae Ducatus [. . .], t.1 (Manuscript), ca 1750. Biblioteka Uniwersytecka Wrocław, Odział Rękopisów, sygn.R550. Available online: https://www.bibliotekacyfrowa.pl/dlibra/publication/7114/edition/14302/content (accessed on 20 January 2021).
84. Kleiner, S. Vera et Accurata Delineatio Omnium Templorum et Coenobiorum quae Tam in Caesarea Urbe ac Sede Vienna, Austriae; part 1–4; Augsburg 1724–1737. Available online: https://www.e-rara.ch/zut/content/zoom/201865 (accessed on 22 January 2021).
85. Falda, G.-B. Ville e Giardini di Roma; Roma. Available online: https://archive.org/stream/gri_33125008634194#page/n5/mode/2up (accessed on 10 January 2021).
86. Rossi, G.G. Li Giardini di Roma; Roma 1683. Available online: https://polona.pl/item/li-giardini-di-roma-con-le-loro-piante-alzate-e-vedute-in-prospettiva,NTg1MTk5MQ/16/#info:metadata (accessed on 10 January 2021).
87. Diesel, M. Erlustierende Augenweide in Vorstellung herrlicher Gärten und Lustgebäude; Augsburg 1717. Available online: https://bildsuche.digitale-sammlungen.de/index.html?c=viewer&bandnummer=bsb00073534&pimage=7&v=2p&nav=&l=de (accessed on 10 January 2021).
88. Blondel, J.-F. *De la Distribution des Maisons de Plaisance et de la Décoration des Edifices en Général*; Charles-Antoine Jombert: Paris, France, 1737; Volume 1–2, pp. 1737–1738. Available online: https://gallica.bnf.fr/ark:/12148/btv1b8624593q.image (accessed on 20 January 2021).
89. Le Rouge, G.-L. Jardins Anglo-Chinois à la Mode; Cahiers 1–21, Paris 1775–1789. Available online: https://gallica.bnf.fr/html/und/images/les-jardins-anglo-chinois-de-georges-louis-lerouge?mode=desktop (accessed on 20 January 2021).
90. Czartoryska, I. Myśli Różne o Zakładaniu Ogrodów; Wrocław 1805. Available online: https://polona.pl/item/mysli-rozne-o-sposobie-zakladania-ogrodow,MTIzMzk2OQ/2/#info:metadata (accessed on 20 January 2021).
91. Hirschfeld, C.C.L. Theorie der Gartenkunst; Bd. 1–5, Leipzig 1779–1785. Available online: https://digi.ub.uni-heidelberg.de/diglit/hirschfeld1779 (accessed on 20 January 2021).
92. de Caus, S. Hortus Palatinus: A Friderico Rege Boemiae Electore Palatino Heidelbergae Exstructu; Frankfurt a. M., 1620. Available online: https://gallica.bnf.fr/ark:/12148/bpt6k15122355.image (accessed on 10 January 2021).
93. von Pückler-Muskau, H. Andeutungen über Landschaftsgärtnerei, Verbunden mit der Beschreibung Ihrer Praktischen An-wendung in Muskau; Stuttgart, 1834. Available online: https://digi.ub.uni-heidelberg.de/diglit/pueckler1834a (accessed on 20 January 2021).
94. Żaba, A. Drawing of F.B. Werner and geometry. *J. Pol. Soc. Geom. Eng. Graph.* **2016**, *28*, 63–70. Available online: http://yadda.icm.edu.pl/yadda/element/bwmeta1.element.baztech-c63e043f-ed01-4b50-b416-b1bfab6d4fec (accessed on 10 January 2021).
95. Sroczyńska, K. *Zygmunt Vogel, Rysownik Gabinetowy Stanisława Augusta, Zakład Narodowy im*; Ossolińskich: Wrocław-Warsaw-Kraków, Poland, 1969.
96. MAPIRE—Historical Maps Online. Available online: http://mapire.eu (accessed on 20 January 2021).
97. Zachariasz, A. Przydatność Archiwalnych Źródeł Kartograficznych dla Współczesnych Badań Krajobrazowych. Prace Komisji Krajobrazu Kulturowego. *Pol. Tow. Geogr.* **2012**, *16*, 63–83. Available online: http://krajobraz.kulturowy.us.edu.pl/publikacje.artykuly/16.kartografia/4-zachariasz.pdf (accessed on 5 January 2021).
98. Pozzana, M. *Giardini Storici. Principi e Techniche della Conservation*; Alinea Città: Florence, Italy, 1996.
99. Mosser, M. The Impossible Quest of the Past. In *The Architecture of Western Gardens*; Mosser, M., Teyssot, G., Eds.; MIT Press: Cambridge, UK, 1991; pp. 525–529.
100. Bronwyn, H. Innovation in Conservation, a Timeline History of Australia ICOMOS and the Burra Charter; Sydney 2015. Available online: https://www.researchgate.net/publication/316924146_Innovation_in_Conservation_a_Timeline_History_of_Australia_ICOMOS_and_the_Burra_Charter (accessed on 22 January 2021).
101. Heinich, N. *La Fabrique du Patrimoine. De la Cathédrale à la Petite Cuillère*; Éditions de la Maison des Sciences de L'Homme: Paris, France, 2009.
102. Obad Šćitaroci, M.; Marić, M.; Vahtar-Jurković, K.; Knežević, K.R. Revitalisation of Historic Gardens—Sustainable Models of Renewal. In *Cultural Urban Heritage. Urban Book Series*; Springer: Cham, Switzerland, 2019; pp. 423–441.
103. Cabral, I.; Keim, J.; Engelmann, R.; Kraemer, R.; Siebert, J.; Bonn, A. Ecosystem services of allotment and community gardens: A Leipzig, Germany case study. *Urban For. Urban Green.* **2017**, *23*, 44–53. Available online: http://dx.doi.org/10.1016/j.ufug.2017.02.008 (accessed on 10 January 2021). [CrossRef]
104. ICOMOS International "Journeys to Authenticity" Initiative. Available online: http://www.icomos.pt/index.php/2-inicial/160-icomos-international-journeys-to-authenticity-initiative (accessed on 8 September 2020).
105. Report on the Scientific Council meeting in Marrakesh—14 October 2019. Available online: https://www.icomos.org/en/about-icomos/governance/general-information-about-the-general-assembly/list-of-general-assemblies/annual-general-assembly-2019/69712-report-on-the-2019-scientic-council-meeting-in-marrakesh-october-2019 (accessed on 10 January 2021).

Article

Baroque Origins of the Greenery of Urban Interiors in Lower Silesia and the Border Areas of the Former Neumark and Lusatia

Bogna Ludwig

Department of Architecture Conservation and Restoration of Cultural Landscape, Faculty of Architecture, Wroclaw University of Science and Technology, 50-370 Wroclaw, Poland; bogna.ludwig@pwr.edu.pl

Abstract: The article is the first attempt to gather information on the beginnings of using green elements in urban compositions in Lower Silesia and border areas, in the former Neumark and Lusatia. It presents Baroque urban arrangements with the use of green ground floors, tree espaliers and avenues, from the earliest ones—occurring in the aftermath of the Thirty Years' War—and the solutions applied in private municipalities in the Habsburg, Wettin, and Hohenzollern states, which were recovering from war damage, to urban developments at the end of that period, in the areas already under Prussian rule and its strict regulations. A comparison with the achievements of European urban planning in this field allows us to trace the paths of inspiration, but also to uncover some innovative achievements.

Keywords: urban baroque greenery; 17th–18th century; allée; Lower Silesia; preservation and renewal of heritage

Citation: Ludwig, B. Baroque Origins of the Greenery of Urban Interiors in Lower Silesia and the Border Areas of the Former Neumark and Lusatia. *Sustainability* **2021**, *13*, 2623. https://doi.org/10.3390/su13052623

Academic Editor: Steve Kardinal Jusuf

Received: 21 January 2021
Accepted: 19 February 2021
Published: 1 March 2021

1. Introduction

Although there were no large or dynamically developing urban centers in Lower Silesia during the Baroque period, there were changes in the appearance and methods of design of architecture, which largely influenced the appearance of municipalities, changes which were based on a modern view of the relationship between public space and buildings. Greenery complexes were among the new components introduced to Baroque spatial arrangements of urban interiors. These urban planning activities have rarely been noticed by researchers, as a result of which no comprehensive scientific study has yet been devoted to them yet.

Greenery was only introduced into urban planning arrangements when the residential municipalities, still autonomous of the Habsburg State, were transformed in the Duchies ruled by the Houses of Piast and Württemberg as well as in Albrecht von Wallenstein's estates. The investments of Louise of Anhalt-Dessau (1631–1680), Christian Ulrich Württemberg (1652–1704), and his wife Anna Elisabeth of Anhalt-Bernburg (1647–1680) were of particular importance. In the following period, new small private urban centers were established in which the owners of the border lands received religious refugees from Silesia, Bohemia, and Moravia; in such municipalities, tree avenues made up the center or one of the most important elements of the composition. The initiatives of Count Balthasar Erdmann von Promnitz (1656–1703), Konrad von Troschke (1671–1728), and then the extensive actions of Nikolaus Ludwig von Zinzendorf (1700–1760), resulted from economic reasons, but they also clearly had an ideological basis. The last villages with Baroque layout and greenery compositions were created in the Prussian state after the borders had been shifted. At the same time, according to the new principles of fortification design, tree espaliers became a component of fortification systems and also served to protect the roads, which contributed to their further dissemination as a composition element in urban planning.

The composition of greenery as a new urban solution was a trend that municipality owners took over from other regions of Europe. For this purpose, they hired suitable contractors. This phenomenon was initially clearly related to the descent and connections of ducal families in Lower Silesia. In the second period, the planning solutions were modeled primarily on those that were being developed in the immediate vicinity. Inspirations were also drawn during trips to remote European centers.

2. Materials and Methods. State of the Art and Sources

The study was based on a typical research method used in the humanities. Based on a review of the scientific literature on baroque urban greenery in Europe, the research background was characterized. The current state of knowledge was presented, and in some cases an attempt was made to supplement it using the analysis of sources. A review of scientific literature concerning the chosen subject of research has been made. The identification of urban greenery systems in Lower Silesia and neighboring areas created in the Baroque period was based on a meticulous analysis of available archival documents—cartographic, iconographic, and written documents—as well as chronicles and topographic descriptions. The few preserved greenery complexes in the discussed area arranged in the Baroque period were examined. The historical background of their creation is presented; the founders and—where possible—designers are indicated, with an attempt to trace the origin of inspiration. Their further development is briefly characterized. The description and formal analysis of the presented urban greenery structures was based on source information and the remains present in the urban layout. In this way, the results of the research were formulated in a chronological and typological sequence of urban greenery in Lower Silesia and the border areas of the former New March and Lusatia in the Baroque period. This allowed comparisons to be made with European examples, indicating the typicality of solutions, the use of selected patterns, and—in some cases—innovation.

Research on the history of greenery design in Europe [1] shows the genesis and dissemination of solutions from modern times for introducing greenery composition into urban interiors. In this respect, it is important to indicate why and how the ideas of designing greenery systems, initially only tree espaliers and avenues, used in gardens and open landscapes, were transferred to inner municipality structures. The juxtaposition of the emerging pan-European concepts in this respect allows us to refer to them realizations from the presented region.

Baroque urban planning in Lower Silesia and the border areas is a rarely studied issue. Therefore, it is difficult to expect research on selected, quite detailed problems, such as the occurrence and the way of shaping greenery in public interiors. Studies of the historical spatial transformations of individual municipalities—from the oldest publications from the turn of the 20th century [2,3] to contemporary ones [4–6]—reveal the characteristics and sometimes also the landscape role of Baroque transformations. These publications also include information on the introduction of greenery systems to urban design. However, the very phenomenon of the greenery of urban interiors occurring for the first time in the Baroque period has not been properly noticed. This issue is partially discussed in an article devoted to the history of gardens from that period, which were also used as types of parks established at the settlements of the Moravian brothers in Lusatia, Hessen and Lower Silesia [7]; yet, the aforementioned text does not characterize the phenomenon of urban interior design with greenery systems. Similarly, the book by the same authors mainly mentions the greenery of gardens and parks, not urban greenery [8]. Historical research, in which the landscape role of the avenue is discussed, concerns the later assumptions and usually palace-park complexes [9]. In recent years, there have been fairly extensive studies on the natural qualities of avenues, which is a completely separate theme.

The analysis of local sources makes it possible to trace when and in what forms the first elements of greenery that shape urban compositions in Lower Silesia and the border areas of the Neumark and Lusatia emerged. The most precise information on this subject is provided by iconographic representations, made shortly after their creation, such as drawings by F.B. Wernher [10–16]. Cartographic records coming only from the 19th century are a valuable supplement of date [17–22]. Notes from Baroque chronicles written by F. Lucae [23] and the already mentioned F.B. Wernher [11–13], as well as from

18th and 19th century historical and topographical descriptions of Prussian Silesia by F.A. Zimermann [24] and J.G. Meissner [25], as well as descriptions of the March of Brandenburg by F.W.A. Bratring [26] and Saxony by F.A. Schumann and A. Schiffner [27] are important sources of knowledge.

From these juxtapositions, one can read out what patterns and how quickly they were adopted in the discussed area, answer the question of whether own solutions were developed and to what extent this new form of space arrangement using greenery composition was widespread.

3. Results

3.1. Avenues and Tree Espaliers—The Beginnings of Urban Green Space in European Municipalities

Since the end of the 17th century, trees have been consciously introduced as compositional elements, first in the form of tree espaliers and then allées. In ducal municipalities, the tree espaliers began to mark important spatial connections between the ruler's seat and the temple. Then, they became widespread in the municipalities possessed by the landed aristocracy as well. The tree espaliers began to be used as a composition element during the Renaissance. Allées appeared in garden arrangements. Initially, the straight paths were covered with pergolas overgrown with vines to provide shade (A1 in Appendix A) [28]. Such solutions were already applied in the Middle Ages. The name espalier (Italian: *spalliera*) is derived from that period. Alberti in the Renaissance villa Quaracchi Giovani Ruccelai (1459) divided the grove—*boschetto*—with straight paths [29,30]. This method of design quickly spread in the sixteenth century and became the principle of shaping the gardens of the Italian country villas (Villa Medcici in Castello, in Pratolino, the Boboli Gardens by the Pitti Palace, the villa in Poggio a Caiano, Poggio Reale near Naples) and the gardens by the residences all over Western Europe, especially the hunting parks (Blois, Gaillon, Hampton Court). It was also then that the term allée was introduced to name such paths (A2 in Appendix A) [31,32]. Wooden avenues were shaped outside the garden systems as paths leading to them (Chenoncaeux, Fointenbleau, Blois, Morienmont, Aranjues) [1] (pp. 18–19). Palladio in his treatise recommended providing country roads with tree espaliers giving shade and constituting greenery as a respite for eyes (A3 in Appendix A) [33] (pp. 262, 266). He also gave examples of the few such roads to Villa Cicogna and Villa Quinto in its area.

At the same time, rows of trees planted over moats and canals began to become a decorative element of the municipality. At the end of the 16th century, Dutch municipalities such as Bruges, Gouda, Harlem, The Hague, Amersfoort, and Kampen had wooded canal piers [34–37]. In The Hague, at the inner lake, Hofvijver, which was regulated by a rectangular basin, tree rows and greenery systems appeared as early as in the 14th century [38,39]. The wooded area of Lange Voorhout was incorporated into the municipality at the beginning of the 15th century. In 1536, during his visit, Emperor Charles V ordered the planting of an avenue of linden trees, which would connect the adjacent gardens in this area [40–42]. Trees were also planted in the 1670s around the fortifications in Antwerp, and at the end of the 16th century also in Italian municipalities of Florence, Siena, Piacenza, Lucca, and Padua. The enlargement of municipalities also boosted the appearance of groups of trees. In Amsterdam, tree-lined canals were introduced in place of old fortifications after they had been shifted to a new line in 1585 and once again in 1610 [43]. In Leiden, such changes occurred slightly later, and in Utrecht in 1660 [1] (pp. 44–45). In the 17th century, the custom became widespread and tree-lined canals were created in many Dutch municipalities.

In accordance with Palladio's recommendations, roadsides were lined with trees from the end of the 16th century. The suburban roads of Rome have mostly turned into avenues over several decades [44,45]. In France, already in 1553 and 1575, kings Henry II and Henry III ordered the royal roads to be tree-lined. However, it was only J.B. Colbert's regulations of the 1650s, which applied the 1601 Edict of Sully, that decided that this principle was fully implemented [1] (p. 19) (which, for example, was clearly visible around Reims [46]). Outside Italy, roads leading to residences were also designed in the form of avenues. The earliest example of which is the tract to Hellbrunn from Salzburg, created in 1615–1619 [47,48]. In the 17th and 18th century, avenues leading to palaces became the norm (e.g., from the more famous Fürstenallee in North Westphalia to the hunting lodge Oesterholz, 1725, Poppelsdorferallee near Bonn to the palace of the same name, around 1730, Wilhelmshöher Allee in Kassel, from 1767, and Laxenburger Allee and Schönbrunner Allee near Vienna, 1741).

Social changes in big municipalities—the settlement of courtiers and nobles, as well as the growth of a rich patrician group—changed the habits and the manner of having entertainment. Horsemanship, outdoor games, such as tennis, golf, various ball games, pall-mall (paille-maille), and shooting galleries, required green areas. Suburban recreation areas began to emerge; in the Renaissance, suburban gardens (e.g., in the neighborhoods of Florence, Rome, Naples, Paris, London, Bruges, and Antwerp), in the Baroque were replaced by promenades and suburban avenues and boulevards. Special areas for pall-mall appeared from the end of the 16th century in Paris, The Hague, Utrecht, and London. Then, they were often used as walking promenades and as such were copied in other municipalities in the 17th and 18th centuries. Riding avenues, the route over the Arn near Florence and the Course la Reine on the Seine, modeled on it in Paris (1616), inspired further routes in the vicinity of Marseille and Aix, but also outside France near Madrid or London [1] (pp. 32–37).

The first urban interiors designed using tree espaliers of trees were also realized. In Aix en Provence, the square of the new district of Villeneuveve was lined with trees in 1580. In Willemstadt and Klundert, the new fortress municipalities founded in the 1580s in North Brabant by Wilhelm of Orange, trees surrounded some squares and streets [1] (pp. 22–26) [49,50]. Created in 1630, Covent Garden Piazza in expanding London gave rise to squares as centers of aristocratic districts—Lincolns Inn, Bloomsbury Square, St. James Sq. in 1650–1660—which were initially only lined with by tree espaliers and lawns. Others appeared after the great municipality fire (1666) and became a model for other English municipalities; e.g., Warwick or Bristol. In the second half of the 17th century, walking avenues were also created in English municipalities; leading to the cathedral in London, in Bristol, Bath, Wells, and many provincial centers [1] (pp. 38–51). In Rome, from the beginning of the seventeenth century, elm tree avenues were introduced at the passageways: connecting the basilicas of Santa Croce and San Giovanni and further on Santa Maria Maggiore, the Arch of Constantine with the church of San Gregorio al Celio, as well as at Campo Vaccine on the site of the cattle market at the Roman Forum recreating the Via Sacra between the arches of Titus and Sever (A4 in Appendix A) [51]. Within the precincts of Paris there were established the royal gardens of Thulleries (created from 1564) and Luxembourg (completed in 1612), in the vicinity of Place Royale (later Place des Vosges, also formed until 1612) along with the gardens of Palais Royale (in 1633), and then the boulevards created on the site of the fortifications (Les Grands Boulevards, 1668–1705). However, public greenery appeared inside the municipality sporadically, only in the form of the greenery of Vosges Square in 1680. Further expansion of the metropolis connected it with suburban green areas such as Les Invalides Square (1670–1676), Jardin des Plantes (1626, as a public garden since 1640), or the Champ de Mars (1751) [52]. The tree arrangements in the urban composition did not gain a significant share in the Baroque in France. Even in the second half of the 18th century, when e.g., the extension of Rouen was planned, the project made by M. le Crapentier (1758), which included avenues and squares surrounded by tree espaliers, was mostly not realized; some of the works started only at the very end of the 18th century. Similarly, two proposals for the extension of Lyon (1768 and 1769) containing analogue concepts were rejected [1] (p. 66).

It was different in the expanded and established municipalities in Prussia and German countries, where Dutch examples of the use of tree espaliers but also French garden solutions were applicated in shaping urban layouts. The Hague Voorhout became an inspiration for an avenue, initially simply called Erste Straße, which was created on the initiative of the Great Elector Frederick William I in Berlin. It was planted as early as 1647 according to the Dutch model, with lime trees (hence from 1734 called Unter den Linden) [53]. The avenues formed the urban backbone of the New Municipality in Dresden; first, the main axis of the Hauptstrasse (1687) was formed and then extended along the Koenigsstrasse (1722–1732). In Frankfurt, in 1712 the main square of the New Municipality—Rossmarkt—was, on the order of the municipality council, provided with greenery around the fountain and a chestnut allée extended therefrom along the frontage of new houses [54]. The greenery compositions had the greatest share in the planning of the new residential municipality of Charles III Wilhelm von Baden-Durlach—Karlsruhe—in 1715, often regarded as a manifestation of Classicist urban planning. This architectural concept was based on a pattern of radial tracts spreading from the palace (or in fact the tower next to it), formed as the streets continued by the garden avenues. Similarly, one of the two axes shaping the composition of St. Petersburg (founded in 1703)—Nevsky Prospect—featured tree espaliers [55]. The avenues also became the basis of the layout of municipalities and districts for Huguenot refugees, as was the case Potsdam. Potsdam became a residential municipality from around 1675, intensively populated by Huguenots emigrating from France after the Fontainebleau and Potsdam Edict from 1685. It developed along the allée stretching from the elector's hunting lodge into an open area. The Huguenots' refugee districts built near the capital municipalities of Erlangen, Kassel, Magdeburg, and Plön were similarly planned. For example, the new municipality in Erlangen, designed by Johann Moritz Richter, builder of margrave Christian Ernest Hohenzollern-Bayreuth (1686, 1706), together with the construction of the palace from 1700 onwards, took on the layout of a municipality-residence with a palace-garden composition axis. The avenues also surrounded and connected new districts of a different character, such as the Herrenhauser—residential district of Hanover. Chestnut promenades connected the expanding Bayreuth (1725) and Ansbach (1737) [56,57]. Tree espaliers also began to be used to emphasize the representativeness of the site, as in Manheim around the monument on the Pardeplatz designed in 1712 (made in 1743) to commemorate the victory in the war of succession in the Palatinate (1688–1697).

The problem of trees was treated somewhat differently by Frederick William I (1688–1740) and then Frederick II the Great (1712–1786), also taking into account their economic significance. It was recommended to plant fruit trees, mulberries for silkworm rearing and willows for wicker production. In time, state orders were issued, initially covering the protection of planting, and subsequently orders for the introduction of avenues of trees (A5 in Appendix A) [58,59]. In 1742, the planting of fruit trees and mulberries was ordered, which also covered Silesia, conquered at that time [60–63].

3.2. Greenery of Urban Ducal Residences in Lower Silesia

As in Western Europe, the initial introduction of greenery into municipalities in Lower Silesia was associated with the transformation of the ducal urban seats, while on their outskirts with the formation of links between municipalities and neighboring and suburban palaces. In modern times, greenery compositions were gaining in importance in residential layouts, initially only for recreation and decoration, then also for composition.

The visual impact of residences on society began to become increasingly important in modern times. For this purpose, different means were made use of than those in the Middle Ages, where monumentality and inaccessibility were the dominant features of a ruler's residence or a feudal lord. In modern concepts, the criterion of representativeness expressed through classical order architecture but also through the surrounding greenery came to the fore. In addition to the architectural changes in line with the current trends, it was the garden facilities that gave prestige to the ruler's seat, as evidenced by the castle gardens in Brieg (currently Brzeg) and Glogau (Głogów). The Brieg castle was transformed and thoroughly rebuilt during the times of Duke Frederick II (1480–1547) and his son George II (1523–1586), and thus changed from a Gothic defense building into a magnificent Renaissance residence [23] (pp. 1363–1364) [64] (pp. 2–4) [65] (pp. 25, 28–40) [66]. From the foundation of George II and his wife in 1554–1560, a three-story building of the entrance gate, the most important investment shaping the appearance of the castle from the municipality side and presenting a rich iconographic program, was erected between the south wing and St. Hedwig's collegiate chapel [67]. At the same time, the collegiate church was transformed into a mausoleum of dukes. Italian architects Jakob and Franziskus Pahr were employed in the reconstruction of the castle. The continuators were Bernard and Peter Niuron, who also fortified the residence at the end of the century. The surroundings were also changed to shape the castle's representative gardens. To the north of the castle, an orchard was planted along the walls, which incorporated the former Dominican monastery grounds, with a hill (Sperling Berg) [68] (pp. 77–78) (A6 in Appendix A). On the east side, a garden (*Lustgarten*) was arranged in front of the windows, covered with a wall on the municipality side Figure 1a.

Similar transformations were made in the Glogau castle. At the turn of the 15th and 16th centuries, Glogau was inhabited by Prince John Albert, and then by Sigismund Jagiellon (both after Kings of Poland and Grand Dukes of Lithuania), who began rebuilding the castle, which had been ruined during the siege by Matthias Corvinus' army. Gardens were then arranged next to the castle, while behind the bailey and the second line of moats and walls, an entrance square was created [69].

In the Baroque era, residences and gardens began to be composed in a new way. Activities were undertaken both in the ducal residences and the former manors of the landed gentry which were rebuilt into palaces. In exceptional cases, they were matched with the municipality layout, as was the case in Juliusburg (Dobroszyce) or Festenberg (Twardogóra). After the Thirty Years' War, not only different Baroque architectural and ornamental forms were used, but also the approach to space arrangement changed. Compositions with a dominant axis were introduced. In representative forms, the foreground of the residence was designed. Extensive garden arrangements with pavilions, orangeries and loggias were created.

The first gardens of this type were established by the order of Louise of Anhalt in the 1670s, in the Duchess's seats in Brieg, Ohlau (Oława), and Wohlau (Wołów), which were being rebuilt at that time. As Baroque chronicler Lucae noted, these gardens were to provide a beautiful vista [23] (p. 1374). In all castles, the Duchess introduced terraces and vista points to the greenery of the gardens. In Brieg and Wohlau, it was probably mainly about views from the palace windows to the garden, but in Ohlau the front gardens also provided an attractive view of the new residence from the municipality side. The green areas thus became the backdrop for the ducal residence.

In Brieg, the transformation of the residence began as early as during the times of Duke George III (1611–1664). In 1654, George took over the Brieg Duchy [70]. Already in 1656, he started the reconstruction of the Renaissance castle in Brieg. One of the first investments was to clean up the foreground on the municipality side. An openwork fence was erected to separate the courtyard and the castle garden from the municipality square (Topffmarkt), to which a gate with wickets, situated in front of the gate tower, led, which was shaped in the Palladian form, being a kind of triumphal arch [23] (p. 1363) [11] (pp. 297–301). On the western side, behind the fence, partly in front of St. Hedwig's Castle Chapel, in 1658, there were farm buildings and a riding school, forge, and stable, forming a screen around the castle. The front garden (*Lustgarten*) was redesigned. In the following year, the castle façades and all towers were renovated after the Lions' Tower, rebuilt in 1649, was destroyed by thunder. Already during the reign of the next Duke Christian, garden buildings were erected in the orchard; a loggia with an orangery, a birdhouse and a shooting range (A7 in Appendix A) [8] (pp. 156–157). The castle building was rebuilt to open from the east side to the garden with arcades [23] (pp. 1364, 1367, 1378, 1380) [64] (pp. 23–25).

Further Baroque changes were ordered in 1673 by Duchess Louise of Anhalt, widow of Duke Christian, George's brother. The bay window on the eastern façade of the castle was then extended; it was crowned with a terrace and enlarged while its windows were made uniform, and a new arcade loggia was constructed on the municipality and garden side. In this way, the Brieg Castle received not only a representative entrance square on the municipality side and a decorative front garden but also a vantage point Figure 1b.

(a) (b)

Figure 1. (**a**) The ducal residence in Brieg. A fragment of the Brieg plan, probably from the period between 1618 and 1633 [71]; (**b**) The ducal castle in Brieg around 1750 in a drawing by F.B. Wernher [11] (p. 297).

As a regent, the Duchess held power in the Liegnitz-Brieg (Legnica-Brzeg) Duchy in 1672–1675 because her son George IV William (1660–1675) was underage. After the death of George William, who reigned for only a few months, the duchy came under direct rule of the Habsburgs and the Duchess settled in Ohlau. The Brieg castle used by Austrian officials and visiting dignitaries was partially neglected [68] (p. 82). The Jesuits soon located their seat next to it and occupied most of the former Dominican monastery grounds, which was to become the castle garden. In 1741, the Brieg residence was shelled and burnt down by the Prussian army. The façade of the entrance building and part of the ground floor of the eastern wing survived [64] (pp. 16–17). It was rebuilt into an inn and warehouses (1743–1755), while its architectural form was no longer taken care of.

In the modestly presented urban development of Ohlau, the castle's representative foreground was also taken care of. An original layout was created here, consistent with the municipality's location plan, which included a Baroque urban interior combining two areas of urban and residential function. In 1654, after the death of Duke Georg Rudolf, the lands of the Duchy of Liegnitz were divided among the sons of his brother Johannes Christian, Duke of Brieg [70]. The youngest of them, Christian (1618–1672), received Wohlau and Ohlau. He had Carlo Rossi from Como start the reconstruction of the dilapidated Ohlau residence in 1659 [71]. The works under the direction of the Italian architect lasted until 1680, well into the time of Louise of Anhalt [23] (pp. 1405–1406), whom the emperor granted the Ohlau and Wohlau lands for life. At the same time, the castle changed the defensive system of the roundel into an earth bastion. Rossi modernized the residence itself [72] (pp. 43–53) [73] (p. 383). In the years 1659–1673, he erected or converted the former building, which was added to the residential section on the east side (now a church stands there), into a two-story pavilion in early Baroque forms topped with a terrace with figures [11] (p. 398–400). This part of the castle was henceforth called *Christianbau*. From 1655, a chapel on the northern side of the courtyard began also to be rebuilt and a garden behind the palace. In the further 1673–1680, an extension of the castle, a four-story part called Duchess Louise's palace (*Louisenbau*), was created, which opened from the courtyard with an arcade loggia, and was decorated with Baroque details on the front complete with a richly decorated portal depicting the Duchess's coat of arms [23] (p. 1406). In front of the residence, the foreground was cleaned; the buildings were demolished, the bridge was moved, leaving the moat, the castle fortifications on the municipality side were removed, and a garden was established in their place. A decorative fence with stone posts along the palace façade was introduced, separating the green belt from the moat on both sides of the bridge, Figure 2 [11] (pp. 267, 396–397, 400) [74] (A8 in Appendix A) [75–78]. A drawbridge, preceded by stone pillars-pylons, leading to the main entrance portal to the inner courtyard of the mansion was built in the middle of it (A9 in Appendix A). The square's compositional axis was arranged, directed at the portal on the palace's façade, which was deliberately moved from the building's axis of symmetry to the east and emphasized by the use of double windows and the emerging castle tower behind it. This axis was marked out on the axis of the location municipality, i.e., in the middle of the market quarter facing the Castle Square, Figure 3. Its continuation in the open landscape was the main avenue of gardens behind the moat and bastion fortifications, highlighted by a pavilion in the middle [11] (p. 397) [23] (p. 1402). In this way, a very coherent and homogeneous composition of this urban layout was created, which obliterated the various origins of the elements of the location municipality and the medieval castle being transformed into a palace.

Figure 2. The southern elevation of Ohlau Castle on a drawing by F.B. Wernher from around 1730 [11] (p. 267).

Figure 3. A view of castle buildings and gardens in a drawing from the times of James Louis Sobieski [78].

In 1691, Prince James Louis Sobieski settled in the castle in Ohlau to live there for over forty years. After his 1734 move to Żółkiew, the building began to deteriorate. It was taken over by the imperial court after Sobieski's death (1737) and became the residence of imperial administration officials. After Silesia had been incorporated into Prussia, a military lazaret was installed in the castle in 1744 and a field bakery in 1761. The moats were eliminated whereas the gardens were expanded. Rows of mulberry trees were planted.

From 1764, next to the royal office, there was a silk spinning mill in the castle. In the years 1833–1834, a part of the castle (*Chrystianenbau*) was demolished and a Catholic church designed by Karl Friedrich Schinkel was erected in its place; after a 1938 fire, a modernist tower was added to it. In Louise's building, a school was built, next to which a small barracks guardhouse was erected and at the beginning of the 20th century a water tower, thus degrading the completely representative character of the castle complex [79]. The castle square in the second half of the 19th century was turned into a green square with a monument to the Victory in the War of 1870 (Siegesdenkmal).

In Wohlau, ravaged by the Thirty Years' War, the castle was repaired and decorated in a new Baroque style. A representative bridge leading to the mansion over narrow moats was also built [23] (pp. 1162–1163) [80]. It is very probable that already then, when demolishing the serviced buildings of the castle and the north-western corner of the market quarter, an entrance square was marked out from the municipality side. Along with these regulations, a modest green ground floor in front of the castle façade was perhaps introduced. They certainly existed already after the reconstruction ordered by the managers of the Wohlau camera, located in the residence, when after the death of the Duchess (1680) the land came under the direct rule of the imperial court [11] (pp. 562, 569).

In Bernstadt (Bierutów), like in Ohlau, during the times of Christian Ulrich of Württemberg (1652–1704), front gardens were established on the foreground of the duke's seat after the land had been cleared. Bernstadt from one of the subordinate urban centers in the Duchy of Oels (Oleśnica) changed in modern times into the seat of the side lines of the ruling family. The owners of the residence and the municipality tried to give it attractive external features, raising the importance of their family. As soon as Henry II (1507–1548), the third of the House of Poděbrady, took the reins of power, he chose the Bernstadt castle as his residence. In the years 1534–1540, he carried out thorough reconstruction and gave the castle Renaissance forms. In 1603, a fire wreaked havoc with the castle and the whole municipality. The eastern wing and tower burnt down, and reconstruction lasted for the next few decades [81] (pp. 516–519) [82] (pp. 22–30) [73] (pp. 530–533). In spite of that, from 1618, Duke Henry Wenceslaus of Poděbrady lived there (1592–1639, Duke of Bernstadt in the years 1617–1639, and in the years 1629–1639 the General Governor of Silesia). The Bernstadt residence was a simple two-winged building (with southern and eastern buildings) with a preserved Gothic tower at the end of the eastern wing. Three-story arcade porches were constructed at the wings, giving it a Renaissance character, and in 1622, the roof on the tower was replaced by an early Baroque glorious helmet. At the same time, the municipality, which gained the status of a ducal capital, was changing intensively and was expanding despite fires (1603, 1659). After the Württembergs took over the Duchy of Oels in 1647, Bernstadt was, after the fire of 1659, an insignificant ducal seat [80,81] (p. 467). However, in 1673, when the duchy was divided, it became the residence of Silvius Nimrod's second son, Christian Ulrich. The eldest son, Silvius Frederick (1651–1697), received the Oels part, while the youngest, Julius, the Juliusburg part, which was ruled by Elizabeth (1625–1686), as he was a minor until his death [83].

Christian Ulrich resided in Bernstadt until the death of the oldest of his brothers, Silvius Frederick, in 1697, when he inherited Oels. During this period, he took steps to modernize the castle. He expanded it by constructing one story more. He also tried to raise its representativeness by changing the form of the avant-corps of the south wing, introducing there a Baroque portal decorated with the family coat of arms, while at the same time renovating the tower and crowning it with a helmet (1679–1680) [81] (pp. 469, 521–522). On the municipality's side, in the wall closing the castle courtyard, a decorative gate in the form of a palladiana was inserted around 1680, and in front of it a bridge over the moat preceded by a small entrance square (A10 in Appendix A). A larger viewing foreground was probably to be ensured by replacing the last quarter adjacent to the castle with a riding arena, Figure 4 [13] (pp. 688–695). Gardens were to be an additional decoration and attraction; a small Lustgarten with an orangery on the northern side of the castle, a castle garden on the southern side, and extensive gardens set up behind the municipality walls in the immediate vicinity of the residence [81] (p. 521). These were realized under the close supervision and active collaboration of the Duke and his gardener Georg Herbst, who described the achievements in a work published in Oels [84]. The main entrance of the castle gate was located on a long viewing axis opening from the market square, via Breslauerstrasse, a view of the castle tower. From another point, the coat of arms portal of the main wing of the castle is visible through a side passage of the gate [85]. A line of trees was introduced parallel to the wall covering the garden. (Judging by Wernher's drawing, these were linden trees.) The espalier ran along the road connecting the gate with the church, or in fact the side gate in the church wall, which the Württemberg family used to go to the Duke's lodge (founded by Duke Christian Ulrich in 1679, as confirmed by the coat of arms of Württemberg and his wife Anna Elizabeth of Anhalt-Bernburg).

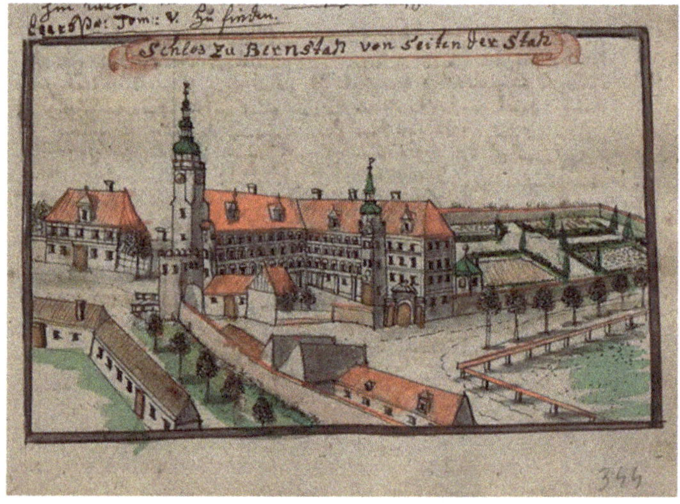

Figure 4. Bernstadt Castle illustrated by F.B. Wernher [13] (p. 691).

From the Baroque arrangement, only the portal of the entrance gate and the portal of the main castle wing have survived to this day; there, the coats of arms commemorate the founders. In the panorama and views, the presence of the castle is still signalled by the tower with its magnificent Baroque coat of arms. In the foreground of the residence on the municipality side there is still greenery, no longer a regular line. A similar but even more modest arrangement of the entrance to the castle from the municipality side was realized in Praisnitz (Prusice), where only a short line of trees preceded the elevation of the residence [11] (p. 651).

The gardens surroundings of the residences founded by the princes of Oels were the best witnesses of the importance of greenery in functioning of their courts. All three successive wives of Duke Christian Ulrich received new palace buildings with extensive gardens. Maria Sybillia received the palace in Sybillenort (Szczodre) and Sophia Wilhelmina, the palace in Wilhelminenort (Brzozowiec). In Oels, the gardens were established only for the fourth wife of Duke Christian Ulrich—Sophia of Mecklenburg at the beginning of the eighteenth century. In addition, his sister-in-law, Silvius Frederick's wife—Eleonore Charlotte of Württemberg-Montbéliard (1656–1743)—arranged a garden in Festenberg surrounded by rows of trees on the outskirts of the municipality [13] (pp. 740–741). The impossibility of introducing green complexes in the immediate vicinity of the residence was compensated for by creating separate suburban gardens. In Liegnitz, the gardens of Duchess Anna Sophia of Mecklenburg were created at the Sophienthal manor house (1657) and behind the Goldberger (Złotoryjska) Gate (1672) [23] (p. 1199). In Bernstadt, apart from the gardens within the municipality walls to the north and south, a large garden complex was also set up outside the municipality walls in the immediate vicinity of the castle.

The compositional significance of trees in the urban layout was planned to be used in Silesia probably already before. Perhaps already in the first half of the 17th century, it was intended to link the layout of the Sagan (Żagań) palace with the Jesuit church, whose guardian was the duke, with a properly designed composition of greenery on a vast area obtained after the demolition of a part of the municipality buildings.

In 1549, the land of Sagan came directly under Habsburg rule. During the campaign of the Thirty Years' War, in 1627, Albrecht von Wallenstein appeared as one of the main commanders of the imperial army in Silesia. In 1628, the Emperor, who was constantly strengthening his court position, handed over the Duchy of Sagan as compensation for outstanding financial obligations [86] (pp. 59–68). Wallenstein started the reconstruction of the municipality destroyed during the warfare. In this almost completely Protestant center, he decided to create a center of Catholicism and a magnificent seat of power. The first manor house to be built was located at the New Market Square (present Słowiański Square). At the same time, Wallenstein started to build a palace in the place of a castle demolished for this purpose [86] (pp. 59–68). The residence, which combines defensive and representative functions, was built based on the designs of Vincenzo Boccacci, the duke's architect. The four-winged form with bastions in the corners surrounded by a moat referred to the Gothic castle, reproducing the Renaissance solutions of the Italian defense residences from the 16th century—*palazzo in fortezza*. The construction of the palace began in the spring of 1630. The ground floor and walls of the first floor of the building were erected on a three-meter-high platform. At the same time, the new duke expelled the Lutheran predecessors. He assigned an abandoned Franciscan monastery to the Jesuits. In 1632, the reconstruction of the municipality was disturbed by the war. Sagan was captured twice by General Arnim and the Swedish army. However, at the end of the year, Wallenstein regained Silesia. In 1633, the duke, who resided in Sagan, planned an extensive construction of a Jesuit college and a convent for a hundred alumni. The project was also developed by Vincenzo Boccaccio. The complex was to consist of two-story buildings added to the former Franciscan church, grouped around four inner courtyards. Only part of the work was carried out until Wallenstein suddenly died in 1634 [86] (pp. 82, 87–89).

In front of the palace in Sagan, which was built as a modern residence under Wallenstein, it was decided to create a forecourt by deliberately demolishing the buildings. The result was a large square which, together with the articulation of the palace's façade, was to create an urban system that was subordinated to typical Baroque rules. In order to provide a suitable perspective for its seat, Wallenstein ordered the demolition of over 70 houses, which was largely carried out [86] (pp. 63–68) [87,88]. Within this space, two composition axes were arranged, one facing the western façade, the other facing the northern, shorter and linked to the façade of the opposite tenement. The latter was overshadowed at the end of the 18th century by the palace of the duke's administration (Herzogliches Kammergebäude), a classicist building erected for the General Plenipotentiary of the Sagan Duchy (now the district court). The longer composition axis ran in the middle of an elongated square to the north of the municipality walls (currently occupied by the tower apartment buildings), created after the demolition. It was a representative entrance to the residence from the side of the square. The spacious square north of the mansion, obtained by removing four quarters of the buildings connecting with the New Market Square, was not finally arranged [5]. Perhaps there were plans to completely clean up the space between the palace and the Jesuit college founded by Wallenstein in the former Franciscan monastery stretching from the square along the municipality walls to the north, Figure 5. Undoubtedly, these were one of the broader plans to transform the municipality into a representative space in front of the residence. The open square had dimensions of almost 130 by 150 m, and the farthest perspective view of the palace opened from the intersection of the streets behind the New Market (currently at the intersection of Słowackiego and Teatralna Streets), i.e., at a distance of about 350 m. This allowed for the creation of exceptionally monumental perspective compositions. Such extensive areas were probably planned as compositions with the use of greenery systems, tree espaliers and garden parterre, as it was done in the immediate vicinity of the palace at the turn of the 18th and 19th centuries. However, there is no design evidence or records of this. Indirectly, such plans may be evidenced by the care provided to the chateau garden surrounded by avenues with a zoo and a pheasantry garden [86] (p. 60).

Figure 5. A fragment of the Sagan panorama showing a Jesuit palace and monastery with a church in the mid-18th century [87].

3.3. *Tree Espaliers in Municipalities and Their Periphery*

The green areas introduced into the municipality limits had the character of front garden parterres and tree espaliers, such as those in front of the residence in Bernstadt or Prusnitz, which were beginning to become an increasingly important compositional element. In Bernstadt, as it was probably mentioned in the 1680s, a line of linden trees protected the wall of the castle garden from the side of the municipality and marked the road serving the princely passage from the gate to the church. In a little town, Auras (Uraz), the line planted along the façade of a yuft factory founded in 1722 by Philipp Wilhelm Luther [89,90] and the accompanying residential buildings—the frontage of the market square and the street—marked the road to the bridge and the castle portal, Figure 6 [10] (p. 347). Wroclaw municipality-dweller Luther was granted an imperial patent for exclusive production of yuft in Silesia [91] and appropriate privileges for imported workers. He also took care of the hospital and suitable flats for them.

Figure 6. Auras in the middle of the 18th century in a drawing by F.B. Wernher [10] (p. 347).

Similarly to other European countries, tree espaliers and avenues were introduced in Silesia as early as in the 17th century. The lime avenue in Leubus (Lubiąż) was described as early as the end of the 17th century, so it had to be planted at least several dozen years earlier [23] (p. 1167). In Dyhernfurth (Brzeg Dolny), probably as early as 1666 or with the foundation of the Stations of the Cross after 1701, an allée connecting the residence with the sanctuary was planted [3] (pp. 47–50, 90–94) [4] (p. 75). Trees also lined roads—the old road to Wahren (Warzyń) and a new route from the village to the chapel and further towards the Odra River (led along the border of the medieval location department) [9] (pp. 410–411) [8] (p. 152). In Ohlau, a garden avenue was created in the 1670s, with walls in the form of a hedge or trellis, which was an extension of the composition axis of the residence behind the moat in the suburban gardens [77]. In Juliusburg, a palace and church garden and a road from the manor house were planted in rows [13] (pp. 700, 703). Regularly planted trees appeared on the fortifications of the Liegnitz (Legnica) castle [11] (pp. 7, 89). Avenues spread in Silesia, and lines of trees were created in the municipalities and around their walls [25]. Both smaller Silesian municipalities, such as Haynau, Ghurau, Kant, Nimptsch, Prochowitz, Steinau, Neumarkt, Goldberg and Landeshut (Chojnów, Góra, Kąty, Niemcza, Prochowice, Ścinawa, Środa, Złotoryja, and Kamienna Góra) [9] (pp. 559–560, 484–485) [11] (pp. 206–207, 208–209, 396–397) [14] (pp. 194–195) or much bigger such as Wolau, Ohlau, Bernstadt, Sagan, Oels [11] (pp. 396–397, 562, 567) [13] (pp. 688, 694–695) [87] or Schweidnitz (Świdnica) [15] (pp. 343–344) (A11 in Appendix A) [92]. They were surrounded by rings of fruit trees. Fortifications in Breslau (Wrocław) were provided with plantings on the southern side of the municipality [9] (pp. 196–199).

Even more widespread use was made of tree espaliers and avenues in the open landscape. They were commonly used in the complexes of residences, as shown in Wernher's drawings [9] (pp. 390–589) [11] (pp. 373–619) [12,13] (pp. 252–449, 662–681) [14]. They were used as a kind of shielding and decoration of monotonous walls or railroad fences and manor buildings of both extensive palace buildings and small manors. Such a line was planted in the aforementioned palace in Sybillenort. They can be seen in illustrations of numerous manor and palace complexes. Roads leading to palaces and manor houses were also transformed into avenues.

The tree espaliers were planted along the roads leading to the churches of Peace in Schweidnitz, Jauer (Jawor) (A12 in Appendix A) and Glogau [16] (p. 27) [93]. Three oak allées, of which fragments of one have survived to this day, leading to the gates in the wall surrounding the property of the Evangelical parish in Schweidnitz were built together with its construction, which is shown on plans from before the mid-18th century [94]. An avenue also led to the Church of Grace in Sagan [95]. All those suburban Evangelical churches of Peace and Grace were accompanied by cemeteries and schools and surrounded by lines of trees. Then, already in Prussian times, roads to the churches and their fences were provided with plantings. Similarly, trees were planted along the church walls, and the elevations of newly established Lutheran high schools, e.g., in Hirschberg (Jelenia Góra) [96], as shown in Wernher's drawings, which was in line with the state recommendations concerning the plantings surrounding schools and temples.

All main roads leading from Breslau began to be lined with trees already before the middle of the 18th century (A13 in Appendix A) [97–100]. At the turn of the 18th and 19th century, they transformed into highroads (*Kunststrassen*, *Chausseen*) 100 as well as many main roads in the whole region [101]. The Prussian national orders recommending the planting of trees decided that most of them took on the character of an avenue. However, the avenues did not become part of the composition of dense urban complexes of municipalities expanding over time. The avenue in Tschepin (Szczepin), shown on plans from the end of the 18th century, was not preserved in Breslau after it was incorporated into the urban layout [102], nor the famous Scheitniger (Szczytnicka) Avenue—a poplar avenue (virginischen Pappeln, Populus deltoides), planted in 1790, leading to the park founded by Friedrich Ludwig von Hohenlohe-Ingelfingen [103].

3.4. Avenues in Squares and Streets of Lower Silesian-Lusatian Border Municipalities

At first, in the neighborhood of Silesia, avenues started to play an important role in the layout and shaping of green urban interiors of municipalities. At that time, on the borderland of Lusatia and the Neumark, which lay beyond the borders of the Habsburg state, there were settlements related to the influx of religious fugitives, mainly from Silesia, valuable craftsmen, especially weavers. Many of the centers, which were established at that time as municipalities, lost their municipal rights quite quickly. Their beginnings, however, and thus their spatial organization, are connected with the process of founding the city.

In the medieval village of Trebschen (Trzebiechów), first mentioned in 1308 and situated in the Neumark, which in 1701 was incorporated into the Prussian state, a village that developed around the local Renaissance manor house belonging to the Troschke family from Züllichau (Sulechów) (A14 in Appendix A) [104] (p. 344) a Lutheran border church was built in 1654. Religious fugitives, among them weavers, began to come here. In 1680, a new large temple was erected [105]. The expanding village was granted municipality rights by a royal decree from 1707, and its then owner Konrad von Troschke named it after the first Prussian king Frederick I Hohenzollern (1657–1713)—Friedrichshuld (Frederick's Grace) [26,106].

The new buildings (in 1719, it was 25 houses) were erected along an avenue planted with double rows of lime trees, which connected the palace and the church (now Lipowa Street), Figure 7. The street was perhaps founded as early as 1680, and it certainly existed after 1707. It was designated as a composition axis connecting the entrance to the palace with the dominant of the church tower. The central part formed a kind of elongated square in the greenery, which was the center of the municipality. It was complemented by park avenues stretching north of the landowner's seat. The Baroque layout of the village is well readable today, although in 1823 a new building designed by K.F. Schinkel and his collaborators was erected on the site of the previous Evangelical temple, and the municipality houses were replaced after the fire of 1830, while the palace was rebuilt in the 1870s and 1880s, when Trebschen lost its municipal rights (1870) [9]. Linden alley was undoubtedly modelled largely on the Dresden Hauptstrasse as the closest example, probably well known to the Saxon courtier. However, the most important model was probably Berlin's Uner den Linden, as evidenced by the patron of the municipality, the first king in Prussia, mentioned in the name.

Figure 7. Aleja Lipowa (Linden Allée) in Trzebiechowo. Fot. Piotr Frąszczak, www.polska-org.pl (accessed on 27 October 2020).

In 1679, the old settlement of Dziadoszanie—Halbau (Iłowa) was granted municipality rights from the Elector of Saxony John George II (1613–1680) due to the request by Countess Helga Margaret von Friesen. The rights were valid until 1830, when the town was degraded to a market municipality (Marktflecken) [27,107]. Its inhabitants were refugees from Silesia, who founded the first Evangelical parish in 1668 [108] (p. 836). The village developed in the hands of a new (since 1682) owner, Count Balthasar Erdmann von Promnitz, owner of Pless and Sorau (Pszczyna, Żary) [104] (pp. 101–102) [109], in the early 18th century. At that time, a rectangular market square was built up, with streets running out of corners, situated to the southeast of the land property. The erection of an evangelical parish church in 1720, designed by Giulio Simonetti [110,111], began after a fire in 1725, rebuilt and completed shortly before the next municipality fire, in 1749. At the same time, the Renaissance manor house from 1626 was converted and extended with a Baroque wing, giving it a palace-like form [108] (p. 836). Both buildings of the church and the palace, surmounted with towers protruding in front of the façades, were connected by a viewing axis, which in time transformed into a street surrounded by frontages [A15] [24,27]. At that time, around 1720, a second axial layout was also formed in Lindenstrasse (now K. Pułaskiego Street), located north of the first one, leading from the residence to the east (at its end in the 19th century a parish house was built up) [112–114] Figure 8 [20]. It was an extended avenue forming a kind of square planted with lime trees. In 1816, it was already an impressive avenue, so it must have been built several dozen years earlier [27]. The choice of trees indicates the Berlin pattern. However, it should be remembered that the founder knew the assumptions originating in Italy, France and the Netherlands [104] (p. 101). Greenery filling was not used in the case of the main urban interior—a marketplace, which was traditionally shaped like a square with streets leading from its corners; a second urban interior was added, probably related from the beginning to the area belonging to the Protestant community, where the school and parish buildings were erected [108] (p. 836).

Figure 8. Halbau on a topographic plan from 1939 [113].

Within the borders of Silesia, which was taken over by the Prussian state, the first example of the introduction of urban greenery was its implementation of a new municipality established at that time—Neusalz (Nowa Sól). A significant economic problem of Lower Silesia was the lack of deposits of rock salt. On the initiative of Emperor Ferdinand I (1503–1564), attempts were made to become independent from the export of this product from Poland. As a result, the settlement of Neusalzburg (Neusalz) was established, where sea salt breweries began to operate from 1553, and which finally became a municipality in 1743 from the edict of King Frederick II of Prussia [2] (pp. 3–6, 29, 43 in, 136) [6] (pp. 8–17).

Neusalz, which initially developed as a settlement on the Oder River, was later given a Baroque irregular layout with the center in the vicinity of the salworks and the building of the salt office (later the town hall). Simultaneously with the granting of municipality rights, King Frederick II gave permission for religious freedom, thus allowing the settlement of the Moravian Brothers (Herrnhuter). On the route from Breslau to Grünberg (Zielona Góra), along the section of the road connecting Neustädtel (Nowe Miasteczko) and Wartenberg (Otyń), which was created with the formation of post riders, a colony was established, equipped with a house of prayer, hospital, school, and pharmacy [2] (pp. 46, 49–52, 58) [6] (p. 19). Residential buildings, identical story houses, stood on both sides of an avenue planted with chestnut trees, Figure 9 [2] (p. 52) [14] (pp. 172–173). The choice of these trees, considered to be very ornamental, associated with superfluousness, is evidence of a desire to accentuate wealth and emphasize prestige. This shows a connection with Huguenot districts in German municipalities, but also with very representative projects, e.g., in Bayreuth and Ansbach.

Figure 9. Neusaltz's perspective plan from the mid-18th century [14] (pp. 172–173).

After the 1759 fire during the Third Silesian War, it was rebuilt again according to strict rules from 1765 [2] (pp. 765, 73–75) [115]. The avenue existed in the 19th century, with the last trees being removed at the beginning of the 20th century [116], and its memory was immortalized by the name Kastanienallee (now Wrocławska Street).

The introduction of greenery into the center of the layouts has become a principle in the settlements of the Moravian Brothers. The Moravian (Czech) brothers under the care of Nikolaus von Zinzendorf received permission to settle in Upper Lusatia in Herrnhut in 1722 and then in Hesse in Herrnhaag in 1738 (A16 in Appendix A), in Niesky in 1742, and in 1756 in Kleinwelka near Bautzen. In Silesia, Frederick II allowed the construction of settlements in Neusaltz 1742, Gnadenberg (Godnów, currently in Kruszyn, near Bolesławiec), and Gnadenfrei near Reichenbach (Piława Górna, Dzierżoniów). Both settlements began to be built from 1743. Much later, in 1772, the Brothers also settled in Gnadenfeld (Pawłowiczki) in Upper Silesia [117–119]. Around the square with the house of prayer, their first settlement Herrnhut (1722) and another Herrnhaag (1738) were built, although they did not have municipality rights, but had a regular layout [7,8] (pp. 194–198).

Although the architecture of the main buildings of these settlements was designed by professionals such as architect Siegmund August von Gersdorff (1702–1777), a member of the community since 1743 [120] (pp. 109–112), their layout, according to recorded tradition, was the result of the ideas of the first spiritual leaders of the community. It is known that the Herrenhag settlement plan was drawn up by the then bishop of the community, Christian David Nitschmann (1696–1772) [121]. The concepts stemmed from the implementation of biblical principles considered to be the basis for the functioning of the community into not only European, but also American urban planning solutions known to the organizers.

In Herrnhut, the central square was occupied by a large building of the community house of prayer, the so-called Great Hall. In front of the house of prayer—opposite the main axis of the complex marked outside the housing estate by an allée—a small front garden was made [122]. In Herrnhaag and Niesky, the houses of prayer were erected in frontages of squares, and their area was designated for extensive greenery, where rows of trees, connected by fences surrounded green parterres [123,124], Figure 10. In Herrnhaag, in the middle, a pavilion for a well was erected [125]. In Gnadenfrei Figure 11, Gnadenberg Figure 12, and Kleinwelka, initially the square was only decorated with trees, but as early as the beginning of the 19th century it was also surrounded by hedges [126]. This connects them not only with the history of the introduction of avenues in European municipalities, but also with the establishment of squares formed in a manner of a garden with the use of one-story greenery, the origins of which appeared in English municipalities and were a new element on the continent.

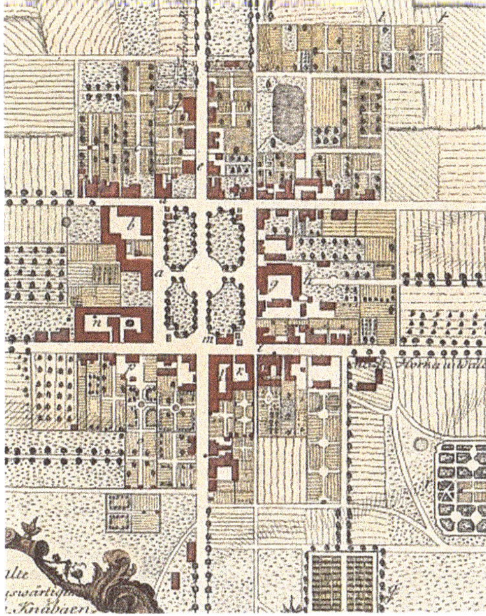

Figure 10. The Herrhaag plan from the time of its creation. Extract from the plan [122].

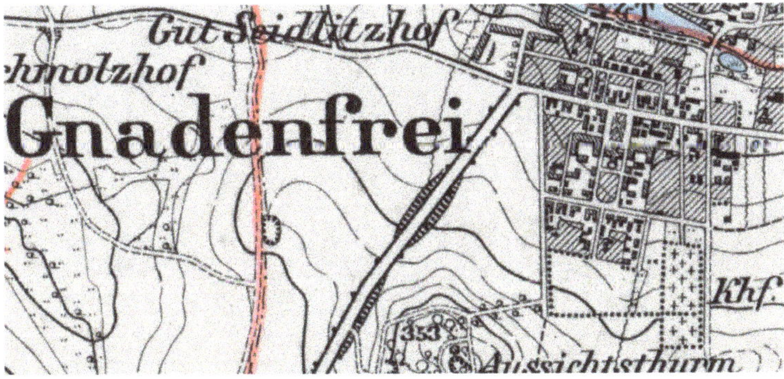

Figure 11. Gnadenfrei on a topographic plan from the early 20th century. Fragment of the plan [19].

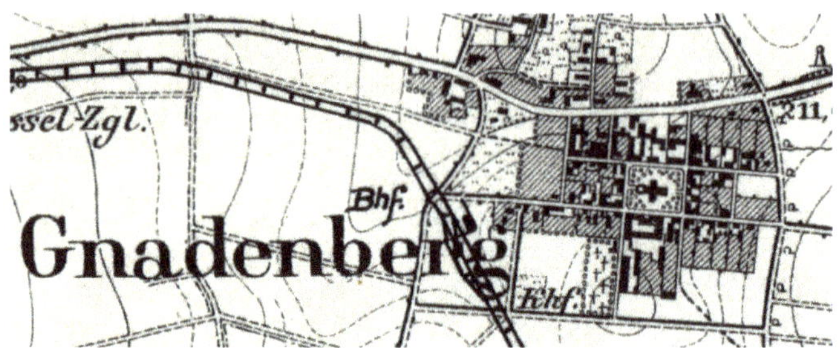

Figure 12. Gnadenberg of the topographical plan from the end of the 19th century. Fragment of the plan [17].

4. Discussion

In Lower Silesia, we associate urban greenery with the urban planning of the 19th century. However, it should be remembered that the origins of the use of gardens and trees in urban composition date back to modern times. Initially, in the municipalities of Lower Silesia and its borders, the only greenery was introduced by closed gardens. Decorative trees and shrubs were only visible from the windows of the palaces, the routes and loggias designed within them, and through the rails of the fences, as in front of the Brieg seat of George III, Christian and Louise of Anhalt. These were still largely viewing openings created in the Renaissance style. However, by this time, the role of greenery in forming the overall appearance of the residence was probably already beginning to be taken into account, which was an innovation in the design approach. In the process of shaping the foreground of the Baroque municipality residences, starting from the earliest examples created just after the Thirty Years' War in the middle of the 1670s in Ohlau and Wohlau, greenery compositions in the form of garden parterres became an element of entrance squares. The same was probably also planned for the unimplemented arrangements in Sagan. Slightly later, lines of trees were used to emphasize the façade and the procession road in front of the ducal seat in Bernstadt. In all these layouts, a long compositional axis was used—the most visible in the case of the Ohlau residence—leading to the palace portal through the middle of the garden plots, marked by the use of small architecture in the form of decorative bridges adorned with obelisks. At the back of the residences in Ohlau and Bernstadt, the axis was continued in the palace gardens, through an avenue with a trellis or a line of trees. In this way, a typical baroque solution of the palace layout was introduced, which was only beginning to be popular at that time. Moreover, the method of using greenery confirmed the knowledge of the most up-to-date patterns in the design of this type of complexes. This is clearly related to the founders, their education, and their connections. Louise of Anhalt and the dukes of both Württemberg lines belonged to the group of enlightened people and were widely related to European—German and French—aristocratic families. Scarce information concerning the designers of urban layouts—such as Vincenzo Boccaccio in Sagan, Carlo Rossi of Como in the case of the Ohlau residence, or Georg Herbst in the case of the gardens in Bernstadt [127]—indicates that only local artists were employed, which speaks all the more in favor of the importance of investors in the choice of design solutions. In these concepts, one can clearly trace the familiarity with French models, especially Versailles and the residences in Paris, which were being built at that time, and German models, those applied in the districts of Berlin or Dresden.

Alleys and tree rows had already become common in Lower Silesia and neighboring areas by the middle of the 18th century. In the urban layout, the lines surrounding the municipalities were incorporated into the fortification systems and they screened roads leading out of the cities. The avenues, which initially had a largely purely utilitarian significance, became the basic compositional skeleton of the new suburbs; the same happened in the intensively developing European cities of the time. In the open landscape, they were used to cover and accentuate roads, introducing optical partitions and Baroque viewing axes.

They were introduced into spatial arrangements creating perspective axes and shaping views and panoramas. In this way, the importance of the architecturally modest suburban Protestant churches of Peace and Grace and the accompanying schools was emphasized. The use of lines of trees was a kind of spiritual elevation. It also indicated ideological ties. The adoption of Dutch solutions through their use in German urban designs and districts for Huguenot refugees lent them an ideological sense in Lower Silesia. It signaled the Protestant affiliation of the emerging urban complexes. Tree espaliers and avenues were widespread and used for both utilitarian and aesthetic reasons; they also found application in urban interiors. In this case, too, they had not only an aesthetic significance but also a propaganda dimension. The tree espaliers in Auraz and the chestnut avenue in the district of the Moravian Brothers in Neusaltz highlighted the frontage of the buildings. At the same time, they indicated the area of the city inhabited by a separate Protestant community.

Avenues extended to elongated to green squares, they were the first to appear not in Silesia, where no new municipalities were created in the 18th century and no large urban districts developed, but in the border area of Lusatia and the Neumark at the beginning of the 18th century, in a period of cultural and political transition, initially still within the borders of the Wettin state, and later under Prussian rule, in close connection with the settlement of religious fugitives. In the case of these settlements, the existing arrangements based on traditional solutions with a rectangular square and streets running out of its corners were planned or adapted for expansion, but also Baroque axial systems connecting the residences with the square or the most important buildings of the villages were introduced. A specific novelty was the introduction of avenues planted with lime trees forming squares in Halbau and Friedrichshuld. The layout of new German urban districts for Huguenots and other religious refugees was modeled on the Berlin Dorotheenstadt and Friedrichsstadt with the Unter den Linden avenue. The form and prominence in the urban layout give them the expression of a green urban space, which is comparable to those shaped in the aforementioned cities in northern Brabant. In the municipalities founded by the Moravian brothers, green central squares were designed, in which not only trees were planted, but also lawns surrounded by fences were established. In this case, links can undoubtedly be made to the solutions used in cities and neighborhoods for religious exiles, starting with the first Renaissance layouts, through examples from various German countries. However, the way greenery was introduced into the interiors of the central squares also links them to the settlements of Protestant exiles in the neighboring areas mentioned in the article. The form of lines and green parterres, forming a kind of square—in principle not yet used on the continent (with the modest exception of the Vosges Square)—points to English inspiration. The fact that one of the communities was founded in London while Zinzendorf had contacts with and even made trips to that city would confirm the close links with that country. They thus become one of the earliest examples of the use of squares in European urban planning outside England.

5. Conclusions

These greenery compositions appeared in urban development in the region of Lower Silesia with some delay as compared to their popularization in western Europe, although the stages of their introduction are similar. The use of greenery in the layout of ducal urban residences began at the same time as this innovation spread across the continent, i.e., decades after the first similar developments. Even later, in relation to the earliest examples, tree rows were planted in Lower Silesia and border areas. The introduction of greenery into urban interiors was initially on a very limited scale. Green squares and inner municipality avenues appeared in the region later than in the Netherlands, France, Prussia, or other German countries. Patterns were undoubtedly drawn from these regions. It is worth remembering, however, that these were not only imitations of the activities of the Saxon and Prussian courts, but also, in part, were inspired by the solutions already applied in Lower Silesia. The ideas of the investors of the dukes, then the high courtiers, were accepted by religious communities. Thanks to this, they probably became a kind of distinguishing feature of bourgeois culture, and utilitarian reasons decided to popularize avenues in the Prussian state.

Over time, the avenues became a distinctive feature of the landscape, preserved to this day in the areas of the former Neumark and Lusatia. In Lower Silesia, since the beginning of the 19th century, green promenades have been commonly created in areas after the demolition of fortifications (Breslau, Liegnitz, Glogau, Schwiednitz, Neisse but also, e.g., much smaller Bunzlau/Bolesławiec, Lüben/Lubin). The squares became popular in the 19th century and the beginning of the next one; in Breslau (A17 in Appendix A) and in all larger municipalities in newly planned squares in developing downtown districts (A18 in Appendix A). The tree espaliers were also planted within municipality centers along the market and street frontages. Such popularity and prestigious importance of the urban greenery had undoubtedly much to do with Baroque traditions.

Baroque urban solutions have left their mark on the urban layout to this day. Unfortunately, the arrangement of greenery spaces has largely disappeared. In residential municipalities, the compositions have undergone further transformations as in Oława or elimination as in Bierutów. Today, greenery arrangements are sometimes created, partly inspired by Baroque front gardens, for example in Brzeg. The Baroque avenues were removed from Uraz (along with the elimination of the whole municipality layout) and Nowa Sól after the expansion of the communication route. Of the municipalities founded by the Moravian Brothers, Herrnhaag existed for only a few decades and disappeared with the emigration of the community in 1753, the urban complex in Godnów was completely destroyed during the Second World War. This makes it all the more valuable for the present day to see the few arrangements that have survived to this day [128]. In Germany, the greenery complexes in Herrnhut and Niesky were revalorized and restored, while the lime trees were supplemented. In Trzebiechów, the composition of greenery with double lines (with trees evidenced in the 19th and 20th century) is largely preserved. Although in Iłowa the trees of the avenue have been mostly removed, the spatial arrangement has survived until today. The square in Piława Górna has turned into a modern green square without the rows of lime trees once surrounding it. It is worth understanding what the spatial impact was; these specific Baroque urban arrangements are worth protecting and revalorizing.

Funding: This research received no external funding

Institutional Review Board Statement: Not applicable.

Informed Consent Statement: Not applicable.

Data Availability Statement: Not applicable.

Conflicts of Interest: The authors declare no conflict of interest.

Appendix A. Details and Data Supplemental to the Main Text

A1 Bocacio characterized: "per lo mezzo in assai parti vie ampissime; tutte diritte come strale e coperte di pergolatidi viti" [28]

A2 "Ambulacrum siue ambulatio, une allee", in a book for children from 1536) [31,32]

A3 "bellissima uista una strada diritta, ampia e polita... con gli arbori, iquali essendopiantati dall'una, e dall'altra parte con la uerdura allegrano glianimi nostri, e con l'ombra ne fanno commodo grandissimo" "Le vie fuori della Città si deuono far ampie, commode, con arbori d'amendue le parti; da i quali i uiandanti l'estate siano difesi dall'ardor del sole, e prendano gli occhi loro qualche ricreatione per la uerdura") [33] (pp. 262, 266)

A4 Some of these were removed as early as the eighteenth century.

A5 Edict of 1731, renewed in 1743 and rewritten in 1746, covered the protection of willow, mulberry and lime tree avenues and others. The edict recommended planting of forests with rows of lime trees [58,59].

A6 Already recorded in the 1603 urbarium.

A7 Then moved to the other side of the Oder River [8] (pp. 156–157).

A8 Perfectly representing the spatial relations due to the triangulation measurement points marked, the situation plan of the royal castle). From the municipality side, a square was formed along the entire palace façade—currently Castle Square [11] (pp. 267, 397, 400).

A9 It might be of interest to know how much trouble Wernher had with designing this entrance correctly. In each drawing, the portal and the bridge that precedes it is in a different place of the wing of Duchess Louise.

A10 The moat was probably eliminated in the 18th century.

A11 In 1769, it was already 560 trees [92].

A12 The alley is now reconstructed.

A13 Roads leading from Breslau to Ohlau via Tschechnitz (Siechnice), to Brieg, to Grottkau (Grodków), to Strehlen (Strzelin), to Schweidnitz via Klettendorf (Klecina) and Klein Tinz (Tyniec Mały), to Kant via Gräbschen (Grabiszyn), to Striegau (Strzegom) via Mochbern (Muchobór), to Neumarkt, Liegnitz and Glaogau via Lissa (Leśnica), to Auras and further to Trebnitz (Trzebnica) via Rosenthal (Różanka) and to Oels via Hundsfeld (Psie Pole).

A14 It was Konrad von Troschke (1638-1702)—between 1688-1691, leaseholder of the castle in Schwiebus (Świebodzin) and royal official—and his son Konrad (1671-1728), the last of the family) [104] (p. 344).

A15 In the second half of the 18th century, there were fewer than 500 inhabitants. Probably only the area around the marketplace, a place in the form of a square, later a new square, remained built-up at that time [24,27].

A16 The Herrnhuter temporarily expelled from Saxony settled in neighboring Hesse.

A17 The square: Am Wäldchen now Pomorska Street, Tauentzien Pl./T. Kościuszki Square, Matthiasplatz/Św. Maciej Sq., Kaiser Wilhelm Platz/Powstańców Roundabout.

A18 F.e in Liegnitz: Fridrichs Pl./pl. Słowiański, Bilze Pl./Skwer Orląt Lwowskich; in Waldenburg/Wałbrzych: Hermannsplatz/ Konstytucji 3 Maja Sq.; in Glogau: Wilhelm Pl. Currently the square of Rev. Z. Kutzan; in Hirschnerg; Wilhelms Pl./T. Kościuszki Sq., in Neisse: Vickoria Pl./Kopernika Sq.; in Oppeln/Opole: Friedrichs Pl./Daszyńskiego Sq.

Source of Figures

Biblioteka Uniwersytetu Wrocławskiego, Oddział Rękopisów. Wydział Geografii i Studiów Regionalnych Uniwersytetu Warszawskiego. Wydział Nauk Geograficznych Uniwersytetu Łódzkiego—all in Public Domain.

Sächsische Landesbibliothek-Staats- und Universitätsbibliothek Dresden SLUB/Deutsche Fotothek—Creative Commons Attribution-ShareAlike 4.0 International License.

References

1. Lawrence, H.W. *City Trees: A Historical Geography from the Renaissance through the Nineteenth Century*; University of Virginia Press: Charlottesville, VA, USA; London, UK, 2006; ISBN 978-0813928005.
2. Bronisch, P. *Geschichte von Neusalz an der Oder*; Siltz: Neusalz, German, 1893.
3. Herda, M. *Dyhernfurth*; Schlesische Vorzeitung: Wohlau, Germany, 1913.
4. Kaszuba, E.; Ziątkowski, L. *Historia Brzegu Dolnego*; Korab: Brzeg Dolny/Wrocław, Poland, 1998; ISBN 83-910321-0-8.
5. Kowalski, S. Kompleks urbanistyczny Żagania jako element dziedzictwa kulturowego. In *Dziedzictwo Artystyczne Żagania*, Czechowicz, B., Konopnicka, M., Eds.; Oficyna Wydaw. ATUT- Wrocławskie Wydaw; Oświatowe: Wrocław/Zielona Góra/Żagań, Poland, 2006; pp. 17–25; ISBN 8374320435.
6. Andrzejewski, T.; Gącarzewicz, M.; Sobkowicz, R. *Dzieje Nowej Soli. Wybór Źródeł i Materiałów*; Muzeum Miejskie w Nowej Soli: Nowa Sól, Poland, 2006; ISBN 83-919139.
7. Jagiełło, M.; Brzezowski, W.; Wowrzeczka, B. Gardens and parks in the settlments of the unity of the Brethren in Lusatia, Hesse and Silesia in the XVIII century (Herrnhut, Herrnhaag, Niesky, Nowa Sól, Godnów, Piława Górna). *Czas. Tech. Archit.* **2016**, 15–34. [CrossRef]
8. Brzezowski, W.; Jagiełło, M. *Ogrody na Śląsku. Barok*; Oficyna Wydawnicza Politechniki Wrocławskiej: Wrocław, Poland, 2017; Volume 2; ISBN 9788374939522.
9. Szymańska-Dereń, M. Trzebiechów. In *Zamki, Dwory i Pałace Województwa Lubuskiego*, 2nd ed.; Bielinis-Kopeć, B., Ed.; Wojewódzki Urząd Ochrony Zabytków: Zielona Góra, Poland, 2008; pp. 417–419; ISBN 978-83-924669-3-2.

10. Wernher, F.B. *Topographia Oder Prodromus Delineati Silesiae Ducatus*; Manuscript; Biblioteka Uniwersytetu Wrocławskiego (BUWr) Oddział Rękopisów, Biblioteka Cyfrowa Uniwersytetu Wrocławskiego: Wrocław, Poland, 1750–1800; Volume 1. Available online: Oai:bibliotekacyfrowa.pl:14302 (accessed on 27 October 2020).
11. Wernher, F.B. *Topographia Oder Prodromus Delineati Principatus Lignicensis Bregensis, et Wolaviensis [...]*; Manuscript; BUWr Oddział Rękopisów, Biblioteka Cyfrowa Uniwersytetu Wrocławskiego: Wrocław, Poland, 1750–1800; Volume 2. Available online: Oai:bibliotekacyfrowa.pl:14652 (accessed on 27 October 2020).
12. Wernher, F.B. *Silesia in Compendio Seu Topographia das ist Praesentatio und Beschreibung des Herzogthums Schlesiens [...] Pars I*; Manuscript; BUWr Oddział Rękopisów, Biblioteka Cyfrowa Uniwersytetu Wrocławskiego: Wrocław, Poland, 1750–1800; Volume 3. Available online: Oai:bibliotekacyfrowa.pl:15414 (accessed on 27 October 2020).
13. Wernher, F.B. *Topographia Seu Compendium Silesiae, Pars II [. . .]*; Manuscript; BUWr Oddział Rękopisów, Biblioteka Cyfrowa Uniwersytetu Wrocławskiego: Wrocław, Poland, 1750–1800; Volume 4. Available online: oai:bibliotekacyfrowa.pl:16339 (accessed on 27 October 2020).
14. Wernher, F.B. *Topographia Seu Compendium Silesiae, Pars V [...]*; Manuscript; BUWr Oddział Rękopisów, Biblioteka Cyfrowa Uniwersytetu Wrocławskiego: Wrocław, Poland, 1750–1800; Volume 5. Available online: Oai:bibliotekacyfrowa.pl:16345 (accessed on 27 October 2020).
15. Wernher, F.B. *Topographia Silesiae, Münsterberg-Frankenstein, Schweidnitz, Freie Standesherrschaften*; Manuscript; Rep. 135, Nr. 526-2; Geheimes Staatsarchiv Preußischer Kulturbesitz XVII. HA: Berlin, Germany, 1750–1800; Volume 3.
16. Wernher, F.B. *Scenographia Urbium Silesiae*; Hofmann: Nürnberg, 1737–1752; BUWr Oddział Starych Druków: Wrocław, Poland. Available online: Oai:bibliotekacyfrowa.pl:15058 (accessed on 27 October 2020).
17. *Bunzlau 2758 [New Nr 4759], Topographische Karte (Messtischblatt) Ostdeutschland 1: 25,000*; Map; Wydział Geografii i Studiów Regionalnych Uniwersytetu Warszawskiego; Königlich-Preussische Landesaufnahme: Berlin, Germany, 1888. Available online: http://maps.mapywig.org/m/German_maps/series/025K_TK25/4759_(2758)_Bunzlau_1888_UW.jpg (accessed on 27 October 2020).
18. *Trebschen 2192 [New Nr 3960], Topographische Karte (Messtischblatt) Ostdeutschland 1: 25,000*; Map; BUWr; Reichsamt für Landesaufnahme: Berlin, Germany, 1896. Available online: https://www.bibliotekacyfrowa.pl/dlibra/publication/33891/edition/40950 (accessed on 27 October 2020).
19. *Gnadenfrei 3135 [New Nr 5366], Topographische Karte (Messtischblatt) Ostdeutschland 1: 25,000*; Wydział Nauk Geograficznych Uniwersytetu Łódzkiego, Biblioteka WNG; Königlich-Preussische Landesaufnahme: Berlin, Germany, 1900. Available online: http://maps.mapywig.org/m/German_maps/series/025K_TK25/5366_(3135)_Gnadefrei_1900_BWNG_UL.jpg (accessed on 27 October 2020).
20. *Halbau 2552 [New Nr 4457], Topographische Karte (Messtischblatt) Ostdeutschland 1: 25,000*; Sächsische Landesbibliothek—Staats- und Universitätsbibliothek Dresden (SLUB); Königlich-Preussische Landesaufnahme: Berlin, Germany, 1905. Available online: http://www.deutschefotothek.de/documents/obj/71054713 (accessed on 27 October 2020).
21. *Auras 2766 [New Nr 4767] Topographische Karte (Messtischblatt) Ostdeutschland 1: 25,000*; Map; Reichsamt für Landesaufnahme: Berlin, Germany, 1916. Available online: https://www.bibliotekacyfrowa.pl/dlibra/publication/33753/edition/44017/content (accessed on 27 October 2020).
22. *Neusalz 2335 [New Nr 4160] Topographische Karte (Messtischblatt) Ostdeutschland 1: 25,000*; Map; Reichsamt für Landesaufnahme: Berlin, Germany, 1922. Available online: http://maps.mapywig.org/m/German_maps/series/025K_TK25/4160_(2335)_Neusalt_1922_UW.jpg (accessed on 27 October 2020).
23. Lucae, F. *Schlesiens Curiose Denckwürdigkeiten Oder Vollkommene Chronica von Ober- und Nieder-Schlesien*; Knoch: Franckfurt, Germany, 1689; Volume 2.
24. Zimmermann, F.A. *Beyträge zur Beschreibung von Schlesien*; Bd. 7; Tramp Press: Brieg, Prussia, 1796; p. 100.
25. Meissner, J.G. *Kurze Beschreibung von Schlesien [...]*, 3rd ed.; 1795; Linsner: Bunzlau, Prussia, 1805; p. 111.
26. Bratring, F.W. *Statistisch-Topographische Beschreibung der Gesammten Mark Brandenburg*; Maurer: Berlin, Prussia, 1809; Volume 3, p. 329.
27. Schumann, F.A.; Schiffner, A. *Vollständiges Staats- Post- und Zeitungslexikon von Sachsen*; Schumann: Zwickau, Prussia, 1816; Volume 3, pp. 670–671.
28. Boccacio, G. *Decameron, (1470 ca)*; Laterza: Bari, Italy, 1927; p. 184. Available online: https://it.wikisource.org/wiki/Pagina:Boccaccio_-_Decameron_I.djvu/188 (accessed on 27 October 2020).
29. Battisti, E. "Natura artificiosa" to "Natura artificialis". In *The Italian Garden*; Coffin, D.R., Ed.; Dumbarton Oaks: Washigton, DC, USA, 1972; pp. 15–16.
30. MacDougall, E.B. *Fountains, Statues, and Flowers: Studies in Italian Gardens of the Sixteenth and Seventeenth Centuries*; Dumbarton Oaks: Washigton, DC, USA, 1994; p. 97.
31. Estienne, C. *De re Hortensi Libellus*; V. Lugduni: Paris, France, 1536; p. 19.
32. Wimmer, K.A. Alleen—Begriffsbestimmung, Entwicklung, Typen, Baumarten. In *Alleen in Deutschland*; Lehmann, I., Rohde, M., Eds.; Edition Leipzig: Leipzig, Germany, 2006; pp. 14–23.
33. Palladio, A. Delle vie, Cap. Primo. In *I Quatro Libri dell Architettura*; Venezia 1570; repr. Hoepli: Milano, Italy, 1945; pp. 262–266. Available online: https://www.liberliber.it/mediateca/libri/p/palladio/i_quattro_libri_dell_architettura/pdf/palladio_i_quattro_libri_dell_a.pdf (accessed on 27 October 2020).

34. Braun, G.; Hogenberg, F. Brugae [Brugge]. In *Flandricarum Urbium Ornamenta*; Map; Bibliotheque Nationale de France, Gallica: Cologne, France, 1572; p. 625. Available online: http://gallica.bnf.fr/ark:/12148/btv1b550045931.r=bruges.langEN (accessed on 27 October 2020).
35. Braun, G.; Hogenberg, F. *Harlemum, Sive ut Ha: Barlan Her:lemum Urbs Hollandia Famosa*; The Barry Lawrence Ruderman Map Collection: La Jolla, CA, USA, 1575. Available online: https://exhibits.stanford.edu/ruderman/catalog/tt323th0524 (accessed on 27 October 2020).
36. Braun, G.; Hogenberg, F. *Gouda, Elegantiss Hollandiae Opp. ad Isalam Amnem, ubi Goudam flu. à Quo Oppidum Nomen Habet, Absorbet 1585*; The Allard Pierson in Focus: Amsterdam, The Netherlands, 1588. Available online: https://hdl.handle.net/11245/3.163 (accessed on 27 October 2020).
37. Braun, G.; Hogenberg, F. *Campia. Campen*; Zuiderzeemuseum Enkhuizen: Enkhuizen, The Netherlands, 1589–1635. Available online: https://www.zuiderzeecollectie.nl/object/collect/Zuiderzee_museum-3041 (accessed on 27 October 2020).
38. *Kaart Met Stadsmuren Voor Den Haag*; Holland Nationaal Archief: The Hague, The Netherlands, 1530. Available online: https://geschiedenisvanzuidholland.nl/collecties/kaart-met-stadsmuren-voor-den-haag-1530- (accessed on 27 October 2020).
39. Londerseel van, J. *Lange Voorhout, Vogelvlucht*; engraving; Rijksdienst voor het Cultureel Erfgoed: Amersfoort, The Netherlands, 1614. Available online: https://beeldbank.cultureelerfgoed.nl/rce-mediabank/detail/4bdf0d89-bd02-71fe-92d6-be6309b8c39a/media/98b4d8bf-207a-f1d8-bb1b-91e968f23e69?mode=detail&view=horizontal&q=20085921&rows=1&page=1&sort=order_s_objectnummer%20asc (accessed on 27 October 2020).
40. Blaeu, J. Het Voorhout Haga Comitis Vulgo's Graven-Hage. In *Blaeu's Toonneel der Steden*; Delft University of Technology: Delft, The Netherlands, 1649. Available online: https://repository.tudelft.nl/view/MMP/uuid:4985aa39-66cd-4649-97c9-5f8e2f383784 (accessed on 27 October 2020).
41. Mourik van, P.; Veen van der, G. *Langs Haagse Bomen*; Eburon Uitgeverij, B.V.: The Hague, The Netherlands, 2013; p. 17; ISBN 9059727320.
42. Lange Voorhout. In *Het Lange Voorhout: Monumenten, Mensen en Macht*; Wijsenbeek-Olthuis, T. (Ed.) Geschiedkundige Vereniging Die Haghe: Zwolle/Waanders/Den Haag, The Nederlands, 1998; ISBN 90-400-9244-3. Available online: https://nl.wikipedia.org/wiki/Lange_Voorhout (accessed on 27 October 2019).
43. Berckenrode, B.F. *Plattegrond van Amsterdam (Blad Middenboven)*; Rijksprentenkabinet, Rijksmuseum: Amsterdam, The Netherlands, 1625. Available online: https://www.rijksmuseum.nl/en/collection/RP-P-1892-A-17491A (accessed on 27 October 2020).
44. Braun, G.; Hogenberg, F. *Civitates Orbis Terrarum, Liber Primus*; Coloniae Agrippinae. Map; BUWr Oddział Zbiorów Kartograficznych: Wrocław, Poland, 1612; pp. 46–47. Available online: https://www.bibliotekacyfrowa.pl/dlibra/publication/31483/edition/110258/content (accessed on 27 October 2020).
45. Blaeu, W.J. Roma (1663). In *Het Nieuw Stede Boek van Italie, Ofte Naauwkeurige Beschrijving van Alle Deszelfs Steden, Paleyzen, Kerken &c., Nevens de Land-Kaarten van Alle Deszelfs Provincien*; map; Mortier, C.: Amsterdam, The Netherlands; Utrecht University Library: Utrecht, The Netherlands, 1704; p. 4A. Available online: https://objects.library.uu.nl/reader/viewer.php?obj=1874-36121&pagenum=9&lan=en (accessed on 27 October 2020).
46. Merian, M. Reims En Champagne. In *Topographia Galliae*; Map; Merian: Franckfurt/Koblenz, Germany, 1655; p. 24C. Available online: LandesbibliothekszentrumRheinland-Pfalz2011pdf (accessed on 27 October 2020).
47. Schaber, W. Schloss Hellbrunn und die Hellbrunner Allee. In *400 Jahre Hellbrunner Allee*; Gföllner, K., Ed.; Samson Druck: Salzburg, Germany, 2015; pp. 7–13; ISBN 978-3-85015-282-2. Available online: https://www.salzburg.gv.at/kultur_/Documents/Publikationen/400_Jahre_Hellbrunner_Allee_online_2.pdf (accessed on 27 November 2020).
48. Schlager, G. Salzburger Baumschutzverordnung 1992—Wunsch und Wirklichkeit. *Nat. Landsch.* **1993**, *68*, 302–307.
49. *Clundert, [ca 1632]*; Blaeu, J.: Amsterdam, The Netherlands, 1649–1652; Map. Rijksprentenkabinet, Rijksmuseum: Amsterdam, The Netherlands. Available online: https://www.rijksmuseum.nl/en/collection/RP-P-AO-14-61-1 (accessed on 3 December 2020).
50. *Willemstadt, 1632*; Blaeu, J.: Amsterdam, The Netherlands, 1649–1652; Map. Rijksprentenkabinet, Rijksmuseum: Amsterdam, The Netherlands. Available online: https://www.rijksmuseum.nl/en/collection/RP-P-AO-14-62 (accessed on 3 December 2020).
51. Falda, G.B. *Novissima et Accuratissima Delineatio Romae Veteris et Novae*; map; Feuille de la, J.: Amsterdam, The Netherlands; The Getty Research Institute: Los Angeles, CA, USA, 1690. Available online: https://rosettaapp.getty.edu/delivery/DeliveryManagerServlet?dps_pid=FL2115639 (accessed on 3 December 2020).
52. de Vaugondy, G.R.; Delamarche, C.F. *Plan Geometral De Paris Et De Ses Fauxbourgs*; map; A Paris: Chez Delamarche, Géog., Rue du Foin St. Jacques, au Collège de Mtre; Gervais: Paris, France, 1797. Available online: https://collections.leventhalmap.org/search/commonwealth:6w924q228 (accessed on 27 November 2020).
53. Waltern, J.F. *Die Konigl. Preus. u. Churf. Brandenburg. Residenz-Stadt Berlin*; Homann: Nuremberg; [1740 ca]; map. Zentral- und Landesbibliothek Berlin: Berlin, Germany. Available online: https://digital.zlb.de/viewer/toc/15453578/1/-/ (accessed on 27 November 2020).
54. Faber, J.H. *Topographische, Politische und Historische Beschreibung der Reichs- Wahl- und Handelsstadt Frankfurt am Mayn*; Jäger: Frankfurt, Germany, 1788; Volume 1, pp. 25–61.
55. Seutter, M. *Topographia Sedis Imperatoriæ Moscovitarum Petropolis Anno 1744 Jam Publici Juris Facta*; Map; BnF Gallica: Augsburg, Germany, 1744. Available online: https://gallica.bnf.fr/ark:/12148/btv1b530395167 (accessed on 27 November 2020).
56. Fischer, J.B. *Geschichte und Beschreibung der Markgräfl.-Brandenb. Haupt- & Res. Stadt Anspach*; Fischer: Neu-Anspach, Germany, 1786; p. 32.

57. Heinritz, J.G. *Geschichte der Stadt Bayreuth (in drei Teilen)*; Bayreuth, 1823; Reed. Stadtarchiv Bayreuth: Bayreuth, Germany, 2019; Volume 1, pp. 56–57. Available online: https://www.bayreuth.de/wp-content/uploads/2019/02/Heinritz_Geschichte-der-Stadt-Bayreuth.pdf (accessed on 27 November 2020).
58. Decret No. 96 Berlin 26 Nov. 1746. In *Novum Corpus Constitutionum Prussico-Brandenburgensium*; Königl. Preußische Acad. der Wiss.: Berlin, Germany, 1771; Volume 4, p. 614. Available online: https://www.bayreuth.de/rathaus-buergerservice/stadtverwaltung/zahlen-fakten-2/stadtgeschichte/bayreuther-stadtgeschichte/ (accessed on 27 November 2020).
59. Gottlieb Beckmann, J. Edickt wegen Satzung Weiden und Linden. Ed. 10 Juli 1743. In *Beyträge zur Verbesserung der Forstwissenschaft*; Beckmann, J.G., Ed.; Bd. 3; Stossel: Chemnitz, Germany, 1763; p. 183.
60. Edict XXXII, Berlin 13 November 1742. Edict XXXII, Berlin 13 November 1742. Corpus Constitutionum Marchicarum, Continuatio II. Derer in der Chur- und Marck Brandenburg, Auch Incorporirten Landen, Ergangenen Edicten, Mandaten, Rescripten, von 1741. Biß 1744. Inclusive. 1744, p. 83. Available online: https://web-archiv.staatsbibliothek-berlin.de/altedrucke.staatsbibliothek-berlin.de/Rechtsquellen/CCMC2/start.html?image=06351&size=1 (accessed on 27 November 2020).
61. Pfeiffer von, J.F. *Vermischte Verbeßrungsvorschläge und freie Gedanken über Verschiedene, den Nahrungszustand, die Bevölkerung und Staatswirtschaft der Deutschen, betreffende Gegenstände*; In der Eßlingerischen Buchhandlung: Frankfurt, Germany, 1778; Volume 1, pp. 60–70.
62. Verordnung wegen der Maulber-Baume […]. 16 August 1765. In *Novum Corpus*; Königl. Preußische Acad. der Wiss.: Berlin, Germany, 1766; Volume 3, p. 1008.
63. Sturm, L. *Geschichte der Stadt Goldberg in Schlesien*; Sturm: Goldberg, Prussia, 1888; p. 409.
64. Kunz, H. *Das Schloss der Piasten zum Briege*; Bänder: Brieg, Prussia, 1885.
65. Bimler, K. *Das Piastenschloss zu Brieg*; Maruschke und Berendt: Breslau, Germany, 1934.
66. Zlat, M. *Zamek piastowski w Brzegu*; Instytut Śląski w Opolu: Opole, Poland, 1988.
67. Zlat, M. Brama zamkowa w Brzegu. *Biul. Hist. Szt.* **1962**, *24*, 284–322.
68. Schönwälder, K.F. *Geschichtliche Ortsnachrichten von Brieg und Seiner Umgebung*; Kern: Brieg, Prussia, 1847; Volume 2.
69. *Grund Oder Abris der Stad Grosglogau*; [1618–1632]; map. BUWr Oddział Zbiorów Kartograficznych: Wrocław, Poland. Available online: Oai:bibliotekacyfrowa.pl:37485 (accessed on 27 October 2020).
70. Boras, Z. *Książęta Piastowscy Śląska*; Wyd. Śląsk: Wrocław, Poland, 1972; p. 456.
71. *Aigentliche Vorbildung der Vestung Brieg Wie Selbte Gelegen und Mit Ihren Bollwercken Versehen*; BUWr Oddział Zbiorów Kartograficznych: Wrocław, Poland. Available online: http://www.bibliotekacyfrowa.pl/dlibra/publication/28379?tab=1 (accessed on 27 November 2020).
72. Bimler, K. *Das Piastenschloss zu Ohlau*; Maruschke und Berendt: Breslau, Germany, 1936.
73. Lutsch, H. *Verzeichnis der Kunstdenkmäler der Provinz Schlesien*; Korn: Breslau, Prussia, 1889; Volume 2.
74. Eysymontt, R. Architektura miasta Oławy. In *Oława*; Eysymontt, R., Ed.; Institute of Archeology and Ethnology of the Polish Academy of Sciences: Wrocław, Poland, 2015; Volume 4, pp. 14–18.
75. Geissler, K.G. *Situations Plan F. Von dem Königs Schloss zu Ohlau*; Eysymontt, R., Atlas Historyczny Miast Polskich, Czaja, R., Śląsk, Młynarska-Kaletynowa, M., Eds.; Institute of Archeology and Ethnology of the Polish Academy of Sciences: Wrocław, Poland, 2015; Volume 4.
76. Säbisch, V. *Plan of Bastion Fortifications of Oława*; Eysymontt, R., Atlas Historyczny Miast Polskich, Czaja, R., Śląsk, Młynarska-Kaletynowa, M., Eds.; Institute of Archeology and Ethnology of the Polish Academy of Sciences: Wrocław, Poland, 2015; Volume 4, p. 44.
77. *Ohlau. Gezeichnet von Hauptmann Chalaupka. 1 gez. Blatt (Kopie)*; Atlas Historyczny Miast Polskich; Eysymontt, R.; Czaja, R.; Śląsk; Młynarska-Kaletynowa, M. (Eds.) Institute of Archeology and Ethnology of the Polish Academy of Sciences: Wrocław, Poland, 2015; Volume 4, p. 20.
78. Wernher, F.B.; Engelbrecht, M. *Arx Olavie in Silesia In Views of Cistercian Monasteries and Palaces of Lower Silesia*; [Augsburg?]; BUWr Oddział Starych Druków: Wrocław, Poland, 1739; p. 38. Available online: https://www.bibliotekacyfrowa.pl/dlibra/publication/10927/edition/69639/content (accessed on 27 November 2020).
79. Wissmach, E. *Die Stadt Ohlau und ihre Umgebung*; Magistr. der Stadt Ohlau: Ohlau, Germany, 1929; pp. 22–24.
80. Eysymontt, R. Rozwój przestrzenny, architektura, sztuka. In *Wołów. Zarys Monografii Miasta*; Kościk, E., Ed.; Silesia: Wrocław/Wołów, Poland, 2002; pp. 154–198; ISBN 83-85689-99-0.
81. Sinapius, J. *Olsnographia oder Eigentliche Beschreibung des Oelsnischen Fürstenthums In Nieder-Schlesien welche in zwey Haupt-Theilen so wohl insgemein Dessen Rahmen, Situation, Regenten Religions-Zustand, [...] Ausgefertiget von Johanne Sinapio, [...]. T.1*; Brandenburger: Leipzig, Germany, 1706; Volume 2, pp. 516–519.
82. Bimler, K. *Die Schlesischen Massiven Wehrbauten, Fürstentum Oels—Wohlau*; Kommission Heydebrand: Breslau, Germany, 1942; Volume 3, pp. 22–30.
83. Schulze, H. *Die Succession im Fürstenthum Oels Beim Abgange der Ältem Linie des Hauses[…]*; Korn: Breslau, Prussia, 1868; pp. 6–7.
84. Herbst, G. Des Schlesischen Gärtners Lustiger Spatziergang. 1692. Available online: http://diglib.hab.de/wdb.php?dir=drucke/oe-283&distype=thumbs (accessed on 27 November 2020).
85. Jenderich, C. *Mahre eigentilische Abbildung […] Residenz in Bernstadt*; Bernstadt, Germany, 1692; p. 414.
86. Heinirch, A. *Wallenstein als Herzog von Sagan*; Goerlich & Coch: Breslau, Germany, 1896.

87. Wernher, F.B. *Das Fürstenthum Sagan in Nieder Schlesien*; BUWr Oddział Zbiorów Kartograficznych: Wrocław, Poland, 1750. Available online: Oai:bibliotekacyfrowa.pl:37488 (accessed on 27 November 2020).
88. Leipelt, A. *Geschichte der Stadt und des Herzogthums Sagan...*; Rauertsch: Sorau, Prussia, 1853; pp. 128–129.
89. Kanold, J. *Sammlung von Natur- und Medicin-, Wie Auch Hierzu Gehörigen Kunst- und Literatur Geschichten*; Richter: Leipzig, Germany, 1726; pp. 747–753.
90. Gierowski, J. Miasta, manufaktury. In *Historia Śląska*; Part 3; Maleczyński, K., Ed.; Ossolineum: Wrocław/Warszawa/Kraków, Poland, 1963; Volume 1, p. 214.
91. *Sammlung Der Wichtigsten Und Nöthigsten, Bisher Aber Noch Nicht Herausgegebenen Käyser- Und Königlichen Auch Hertzoglichen Privilegien, Statuten, Rescripten und Pragmatischen Sanctionen des Landes Schlesien: Aus den nach und nach publicirten Königlichen Ober-Amts-Patenten, wie auch aus den wichtigsten Landes-Cantzeleyen und Archiven. Bd. 2*; Habsburg Empire: Breslau, Poland, 1739; Volume 2, pp. 423–424.
92. Schmidt, F.J. *Geschichte der Stadt Schweidnitz*; Heege: Schweidnitz, Prussia, 1846; Volume 2, p. 300.
93. Wernher, F.B.; Merz, J.G. *Prospectus Templi Evangelici Extra Urbem Iauriam [...]*; engraving; Österreichische Nationalbibliothek: Wien, Austria, 1735. Available online: Oai:baa.onb.at:19191100 (accessed on 27 November 2020).
94. Festungsplan von Schweidnitz. In *Befestigungsentwürfe für die Stadt Schweidnitz*; Walrave, G.C.; Sers, L.P.; Enbers, W.; Foris, F.A.; Dariez (Eds.) Staatbibliothek Berlin: Berlin, Germany, 1743–1748.
95. Wernher, F.B.; Merz, J.G. *Prospectus Templi Evangelici Extra Urbem Sanagum [...]*; Österreichische Nationalbibliothek: Wien, Austria, 1735; Available online: Oai:baa.onb.at:19252904 (accessed on 27 November 2020).
96. Wernher, F.B.; Merz, J.G. *Prospectus Templi Evangelici Extra Urbem Hirschbergam in Silesia Inferiore, Cui Nomen ad Crucem Christi. Prospect der Evangelischen Kirche vor der Stadt Hirschberg, in Nider Schlesien, genandt zum Creutz-Christi. 1709*; Österreichische Nationalbibliothek: Wien, Austria, 1735; Available online: Oai:baa.onb.at:13889693 (accessed on 27 November 2020).
97. *Plan de la Bataille de Breslau 1757*; Map; Hond: Paris, France, 1758. Available online: https://polska-org.pl/915320,foto.html?idEntity=554891 (accessed on 27 November 2020).
98. Bataille De Breslau 22. Nov. 1757. In *Collection de Quarante Deux Planes de Batailles, Sieges et Affaires les Plus Memorables de la Guerre de Sept Ans [...]*; map; Roesch, J.F. (Ed.) Jaeger: Franckfurt, Germany, 1790. Available online: https://polska-org.pl/831827,foto.html?idEntity=554891 (accessed on 27 November 2020).
99. Plan vor den Gegend und den Draeffen Welche un die Stadt Breslau Bis Lissa Hin Ligen, 1756, map. Available online: https://polska-org.pl/632755,foto.html?idEntity=3571227 (accessed on 27 November 2020).
100. Breslau, 270, Topographische Karte, Pruss.Generalstab, 1826–1831, Hosse: [Breslau] 1832, with Changes from 1856, map. Available online: https://polska-org.pl/7902794,foto.html?idEntity=3571227 (accessed on 27 November 2020).
101. Breslauer Regierungs Bezirk. 1828–1831. Available online: oai:kpbc.umk.pl:171533 (accessed on 27 November 2020).
102. *Grundriss der Königl. Prusisch. und Herzogl. Schlesischen Haupt-Stadt Bresslau*; map; BUWr Oddział Zbiorów Kartograficznych: Wrocław, Poland, 1750. Available online: Oai:bibliotekacyfrowa.pl:42002 (accessed on 27 November 2020).
103. Herold, J. *Von Scheitnig und Seinem Parke, dem "Fürstengarten": Eine Ortsgeschichtliche Darstellung*; Trewendt & Granier: Breslau, Germany, 1913; p. 13.
104. Sinapius, J. *Schlesische Curiositäten Erste Vorstellung, Darinnen die Ansehnlichen Geschlechter Des Schlesischen Adels*; Sinapius: Leipzig, Germany, 1720; p. 344.
105. Foerster, A. *Geschichtliches von den Dörfern des Grünberger Kreises*; Weller: Grünberg, Germany, 1905; p. 207.
106. Hinz, B. *Die Schöppenbücher der Mark Brandenburg: Besonders des Kreises Züllichau -Schwiebus*; Gerd, H., Ed.; De Gruyter: Berlin, Germany, 1964; pp. 88, 153.
107. Worbs, J.G. *Geschichte des Herzogthums Sagan*; Frommann: Züllichau/Sagan, Prussia, 1795; pp. 90, 395.
108. Knie, J.G. *Alphabetisch-Statistisch-Topographische Übersicht der Dörfer, Flecken, Städte und andern Orte der Königl. Preuß. Provinz Schlesien*; Graß, Barth und Co.: Breslau, Prussia, 1845.
109. Sinapius, J. *Des Schlesischen Adels Anderer Theil, Oder Fortsetzung Schlesischer Curiositäten: Darinnen Die Gräflichen, Freyherrlichen und Adelichen Geschlechter, So Wohl Schlesischer Extraction, Als auch Die aus Andern Königreichen und Ländern in Schlesien Kommen, Und Entweder Darinnen Noch Floriren, Oder Bereits Ausgangen, Jn Völligen Abrisse Dargestellet Werden, Nebst Einer Nöthigen Vorrede und Register*; Rohrlach: Leipzig, Germany; Breslau, Poland, 1728; p. 178.
110. Simonetti, Giulio. In *Allgemeines Lexikon der Bildenden Künstler von der Antike bis zur Gegenwart. Begründet von Ulrich Thieme und Felix Becker*; Vollmer, H. (Ed.) Siemering–Stephens; Seemann, E.A.: Leipzig, Germany, 1937; Volume 31, p. 73.
111. Kalinowski, K. *Architektura Doby Baroku na Śląsku*; Państwowe Wydawnictwo Naukowe: Warszawa, Poland, 1977; p. 132.
112. Jansen, H. *Katasterkarte Halbau*; Map; Technische Universität Berlin: Berlin, Germany, 1907. Available online: https://architekturmuseum.ub.tu-berlin.de/index.php?p=79&POS=2 (accessed on 27 November 2020).
113. *Halbau 2552 [New Nr 4457], Topographische Karte (Messtischblatt) Ostdeutschland 1: 25,000*; Königlich-Preussische Landesaufnahme: Berlin, Germany, 1939. Available online: https://www.bibliotekacyfrowa.pl/dlibra/publication/33832/edition/42477 (accessed on 27 October 2020).
114. Lindenstrasse, Halbau. 1900. Available online: https://polska-org.pl/8185306,foto.html?idEntity=599261 (accessed on 27 November 2020).

115. Zollner, J.F. Briefe über Schlesien, Krakau, Wieliczka und die Grafschaft Glätz aufeiner Reise im Jahr 1791, Berlin, 1792. Konopnicka, M. tr. In *Źródła i Materiały do Dziejów Środkowego Nadodrza*; Bartkiewicz, K., Ed.; Pedagogical University of Tadeusz Kotarbiński in Zielona Góra: Zielona Góra, Poland, 1996; pp. 103–104.
116. *Grüß aus Neusalz, Bresslauer Strasse, Postcard*; Bohm: Neusalz, Germany, 1900. Available online: https://polska-org.pl/912284,foto.html?idEntity=596123 (accessed on 27 November 2020).
117. Mettele, G. *Weltbürgertum oder Gottesreich: Die Herrnhuter Brüdergemeine Als Globale Gemeinschaft 1727–1857*; Vandenhoeck & Ruprecht: Göttingen, Germany, 2009; pp. 49–54; ISBN 978-3-525-36844-2.
118. Karczyńska, H. *Odnowiona Jednota Braterska w XVIII-XX w.*; Semper: Warszawa, Poland, 2012.
119. Battek, M.J. Ansiedlung der Unitäts-Brüder in Schlesien. Available online: https://docplayer.org/50866439-Ansiedlung-der-unitaets-brueder-in-schlesien-und-ihre-spuren.html (accessed on 27 November 2020).
120. Lewis, M.J. *City of Refuge: Separatists and Utopian Town Planning*; Princeton University Press: Princeton, NJ, USA; Oxford, UK, 2016; pp. 22–113; ISBN 13 9780691171814.
121. Neudecker, C.G. *Geschichte des Evangelischen Protestantismus in Deutschland*; K.F. Köhler: Leipzig, Prussia, 1845; Volume 2, pp. 646–655.
122. Keyl, M.; Krause, I.G. *Grundriss und Prospecte der Drey Evangelischer Bruder Gemein Orte in Marggraffthum Ober Lausitz, Herrnhut, Nisky, Klein-Welke: [gewidmet:] Reichsgrafen zu Reuss Heinrich den XXVIII,...*; Sächsische Landesbibliothek—Staats- und Universitätsbibliothek Dresden (SLUB), Kartensammlung: Dresden, Germany, 1782. Available online: http://www.deutschefotothek.de/documents/obj/70400307 (accessed on 27 November 2020).
123. Reuter, P.C. *Herrenhaag*; Moravian Archive: Winston-Salem, NC, USA, 1760. Available online: https://www.researchgate.net/figure/P-C-G-Reuter-Herrnhag-nd-Courtesy-of-the-Moravian-Archives-Winston-Salem-North_fig4_265839817 (accessed on 27 November 2020).
124. Reuter, P.C. *Nieska*; Moravian Archive: Winston-Salem, NC, USA, 1760. Available online: https://www.researchgate.net/figure/P-C-G-Reuter-Nieska-1760-Courtesy-of-the-Moravian-Archives-Winston-Salem-North_fig5_265839817 (accessed on 27 November 2020).
125. Brandt, A.L. *Büdingen Herrnhaag*; SLUB: Dresden, Germany, 1755. Available online: https://www.slub-dresden.de/ueber-uns/buchmuseum/ausstellungen-fuehrungen/archiv-der-ausstellungen/ausstellungen-2010/die-welt-in-herrnhut/herrnhut-impressionen/ (accessed on 27 November 2020).
126. Brandt, A.L. *Gnadenfrei*; SLUB: Dresden, Germany, 1755. Available online: https://www.slub-dresden.de/ueber-uns/buchmuseum/ausstellungen-fuehrungen/archiv-der-ausstellungen/ausstellungen-2010/die-welt-in-herrnhut/herrnhut-impressionen/ (accessed on 27 November 2020).
127. Brzezowski, W. Ślązacy a teoria sztuki ogrodniczej od XVI do XVIII w. w świetle zbiorów Biblioteki Uniwersyteckiej we Wrocławiu, The Silesians and theories of garden art from the 16th to the 18th century with new research from the Wrocław University Library collection. *Architectus* **2013**, *33*, 5–6. [CrossRef]
128. Szilágyi, K.; Lahmer, C.; Szabó, K. Allées in Landscape Architecture and Garden. Art—Types, Preservation, and Renewal of the Living. Heritage of Baroque Allées in Hungary. *Land* **2020**, *9*, 283. [CrossRef]

Article

The Greenery of Early Modernist Housing Estates: The 1919–1927 Wałbrzych Agglomeration

Bogna Ludwig

Department of Architecture Conservation and Restoration of Cultural Landscape, Faculty of Architecture Wrocław, University of Science and Technology, 50-317 Wrocław, Poland; bogna.ludwig@pwr.edu.pl

Abstract: Using the Wałbrzych agglomeration housing estates—once the most important mining and industrial region in Lower Silesia—as an example, this article illustrates the specific significance of the design of green spaces, including urban layouts, and the issue of protecting unique trees and green spaces in the concepts of estates from the early modernism period after the First World War in the years 1919–1927. This article tries to deepen the knowledge on the origins of the design solutions of public and private greenery systems while considering natural, landscape, and social needs. This study complements the information gathered so far in the field of forming green areas in modernist housing estates and highlights the importance of this issue in complex urban design. The Wałbrzych housing settlements are crucial because they were among the first of their kind, not only in Lower Silesia but also in the whole of the Weimar Republic. Based on literature and source studies, it was possible to reconstruct design ideas concerning the composition of green areas in most housing estates in the discussed area. The most interesting ones were presented and broken down into the landscape-related and functional aspects of the use of greenery in housing estates. This made it possible to select specific solutions applied by designers in order to indicate sources of inspiration and theoretically developed rules which then and now seem to be extremely adequate.

Keywords: greenery; housing estates; Lower Silesia; preservation and renewal of heritage

Citation: Ludwig, B. The Greenery of Early Modernist Housing Estates: The 1919–1927 Wałbrzych Agglomeration. *Sustainability* **2021**, *13*, 3921. https://doi.org/10.3390/su13073921

Academic Editor: Jan K. Kazak

Received: 22 February 2021
Accepted: 29 March 2021
Published: 1 April 2021

1. Introduction

Modernism is, at times, primarily associated with functionalism and a rationalist approach to design. The first period of this trend remained deeply tied to the concepts that first appeared in the early 20th century. In Germany, the Art Nouveau ideas initiated by Camillo Sitte [1–3], the activities of the first heritage protection organizations (Heimatschutzbewegung) [4], and then Howard's widely discussed theory [5] laid the foundations for design concepts in the first decade after the World War. Modernism provided additional sources of inspiration and indicated new formal interpretations, both urban and architectural [6–8]. The postulation to protect the landscape and its natural and cultural values became one of the most important ideological foundations in design. Regionalism of the early modernism period was part and parcel of the landscape and developed the existing compositional rules. Arranging greenery in urban layouts was an important issue. Its special social role was also recognized.

The architecture and urban planning of early modernist housing estates is already a partly researched issue in contrast to the concept of urban greenery of the greenbelt type, which was also developing based on Howard's visions in the late 1920s and 1930s. However, problem of design greenery—so important for these housing estates—has hardly been taken up (apart from the ecological and conservation aspects [9]) as a leading subject of research [10–13]. Only the issue of employee gardens (Kleingarten) has received more attention [14–17]. In a comprehensive study of the housing estates of the Wałbrzych agglomeration in the interwar period, only the issue of the history of design green spaces in residential complexes in the early modernism period was signaled [18] (pp. 584–586). The topic is worth discussing due to the pioneering significance of the housing estate projects in the Wałbrzych (Germ. Waldenburg) mining and industrial district.

By using the Waldenburg agglomeration housing estates an example, this article illustrates the specific significance of the design of green spaces, including urban layouts, and the issue of protecting unique trees and green spaces in the concepts of estates from the early modernism period after the First World War in the years 1919–1927 (see Appendix A, A1). The role of both tradition and innovative research-based approaches to these issues is indicated.

The Waldenburg housing settlements are crucial because they were among the first of their kind, not only in Lower Silesia but also in the whole of the Weimar Republic. The concepts developed during the design process soon spread beyond the borders of the country. This also applied to the development of the principles and methods of composing the greenery of these settlements. Reference was made to tradition, and new or significantly altered forms were introduced. Relevant literature and source studies made it possible to reconstruct design ideas concerning the composition of green areas in most housing estates in the discussed area. The most interesting ones were presented and broken down into those landscape-related and functional aspects of the use of greenery in housing estates. This allowed for the selection of specific solutions applied by designers, the indication of sources of inspiration, and theoretically developed rules, which seem to be extremely adequate both then and now. In addition to the already known and popular forms, the designers used innovative greenery layouts, green street bays, and complexes of communal employee gardens, sometimes also merging into compositions with public greenery. Particularly important was their attitude to the heritage of the past—valuable trees and existing avenues, as well as natural greenery like the surrounding forests, watercourses, and ravines. Greenery systems were used there as both elements of urban composition and as recreational facilities. Carefully considered decisions were successful in that the greenery in the early modernist housing settlements in the Wałbrzych agglomeration still belonged to one of the better designed layouts and could still be a model for contemporary developers of urban plans for residential complexes.

This article tries to deepen the knowledge on the beginnings of solutions in the design of green systems, both public and private, by analyzing this selected object of research. It tries to trace the methods of development and implementation of housing estate greenery systems while considering natural, landscape, and social needs. In this way, it complements the information collected to date in the field of shaping green areas in modernist housing estates, thus indicating the importance of this issue in complex urban planning, as well as in relation to historical natural specimens and existing alley complexes like the greenery of watercourses and ravines.

To this end, the basic premises for the emergence of new housing concepts in Germany in the early 20th century were characterized. The significance of housing estates in the Walbrzych region is discussed against the background of the history of housing estates in the Weimar Republic. In the main part of the paper, three separate categories of shaping green areas in residential complexes of the Walbrzych agglomeration are presented: (a) the landscape, (b) the composition of the role mainly played by public greenery, and (c) the socioeconomic category related to the planning of home and working gardens. The discussion pointed to the special treatment of greenery design in the planning of modernist housing estates and a comprehensive approach to this issue.

2. Materials and Methods

The study was based on the research method used in the humanities that required an analysis of the background of the phenomenon; a study of available sources, research, and analysis of the subject matter; and an attempt to create a synthetic comparative characteristic. The historical background was characterized in relation to the problem of the architectural and urban design of housing estates of a selected period in Germany and Silesia. A full review of the scientific literature on the selected research subject was carried out. All available archival documents—designs, maps, drawings, photographs, and written reports preserved in the State Archives in Wrocław (Archiwum Państwowe we Wrocławiu (APW)) and its branch in Kamieniec Ząbkowicki (APW Kamieniec), as well as the collection of the Porcelain Museum in Wałbrzych—were analyzed. Local press (especially regarding those stored in the collection of the Walbrzych Museum) and professional press (especially the Schlesisches Heim magazine) queries concerning the aspect of design and implementing greenery systems in the housing estates of the Walbrzych region were carried out. Field research was carried out, and up-to-date reports on the state of preservation of the presented greenery in the housing estates were found. The study of urban planning and architecture of the Wałbrzych agglomeration settlements had been carried out by the author since 2000. In recent years, this was supplemented by a search for archival documents from private collections (mainly iconographic) made available on websites, as well as subsequent local inspections of the changing resources. In 2020, materials relating to the issue of greenery design in the surveyed housing estates were completely gathered. Based on the study of the available scientific literature and sources, it was possible to reconstruct design ideas concerning the composition of green areas in most estates of the selected period in the discussed area. This article presents the most interesting of them, grouping them according to the landscape and functional aspects of the use of greenery in housing estates. This made it possible to select specific solutions applied by designers and indicate sources of inspiration and adopted rules reported in the professional press.

3. Ideological and Historical Background

3.1. Early Modernist Housing Estates—From Art Nouveau to Regionalism

In the period before and after World War I, the immediate environment began to play an increasingly important role for the Art Nouveau artists. Criticism of the artificiality of classical and historical patterns in relation to the designs of the historicist period also began to include the inspirations of early Art Nouveau derived from Gothic and Far Eastern art. In Germany, this manifested itself in the form of a mental and cultural movement. Its slogan became the protection of national heritage (Heimatschutzbewegung). Breslau became the seat of the organization Schlesischer Bund für Heimatschutz. Its founder Hans Poelzig and Max Berg, an important activist, were teachers and mentors for younger architects [19,20]. Thanks to their involvement, the design trend spread in architectural literature and academic teaching. The interest of Art Nouveau artists in the Middle Ages and their fascination with nature led them to perceive the value of historic buildings not only in their individual form but also in combination with their surroundings. The emerging principles and institutions of conservatory protection also drew attention to the issues of landscape protection together with its natural values. These problems were included in theoretical and design architectural considerations, in which the planning of green spaces came to the fore.

Ebenezer Howard's book played a special role in integrating greenery arrangement issues into urban design [21,22]. In the spirit of Art Nouveau, Ebenezer Howard tried to develop principles combining the advantages of urban and rural life. The solution was found in the decentralization of cities by creating a network of self-sufficient garden-cities surrounding the superior center. These postulations were partly put into practice in Lower Silesia by building villa districts near cities, e.g., Wrocław (Breslau, Kleinburg, 1872–1900), then suburban garden settlements and districts in Wrocław like Gartenstadt Bischofswalde (1908–1911) and Gartenstadt Carlowitz (1911), and other cities like Neustadt Waldenburg (1904), Gartenstadt Neisse (1911), and Gartenvorstadt Liegnitz (1911) [23–27].

The competition for the development of Great Berlin in 1910, the international urban planning exhibition taking place at the same time (Berlin, Germany, 1910)—as well as the Town Planning Conference in London organized by the Royal Institute of British Architects (1910) [1] (p. 57)—provided an opportunity to lay down final principles for the planning of modern cities. Among the new ideas were polycentric town planning, the separation of new development by green belts and wedges of greenery, and the zoning of urban development, all of them referred to Ebenezer Howard's idea of the "garden city".

The Green Manifesto (1918) of Leberecht Migge [28], already a well-known landscape architect at that time, became the last decisive element for the formation of a program for improving the functionality of estates of the interwar period in Germany. Migge advocated solving the social and economic problems of the German people by creating gardens, parks, and (first of all) small vegetable gardens (200–400 m^2) to supplement food and to improve the climatic qualities of residential areas [28,29] (pp. 8–20).

3.2. Attempts to Solve the Housing Problem in the Weimar Republic, Lower Silesia, and Lower Silesian Coal District after World War I

After the First World War, the Weimar Republic created a favorable situation for the development of the theory and practice of housing estates design. On the one hand, the housing shortage due to industrial development and war damage was desperate; on the other hand, a favorable political climate was created for attempts to solve this problem. Power passed into the hands of social democrats. Their program included the improvement of living conditions of workers and civil servants. Apart from efforts to improve working conditions, the housing program became an equally important element of the government's social policy [30]. Immediately after the war, on the basis of the idea propagated in Germany by Leberecht Migge and Hermann Muthesius [31], the authorities developed the concept of housing development that consisted of erecting Kleinsiedlung—small housing estates with home gardens [32–34]. Legal tools that allowed for the launching of economic and social mechanisms were applied (see Appendix A, A2). In 1918, the first communal housing cooperatives (gemeinnützige Baugesellschaften) were established [30].

In July 1919, the company Schlesisches Heim (Schlesische Heimstätte) (see Appendix A, A3) was founded in Silesia [18] (pp. 243–246) [35–37]. Its aim was to erect cheap, small houses for workers and civil servants in suburban settlements and to build rural settlements near towns and villages. The urban and architectural program of these settlements was initially run exclusively by Ernst May (A4). He also outlined a general vision of these developments [38–41]. In his vision, the city of the future was to have a center with high dense buildings, on the periphery of which small settlements of detached houses would be built and then connected to the main center by efficient public transport. The urban and architectural concepts were to draw inspiration from the layout and development of the Silesian countryside. May proposed complexes of scattered, street-based, and oval-village-shaped settlements concentrated around a village green (Streusiedlung, Strasensiedlung, and Angersiedlung), depending on the local landscape conditions and needs [42]. Apart from flower and vegetable gardens, the buildings were to be surrounded by greenery planted around squares and streets. With time, however, the activities of the Schlesische Heimstätte company dominated the housing market in the construction of suburban settlements. It operated in 22 town districts, and it cooperated with the administration of towns and municipalities (especially suburban ones) and with local companies and societies [42,43].

The region of the Lower Silesian Coal District (Niederschlesischen Steinkohlbezirk), and Waldenburg itself (now Wałbrzych), the second largest city in Lower Silesia, was suffering from the greatest housing shortage on a national scale [43] (p. 74) and was the first, in a sense, experimental area of May and the Schlesische Heimstätte company's activity. In a short period of time, thanks to the efforts of the administration, party, and trade union activists, a mining housing fund (Treuhandstelle für Bergmannswohnstätten in Niederschlesischen Steinkohlbezirk) was established with its seat in Salzbrunn (Szczawno Zdrój) [43,44] (p. 91) [45,46], which cooperated with construction companies and communal cooperatives to facilitate the financing of investments. In initial cooperation with local mining and industrial tycoons, first of all with Fürsten v. Pless Consolidierte Fürstensteiner Gruben and later with communal companies, the Silesian organization Schlesische Heimstätte erected more than twenty housing estates between 1919 and 1924 in Waldenburg, Gottesberg (now Boguszów)—two towns of the mining district and several neighboring communes. Land for the construction of housing settlements was purchased and 815 flats were erected, financed from the funds of the district and the state (5.8 million marks from the state mortgage and 1.9 million from other sources) (see Appendix A, A5) [43] (p. 95). In the following years, the acceleration of works on the erection of communal settlements was significantly influenced by the Act of 1924 on rents [30] (p.251) that, by increasing the fee for mortgage charges, provided an additional source for the construction of new flats. In the years 1924–1927, a further 1353 residential premises were built in the Waldenburg area.

4. Results

4.1. Greenery in Panoramas and Views from Housing Settlements in the Waldenburg Area

The hilly and mountainous terrain of the Waldenburg area made it possible to create special landscape effects. Housing estates were built at the foot of towering peaks—small ones like Wilhelmshöhe (Gediminas Hill) and quite impressive mountains like Schwarze-Berg (853 m above mean see level, Borowa) or Hochwald (850 m, Chełmiec)—and on the edge of old villages and developing mining towns (usually neighboring with vast areas of private forests that were mostly owned by the Hochbergs but sometimes owned by the state. The artists were perfectly aware of the possibility of exploiting the landscape values of the area. Ernst May knew the surrounding mountains and villages and made sketches of them in his sketchbook (see Appendix A, A6). Most of these settlements were created as special landscape complexes. They were created as separate units that were carefully integrated into their environment. This was manifested in the conscious and extremely careful shaping of the panoramas of the complexes. This problem was raised, among others, by May himself when discussing the issues of housing estates design. He emphasized the necessity of subordinating an urban complex to a dominant feature. In most cases, these were already existing church towers in neighboring villages. This is how the housing estate in Gottesberg was designed. In the settlement Stadtparksiedlung (Gaj), a chapel was to be built as a dominant element. The housing estate on Sandberg (Piaskowa Góra) in Salzbrunn was subordinated to the dominant of the Catholic church, and the housing estate in Dittersbach (Podgórze) near Waldenburg was subordinated to the dominant elements of the Lutheran church. The designed housing estates were adapted to the existing development method in the neighborhood, its intensity, and the height of the buildings; they received similarly shaped composition nodes and a dilution of the building fabric in the form of squares and street extensions. This created an analogous urban pattern that imitated the spontaneity of the settlements' development in the landscape. The settlements built by or in cooperation with Schlesische Heimstätte show that the best results were achieved when the topography was preserved and followed. Small slopes, escarpments, or depressions gave the settlements a natural and picturesque landscape. The location of settlements on hills or slopes gave views of the surrounding panorama.

Hermann Jansen (see Appendix A, A7) took particular care to exploit these opportunities [47,48]. When, after the First World War, the city authorities of Waldenburg independently made another attempt (after the foundation of Neustadt; Nowe Miasto) to build a large residential district, they asked this well-known Berlin academic for a design. In the arrangement of Hartebusch Siedlung in Waldenburg, Jansen repeated the principles of spatial organization used in the Berlin housing estates of his design; see Figure 1.

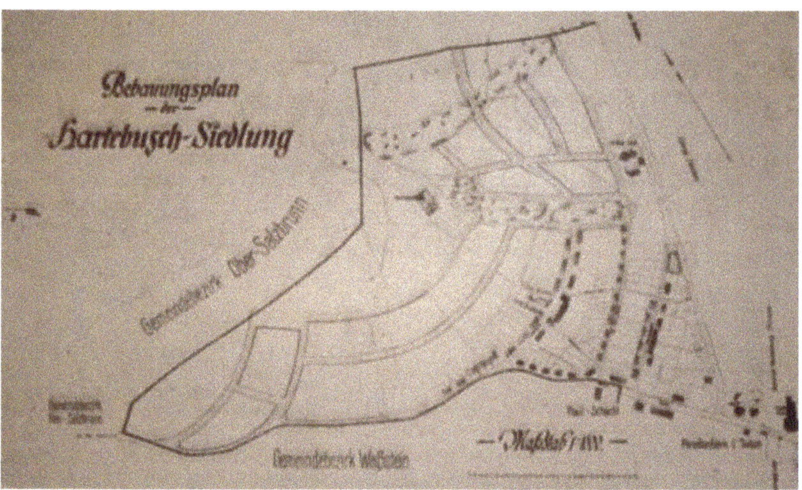

Figure 1. The project of the housing estate Hartebusch (part of Stary Zdroj in Wałbrzych). Original version developed by Hermann Jansen. From [44] (p. 91) (Public domain), https://www.dbc.wroc.pl/dlibra/publication/3031?language=pl (accessed on 7 January 2021).

However, the specific topographical conditions dictated special solutions. Due to a shortage of land in the neighborhood of Waldenburg, it was decided to locate the housing estate on this exceptionally steep slope. This imposed a kind of terraced layout on the whole settlement; see Figure 2.

Figure 2. The housing estate in Hartebusch Siedlung (Stary Zdroj in Waldenburg) at the time of construction. Adapted with permission from collection of Historical Department of the Museum in Wałbrzych (Muzeum w Wałbrzychu, Oddział Historyczny) 2505.

Along the main road, the estate was enclosed by a row of multi-family buildings. Along the main streets, the composition nodes were regularly arranged in the form of squares and terraces connected with the largest three-story multi-family houses, from which one could admire the panorama of Waldenburg and Altwasser (Stary Zdrój, Jastrzębie-Zdrój, Poland) [18] (p. 247–264).

Perhaps it was this successful settlement that inspired Ernst Pietrusky (see Appendix A, A8) [49–51] to shape a similarly situated settlement in Nieder Hermsdorf (Sobięcin Dolny); see Figure 3a. Unfortunately, most of the planned viewpoints were not completed. The project of Ernst Pietrusky's housing estate in Nieder Hermsdorf was selected as a result of a contest announced in the magazine *Schlesisches Heim* [37] (pp. 8,68,111) [52–55]. The architect based his conception on a strict representation of the terrain's shape; the irregular lines of the streets were running horizontally. Probably this naturalness and the use of a high building density for a housing estate of this type were decisive for the choice of this solution. As in the case of the settlement on the slopes of Wilhelmshöhe in Hartebusch Siedlung, squares were to play particularly important role in the layout of the Nieder Hermsdorf housing estate, including appropriately placed squares at the end of the streets in the form of viewing terraces opening onto the valley of the stream flowing through the village.

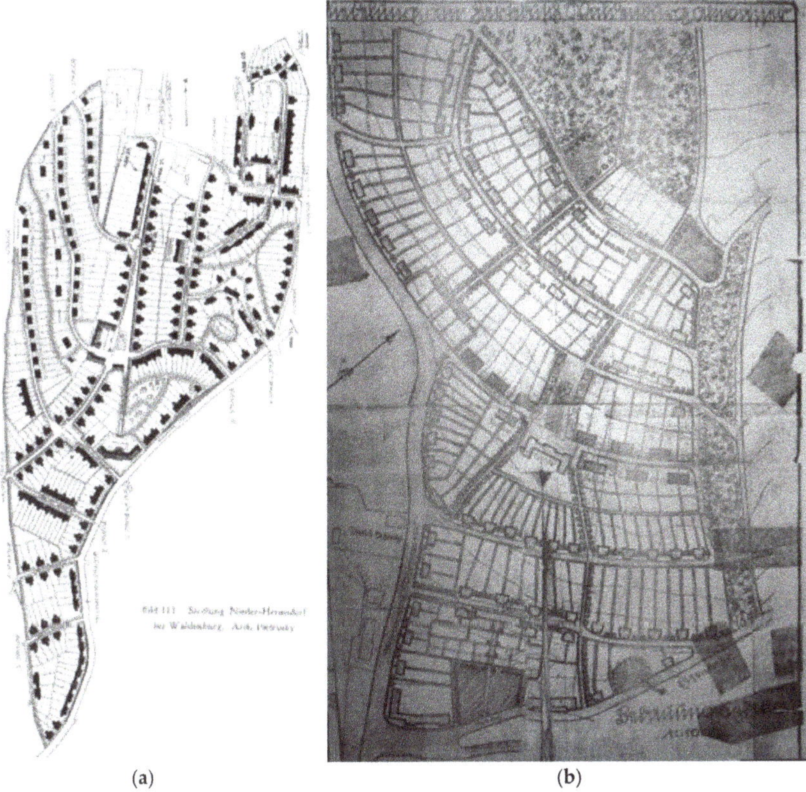

(a) (b)

Figure 3. (**a**) The project of the housing estate in Nieder Hermsdorf (Sobięcin Dolny) developed by Ernst Pietrusky; from [37] (Public domain), https://www.bibliotekacyfrowa.pl/publication/37009 (accessed on 7 January 2021). (**b**) The project of the zoning of the settlement Hartebusch on the slope of Wilhelmshöhe (Stary Zdrój) made in the municipal office of Waldenburg. Reproduction without scale. Adapted with permission from collection State Archives in Wrocław, Branch in Kamieniec Ząbkowicki, Acta specialia des Magistrat zu Waldenburg, No. 21, p. 6, Siedlung am Hartebusch, Waldenburg Altwasser, Stadtbaurat, 1:1000, no year. (after 1920).

The early modernist housing estates of the Waldenburg agglomeration were formed as unique landscape complexes that were carefully inscribed into the mountainous landscape of the region. Subordinated to selected dominants, with carefully shaped accents in height and volume, they revealed themselves in interesting panoramas and views. They were enclosed by greenery that separated them from the neighboring areas. Viewing points were planned within the residential complexes, where views of old villages, mountains and hills were given between the houses and rows of trees. Inside the housing estates, green alleys and gardens surrounded the buildings, separated the sub-units, and contributed to the compositional skeleton.

4.2. Greenery in the Settlements: Squares, Green Spaces, Tree Rows and Alleys

In the design of this period, the role of public greenery was appreciated, both in larger areas in the form of parks and huge squares and in the form of street rows.

On the Hartebusch estate, Hermann Jansen, in his typical fashion, used green strips to separate the different levels of the housing to accentuate the main public spaces and to emphasis the composition. Between the two main streets, which run horizontally from side to side, a green public area with squares and playgrounds was laid out in the middle of the estate. They constitute the skeleton of the settlement, which is emphasized by the grouping of multi-family buildings and the creation of a green space between them, Eichenplatz (i.e., Oak Square, now Zawisza Czarny Square), where the architect decided to preserve the centuries-old oak trees. Below the two main streets, two levels of parallel streets were designed, and above the main streets, one was designed. To the north, at the end of the streets, Jansen planned to leave a green belt in a shallow ravine, subordinated to the compositional axis created by the water tower visible over the housing estate. On the other side of this green space on the north-eastern slope of the hill, he designed the second part of the housing estate in a similar terraced arrangement.

Another version of the project was prepared by Otto Rogge (A9]) (Figure 3b), a building advisor at the municipal office [18] (pp. 18, 247 et seq.) [56]. He took even more care to preserve the existing valuable trees and arrange new plantings along the estate streets. The lower main street of the estate, the original village road, is lined with rows of lime trees; see Figures 2 and 4. In the highest part of the settlement, there is a large square called Buchenplatz (i.e., Beech Square or Gediminas Square), where beech trees are protected. All squares and plazas of the settlement were connected to extremely steep pedestrian walkways that are, in some places, arranged as stairs running up and down the hill. At most of them, tree rows were designed to highlight their course. The architectural lines were also planned to be highlighted by connecting the free-standing buildings with rows of trees. From the first stage of construction, tree rows were planted along the streets; for this purpose, specific species were chosen, e.g., the lime alley was supplemented with faster-growing chestnut trees. The centuries-old oak trees were preserved, but the beech trees were unfortunately partly cut down.

In the arrangement of the greenery, familiar forms of rows, avenues, and squares were used, sometimes extending like a small park. The special role assigned to the greenery was different. In the final effect, the housing estate gained a system of greenery, co-creating the basic urban composition. Successive lines of buildings on the slope were separated by strips of greenery that formed a background for them. The rows emphasized the rhythm of the urban structure. The green spaces stretching down the slope and continued by the trees on the steep pedestrian approaches were used to emphasize the main compositional axes of the urban complex.

Ernst May had a slightly different approach to the design of the green spaces in the neighborhood when he planned the housing settlement—Stadparksiedlung [18] (pp. 280–290); see Figure 5.

Figure 4. The housing estate in Stary Zdrój in Wałbrzych (Hartebusch Siedlung). The lower main street (Żeromskiego Street) of the estate (the original village road) lined with rows of trees. Photo by the author in 2009.

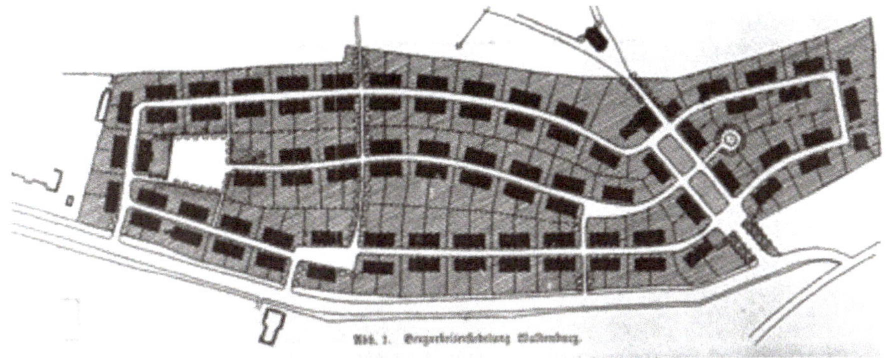

Figure 5. The project of the settlement Stadtparksiedlung in Waldenburg (Gaj in Wałbrzych) developed by Ernst May. Adapted from ref. [45] (p. 3). (public domain) in collection of Wrocław University Bibilotheca.

It was to be a residential complex reminiscent of the villages near Waldenburg, with buildings appearing to be spontaneously developed. The greenery introduced into the layout had to emphasize this character by creating groups of trees and bushes reminiscent of those spontaneously growing by the roads and main squares of the village. The concept of the settlement resembled the construction of Neustadt or the settlement Hartebusch—the creation of a completely independent urban organism. The land to be developed was located behind a mountain railway viaduct that was far from the town and surrounded by forests on all sides. The solution proposed by Ernst May took the shape of the land into consideration (levelling was kept to a minimum). The two main roads run inside the area along a closed lenticular circumference. Small, semi-detached, single-family houses were designed on both sides of the streets. The plots marked out on both sides have a variety of shapes. The main square, as well as several small squares inside the settlements, was planned. The complex was made to have charming alleys. There is no rigidity, there is variety, and each section has a different curve due to the terrain. The narrow streets are winding, almost like country paths. Diverse groupings of buildings were created around squares, and accents and compositional nodes were created despite the use of identical architectural objects. The complex is an example of implementation of the "garden settlement" concept in the form of a village-like unit with greenery that imitates groups of natural vegetation. The idea was based on the creation of a small semi-rural center—a square with a school and a chapel in the north-eastern part (from the town side), closed on two sides with two-story frontages, with passages in the that which could satisfy the needs of the inhabitants. It also had its advantages in terms of the landscape. The introduction of the square created a hierarchy of space and gave independence to the settlement that was thus gaining a rank equivalent to the neighboring industrialized villages [45,57,58]. Both the main square and all small squares of the settlement were filled with greenery, though some of them only with low levels of such. However, the basic greenery in the settlement was planned to be private fruit and vegetable gardens with flowers, thus creating a rural landscape. The greenery complemented the buildings, and together they formed a housing unit that very closely copied the appearance of the neighboring medieval villages.

It was also possible to combine these two ways of using greenery of traditional composed forms and imitation wild vegetation. In the unusually careful design of the Nieder Hermsdorf settlement (Figure 3a), Pietrusky was inspired by both the propositions by Jansen and some of May's ideas in the design of the greenery. The concept was based on keeping a road connecting an orphanage (which had been erected in the former village at the end of the 19th century) as the main street, with a folk house designed in the southern end of the settlement on a hill, to which a line of stairs in greenery would lead; this settlement was to be preceded by a triangular square that constituted the settlement's main square [37,55] (p. 68). Ernst Pietrusky planned three-story buildings with commercial premises on the ground floor. Thus, the square was to constitute the functional and spatial center of the settlement. The remaining streets were laid out along curve lines in accordance with the slope's contours, crossed with streets running down the slope of the terrain [54]. The irregular course of the streets introduced diversity and a high individualization of space, despite the use of typical buildings. The use of small squares was to highlight the rank of selected objects, and, in the final version, it was used to create distinctive places where the life of the settlement would concentrate.

The housing settlement was conceived as a compact and closed urban layout whose composition was based on emphasizing the main axis, within which the center of the housing settlement—clearly marked and emphasized with the use of higher buildings—was developed. The streets and narrow pedestrian passages were to be planted with tree rows. Numerous green squares enriched the layout. In the eastern part of the settlement, at the foot of the slope, a rectangular elongated square was planned with a small square connected to it at the corner, and an irregular square in the bend of the street was planned to be above it. In their vicinity, at the crossroads of the streets, a small rectangular square was to be the foreground of the kindergarten and school building. In the original version, another large rectangular square with a bay was planned at the south-western end of the settlement, as were squares at the ends of the streets on the western side—one of them with a viewing terrace. There were also to be two large squares on the settlement, a kind of small park sheltered by a screen of trees. They were located within the quarters of single-family housing on large plots extending the public greenery with private gardens; see Figure 6.

The larger of the squares was adjacent to the steps leading up to the folk house on the settlement's main traffic route. The project was not fully implemented. However, during construction, the greenery in the settlement was introduced from the beginning in accordance with the plans of project. Lime, maple, and chestnut trees were planted along the streets. Some of them also appeared in private front gardens introduced here [59]. As in Hartebusch Siedlung, a lime-poplar alley [60], which ran in the neighborhood and was later supplemented with maples and old pines on the top of the hill, was preserved. A centuries-old yew tree at the northern end of the settlement was also treated with due respect. In 1925, the 19th century park (Volks Park) in the north of the settlement was redesigned. Within several years, the settlement became a green enclave, which, of course, did not reduce the health problems of its residents. From the very beginning, the biggest inconvenience for the housing settlement was the extremely close vicinity of the coking plant and chemical works. Already in the 1920s, the planners working on this housing estate addressed queries to the local authorities about the acceptable rates of air pollution in the residential areas due to the proximity of the coking plant and the chemical works [61]. It is possible that the Nieder Hermsdorf settlement was one of the first places where the issues of industrial pollution started to be addressed.

Figure 6. View of Kresowa Street in Sobięcin (Nieder Hermsdorf). Photo by the author in 2010.

The greenery system of the Nieder Hermsdorf estate co-created the compositional skeleton of the residential complex. It marked the main compositional axis and emphasized the most important urban interiors. The park with a garden restaurant, amphitheater, and music pavilion, as well as spacious green squares, provided areas for rest and recreation. Green front gardens in front of the service buildings and kitchen gardens created a compact green environment for development. Particularly interesting was the incorporation of private gardens into an overall ensemble with green areas accessible to the public and decorated with lines of footpaths. This created an impression of the naturalness of the greenery planted on the estate. Another important aspect was the respect for the existing trees and the park complex, as was the case on the Hartebusch estate.

Similarly shaped green squares, parks, streets, and alleys were designed in other housing settlements built in the region of Waldenburg in cooperation with Schlesische Heimstätte. Squares, some of which were to be planted with trees and others of which were only to be planted with low greenery, were an integral part of the layout. In the settlement planned as an extension of Gottesberg (1919), Ernst May, apart from the square functioning as a market, also designed a triangular green square in the middle of the development surrounded by single-family terraced houses [62]. In the competition-winning design of the housing settlement in Neu Salzbrun (Nowe Szczawno) (1920), Theo Effenberger, already a well-known designer from Wrocław and the author of the housing settlement in Pöpelwitz (Popowice in Wrocław) (1919) (A10]) that was under construction at that time [63,64], introduced an irregular square surrounded by compact building frontages at the end of the axis of the layout—a street emerging from the former communication junction in front of the railway viaduct [18] (p. 357) [65]. In the second of the settlements planned in Rothenbach (Gorce) for the Schlesische Kohlen und Kokswerke company, May and his teammate Bussmann proposed a square in the center with a semi-circular outline, a school, and a church (or folk house?), surrounded with a park (1921) [66]. These complexes were not realized. In Weiβstein (Biały Kamień in Wałbrzych), the designers chose solutions with squares in the center of the housing settlements divided into workers' gardens (1924) (see Figure 7) [18] (pp. 325–337).

Figure 7. View of housing estate of Weiβstein (Biały Kamień) in early 1920s. Adapted with permission from collection of Historical Department of the Museum in Wałbrzych (Muzeum w Wałbrzychu, Oddział Historyczny) 2881.

Squares were important green elements used in early modernist housing estates. They took a different form than those from the 19th and early 20th centuries. They often had an outline of an irregular triangle or were not of any size at all. They accompanied not only important service buildings but also often diversified residential developments, including single-family ones. This type of solution imitated natural groups of vegetation.

Tree rows were equally important for composition. In the case of the housing settlement in Sandberg (1918–1919), rows of lime trees and chestnut trees on both sides of the street constituting the basis of the complex emphasized its character; see Figure 8. Using a cul-de-sac, which resembles an agricultural homestead, it skillfully imitated the development of a typical mountain village stretching along the green banks of a creek [67,68]. The housing settlement in Gottesberg was enriched by street bays with rows of lime, maple, and chestnut trees; see Figure 9. In the above-mentioned Effenberger's project, the compositional skeleton was to be the ring street (Ringstrasse), with a one-sided development on the outer side of the housing complex surrounded by a tree row, from which the Salzbach (Szczawnik) valley would be visible; on the other side, the housing settlement on the Hochwald Mountain would be visible. On both housing settlements in Dittersbach (Podgórze in Wałbrzych)–Neuhäuser Siedlung and Melchior Siedlung (1920), which was designed by Ernst May and adapted in the municipal office by Daehmel [18] (pp. 291–311) [69–73]—one could enjoy the views of grand alleys. At the Melchior shaft (after the war, it was renamed the Mieszko shaft), these were lime alleys on both sides of the railway tracks (Figure 10), while at the second settlement, the old lime alley (Neuehauser Allee) was preserved, running from the village near the Catholic church up the slope towards the castle and the manor in Neuhaus (Nowy Dwór) (Figure 11). Like in Nieder Hermsdorf, front gardens separated by rail fences and rhythmically planted trees appeared in front of the houses in both these settlements. They were also created in subsequent small settlements in Konradsthal, Nieder Salzbrunn, and Adelsbach near Waldenburg (Konradów, Szczawienko, and Struga) (c. 1923–1925) [18] (pp. 358–377).

Figure 8. View of settlement on Sandberg (Piaskowa Góra) and Salzbrunn (Szczawno) in the interwar period. Adapted with permission from collection of Historical Department of the Museum in Wałbrzych (Muzeum w Wałbrzychu, Oddział Historyczny) 2877.

Figure 9. View of housing estate in Gottesberg (Boguszów). Ulica 1 Maja (Street of 1 May) with rows of lime trees. Photo by the author in 2015.

Figure 10. View of Melchior Siedlung in Dittersbach (Podgórze in Wałbrzych) in early 1920s. Adapted with permission from collection of Historical Department of the Museum in Wałbrzych (Muzeum w Wałbrzychu, Oddział Historyczny) 2877.

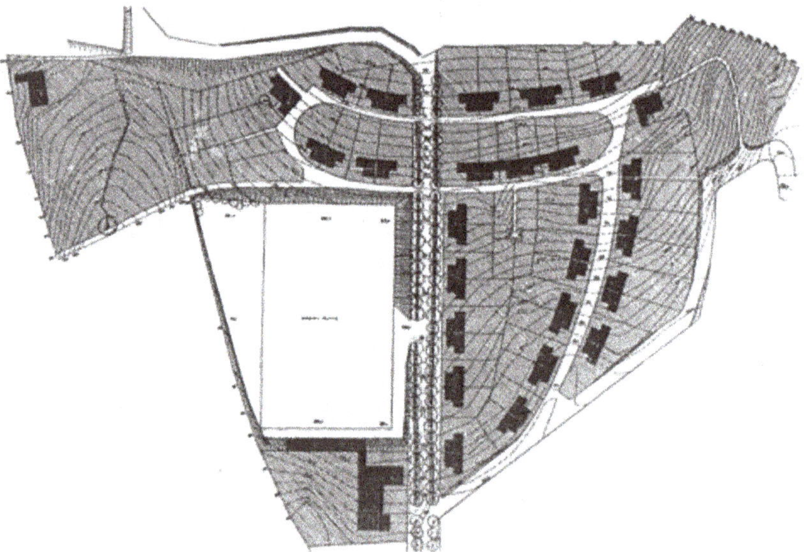

Figure 11. The project of Neuhäuser Siedlung in Dittersbach (Podgórze in Wałbrzych) developed by Ernst May. Adapted from [70] (public domain).

Avenues and rows of trees, both the preserved and carefully protected 19th-century complexes and the new ones, constituted an important element of the urban composition of the residential complexes under discussion. In this case, as in the design of the squares, in addition to continuing the form of a straight avenue and row and a regular roundabout, planting in the form of short rows, which were non-rhythmic in character and imitated natural trees, was sometimes used.

The preservation of some green spaces within the designed housing settlements was also required for reasons of economy and knowledge of the influence of physiography—the topography of the terrain, the presence of watercourses, sun exposure, problems with cold air flow, and ventilation. This could already be seen in the earliest post-war project by Jansen, where the architect left green ravines that were difficult to level and are threatened by cold air flow. Similarly, he approached May's physiographic issues with great emphasis, probably largely influenced by Raymond Unwin. In mountainous areas, the consideration of rules was particularly important, which he highlighted in his articles. A characteristic feature of his projects was the precise adaptation to topographic and physiographic conditions. Taking advantage of the terrain's shape in addition to improving the microclimatic values, significant savings on levelling works during construction usually resulted in a smooth curvilinear line in the urban composition [66]. Consequently, the settlements had small interiors, and the inhabitants enjoyed various views from successive places along the streets. In the Waldenburg region, the settlements built in this period, usually in difficult-to-develop areas with steep slopes, were given a layout based on running streets parallel to the slope contour. In the case of the Dittersbach housing settlements, the main streets were laid out in wide arches that formed serpentines and horseshoe shapes surrounding green spaces. In May's designs (Stadparksiedlung), attention was drawn to the extremely precise mapping of the terrain along the streets, which were thus laid out along a winding line. A completely new theoretical problem was the design of settlements on steep slopes (Rothenbach) [66]. May drew on limited design experience to solve housing estates in mountainous areas. There were no patterns of regular village buildings, so he drew some inspirations from the towns of the Sudeten foothills, e.g., Silberberg (Srebrna Góra) [73] and probably also from Gottesberg (which Ernst May known from the time of design). He developed a layout following the contours of the hill, with farm buildings at the gables of the houses or in a separate line at the back of the plot.

In the design of early modernist housing estates around of Walbrzych (Figure 12), public greenery was of particular importance. It was arranged in a way that imitated the forms existing in the landscape. Old alleys were preserved, and the design of this form of greenery, clearly visible in the landscape, was continued. This emphasized the traditional communication and cadastral structure. Tree rows served to strengthen the rhythm of the buildings. Larger greenery complexes marked the axes and composition nodes of these urban layouts. At the same time, many elements imitating natural groups of roadside and village greens were introduced. They gave the whole arrangement a character of spontaneity.

4.3. Employee Gardens, Tradition, and Design

Another particularly important issue was the provision of housing settlements with home gardens (Kleingarten) or workers' gardens (Schreibergärten) according to sociological assumptions and government guidelines (Figures 13 and 14). This was an idea developed based on government recommendations of the time, but it had an old tradition in the Waldenburg district. Initially, when the initiative in housing matters belonged to the owners of mines and factories, as well as the railway authorities that wanted to provide manpower in their plants, patronage settlements were established. These mostly small complexes, consisting of several houses, were equipped with a common area, a playground, small social facilities and workers' gardens. The entire housing settlement area was divided into small plots. Wooden sheds, pigsties, and dovecotes were built among them. Additionally, in the case of municipal or communal multi-family houses, at least some of the flats were equipped with workers' gardens, as was the case in the new districts of Waldenburg–Neustadt (1904) or the new part of Weiβstein (1904), where the uninvested lands were allocated for this purpose for a short period of time (Figure 15).

Figure 12. View of Neuhäuser Siedlung (behind the railway overpass) in Dittersbach (Podgórze in Wałbrzych) in the interwar period. Adapted with permission from collection of Historical Department of the Museum in Wałbrzych (Muzeum w Wałbrzychu, Oddział Historyczny) 3085.

Figure 13. Workers' gardens around residential buildings in Hartebusch Siedlung shortly after construction. Adapted with permission fFrom collection of Historical Department of the Porcelain Museum (Muzeum Porcelany w Wałbrzychu, Oddział Historyczny) 2507.

Figure 14. Adapted with permission from collection of Historical Department of the Museum in Wałbrzych (Muzeum w Wałbrzychu, Oddział Historyczny) 2507.

(**a**) (**b**)

Figure 15. (**a**) Fragment of the project of the housing estate in Sandberg. Reproduction without scale. Adapted with permission from collection of Communal Archives in Szczawno (Archiwum gminy Szczawno), 65/103, s. 62, Errichtung von 25 Bergmannshaeusern, Lageplan Klein Siedlung Ober-Salzbrunn, M. 1:1000, der Ausführende Wilhelm Kahmann, Bad Salzbrunn 4 July 1920. (**b**) Central square with workers' gardens in Friedenhof Siedlung in Weiβstein in archival photography from the mid-twenties of the 20th century. Adapted from [43] (p. 98) (public domain) https://dbc.wroc.pl/dlibra/publication/17186/edition/15270 (accessed on 7 January 2021).

To a large extent, the functional solutions of the settlements had to be derived from the layout of rural houses, as the population was usually comprised of first-generation migrants from the countryside. The idea of a house with a garden for each family, derived from Howard's theory and implemented by Unwin (in whose studio May worked during his internship in England), was a response to social needs. In earlier periods, the need to cultivate at least a piece of land was secured by providing each flat with a workers' garden. In the projects of the Schlesische Heimstätte studio, both concepts were combined. In the early modernist housing settlements, gardens could have the character of agricultural facilities with a considerable area of 1500–2500 m^2—as in the case of the housing settlement on Sandberg, which initially (similarly to the housing settlement in Goldschmiden (Złotniki in Wrocław)) was to be mainly rented to agricultural workers. In Gottesberg, each flat that was located in the first row of two-story single and semi-detached buildings with single-story connectors was provided with a large garden (450 m^2) behind the buildings and with sheds in the connectors between the houses [74]. In Stadtparksiedlung, gardens varied in size (200–500 m^2), and in the Nieder Sobięcin settlement, the differences were more significant depending on the type and location of buildings (130–350 m^2). Most often, the size of plots exceeded 200 m^2, as in the settlements in Rothenbach (200–300 m^2) or Dittersbach–Neuhäuser Siedlung (200–300 m^2). The smallest were plotted for multi-family housing settlements in Melchior Siedlung (100–150 m^2) and Gottesberg (70 m^2), while in Weiβstein, they slightly exceeded 50 m^2. The plots were carefully planned and enclosed with fences. This created checkered or fan-shaped green arrangements. In the gardens, the outbuildings—that were designed together with the remaining buildings in many housing settlements (e.g., in Dittersbach, Weiβstein, and Konradsthal)—were erected. Vegetables and flowers were grown. In most settlements, poultry and even domestic animals were also kept. However, in some cases, it was strictly forbidden even to have dovecotes (Nieder Hermsdorf). The attachment to this form of contact with nature and relaxation after working hours is evidenced by the fact that the probably planned green spaces in the housing settlement Hartebusch, as in all previous Jansen projects, were parceled and changed into gardens; the same happened at the end of the 1920s and 1930s (in the Waldenburg agglomeration, only two pre-war housing settlements had public green spaces).

Attempts were also made to ensure an attractive appearance, a good level of care of gardens, and an increase in production efficiency. These goals were achieved by establishing companies that promoted knowledge in the field of gardening and publishing magazines addressed to users of small home and working gardens [16,75].

Home and working gardens became the most important characteristics of the housing estates of that period. They provided for the needs of the population from the countryside who were accustomed to cultivating the land and keeping domestic animals. Such facilities also sought to establish social habits. The need to ensure stress-relieving recreation for the miners, which had been fulfilled by visits to taverns or handicrafts (e.g., hosiery making in the Waldenburg region), could be replaced by gardening. It also had an economic aspect, simply supplementing the inhabitants' diet and raising the general level of nutrition, which turned out to be particularly important in the period of the "great crisis." It was also important to transfer some of the work of cleaning and developing the estate to the residents. This could have decisively accelerated the process and rapidly improved the natural qualities of the housing area.

5. Discussion

The specific location of the Waldenburg agglomeration settlements in the mountains imposed unusual solutions, but their creators—prominent German architects of that time like Hermann Jansen, Ernst May, Theo Effenberger, and Ernst Pietrusky or architects associated only with this city like Oscar Rogge—introduced certain principles that were later adopted as the basis for planning housing complexes. This basis also referred to the design of green spaces in housing settlements. Greenery systems were used both as elements of urban composition and as recreational facilities. In the new housing settlements, along with the regulation of streets, carefully considered planting was introduced—mainly rows of lime, maple, and chestnut trees. Traditional solutions of roads in the form of alleys, an early modern idea that was popularized in Germany in the urban planning handbook by Joseph Stübben at the turn of the 19th and 20th centuries, were preserved. The idea of the square as an element that highlights compositional nodes was also taken up. This form of introducing urban greenery, developed at the turn of the 17th and 18th centuries, started to appear in plans of new residential districts of Silesian cities from the beginning of the 19th century and gained popularity in the Art Nouveau era. In early modernist solutions, this form took a slightly different form. In Jansen's design, it took the shape of small park complexes (perhaps due to the need to preserve the old trees). In the concepts of Ernst May and other architects associated with the Schlesische Heimstätte, squares were transformed into green squares at street intersections and much larger central housing settlement squares where the most important services were concentrated. From these skillfully used traditional forms, the designers created a complete housing estate green system. It became an element of the entire urban concept.

Just as the architecture and urban planning of these housing complexes were based on studies of the local development, the same approach was followed for the shaping of green spaces. The studies respected the local natural and cultural landscape. Inspiration was sought in the immediate surroundings of the designed area, and attempts were made to interpret the rules governing nature and the local community. In this way, the settlements gained coherence with their surroundings and an effect of spontaneity in the urban arrangements. Care was given to monuments of nature, e.g., oaks and beeches in Hartebusch Siedlung, yew trees in Nieder Hermsdorf, and the historic alley leading to Neuhaus. The designs of greenery layouts skillfully applied the principle of quotation, i.e., the literal imitation of traditional solutions and prudent stylization (e.g., in urban planning, it was the use of an oval center and in architecture, it was the use of semi-circular passages, dormers of the "oeil de boeuf" type, classical cornices, etc.) in the form of a line of village buildings along a stream (Sandberg), a kind of a village green, a green square analogous to a common village grassland (Stadtparksiedlung), a manor courtyard (the housing settlements in Weiβstein), rural roads with scattered buildings interwoven with short rows of trees, and tree-lined street bays (the housing settlements in Gottesberg and Dittersbach). New forms were introduced based on distinguished, traditional compositional rules (in urban planning, this means a layout with a dominant height and space, as well as a different choice of building density; in architecture, this means diversified proportions of window sizes and shapes), such as green squares at the intersections of streets and pedestrian alleys (Stadtparksiedlung) or in the middle with Unwinian cul-de-sacs (Sandberg), as well as attractive semi-circle or horseshoe layouts of buildings surrounded by greenery that were inspired by English baroque and classicist crescents (the Dittersbach housing settlements and the unrealized settlement on Gleisberg in Waldenburg–Parkowa Góra in Wałbrzych). Careful attention was also paid to the arrangement of household and workers' gardens. Their attractiveness in the landscape was determined by the skillful division of land (which mirrored the parceling out of the villages surrounding Waldenburg) and well-designed garden architecture that replicated regional forms used in residential buildings, and it was sometimes enriched with traditional woodcarving decorations and well-tended crops (for this purpose, an advisory magazine was published). Home and working gardens were of particular social and economic importance. Their creation was part of a deliberate program to create a lifestyle for the residents of the estate, based largely on tradition and habit.

In the early modernist housing estates of the Walbrzych region, the designed greenery layouts were considered not a supplementary decorative element but one of the most important components of a complex urban and architectural design. These were not the first applications of greenery systems in this character. Already in the designs of neo-Baroque and (especially) Art Nouveau districts, greenery compositions constituted an important factor shaping the space. However, it was not until the housing estates of early Modernism (among which the Waldenburg estates are among the first) that the shaping of greenery was so comprehensively included in the functional, ideological, and formal concepts and the social program. Greenery compositions arranged main squares and streets of residential complexes. They fundamentally shaped the landscape character of entire complexes. Parks and squares constituted recreational background, where entertainment and, in time, sports facilities were located. The cultivation of employee and home gardens became a way of spending free time. The culture of spending time outdoors in green surroundings was promoted.

6. Conclusions

The greenery in early modernist housing estates in the Waldenburg agglomeration has survived to the present day. Public green areas, maintained and partly replaced, are attractive recreational areas of the housing estates (See Appendix A, A11). They are also objects of interest and pride for the inhabitants [76–78]. Attractive panoramas and views of housing estates and districts of Walbrzych and neighboring towns are admired because they have been revalorized in recent years through the restorations of viewpoints and the shaping of chaotically overgrown vegetation. Home and working gardens are in constant use. Their presence makes the housing offer of the still-exploited modernist housing estates around Waldenburg particularly attractive.

The green system of these housing estates is still one of the best-designed systems and may still be a model for contemporary developers of urban plans for residential complexes. The introduction of avenues and espaliers and the arrangement of green squares and plazas are worthy of imitation or, at least, of inspiration. Additionally in terms of solving household and working gardens (which are probably becoming popular again after a period of certain loss of attractiveness), the ideas from the 1920s implemented in this area may still prove valuable. The most valuable, however, is probably the very approach to solving design issues while considering broad social needs. This was based on both the implementation of modern ideas in architectural and urban planning concepts, including the planning of green areas in housing estates, and on respect for tradition in the customs and functioning of local communities. In the city landscape and in the urban interiors of new settlements, a sense of the introduction of new values was created, but this was in line with the principles that had long been present in the area.

Funding: This research received no external funding.

Institutional Review Board Statement: Not applicable.

Informed Consent Statement: Not applicable.

Data Availability Statement: The study did not report any data.

Conflicts of Interest: The author declares no conflict of interest.

Appendix A

A1: Apart from the first effects of the economic crisis in Germany, a fundamental change in the design of housing settlements was brought about by the centralization process in the operation of housing companies, which began in this very year. The laws on housing issues have also been fundamentally changed.

A2: A movement supported by the state since 1919, the Kleingarten Law of 1919 and 1920, and the Heimstätterecht Law of 1920. The laws of 1920 and two of 1924 recommended the establishment of settlements with large gardens (Heimstättesiedlung) and gave rise to organizations for such investments (Heimstätte).

A3: The Schlesisches Heim (Schlesische Heimstätte) with the seat in Breslau was founded on 28 July 1919 as the fourth such regional organization. The first was established in Westphalia in 1918, and the last was established in 1925 in Upper Silesia.

A4: Ernst May (1886–1970) studied in England, where he later served his apprenticeship in the studio of Raymond Unwin, and then he studied in Munich. After the First World War, in which he actively participated, he started working in Breslau. He became the technical director of the Schlesisches Landgesellschaft and later of the Schlesische Heimstätte. He took part in the competition for the development plan of Breslau, promoting the idea of satellite settlements. In 1925, he became the head of the municipal building office in his hometown Frankfurt am Main. He drew up a development plan, and, until 1930, he designed or participated in the design of successive housing settlements of the so-called "New Frankfurt." He left for the USSR, where he led a group of architects working on concepts for large housing settlements in Moscow and new cities (Magnitogorsk, Stalinsk, and Nizhni-Tagil). In 1934, he emigrated to Africa. There, he developed both urban and architectural projects for monumental buildings. In 1954, he returned to Germany to Hamburg.

A5: New dwellings built in the Waldenburg agglomeration: in 1924, there were 171 (83 private and 88 communal); in 1925, there were 350 (143 private and 207 communal); in 1926, there were 416 (174 private and 242 communal); and in the middle of 1927, there were 370 (140 private and 230 communal).

A6: Particularly among the notes drawn during the honeymoon trip around Waldenburg. Note in the collection of the Museum in Frankfurt, exhibited during the Ernst May exhibition in the Museum of Architecture in Wrocław.

A7: Hermann Jansen (1869–1945)—a lecturer at the Berlin and Stuttgart universities of applied sciences, specialist in urban planning and housing design, member of the Berlin city council, official in the ministry, and member of the Prussian Academy of Arts from 1918—was the author of numerous urban general and detailed designs. Probably the most important of them was a general plan for Berlin (first prize in a competition organized by the city council in 1910), followed by designs for Berlin districts (Treptow in 1914, Western Berlin in 1917, and Charlottenburg in 1919) and other German towns (Cologne, Leipzig, Schleswig, Nürnberg, Brandenburg, Wiesbaden, and the Enden–Friesland housing settlement in 1915). He also prepared numerous housing settlement projects for Pomeranian towns (Koszalin, Stargard, and Szczecin) and Silesian towns (now: Nysa, Brzeg, Środa,

Złotoryja, Świdnica, Bierutów, Namysłów, and Ząbkowice). He developed projects of districts in Łódź, Riga, Budapest, Constantinople, and Ankara. He commented on small housing settlements (1909), as well as blocks of flats (1910 and 1917) and large cooperative settlements (1910). He was also the author of large architectural buildings such as the Great Market Hall in Berlin (1915).

A8: Ernst Pietrusky, after studying in Breslau, was active during the last years of the war in Goldap (Gołdapia), starting his architectural career with a project for a housing settlement in Nieder Hermsdorf. Then, he was engaged as an architect and town planner in Waldenburg and its surroundings. He designed buildings in modernist style, e.g., the Labor Office and a school in Weiβstein. He also designed several other schools in Lower Silesia, e.g., Hans-Schemm-Schule in Glatz (Kłodzko) and the airport in Gandau (Gądów in Wrocław).

A9: Otto Rogge (1866–1955): in 1910, he became the town building master of Waldenburg. Then, in 1924, he became the town building advisor. He prepared numerous urban and architectural projects in Waldenburg. He planned the expansion of districts and housing settlements, e.g., on the slopes of Gleisberg and Neustadt (from 1927). He designed many public buildings—the Evangelical Lutheran church, numerous schools (e.g., the Catholic school in Altwasser, the Municipal Female Vocational and Trade School, and the comprehensive secondary school in Neustadt (1928)), the sports stadium with a swimming pool in Neustadt (1926), the management board building of the Municipal and District Health Fund (1930), and the communal cemetery in Altwasser.

A10: Theo Effenberger (1882–1968) studied in Breslau and later in Darmstadt. After working in Magdeburg and Augsburg, he returned to Breslau in 1907. In 1910, he was one of the co-founders of the Silesian Association for the Protection of Heritage (Heimatschutzbewegung). From 1919, he mainly worked for Siedlungsgesellschaft Breslau A.G., designing several housing settlements and housing complexes, including the aforementioned Pöpelwitz (1919–1927) and Westend and Viehweide (1925–1929).

A11: This—a separate problem, also in terms of preserving the landscape and functional values of greenery—concerns housing estates whose urban structures have been destroyed by a transit route. This was the case for two of the complexes in question—in Melchior Siedlung in the 1970s and 1980s, as well as in the other Hartebusch Siedlung at the present time.

References

1. Piccinato, G. *La Costruzione Dell'urbanistica;* Officina Edizioni: Roma, Italy, 1974.
2. Collins, G.R.; Crasemann Collins, C. *Camillo Sitte and the Birth of Modern City Planning;* Phaidon Press: New York, NY, USA, 1965.
3. Wilhelm, K. Städtebautheorie als Kulturtheorie, Camillo Sittes „Der Städtebau nach seinen künstlerischen Grundsätzen". In *Cultural Turn. Zur Geschichte der Kulturwissenschaften;* Musner, L., Wunberg, G., Lutter, C., Eds.; Turia & Kant: Vienna, Austria, 2001; pp. 89–109.
4. Speitkamp, W. *Die Verwaltung der Geschichte: Denkmalpflege und Staat in Deutschland 1871–1933;* Vandenhoeck & Ruprecht: Göttingen, Germany, 1996; pp. 36–65.
5. Castex, J.; Depaule, J.C.; Panerai, P. *Formes Urbaines: De l'ilot a' la Barre;* Dunod: Paris, France, 1977; p. 45, et. seq.
6. Posener, J. *Berlin auf dem Wege zu Einer Neuen Architektur, Das Zeitalter Wilhelms II;* Prestel Verlag: München, Germany, 1979.
7. Heinrich, E.; Ernst, W. *Berlin und Seine Bauten, Volume 2, Rechtsgrundlagen und Stadtentwicklung;* Ernst, W. & Sohn: Berlin, Germany, 1964.
8. Barnstone, D.A. Style Debates in Early 20th-Century German Architectural Discourse. *Archit. Hist.* **2018**, *6*, 17. [CrossRef]
9. Battisti, L.; Pille, L.; Wachtel, T.; Larcher, F.; Säumel, I. Residential Greenery: State of the Art and Health-Related Ecosystem Services and Disservices in the City of Berlin. *Sustainability* **2019**, *11*, 1815. [CrossRef]
10. Jackisch, B.A. The Nature of Berlin: Green Space and Visions of a New German Capital, 1900–45. *Cent. Eur. Hist.* **2014**, *47*, 307–333. [CrossRef]
11. Zachariasz, A. Modernizm w kształtowaniu ogrodów i terenów zieleni—Przyczynek do historii architektury krajobrazu w Polsce. In *Architektura Pierwszej Połowy XX w. i Jej Ochrona w Gdyni i w Europie;* Sołtysik, M.J., Hirsch, R., Eds.; Urząd Miasta: Gdynia, Poland, 2011; pp. 143–150. ISBN 978-83-907114-9-2.
12. Walker, M. Zieleń w osiedlach i koloniach robotniczych. In *Historyczne Osiedla Robotnicze;* Bożek, G., Ed.; Śląskie Centrum Dziedzictwa Kulturowego: Katowice, Poland, 2005; pp. 70–76.
13. Andrzejewska, A. Modernistyczni mistrzowie urbanistyki jako prekursorzy współczesnego trendu "zielonej infrastruktury" i "urban garden" w kontekście zachodzących procesów przestrzennych. *Architectus* **2019**, *60*, 53–65.

14. Hilbrandt, H. Everyday urbanism and the everyday state. Negotiating habitat in allotment gardens in Berlin. *Urban Stud.* **2019**, *56*, 352–367. [CrossRef]
15. Lieske, H. Self-Sufficiency in Suburban Home Gardens? On the History and Prospects of the Idea of Food Production in German Home Gardens. In Proceedings of the II International Conference in Landscape and Urban Horticulture, Bologna, Italy, 9–13 June 2009; Gianquinto, G.P., Orsini, F., Eds.; Humana Press: Totowa, NJ, USA, 2010; Volume 881, pp. 807–812.
16. Urbanik, J. Krajobraz nowoczesnych osiedli okresu międzywojennego w Niemczech w dziełach Leberechta Migge. In *Architektura Pierwszej Połowy XX w. i Jej Ochrona w Gdyni i w Europie*; Sołtysik, M.J., Hirsch, R., Eds.; Urząd Miasta: Gdynia, Poland, 2011; pp. 35–42. ISBN 978-83-907114-9-2.
17. Gryniewicz-Balińska, K. Historyczne ogrody działkowe Wrocławia jako element planowego systemu zieleni miejskiej. *Kwart. Archit. Urban.* **2015**, *60*, 45–56.
18. Ludwig, B. *Osiedla Mieszkaniowe w Krajobrazie Wałbrzyskiego Okręgu Górniczo-Przemysłowego (1850–1945)*; Oficyna Wydawn; Politechniki Wrocławskiej: Wrocław, Poland, 2010; pp. 584–586. ISBN 978-83-7493-573-9.
19. Dobesz, J. Hans Poelzig i Wrocławska Akademia Sztuki. In *Ten Wspaniały Wrocławski Modernism*; Loose, S., Ed.; VIA: Wrocław, Poland, 1991; pp. 14–26.
20. Niemczyk, E. Architektura Wrocławia pocz. XX w. a problem kontynuowania tradycji miejscowych. In *Wątki Kulturowe i Symboliczne w Architekturze Współczesnej*; Oficyna Wydawn; Politechniki Wrocławskiej: Wrocław, Poland, 1980; pp. 61–74.
21. Howard, E. *To-Morrow. A Peaceful Path to Real Reform*; Swan Sonnenschein & Co.: London, UK, 1898.
22. Howard, E. *Garden Cities of To-Morrow*; Swan Sonnenschein & Co.: London, UK, 1902.
23. Tomaszewicz, A. Osiedle Borek—historia i architektura (1871–1914). In *Przedmieście Świdnickie we Wrocławiu*; Okólska, H., Górska, H., Eds.; Muzeum Miejskie Wrocławia Wydawnictwo GAJT: Wrocław, Poland, 2012; pp. 98–110.
24. Kononowicz, W. Urbanistyczny rozwój Wrocławia od połowy XIX wieku do II wojny światowej. In *Wrocław*; Atlas Historyczny Miast Polskich; Śląsk, Issue 13; Instytut Archeologii i Etnologii PAN: Wrocław, Poland, 2017; Volume IV, pp. 15–22. ISBN 978-83-63760-82-3.
25. Ludwig, B. Nowe Miasto w Wałbrzychu jako przykład przenikania idei secesyjnych do urbanistyki Josepha Stübbena. *Kwart. Archit. Urban.* **2009**, *54*, 66–77.
26. *Neisse: Ein Führer durch die Stadt und ihre Geschichte*; F. Bär: Neisse, Germany, 1922; p. 37.
27. Störtkuhl, B. *Modernizm w Legnicy*; Muzeum Architektury we Wrocławiu: Wrocław, Poland, 2007; pp. 12, 60.
28. Migge, L. *Jedermann Selbstversorger*; Lugen Dederichs: Jena, Germany, 1918.
29. Haney, D.H. Leberecht Migge's "Green Manifesto": Envisioning a Revolution of Gardens. *Landsc. J.* **2007**, *26*, 201–218. [CrossRef]
30. Tafuri, M. Sozialpolitik e citta' nella Germania di Weimar. In *La Sfera e il Labirinto*; Tafuri, M., Ed.; Einaudi: Turin, Italy, 1980; pp. 243–293.
31. Muthesius, H. *Kleinhaus und Kleinsiedlung*; F. Bruckmann: München, Germany, 1918.
32. Siedlung. *Der Grosse Brockhaus*; Brockhaus: Leipzig, Germany, 1934; Volume VII, p. 382.
33. Damschke, A. Heimstätterecht. In *Wasmuth Lexikon der Baukunst*; Wasmuth: Berlin, Germany, 1931; Volume III, p. 82.
34. Lemmer, L. Kleingarten. In *Wasmuth Lexikon der Baukunst*; Wasmuth: Berlin, Germany, 1931; Volume III, p. 377.
35. Pauly, 10 Jahre Wohnungsfürsorgegesellschaften. *Schles. Heim* **1928**, *9*, 1.
36. May, E. Leiter und Bauabteilung des Schlesisches Heim u. Schlesischen Landgesellschaft, Sielungspläne. *Schles. Heim* **1920**, *1*, 7.
37. Effenberger, T. *Siedlung und Stadtplanung in Schlesien. Volume 2, Ober u. Niederschlesien*; Grass, Barth: Breslau, Germany, 1926; p. 9.
38. *Ernst May 1886–1970*; Quiring, C.; et al. (Eds.) Muzeum Architektury: Wrocław, Poland, 2012; ISBN 9788389262660.
39. Kononowicz, W. May Ernst. In *Encyklopedia Wrocławia*; Harasimowicz, J., Ed.; Wydawnictwo Dolnośląskie: Wrocław, Poland, 2000; p. 493.
40. May, E. Siedlungspläne. *Schles. Heim* **1920**, *1*, 7–10.
41. May, E. Klein Wohnungs Typen. *Schles. Heim* **1920**, *1*, 14–17.
42. Ludwig, B. Ochrona wartości krajobrazowych w projektach osiedli autorstwa spółki „Schlesisches Heim" pod kierunkiem Ernsta Maya na Dolnym Śląsku. *Kwart. Archit. Urban.* **2011**, *56*, 3–22.
43. Ohle, K. *Der Kreis Waldenburg im Nieder-schlesischen Industriegebiet in Vergangenheit und Gegenwart*; Korn: Breslau, Germany, 1927; p. 95.
44. Rogge, O. Bau, Wohnungs-und Siedlungswesen. In *Waldenburg*; Stein, E., Ed.; Monographien deutscher Städte; Deutscher Kommunal-Verlag: Berlin-Friedenau, Germany, 1925; Volume XVI, pp. 93–95.
45. May, E. Eine Bergarbeitersiedlung für Waldenburg. *Schles. Heim* **1920**, *1*, 3.
46. Treuhandstelle für Bergmannswohnstätten in Niederschlesischen Steinkohlbezirk. *Schles. Heim* **1920**, *1*, 1.
47. Hermann, J. *Allgemeines Lexikon der Bildenden Künstler*; Becker,, T., Ed.; E.A. Seemann: Leipzig, Germany, 1967; Volume XVIII, p. 398.
48. Hermann, J. *Wasmuths Lexikon des Baukunst*; Wasmuth: Berlin, Germany, 1931; Volume III, p. 272.
49. Stahl, F. *Ernst Pietrusky*; Ernst Hübsch Verlag: Berlin, Germany; Vienna, Austria, 1927.
50. Rogge, O. *Neubauten im Waldenburger Industriegebiet*; Deutsche Architektur-Bücherei G.M.B.H.: Berlin, Germany, 1930; p. 9.
51. Dobesz, J. *Wrocławska Architektura Spod Znaku Swastyki na Tle Budownictwa III Rzeszy*; Oficyna Wydawn; Politechniki Wrocławskiej: Wrocław, Poland, 1999; pp. 70–72, 105–106.
52. Schlesischer Wettbewerb. *Schles. Heim* **1920**, *1*, 19.
53. Protokoll der Sitzung des Preisgerichts ... Siedlung Nieder Hermsdorf. *Schles. Heim* **1920**, *1*, 14.
54. State Archives in Wrocław, Branch in Kamieniec Ząbkowicki (Kamieniec), Files of Sobięcin, 545–552.

55. Pietrusky, E. Bergmannsiedlung Nieder-Hermsdorf, district Waldenburg. *Schles. Heim* **1925**, *6*, 379, et seq..
56. Grygiel, D. Pro Publico Bono. "Aktywność Państwa i Samorządu Miejskiego w Realizacji Obiektów Użyteczności Publicznej na Terenie Wałbrzycha w Latach 1856–1945. Ph.D. Thesis, University of Wrocław, Politechnika Wrocławska, Wrocław, Poland, 2019.
57. May, E. Klein Wohnungs Typen. *Schles. Heim* **1920**, *1*, 14.
58. State Archives in Wrocław, Branch in Kamieniec Ząbkowicki, Acta specialia der Magistrat zu Waldenburg, [117], Aufstellung des Bebauungsplanes für die Siedlung am Kreiskrankenhaus (Stadtparksiedlung).
59. Postcard of Hermsdorf from 1932. Available online: https://polska-org.pl/4068894,foto.html?idEntity=565166 (accessed on 7 January 2021).
60. Photo of Hermsdorf form 1911. Available online: https://polska-org.pl/3869460,foto.html?idEntity=547753 (accessed on 7 January 2021).
61. State Archives in Wrocław, I/16482, Schlesisches Heimstätte, 1922–1925 , p. 88.
62. May, E. Siedlungspläne. *Schles. Heim* **1920**, *3*, 8–11.
63. Effenberger, T. *Allgemeines Lexikon der Bildenden Künstler des XX Jahrhunderts*; Vollmer, H., Ed.; E.A. Seemann: Berlin, Germany, 1931; Volume 2, p. 13.
64. Nielsen, C. *Theo Effenberger 1882–1968. Architect in Wroclaw and Berlin*; Hänsel-Hohenhausen: Egelsbach, Germany, 1999.
65. Ideen Wettbewerb. *Schles. Heim* **1920**, *1*, 18.
66. May, E. Besiedlungspläne für hügeliges Gelände. *Schles. Heim* **1921**, *2*, 353.
67. Krug. Die Bautätigkeit der Schlesischen Landgesellschaft in Kleinsiedlungssachen im J.1919. *Schles. Heim* **1920**, *1*, 11.
68. May, E. Kleinsiedlung Ober Salzbrunn. *Schles. Heim* **1920**, *1*, 9.
69. May, E. Typen für Mehrgeschossige Kleinwohnungsbauten. *Schles. Heim* **1920**, *1*, 9–12.
70. May, E. Kleinsiedlung Dittersbach. *Schles. Heim* **1920**, *1*, 9.
71. State Archives in Wrocław, Branch in Kamieniec Ząbkowicki, Akta m. Wałbrzych, 1955, Melchior Siedlung.
72. State Archives in Wrocław, Branch in Kamieniec Ząbkowicki, Akta gminy Wałbrzych Główny, 390, s. 2, 3.
73. May, E. Zweigeschossige Vier-und Sechsfamilien Wohnhäuser. *Schles. Heim* **1922**, *3*, 37–42.
74. State Archives in Wrocław, Branch in Kamieniec Ząbkowicki, Spezial Acten Polizei Verwaltung zu Gottesberg, Bausachen, Schuetzenstr. 9.
75. *Siedlungs-Wirtschaft: Mitteilungen der Siedlerschule Worpswede. Die Grüne Illustrierte* **1923–1928**, *Volume 1–6*. (Supplemente in *Schlesisches Heim*).
76. Wałbrzych: Na Sobięcinie Rośnie Jedno z Najstarszych Drzew w Polsce. Available online: https://walbrzych.naszemiasto.pl/walbrzych-na-sobiecinie-rosnie-jedno-z-najstarszych-drzew-w/ar/c8-5132515 (accessed on 10 February 2021).
77. Park im. T. Kościuszki w Wałbrzychu. Available online: http://www.pomniki-przyrody.pl/?p=4497 (accessed on 10 February 2021).
78. Pomniki Przyrody w Wałbrzychu—Sobięcinie. Available online: http://www.pomniki-przyrody.pl/?p=4567 (accessed on 10 February 2021).

Article

Practical Functioning of a Sustainable Urban Complex with a Park—The Case Study of Stavros Niarchos Foundation Cultural Center in Athens

Beata Makowska

Faculty of Architecture, Cracow University of Technology CUT, Warszawska 24, 31-155 Krakow, Poland; bmakowska@pk.edu.pl

Abstract: Intensive urban development has created a shortage of urban green areas. The need to economically plan and use urban green spaces has fueled the redefinition of public spaces and parks so as to provide the residents with both recreation and relaxation facilities, as well as a forum for contact with culture. This paper discusses the case of the Stavros Niarchos Foundation Cultural Center (SNFCC) in the Kallithea district on the outskirts of Athens, near the Mediterranean Sea. It fills a gap in the research on the aspects of the practical functioning of such facilities. The methodology used in the research included an analysis of the literature, the SNFCC's reports, and an in situ survey. The cultural center hosts a number of events aimed at promoting Greece's natural and cultural heritage. The paper includes a detailed analysis of the events organized by the SNFCC in the period 2017–2020 and their immense impact on residents. The aim of the study is to show that the creation of the SNFCC with the park areas has functioned as a factor contributing to the improvement of the quality of urban space and the quality of life of the city's inhabitants. The paper's conclusions indicate that the sustainable SNFCC project, which fulfils the urban ecology criteria, has been very well received by the visitors—citizens and tourists alike. A program-centered innovation introduced by the SN Park has added great value to their lives. The project contributes to economic and cultural growth, as well as the protection and promotion of heritage.

Keywords: urban park; sustainable cultural complex; cultural heritage

Citation: Makowska, B. Practical Functioning of a Sustainable Urban Complex with a Park—The Case Study of Stavros Niarchos Foundation Cultural Center in Athens. *Sustainability* **2021**, *13*, 5071. https://doi.org/10.3390/su13095071

Academic Editors: András Reith and Jan K. Kazak

Received: 8 March 2021
Accepted: 27 April 2021
Published: 30 April 2021

1. Introduction

Rapid urbanization is one of the main drivers of environmental change. Therefore, flexible and sustainable development of contemporary cities is a key element, which can provide a better place for living and recreation [1]. As the space for green areas becomes ever scarcer in the cities, it has become necessary to introduce an innovative approach to the design of parks [2,3], whose impact on the improvement on the living conditions of residents has now been scientifically recognized [4]. Parks ensure a more balanced relationship to the natural environment in the urban context and improve both the quality of life of the city's residents and the quality of urban space [5,6]. Their aesthetic values boost the attractiveness of the working spaces, residential areas, and recreation sites. Parks may also encourage a variety of social activities—neighborhood meetings, direct interpersonal contacts, or the promotion of active and healthy lifestyles. Some recent implementations have shown that parks may attract tourists thanks to their avant-garde solutions and become major components affecting the image of cities, ensuring their profitability [7,8]. Moreover, they may also facilitate the integration of urban spaces, improving their quality and contributing to the activization of their surroundings [9,10]. The 21st century necessitates, once again, the unification of physical urban structures—buildings, spaces, streets, and green areas—with the cultural, social, and economic aspects of city life [11]. The new projects introducing greenery to crowded cities create spaces that are friendly, safe, and open to all residents. Architectural forms and parks linked to the cultural heritage of a specific region [12,13], as well as to its topography and natural elements, are extremely important for sustainable development and have an immense influence on the residents [14].

One excellent example of such a case is the Stavros Niarchos Foundation Cultural Center (SNFCC) project analyzed in this paper. The project allowed for the reclamation of a degraded area. It also merged adjacent spaces in Athens into a coherent urban structure, preventing their fragmentation. The new buildings and the park have blended well into the urban landscape which may be admired by visitors thanks to the observation decks and sightseeing platforms. The links created with the existing vegetation and culture promote the socialization of residents and strengthen their sense of belonging to the place. This innovative approach to the creation of recreational and educational space may be referred to as 'urban fabric remodellation'. One of the main goals of the SNFCC's interdisciplinary project was to reconnect the residential area with the sea, separated by the highway, and create a new type of open public area within the park. The site has functional linkages with sport facilities in the neighborhood (the Kallithea Municipal Swimming Pool, among others) and compositional reference to the Acropolis, which emphasizes the north–south axis towards the sea.

The initial idea of SNFCC project (proj. Renzo Piano Building Workshop, 2012–2016) was the creation of a hill and buildings as a—topos—(in Greek: tópos koinós), a significant place which 'cannot be created, but rather becomes' [15]. For this purpose, Renzo Piano designed a park stretching over the rooftops of the associated cultural facilities, thus creating a special place that establishes a spatial dialogue with the Acropolis hill. The complex consists of the National Library of Greece (NLG) and the Greek National Opera (GNO), the Agora between them, and the Park. Renzo Piano has created an urban square surrounded by the most important buildings for culture—the library and the opera house—the 'two pillars of civilization' [15]. The buildings are translucent; thus, their function is visible from the outside. Their facades emphasize the ecological building performance. Characteristic for their structures is the lightness of the spaces and their transparency. The books at the National Library of Greece, the performances at the Greek National Opera, the cultural events, and the exhibitions taking place at the SNFCC all provide an insight into the Greek culture. It is a 'microcosm of the country' [15]. The interdependence of culture and nature within this interdisciplinary project has contributed to the popularizing of local heritage. The square (Agora) between the buildings with the canal open to the city creates an interesting multi-functional space for the organization of events (Figure 1). The agora has always had an important place in Greek culture.

The SNFCC in Athens was chosen as a case study as one of the best examples of a sustainable approach to design and management which enhances Greece's image internationally. The relevance of the SNFCC project has been confirmed by the award of the Platinum LEED certification—the highest rating possible in terms of green design and construction standards. This is the only public building complex in a park of this size and complexity in Europe [15]. It has exceptional environmental standards and has become a landmark of this Kallithea area [16].

The SNFCC and the other buildings of major cultural institutions with the LEED Platinum have radically changed our perception of the public space and the park, e.g., Vestas Technology and Development Center in Århus, Denmark; Water and Life Museums in Hemet, USA; Park Ventures in Bangkok, Thailand. Those projects help cities to improve their international competitiveness and attract foreign visitors.

The SNFCC structures and their resistant materials are environmentally sustainable and enable the minimal waste of water and energy resources. The designers used, for instance, trees formerly growing in this area, local materials, and those which came from the demolition of old buildings. Many local workers worked there, which was important in the context of the financial crisis related to Greece's debt. One great advantage of this project was also the flexibility in the designers' approach. They made many adjustments during the construction period in order to improve the quality of the complex. One example was the reduction in the canopy size and its construction costs (the total cost of the SNFCC reached EUR 630 million).

(a)

(b)

Figure 1. The SNFCC: (**a**) the Greek National Opera (GNO) and the National Library of Greece (NLG); (**b**) the Agora, photos by B. Makowska, 2019.

This paper fills a gap in the research on the details of some aspects of the practical functioning of the SNFCC. The study aims at filling some of the gaps in the existing research concerning the project and its relevance to the actual functioning of sustainable development in practice. It covers the issues related to the building structure (including a comparison of the canopy panels' efficiency against the mean temperatures, the number of daylight hours and sunshine periods in Athens) as well as the urban ecology and education programs carried out at the site. The study comprises a detailed analysis of the number and category of events organized by the SNFCC in the period 2017–2020. Based on the data available, the numbers of participants were identified, including those who took part in the events remotely in 2020, as a result of the difficult circumstances caused by the COVID-19 pandemic. The obtained results emphasize the importance of the project, its relevance to the residents of Athens and the immense impact scale. They may also popularize an excellent model of sustainable management, which can inspire other cultural and educational complexes.

2. Materials and Methods

The methodology used in the study included an analysis of the printed literature and online sources. It relied on the latest scientific publications concerning the SNFCC project and an analysis of the practical implementation of sustainability in urban design. The analyzed sources included, among others, the SNFCC's Reports [17], the Impact Detailed Report prepared by The Boston Consulting Group [18], and the SNFCC's monthly Booklets [19]. Additionally, a variety of reports concerning Athens were studied: the city population censuses, the climate, the greenery [20], and other documents which were associated with the protection of cultural and natural heritage of the city.

One important publication concerning the SNFCC is the book written by Victoria Newhouse in 2017 entitled '*Chaos and Culture: Renzo Piano Building Workshop and the Stavros Niarchos Foundation Cultural Center in Athens*'. The basic information is also available at the SNFCC's official Internet pages [. This study contains an analysis of publicly available reports published on that website. Many scientific papers raise the problem of the project's adaptation to sustainable design [21] and the use of innovations [22], the protection against flooding and earthquake resistance, including special columns in a base-isolation system as an element of seismic protection. Scientists have also undertaken research on the potential introduction of tickets to the Park based on the surveys among residents [23]. Other interesting studies concerned the improvement of the complex's accessibility by public transport from the city center and connecting the sites by building bikeways and pedestrian paths [24]—unfortunately, problems with soil erosion did not allow for the construction of an underground line which would connect the SNFCC with the center of Athens.

This paper is also based on in situ survey carried out in 2019. On that occasion, photographic documentation of the park and the architectural structures—including their interiors—was prepared during an excellent guided tour.

Thanks to the accessibility of data, this paper includes a chart illustrating the efficiency of the canopy panels in the period 2017–2019 as compared to the mean temperature throughout the year, the number of daylight hours, and sun exposure in Athens. Furthermore, an analysis of the changing number of events organized at the SNFCC and the visitors in the period 2017–2020 has been included. For this purpose, official reports were used, supported by a detailed analysis of the programming booklets prepared for every month. The results concerning the proportions between the specific groups of events organized at the SNFCC have been presented in tables and in charts. The charts have been contrasted against the programming assumptions of the SNF foundation regarding the financing of specific fields. The aim was to emphasize the transparency of the foundation's objectives and the practical implementation of its declared goals. The charts representing the proportions between all the events organized at the site allow us to draw conclusions regarding those that are in highest demand and most popular among residents.

3. The Studied Area of the Stavros Niarchos Park and the District of Kallithea

3.1. The Stavros Niarchos Park

The Stavros Niarchos Park (SN Park) is an integral part of the SNFCC which covers 85% of site. It was designed by Deborah Nevins & Associates according to Renzo Piano's scheme which emphasizes the movement toward the sea. It combines Italian, French, and English forms [25]. The park located on the building's roofs and the surrounding area offers spectacular panoramic views of Athens (Figure 2)—the sea and Faliro Bay, the mountains, Acropolis, and the city. According to Nevins, the park is a repository of the botanical culture of the region—in the same way as the library is 'a repository of Greek and Mediterranean culture' [15]. The project emphasizes the urgency of preserving indigenous endangered species. It also broadens knowledge about them and popularizes them in future public park design. Little-known Greek species were introduced there and used in a public park for the first time. The plants were adopted in order to cool down the paved area for public activities.

In this project, the key challenge involved the climatic extremes—especially between June and September—high temperatures with little rain on the one hand, and floods caused by downpours on the other. Such conditions created a hostile environment for plants. Another problem at the Park was the absence of a steady and reliable source of water and the inability to use the city sewage system, which was already operating at full capacity. That is why Nevins chose drought-resistant plants which were irrigated by roots. The second source of irrigation water is a canal filled with seawater, which is desalinated before use. The canal also serves as runoff in case of flood. Implemented at the SN Park, water management has significant impact on the ecology, e.g., use of rainwater, use of plants with lower irrigation requirements, replacement of surface irrigation with dripping irrigation, and the reuse of wastewater.

The Stavros Niarchos Foundation's aim was to create a flexible and freely accessible place, which would be open for easy programmatic use. Nevins designed a park which was inspired by the Greek landscape and reflected the soul of Greece—with characteristic olive trees, pines, native herbs, water, and stone. All the plants are low maintenance because they are drought-tolerant and wind-resistant (there are about 1400 trees and 310,000 shrubs and perennials). The weight was also an important factor in selecting plants located on buildings' roofs. Characteristic for Greek landscapes is the prevalence of iconic olive trees.

Nevins used them as focal points in the garden. One of the disadvantages of introducing olive trees in the Park is their slow growth—they may never grow tall enough to provide protection from the sun. It was also difficult to import and transport mature trees from abroad [25]. Nevins also had a problem with the soil, which needed careful preparation for successful planting, because the park is located on a hill.

(a)

(b)

Figure 2. Views from the terraces of the SNFCC's buildings: (**a**) view towards the north-east; (**b**) view towards the north; photos by B. Makowska, 2019.

The Great Lawn in the Park is a central place for recreation, concerts, festivals, films, and sport events. In the park, there is also a splash pool, labyrinth ('a meditation tool'), several children's playgrounds, and a green area arranged geometrically, mainly using endemic plants (Figure 3).

(a)

(b)

Figure 3. The Stavros Niarchos Park in Athens: (**a**) endemic plants; (**b**) shrubs and trees; photos by B. Makowska, 2019.

Nevins cited local agricultural forms (hedgerows and groves) with an ordered grid of orchards—arranged in straight and parallel rows (Figure 4).

1 Great Lawn
2 Pine Grove
3 Southern Walks (on the roof)
4 Labyrinth
5 Outdoor Gym
6 Running Track
7 Western Walks
8 Playground
9 Café
10 Splash Pool
11 Sound Garden
12 Mediterranean Garden
13 National Library of Greece
14 Agora
15 Greek National Opera
16 Lighthouse
17 Panoramic Steps
18 Canal
19 Esplanade
20 Small Agora
21 Outdoor Parking Lot

Figure 4. The map of the SNFCC (the green color marks the area of the SN Park and sport facilities on the north-west side, i.a., the Kallithea Municipal Swimming Pool)—based on: OpenStreetMap.

3.2. The Stavros Niarchos Park Versus Greenery in Athens

In 2016, there were only 9059 ha of green areas in Athens, which accounted for as little as 13.8% of the total area of the city (only green areas with a surface area in excess of 0.25 ha were taken into account) [20]. The estimated surface area of the city is 65,594 ha. The southern and south-western parts of the city have particularly little green areas, which is why the creation of the Stavros Niarchos Park is so important for the Kallithea residents (Figure 5).

It has a surface area of about 210,000 m^2. Even though in the context of the entire city it accounts for a small percentage of green areas (2.3%), the role of the park with its innovative and diverse educational, cultural, and recreational benefits cannot be underestimated. A highly flexible approach to adjusting the programming to the current issues is a major advantage of the institution. In 2018, it was estimated that there were 3.99 m^2 of park areas, 1.64 m^2 of forest and woodland, and 0.14 m^2 of maintained grass per capita. In total, this adds up to 5.77 m^2 per inhabitant [26]. Compared to other cities, this value is, unfortunately, quite low. Although there are differences in the total amount of urban green spaces in the cities in different datasets [27], Athens fares badly in many of them. According to the research carried out in 2017 in 43 European capitals, Athens ranks last from the perspective of urban green areas [28]. By contrast, in London, urban greenery is distributed more evenly throughout the city and is much more abundant (Figure 6)—it takes up ca. 40% of the urban area [29], which is triple the percentage recorded in Athens. In 2015, green urban spaces in larger cities in Poland (the total share of forests and woodlands, parks, green enclaves and urban greenery in residential areas) corresponded to 46% of the city area in the case of Katowice, 19% in the case of Warsaw, 18.5% in Poznań, 12% in Wrocław, with Krakow tailing with only 9% [30].

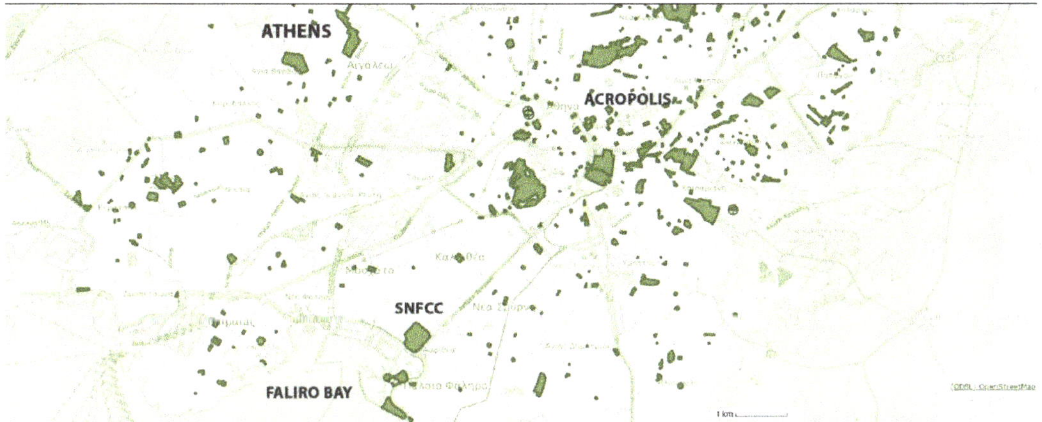

Figure 5. Greenery in Athens—based on: OpenStreetMap https://query2map.toolforge.org/queryinmap.php?name=*&key=leisure&value=park&types=points-areas&BBOX=23.5387,37.8909,23.9387,38.0909 (accessed on 2 September 2020).

(**a**) (**b**)

Figure 6. A comparison of green spaces: (**a**) in Athens; (**b**) in London—based on: OpenStreetMap https://query2map.toolforge.org/queryinmap.php?name=*&key=leisure&value=park&types=points-areas&BBOX=-0.32766,51.40732,0.07234,51.60732 (accessed on 2 September 2020).

According to the Impact Detailed Report FROM the Boston Consulting Group, the SN Park has 'doubled the green surface per capita in the area (including Kallithea, Moshato, Paleo Faliro and Nea Smirni, excluding cemeteries, football stadiums), bringing with it distinct health benefits (exercise and clean air)' [18]. It has already enhanced the local ecosystem and biodiversity. The SN Park supports biodiversity by plant selection, which creates a sustainable ecosystem, implementing green roofs which integrate the buildings with the Park (thermal and noise insulation of the opera and the library, rainwater collecting), using natural and recycled materials in pathways, playgrounds, etc. The benefits are already there: birds and bees have come back to the Kallithea area. The SN Park's vegetation has a positive impact on the micro-climate and decreases the temperature in the neighboring areas.

3.3. The District of Kallithea—Population and History

The district Kallithea is situated near the Faliro Bay, and its name means a 'beautiful view' (in Greek: Καλλιθέα). It is located in the south-western part of Athens (Figure 7).

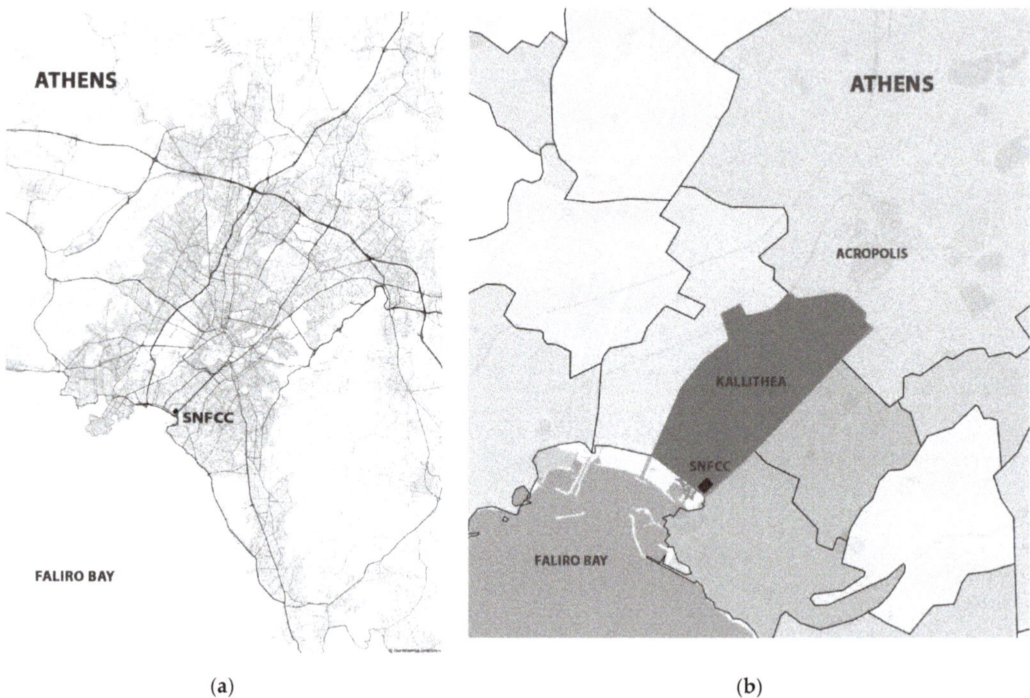

(a) (b)

Figure 7. The location of the SNFCC: (**a**) on the map of Athens—based on: OpenStreetMap https://wiki.openstreetmap.org/w/images/a/ad/Athens.pdf (accessed on 2 September 2020); (**b**) in the Kallithea district in Athens.

Before the SNFCC was created in the Kallithea district, there were no open green spaces there, except the Park near the Church of Metamorphosis Sotiros (Ieros Naos Metamorfosis Sotiros), the small green area near the corner of Praxiitelous and Agisilaos streets, and the tiny playground for children, Paidiki Xara (Παιδική Χαρά), located between Platanos and Achilleos streets. Nowadays, SN Park is the key and most important green area of the district. In order to present the researched issues in a broader context, an analysis of the fluctuations in the number of residents and the population density in the Kallithea district where the SNFCC is located was carried out for the period 1981–2020. For this purpose, the data available from the General Secretariat of National Statistical Service of Greece, i.e., population censuses, were used [31]. Kallithea, with an area of 4.75 km^2, is now among the most densely populated districts in Athens [32]. In the years 1928–1940, the density in this district was estimated at 5–10 inhabitants/km^2; in the years 1951–1961, between 10 and 15 inhabitants/km^2; in the year 1971, at over 20 inhabitants/km^2; and in the period 1981–2001, the density ratio reached almost 25 inhabitants/km^2 [33]. The density slightly decreased in the year 2011, it reached 21.188 inhabitants/km^2 (Table 1).

Table 1. Population and density of the Kallithea district [1].

Year	Population of the Kallithea District	Density (Habitants/km^2)
1981	117,319	24,699
1991 [2]	116,731	24,575
2001 [3]	115,150	24,242
2011 [4]	100,641	21,188
2020 [5]	100,641	21,188

[1] Source: ELSTAT, Population censuses, https://panorama.statistics.gr/ (accessed on 2 September 2020); [2] According to the population census from 17 March 1991; [3] According to the population census from 18 March 2001; [4] According to the population census from 16 March 2011; [5] Source: https://worldpopulationreview.com/countries/cities/Greece (accessed on 2 September 2020).

It is estimated that more than 250,000 people live in Kallithea or are passing by (among them the SNFCC visitors) or working there on a daily basis. The fluctuating population density of the Kallithea district is related to its history. This district was established by the Euthymios Kehagias' Building Company in 1884 [34]. At that time, many houses there were designed by the German architect Ernst Ziller (1837–1923). At the turn of the 19th and 20th centuries, the areas at the Faliro Bay were a popular recreation site among residents of the capital (Figure 8). Water (the Faliro Bay) and greenery (the SN Park) are key elements of creating a good quality housing environment in the nearby Kallithea (Figure 8).

(a)

(b)

Figure 8. Faliro Bay in Athens: (**a**) a postcard from July 1901, source: https://twitter.com/CCavafy/status/1020702785262817280/photo/1 (accessed on 2 September 2020); (**b**) a postcard from about 1900, source: https://www.pinterest.it/pin/346495765057211224/ (accessed on 2 September 2020).

In that period, the number of residents of the district gradually increased. This process was fueled, among other things, by the construction of sports facilities and the organization of the Olympic games—the Summer Olympics took place at the Kallithea Skopefterion in 1896 [35]. Unfortunately, later, the construction of a motorway cut this part of the shore off from the city, decreasing the district's attractiveness. The SNFCC project restores the link between the residential areas and the sea, making it a popular recreation site, just like at the turn of the 19th and 20th centuries.

The Kallithea district has also attracted numerous immigrant residents—'Kallithea has a variety of population groups that settled here from the early 19th century up to nowadays. Most of them are expatriate—refugees who return to Greece from Asia Minor and the coastal areas of the Black Sea as well as expatriate and immigrants from Russia, Czech Republic and other eastern European countries. All these groups have a quite diverse cultural background that gives Kallithea a quite multicultural life' [36]. In the 20th century, Kallithea became a major center of artistic and intellectual activity. Many renowned writers, poets, painters, and sculptors have lived there.

4. 'Sustainability' of the SNFCC's Structure—The Efficiency of the Canopy Panels

The sustainable approach to design can be observed in the structure of the GNO and NLG buildings, as well as in introducing Ecosystem Services which moderate natural phenomena, i.a., flood control, water purification and climate regulation. All systems at the SNFCC have been designed to save energy, such as heating, air conditioning, or lighting. One of them is the innovative ferro-cement construction which was used to create the canopy above the Opera building. It supports technologically advanced solar panels. The canopy panels with 87,000 square feet of photovoltaic cells are expected to meet 15% of the facility's demand for electricity [15]. According to information from the SNFCC's official page, the canopy can cover 100% of energy needs—although only at certain times (e.g., depending on the number of visitors) [19]. The technologies applied allowed for the integration of the form and materials. The handmade methods of construction required sophisticated analysis within a millimeter of tolerance.

To better illustrate the functioning of the sustainable project in practice, this paper includes a comparison of the canopy panels' effectiveness in particular months. Based on the data available on the SNFCC website, a comparative analysis of the panels' efficiency in the period 2017–2019 was carried out (Figure 9). The paper compares the range of efficiency of the panels against the mean temperatures in Athens [37] during the year (Figure 10). A clear dependence between these factors is observable in the similarities of the charts onto which the data have been plotted.

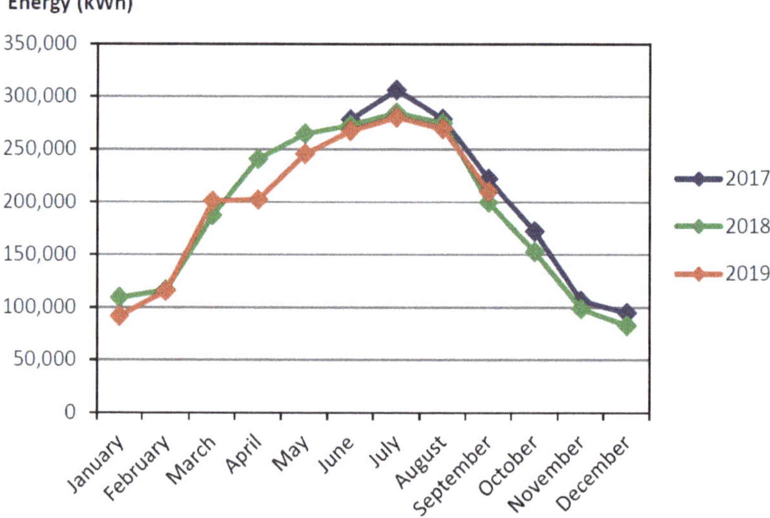

Figure 9. Energy (kWh) from the canopy panels during each month in the years 2017–2019—based on: https://www.snfcc.org/en/sustainability-hub (accessed on 2 September 2020).

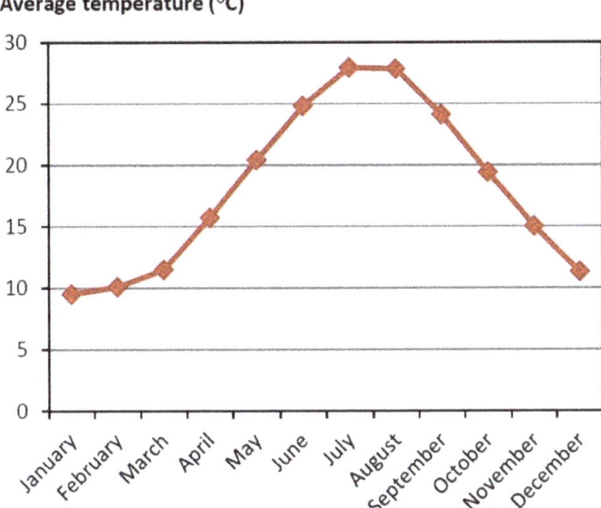

Figure 10. Average temperature in Athens during each month—based on: https://en.climate-data.org/europe/greece/athens/athens-7/ (accessed on 2 September 2020) [37].

Likewise, the chart illustrating the changes to the length of the daylight hours and sunshine hours throughout the year (Figure 11) shows similarity to the chart presenting the efficiency of solar panels, which is closely related to sunshine hours [38]. In Athens, June is the month with the longest days (with average daylight of 14.8 h), and December is the month with the shortest days (with average daylight of 9.6 h). July and August are the months with the most sunshine (with average sunshine time of 12 h), and January is the month with the least sunshine (with an average of 5.6 h).

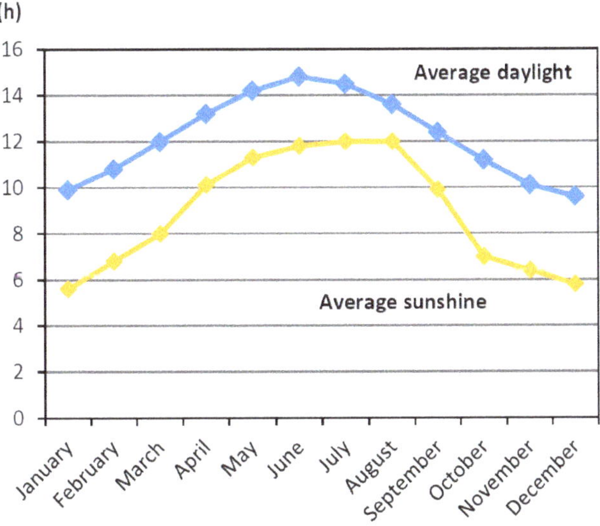

Figure 11. Average daylight hours and sunshine hours in Athens—based on: https://www.weather-atlas.com/en/greece/athens-climate#daylight_sunshine (accessed on 2 September 2020).

The foregoing charts show a clear relationship between the efficiency of the canopy panels and the mean temperature and the length of the sunshine hours throughout the year. Given the global trend of raising temperatures, one should expect that the efficiency of the panels will proportionally increase over the coming years. The increase in temperatures is particularly noticeable in Europe. According to the Copernicus Climate Change Service, December 2019 in Europe was more than 3.2 °C warmer than the December average for 1981–2010 (globally, the temperature was 0.7 °C higher) [39].

The SNFCC is characterized by transparency in terms of analyzing and presenting the data on sustainable design and approach. Its official website in 2017 disclosed the following data:

- The photovoltaic panels on the Energy Canopy were activated in May, meeting 27.98% of the SNFCC energy consumption needs from June to December 2017 [40];
- 'The operation of critical systems, such as air conditioning, were optimized to minimize power consumption. Natural gas consumption was reduced by 21% (comparison periods: May to December 2016 and 2017)' [17];
- 'Thanks to proper planning and streamlining, a 23% reduction in irrigation consumption was achieved' [19] (comparison periods: April to September 2016 and 2017).
- 'SNFCC's energy efficiency initiatives result in a 40% energy reduction compared to what a similar building complex would consume without their implementation' [40];
- 'Energy efficiency initiatives contribute to annual energy savings of 7.4 GWh, equivalent to ~2750 tons of CO_2 and ~EUR 0.6 million saved annually' [41].

5. Results

5.1. The Number of SNFCC's Visitors

This paper contains an analysis of the changing number of visitors to the SNFCC in the period 2017–2020. The data on the number of visitors presented in the table (Table 2) come from the official SNFCC website [42] and have been supplemented with the mean values calculated for the three years and the mean number of visitors per day. The estimated mean number of SNFCC visitors in the period 2017–2019 was more than 13,000 per day. The number of the SNFCC visitors is variable and depends on the season of the year, the day, and the events being organized. The site attracted the largest crowds during weekends (especially Saturdays and Sundays at ca. 18:00–19:00; the place was also slightly busier at lunchtime, ca. 13:00) as well as during the summer months and on the days of major events. The SNFCC provides real-time information on the visitor concentration levels every day. This allowed the complex to continue functioning in 2020 albeit in a limited way.

Table 2. Number of visitors in the SNFCC and average number of visitors per day.

Year	The SNFCC's Visitors	Estimate Average Number of Visitors Per Day
2017 [1]	3,100,000	3,100,000: 365 days = 8493
2018 [2]	5,300,000	5,300,000: 365 days = 14,521
2019 [3]	6,300,000	6,300,000: 365 days = 17,260
average 2017–2019	4,900,000	13,425

[1] From June to December 2017; [2] From January to December 2018 [42]; [3] From January to December 2019.

In the period 2017–2019, Christmas marked the peak in the number of guests at the SNFCC. For instance, at Christmas in December 2018, over 700 thousand visitors were recorded; on 30 November 2019, over 40 thousand visitors watched 'the Christmas World' celebration. As many as 34 thousand visitors welcomed 2020. A comparison of the number of residents of the Kallithea district with the number of the SNFCC visitors shows the broad scope of impact of the site, which has become a major recreation venue and an important tourist attraction, known to a broader public rather than just locally. Provided in the SN Park, Cultural Ecosystem Service (CES) benefits (e.g., recreation, physical and mental health, aesthetic enjoyment) can be essential for human wellbeing and can foster social cohesion. The growing number of visitors was certainly partially spurred by the strategic approach of the SNFCC administration. They have created a new paradigm of a public space—free and open to everyone, without any kind of barriers.

5.2. 'Sustainability' of the SNFCC's Program—The Number of Events and Their Role in Popularizing Cultural and Natural Heritage of Greece

The SNFCC interdisciplinary project plays an important role in popularizing the cultural and natural heritage of Greece. It also offers green education—especially for children and youth [43]. The cultural center was built using funds from the exclusive grant donated by the Stavros Niarchos Foundation. Most of events organized there are free for the public and open to everybody, thanks to an exclusive grant by this foundation. The educational activities and learning opportunities are addressed to people of all ages, especially those who are experiencing financial and social difficulties, thus enabling them to have access to culture—'SNF is dedicated to fostering the vibrancy of civic life, from widely shared access to stellar arts programming, to open dialogue across divides' [44]. Its principle 'focuses on providing cultural stimuli and opportunities for entertainment and learning to people of all ages, while removing economic, social and other barriers to public access to culture' [45].

The goal of the SNFCC program is to deepen the bond between education, society, and creativity. The aim is also to develop a creative society in order to promote the pursuit of a variety of careers in Greece. The courses offered at the center give young and older people an opportunity to access education and learn new competences which are coveted in the job market and useful in daily life. As a result, the center stimulates the engagement of residents in the economy. The SNFCC's educational programs (including art, architecture, nature, and technology workshops) also encourage participants to broaden their range of interests and passions, as well as discover their potential. The program also helps to increase Greece's cultural index—as many as 63% of inhabitants rank 'low', and only 4% rank 'high' in terms of cultural participation [18].

In order to discuss the sustainable and flexible approach assumed in the SNFCC program ('sustainable' program), this paper offers a detailed analysis of the number of events organized at the site divided into three major categories: Arts and Culture, Education, and Sport and Wellbeing/Health. The first category is further divided into the following subgroups: music/concerts, cinema, theatre/performances, exhibitions, dance performances and festival/carnival/conference/forum. A variety of courses are offered within each of the subgroups (Table 3).

An integral part of the SNFCC's mission is the promotion of healthy lifestyles. Among the sports facilities, there are: 200 m of linear running track, 300 m of elliptical running track and 2 km of cycle routes (including Evripidou St.). One particularly noteworthy activity is the daily football training sessions organized in the afternoons and in the evenings which—as in the case of other cities—are offered as a measure of preventing criminal activity among the youth. These events are very popular.

The analysis of the events organized at the SNFCC has been carried out based on the programming prepared for every month and published on the center's official website. The set refers to all events in the aggregate (Table 4), taking account of cyclical and periodical activities (e.g., sport activities organized 2–6 times per week, several times a day) and—in the case of online activities—those viewed multiple times. The analysis covers the period from 2017 until the end of 2020. Unfortunately, in 2020, the coronavirus pandemic upended the functioning of many cultural facilities and parks [46]. From March to June 2020, many events organized by the SNFCC were cancelled due to the pandemic restrictions (lockdown). Fortunately, the SN Park remained open to the public during 2020 with some limitations. Many events—as is the case with many other cultural facilities—are now being organized via the Internet (data updated as of 31 December 2020).

Table 3. Selection of courses offered by the SNFCC.

Arts and Culture	Education	Sport and Wellbeing/Health
• Contemporary art exhibitions; • Different genres and styles of music concerts; • Theatre performances; • Ballet performances; • Park Your Cinema and Park Your Cinema Kids; • Festivals; • Conferences.	• Courses for mothers to inspire their kids' reading; • Programs for students of different ages; • Literature programs designed for the elderly; • Dialogues—series of lectures which delve into basic themes of the Foundations: Arts and Culture, Education, Health and Sports, and Social Welfare; • Social Ballroom is a combination of dance workshops and parties; • Art studio; • Meetings for bibliophiles; • Lectures and workshops on botanical and environmental topics, e.g., tree recognition, Organic Agriculture and Food Security, Experiential workshops for the arts and sustainable development, We make our own compost—we select the suitable soil for our plants, Our Land, The Plants that Care for Us: Spring, A Seed Travels; • Vegetables and Herbs; • Lifelong Learning—courses for elderly; • Cross-cultural Choir; • Panel discussions.	• Gardening courses for the family; • Athletic classes and facilities; • Playgrounds for children; • Daily sports classes and activities; • Walk in the Park, Sundays in the Park; • European Mobility Week; • Team Playing.

Table 4. Number of all events in the years 2017–2020.

Year	2017	2018	2019	2020
Music, concert	56	62	68	25
Cinema (Park Your Cinema and Park Your Cinema Kids)	44	45	24	6
Theatre, performance, spectacle	125	217	245	45
Exhibitions	7	12	20	13
Dance—performance	10	9	23	2
Festival, carnival, conference, forum	**6**	**5**	**11**	**4**
Talks/learning, lecture, reading club, reading seminar, course and workshop, etc.	**774**	**1431**	**904**	**342**
Sport and wellbeing, walk in the park, virtual tour in the park, gardening for the family in the Park, etc.	**1452**	**2547**	**3124**	**978**
Total events (with total number of recurring events)	**2474**	**4328**	**4419**	**1415**

Based on data from 2019, it is estimated that in 2020 there were 336,015 visitors—about 308,360 visitors in the period between January and 13 March 2020 attending the 'live' events and about 27,655 viewers at home via the Internet (Table 5). This figure corresponds to as little as 34.9% of the total number of visitors, and to 37.8% as compared to the mean value for the period 2017–2019. The table below (Table 6) contains detailed data on all the events prepared by the SNFCC and released on the Internet (data accessed: 1 January 2021, the updated data are given in brackets).

Table 5. Number of events and number of participants.

Year	2017	2018	2019	2020 [1]
Number of events	2474	4328	4419	1415
Number of participants [2]	258,973	713,667	963,000	~336,015 [3]

[1] From January to 13 March 2020 (from 13 March to 21 June the SNFCC was closed), from 22 June to December 2020 the SNFCC functions were mainly online; [2] Source: https://www.snfcc.org/en/news/stavros-niarchos-foundation-cultural-center-three-successful-years-63-million-visitors-2019 (accessed on 2 September 2020); [3] Estimated data made on the basis of 2019.

Table 6. List of all videos watched on the Internet in 2020.

Event	Number of Films on the Internet	Number of Viewers
Music, concerts	22 films (29 December 2020)	6109 views
Cinema	0	0
Theater and performances	6 films (4 December 2020)	1696 views
Exhibitions	0	0
Dance	0	0
Festival, carnival, conference	0	0
Talks/learning, lectures, reading club, reading, etc.	Reading for kids: 43 films (22 December 2020) 76 films (18 December 2020) Educational program for kids: 88 films (18 December 2020)	 2786 views 2333 views 4621 views
Sport and wellbeing, walk in the park, virtual tour in the park	Tour of the SNFCC: 8 films (5 November 2020) walk in the Park: 3 films (3 November 2020) Sport [1]: 57 films (9 November 2020)	 898 views 550 views 8662 views [1]
Total	303	27,655

[1] Detailed information in Table 7.

Table 7. List of sports videos watched on the Internet in 2020.

Year: 2020	Yoga	Pilates	Tai Chi	Fit Kids	Exercise and Health	Total
Number views	2270	3268	991	1009	1124	8662
Number of films	12	15	6	12	12	57

Table 7 contains detailed data on sports activities released for the viewers to watch on the Internet (data accessed: 1 January 2021, updated as of 9 November 2020). The most frequently viewed events were sports and concerts.

The survey results can serve as a basis for the following conclusions:

- The proportions of grants donated by the SNF (Figure 12) in the various areas (Art and Culture—30%, Education—32% and Health and Social Welfare—the last two areas account for 38% jointly) are comparable to the proportions between the three categories of events organized by the SNFCC over the years (Figure 15): Art and Culture (from 28.7% to 36.1%), Education (from 21% to 34.5%) and Sport/Wellbeing/Health (between 29.7% and 42.9%). This confirms the consistent policy of the foundation and a conscientious approach to full transparency in financial management [47];
- The obtained data on all events and the proportions between the three distinguished categories (Figures 13 and 16) highlight those which enjoy the most popularity and provide a response to the needs of the community. The particularly high number of sports events (in 2019, they accounted for 70.5%, while in 2020 for 69.1% of all events) confirms the growing interest in the facility and the increased frequency of visits to the park. Higher demand for educational services confirms the immense contribution by the SNFCC in terms of extending the residents' knowledge and inspiring their

interests. This applies, for instance, to the assistance targeted at seniors (computer training, activities for persons experiencing problems with memory, etc.). The SNFCC project was designed to achieve a broad, lasting, and positive social impact. Its diverse cultural, educational, and sports programs have particularly focused on the needs of the young and the elderly people as a way of ensuring public welfare [48,49]. Such activities are highly relevant because of the effects of the crisis in Greece;

- The creation of the SNFCC with the Park is a factor contributing to the improvement of the quality of residents' life (health and wellbeing); for instance, by offering sports activities, which have a positive effect on physical and mental health, general fitness, mobility, and functionality. The prevalence of sports events in the scheme (Figure 14) shows that many residents have been persuaded to adopt a healthy lifestyle and engage in regular exercise;
- The social and cultural aspects of the sustainable urban improvements are fulfilled during practical functioning of the SNFCC. Its program focusses on providing cultural and ecological stimuli and opportunities for learning to everybody;
- Unfortunately, the COVID-19 pandemic has changed the way in which the citizens use the SNFCC complex. The gradual increase in the number of the events organized by the SNFCC in the period 2017–2019 was unfortunately followed by a drop in 2020—the number of events corresponded to only 37.8% of the average value for 2017–2019. The lectures organized in 2020 touched upon pertinent topics and aimed at helping the residents whose mental health [50] had deteriorated as a result of the pandemic, i.e., 'Challenges in an ever-changing landscape', 'The limitation of our social life', 'Mentally resistant during quarantine'. Owing to the flexible approach, program modification, and its releases on the Internet, the SNFCC has received subsequent awards in 2020 for its operations.

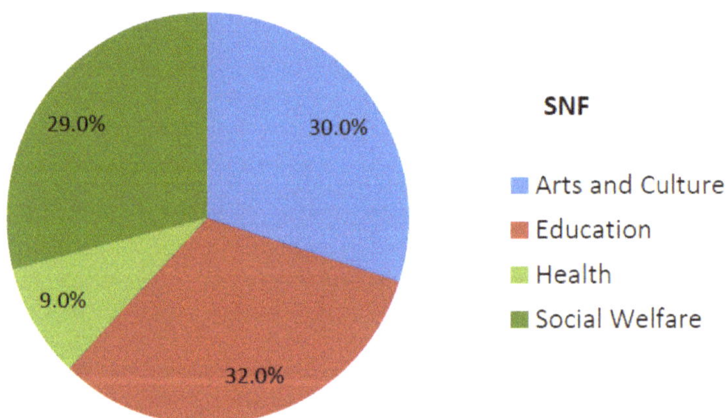

Figure 12. The Stavros Niarchos Foundation's grants in different fields, source: https://www.slideshare.net/Europeana/presentation-it-europeana-cloud-plenary18314 (accessed on 3 April 2014).

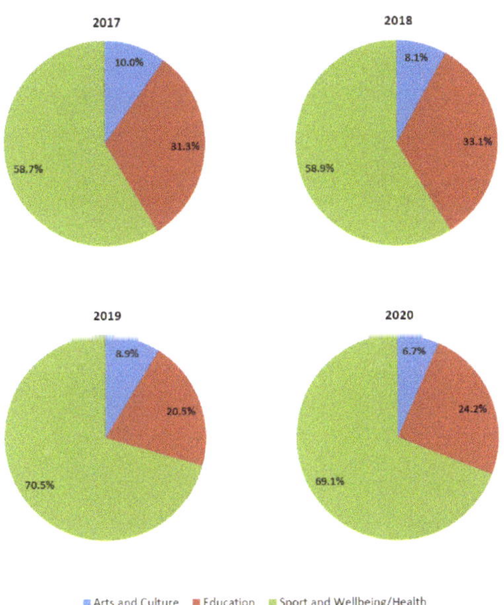

Figure 13. All events—the relationship between 3 groups of events: Art and Culture, Education, Sport and Wellbeing/Health.

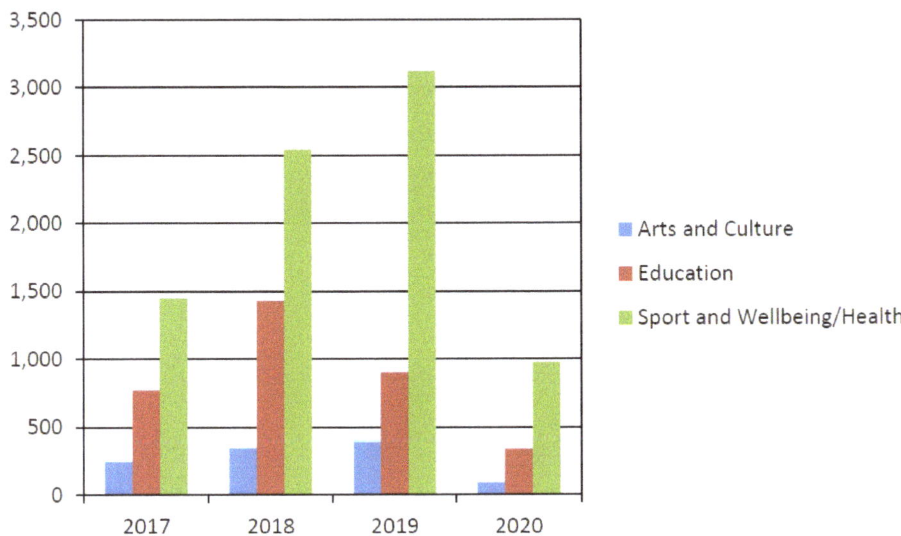

Figure 14. Number of 3 groups of all events: Art and Culture, Education, Sport and Wellbeing/Health during 2017–2020.

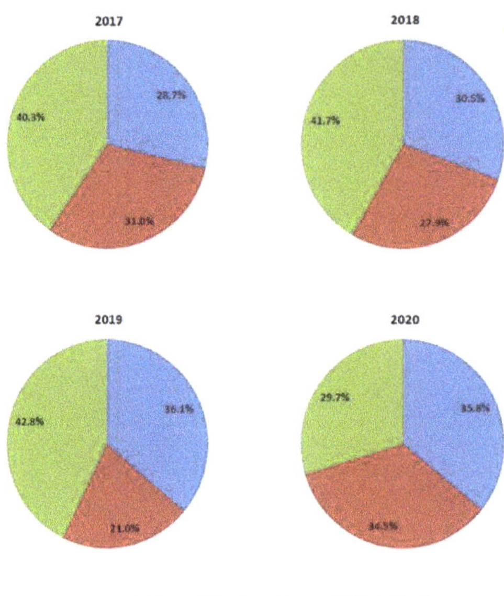

Figure 15. Event units—the relationship between 3 groups of events: Art and Culture, Education, Sport and Wellbeing/Health.

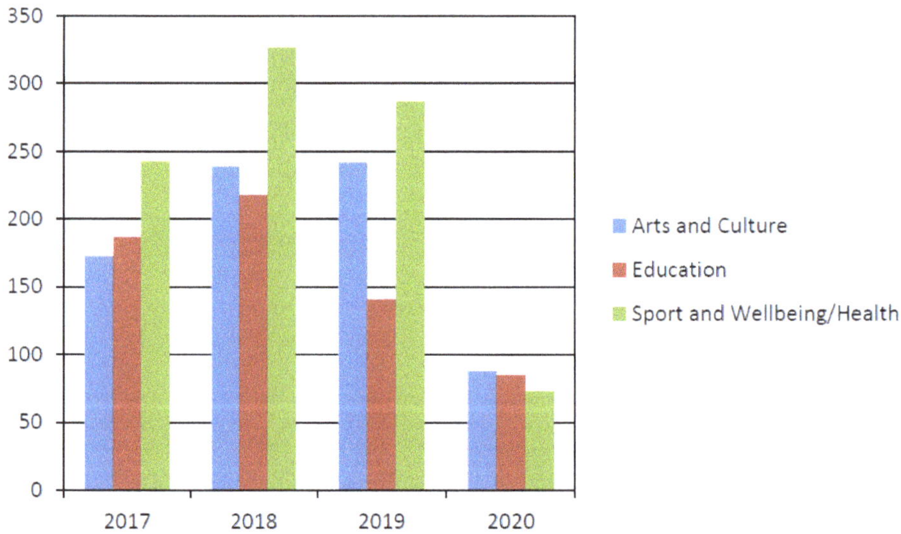

Figure 16. Number of 3 groups of event units: Art and Culture, Education, Sport and Wellbeing/Health during 2017–2020.

6. Discussion

This paper fills a gap in the research on the details of the SNFCC's practical functioning. Its contribution fills some of the gaps in the existing research on the importance of sustainable design from the perspective of practical functioning of sustainable facilities, focusing on the example of the canopy panels as well as the sustainable urban ecology program. The reports on these matters for specific years (except for 2020) are available on the official SNFCC website. However, there are no aggregate studies for the period

2017–2019 or for 2020. The presentation and comparison of the figures regarding the events organized by the SNFCC illustrates the immense scale of the facility's impact on the visitors. The aim of the study was to show that the creation of the SNFCC with the park areas is a factor contributing to the improvement of the quality of urban space and the quality of life of nearby residents. Program-centered innovation introduced by the SN Park is highly valuable to them. The project contributes to economic and cultural development as well as to the protection and promotion of heritage.

The SNFCC in Athens was chosen as a case study as one of the best examples of a sustainable approach to design and management which enhances Greece's image internationally. The program of the buildings is constantly changing according to users' preferences and requirements, moving them to a new complex era [51–53]. In January 2016, it was expected to be a landmark of inclusion which would enrich the area in terms of Greece's cultural and educational legacy and improve the quality of life for the local community. The results of the study fully confirm the expectations regarding the institution. Before the project was made available for use, the investors assumed that the annual number of SNFCC visitors would reach 700,000 (the Operational Plan). The actual number of visitors and participants of events organized at the site has significantly exceeded these original expectations (average 4,900,000 visitors per year from 2017 to 2019). Undoubtedly, one of the key reasons has been the extensive, continuously updated, and modified cultural–educational–recreational aspects, mostly involving events accessible to the public free of charge. The research substantiates a strong conclusion that the collective benefits from the SNFCC operations are huge and will pay dividends in the future. The extensive, 'sustainable' programming gives participants a chance for development and improves their chances of landing a good job. Importantly, it also provides much needed support during the difficult period of the pandemic. The results of this research confirm the importance of the building and its significant impact range.

Since 2018, the SNFCC has conducted many public surveys among Attica residents and visitors. The purpose has been the periodical evaluation of the SNFCC's public image and the quality of service. In 2019, QED market research in Greece (QED) conducted a survey among 3000 people—93% of the residents of Attica were satisfied by the services provided to visitors, and 99% of them had a positive view of the SNFCC, and as many as 91% of them considered the SNFCC to be an open and accessible place, which improved the quality of residents' lives and enriched the cultural and natural landscape. The annual survey was carried out through telephone interviews, as well as with visitors to the SNFCC's site. The mentioned survey also indicated that one in two young residents of Attica (over the age of 18) had already visited this place at least once. People selected during research declared that they had visited the SNFCC on average five times [42]. One in two respondents said that they had been there at least once in 2018, while 99% positively evaluated the venue and its cleaning services [42]. The most frequent visitors were families and people aged between 18 and 44, who were well-educated and predominantly residents of the southern suburbs [47]. The QED survey in 2018 showed that 50% of the visitors were Athenians (which gives the sum of 2.650 million people among 5.3 million visitors in 2018) [54]. The SNFCC's important aim was fulfilled: creation of a place where the residents of Attica can celebrate important occasions, which can help to build a solid relationship with the local community [55]. The other aim promoting this place as a landmark of contemporary Athens abroad was also fulfilled. According to the survey, tourists in Athens considered the SNFCC a key cultural destination [50]. It became the second most visited place in Greece after the Acropolis [56]. Thanks to collaborations with tourism organization, the numbers of visitors have increased, especially in the summer months. According to the SNFCC's visitor statistics in 2019 (by comparison to 2018), there was a 47% growth rate in June and 43% in July. Unfortunately, in the SNFCC's Reports, there is no distinction between the proportion of visitors coming from: Kallithea district, other areas in Athens, elsewhere in Greece, or other countries. According to the survey in 2017, 18,027 children and 22,500 adults participated in the free 'Sports and Wellness' activities organized at the

SNFCC. In 2017, 6378 adults participated in educational courses on architecture, nature, the arts, and science and technology, including computer courses developing digital skills for users aged over 65 (4060 persons participated in such courses in 2017). The residents of Attica especially benefit from the SNFCC's cultural and educational programs. According to the survey carried out by the SNFCC in 2019, 50% of the residents have already visited this place and over 90% of Attica residents believe that it improves the quality of life and cultural standards of the city [19]. The research approach adopted for the purposes of this paper differed from the resident survey methodology. The study prepared by the author involved a detailed analysis of the events organized at the SNFCC and the number of visitors. The data on the solar panels' efficiency have been extracted from the reports drafted by the SNFCC. However, the data were presented in a different way (one chart illustrated the figures for the period 2017–2019) compared against the mean temperatures, the daylight hours, and sunshine exposure in Athens.

Certain limitations have hindered the research; for instance, the availability of information and the assumption of a certain degree of generalization of data on the number of the SNFCC visitors and the events organized online. This may have resulted in certain differences in terms of figures collected by the author (Table 5) as compared to the official data (Table 8) from the SNFCC report from the end of 2019. Unfortunately, there is no analogous information about the number of events in 2020.

Table 8. Number of events and activities [1].

Year	2017	2018	2019
Number of events and activities	2942	3743	3608

[1] Source: https://www.snfcc.org/en/news/stavros-niarchos-foundation-cultural-center-three-successful-years-63-million-visitors-2019 (accessed on 2 September 2020) [43].

The classification of events organized at the SNFCC is not unequivocal. In the classification prepared by the author, conferences, congresses, and festivals were allocated to the Arts and Culture group (rather than to Education), while activities in the garden and the walks were classified to Sport and Wellbeing/Health (rather than as workshops, because they involve physical activity outdoors and 'dynamic' contact with nature). By contrast, 'closed' events organized for schools were not included, because information on such activities is not publicly available. Moreover, there is also no information on the cancellation of planned events publicized in the monthly booklets for the period 2017–2019 (in 2020, during the pandemic, the cancellations were announced); for the purposes of the study, it was assumed that these events did take place. However, this study did not include the guided tours, which are also highly popular: more than 1800 free guided tours were organized in Greek and in English (mainly dedicated to foreign tourists) in 2017. Additionally, one source could not be found online: the Booklet from January 2017—the lack of the data for one month constitutes 2.8% of the total amount of 36 months which were analyzed. Insofar as the number of SNFCC visitors was concerned, the publicly available data may be generalized—the visitors do not necessarily engage in the events organized at the site and cannot be included for the purpose of the analysis.

Regarding the remaining limitations, one should mention:

- The lack of updated data on the population of the Kallithea district in 2020—the data from 10 years ago are still being used (from the last the population census from 16 March 2011);
- No detailed information on the efficiency of the panels from October 2019 to December 2020 or on the number of persons visiting the site in 2020 could be obtained from the SNFCC;
- The limited availability of information in the English language.

The role of a contemporary urban park has undergone a major transformation. Intended to serve as a refuge from city noise in the past, urban parks are now being designed to ensure the fullest possible integration with the city, and aspire to engage the residents by providing them with cultural and recreational benefits [57,58]. This role is particularly pertinent given the diminishing urban green areas and the difficult epidemiological situation. The authors have already conducted research focused on the problem of deficits of green areas in cities. The publications have touched upon the search for the new forms and technologies applied in urban gardens occupying a limited space, including green walls and roofs [59,60]. The innovative parks integrate the urban tissue, improve its quality, and contribute to the revival of the adjacent areas. Often—as in the case of the SNFCC—they comprehensively combine a variety of auxiliary functions in their vast programming offers. The introduction of such an innovative cultural–educational program at the SNFCC has significantly changed our perception of an urban park. It has drawn our attention to the social issues and the necessity to ensure good living conditions to the residents of the local community. The SNF grants and public funding have been allocated to meet the needs of regular people (including those facing exclusion as a result of their lower social status or country of origin) and are not used to generate profits for private institutions and investors.

The COVID-19 pandemic has already changed the perspective and the way in which we use our public spaces and parks. In central and eastern Europe, the number of visitors is declining. The visitor volume decreased to as little as 40–50% of the 2018–2019 numbers [46]. The number of persons visiting the SNFCC in 2020 (including the 'virtual' visits) corresponded to ca. 52% as compared to the mean numbers of 2017–2019. The findings of this study pave the way for further investigations. The problem of sustainable development [61] and the adjustment of programming by cultural institutions and parks to the dynamic situation is important, and thus further research is planned in this area. The SNFCC and the other buildings of major cultural institutions with the LEED Platinum have radically changed our perception of the public space and the park, e.g., Vestas Technology and Development Center in Århus, Denmark; Museum of Water and Life in Hemet, USA; Park Ventures in Bangkok, Thailand. Those projects help cities to improve their international competitiveness and attract foreign visitors [58].

7. Conclusions

The research results justify the following conclusions:

- The SNFCC interdisciplinary project plays an important role in popularizing the cultural and natural heritage of Greece;
- The fact that the SNFCC project has met high ecological standards makes it an endeavor of European relevance. It is one of the leading examples of environmental sustainability across three levels: design and construction, operation, footprint in the local ecosystem and social welfare ('sustainable' program). The best practices in terms of environmental standards implemented at the SNFCC guarantee the sustainability of its operations. The ecological aspects of sustainability performed in the SNFCC contribute to stability of the climate in the neighborhood, improvement in the quality of air, and the renewal of biodiversity. Tangible benefits on the local ecosystem are already apparent in the Park. The Kallithea district has become a more inclusive, safe, and sustainable settlement for inhabitants;
- The sustainable project of the SNFCC, which fulfils the urban ecology criteria, has been very well received by the visitors—citizens and tourists alike. The sustainability of this complex extends beyond the structure of the buildings and the SN Park. The SNFCC's practical functioning thanks to the 'sustainable programme' contributes to other benefits such as the health and wellbeing of residents, quality and accessibility of education, and reduction in inequalities [56]. It offers green and cultural education to all visitors, especially children and youth;
- The development of high-quality buildings and the SN Park is a factor contributing to the improvement of the quality of urban space and the quality of residents' lives. It also contributes to the increased value of the neighboring areas in the Kallithea district and influences a property boom [53]. The sports activities organized at the

site contribute to the improvement of public health and help to promote a healthy lifestyle;

- The SN Park continued to operate during the pandemic in 2020, which was a crucial element for citizens during this difficult time;
- The SN Park is an approach to create a deep link between nature and society according to urban political ecology. In this social and ecological process, local plants became a second nature based on Greek landscapes;
- Introduced Ecosystem Services in the SNFCC project moderate natural phenomena (flood control, water purification, and climate regulation);
- Provided in the SN Park are Cultural Ecosystem Services benefits (e.g., recreation, physical and mental health, aesthetic enjoyment), which may contribute to fostering social cohesion. They can be essential for human wellbeing and can contribute to a sense of place (topos). In the SN Park, there are a few urban spaces indicated on the map (Figure 4) which play an important role for the local community.

Funding: This research received no external funding.

Institutional Review Board Statement: Not applicable.

Informed Consent Statement: Not applicable.

Data Availability Statement: Data are available in publicly accessible repositories of the SNFCC. They are all shown in the 'References' section.

Acknowledgments: The author would like to thank the anonymous reviewers for their thorough work with the manuscript and for providing constructive and insightful comments on this paper.

Conflicts of Interest: The author declares no conflict of interest.

References

1. Ruble, B.A. The Challenges of the 21st Century City. Urban Sustainability Laboratory. Wilson Center. Available online: www.wilsoncenter.org/publication/the-challenges-the-21st-century-city (accessed on 6 December 2012).
2. Papageorgiou, M.; Gemenetzi, G. Setting the grounds for the Green Infrastructure in the metropolitan areas of Athens and Thessaloniki: The role of green space. *Eur. J. Environ. Sci.* **2018**, *8*, 82–92. Available online: https://www.researchgate.net/publication/325897837_Setting_the_grounds_for_the_Green_Infrastructure_in_the_metropolitan_areas_of_Athens_and_Thessaloniki_the_role_of_green_space (accessed on 12 October 2020). [CrossRef]
3. Cranz, G.; Boland, M. Defining the Sustainable Park: A Fifth Model for Urban Parks. *Landsc. J.* **2004**, *23*, 102–120. Available online: https://www.researchgate.net/publication/250231365_Defining_the_Sustainable_Park_A_Fifth_Model_for_Urban_Parks (accessed on 12 October 2020). [CrossRef]
4. Nielsen, A.B.; van den Bosch, M.; Maruthaveeran, S.; van den Bosch, C.K. Species richness in urban parks and its drivers: A review of empirical evidence. *Urban Ecosyst.* **2013**, *17*, 305–327. Available online: https://www.researchgate.net/publication/257671090_Species_richness_in_urban_parks_and_its_drivers_A_review_of_empirical_evidence (accessed on 12 September 2020). [CrossRef]
5. Benedict, M.A.; McMahon, E.T. Green Infrastructure: Smart Conservation for 21st Century. *Renew. Resour. J.* **2002**, *20*, 12–17. Available online: https://www.merseyforest.org.uk/files/documents/1365/2002+Green+Infrastructure+Smart+Conservation+for+the+21st+Century.pdf (accessed on 12 September 2020).
6. Smets, J.; De Blust, G.; Verheyden, W.; Wanner, S.; Van Acker, M.; Turkelboom, F. Starting a Participative Approach to Develop Local Green Infrastructure: From Boundary Concept to Collective Action. *Sustainability* **2020**, *12*, 10107. Available online: https://www.mdpi.com/2071-1050/12/23/10107 (accessed on 12 September 2020).
7. Drapella-Hermansdorfer, A. Współczesny Park Miejski w Europie. Available online: https://docplayer.pl/3164332-Wspolczesny-park-miejski-w-europie.html (accessed on 12 October 2020).
8. Wolski, P. Współczesny park miejski w Europie. In *Wizja Rozwoju Wojewódzkiego Parku Kultury i Wypoczynku im. Gen. Jerzego Ziętka. Materiały Konferencyjne*; Urząd Marszałkowski Województwa Śląskiego: Katowice, Poland, 2006; pp. 1–9. Available online: https://slaskie.pl/images/wpkiw/pw_wpme.pdf (accessed on 12 October 2020).
9. Zachariasz, A. Współczesne kierunki i tendencje w projektowaniu parków publicznych. *Nauka Przyr. Technol.* **2009**, *3*, 1–10.
10. Zachariasz, A. *Zieleń jako Współczesny Czynnik Miastotwórczy ze Szczególnym Uwzględnieniem Roli Parków Publicznych*; Wydawnictwo Politechniki Krakowskiej: Kraków, Poland, 2006. Available online: https://www.researchgate.net/publication/332080375_Zielen_jako_wspolczesny_czynnik_miastotworczy_ze_szczegolnym_uwzglednieniem_roli_parkow_publicznych (accessed on 2 September 2020).

11. Zachariasz, A. A Park in a Small Town. On the Value of Place and Maintenance of Tradition. In *7ULAR: Middle-Sized Cities of Tomorrow*; Juzwa, N., Sulimowska-Ociepka, A., Eds.; Wydawnictwo Politechniki Śląskiej: Gliwice, Poland, 2013; pp. 209–215. Available online: https://www.researchgate.net/publication/343006692_A_PARK_IN_A_SMALL_TOWN_ON_THE_VALUE_OF_PLACE_AND_MAINTENANCE_OF_TRADITION (accessed on 21 September 2020).
12. Makowska, B. Ogrody przy muzeach w Paryżu/The gardens near museums in Paris. *Czas. Tech.* **2010**, *5-A*, 179–188.
13. Makowska, B. Parki rzeźb w Norwegii/The Sculpture Parks in Norway. *Czas. Tech.* **2012**, *2-A*, 203–208. Available online: https://repozytorium.biblos.pk.edu.pl/redo/resources/31199/file/suwFiles/MakowskaB_ParkiRzezb.pdf (accessed on 2 September 2020).
14. Schneider-Skalska, G. Witruwiusz—prekursor projektowania zrównoważonego. *Czas. Tech.* **2009**, *1-A*, 127–131. Available online: https://repozytorium.biblos.pk.edu.pl/redo/resources/33761/file/suwFiles/SchneiderSkalskaG_WitruwiuszPrekursor.pdf (accessed on 2 September 2020).
15. Newhouse, V. *Chaos and Culture: Renzo Piano Building Workshop and the Stavros Niarchos Foundation Cultural Center in Athens*; Monacelli Press: New York, NY, USA, 2017.
16. Awards. Available online: https://www.snfcc.org/en/awards (accessed on 12 September 2020).
17. Stavros Niarchos Foundation Cultural Center 2017 Report. Available online: https://www.snfcc.org/sites/default/files/sitefiles_2019-07/apologismos_eng_1.7.19.pdf (accessed on 12 September 2020).
18. SNFCC Impact Study Report for SNF Board of Directors. Available online: https://www.snf.org/media/4960627/SNFCC-Impact-Study-Detailed-Report-EN.pdf (accessed on 29 January 2020).
19. Stavros Niarchos Foudation Cultural Center. Available online: https://www.snfcc.org (accessed on 12 September 2020).
20. Pafi, M.; Siragusa, A.; Ferri, S.; Halkia, M. *Measuring the Accessibility of Urban Green Areas: A Comparison of the Green ESM with Other Datasets in Four European Cities*; JRC Technical Reports; Publications Office of the European Union: Luxembourg, 2016. Available online: https://publications.jrc.ec.europa.eu/repository/bitstream/JRC102525/190916_siragusa_%20jrc_techrep_accessibility_online.pdf (accessed on 12 September 2020).
21. Kalfa, F.; Kalogirou, N. Quality through Sustainable Practices during the Design and Construction Phase—The case of the SNFCC. *Procedia Environ. Sci.* **2017**, *38*, 781–788. [CrossRef]
22. Asprogerakas, E. Strategic Planning and Urban Development in Athens. The Current attempt for Reformation and future challenges. In Proceedings of the Sustainable Urban Planning and Design Symposium, Nicosia, Cyprus, 13 May 2016. Available online: https://www.researchgate.net/publication/320695579_Strategic_Planning_and_Urban_Development_in_Athens_The_Current_attempt_for_Reformation_and_future_challenges (accessed on 8 October 2020).
23. Sardianou, E.; Leonti, A. Applying the Contingent Valuation Method in Assessing Urban Parks: The Case of Niarchos in Greece. *IOP Conf. Ser. Earth. Environ. Sci.* **2019**, *362*, 1–6. Available online: https://www.researchgate.net/publication/337289250_Applying_The_Contingent_Valuation_Method_in_Assessing_Urban_Parks_the_Case_of_Niarchos_in_Greece (accessed on 12 October 2020). [CrossRef]
24. Bakogiannis, E.; Kyriakidis, C.; Siti, M.; Floropoulou, E. Reconsidering Sustainable Mobility Patterns in Cultural Route Planning: Andreas Syngrou Avenue, Greece. *Heritage* **2019**, *2*, 104. [CrossRef]
25. Deborah Nevins & Associates. Available online: https://www.dnalandscape.com/athens.html (accessed on 24 August 2020).
26. Clark, D. Green Space per Inhabitant in the City of Athens in Greece in 2018. Statista. Available online: https://www.statista.com/statistics/860773/green-areas-per-inhabitant-in-athens-in-greece (accessed on 4 September 2020).
27. Feltynowski, M.; Kronenberg, J.; Bergier, T.; Kabisch, N.; Łaszkiewicz, E.; Strohbach, M.W. Challenges of urban green space management in the face of using inadequate data. *Urban For. Urban Green.* **2018**, *31*, 56–66. Available online: https://www.sciencedirect.com/science/article/abs/pii/S1618866717304569 (accessed on 12 October 2020). [CrossRef]
28. Forde, T. Satellite Images Ranks Europe's Greenest (and Not so Green) Cities. Available online: https://www.archdaily.com/883707/satellite-images-ranks-europes-greenest-and-not-so-green-cities (accessed on 20 September 2020).
29. London 'Greenest City' in Europe. Available online: https://www.edie.net/news/6/London--greenest-city--in-Europe-/ (accessed on 20 September 2020).
30. Prajsner, A. Które z Naszych Miast jest Najbardziej "Zielone"? Available online: https://forsal.pl/artykuly/1027174,ktore-z-naszych-miast-jest-najbardziej-zielone.html (accessed on 20 September 2020).
31. Dímos Kallithéas. Municipality in Attica. Available online: https://citypopulation.de/en/greece/attiki/4801__d%C3%ADmos_kallith%C3%A9as/ (accessed on 12 September 2020).
32. Kallithea. Euromed. Available online: http://www.reseau-euromed.org/en/ville-membre/kallithea-2/ (accessed on 12 September 2020).
33. Wikipedia. Population Density in Athens. Available online: https://en.wikipedia.org/wiki/File:Population_Density_in_Athens.PNG (accessed on 12 September 2020).
34. Daniels, E. The Area I Live in: Kallithea. Available online: https://slideplayer.com/slide/13793946/ (accessed on 8 October 2020).
35. Wikipedia. 1896 Summer Olympics. Available online: https://en.wikipedia.org/wiki/1896_Summer_Olympics (accessed on 12 September 2020).
36. Europa dla Obywateli. Available online: https://europadlaobywateli.pl/partnerzy/municipality-of-kallithea// (accessed on 12 September 2020).

37. Ateny Klimat (Grecja). Available online: https://pl.climate-data.org/europa/grecja/ateny/ateny-7/ (accessed on 12 September 2020).
38. Weather Atlas. Monthly Weather Forecast and Climate Athens, Greece. Available online: https://www.weather-atlas.com/en/greece/athens-climate#daylight_sunshine (accessed on 12 September 2020).
39. Surface Air Temperature for December 2019, Implemented by ECMWF, Copernicus 2019. Available online: https://climate.copernicus.eu/surface-air-temperature-december-2019 (accessed on 12 September 2020).
40. SNFCC. Energy. Available online: https://www.snfcc.org/en/energy (accessed on 8 September 2020).
41. SNFCC. Impact Study Report for SNF Board of Directors. Available online: https://www.snf.org/media/4963854/SNFCC-Impact-Study-Report-FINAL-EN.pdf (accessed on 8 September 2020).
42. Stavros Niarchos Foundation Cultural Center SNFCC. Three Successful Years—6.3 Million Visitors in 2019. Available online: https://www.snfcc.org/en/news/stavros-niarchos-foundation-cultural-center-three-successful-years-63-million-visitors-2019 (accessed on 12 September 2020).
43. SNFCC. Monthly Booklets. Available online: https://www.snfcc.org/en/monthlyevents (accessed on 8 September 2020).
44. An Innovative Collaboration Reimagines and Reactivates Public Space for a New Era, from New York to Athens. Available online: https://www.snf.org/en/newsroom/news/2020/12/an-innovative-collaboration-reimagines-and-reactivates-public-space-for-a-new-era (accessed on 3 December 2020).
45. SNFCC. Culture. Available online: https://www.snfcc.org/en/snfcc/our-work/culture (accessed on 8 September 2020).
46. Hodor, K.; Przybylak, Ł.; Kuśmierski, J.; Wilkosz-Mamcarczyk, M. Identification and Analysis of Problems in Selected European Historic Gardens during the COVID-19 Pandemic. *Sustainability* **2021**, *13*, 1332. Available online: https://www.mdpi.com/2071-1050/13/3/1332 (accessed on 1 March 2021). [CrossRef]
47. SNFCC. Close Collaboration between the SNFCC and the Stavros Niarchos Foundation. Available online: https://www.snfcc.org/en/news/close-collaboration-between-snfcc-and-stavros-niarchos-foundation/1240 (accessed on 8 September 2020).
48. Visitors to SNFCC Reach 5.3 Million in 2018. Available online: https://www.ekathimerini.com/238024/article/ekathimerini/life/visitors-to-snfcc-reach-53-million-in-2018 (accessed on 8 September 2020).
49. The Stavros Niarchos Foundation. Available online: https://www.efc.be/member-post/stavros-niarchos-foundation-2/ (accessed on 8 September 2020).
50. SNFCC. Pandemic: Challenges in an Ever-Changing Landscape. Available online: https://www.snfcc.org/en/PandemicLectures (accessed on 12 September 2020).
51. Oberfrancová, L.; Legény, J.; Špaček, R. Critical Thinking in Teaching Sustainable Architecture. *World Trans. Eng. Technol. Educ.* **2019**, *17*, 127–133. Available online: http://www.wiete.com.au/journals/WTE&TE/Pages/Vol.17,%20No.2%20(2019)/01-Spacek-R.pdf (accessed on 18 January 2021).
52. Stavros Niarchos Foundation Cultural Center. Presentation it European Cloud Plenary 18 March 2014. Available online: https://www.slideshare.net/Europeana/presentation-it-europeana-cloud-plenary18314 (accessed on 8 September 2020).
53. Gkinni, Z.; Tsaroucha, C.; Sarris, N. The National Library of Greece: Moving into a new era. In Proceedings of the World Library and Information Congress, 83rd IFLA General Conference and Assembly, Wrocław, Poland, 19–25 August 2017. Available online: https://www.researchgate.net/publication/340447525_The_National_Library_of_Greece_moving_into_a_new_era (accessed on 8 September 2020).
54. Kampouris, N. Over Five Million Visit Athens' Stavros Niarchos Cultural Center in 2018. Available online: https://greekreporter.com/2019/02/25/over-five-million-visit-athens-stavros-niarchos-cultural-center-in-2018/ (accessed on 18 January 2021).
55. Greek City Time. Over 6.3 Million People Visit Stavros Niarchos Foundation Cultural Centre in 2019. Available online: https://greekcitytimes.com/2020/01/30/over-6-3-million-people-visit-stavros-niarchos-foundation-cultural-centre-in-2019/ (accessed on 18 January 2021).
56. Andriopoulou, E. The Stavros Niarchos Foundation Cultural Center: A Case Study of a Public-Private Partnership for the Benefit of Society. Available online: https://www.wtflucerne.org/snfcc-as-ppp-success-story (accessed on 18 January 2021).
57. Markou, M. Renovation Projects at Faliro Bay. Available online: https://www.athenssocialatlas.gr/en/article/faliro-bay/ (accessed on 12 October 2020).
58. Fotiadi, I. The SNFCC in Athens Local Improvement Rekindles a District. Goethe Institut. Available online: https://www.goethe.de/en/kul/ges/eu2/rhr/21017619.html (accessed on 8 September 2020).
59. Makowska, B. Green elevations and roofs of the buildings (the 5th elevation) in 21st century cities. In *Future of the Cities—Cities of the Future*; Gyurkovich, J., Wójcik, A., Eds.; Wydawnictwo Politechniki Krakowskiej: Kraków, Poland, 2014; pp. 29–50.
60. Makowska, B. Ogrody i Parki w Miastach Hiszpanii/Gardens and Parks in Spain Cities. *Czas. Tech.* **2012**, *6*, 235–243. Available online: https://repozytorium.biblos.pk.edu.pl/redo/resources/30944/file/suwFiles/CzasopismoTechniczne_z.19.Architektura_z.6-A.pdf (accessed on 2 September 2020).
61. How Many Buildings in the World are LEED Platinum Certified? Design Ideas for the Built World. Available online: https://caddetailsblog.com/post/how-many-buildings-in-the-world-have-become-leeds-platinum-certified (accessed on 12 October 2020).

Article

Evolution of the Concept of Sensory Gardens in the Generally Accessible Space of a Large City: Analysis of Multiple Cases from Kraków (Poland) Using the Therapeutic Space Attribute Rating Method

Izabela Krzeptowska-Moszkowicz, Łukasz Moszkowicz and Karolina Porada *

Faculty of Architecture, Cracow University of Technology, 31-155 Krakow, Poland; ikrzepto@pk.edu.pl (I.K.-M.); lmoszkowicz@pk.edu.pl (Ł.M.)
* Correspondence: kporada@pk.edu.pl

Abstract: This paper presents a study on public gardens with sensory features located in Kraków (Poland). Data for the analysis of the facilities were obtained during site visits using observations. The paper uses a research method for the analysis of therapeutic outdoor areas in cities based on the evaluation of their attributes. This method makes it possible to characterise features of objects as well as their value. It is a practical tool, which enables an in-depth analysis of public spaces. The study showed that public gardens with sensory features located in Kraków have significant deficiencies, which make it impossible to fully exploit the potential of the sensory space.

Keywords: therapeutic space; sensory garden; Kraków; public space; rating method

Citation: Krzeptowska-Moszkowicz, I.; Moszkowicz, Ł.; Porada, K. Evolution of the Concept of Sensory Gardens in the Generally Accessible Space of a Large City: Analysis of Multiple Cases from Kraków (Poland) Using the Therapeutic Space Attribute Rating Method. *Sustainability* **2021**, *13*, 5904. https://doi.org/10.3390/su13115904

Academic Editors: Katarzyna Hodor, Magdalena Wilkosz-Mamcarczyk and Jan K. Kazak

Received: 30 April 2021
Accepted: 17 May 2021
Published: 24 May 2021

1. Introduction

Sensory gardens are typically described as a type of therapeutic garden, e.g., in the cross-sectional work by Winterbottom and Wagenfeld [1]. These types of gardens are created in public spaces and sometimes form part of therapy and healthcare centres. This trend grew even more when it was demonstrated that the view of greenery and the presence in a green space have a positive impact on patient health [2]. Initially, sensory gardens were designed primarily with blind people in mind (the oldest included the touch- and smell-focused garden at the John J. Tyler Arboretum, in Lima (1949, PA, USA), or the fragrance garden at the Cambridge University Botanical Garden (1960, Cambridge, UK)), but over time they were tailored to all visitors, regardless of ability or disability, which was labelled universal design [1]. Few detailed studies about them have been published, and the existing literature focuses mostly on cases from specific countries. Such works include those of Hussein [3], which focused on therapeutic sensory gardens accompanying educational facilities for the disabled in Great Britain. Gonzales and Kirkevold discussed the problem of therapeutic gardens accompanying care homes in Norway [4]. An analysis of various therapeutic sensory gardens in Lithuania was presented by Balode [5].

Contemporary sensory gardens are often intended to stimulate various human senses. Apart from the five basic senses—touch, taste, smell, hearing, and vision—the holistic sensory systems of the human body have also become a focus. Winterbottom and Wagenfeld argue that the designs of therapeutic gardens should focus on the relationship between gardens and movement, balance, and touch [1]. This appears to be of particular significance in sensory healing gardens.

The Sensory Trust (from the United Kingdom) defines a sensory garden as 'a self-contained area that concentrates a wide range of sensory experiences. Such an area, if designed well, provides a valuable resource for a wide range of uses, from education to recreation' [6]. This definition does not categorise stimuli and does not focus on how many and which senses should be used to receive them, focusing instead on an area's potential to generate stimuli. With this definition, sensory gardens can have a variety of uses, the therapeutic use not being the primary one. It better characterises sensory gardens designed in public spaces, i.e., where such gardens accompany various places and institutions (e.g., at housing estates, at schools, as parts of public parks and gardens) and have a diverse set of users.

Hussein (2009) noted that sensory gardens should be designed so as to be perceived and experienced from up close, using all of one's senses [7]. This type of garden allows the user to develop a close and highly individualised relationship with each of its elements. The senses can be influenced by various elements, such as vegetation, paving, or outdoor furniture [1]—especially sound, tactile, and kinetic elements such as reliefs or musical instruments. Water is also an essential element in all therapeutic gardens. Its basic property is to feed the soil and the plants. Winterbottom and Wagenfeld point out that a good practice in therapeutic gardens is to establish bioswales and raingardens, which not only support rainwater management but also bring birds, butterflies, and dragonflies into the garden [1]. Water features can also have sensory significance. The sound of water in a stream, fountain, or cascade is a factor that affects the sense of hearing.

Polish-language literature features few works on sensory gardens, which could be associated with their being relatively new. Our search of academic literature indicates that interest in the topic of sensory gardens in Poland emerged after 2000, and one of the first authors was Pawłowska [8]. In publications from the period 2010–2020, a division can be seen between literature describing Polish case studies and the problem of using sensory gardens for the rehabilitation of users with visual impairment. A study of various Polish therapeutic garden cases, including sensory gardens, can be found in publications by Dudkiewicz et al. [9], Górska-Kłęk [10], or Trojanowska [11]. Studies of sensory gardens as spaces for the visually impaired were published by Dąbski & Dudkiewicz [12], Trojanowska [13], Woźny & Lauda [14], and Zajadacz & Lubarska [15], who analysed selected Polish cases.

As for generally available spaces, sensory gardens were found to have an even poorer representation in Polish-language literature. The idea of aligning sensory spaces with the structure of therapeutic parks was proposed by Trojanowska [11], who argued that multi-sensory stimulation is the fundamental purpose of such parks and is associated with creating separate spaces dedicated to sensory impacts. A multiple case study of small sensory gardens located in the most densely developed parts of New York was presented by Pawłowska [8]. This study underscores their role as spaces that positively influence human wellbeing, especially of individuals who work in downtown areas of large cities, who are the typical users of such gardens. Sensory gardens in cities can thus play the role of restorative gardens, i.e., places that offer conditions favourable to reattaining the body's internal balance (homeostasis), which can improve emotional and psychological wellbeing, for instance by reducing stress [16]. Krzeptowska-Moszkowicz et al. [17] discussed creating sensory gardens that would be friendly both to humans and animals in large cities, as this could aid in enhancing the scope of positive stimuli that affect the human senses.

2. Evolution of the Idea of Sensory Gardens in Poland

Sensory gardens are closely tied with the human senses but can have different definitions, as our perspective on this garden type is evolving. One of the more often-quoted definitions in Polish-language literature states that 'mianem ogrodu sensorycznego określa się kompozycję tak zaprojektowaną, aby bodźce pozawzrokowe były użyte celowo i to w większym natężeniu niż zwykle' (a sensory garden is a composition designed so that non-visual stimuli are used intentionally and at a much greater intensity than usual) [8]. It classifies sensory gardens as layouts addressed primarily to the blind, which does not exclude their use by all visitors. This is a view that stems from the original perceptions of these gardens as places primarily intended for the visually impaired.

The construction of sensory gardens in Poland began only two decades ago. In 2008, Pawłowska wrote that 'the idea of sensory gardens was not yet popular in Poland' [8].

Design work on sensory gardens was also viewed as a certain form of experimentation. In the first decade of the twenty-first century, this idea was already present in Poland, and conceptual designs were created at universities—e.g., at the Cracow University of Technology, some students prepared concepts for sensory gardens as part of their diploma theses under Pawłowska's guidance [18], although completed projects with such gardens were far from common, especially in public spaces.

The first projects to focus on generally accessible spaces in Poland were sites near national park headquarters and as parts of arboretums. They included: the Environmental Garden of the Senses near the Mouth of the Warta National Park Directorate Building, which had educational features targeting children and young people (2000), the Garden of the Senses near the Babia Góra National Park headquarters building in Zawoja (2007), with amenities for the disabled, including the blind, and the Sensory Garden in the Arboretum and Physiography Facility in Bolestraszyce (2008), which was also adapted to the needs of the disabled, including blind people. During this initial period, the first generally accessible sensory gardens in Poland were located outside of large cities. They were also linked with educational facilities. Only during the second decade of the twenty-first century did such gardens begin to appear in cities, associated with functions that were not educational. Their scope of use was also extended, which brought them closer to the definition of sensory gardens proposed by the Sensory Trust. In Kraków, the establishment of gardens with sensory features and located in public spaces has only been practised fora couple of years, which makes them relatively new facilities, dating from the period between 2013 and 2020, but more are being planned.

Large cities, as areas with a mosaic of places with a diverse range of functions, are dedicated to specific forms of activity. We can thus ask whether sensory gardens are built in spaces with different uses, or whether they are associated with a specific type of urban space. In this paper, we analyse the case of Kraków, one of Poland's largest cities. The objective of this paper was also to characterise and assess gardens with sensory features and located in Kraków as therapeutic spaces, with the use of a specific analytical tool.

3. Object of Study

Kraków is Poland's second-largest city, the largest being Warsaw. Its area is around 327 km^2, and its population is around 780 thousand [19]. Kraków is a geomorphologically varied city. Various geographic and phytogeographic regions come into contact here [20].

The city's urban greenery system consists of natural greenery that comprises 7.88% of the city's area, with forests making up 6.04%. Apart from forests, these areas also include complexes of aquatic and epilithic plants, as well as naturally growing plants. Urban greenery also includes semi-natural plant complexes such as meadows and pastures, which cover 7.39% of the city's area. There is also anthropogenic plant cover, which constitutes 36% of the city, and includes: fields, orchards, allotment gardens, and ruderal plants. This type of greenery also accompanies developed areas and household gardens and amounts to 29.11% of Kraków. Landscaped greenery covers only 13.47% of the city's area and includes public parks, fortress greenery, cemetery greenery, greenery accompanying sports grounds, streets, Jordan gardens (in Poland, this name is given to green areas established mainly in cities, intended for children and young people), and housing estates [21].

Based on the Polish-language literature and our own investigation, we determined that out of all Polish cities, Kraków had the most publicly available gardens with sensory features. Most of these gardens have never been analysed or mentioned in the Polish-language literature, as they were built relatively recently. The following gardens of this type can be found in Kraków (list by date of completion—garden area measurements were performed based on orthophotomaps: the area values are therefore estimates), the location is shown on the map (Figure 1):

(A) 'Zapachowo'—fragrance garden in the S. Lem Educational Park (2013)—ca. 742 m^2;
(B) Sensory Garden at the Piaski Nowe housing estate (2019)—ca. 903 m^2 (Figure 2);

(C) Garden with sensory features near the J. Czapski Musuem (2018–2019)—ca. 260 m^2 (Figure 3);
(D) Sensory garden with a sensory path in Tysiąclecia Park (2020)—ca. 1840 m^2;
(E) Sensory garden in Reduta Park (2020)—ca. 2385 m^2;
(F) Playground with sensory features in Jordan Park (built gradually, mostly in recent years)—ca. 2800 m^2.

Figure 1. Placement of existing gardens with sensory features (A–F) in Kraków. City parks have been marked in green.

Figure 2. Sensory Garden at the Piaski Nowe housing estate, phot. Krzeptowska-Moszkowicz I., 2021.

Figure 3. Garden with sensory features near the J. Czapski Musuem, phot. Porada K., 2020.

The existing sensory gardens in Kraków are small spaces within the urban tissue. The first garden of this type in the city was located near an educational building, which aligned with the initial trend in Poland. Gardens in different settings, e.g., in city parks, started to be built later. In our study we used material procured during visits to each site, which were conducted between 2019 and 2021, during spring and summer periods (due to vegetation). These were single visits, lasting several hours each, during which we conducted an inventory of sensory elements. The visits also allowed us to obtain specific data used to characterise each garden and to collect photographic material that can be of aid in analysing and illustrating specific problems (Figures 2 and 3).

4. Research Method

To determine the spaces where gardens with sensory features are located, we analysed the setting of each of their locations. We did this during on-site analyses.

Afterwards, we used a modified version of Trojanowska's method [22], assuming that essential values of therapeutic spaces—in the form of green areas—are comparable. This method was used to analyse public areas in cities, which is another argument in favour of using it to assess the generally accessible gardens with sensory features under study. The method is based on attribute analysis. The attribute was characterised as 'a feature of space or the presence of a type of equipment' [22]. The method assumes that as the number of attributes present in an area increases, so does its therapeutic potential.

Trojanowska used this method to analyse the therapeutic potential of parks, which is why it was necessary to adapt it to highly specific features, namely sensory gardens.

Well-designed sensory gardens should be characterised by the following distinctive characteristics [15]:

- they should be designed with the intent to stimulate human senses;
- they should form a complete whole, isolated from the surroundings;
- they should affect all the senses;
- they should focus on non-visual stimulation;
- apart from sense-stimulating plants, they should also feature other elements that stimulate the senses.

Additional elements can be added to this list:

- they should be animal-friendly, as the presence of animals increases the scope of positive stimuli [17];
- they should be equipped with water features, because of their sonic properties and their importance for plants and animals;
- they should have dedicated use indications that facilitate experiencing the garden from up close [17];

It was necessary to account for the characteristics presented above in the attribute set used to rate the garden. This was possible as the method was not designed as a finite tool, allowing for the addition of attributes.

After analysing the attributes listed by Trojanowska and comparing them with the characteristics of sensory gardens, we found that there was a significant difference concerning one element. Sensory gardens are spaces that are isolated from the surroundings, as experiencing them is oriented inwards, towards perception from up close, while large green areas such as parks are mainly focused outwards, on perceiving distant vistas, including those from outside of a given park's space. This was accounted for in the attribute set, with only a small number of attributes being removed, as we decided that their use in rating sensory gardens was not justified. The remaining attributes, as used by Trojanowska, remained unchanged, with some being combined, as it was possible to rate them together. These included 'elements that indirectly affect the comfort of use of the space: access to food, drink, toilets and others'. We did not find it justified to rate each of these elements separately in the case of sensory gardens, as they all determine user comfort and are often provided in a set (e.g., by the coffee shop in the garden of the J. Czapski Museum, as well as in Reduta Park).

Trojanowska segregated the attributes she had selected by assigning them to design stages—the same approach was used here, with the following stages: functional programme, functio-spatial structure, design of internal spaces, architectural form and place-making. We also accounted for sustainability criteria.

This method allows for characterising a garden's features and rating its value. It is a useful analytical tool, as it allows for identifying a garden's specific weaknesses as a therapeutic space. The less attributes, the lower the therapeutic functionality of such a space, and the user can derive less benefits from being in it.

5. Results

5.1. Location of Kraków's Gardens with Sensory Features and Their Users

Based on an analysis of the surroundings of Kraków's gardens with sensory features, we isolated the following urban activity zones (Table 1).

Table 1. Overview of Kraków's gardens with sensory features and an analysis of their surroundings with urban activity zones isolated.

Garden with Sensory Features	General Overview	Location Overview	Use, Users	Urban Activity Zone
(A) in the Educational S. Lem Park	The garden has a freeform composition and consists of a path that meanders between fragrant plants	The garden is a part of a sensory educational park that familiarises users with the laws of physics and that features several dozen installations for personal experimentation	Learning through play, primarily for children, the youth, and adults	Educational zone
(B) accompanying the J. Czapski Museum	The garden has a geometric layout and consists of a large, square lawn and a path that runs around it and leads to the museum building. There are tall pots, mostly with fragrant plants and herbs, along the path	The garden is located in the city centre and belongs to the J. Czapski Museum grounds It was designed as a place intended for the museum's supplementary events. The building's outer wall is used as a screen for film screenings, and there is also a coffee shop in the garden. There is another Museum nearby, named after E. Hutten-Czapski	Cultural activity intended for visitors to the city and its residents	Cultural and tourist zone

Table 1. *Cont.*

Garden with Sensory Features	General Overview	Location Overview	Use, Users	Urban Activity Zone
(C) at the Nowe Piaski housing estate	It was designed with central paths extending from a central section and leading to various garden interiors. It has a wealth of plants and small meadows covered with grass and surrounded by greenery	The garden is located on a typical block housing estate distinctive for large Polish cities. The garden is surrounded by housing blocks, streets, and parking lots	Local communal activity, it is also used as a meeting space and to stimulate activity among seniors	Urban housing estate resident activity zone
(D) in Tysiąclecia Park	The garden consists of several sections, its composition is part-freeform and part-geometric.	The garden is located in the centre of a large city park	Outdoor exercise and rest among greenery for local citizens	Recreational zone in an urban green area
(E) in Reduta Park	The garden is located on a slope; most of it is geometric. It features large, narrow beds with fruit bushes and low beds mostly filled with fragrant herbs. The garden features a coffee shop with a roof with a gravel surface and low planters with fragrant plants. The roof can be reached via stairs.	The garden is located in a large city park	Outdoor exercise and rest among greenery for local citizens	Recreational zone in an urban green area
(F) in H. Jordan Park	The garden was built gradually and consists of a sand garden, a section of wooden play equipment for children, a fountain with jets in the pavement surface, and a green labyrinth.	The garden is located in a large park which has been dedicated to children's and youth activity from its inception.	Stimulating children and the youth to lead an active and healthy lifestyle	A zone dedicated to sports activity among children and the youth

The listing above shows that the gardens under analysis are not assigned to any single specific urban activity zone, but instead are located in very different zones within the urban tissue, and as such are dedicated to different user groups. Each of the gardens has specific intended users, ranging from the local community to the general population of the city and tourists. These gardens are therefore directed towards specific users and there is a certain specialisation in this regard. However, when looking at them holistically, it can be said that they are areas of activity for every age group—children, youth, adults, and seniors.

None of the gardens under analysis had any distinguishing features that would suggest an intention of catering to the blind or visually impaired (like the first sensory gardens to be built in Poland and around the world). They lacked dedicated amenities for these groups. It was probably assumed that persons with visual impairments were not intended to be the main users of these gardens.

5.2. Sensory Stimulation in Kraków's Sensory Gardens

While accounting for the evolution of the idea of sensory gardens and the projects built in Kraków, it was found that the gardens with sensory features that existed there at the time of this study were based primarily on the five basic human senses (Table 2). There were very few elements that stimulated other senses.

Table 2. Listing of sensory stimuli in Kraków's gardens with sensory features.

Garden with Sensory Features	Taste	Touch	Sight	Hearing	Smell	Others
(A) in the Educational S. Lem Park	None	There is a sensory path near the garden—tactile sensations in feet	Flowers, greenery, plants	People laughing, birds singing	Large areas of fragrant herbs	There is a plant-based labyrinth near the garden
(B) accompanying the J. Czapski Museum	Fruits	The touch of grass, plants, sensory path—tactile sensations in feet	Colours of flowers, fruits, narrow mirrors that reflect plants	Insect buzzing	Fragrant herbs and flowers	Wigwam from willow branches, isolated meadows with grass, a small sensory path
(C) at the Nowe Piaski housing estate	The taste of herbs, nearby coffee shop	Ability to touch plants with different textures	Green and white—the plants and the surroundings	Insect buzzing	Fragrant herbs and flowers	None
(D) in Tysiąclecia Park	None	The touch of perennial plants, sensory path—tactile sensations in feet	Colours of flowers	The sound of water (a fountain)	Bitterling, the smell of herbs and flowers	Elements in the form of a line net
(E) in Reduta Park	Fruits	None	Colours of flowers	Insect buzzing	Smell of herbs	Observation deck with gravel surface
(F) in H. Jordan Park	None	Touch of sand, water, and wood	Green of plants, clear water	Sound of steps on wooden surfaces, birds singing, the sound of water	None	Labyrinth made of plants

5.3. Garden Attribute Analysis

We performed our analysis using the methodology prepared by Trojanowska [22]. Initially, we described the form the attributes took in each garden. To rate the presence of attributes (using numbers), we used the following scale:

- 1—the attribute was present, and its potential was being used well,
- 0.5—the attribute was present, but it was not used to its full potential,
- 0—the attribute was not present.

The assessment of each object was based on observations. Here we used an enhanced rating (by adding the 0.5 value) so that we could assess whether an attribute's potential was fully introduced into the garden's space (Tables 3 and 4). The occurrence of the attributes characterising the therapeutic spaces in sensory gardens in Krakow is also presented as a percentage in the graphs (Figure 4).

Table 3. Analysis of the attributes of Kraków's gardens with sensory features—the first group of attributes.

		The First Group of Attributes					
		(A)	(B)	(C)	(D)	(E)	(F)
Functional Programme	**Attribute**						
(a) Enabling physical and psychological regeneration	Places for rest that facilitate experiencing the surroundings from up close	YES/NO (0.5) too few benches, but they are deep in the garden	YES (1) many seats in various parts of the garden, there is also a small meadow with sunbeds	YES/NO (0.5) there are seats on a platform, but no isolated places or places near plants	YES/NO (0.5) there are seats in various places, but usually along the main path	YES/NO (0.5) too few benches, only in selected places and arranged linearly	YES/NO (0.5) too few benches
	Isolation from the urban environment, noise, smells, and the pressure of time and fast living	YES (1) the garden is at the edge of a science garden and faces the park	YES (1) densely placed plants partially insulate from streets, but there are also places deeper in the garden	YES (1) the garden is located in an interior courtyard	NO (0) a busy path runs through the garden, no plants along the edges	YES/NO (0.5) frequented paths run through the garden, and there is a playground nearby, but the park itself is isolated from busy streets	YES/NO (0.5) the garden is near the park's outer fence, but there are bushes and trees here
	Ability to easily observe animals or people	YES/NO (0.5) insects—there are taller bushes, but many plants are close to the soil surface	YES (1) insects and birds—there are both taller bushes and perennials, as well as bench-side plants	YES (1) insects, as plants are in taller pots	YES/NO (0.5) there are taller perennials, but most plants are short	NO (0)—the benches are placed in a way that prevents a comfortable observation of the nearby park area	YES (1) ability to observe children at play, or birds
(b) Facilitating social contact	Ability to meet as a group	YES/NO (0.5) there is a small space with three benches, but the main path crosses through it, which can make conversing difficult	YES (1) there is a meeting space; benches are often placed in pairs and at an angle, facilitating eye contact and conversation	YES (1) there is a space for meetings—a terrace with chairs and tables that can be rearranged	YES/NO (0.5) meetings are possible, but usually for two people—benches and chairs are placed in pairs	YES/NO (0.5) ability to meet at a coffee shop, but the sensory spaces do not facilitate this	YES (1) the entire space allows group play
(c) Facilitating physical activity	Places for play and recreation	YES (1) a large lawn, with equipment for performing experiments nearby	YES (1) a meadow in a garden with sunbeds and a wigwam made of living willow branches for children	YES/NO (0.5) there is a lawn, but due to the proximity of a Museum it is not used for physical activity	YES (1) features with a line net suspended from metal frames; extensive lawns nearby	YES/NO (0.5) there are lawns, but the site is on a slope, which excludes several activity types	YES (1) the entire area facilitates play for children and their caretakers
	Place dedicated to gardening classes or hortitherapy	NO (0)—absent	YES (1) a large table for classes in the central part	NO (0)—absent	NO (0)—absent	NO (0)—absent	NO (0)—absent

Table 3. *Cont.*

		The First Group of Attributes					
		(A)	(B)	(C)	(D)	(E)	(F)
Functional Programme	Attribute						
(d) Meeting essential user needs	Safety in the garden space	YES (1) the garden is located in a larger, fenced science garden	YES (1) the garden is fenced	YES (1) the garden is close to the Museum, in its internal courtyard	NO (0) lack of isolation from the remainder of the park	NO (0) lack of isolation from the remainder of the park	YES (1) the site is fenced with a low fence
	Safety in direct contact with plants	YES (0.5) the plants are safe, but there are also roses	YES/NO (0.5) most plants are safe, but there are also prickly plants	YES (1) safe plants were used, mainly herbs	YES/NO (0.5) there are rose bushes around a fountain	YES/NO (0.5) safe plants were used, but there are gooseberry bushes	YES (1) the labyrinth is made from safe plants
	Seating or shelter	YES/NO (0.5) there is shelter under tree canopies	YES (1) shelter under a sail suspended above the garden's central space	YES (1) shelter inside a coffee shop that opens towards the garden	YES (0.5) there is shelter under tree canopies	YES (1) shelter inside a coffee shop	YES/NO (0.5) there is shelter under tree canopies
	Sunny and shaded places	YES/NO (0.5) there are only sunny spaces in the garden	YES (1)	YES (1)	YES (1)	YES/NO (0.5) there are only sunny spaces in the garden	YES (1)
	Amenities for the disabled	NO (0) a gravel path with solid strips, but turns are at an angle that make it impossible for wheelchair-bound persons to make them	YES/NO (0.5) wide paths, but no amenities for the blind	YES/NO (0.5) wide paths, but no dedicated amenities for the blind	YES/NO (0.5) wide, even paths, but no dedicated amenities for the blind	NO (0) the garden is on a slope, the paths along bushes are narrow and dead-ended, which hinders wheelchair movement, stairs to the observation deck	NO (0) absent
	Elements that indirectly affect comfort of use: access to food and drink, toilets, and others	YES (1) they are present in the park	YES/NO (0.5) they are absent in the park; there are bicycle stands and waste bins	YES (1) there is a coffee shop that opens towards the garden	YES/NO (0.5) there are bicycle stands	YES (1) there is a coffee shop with restrooms	YES (1) they are present in the park
(e) Cognitive support	Features that facilitate education in the garden	YES (1) plaques with plant names	YES (1) information boards that also feature plant names	NO (0)	NO (0) no plaques	NO (0) no plaques	YES/NO (0.5) learning through play facilitated by playground equipment

Table 4. Analysis of the attributes of Kraków's gardens with sensory features—the second group of attributes.

		The Second Group of Attributes					
		(A)	**(B)**	**(C)**	**(D)**	**(E)**	**(F)**
	Attribute						
Functio-spatial structure	Isolation of the garden from its surroundings, creating a separate, intimate space	YES/NO (0.5)	YES (1)	YES (1)	NO (0)	NO (0)	YES (1)
	Siting in a place that retains fragrances and sounds inside the garden	YES (1) there is a hill near the garden that shelters it from one side	YES (1) plants located around the outer rim of the garden shelter it from draughts	YES (1) the garden is surrounded by solid fences and walls	YES/NO (0.5) the garden is in a depressed area, but has no buffer greenery to protect it from draughts	YES/NO (0.5) only partially protected	NO (0)
Internal space and architectural form design	Garden complexity, presence of various garden interiors, proper path system	YES/NO (0.5) interesting path course, no interiors for longer stays	YES (1) the garden is divided into interiors where one can sit and stay longer	YES/NO (0.5) the layout is simple, the path encircles a lawn, but plant pots are only on one side	YES (1) the garden has varied interiors and each has rest spaces	YES/NO (0.5) there are diverse interiors but no paths, some paths have dead ends, little room to sit among the plants	YES/NO (0.5) the path runs through the garden's centre, which allows easy access to its every part; there are two entrances
	Legibility of composition	YES/NO (0.5) the composition is not entirely legible, there are dead-ended side paths	YES (1) a clear central path with side paths	YES (1) legible, geometric composition	YES/NO (0.5) consists of several sections, which appear isolated from each other	YES/NO (0.5) consists of several sections which do not appear coherent	YES/NO (0.5) consists of several sections with poor compositional linkages
	Presence of water, especially water in motion	NO (0)	NO (0)	NO (0)	YES (1) there is a fountain	NO (0)	YES (1) there is a fountain in the form of water jets
	Plant sensory impact on each of the senses	YES/NO (0.5) mostly a scent-based garden	YES (1) each of the five senses is highlighted in a separate part of the garden	YES (1) plants in pots provide a diverse range of sensations, they can also be tasted	YES/NO (0.5) various plants that induce sensory experiences, no experiences based on taste	YES/NO (0.5) numerous fragrant plants and fruit-bearing bushes, one can hear the rustle of grass and the sound of gravel, but touch is not stimulated	NO (0)
	Intensity of plant sensory impact (e.g., diversity of species, large spaces, elevated beds)	YES (1) large patches of fragrant plants situated along paths	YES (1) wealth of species and strains; plants also placed in pots built into benches	YES (1) plants elevated and placed in large pots	YES (1) sensory path surrounded by sensory active plants	YES/NO (0.5) presence of fragrant plants in isolated areas, but difficult to reach and without seats	NO (0) absent

Table 4. *Cont.*

	The Second Group of Attributes						
		(A)	(B)	(C)	(D)	(E)	(F)
	Attribute						
	Other sensory active elements (e.g., labyrinth, sensory path)	YES (1) sensory path and labyrinths near the garden	YES (1) small sensory path	NO (0) absent	YES (1) sensory path	YES/NO (0.5) observation deck on roof, with a gravel surface, but only accessible via stairs	YES (1) labyrinth
Placemaking	Ability to personalise the space	NO (0)	NO (0)	NO (0)	NO (0)	NO (0)	NO (0)
	Ability to animate the space	YES/NO (0.5) lessons for children, the youth, or adults in the open	YES (1) a green meadow and a tent from living willow branches form a place for play and recreation	YES (1) multimedia presentations, films on the museum wall (chairs laid out on the lawn)	YES/NO (0.5) small tables where one can play board games or chess	NO (0)	YES (1) potential to use one's imagination and engage in various forms of play
	Artistic creations	NO (0)	YES (1) white cubes that act as seats	YES (1) the Museum wall has an artistic expression	NO (0)	NO (0)	NO (0)
	Special indications for use that facilitate experiencing the garden from up close	NO (0)	YES (1) one can taste fruits	YES (1) one can taste herbs	NO (0)	YES (1) one can taste fruits	YES (1) one can directly use all the equipment, including the fountain
Sustainability criteria	Biodiversity preservation: use of domestic plant species and plants attractive to various groups of animals, creating habitats for animals	YES/NO (0.5) some plants are attractive to butterflies or other insects	YES (1) introduction of plants attractive to the hymenoptera species, plants with fruit for birds; bird habitats	YES/NO (0.5) intentional use of species attractive to species of hymenoptera	YES/NO (0.5) some plants are attractive to butterflies or other insects	YES/NO (0.5) some plants are attractive to insects	NO (0)
	Sustainable water management, e.g., stormwater collection and use	NO (0)	NO (0)	NO (0)	NO (0)	NO (0)	NO (0)
	Natural energy sources	NOT APPLICABLE—no electrical appliances	NOT APPLICABLE—no electrical appliances	NO (0)	NO (0)	NO (0)	NO (0)
	Natural garden maintenance methods	No data	No data	YES (1)	No data	No data	No data

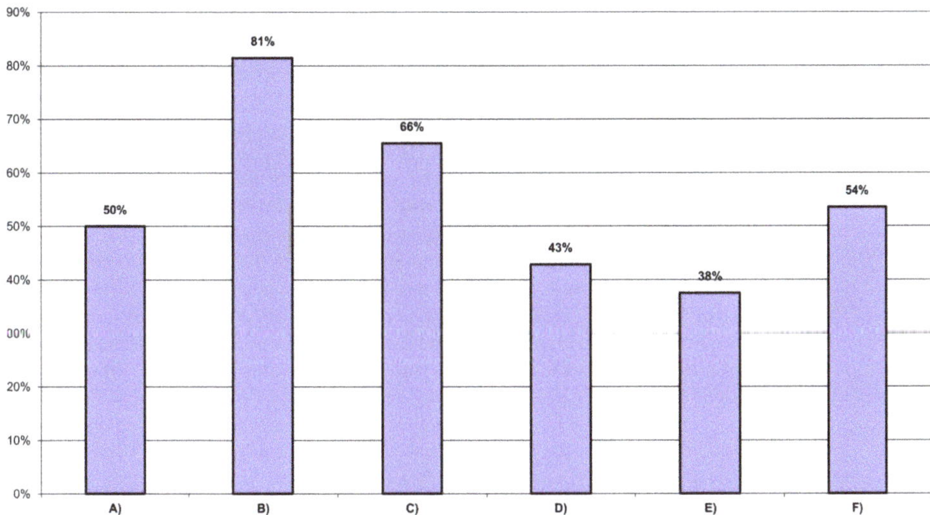

Figure 4. Presence of attributes that characterise the therapeutic spaces in each garden with sensory features in Kraków, in percentages. The letters A–F on the horizontal axis represent the gardens analysed; the description follows Figure 1.

The greatest number of attributes was found in the sensory garden at the Piaski Nowe housing estate. It was not given a perfect rating, as it did not feature any water features, including those based on stormwater. One of the least demanding solutions of this type could be a fountain for birds or for insects, with water obtained from waste, while more advanced water features include fountains, cascades, or flowing water. Following the principle that the more attributes a garden has, the better it is at performing its therapeutic function, it can be stated it was designed very well for this type of space. The deficiencies that were identified during assessment were not significant and making up for them is not impossible. We found it could be done without major changes to the garden's structure.

The least amount of points was given to two gardens that were recently built in city parks: Tysiąclecia Park and Reduta Park, as they had the lowest number of attributes. The most serious deficiency in both of them was that the sensory space was not isolated from the remainder of the parks. Isolation is a crucial characteristic of sensory gardens, and not providing it results in the gardens not having the atmosphere necessary to perceive the garden from up close and not receiving the resultant psychological regeneration. During a site visit we observed that dogs were allowed into these spaces and that people who crossed the park space were moving in a hurry. This indicates that these parks are often used as shortcuts to other destinations and visitors have no intention of using the therapeutic garden space.

In the case of the second garden, in Reduta Park, its designers used many bold and non-standard ideas (large areas covered by plants with edible fruit, a terrace with noisy gravel, or grass surrounding benches), yet their implementation, considering the failure to meet essential requirements of sensory gardens as therapeutic spaces, prevented a positive assessment.

The playground in H. Jordan Park has many features of a sensory garden. This garden could provide better sensory stimulation, both to children that visit the playground and to their caretakers, if it featured beds planted with stimulating plants.

Both gardens with sensory features that possessed intentionally placed elements that are friendly to pollinating insects, mainly those from the order hymenoptera and butterflies (the garden at the Piaski Nowe housing estate and near the J. Czapski Museum) were found to be places where pollinating insects visited flowering plants. Other sensory gardens (in the S. Lem Park, Tysiąclecia Park, and Reduta Park), despite not being intentionally designed as animal-friendly, fulfilled a similar role due to having large areas with blooming plants.

6. Discussion

The analytical tool based on Trojanowska's method [22] which we used to study Kraków's gardens with sensory features yielded results that are easy to list and allow for a quick comparison of each garden, combined with an identification of their weaknesses. This is due to the method employing a very good listing of therapeutic space attributes, which introduces a certain standardisation when studying structures of this type. This method also allows the list of attributes to be extended as the concept of therapeutic gardens evolves and develops, in addition to offering a better understanding of their potential while maintaining its value as an analytical tool. The potential of sensory gardens, both in generally accessible spaces and in healthcare and therapeutic facilities, has yet to be fully explored, with additional studies providing greater insight into their value and practical use [7].

However, it should be noted that this method was originally designed for studying public spaces. It may not be well suited to the study of gardens with sensory features that were designed to accompany healthcare and therapeutic facilities, as such gardens are prepared for highly specific users, who often suffer from health disorders and have special needs. When one designs a garden to accompany such an institution, the participation of their future users should be greater. Hussein argued that it is necessary to account for the therapeutic needs of specific patients, which is crucial to a positive outcome of their therapy [3,7], and Bengsston noted that such therapeutic gardens should be designed to offer patients a choice of activity and a sensory perception of the garden adapted to the level that they are currently capable of [23]. This means that such gardens should have a different structure than those located in public places.

The fifth group of attributes, namely accounting for sustainability precepts, was rated the lowest in the gardens under analysis. The first attribute, which concerns supporting biodiversity in sensory gardens, was found to be present only partially. The Polish-language literature highlights the significance of animal presence in therapeutic gardens [24]. It can even be suggested that it should be one of the essential characteristics of a sensory garden, as it considerably enhances the scope of sensory stimuli [17]. The presence of flowering and fruit-bearing plants in Kraków's gardens with sensory features supports urban fauna, especially insects and birds, which was visible in those gardens where specific plant species had been introduced (the garden in plant pots by the J. Czapski Museum and the garden at the Piaski Nowe housing estate). This is crucial in Kraków city parks, as beds with blooming plants are a rarity there. Furthermore, the lawns are intensively mowed, which is not conducive to plant species variety [25]. The use of ornamental plants aids in maintaining pollinator diversity, but only in the condition that the flowers have properties beneficial to them. Gardens intentionally designed as bee- or butterfly-friendly provide the best effect due to the use of species that are appropriate for them [26].

Sensory gardens have not been built in every urban activity zone in Kraków that could accommodate them. Other large cities around the world feature spaces for employees. New York provides the best practices in this regard, with its two pocket gardens: GreenAcre Park and Paley Park [8]. There are also cases of external green spaces accompanying places of employment, such as office gardens, which are sometimes built on rooftops and are dedicated solely to a given building's employees, of which London is a good example [27]. There are studies that show how crucial green surroundings are to office workers. They found that windows with a view of greenery and indoor plants visibly reduced workplace stress [28]. The ability to spend one's lunchbreak or free time in a sensory garden space would be all the more beneficial. Of the gardens under analysis, the garden near the J. Czapski Museum is the closest to this concept, although it is not a typical garden dedicated to employees. It was constructed in the city centre and includes a coffee shop terrace that is a part of the Museum, which allows visitors to spend more time in a sensory environment.

7. Conclusions

The method we used in this study allowed us to critically assess these complexes, which can enable the introduction of specific changes, as they can be easily identified as attributes. Thus, the method can be used both to study completed gardens and earlier. during the design phase. It can therefore be used in practice. It can be said that there is a need to apply it, as demonstrated by the latest gardens with sensory features that have been built in Kraków, and which have visible, essential deficiencies that prevent making use of their full potential as sensory spaces.

The method is not only a practical tool but also he proposal of an academic method (an analytical procedure that can be used in investigations) that allows for an in-depth analysis of such gardens.

In Poland, a country that has only recently began introducing sensory gardens into public spaces, the evolution of this concept has started to develop rapidly, and Kraków's gardens are not copies of some specific scheme. They include a variety of solutions in terms of layout and sensory stimulus sources.

Author Contributions: Investigation: I.K.-M., Ł.M., K.P.; Methodology: I.K.-M., Ł.M.; Visualization: K.P., Ł.M.; Writing—original draft: I.K.-M.; Writing—review & editing: K.P. All authors have read and agreed to the published version of the manuscript.

Funding: This research received no external funding.

Institutional Review Board Statement: Not applicable.

Informed Consent Statement: Not applicable.

Data Availability Statement: The study did not report any data.

Conflicts of Interest: The authors declare no conflict of interest.

References

1. Winterbottom, D.; Wagenfeld, A. *Therapeutic Gardens: Design for Healing Spaces*; Timber Press: Portland, OR, USA, 2015.
2. Ulrich, R.S. Health Benefits for Gardens in Hospitals. Ph.D. Thesis, Texas A & M University, College State, TX, USA, 2002.
3. Hussein, H. Using the sensory garden as a tool to enhance the educational development and social interaction of children with special needs. *Support Learn.* **2010**, *25*, 25–31. [CrossRef]
4. Gonzalez, M.T.; Kirkevold, M. Clinical use of sensory gardens and outdoor environment in Norvegian nursing homes: A cross-sectional e-mail survey. *Issues Ment. Health Nurs.* **2015**, *36*, 35–43. [CrossRef] [PubMed]
5. Balode, L. The design guidelines for therapeutic sensory gardens. *Res. Rural Dev.* **2013**, *2*, 114–119.
6. Sensory Trust. Available online: Sensorytrust.org.uk/information/factsheets/sensory-garden-1.html (accessed on 21 January 2021).
7. Hussein, H. Sensory Garden in Special Schools: The issues, design and use. *J. Des. Built Environ.* **2009**, *5*, 77–95.
8. Pawłowska, K. Ogród sensoryczny. In *Dźwięk w Krajobrazie Jako Przedmiot Badań Interdyscyplinarnych. Prace Komisji Krajobrazu Kulturowego PTG*; Bernat, S., Ed.; Wydawnictwo Polihymnia: Lublin, Poland, 2008; Volume XI, pp. 143–152.
9. Dudkiewicz, M.; Marcinek, B.; Tkaczyk, A. Idea ogrodu sensorycznego w koncepcji zagospodarowania atrium przy Szpitalu Klinicznym nr 4 w Lublinie. *Acta Sci. Pol. Archit.* **2014**, *13*, 71–77.
10. Górska-Kłęk, L. *ABC Zielonej Opieki. Seria „Biblioteka Nestora", t. VIII*; Dolnośląski Ośrodek Polityki Społecznej: Wrocław, Poland, 2016; p. 151.
11. Trojanowska, M. *Parki i Ogrody Terapeutyczne*; Wydawnictwo Naukowe PWN SA: Warszawa, Poland, 2017; p. 231.
12. Dąbski, M.; Dudkiewicz, M. Przystosowanie ogrodu dla niewidomego użytkownika na przykładzie ogrodów sensorycznych w Bolestraszycach, Bucharzewie i Powsinie. *Teka Kom. Archit. Urban. Studiów Kraj.* **2010**, *6*, 7–17.
13. Trojanowska, M. Sensory gardens inclusively designed for visually impaired users. *PhD Interdiscip. J.* **2014**, *1*, 30–317.
14. Woźny, A.; Lauda, A. Ogrody dla osób z dysfunkcją wzroku w świetle ich oczekiwań. *Archit. Kraj.* **2014**, *4*, 65–71.
15. Zajadacz, A.; Lubarska, A. Sensory gardens as places for outdoor recreation adapted to the needs of people with visual impairments. *Studia Perieget.* **2020**, *2*, 25–43. [CrossRef]
16. Westphal, J.M. Hype, Hyperbole, and Health: Therapeutic site design. In *Urban Lifestyles: Spaces, Places People*; Benson, J.F., Rowe, M.H., Eds.; Brookfield A. A. Balkema: Rotterdam, The Netherlands, 2000.
17. Krzeptowska-Moszkowicz, I.; Moszkowicz, Ł.; Porada, K. Znaczenie miejskich ogrodów sensorycznych o cechach przyjaznych organizmom rodzimym, na przykładzie dwóch przypadków z terenu dużych miast europejskich: Krakowa i Londynu. In *Integracja Sztuki i Techniki w Architekturze i Urbanistyce*; Wydawnictwa Uczelniane Uniwersytetu Technologiczno-Przyrodniczego: Bydgoszcz, Poland, 2020; pp. 61–69.

18. Kuczko, D. *Ogrody sensoryczne na Błoniach Niepołomickich*; Cracow University of Technology: Kraków, Poland, 2005.
19. Kraków w Liczbach. 2019, Urząd Miasta Krakowa. Available online: https://www.bip.krakow.pl/?mmi=6353 (accessed on 21 January 2021).
20. Kondracki, J. *Geografia Regionalna Polski*; Polskie Wydawnictwo Naukowe: Warszawa, Poland, 2001; p. 441.
21. Rackiewicz, I. *Diagnoza Stanu Środowiska Miasta Krakowa*; Urząd Miasta Krakowa: Kraków, Poland, 2012.
22. Trojanowska, M. The Universal Pattern of Design for Therapeutic Parks. Methods of Use. *Int. J. Sci. Eng. Res.* **2018**, *9*, 1410–1413.
23. Bengtsson, A.; Grahn, P. Outdoor environments in healthcare settings: A quality evaluation tool for use in designing healthcare gardens. *Urban For. Urban Green.* **2012**, *13*, 878–891. [CrossRef]
24. Nowacki, J.; Bunalski, M.; Sienkiewicz, P. Fauna jako istotny element hortiterapii. In *Hortiterapia Jako Element Wspomagający Leczenie Tradycyjne*; Krzymińska, A., Ed.; Rhytmos: Poznań, Poland, 2012; pp. 117–144.
25. Moszkowicz, Ł.; Krzeptowska-Moszkowicz, I. Impact of the public parks location in the city on the richness and diversity of herbaceous vascular plants on the example of Krakow Southern Poland. In Proceedings of the Plants in Urban Areas and Landscape, Nitra, Slovakia, 7–8 November 2019. [CrossRef]
26. Willmer, P. *Pollination and Floral Ecology*; Princeton University Press: Princeton, NJ, USA; Oxford, UK, 2011; p. 778.
27. Chance, H.; Winterbottom, D.; Bell, J.; Wagenfeld, A. Gardens for Well-Being in Workplace Environments. In Proceedings of the 3rd European Conference on Design4Health, Sheffield, UK, 13–16 July 2015. Available online: https://research.shu.ac.uk/design4health/wp-content/uploads/2015/07/D4H_Chance_et_al.pdf (accessed on 1 September 2020).
28. Chang, C.Y.; Chen, P.K. Human Response to window views and indoor plants in the workplace. *HortScience* **2005**, *40*, 1354–1359. [CrossRef]

sustainability

Article

Identification and Analysis of Problems in Selected European Historic Gardens during the COVID-19 Pandemic

Katarzyna Hodor [1,*], Łukasz Przybylak [2], Jacek Kuśmierski [3] and Magdalena Wilkosz-Mamcarczyk [4]

1 Chair of Landscape Architecture, Faculty of Architecture, Cracow University of Technology CUT, Warszawska 24, 31-155 Kraków, Poland
2 European Route of Historic Gardens, Plaça de la Vila 1, 17310 Lloret del Mar, Spain; vicepresident@europeanhistoricgardens.eu
3 Museum of King John III's Palace at Wilanów, Stanisława Kostki Potockiego 10/16, 02-958 Warsaw, Poland; jkusmierski@muzeum-wilanow.pl
4 Department of Land Management and Landscape Architecture, Faculty of Environmental Engineering and Land Surveying, University of Agriculture in Kraków, Balicka 253 c, 30-198 Kraków, Poland; magdalena.wilkosz-mamcarczyk@urk.edu.pl
* Correspondence: khodor@pk.edu.pl; Tel.: +48-608396220

Abstract: The paper is based on a survey and investigates the functioning of historic gardens during the pandemic. The authors collected and analysed information on the impact of the pandemic on the behaviour of visitors, maintenance, and condition of cultural heritage assets, European historic gardens. Four aspects were considered particularly carefully: the situation of gardens during the COVID-19 pandemic, maintenance and care in gardens, virtual activity and communication, and financial consequences. The authors determined the conditions of the gardens and the problems they faced based on a survey completed by 23 managers of 31 historic gardens from June to August 2020 and then proposed a diagnosis. The paper presents the survey results. In general, visitor volumes tended to drop in 2020, which significantly affected gardens' financial standing and contributed to workforce reductions. The garden condition and treatments were affected, as well. Reduced visitor volumes resulted in positive environmental changes. Among them were ecological succession, the stability of landscaped plants, increase in vegetation, improved biodiversity in the ground cover, and enhanced animal presence. Additional safety measures were implemented after the gardens were reopened to the public during the pandemic, mostly social distancing, and obligatory face masks. Less than half of the gardens had contingency plans, and 25% of the respondents were working to develop one. The analyses provided foundations to start working on a universal emergency strategy similar to procedures used for years for permanent collections at museums. Note that, being open public spaces and live museums, historic gardens were the first places reopened after the lockdown. Recommendations based on the study can contribute to the future safe functioning of historic gardens in other similar crises. The guidelines offer instructions, advice, and recommendations that form foundations of the development of a universal management model facilitating the preservation of historic gardens in good condition while exploiting their ecological potential.

Keywords: COVID-19; historic parks; historic gardens; landscape architecture; cultural heritage; green cultural heritage; green area

Citation: Hodor, K.; Przybylak, Ł.; Kuśmierski, J.; Wilkosz-Mamcarczyk, M. Identification and Analysis of Problems in Selected European Historic Gardens during the COVID-19 Pandemic. *Sustainability* **2021**, *13*, 1332. https://doi.org/10.3390/su13031332

Received: 19 December 2020
Accepted: 25 January 2021
Published: 27 January 2021

1. Introduction

Today, the term 'historical object' applies to every object important for the cultural heritage and development of culture because of its historical, scientific, or artistic value [1]. The category includes historic gardens, which were appreciated numerous times by many international organisations, also by being included on the UNESCO World Heritage List [2]. Moreover, the 1964 Venice Charter defines a historic monument much more broadly than as just an individual architectural work and includes a rural or urban setting as well. The

1981 Florence Charter outlines procedures and describes protection of historic gardens. It defines the historic garden as 'an architectural and horticultural composition of interest to the public from the historical or artistic point of view' [3]. It points out that a historic garden is an architectural composition but also a living vegetal fabric and proposes to consider it a natural monument and apply the principles of the Venice Charter to historic gardens [4]. Many European countries became interested in monument protection in the 19th century. France established the Commission of Historic Monuments (Commission des Monuments Historiques), the United Kingdom started The National Trust, and Belgium, the Royal Committee for Monuments and Sites (Commission Royale des Monuments et Sites). Changes in the perception of historic green sites that commenced in the 19th century led to the intentional protection of the gardens [5].

Historic gardens were first studied in the 16th century in treaties on the construction and maintenance of gardens. The art of garden design and care was considered science in the 18th century [6]. Today, historic gardens are appreciated for their social, aesthetic, environmental, cultural, architectural, and perceptual qualities. Research on historic gardens involves analysis of the landscape and its complexity combined with landscape valuation to help users realise the import of heritage they represent [7].

The investigated objects—historic gardens in Europe—have clear aesthetic, natural, and cultural qualities. Despite differences in the scale, origin, and location, they constitute a group of unique heritage assets, increasingly often considered museums. Protection and conservation of historic gardens have been supplemented with the art of remembrance and curation, which resulted in the musealisation of the gardens. It is reflected in various forms of nature and monument protection intended to preserve individual plant specimens, garden compositions, or even whole landscapes in an unaltered form [8,9]. Their conservators employ detailed strategies with protection and conservation plans, directions for restoration, procedures for making them available to the public, and even an exhibition plan. Gardens with works of nature and man became open-air museums [10]. Their growing popularity gave rise to 'garden tourism' as a highly specialised and rather profitable branch of cultural tourism [11]. It is important to appreciate their social and ecological roles in the landscape and their uniqueness related to cultural heritage. Historic gardens are a crucial part of cultural heritage. They remain a memento of the centuries past with their invariable style and character reflecting their times despite changes in urban and rural environments over time [12–14].

Experts focus on so-called process conservation today. It involves regular shaping, maintenance, change management, and resource management in individual time intervals. As regards physical operations, reconstruction of former gardens, in particular, they propose conservation through documentation and preservation of value. The cultural approach to the gardens helps expand the knowledge of history and sources of inspiration for art, religion, and philosophy [15]. The heritage of historic gardens is perceived considering the preservation indicators and its values. Experts look into the historical value, presence of plants from the initial design, historic buildings, and park furniture and small structures. The artistic qualities, garden's design, shape, and colour matter as well [6]. Being a living tissue, historic gardens need to be conserved and their heritage, protected. The care for the gardens, which are also artistic objects, and their conservation is the expression of the culture of our societies [16,17] Otherwise, they may be degraded, and the damage can be irreversible.

Management is a key component of the effort to protect historic gardens as heritage sites. Sufficient funding for historic gardens in the current COVID-19 crisis may contribute to a thriving urban environment in the future. Contact with history and nature promotes the identity of people and their aesthetic experience after a long quarantine. Well-organised museums, historic gardens, and botanic gardens may play an important role in research, education, and environmental protection [18]. Therefore, it is necessary to complement research on historic gardens with methods specific to social sciences, economics, and resource assessment, tourism research, and urban heritage studies [4,19]. The current

COVID-19 epidemic led to crises in many industries [20], also in tourism. Lifestyle, tourist behaviour, and travel preferences changed [21,22]. The slump in international tourism and at the local level [23] affected the functioning of gardens significantly. The usual forms of activity and recreation in gardens were interrupted, which affected the well-being of regular garden users [24]. Some organisations took action (in line with the limitations in place in their respective countries at the time) to develop documents with hints and advice regarding the management of parks and sites so that they could become safe spaces for leisure with nature and in conformity with all necessary safety principles, such as Comité des Parcs et Jardins de France, La Demeure Historique, Greenspace Scotland, or The Fields in Trust.

The COVID-19 outbreak in late 2019 and early 2020 intensified all known groups of management and restoration challenges and merged them into one global factor that affected all garden heritage sites. The scale of the new threat to cultural heritage, including historic gardens, can be compared to the period of 1939–1945 when the global military conflict endangered the future of cultural civilisation assets. The Second World War was a trying time for historic sites. Many cities were destroyed, with parks and gardens along with them [25,26].

This completely new 'super factor' brought to view the need to pay historic gardens the attention similar to the one reserved for works of art and architecture. Museum items are often secured with evacuation procedures for collections [27,28]. The buildings often have emergency plans and conservation monitoring systems that keep track of environmental data and help with prevention and rescue activities. They also have systems preventing damage by vandalism, theft, or fire. [29–31]. Historic gardens do not have such precautions. This gigantic emergency-response gap calls for universal procedural models that can be adapted for small sites and large landscape complexes facing a crisis. It should be a set of management practices determined first and foremost by the style and specific character of the garden composition that will help keep it in good condition in the face of a similar crisis. Each institution in charge of a historic garden became a unique case study when the pandemic broke out. Its experience can either be used in a universal management model or as a caution against decisions that could lead to degradation of a monument with time.

2. Study Area and Objectives

The analysis focuses on historic gardens situated in Europe. The authors addressed associations of historic gardens and people in charge of individual gardens. The survey was addressed to managers of historic gardens. We followed two tracks to encourage them to complete the questionnaires on an online platform. The first one was to contact the gardens directly (requests sent to the managers, also members of the European Network of Historic Gardens: 31 members from Germany, Georgia, Italy, Poland, Portugal, and Spain). The other way was to post questionnaires to social media and open-access websites (such as the newsletter of the Dutch Castles, Historic Country Houses & Rural Estates Foundation sKBL (stichting Kastelen Buitenplaasten Landgoederen), ERHG (European Route of Historic Gardens), Landscape Conference in Kraków, and in industry-specific groups on LinkedIn). The survey is estimated to have reached 56 respondents apart from the open-access availability mentioned above.

Survey questionnaires were completed by garden managers from Poland, Germany, the Netherlands, Portugal, Spain, United Kingdom, Sweden, and Italy (Table 1, Figure 1).

Table 1. List of respondents.

No.	Country	Place	Garden Name
1.	Spain	Aranjuez	Jardín del Príncipe
2.			Jardín de la Isla
3.		Cambrils	Parc Samà
4.		Grenada	Alhambra
5.			Generalife
6.		Lloret de Mar	Jardín botánico Santa Clotilde
7.	The Netherlands	Apeldoorn	Paleis Het Loo
8.	Germany	Dresden	Großer Gärten
9.		Hanower	Herrenhäuser Gärten
10.		Heidenau	Barockgarten Großsedlitz
11.		Müglitztal	Schloss Weesenstein
12.		Pillnitz	Schloss und Park Pillnitz
13.	Poland	Gdańsk	Park Oliwski im. Adama Mickiewicza
14.		Kozłówka	Pałac Zomoyskich w Kozłówce
15.		Kraków	Zamek Królewski na Wawelu
16.		Łańcut	Zamek w Łańcucie
17.		Nieborów	Pałac w Nieborowie i romantyczny park w Arkadii
18.		Pieskowa Skała	Zamek w Pieskowej Skale
19.		Rogalin	Pałac w Rogalinie
20.		Warszawa	Zamek Królewski w Warszawie
21.			Pałac w Wilanowie
22.			Park im. Stefana Żeromskiego
23.			Muzeum Łazienki Królewskie
24.		Wrocław	Pałac Kornów w Pawłowicach
25.	Portugal	Queluz	Palácio Nacional e Jardins de Queluz
26.		Sintra	Parque e Palácio de Monserrate
27.			Parque e Palácio National da Pena
28.	Sweden	Mölndal	Gunnebo House and Garden
30.	United Kingdom	Waddesdon	National Trust Waddesdon Manor
31.		Wisley	Royal Horticultural Society's Wisley Garden
32.	Italy	Florence	Giardino di Boboli

Figure 1. Map of Europe with countries where the participating gardens are located 1. Jardín del Príncipe (UNESCO Aranjuez Cultural Landscape) 2. Jardín de la Isla 3. Parc Samà 4. Alhambra (UNESCO Alhambra, Generalife and Albayzín, Granada), 5. Generalife (UNESCO Alhambra, Generalife and Albayzín, Granada), 6. Jardín botánico Santa Clotilde 7. Paleis Het Loo 8. Großer Gärten 9. Herrenhäuser Gärten 10. Barockgarten Großsedlitz 11. Schloss Weesenstein 12. Schloss und Park Pillnitz 13. Park Oliwski im. Adama Mickiewicza 14. Pałac Zomoyskich w Kozłówce 15. Zamek Królewski na Wawelu (UNESCO Historic Centre of Cracow), 16. Zamek w Łańcucie 17. Pałac w Nieborowie i romantyczny park w Arkadii 18. Zamek w Pieskowej Skale 19. Pałac w Rogalinie 20. Zamek Królewski w Warszawie (UNESCO Historic Centre of Warsaw) 21. Pałac w Wilanowie 22. Park im. Stefana Żeromskiego 23. Muzeum Łazienki Królewskie 24. Pałac Kornów w Pawłowicach 25. Palácio Nacional e Jardins de Queluz 26. Parque e Palácio de Monserrate (UNESCO Cultural Landscape of Sintra) 27. Parque e Palácio National da Pena 28. National Trust Waddesdon Manor 29. Royal Horticultural Society's Wisley Garden 30. Giardino di Boboli (UNESCO Medici Villas and Gardens in Tuscany) 31. Gunnebo House and Garden.

All the participating gardens are protected and listed as national monuments or included on the UNESCO World Heritage List (individually or as part of a site). They represent almost all garden style groups: Islamic with late Medieval components, Medieval, Renaissance, Baroque, Landscape, Romantic, neo-style, and Modernist. The objects of interest are public spaces situated in city centres (8), in cities or towns (2), on the outskirts (10), and in rural areas (11).

This paper aims to present an initial estimate of historic gardens affected by lockdown and the steps they have taken to continue their activity. The research includes the identification and listing of main landscape, financial, and managerial issues in public spaces that are historic gardens. The particular focus was the impact of the COVID-19 pandemic on the preservation, maintenance, and condition of gardens as cultural heritage assets. The responses allowed the authors to develop an outline of recommendations for a universal managerial model that should offer procedural guidelines for if the pandemic continues and other similar emergencies. The authors have based their multifaceted assessment of the influence of the emergency on the management of historic gardens' resources on financial activities, virtual actions, care treatments, and human resources initiatives.

3. Methodology

Management challenges that historic gardens face today can be classified into six groups. These are methodological, regulatory, technological, administrative, social, and environmental challenges that affect what model of management is implemented in a facility [32]. Although the origins of the challenges are often global, their intensity that determines the condition and maintenance methods for a historic garden, as well as mitigation actions, are evaluated at a much smaller scale, often a region or local environment of an individual garden.

To identify the scale of impact of the COVID-19 pandemic and relevant restrictions on the stability of conservation and social position of historic gardens, questions needed to be based on management methods. The management method is determined by the vertical and horizontal, administrative, and financial structure of a garden. The vertical structure is its measurable geographical reach, topography, and hydrology in the area. The vertical structure includes not only all components of the vegetation but also mobile and immobile garden equipment (including sculptures and fountains). The administrative structure means the mode of ownership, the profile of the institution in charge of the garden, statutory activities based on its resources, and the way the personnel responsible for ongoing maintenance of the garden is managed. Note that the extent of the vertical and horizontal structure should proportionally shape the size of the administrative and financial structures that ensure the maintenance of the existing condition of the garden or its smooth restoration [33].

Survey

The authors identified aspects of historic garden maintenance and used them to develop a set of issues relevant to the basic information on ongoing care, statutory activity, and administrative and financial circumstances. The palette of questions helped collect basic and detailed data from many facilities on the time and nature of the response of the management and the care, conservation, and marketing activities that followed. The combination of interpretation of general information about components of the horizontal and vertical structures of the gardens and responses to in-depth questions regarding their administrative and financial structures facilitated a thorough and innovative attempt to develop universal conservation guidelines for the maintenance and functioning of European historic gardens and parks in the face of the COVID-19 pandemic.

The survey questionnaire 'Garden heritage in the face of the COVID-19 pandemic' was distributed online from 19 June 2020 to 12 August 2020. It contained four categories of questions: the situation of gardens during the COVID-19 pandemic, maintenance, and care in gardens, virtual activity and communication, financial consequences (Figure 2). The questionnaire consisted of 23 questions. Eleven of them were open-ended questions, and twelve were closed-ended. The latter were single-select and multi-select questions depending on its scope. The respondents could offer a comment on some of them to provide more in-depth insight. Open-ended questions focused on issues where qualitative results were desirable to provide a broader view of the impact of the pandemic on historic gardens.

The general inquiries involved the name of the site, which facilitated a more thorough insight into the character of the garden, especially as regards its vertical structure and the area (in hectares) to represent the scale of its horizontal structure. When combined with information about the number of people involved in routine care, changes in team sizes as a result of the outbreak, and identifiable deterioration of the condition of the garden, the general data helped uncover further relationships. The information demonstrated how varied the participating sites were. The question about a strategic document with maintenance standards or restoration plan for the site was an important one. Being an unforeseen emergency, the COVID-19 pandemic could be a trigger to verify the documents as part of a change management process, for example, through the reduction of teams involved in garden maintenance or limitation of the scope of operations.

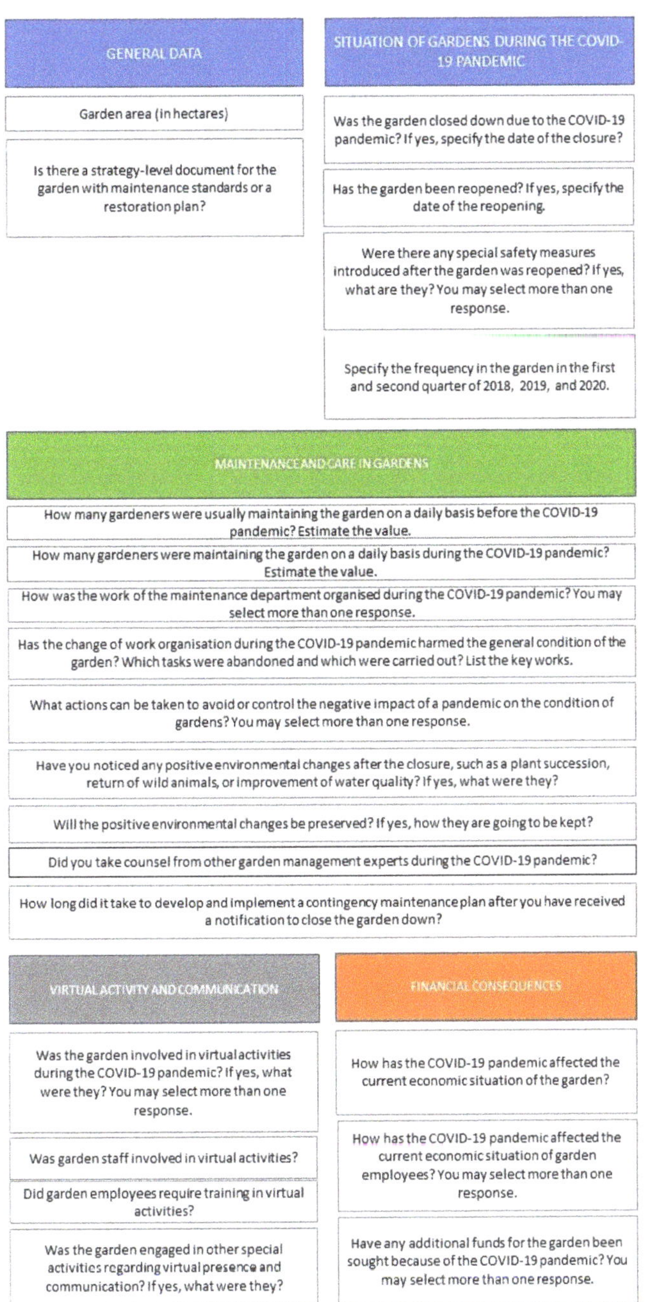

Figure 2. A diagram of survey questions.

The next section looked into the situation of the gardens during the pandemic. It focused on whether or not the garden was closed down, when, if at all, it was reopened, and what precautions were implemented. Its goal was to demonstrate the impact of the pandemic on the accessibility of the sites to the public. Visitor number was one of the factors illustrating the effects of COVID-19 on the financial situation of historic gardens and their social position. To paint a full picture of changes, the respondents were asked to provide visitor number data for the first and second quarters of 2018, 2019, and 2020. Note that the influence of the epidemic on historic gardens is a long-term phenomenon and will require data for the same quarters of 2021 and 2022. Nevertheless, the data for the 31 sites gave insight into social trends regarding the need to visit historic gardens in the initial period after the restrictions on the use of public spaces were lifted.

The third group of questions concerned maintenance and care in the gardens before and after the outbreak. The key matter was to determine the number of people involved in the everyday care of a garden and changes in human resources after health safety regulations were introduced in the initial phase of the pandemic. One question concerned the way work of the maintenance team was organised during the pandemic to determine the most popular management methods used to limit movement and ensure social distancing.

The respondents were further asked what actions could be taken to avoid or control the negative impact of a pandemic on the condition of the gardens. Its objective was to determine whether it was material or intangible aid that is more important to managers in a crisis. They were also asked whether they consulted other experts in garden management during the COVID-19 pandemic and how long it took to develop and implement a contingency plan after they received a notification to close the garden down. The goal was to check how many of the respondents needed additional professional support and how many acted on their own. The reaction time to the pandemic was investigated as well.

The one before the last section of the questionnaire looked into virtual activities and communication. The respondents were asked whether their facilities undertook such actions, if their employees took part in them, and if there was a need to train them in virtual operations. The questions were intended to determine the participation of historic gardens in the virtual world and whether it required an improvement of digital competencies of the personnel. The last, fifth section of the questionnaire handled financial consequences. The respondents declared the impact of the pandemic on the budgets of their sites, the financial situation of the employees, and whether any additional measures were taken to obtain additional funds to preserve jobs. The objective was to find out whether and to what extent the closure of the gardens affected their financial standing, if the crisis financially hurt the employees, and what corrective actions the managers implemented.

4. Results

The responses to the question concerning the date of closure and reopening of the gardens to the public yielded the average duration of the period of about 2.5 months. The knowledge of the period of zero attendance gives some idea of the dynamics of natural succession and general biocoenotic changes in the sites. The closure took place in the first and second quarter of 2020. The visitor number data for the same periods of 2018, 2019, and 2020 yielded general trends in the 'consumption' of gardens (irrespective of the offer of a particular garden) but also an insight into its radical drop due to health restrictions with its far-reaching financial consequences. Complete quarterly data were provided by seven respondents (Santa Clotilde Botanical Garden—Spain, Herrenhausen Garden—Germany, Park of Pena—Portugal, Park of Monserrate—Portugal, National Palace Gardens of Queluz—Portugal, Wilanów Palace Gardens—Poland, and the Royal Łazienki—Poland). The non-standard typical schedule of one of the gardens, Wawel Royal Castle (Poland), allowed it to provide data only for second quarters (Figure 3). Five respondents provided aggregate attendance data for 2018 and 2019 and partial data for 2020 (Figure 4). Over half of the respondents failed to provide relevant data.

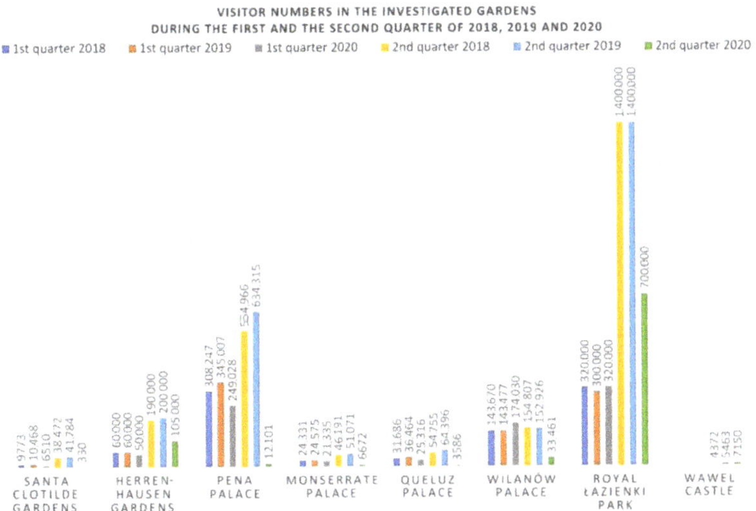

Figure 3. Visitor numbers in the investigated gardens during the first and the second quarter of 2018, 2019, and 2020.

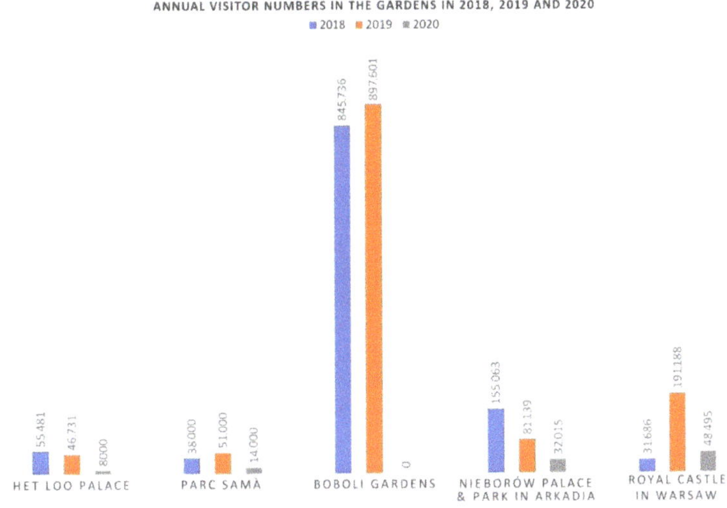

Figure 4. Annual visitor numbers in the gardens in 2018, 2019, and 2020.

The reopening was followed by additional safety measures. The most popular were distancing (25%) and the obligation to wear masks (19%). Next were the reduction in the number of visitors allowed (15%), shorter opening hours (15%), and disinfection of benches and other resting places (12%). The diverse scales of health restrictions implemented in European countries represented by the respondents were reflected in the attendance changes in the second quarter of 2020. For Central and Eastern European countries, a clear trend was found. The visitor volume went down to 40–50% of the 2018–2019 numbers. Respondents from Western Europe provided devastating visitor data for the second quarter of 2020. According to their records, the visitor numbers in West-European tourist destination historic gardens plunged to 0.78–13.06% of 2019 figures. Such a dramatic decrease in visitor volume paints a morbid picture of the financial situation of sites, the primary source of income of which are entrance fees. The consequences are a direct threat to the stability of the garden and the clarity of its historical spatial structure. It also suggests the answer to the question of the financial and personnel impacts of COVID-19. Significant financial losses and employment reductions among those responsible for maintenance will lead to loss of artistic qualities of the garden.

Data on the number of people involved in the everyday care of a garden and changes in human resources after health safety regulations were introduced in the initial phase of the pandemic indicate that 32% of the gardens maintained complete personnel responsible for ongoing maintenance. The typical solution regarding the size of the team responsible for fieldwork was a reduction to 74–50% and 49–25%, for the investigated gardens. This change was most often accompanied by a rotational shift scheme or a reduction in daily hours. An alarming, yet widespread approach (18% of the respondents) was to restrict personnel to 25–0% during the pandemic.

The declared temporary changes at the management level resulted in an evident deterioration of the condition of almost one-fourth of the gardens. The respondents indicated two leading factors when asked about the unfavourable effects of the reduction in the gardening personnel. The first one was unforeseen violent weather events during the closure and declared a decrease in gardener and technician teams. Reduced personnel significantly extended the response time and repairs necessary to restore the condition of the garden. The other factor contributing to the deterioration of the condition of the sites was significant changes in the typical schedule of gardening operations. Postponing of such treatments as spring lawn renovation or formative and rejuvenation pruning of shrubs and trees lowered the aesthetic value of the garden. The prioritisation of maintenance operations in gardens closed to the public results in particular in minimisation of the maintenance of the vertical diversity of the garden structure. The leading compromise in this group was a conscious decision not to plant seasonal plants in part or the whole of the garden. Note that a temporary failure to plant seasonal vegetation does not impact the general condition of a historic garden as long as the knowledge on the location, form, and structure of specific compositions survives. A negative aspect related to the maintenance of the vertical structure is the permanent abandonment of topiaries and uncontrolled growth of voluntary seedlings of bushes and trees. This issue was identified in 14% of the historic gardens, including in those focusing on geometric compositions. In the long term, such an approach will lead to a permanent loss of artistic qualities of the site.

In the context of prioritisation of gardening operations, the respondents indicated prevention of epiphytotic diseases and epizootics. Key maintenance operations carried out by gardeners regardless of the extent of temporary personnel reductions were watering and pest control, including the box tree moth (*Cydalima perspectalis*).

As regards the trends discussed above, note that regardless of the extent of the horizontal and vertical structure of a garden, a personnel reduction to 99–75% did not cause a deterioration in its condition, and preserved the historical clarity of the spatial composition. The high-risk range regarding the safety of maintaining a garden in a non-deteriorated condition was the staff reduction intervals to 74–50% and 49–25%. A reduction to 25–0% was an immediate and direct threat to the stability regardless of the preparation.

The ranges mentioned above pose a threat to the garden, especially when combined with a lack of a strategic document to indicate the general principles of restoration and parameters for the ongoing maintenance. Only less than half of the managers participating in the survey had such a document, and 24% were working on it. The threat referred to above is real for 28% of the sites that did not have the strategic document. Additionally, 69% of the respondents did not take advantage of other experts in garden management during COVID-19. This demonstrates two facts: their high level of self-reliance but also that they were making intuitive decisions according to the current situation, their knowledge and experience. The response time is worth noting, as well. In over half of the facilities, it took a week to develop and implement an emergency plan after the managers received a notification to close the garden down.

The general changes in historic gardens caused by the COVID-19 pandemic involved more than just garden compositions, as can be seen from the survey questions. The respondents identified biocoenotic changes in the gardens caused by changes in the availability to the public and intensity of maintenance operations.

Over half of the respondents (59%) indicated beneficial environmental changes. Four main types of biocoenotic changes emerge from the responses. These are phytocoenotic, zoocoenotic, physicochemical changes in water and air, and other changes. The phytocoenotic changes declared by the respondents were accelerated ecological succession, mostly occurrence of park ground cover species in the vertical structure. Other positive phytocoenotic changes were the stability of designed plant structures that were degraded before due to intensive use by visitors. The phytocoenotic changes are driven by:

- the season when the COVID-19 pandemic broke out,
- a radical drop in attendance,
- limitation of the size of teams dedicated to ongoing maintenance,
- prioritisation of care operations, whereby lawn mowing was most often less frequent.

Beneficent changes in the vertical structure mean generally increased biodiversity of the ground cover and amount to 32% of the identified changes. Zoocoenotic changes were another type of leading biocoenotic effects. It means not only increased activity and presence of birds, small mammals, deer, wild boar, or insects not seen in the garden for a long time but also an increase in their local populations. The main driver of beneficial zoocoenotic changes in the investigated historic gardens was the closure for safety reasons during the mating and breeding seasons of animals.

The global and local reduction in transport and industry activity had an impact on the ecosystems of historic gardens as well. It was reflected in the responses as a noticeable improvement of physicochemical properties of water and air. Another matter worth mentioning is the effect of a lower rate of mechanical damage caused by high visitor volumes or occasional vandalism. It was beneficial for the stability of both vertical and horizontal structures.

The additional question about the possibilities of maintaining the beneficial biocoenotic changes in the historic gardens made abundantly clear the strong relationship between the attendance and the financial situation of the sites. Restriction of the former could help maintain most of the positive environmental changes, but would also reduce the income necessary for stable management and maintenance. Some respondents declared that positive phytocoenotic changes could be maintained by administrative decisions such as modified lawn mowing schedules.

The gardening staff could continue their work in a relatively normal manner as opposed to other divisions in the complex organisational structures of the gardens, such as finance, conservation, or education. This exhibited their educational potential. Almost 36% of the respondents indicated that the personnel was not involved in any virtual activity. This situation could have diverse causes, such as the organisation of work and workload of gardeners or the fact that other divisions were responsible for such activities. The auxiliary question regarding the necessity to improve gardening personnel competencies in customer service and virtual activities demonstrated no demand for such an approach. Sixty-nine percent of the respondents believed their teams did not need additional training and only 14% followed this development path for their employees.

As regards virtual activities, the managers aimed their efforts mostly towards social media (25 responses) where they published posts, pictures, and stories to compensate for the temporary unavailability of their gardens. Content analyses demonstrated that the largest reach and activity of social media users were reached by accounts on the current state of the garden (it was the time of spring and early summer in Europe) or content regarding garden history. It is an important insight into the significant potential of the virtual world for information and awareness-raising regarding the existence, character, and value of garden patrimony. Videos and podcasts were less popular (15 responses) followed by virtual tours (12 responses), which was most likely due to technical limitations.

Some sites expanded their websites (9 responses), shared digital publications (8 responses), and held webinars and online education courses (3 responses). Other solutions included photograph challenges (3 responses) that encouraged users to share their photographs online. Note that only two Polish facilities refrained from online activities.

The financial standing of historic gardens provided important insight into the impact of the pandemic. Most of the respondents indicated they suffered significant losses (72%). Only a small share of the sites recorded minor losses (4%) or were not hit by the pandemic financially (4%). In their comments, the respondents pointed out that the consequences could be long-term and be felt as late as in the autumn or even in the next season in 2021. The deterioration of the situation of historic gardens will have a lasting effect on their condition and restoration progress. It can be rendered even worst by the second wave of the COVID-19 pandemic and escalating economic crisis, which can turn into the worse global recession in decades according to the World Bank [34]. It is bad news for institutions with entry fees, events, and services for local visitors as the sole source of income because it threatens their stability.

The managers are already taking steps to improve the financial situation. One of them is to reduce personnel and outsourcing costs. The first actions taken by the respondents was to implement redundancy programs (7 responses), reduce bonuses (7 responses), or reduce base salaries (6 responses). They further mentioned that third-party service providers also suffered through contract suspension or termination. The saving schemes to alleviate the financial impact resulted in a complete restriction of hiring and outsourcing (8 responses). Employees of ten historic gardens participating in the survey were not affected financially by COVID-19.

Apart from the reduction in personnel costs, the administrators took other actions to secure additional funding for their gardens. Most of them intensified promotion (14 responses). Some sites expanded their commercial offer for visitors (7 responses) and their educational portfolio (4 responses). The administrators decided to apply for government reimbursement of losses from the pandemic (4 responses), but none addressed NGOs or EU institutions. Other solutions included intensified partner collaboration (private business and institutions), which contributed additional means for garden maintenance (8 responses). As many as nine of the respondents took no action to balance their budgets, all of them state-funded.

5. Discussion

The COVID-19 pandemic and the changes it brought inspired many documents with proposed emergency procedures in line with national prevention policies. They focused on the functioning of cities and various domains of everyday life.

One of the general proposals published by the ICOMOS in April 2020 was *Urban Function-Spatial Response Strategy for the Epidemic—A Concise Manual on Urban Emergency Management* by the Urban Heritage Conservation and Sustainable Development Research Team, School of Architecture, Southeast University (SEU), China SEU Key Laboratory of Urban and Architectural Heritage Conservation, Ministry of Education, China, UNESCO Chair in Cultural Resource Management based on Chinese experience from late 2019 and early 2020. It describes the inability of modern Chinese cities to respond to emergencies. It focused on outlining instructions in accordance with the International Health Regulations (2005), Public health preparedness and response (2018), and the World Health Organisation documents (WHO). It was an attempt to adapt at the micro and macro level, focusing on transport, medical facilities, and spatio-functional adjustments, while disregarding landscape heritage structures [35].

Gardens (historical, botanical, and other) have been considered part of the social life for centuries. It was a popular place of respite and recreation, a haven for those seeking refuge from the hustle and bustle of the city [36]. Visiting a garden is a form of tourism and recreation as the garden is perceived as a place that brings people together [37].

In May, a group of researchers conducted analyses of a social-media survey regarding the importance and availability of green sites during the COVID-19 pandemic. The survey spanned four European countries. Its results showed that green areas became a high-priority service of high impact on the health and well-being of the public [38,39]. The authors emphasised the value of urban forests that became the 'critical infrastructure' for the whole urban system [40,41]. As regards publications by managers of historic gardens affected by the COVID-19 pandemic, only the European Route of Historic Gardens, an association of 33 historic gardens published on its website a paper reflecting on the situation [42].

Other research demonstrated that contact with nature (gardens) during a COVID-19 lockdown reduces the rate of reported depression and anxiety and may shield from a negative impact of a quarantine, protecting the mental and physical health through access to green areas and various forms of activities and recreation [26,43,44]. In addition to international publications, individual countries have various response strategies. In France, Comité des Parcs et Jardins de France published special recommendations *Spécial COVID-19—Réouverture et Charte sanitaire* for handling public access to gardens and health rules [45]. The country was divided into green and red regions. Gardens in the first zone were opened on 11 May with general safety principles applied. The gardens in the other zone remain closed to the public. Their managers may, however, address their local prefect with a request for a waiver, specifying the date of opening, financial reasons, and measures that would be taken to conform to national safety regulations. La Demeure Historique published similar recommendations for its members, *COVID-19: Mesures d'accompagnement et de soutien* [46]. It contained a health protection plan and a good safety practice guide.

In the United Kingdom, The Gardens Trust published a short note, *The impact of Covid-19 on parks and gardens* [47], where it listed the most significant consequences of the pandemic such as cancelled events and suspended volunteer schemes. Greenspace Scotland published a guide, *Managing Scotland's parks and greenspaces during COVID-19* [48], with key guidelines for park care, infrastructure maintenance, visitor management and activities, personnel, contractors, and volunteer management, and communication. A similar brochure *Managing Public Parks during Covid-19* was published by Community First Partnership with The National Lottery Heritage Fund, Local Government Association, National Trust, Association of Public Service Excellence the Midlands Parks Forum, and mangers of green sites [49]. The Royal Parks introduced a special policy for park-goers such as closed public toilets, sports facilities, and playgrounds until further notice, opening of several catering kiosks for takeaway only, or suspended cycling in Richmond Park [50]. The Fields in Trust published a guide for managers of green areas, *Management of green spaces during Covid-19*. They do not recommend a complete closure of parks unless absolutely necessary, and no safety measures can be implemented. The recommendations include adaptation of infrastructure and equipment, restoration of staff and volunteer work, opening green areas to visitors, restoration of past activities, and communication [51].

The survey demonstrated a different direction leading to the identification of problems of a specific domain that is historic gardens (museums). The proposed survey approach involving eight European countries provides a broader outlook on problems the research identified. These include the specificity of treatments based on conservation and continuation of restoration schemes. The research shows that 68% of the curators limited gardening operations to minimise the risk of COVID-19.

This approach stemmed from general safety standards implemented in most green public spaces in Europe. The results indicate that 84% of the historic gardens implemented special safety measures, such as imposed social distance, obligatory face masks, limited visitor volume, reduced opening hours, and disinfection of benches and other relaxation areas.

The education and promotion aspect, connected to financial matters, was important for historic gardens, as they are different from green public areas. The virtual effort, including social media, videos, podcasts, virtual tours, website expansion, and online publication was noted for 96.5% of the historic gardens.

The difficult financial situation contributed to reduced gardening teams (18%), which may escalate in future. To set off the further financial strain, 81% of the gardens made an effort to secure funds. The effort involved extra promotion, improved commercial offer, and educational programs.

6. Conclusions

The paper complements research on management in a specific group of historic gardens during crises. On the one hand, the survey results demonstrated weaknesses of garden management (restriction of expenses, personnel reductions). On the other hand, they showed personnel management opportunities to improve the chance of controlling damage to the historic green structure of gardens. It proposes promotion approaches that can help improve the financial standing of the gardens. Thirty percent of the respondents pointed out the need to ensure funds to minimise the impact on the condition of historic gardens. Moreover, 22% of them indicated the need for crisis management, counselling, and training. Twenty-one percent suggested a pandemic manual, and 19%, a platform for exchanging knowledge and experience.

An outline of procedural recommendations for continued pandemic and other similar emergencies is provided in eight key actions to help protect valuable cultural assets such as historic gardens.

Their value for the global patronage related to the landscape was emphasised in the Florence Charter, which determined the general principles for care and conservation of the designed landscape.

1. An indication of the necessity for every historic site to develop a maintenance policy for a crisis and a management-level document superior to it to determine the general principles and scope of restoration (if applicable) of the garden, taking into account treatment priorities, and set parameters for the number and competencies of the personnel responsible for the ongoing maintenance or smooth restoration together with any machinery necessary,
2. The development of a binding emergency management policy to ensure non-deteriorated conditions of the garden and prioritise care operations following the style of the garden or its part,
3. A recommendation to refrain from redundancy schemes concerning personnel responsible for everyday care, while implementing strict health precautions when facing a situation similar to COVID-19,
4. The extension of professional development schemes for gardeners so that they can co-create statutory content when the activity is shifted to the online environment,
5. The establishment of centres for emergency coordination for historic gardens and cultural landscape with ministries of culture (not only for matters related to a pandemic, but also for epiphytotic diseases and epizootics that directly endanger the stability of historic gardens and their components as well as the negative impact of long-term weather events or ecological crises). Simultaneous provision of funds necessary to continue the ongoing treatments in gardens, including to cover salaries of the personnel and contractors responsible for ongoing operations as well as funds necessary to secure interventions and prevention,
6. An indication to develop financial aid schemes by state administration that would allow:

- the provision of means necessary to continue ongoing care in gardens, including to cover salaries of personnel and contractors responsible for ongoing operations,
- the provision of means necessary to make interventions and take preventive actions when epiphytotic diseases or epizootics are confirmed that threaten the condition of the historic garden or its important part (box tree moth, horse-chestnut leaf miner, Dutch elm disease, Verticillium wilt) and devastating effects of long-term weather conditions (droughts) and ecological crisis (contamination of surface water and groundwater).

A recommendation to implement maintenance breaks in historic gardens from March to April (one to two months) for positive biocoenotic changes to take place in what is often the only environmental reservoirs in very urbanised areas. A simultaneous recommendation to provide financial reserves (in facilities administered by state or local cultural institutions) to maintain them in this period (or significant visitor number restrictions from March to April).

Specification of the recommended maximum daily visitor numbers for gardens on the UNESCO World Heritage List, historic monuments, and listed monuments to ensure the safety of their spatial composition and individual components.

Prioritisation of maintenance operations limited due to extraordinary situations, focus on key zones of the garden, coordination of relevant actions in accordance with the profile each site and the possibility to close historic gardens to reconstruct plant and animal resources. These recommendations can contribute to the continued good condition of gardens and exploitation of their ecological potential both in historic gardens and large landscape complexes.

Author Contributions: Conceptualization, K.H. and Ł.P.; methodology, K.H. and Ł.P.; software, J.K.; validation, K.H., Ł.P., J.K. and M.W.-M.; formal analysis, K.H. and Ł.P.; investigation, K.H., Ł.P., J.K.; resources, J.K.; data curation, K.H. and Ł.P.; writing—original draft preparation, K.H., Ł.P., J.K. and M.W.-M.; writing—review and editing, K.H. and M.W.-M.; visualization, K.H., Ł.P., J.K. and M.W.-M.; supervision, K.H. and Ł.P.; project administration, K.H.; funding acquisition, K.H. and M.W.-M. All authors haveread read and agreed to the published version of the manuscript.

Funding: This publication has received financial support from the Polish Ministry of Science and Higher Education under subsidy for science for the FA CUT in 2019 and 20 and Environmental engineering, mining, and energy science 030008-D014/KGPiAK/2021.

Institutional Review Board Statement: Not applicable.

Informed Consent Statement: Not applicable.

Data Availability Statement: Survey data are stored according to GDPR requirements on CUT servers.

Acknowledgments: The authors would like to thank the anonymous reviewers for their thorough work with the manuscript and for providing constructive and insightful comments on this paper.

Conflicts of Interest: The authors declare no conflict of interest.

References

1. Majdecki, L. *Ochrona i konserwacja zabytkowych założeń ogrodowych*; Wydawnictwo Naukowe PWN: Warszawa, Poland, 1993; pp. 18–20.
2. World Heritage List. Available online: http://whc.unesco.org/en/list/ (accessed on 15 August 2020).
3. O'Donnell, P.M. Florence Charter on Historic Gardens (1982). In *Encyclopedia of Global Archaeology*; Springer Nature: New York, USA, 2014; pp. 2812–2817.
4. Athanasiadou, E. Historic Gardens and Parks Worldwide and in Greece: Principles of Acknowledgement, Conservation, Restoration and Management. *Heritage* **2019**, *2*, 2678–2690. [CrossRef]
5. Majdecki, L.; Majdecka-Strzeżek, A. Ochrona i konserwacja zabytkowych założeń ogrodowych. Wydawnictwo Naukowe PWN: Warszawa, Poland, 2019.
6. Carneiro, A.R.S.; da Silva, J.M.; de S. Veras, L.M.; de F. Silva, A. The complexity of historic garden life conservation. Measuring Heritage Conservation Performance at the 6th International Seminar on Urban Conservation, Rome, Italy, 29–31 March 2011; Zancheti, S.M., Similä, K., Eds.; pp. 33–41.
7. Cazzani, A.; Zerbi, C.M.; Brumana, R.; Lobovikov-Katz, A. Raising awareness of the cultural, architectural, and perceptive values of historic gardens and related landscapes: Panoramic cones and multi-temporal data. *Appl. Geomatics* **2020**, 1–34. [CrossRef]
8. Hodor, K. Contemporary threats to historic gardens in Poland. *Czas. Tech* **2016**, *5*, 125–143.
9. Tieskens, K.F.; Schulp, C.J.; Levers, C.; Lieskovský, J.; Kuemmerle, T.; Plieninger, T.; Verburg, P.H. Characterizing European cultural landscapes: Accounting for structure, management intensity and value of agricultural and forest landscapes. *Land Use Policy* **2017**, *62*, 29–39. [CrossRef]
10. Schneider, A. Weimarer Hausgärten von der Klassik bis zur Moderne. *AHA! Misz, Gartengeschichte Gartendenkmalpfl.* **2020**, *6*, 12–25. [CrossRef]

11. Benfield, R. Garden Tourism. In *The SAGE International Encyclopedia of Travel and Tourism*; Lowry, L.L., Ed.; SAGE Publications, Inc: Thousand Oaks, CA, USA, 2017; pp. 165–191.
12. Ahmad, Y. The Scope and Definitionas of Heritage: From Tangible to Intangible. *Int. J. Herit. Stud.* **2006**, *12*, 292–300. [CrossRef]
13. Bell, S.; Fox-Kämper, R.; Keshavarz, N.; Benson, M.; Caputo, S.; Noori, S.; Voigt, A. *Urban Allotment Gardens in Europe*; Routledge: New York, NY, USA, 2016.
14. Hristov, D.; Naumov, N.; Petrova, P. Interpretation in historic gardens: English Heritage perspective. *Tour. Rev.* **2018**, *73*, 199–215. [CrossRef]
15. Kuśmierski, J. *Hortuseum Musealisation of the European Gardens in the Twenty-First Century*; The Sarny Castle Trust: Ścinawka Górna, Poland, 2020.
16. Rohde, M. Historische Gärten Als Kulturaufgabe. In *Historische Gärten und Klimawandel*; Hüttl, R.F., Karen, D., Schneider, B.U., Eds.; De Gruyter Akademie Forschung: Berlin, Germany, 2020; pp. 31–51. Available online: https://www.degruyter.com/view/book/9783110607772/10.1515/9783110607772-005.xml (accessed on 15 January 2021).
17. Harney, M. *Gardens and Landscapes in Historic Building Conservation*; Wiley-Blackwell: Chichester, UK, 2014.
18. Qumsiyeh, M.; Handal, E.; Chang, J.; Abualia, K.; Najajrehm, M.; Abusarhan, M. Role of museums and botanical gardens in ecosystem services in developing countries: Case study and outlook. *Int. J. Environ. Stud.* **2017**, *74*, 340–350. [CrossRef]
19. Funsten, C.; Borsellino, V.; Schimmenti, E. A Systematic Literature Review of Historic Garden Management and Its Economic Aspects. *Sustainability* **2020**, *12*, 10679. [CrossRef]
20. La, V.-P.; Pham, T.-H.; Ho, M.-T.; Nguyen, M.-H.; Nguyen, P.K.-L.; Vuong, T.-T.; Nguyen, H.-K.T.; Tran, T.; Khuc, Q.; Ho, M.-T.; et al. Policy Response, Social Media and Science Journalism for the Sustainability of the Public Health System Amid the COVID-19 Outbreak: The Vietnam Lessons. *Sustainability* **2020**, *12*, 2931. [CrossRef]
21. Wen, J.; Kozak, M.; Yang, S.; Liu, F. COVID-19: Potential effects on Chinese citizens' lifestyle and travel. *Tour. Rev.* **2020**. [CrossRef]
22. Allam, Z.; Jones, D.S. Pandemic stricken cities on lockdown. Where are our planning and design professionals [now, then and into the future]? *Land Use Policy* **2020**, *97*, 104805. [CrossRef] [PubMed]
23. Carr, A. COVID-19, indigenous peoples and tourism: A view from New Zealand. *Tour. Geogr.* **2020**, *22*, 491–502. [CrossRef]
24. Pouso, S.; Borja, Á.; Fleming, L.E.; Gómez-Baggethun, E.; White, M.P.; Uyarra, M.C. Contact with blue-green spaces during the COVID-19 pandemic lockdown beneficial for mental health. *Sci. Total. Environ.* **2021**, *756*, 143984. [CrossRef] [PubMed]
25. Kostof, S. *The City Assembled -The Elements of Urban Form Through History*; Thames & Hudson: London, UK, 1992.
26. Racoń-Leja, K. *Miasto i Wojna*; Wydawnictwo Politechniki Krakowskiej: Kraków, Poland, 2019.
27. Henderson, J. Disasters without planning: Lessons for museums. *Conservator* **1995**, *19*, 52–57. [CrossRef]
28. Caple, C. *Presentive Conservation in Museums*; Routledge Oxford: Abingdon, UK; p. 2012.
29. Oddy, A.; Keene, S. Managing Conservation in Museums. *Stud. Conserv.* **1997**, *42*, 126. [CrossRef]
30. Ferraro, J.; Henderson, J. Identifying Features of Effective Emergency Response Plans. *J. Am. Inst. Conserv.* **2011**, *50*, 35–48. [CrossRef]
31. Szmit-Naud, E. Działania z zakresu konserwacji zapobiegawczej: Audyt konserwatorski-metoda i przykłady. *Acta Univ. Nicolai Copernici Zabytkozn. Konserw.* **2015**, *45*, 311–334.
32. Przybylak, Ł. Models of implementation and maintenance activities implemented in the Wilanów garden in response to contemporary restoration challenges of historical gardens. *Ochr. Zabyt.* **2019**, *2*, 93–119.
33. Karaşah, B.; Var, M. Recreational Functions of Botanical Gardens And Examining Sample of Nezahat Gökyiğit Botanical Garden. *Geography* **2013**, 803–809.
34. World Bank Group, Global Economic Prospects. Available online: https://www.worldbank.org/en/publication/global-economic-prospects (accessed on 15 August 2020).
35. ICOMOS International Conservation Center-Xi'an. Available online: https://www.icomos.ch/wp-content/uploads/2020/03/20200318-ICOMOS-CHINA.pdf (accessed on 15 August 2020).
36. Čakovská, B. Sustainable Garden Tourism in the United Kingdom or What's Behind the Fence? *Acta Hortic. Regiotect.* **2017**, *20*, 49–54. [CrossRef]
37. Connell, J. Managing gardens for visitors in Great Britain: A story of continuity and change. *Tour. Manag.* **2005**, *26*, 185–201. [CrossRef]
38. Uchiyama, Y.; Kohsaka, R. Access and Use of Green Areas during the COVID-19 Pandemic: Green Infrastructure Management in the "New Normal". *Sustainability* **2020**, *12*, 9842. [CrossRef]
39. Xie, J.; Luo, S.; Furuya, K.; Sun, D. Urban Parks as Green Buffers During the COVID-19 Pandemic. *Sustainability* **2020**, *12*, 6751. [CrossRef]
40. Derks, J.; Giessen, L.; Winkel, G. COVID-19-induced visitor boom reveals the importance of forests as critical infrastructure. *For. Policy Econ.* **2020**, *118*, 102253. [CrossRef]
41. Clearing House H2020. Available online: http://clearinghouseproject.eu/2020/07/23/green-spaces-urban-forests-during-the-covid-19-pandemic/ (accessed on 15 August 2020).
42. European Route of Historic Gardens, Historic Gardens and COVID-19: Times for reflection and action. Available online: http://europeanhistoricgardens.eu/en/historic-gardens-context-covid19-reflection-action/ (accessed on 15 August 2020).

43. Dzhambov, A.M.; Lercher, P.; Browning, M.H.E.M.; Stoyanov, D.; Petrova, N.; Novakov, S.; Dimitrova, D.D. Does greenery experienced indoors and outdoors provide an escape and support mental health during the COVID-19 quarantine? *Environ. Res.* **2020**, 110420. [CrossRef]
44. Szopińska, E.; Kazak, J.; Kempa, O.; Rubaszek, J. Spatial Form of Greenery in Strategic Environmental Management in the Context of Urban Adaptation to Climate Change. *Pol. J. Environ. Stud.* **2019**, *28*, 2845–2856.
45. Comité des Parcs et Jardins de France, Spécial COVID-19 - Réouverture et Charte sanitaire. Available online: https://www.parcsetjardins.fr/actualites/manifestations-nationales/special-covid-19-reouverture-et-charte-sanitaire-1657 (accessed on 15 August 2020).
46. La Demeure Historique, COVID-19: Mesures d'accompagnement et de soutien. Available online: https://www.demeure-historique.org/divers/mesures-daccompagnement-et-de-soutien/ (accessed on 15 August 2020).
47. The Gardens Trust, The impact of Covid-19 on parks and gardens. Available online: http://thegardenstrust.org/the-impact-of-covid-19-on-parks-and-gardens/ (accessed on 15 August 2020).
48. Greenspace Scotland, Managing parks and greenspaces during Covid-19. Available online: https://www.greenspacescotland.org.uk/pages/category/covid-19 (accessed on 15 August 2020).
49. Community First Partnership (CFP). Managing Public Parks during Covid-19. Available online: http://www.cfpuk.co.uk/wp-content/uploads/Managing-Public-Parks-During-COVID-19-FINAL-v7.1.pdf (accessed on 15 August 2020).
50. The Royal Parks, The Royal Parks welcomes visitors but warns it's not 'business as usual'. Available online: https://www.royalparks.org.uk/media-centre/press-releases/the-royal-parks-welcomes-visitors-but-warns-its-not-business-as-usual (accessed on 15 August 2020).
51. Fields in Trust, Management of green spaces during Covid-19. Available online: http://www.fieldsintrust.org/knowledge-base/management-of-green-spaces-during-covid-19 (accessed on 15 August 2020).

MDPI
St. Alban-Anlage 66
4052 Basel
Switzerland
Tel. +41 61 683 77 34
Fax +41 61 302 89 18
www.mdpi.com

Sustainability Editorial Office
E-mail: sustainability@mdpi.com
www.mdpi.com/journal/sustainability

www.ingramcontent.com/pod-product-compliance
Lightning Source LLC
LaVergne TN
LVHW071628170726
843515LV00009B/2509

9783036517520